Special Functions

Special functions, which include the trigonometric functions, have been used for centuries. Their role in the solution of differential equations was exploited by Newton and Leibniz, and the subject of special functions has been in continuous development ever since. In just the past thirty years several new special functions and applications have been discovered.

This treatise presents an overview of the area of special functions, focusing primarily on the hypergeometric functions and the associated hypergeometric series. It includes both important historical results and recent developments and shows how these arise from several areas of mathematics and mathematical physics. Particular emphasis is placed on formulas that can be used in computation.

The book begins with a thorough treatment of the gamma and beta functions, which are essential to understanding hypergeometric functions. Later chapters discuss Bessel functions, orthogonal polynomials and transformations, the Selberg integral and its applications, spherical harmonics, q-series, partitions, and Bailey chains.

This clear, authoritative work will be a lasting reference for students and researchers in number theory, algebra, combinatorics, differential equations, mathematical computing, and mathematical physics.

George E. Andrews is Evan Pugh Professor of Mathematics at The Pennsylvania State University.

Richard Askey is Professor of Mathematics at the University of Wisconsin-Madison.

Ranjan Roy is Professor of Mathematics at Beloit College in Wisconsin.

ENCYCLOPEDIA OF MATHEMATICS AND ITS APPLICATIONS

EDITED BY G.-C. ROTA

Editorial Board
B. Doran, M. Ismail, T.-Y. Lam, E. Lutwak

Volume 71

Special Functions

ENCYCLOPEDIA OF MATHEMATICS AND ITS APPLICATIONS

Special Functions

GEORGE E. ANDREWS   RICHARD ASKEY   RANJAN ROY

CAMBRIDGE
UNIVERSITY PRESS

PUBLISHED BY THE PRESS SYNDICATE OF THE UNIVERSITY OF CAMBRIDGE
The Pitt Building, Trumpington Street, Cambridge, United Kingdom

CAMBRIDGE UNIVERSITY PRESS
The Edinburgh Building, Cambridge CB2 2RU, UK
40 West 20th Street, New York, NY 10011-4211, USA
10 Stamford Road, Oakleigh, Melbourne 3166, Australia
Ruiz de Alarcón 13, 28014 Madrid, Spain
Dock House, The Waterfront, Cape Town 8001, South Africa

http://www.cambridge.org

First published 1999
First paperback edition 2000

Printed in the United States of America

Typeset in 10/13 Times Roman in LaTeX [TB]

A catalog record for this book is available from the British Library

Library of Congress Cataloging in Publication data is available

ISBN 0 521 62321 9 hardback
ISBN 0 521 78988 5 paperback

To Leonard Carlitz, Om Prakash Juneja,
and Irwin Kra

Contents

Preface

Paul Turán once remarked that special functions would be more appropriately labeled "useful functions." Because of their remarkable properties, special functions have been used for centuries. For example, since they have numerous applications in astronomy, trigonometric functions have been studied for over a thousand years. Even the series expansions for sine and cosine (and probably the arc tangent) were known to Madhava in the fourteenth century. These series were rediscovered by Newton and Leibniz in the seventeenth century. Since then, the subject of special functions has been continuously developed, with contributions by a host of mathematicians, including Euler, Legendre, Laplace, Gauss, Kummer, Eisenstein, Riemann, and Ramanujan.

In the past thirty years, the discoveries of new special functions and of applications of special functions to new areas of mathematics have initiated a resurgence of interest in this field. These discoveries include work in combinatorics initiated by Schützenberger and Foata. Moreover, in recent years, particular cases of long familiar special functions have been clearly defined and applied as orthogonal polynomials.

As a result of this prolific activity and long history one is pulled different directions when writing a book on special functions. First, there are important results from the past that must be included because they are so useful. Second, there are recent developments that should be brought to the attention of those who could use them. One also would wish to help educate the new generation of mathematicians and scientists so that they can further develop and apply this subject. We have tried to do all this, and to include some of the older results that seem to us to have been overlooked. However, we have slighted some of the very important recent developments because a book that did them justice would have to be much longer. Fortunately, specialized books dealing with some of these developments have recently appeared: Petkovšek, Wilf, and Zeilberger [1996], Macdonald [1995], Heckman and Schlicktkrull [1994], and Vilenkin and Klimyk

xiii

[1992]. Additionally, I. G. Macdonald is writing a new book on his polynomials in several variables and A. N. Kirillov is writing on R-matrix theory.

It is clear that the amount of knowledge about special functions is so great that only a small fraction of it can be included in one book. We have decided to focus primarily on the best understood class of functions, hypergeometric functions, and the associated hypergeometric series. A hypergeometric series is a series Σa_n with a_{n+1}/a_n a rational function of n. Unfortunately, knowledge of these functions is not as widespread as is warranted by their importance and usefulness. Most of the power series treated in calculus are hypergeometric, so some facts about them are well known. However, many mathematicians and scientists who encounter such functions in their work are unaware of the general case that could simplify their work. To them a Bessel function and a parabolic cylinder function are types of functions different from the $3 - j$ or $6 - j$ symbols that arise in quantum angular momentum theory. In fact these are all hypergeometric functions and many of their elementary properties are best understood when considered as such.

Several important facts about hypergeometric series were first found by Euler and an important identity was discovered by Pfaff, one of Gauss's teachers. However, it was Gauss himself who fully recognized their significance and gave a systematic account in two important papers, one of which was published posthumously. One reason for his interest in these functions was that the elementary functions and several other important functions in mathematics are expressible in terms of hypergeometric functions. A half century after Gauss, Riemann developed hypergeometric functions from a different point of view, which made available the basic formulas with a minimum of computation. Another approach to hypergeometric functions using contour integrals was presented by the English mathematician E. W. Barnes in the first decade of this century. Each of these different approaches has its advantages.

Hypergeometric functions have two very significant properties that add to their usefulness: They satisfy certain identities for special values of the function and they have transformation formulas. We present many applications of these properties. For example, in combinatorial analysis hypergeometric identities classify single sums of products of binomial coefficients. Further, quadratic transformations of hypergeometric functions give insight into the relationship (known to Gauss) of elliptic integrals to the arithmetic-geometric mean. The arithmetic-geometric mean has recently been used to compute π to several million decimal places, and earlier it played a pivotal role in Gauss's theory of elliptic functions.

The gamma function and beta integrals dealt with in the first chapter are essential to understanding hypergeometric functions. The gamma function was introduced into mathematics by Euler when he solved the problem of extending the factorial function to all real or complex numbers. He could not have foreseen the extent of its importance in mathematics. There are extensions of gamma and beta functions that are also very important. The text contains a short treatment of Gauss and Jacobi sums, which are finite field analogs of gamma and beta functions. Gauss sums were encountered by Gauss in his work on the constructibility of regular polygons where they arose as "Lagrange resolvents," a concept used by Lagrange to study algebraic equations. Gauss understood the tremendous value of these sums for number theory. We discuss the derivation of Fermat's theorem on primes of the form $4n + 1$ from a formula connecting Gauss and Jacobi sums, which is analogous to Euler's famous formula relating beta integrals with gamma functions.

There are also multidimensional gamma and beta integrals. The first of these was introduced by Dirichlet, though it is really an iterated version of the one-dimensional integral. Genuine multidimensional gamma and beta functions were introduced in the 1930s, by both statisticians and number theorists. In the early 1940s, Atle Selberg found a very important multidimensional beta integral in the course of research in entire functions. However, owing to the Second World War and the fact that the first statement and also the proof appeared in journals that were not widely circulated, knowledge of this integral before the 1980s was restricted to a few people around the world. We present two different evaluations of Selberg's integral as well as some of its uses.

In addition to the above mentioned extensions, there are q-extensions of the gamma function and beta integrals that are very fundamental because they lead to basic hypergeometric functions and series. These are series Σc_n where c_{n+1}/c_n is a rational function of q^n for a fixed parameter q. Here the sum may run over all integers, instead of only nonnegative ones. One important example is the theta function $\sum_{-\infty}^{\infty} q^{n^2} x^n$. This and other similar series were used by Gauss and Jacobi to study elliptic and elliptic modular functions. Series of this sort are very useful in many areas of combinatorial analysis, a fact already glimpsed by Euler and Legendre, and they also arise in some branches of physics. For example, the work of the physicist R. J. Baxter on the Yang–Baxter equation led a group in St. Petersburgh to the notion of a quantum group. Independently, M. Jimbo in Japan was led by a study of Baxter's work to a related structure.

Many basic hypergeometric series (or q-hypergeometric series), both polynomials and infinite series, can be studied using Hopf algebras, which make up quantum groups. Unfortunately, we could not include this very important new approach to basic series. It was also not possible to include results on the multidimensional $U(n)$ generalizations of theorems on basic series, which have been studied extensively in recent years. For some of this work, the reader may refer to Milne [1988] and Milne and Lilly [1995]. We briefly discuss the q-gamma function and some important q-beta integrals; we show that series and products that arise in this theory have applications in number theory, combinatorics, and partition theory. We highlight the method of partition analysis.

P. A. MacMahon, who developed this powerful technique, devoted several chapters to it in his monumental *Combinatory Analysis*, but its significance was not realized until recently.

The theory of special functions with its numerous beautiful formulas is very well suited to an algorithmic approach to mathematics. In the nineteenth century, it was the ideal of Eisenstein and Kronecker to express and develop mathematical results by means of formulas. Before them, this attitude was common and best exemplified in the works of Euler, Jacobi, and sometimes Gauss. In the twentieth century, mathematics moved from this approach toward a more abstract and existential method. In fact, agreeing with Hardy that Ramanujan came 100 years too late, Littlewood once wrote that "the great day of formulae seem to be over" (see Littlewood [1986, p. 95]). However, with the advent of computers and the consequent reemergence of computational mathematics, formulas are now once again playing a larger role in mathematics. We present this book against this background, pointing out that beautiful, interesting, and important formulas have been discovered since Ramanujan's time. These formulas are proving fertile and fruitful; we suggest that the day of formulas may be experiencing a new dawn. Finally, we hope that the reader finds as much pleasure studying the formulas in this book as we have found in explaining them.

We thank the following people for reading and commenting on various chapters during the writing of the book: Bruce Berndt, David and Gregory Chudnovsky, George Gasper, Warren Johnson, and Mizan Rahman. Special thanks to Mourad Ismail for encouragement and many detailed suggestions for the improvement of this book. We are also grateful to Dee Frana and Diane Reppert for preparing the manuscript with precision, humor, and patience.

1

The Gamma and Beta Functions

Euler discovered the gamma function, $\Gamma(x)$, when he extended the domain of the factorial function. Thus $\Gamma(x)$ is a meromorphic function equal to $(x - 1)!$ when x is a positive integer. The gamma function has several representations, but the two most important, found by Euler, represent it as an infinite integral and as a limit of a finite product. We take the second as the definition.

Instead of viewing the beta function as a function, it is more illuminating to think of it as a class of integrals – integrals that can be evaluated in terms of gamma functions. We therefore often refer to beta functions as beta integrals.

In this chapter, we develop some elementary properties of the beta and gamma functions. We give more than one proof for some results. Often, one proof generalizes and others do not. We briefly discuss the finite field analogs of the gamma and beta functions. These are called Gauss and Jacobi sums and are important in number theory. We show how they can be used to prove Fermat's theorem that a prime of the form $4n + 1$ is expressible as a sum of two squares. We also treat a simple multidimensional extension of a beta integral, due to Dirichlet, from which the volume of an n-dimensional ellipsoid can be deduced.

We present an elementary derivation of Stirling's asymptotic formula for $n!$ but give a complex analytic proof of Euler's beautiful reflection formula. However, two real analytic proofs due to Dedekind and Herglotz are included in the exercises. The reflection formula serves to connect the gamma function with the trigonometric functions. The gamma function has simple poles at zero and at the negative integers, whereas $\csc \pi x$ has poles at all the integers. The partial fraction expansions of the logarithmic derivatives of $\Gamma(x)$ motivate us to consider the Hurwitz and Riemann zeta functions. The latter function is of fundamental importance in the theory of distribution of primes. We have included a short discussion of the functional equation satisfied by the Riemann zeta function since it involves the gamma function.

In this chapter we also present Kummer's proof of his result on the Fourier expansion of $\log \Gamma(x)$. This formula is useful in number theory. The proof given uses Dirichlet's integral representations of $\log \Gamma(x)$ and its derivative. Thus, we have included these results of Dirichlet and the related theorems of Gauss.

1.1 The Gamma and Beta Integrals and Functions

The problem of finding a function of a continuous variable x that equals $n!$ when $x = n$, an integer, was investigated by Euler in the late 1720s. This problem was apparently suggested by Daniel Bernoulli and Goldbach. Its solution is contained in Euler's letter of October 13, 1729, to Goldbach. See Fuss [1843, pp. 1–18]. To arrive at Euler's generalization of the factorial, suppose that $x \geq 0$ and $n \geq 0$ are integers. Write

$$x! = \frac{(x+n)!}{(x+1)_n},\tag{1.1.1}$$

where $(a)_n$ denotes the shifted factorial defined by

$$(a)_n = a(a+1)\cdots(a+n-1) \quad \text{for } n > 0, (a)_0 = 1,\tag{1.1.2}$$

and a is any real or complex number. Rewrite (1.1.1) as

$$x! = \frac{n!(n+1)_x}{(x+1)_n} = \frac{n!n^x}{(x+1)_n} \cdot \frac{(n+1)_x}{n^x}.$$

Since

$$\lim_{n\to\infty} \frac{(n+1)_x}{n^x} = 1,$$

we conclude that

$$x! = \lim_{n\to\infty} \frac{n!n^x}{(x+1)_n}.\tag{1.1.3}$$

Observe that, as long as x is a complex number not equal to a negative integer, the limit in (1.1.3) exists, for

$$\frac{n!n^x}{(x+1)_n} = \left(\frac{n}{n+1}\right)^x \prod_{j=1}^{n} \left(1+\frac{x}{j}\right)^{-1} \left(1+\frac{1}{j}\right)^x$$

and

$$\left(1+\frac{x}{j}\right)^{-1} \left(1+\frac{1}{j}\right)^x = 1 + \frac{x(x-1)}{2j^2} + O\left(\frac{1}{j^3}\right).$$

Therefore, the infinite product

$$\prod_{j=1}^{\infty} \left(1 + \frac{x}{j}\right)^{-1} \left(1 + \frac{1}{j}\right)^{x}$$

converges and the limit (1.1.3) exists. (Readers who are unfamiliar with infinite products should consult Appendix A.) Thus we have a function

$$\Pi(x) = \lim_{k \to \infty} \frac{k! k^x}{(x+1)_k} \tag{1.1.4}$$

defined for all complex $x \neq -1, -2, -3, \ldots$ and $\Pi(n) = n!$.

Definition 1.1.1   *For all complex numbers $x \neq 0, -1, -2, \ldots$, the gamma function $\Gamma(x)$ is defined by*

$$\Gamma(x) = \lim_{k \to \infty} \frac{k! k^{x-1}}{(x)_k}. \tag{1.1.5}$$

An immediate consequence of Definition 1.1.1 is

$$\Gamma(x+1) = x\Gamma(x). \tag{1.1.6}$$

Also,

$$\Gamma(n+1) = n! \tag{1.1.7}$$

follows immediately from the above argument or from iteration of (1.1.6) and use of

$$\Gamma(1) = 1. \tag{1.1.8}$$

From (1.1.5) it follows that the gamma function has poles at zero and the negative integers, but $1/\Gamma(x)$ is an entire function with zeros at these points. Every entire function has a product representation; the product representation of $1/\Gamma(x)$ is particularly nice.

Theorem 1.1.2

$$\frac{1}{\Gamma(x)} = xe^{\gamma x} \prod_{n=1}^{\infty} \left\{ \left(1 + \frac{x}{n}\right) e^{-x/n} \right\}, \tag{1.1.9}$$

where γ is Euler's constant given by

$$\gamma = \lim_{n \to \infty} \left(\sum_{k=1}^{n} \frac{1}{k} - \log n \right). \tag{1.1.10}$$

Proof.

$$\frac{1}{\Gamma(x)} = \lim_{n \to \infty} \frac{x(x+1)\cdots(x+n-1)}{n!n^{x-1}}$$

$$= \lim_{n \to \infty} x\left(1+\frac{x}{1}\right)\left(1+\frac{x}{2}\right)\cdots\left(1+\frac{x}{n}\right)e^{-x\log n}$$

$$= \lim_{n \to \infty} xe^{x\left(1+\frac{1}{2}+\cdots+\frac{1}{n}-\log n\right)} \prod_{k=1}^{n} \left\{\left(1+\frac{x}{k}\right)e^{-x/k}\right\}$$

$$= xe^{\gamma x} \prod_{n=1}^{\infty} \left\{\left(1+\frac{x}{n}\right)e^{-x/n}\right\}.$$

The infinite product in (1.1.9) exists because

$$\left(1+\frac{x}{n}\right)e^{-x/n} = \left(1+\frac{x}{n}\right)\left(1-\frac{x}{n}+\frac{x^2}{2n^2}\cdots\right) = 1 - \frac{x^2}{2n^2} + O\left(\frac{1}{n^3}\right),$$

and the factor $e^{-x/n}$ was introduced to make this possible. The limit in (1.1.10) exists because the other limits exist, or its existence can be shown directly. One way to do this is to show that the difference between adjacent expressions under the limit sign decay in a way similar to $1/n^2$. ∎

One may take (1.1.9) as a definition of $\Gamma(x)$ as Weierstrass did, though the formula had been found earlier by Schlömilch and Newman. See Nielsen [1906, p. 10].

Over seventy years before Euler, Wallis [1656] attempted to compute the integral $\int_0^1 \sqrt{1-x^2}dx = \frac{1}{2}\int_{-1}^{+1}(1-x)^{1/2}(1+x)^{1/2}dx$. Since this integral gives the area of a quarter circle, Wallis's aim was to obtain an expression for π. The only integral he could actually evaluate was $\int_0^1 x^p(1-x)^q dx$, where p and q are integers or $q = 0$ and p is rational. He used the value of this integral and some audacious guesswork to suggest that

$$\frac{\pi}{4} = \int_0^1 \sqrt{1-x^2}dx = \frac{1}{4}\lim_{n \to \infty}\left[\frac{2\cdot4\cdot6\cdots2n}{1\cdot3\cdot5\cdots(2n-1)}\cdot\frac{1}{\sqrt{n}}\right]^2 = \Gamma\left(\frac{3}{2}\right)\Gamma\left(\frac{3}{2}\right).$$

$$(1.1.11)$$

Of course, he did not write it as a limit or use the gamma function. Still, this result may have led Euler to consider the relation between the gamma function and integrals of the form $\int_0^1 x^p(1-x)^q dx$ where p and q are not necessarily integers.

Definition 1.1.3   *The beta integral is defined for* Re $x > 0$, Re $y > 0$ *by*

$$B(x, y) = \int_0^1 t^{x-1}(1-t)^{y-1}dt. \tag{1.1.12}$$

One may also speak of the beta function $B(x, y)$, which is obtained from the integral by analytic continuation.

The integral (1.1.12) is symmetric in x and y as may be seen by the change of variables $u = 1 - t$.

Theorem 1.1.4

$$B(x, y) = \frac{\Gamma(x)\Gamma(y)}{\Gamma(x + y)}. \tag{1.1.13}$$

Remark 1.1.1 The essential idea of the proof given below goes back to Euler [1730, 1739] and consists of first setting up a functional relation for the beta function and then iterating the relation. An integral representation for $\Gamma(x)$ is obtained as a byproduct. The functional equation technique is useful for evaluating certain integrals and infinite series; we shall see some of its power in subsequent chapters.

Proof. The functional relation we need is

$$B(x, y) = \frac{x + y}{y} B(x, y + 1). \tag{1.1.14}$$

First note that for $\operatorname{Re} x > 0$ and $\operatorname{Re} y > 0$,

$$B(x, y + 1) = \int_0^1 t^{x-1}(1 - t)(1 - t)^{y-1}dt$$
$$= B(x, y) - B(x + 1, y). \tag{1.1.15}$$

However, integration by parts gives

$$B(x, y + 1) = \left[\frac{1}{x}t^x(1 - t)^y \right]_0^1 + \frac{y}{x} \int_0^1 t^x(1 - t)^{y-1}dt$$
$$= \frac{y}{x}B(x + 1, y). \tag{1.1.16}$$

Combine (1.1.15) and (1.1.16) to get the functional relation (1.1.14). Other proofs of (1.1.14) are given in problems at the end of this chapter. Now iterate (1.1.14) to obtain

$$B(x, y) = \frac{(x + y)(x + y + 1)}{y(y + 1)} B(x, y + 2) = \cdots = \frac{(x + y)_n}{(y)_n} B(x, y + n).$$

Rewrite this relation as

$$B(x, y) = \frac{(x + y)_n}{n!} \frac{n!}{(y)_n} \int_0^n \left(\frac{t}{n} \right)^{x-1} \left(1 - \frac{t}{n} \right)^{y+n-1} \frac{dt}{n}$$
$$= \frac{(x + y)_n}{n! n^{x+y-1}} \frac{n! n^{y-1}}{(y)_n} \int_0^n t^{x-1} \left(1 - \frac{t}{n} \right)^{n+y-1} dt.$$

As $n \to \infty$, the integral tends to $\int_0^\infty t^{x-1} e^{-t} dt$. This may be justified by the Lebesgue dominated convergence theorem. Thus

$$B(x, y) = \frac{\Gamma(y)}{\Gamma(x+y)} \int_0^\infty t^{x-1} e^{-t} dt. \tag{1.1.17}$$

Set $y = 1$ in (1.1.12) and (1.1.17) to get

$$\frac{1}{x} = \int_0^1 t^{x-1} dt = B(x, 1) = \frac{\Gamma(1)}{\Gamma(x+1)} \int_0^\infty t^{x-1} e^{-t} dt.$$

Then (1.1.6) and (1.1.8) imply that $\int_0^\infty t^{x-1} e^{-t} dt = \Gamma(x)$ for $\mathrm{Re}\, x > 0$. Now use this in (1.1.17) to prove the theorem for $\mathrm{Re}\, x > 0$ and $\mathrm{Re}\, y > 0$. The analytic continuation is immediate from the value of this integral, since the gamma function can be analytically continued. ∎

Remark 1.1.2 Euler's argument in [1739] for (1.1.13) used a recurrence relation in x rather than in y. This leads to divergent infinite products and an integral that is zero. He took two such integrals, with y and $y = m$, divided them, and argued that the resulting "vanishing" integrals were the same. These canceled each other when he took the quotient of the two integrals with y and $y = m$. The result was an infinite product that converges and gives the correct answer. Euler's extraordinary intuition guided him to correct results, even when his arguments were as bold as this one.

Earlier, in 1730, Euler had evaluated (1.1.13) by a different method. He expanded $(1-t)^{y-1}$ in a series and integrated term by term. When $y = n + 1$, he stated the value of this sum in product form.

An important consequence of the proof is the following corollary:

Corollary 1.1.5   *For $\mathrm{Re}\, x > 0$*

$$\Gamma(x) = \int_0^\infty t^{x-1} e^{-t} dt. \tag{1.1.18}$$

The above integral for $\Gamma(x)$ is sometimes called the Eulerian integral of the second kind. It is often taken as the definition of $\Gamma(x)$ for $\mathrm{Re}\, x > 0$. The Eulerian integral of the first kind is (1.1.12). Legendre introduced this notation. Legendre's $\Gamma(x)$ is preferred over Gauss's function $\Pi(x)$ given by (1.1.4), because Theorem 1.1.4 does not have as nice a form in terms of $\Pi(x)$. For another reason, see Section 1.10.

The gamma function has poles at zero and at the negative integers. It is easy to use the integral representation (1.1.18) to explicitly represent the poles and the analytic continuation of $\Gamma(x)$:

$$\Gamma(x) = \int_0^1 t^{x-1} e^{-t} dt + \int_1^\infty t^{x-1} e^{-t} dt$$

$$= \sum_{n=0}^\infty \frac{(-1)^n}{(n+x)n!} + \int_1^\infty t^{x-1} e^{-t} dt. \tag{1.1.19}$$

The second function on the right-hand side is an entire function, and the first shows that the poles are as claimed, with $(-1)^n/n!$ being the residue at $x = -n, n = 0, 1, \ldots$.

The beta integral has several useful forms that can be obtained by a change of variables. For example, set $t = s/(s+1)$ in (1.1.12) to obtain the beta integral on a half line,

$$\int_0^\infty \frac{s^{x-1}}{(1+s)^{x+y}} ds = \frac{\Gamma(x)\Gamma(y)}{\Gamma(x+y)}. \tag{1.1.20}$$

Then again, take $t = \sin^2 \theta$ to get

$$\int_0^{\pi/2} \sin^{2x-1} \theta \cos^{2y-1} \theta d\theta = \frac{\Gamma(x)\Gamma(y)}{2\Gamma(x+y)}. \tag{1.1.21}$$

Put $x = y = 1/2$. The result is

$$\frac{[\Gamma(\frac{1}{2})]^2}{2\Gamma(1)} = \frac{\pi}{2},$$

or

$$\Gamma(1/2) = \sqrt{\pi}. \tag{1.1.22}$$

Since this implies $[\Gamma(\frac{3}{2})]^2 = \pi/4$, we have a proof of Wallis's formula (1.1.11). We also have the value of the normal integral

$$\int_{-\infty}^\infty e^{-x^2} dx = 2 \int_0^\infty e^{-x^2} dx = \int_0^\infty t^{-1/2} e^{-t} dt = \Gamma(1/2) = \sqrt{\pi}. \tag{1.1.23}$$

Finally, the substitution $t = (u-a)/(b-a)$ in (1.1.12) gives

$$\int_a^b (b-u)^{x-1}(u-a)^{y-1} du = (b-a)^{x+y-1} B(x, y) = (b-a)^{x+y-1} \frac{\Gamma(x)\Gamma(y)}{\Gamma(x+y)}. \tag{1.1.24}$$

The special case $a = -1, b = 1$ is worth noting as it is often used:

$$\int_{-1}^1 (1+t)^{x-1}(1-t)^{y-1} dt = 2^{x+y-1} \frac{\Gamma(x)\Gamma(y)}{\Gamma(x+y)}. \tag{1.1.25}$$

A useful representation of the analytically continued beta function is

$$B(x, y) = \frac{\Gamma(x)\Gamma(y)}{\Gamma(x + y)} = \frac{(x + y)}{xy} \prod_{n=1}^{\infty} \frac{\left(1 + \frac{x+y}{n}\right)}{\left(1 + \frac{x}{n}\right)\left(1 + \frac{y}{n}\right)}. \qquad (1.1.26)$$

This follows immediately from Theorem 1.1.2. Observe that $B(x, y)$ has poles at x and y equal to zero or negative integers, and it is analytic elsewhere.

As mentioned before, the integral formula for $\Gamma(x)$ is often taken as the definition of the gamma function. One reason is that the gamma function very frequently appears in this form. Moreover, the basic properties of the function can be developed easily from the integral. We have the powerful tools of integration by parts and change of variables that can be applied to integrals. As an example, we give another derivation of Theorem 1.1.4. This proof is also important because it can be applied to obtain the finite field analog of Theorem 1.1.4. In that situation one works with a finite sum instead of an integral.

Poisson [1823] and independently Jacobi [1834] had the idea of starting with an appropriate double integral and evaluating it in two different ways. Thus, since the integrals involved are absolutely convergent,

$$\int_0^{\infty} \int_0^{\infty} t^{x-1} s^{y-1} e^{-(s+t)} ds dt = \int_0^{\infty} t^{x-1} e^{-t} dt \int_0^{\infty} s^{y-1} e^{-s} ds = \Gamma(x)\Gamma(y).$$

Apply the change of variables $s = uv$ and $t = u(1 - v)$ to the double integral, and observe that $0 < u < \infty$ and $0 < v < 1$ when $0 < s, t < \infty$. This change of variables is suggested by first setting $s + t = u$. Computation of the Jacobian gives $dsdt = ududv$ and the double integral is transformed to

$$\int_0^{\infty} e^{-u} u^{x+y-1} du \int_0^1 v^{x-1} (1 - v)^{y-1} dv = \Gamma(x + y) B(x, y).$$

A comparison of two evaluations of the double integral gives the necessary result. This is Jacobi's proof. Poisson's proof is similar except that he applies the change of variables $t = r$ and $s = ur$ to the double integral. In this case the beta integral obtained is on the interval $(0, \infty)$ as in (1.1.20). See Exercise 1.

To complete this section we show how the limit formula for $\Gamma(x)$ can be derived from an integral representation of $\Gamma(x)$. We first prove that when n is an integer ≥ 0 and $\operatorname{Re} x > 0$,

$$\int_0^1 t^{x-1} (1 - t)^n dt = \frac{n!}{x(x + 1) \cdots (x + n)}. \qquad (1.1.27)$$

This is actually a special case of Theorem 1.1.4 but we give a direct proof by induction, in order to avoid circularity in reasoning. Clearly (1.1.27) is true for

$n = 0$, and

$$\int_0^1 t^{x-1}(1-t)^{n+1}dt = \int_0^1 t^{x-1}(1-t)(1-t)^n dt$$

$$= \frac{n!}{(x)_{n+1}} - \frac{n!}{(x+1)_{n+1}}$$

$$= \frac{(n+1)!}{(x)_{n+2}}.$$

This proves (1.1.27) inductively. Now set $t = u/n$ and let $n \to \infty$. By the Lebesgue dominated convergence theorem it follows that

$$\int_0^\infty t^{x-1}e^{-t}dt = \lim_{n\to\infty} \frac{n!n^{x-1}}{(x)_n} \quad \text{for } \operatorname{Re} x > 0.$$

Thus, if we begin with the integral definition for $\Gamma(x)$ then the above formula can be used to extend it to other values of x (i.e., those not equal to $0, -1, -2, \ldots$).

Remark 1.1.3 It is traditional to call the integral (1.1.12) the beta function. A better terminology might call this Euler's first beta integral and call (1.1.20) the second beta integral. We call the integral in Exercise 13 Cauchy's beta integral. We shall study other beta integrals in later chapters, but the common form of these three is $\int_C [\ell_1(t)]^p [\ell_2(t)]^q dt$, where $\ell_1(t)$ and $\ell_2(t)$ are linear functions of t, and C is an appropriate curve. For Euler's first beta integral, the curve consists of a line segment connecting the two zeros; for the second beta integral, it is a half line joining one zero with infinity such that the other zero is not on this line; and for Cauchy's beta integral, it is a line with zeros on opposite sides. See Whittaker and Watson [1940, §12.43] for some examples of beta integrals that contain curves of integration different from those mentioned above. An important one is given in Exercise 54.

1.2 The Euler Reflection Formula

Among the many beautiful formulas involving the gamma function, the Euler reflection formula is particularly significant, as it connects the gamma function with the sine function. In this section, we derive this formula and briefly describe how product and partial fraction expansions for the trigonometric functions can be obtained from it. Euler's formula given in Theorem 1.2.1 shows that, in a sense, the function $1/\Gamma(x)$ is half of the sine function.

Theorem 1.2.1 *Euler's reflection formula:*

$$\Gamma(x)\Gamma(1-x) = \frac{\pi}{\sin \pi x}. \tag{1.2.1}$$

Remark   The proof given here uses contour integration. Since the gamma function is a real variable function in the sense that many of its important characterizations occur within that theory, three real variable proofs are outlined in the Exercises. See Exercises 15, 16, and 26–27.

Since we shall show how some of the theory of trigonometric functions can be derived from (1.2.1), we now state that $\sin x$ is here defined by the series

$$\sin x = \frac{e^{ix} - e^{-ix}}{2i} = x - \frac{x^3}{3!} + \frac{x^5}{5!} - \cdots.$$

The cosine function is defined similarly. It is easy to show from this definition that sine and cosine have period 2π and that $e^{\pi i} = -1$. See Rudin [1976, pp. 182–184].

Proof.   Set $y = 1 - x, 0 < x < 1$ in (1.1.20) to obtain

$$\Gamma(x)\Gamma(1 - x) = \int_0^\infty \frac{t^{x-1}}{1+t} dt. \tag{1.2.2}$$

To compute the integral in (1.2.2), consider the integral

$$\int_C \frac{z^{x-1}}{1 - z} dz,$$

where C consists of two circles about the origin of radii R and ϵ respectively, which are joined along the negative real axis from $-R$ to $-\epsilon$. Move along the outer circle in the counterclockwise direction, and along the inner circle in the clockwise direction. By the residue theorem

$$\int_C \frac{z^{x-1}}{1 - z} dz = -2\pi i, \tag{1.2.3}$$

when z^{x-1} has its principal value. Thus

$$-2\pi i = \int_{-\pi}^\pi \frac{i R^x e^{ix\theta}}{1 - R e^{i\theta}} d\theta + \int_R^\epsilon \frac{t^{x-1} e^{ix\pi}}{1+t} dt + \int_\pi^{-\pi} \frac{i\epsilon^x e^{ix\theta}}{1 - \epsilon e^{i\theta}} d\theta + \int_\epsilon^R \frac{t^{x-1} e^{-ix\pi}}{1+t} dt.$$

Let $R \to \infty$ and $\epsilon \to 0$ so that the first and third integrals tend to zero and the second and fourth combine to give (1.2.1) for $0 < x < 1$. The full result follows by analytic continuation. One could also argue as follows: Equality of (1.2.1) for $0 < x < 1$ implies equality in $0 < \operatorname{Re} x < 1$ by analyticity; for $\operatorname{Re} x = 0, x \neq 0$ by continuity; and then for x shifted by integers using $\Gamma(x + 1) = x\Gamma(x)$ and $\sin(x + \pi) = -\sin x$.   ∎

The next theorem is an immediate consequence of Theorem 1.2.1.

Theorem 1.2.2

$$\sin \pi x = \pi x \prod_{n=1}^\infty \left(1 - \frac{x^2}{n^2}\right), \tag{1.2.4}$$

$$\pi \cot \pi x = \frac{1}{x} + \sum_{n=1}^{\infty}\left(\frac{1}{x+n} + \frac{1}{x-n}\right) = \lim_{n\to\infty}\sum_{k=-n}^{n}\frac{1}{x-k}, \qquad (1.2.5)$$

$$\frac{\pi}{\sin \pi x} = \frac{1}{x} + 2x\sum_{n=1}^{\infty}\frac{(-1)^n}{x^2-n^2} = \lim_{n\to\infty}\sum_{k=-n}^{n}\frac{(-1)^k}{x-k}, \qquad (1.2.6)$$

$$\pi \tan \pi x = \lim_{n\to\infty}\sum_{k=-n}^{n}\frac{1}{k+\frac{1}{2}-x}, \qquad (1.2.7)$$

$$\pi \sec \pi x = \lim_{n\to\infty}\sum_{k=-n}^{n}\frac{(-1)^k}{k+x+\frac{1}{2}}, \qquad (1.2.8)$$

$$\frac{\pi^2}{\sin^2 \pi x} = \sum_{n=-\infty}^{\infty}\frac{1}{(x+n)^2}. \qquad (1.2.9)$$

Proof. Formula (1.2.4) follows from the product formula

$$\frac{1}{\Gamma(x)} = xe^{\gamma x}\prod_{n=1}^{\infty}\left(1+\frac{x}{n}\right)e^{-x/n}$$

proved in the previous section and from Theorem 1.2.1 in the form $\Gamma(x)\Gamma(1-x) = -x\Gamma(x)\Gamma(-x) = \pi/\sin \pi x$.

Formula (1.2.5) is the logarithmic derivative of (1.2.4), and (1.2.6) follows from (1.2.5) since $\csc x = \cot \frac{x}{2} - \cot x$. The two formulas (1.2.7) and (1.2.8) are merely variations of (1.2.5) and (1.2.6). Formula (1.2.9) is the derivative of (1.2.5). ∎

It is worth noting that (1.2.6) follows directly from (1.2.1) without the product formula. We have

$$x \csc \pi x = \int_0^{\infty}\frac{t^{x-1}}{1+t}dt = \int_0^1\frac{t^{x-1}}{1+t}dt + \int_1^{\infty}\frac{t^{x-1}}{1+t}dt$$

$$= \int_0^1\frac{t^{x-1}+t^{-x}}{1+t}dt = \int_0^1(t^{x-1}+t^{-x})\left[\sum_{k=0}^{n}(-1)^k t^k + \frac{(-1)^{n+1}t^{n+1}}{1+t}\right]dt$$

$$= \sum_{k=-n}^{n+1}\frac{(-1)^k}{x-k} + R_n,$$

where

$$|R_n| \le \left|\int_0^1(t^{n+x}+t^{n-x+1})dt\right| \le \frac{1}{n+x+1} + \frac{1}{n-x+2}.$$

Thus (1.2.6) has been derived from (1.2.1).

Before going back to the study of the gamma function we note an important consequence of (1.2.5).

Definition 1.2.3   *The Bernoulli numbers B_n are defined by the power series expansion*

$$\frac{x}{e^x - 1} = \sum_{n=0}^{\infty} B_n \frac{x^n}{n!} = 1 - \frac{x}{2} + \sum_{k=1}^{\infty} B_{2k} \frac{x^{2k}}{(2k)!}. \qquad (1.2.10)$$

It is easy to check that $\frac{x}{e^x-1} + \frac{x}{2}$ is an even function. The first few Bernoulli numbers are $B_1 = -1/2$, $B_2 = 1/6$, $B_4 = -1/30$, $B_6 = 1/42$.

Theorem 1.2.4   *For each positive integer k,*

$$\sum_{n=1}^{\infty} \frac{1}{n^{2k}} = \frac{(-1)^{k+1} 2^{2k-1}}{(2k)!} B_{2k} \pi^{2k}. \qquad (1.2.11)$$

Proof.   By (1.2.10)

$$x \cot x = ix \frac{e^{ix} + e^{-ix}}{e^{ix} - e^{-ix}} = ix + \frac{2ix}{e^{2ix} - 1} = 1 - \sum_{k=1}^{\infty} (-1)^{k+1} B_{2k} \frac{2^{2k} x^{2k}}{(2k)!},$$

and (1.2.5) gives the expansion

$$x \cot x = 1 + 2 \sum_{n=1}^{\infty} \frac{x^2}{x^2 - n^2 \pi^2} = 1 - 2 \sum_{n=1}^{\infty} \sum_{k=1}^{\infty} \frac{x^{2k}}{n^{2k} \pi^{2k}}.$$

Now equate the coefficients of x^{2k} in the two series for $x \cot x$ to complete the proof. ∎

Eisenstein [1847] showed that a theory of trigonometric functions could be systematically developed from the partial fractions expansion of $\cot x$, taking (1.2.5) as a starting point. According to Weil [1976, p. 6] this method provides the simplest proofs of a series of important results on trigonometric functions orginally due to Euler. Eisenstein's actual aim was to provide a theory of elliptic functions along similar lines. A very accessible account of this work and its relation to modern number theory is contained in Weil's book. Weil refers to $\lim_{n \to \infty} \sum_{-n}^{n} a_k$ as Eisenstein summation.

Theorem 1.2.2 shows that series of the form

$$\sum_{-\infty}^{\infty} \frac{1}{(x+n)^k},$$

where k is an integer, are related to trigonometric functions. As we shall see next, the "half series"

$$\sum_{0}^{\infty} \frac{1}{(x+n)^k}$$

bears a similar relationship to the gamma function. In fact, one may start the study of the gamma function with these half series.

Theorem 1.2.5

$$\Gamma'(1) = -\gamma, \tag{1.2.12}$$

$$\frac{\Gamma'(x)}{\Gamma(x)} - \frac{\Gamma'(1)}{\Gamma(1)} = -\sum_{k=0}^{\infty} \left(\frac{1}{x+k} - \frac{1}{k+1} \right), \tag{1.2.13}$$

$$\frac{d^2 \log \Gamma(x)}{dx^2} = \sum_{k=0}^{\infty} \frac{1}{(x+k)^2}. \tag{1.2.14}$$

Proof. Take the logarithmic derivative of the product for $1/\Gamma(x)$. This gives

$$\frac{-\Gamma'(x)}{\Gamma(x)} = \gamma + \sum_{k=1}^{\infty} \left(\frac{1}{x+k-1} - \frac{1}{k} \right).$$

The case $x = 1$ gives (1.2.12). The other two formulas follow immediately. ∎

Corollary 1.2.6 Log $\Gamma(x)$ *is a convex function of x for $x > 0$.*

Proof. The right side of (1.2.14) is obviously positive. ∎

Remark The functional equation (1.1.6) and logarithmic convexity can be used to derive the basic results about the gamma function. See Section 1.9.

We denote $\Gamma'(x)/\Gamma(x)$ by $\psi(x)$. This is sometimes called the digamma function. Gauss proved that $\psi(x)$ can be evaluated by elementary functions when x is a rational number. This result is contained in the next theorem.

Theorem 1.2.7

$$\psi(x+n) = \frac{1}{x} + \frac{1}{x+1} + \cdots + \frac{1}{x+n-1} + \psi(x), \quad n = 1, 2, 3, \ldots, \tag{1.2.15}$$

$$\psi\left(\frac{p}{q}\right) = -\gamma - \frac{\pi}{2} \cot \frac{\pi p}{q} - \log q + 2 \sum_{n=1}^{\lfloor q/2 \rfloor} {}' \cos \frac{2\pi np}{q} \log \left(2 \sin \frac{\pi n}{q} \right), \tag{1.2.16}$$

where $0 < p < q$; $\sum'$ means that when q is even the term with index $n = q/2$ is divided by 2. Here $\lfloor q/2 \rfloor$ denotes the greatest integer in $q/2$.

Proof. The first formula is the logarithmic derivative of

$$\Gamma(x+n) = (x+n-1)(x+n-2) \cdots x \Gamma(x).$$

We derive Gauss's formula (1.2.16) by an argument of Jensen [1915–1916] using roots of unity. Begin with Simpson's dissection [1759]:

If $f(x) = \sum_{n=0}^{\infty} a_n x^n$, then

$$\sum_{n=0}^{\infty} a_{kn+m} x^{kn+m} = \frac{1}{k} \sum_{j=0}^{k-1} w^{-jm} f(w^j x),$$

where $w = e^{2\pi i/k}$ is a primitive kth root of unity. This is a consequence of $\sum_{j=0}^{k-1} w^{jm} = 0, m \not\equiv 0 \pmod{k}$. Now by (1.2.13)

$$\psi(p/q) - \psi(1) = \sum_{n=0}^{\infty} \left(\frac{1}{n+1} - \frac{q}{p+nq} \right)$$

$$= \lim_{t \to 1^-} \sum_{n=0}^{\infty} \left(\frac{1}{n+1} - \frac{q}{p+nq} \right) t^{p+nq} =: \lim_{t \to 1^-} s(t)$$

by Abel's continuity theorem for power series. From the series $-\log(1-t) = \sum_{n=1}^{\infty} t^n/n$, and Simpson's dissection with $\omega = e^{2\pi i/q}$, we get

$$s(t) = -t^{p-q} \log(1 - t^q) + \sum_{n=0}^{q-1} \omega^{-np} \log(1 - \omega^n t)$$

$$= -t^{p-q} \log \frac{1-t^q}{1-t} - (t^{p-q} - 1) \log(1-t) + \sum_{n=1}^{q-1} \omega^{-np} \log(1 - \omega^n t).$$

Let $t \to 1^-$ to get

$$\psi(p/q) = -\gamma - \log q + \sum_{n=1}^{q-1} \omega^{-np} \log(1 - \omega^n).$$

Replace p by $q - p$ and add the two expressions to obtain

$$\psi\left(\frac{p}{q}\right) + \psi\left(\frac{q-p}{q}\right) = -2\gamma - 2\log q + 2\sum_{n=1}^{q-1} \cos\left(\frac{2\pi np}{q}\right) \log(1 - \omega^n).$$

The left side is real, so it is equal to the real part of the right side. Thus

$$\psi\left(\frac{p}{q}\right) + \psi\left(\frac{q-p}{q}\right) = -2\gamma - 2\log q + \sum_{n=1}^{q-1} \cos\frac{2\pi np}{q} \log\left(2 - 2\cos\frac{2\pi n}{q}\right).$$

$$(1.2.17)$$

But

$$\psi(x) - \psi(1-x) = \frac{d}{dx} \log \Gamma(x)\Gamma(1-x) = -\pi \cot \pi x.$$

So

$$\psi(p/q) - \psi(1 - (p/q)) = -\pi \cot \pi p/q. \tag{1.2.18}$$

Add this identity to (1.2.17) to get

$$\psi\left(\frac{p}{q}\right) = -\gamma - \frac{\pi}{2}\cot\frac{\pi p}{q} - \log q + \frac{1}{2}\sum_{n=1}^{q-1}\cos\frac{2\pi np}{q}\log\left(2 - 2\cos\frac{2\pi n}{q}\right).$$

$$\tag{1.2.19}$$

But $\cos 2\pi(q - n)/q = \cos 2\pi n/q$, so the sum can be cut in half, going from 1 to $\lfloor q/2 \rfloor$, where $\lfloor x \rfloor$ denotes the greatest integer in x. Thus

$$\psi\left(\frac{p}{q}\right) = -\gamma - \frac{\pi}{2}\cot\frac{\pi p}{q} - \log q + \sum_{n=1}^{\lfloor q/2 \rfloor}{}'\cos\frac{2\pi np}{q}\log\left(2 - 2\cos\frac{2\pi n}{q}\right)$$

$$= -\gamma - \frac{\pi}{2}\cot\frac{\pi p}{q} - \log q + 2\sum_{n=1}^{\lfloor q/2 \rfloor}{}'\cos\frac{2\pi np}{q}\log\left(2\sin\frac{\pi n}{q}\right). \qquad \blacksquare$$

1.3 The Hurwitz and Riemann Zeta Functions

The half series

$$\zeta(x, s) = \sum_{n=0}^{\infty}\frac{1}{(n + x)^s} \quad \text{for } x > 0, \tag{1.3.1}$$

called the Hurwitz zeta function, is of great interest. We have seen its connection with the gamma function for positive integer values of s in the previous section. Here we view the series essentially as a function of s and give a very brief discussion of how the gamma function comes into the picture.

The case $x = 1$ is called the Riemann zeta function and is denoted by $\zeta(s)$. It plays a very important role in the theory of the distribution of primes. The series converges for $\operatorname{Re} s > 1$ and defines an analytic function in that region. It has a continuation to the whole complex plane with a simple pole at $s = 1$. The analytic continuation of $\zeta(s)$ up to $\operatorname{Re} s > 0$ is not difficult to obtain. Write the series for $\zeta(s)$ as a Stieltjes integral involving $\lfloor x \rfloor$. Thus for $\operatorname{Re} s > 1$

$$\sum_{n=1}^{\infty}\frac{1}{n^s} = 1 + \int_{1+}^{\infty}\frac{d\lfloor x \rfloor}{x^s} = 1 + \left.\frac{\lfloor x \rfloor}{x^s}\right|_1^{\infty} + s\int_1^{\infty}\frac{\lfloor x \rfloor dx}{x^{s+1}}$$

$$= 1 + \frac{1}{s-1} + s\int_1^{\infty}\frac{\lfloor x \rfloor - x}{x^{s+1}}dx.$$

The last integral converges absolutely for $\operatorname{Re} s > 0$ and we have the required continuation. The pole at $s = 1$ has residue 1 and, moreover,

$$\lim_{s \to 1} \left\{ \zeta(s) - \frac{1}{s-1} \right\} = 1 + \int_1^\infty \frac{\lfloor x \rfloor - x}{x^2} dx$$

$$= \lim_{n \to \infty} \left(1 + \int_1^n \frac{\lfloor x \rfloor - x}{x^2} dx \right)$$

$$= \lim_{n \to \infty} \left(\sum_{m=1}^n \frac{1}{m} - \log n \right) = \gamma. \tag{1.3.2}$$

The best way to obtain analytic continuation to the rest of the plane is from the functional relation for the zeta function. We state the result here, since the gamma function is also involved. There are several different proofs of this result and we give a nice one due to Hardy [1922], as well as some others, in the exercises. In Chapter 10 we give yet another proof.

Theorem 1.3.1   *For all complex s,*

$$\pi^{-s/2}\Gamma(s/2)\zeta(s) = \pi^{-((1-s)/2)}\Gamma((1-s)/2)\zeta(1-s). \tag{1.3.3}$$

If $s < 0$, then $1 - s > 1$ and the right side provides the value of $\zeta(s)$. This relation was demonstrated by Euler for integer values of s as well as for $s = 1/2$ and $s = 3/2$. He had proofs for integer values of s, using Abel means. An interesting historical discussion is contained in Hardy [1949, pp. 23–26]. The importance of $\zeta(s)$ as a function of a complex variable in studying the distribution of primes was first recognized by Riemann [1859].

The last section contained the result

$$\zeta(2k) = \frac{(-1)^{k-1}2^{2k-1}}{(2k)!} B_{2k}\pi^{2k}.$$

The following corollary is then easy to prove.

Corollary 1.3.2

$$\zeta(1 - 2k) = \frac{-1}{2k}B_{2k}, \quad \zeta(0) = -\frac{1}{2} \quad and \quad \zeta(-2k) = 0 \quad for \ k = 1, 2, 3, \ldots. \tag{1.3.4}$$

Corollary 1.3.3

$$\zeta'(0) = -\frac{1}{2}\log 2\pi. \tag{1.3.5}$$

Proof.   From the functional equation and the fact that

$$\Gamma\left(\frac{1-s}{2}\right) = \frac{2}{1-s}\Gamma\left(\frac{3-s}{2}\right)$$

we have

$$-\zeta(1-s) = \pi^{-s+1/2}\frac{\Gamma(s/2)}{2\Gamma((3-s)/2)}(s-1)\zeta(s). \qquad (1.3.6)$$

Now (1.3.2) implies that $(s-1)\zeta(s) = 1 + \gamma(s-1) + A(s-1)^2 + \cdots$. So take the logarithmic derivative of (1.3.6) to get

$$\frac{\zeta'(1-s)}{\zeta(1-s)} = \log\pi - \frac{1}{2}\psi\left(\frac{1}{2}s\right) - \frac{1}{2}\psi\left(\frac{3-s}{2}\right) - \frac{\gamma + 2A(s-1) + \cdots}{1 + \gamma(s-1) + \cdots}.$$

Set $s = 1$ and use Gauss's result in Theorem 1.2.7 with $p = 1$ and $q = 2$. This proves the corollary. ∎

There is a generalization of the last corollary to the Hurwitz zeta function $\zeta(x, s)$. A functional equation for this function exists, which would define it for all complex s, but we need only the continuation up to some point to the left of Re $s = 0$. This can be done by using the function $\zeta(s)$. Start with the identity

$$\zeta(x, s) - (\zeta(s) - sx\zeta(s+1)) = x^{-s} + \sum_{n=1}^{\infty} n^{-s}[(1+x/n)^{-s} - (1 - sx/n)].$$

The sum on the right converges for Re $s > -1$, and because $\zeta(s)$ is defined for all s, we have the continuation of $\zeta(x, s)$ to Re $s > -1$.

The following theorem is due to Lerch.

Theorem 1.3.4

$$\left(\frac{\partial\zeta(x, s)}{\partial s}\right)_{s=0} = \log\frac{\Gamma(x)}{\sqrt{2\pi}}. \qquad (1.3.7)$$

Proof. The derivative of the equation $\zeta(x+1, s) = \zeta(x, s) - x^{-s}$ with respect to s at $s = 0$ gives

$$\left(\frac{\partial\zeta(x+1, s)}{\partial s}\right)_{s=0} - \left(\frac{\partial\zeta(x, s)}{\partial s}\right)_{s=0} = \log x. \qquad (1.3.8)$$

For Re $s > 1$,

$$\frac{\partial^2\zeta(x, s)}{\partial x^2} = s(s+1)\sum_{n=0}^{\infty}\frac{1}{(n+x)^{s+2}},$$

so

$$\frac{d^2}{dx^2}\left(\frac{\partial\zeta(x, s)}{\partial s}\right)_{s=0} = \sum_{n=0}^{\infty}\frac{1}{(x+n)^2}. \qquad (1.3.9)$$

Now (1.3.8) and (1.3.9) together with (1.2.14) of Theorem 1.2.5 imply that

$$\left(\frac{\partial \zeta(x,s)}{\partial s}\right)_{s=0} = C + \log \Gamma(x).$$

To determine that the constant $C = -\frac{1}{2}\log 2\pi$, set $x = 1$ and use Corollary 1.3.3. This completes the proof of Lerch's theorem. ■

   For a reference to Lerch's paper and also for a slightly different proof of Theorem 1.3.4, see Weil [1976, p. 60].

1.4 Stirling's Asymptotic Formula

De Moivre [1730] found that $n!$ behaves like $Cn^{n+1/2}e^{-n}$ for large n, where C is some constant. Stirling [1730] determined C to be $\sqrt{2\pi}$; de Moivre then used a result of Stirling to give a proof of this claim. See Tweddle [1988, pp. 9–19]. This formula is extremely useful and it is very likely that the reader has seen applications of it. In this section we give an asymptotic formula for $\Gamma(x)$ for $\mathrm{Re}\, x$ large, when $\mathrm{Im}\, x$ is fixed. First note that $\log \Gamma(x+n+1) = \sum_{k=1}^{n} \log(k+x) + \log \Gamma(x+1)$. We then employ the idea that an integral often gives the dominant part of the sum of a series so that if the integral is subtracted from the series the resulting quantity is of a lower order of magnitude than the original series. (We have already used this idea in Equation (1.3.2) of the preceding section.) In Appendix D we prove the Euler–Maclaurin summation formula, a very precise form of this idea when the function being integrated is smooth. Two fuller accounts of the Euler–Maclaurin summation formula are given by Hardy [1949, pp. 318–348] and by Olver [1974, pp. 279–289].

Theorem 1.4.1

$$\Gamma(x) \sim \sqrt{2\pi}\, x^{x-1/2} e^{-x} \quad as \quad \mathrm{Re}\, x \to \infty.$$

Proof.   Denote the right side of the equation

$$\log \Gamma(x+n) = \sum_{k=1}^{n-1} \log(k+x) + \log \Gamma(x+1)$$

by c_n, so that

$$c_{n+1} - c_n = \log(x+n).$$

By the analogy between the derivative and the finite difference we consider c_n to be approximately the integral of $\log(x+n)$ and set

$$c_n = (n+x)\log(n+x) - (n+x) + d_n.$$

Substitute this in the previous equation to obtain

$$\log(x + n) = (n + 1 + x)\log(n + 1 + x) - (n + x)\log(n + x) + d_{n+1} - d_n - 1.$$

Thus

$$d_{n+1} - d_n = 1 - (n + x + 1)\log\left(1 + \frac{1}{n + x}\right)$$

$$= 1 - (n + x + 1)\left[\frac{1}{n + x} - \frac{1}{2(n + x)^2} + \frac{1}{3(n + x)^3} + \cdots\right]$$

$$= -\frac{1}{2(n + x)} + \frac{1}{6(n + x)^2} + \cdots.$$

Proceeding as before, take

$$d_n = e_n - \frac{1}{2}\log(n + x),$$

and substitute in the previous equation to get

$$e_{n+1} - e_n = \frac{1}{2}\log\left(1 + \frac{1}{n + x}\right) - \frac{1}{2(n + x)} + \frac{1}{6(n + x)^2} + \cdots$$

$$= -\frac{1}{12(n + x)^2} + O\left(\frac{1}{(n + x)^3}\right).$$

Now

$$e_n - e_0 = \sum_{k=0}^{n-1}(e_{k+1} - e_k) = \sum_{k=0}^{n-1}\left[-\frac{1}{12(k + x)^2} + O\left(\frac{1}{(k + x)^3}\right)\right]; \quad (1.4.1)$$

therefore, $\lim_{n\to\infty}(e_n - e_0) = K_1(x)$ exists. Set

$$e_n = K(x) + \frac{1}{12(n + x)} + O\left(\frac{1}{(n + x)^2}\right),$$

where $K(x) = K_1(x) + e_0$. The term $(n + x)^{-1}$ comes from completing the sum in (1.4.1) to infinity and approximating the added sum by an integral. So we can write

$$c_n = (n + x)\log(n + x) - (n + x) - \frac{1}{2}\log(n + x)$$

$$+ \log C(x) + \frac{1}{12(n + x)} + O\left(\frac{1}{(n + x)^2}\right),$$

where $K(x) = \log C(x)$. This implies that

$$\Gamma(x+n) = C(x)(n+x)^{n+x-\frac{1}{2}}\exp\left[-(n+x) + \frac{1}{12(n + x)} + O\left(\frac{1}{(n + x)^2}\right)\right].$$

$$(1.4.2)$$

We claim $C(x)$ is independent of x. By the definition of the gamma function

$$\lim_{n\to\infty} \frac{\Gamma(n+x)}{\Gamma(n+y)} n^{y-x} = \frac{\Gamma(x)}{\Gamma(y)} \lim_{n\to\infty} \frac{(x)_n}{(y)_n} n^{y-x} = \frac{\Gamma(x)}{\Gamma(y)} \cdot \frac{\Gamma(y)}{\Gamma(x)} = 1. \qquad (1.4.3)$$

Now, from (1.4.2) and (1.4.3) we can conclude that

$$1 = \lim_{n\to\infty} n^{-x} \frac{\Gamma(n+x)}{\Gamma(n)} = \frac{C(x)}{C(0)} \lim_{n\to\infty} \left(1 + \frac{x}{n}\right)^n e^{-x} = \frac{C(x)}{C(0)}.$$

Thus $C(x)$ is a constant and

$$\Gamma(x) \sim Cx^{x-1/2} e^{-x} \quad \text{as} \quad \operatorname{Re} x \to \infty.$$

To find C, use Wallis's formula:

$$\sqrt{\pi} = \lim_{n\to\infty} \frac{2^{2n} (n!)^2}{(2n)!} \frac{1}{\sqrt{n}}$$

$$= \lim_{n\to\infty} \frac{2^{2n} C^2 n^{2n+1} e^{-2n+O(\frac{1}{n})}}{C(2n)^{2n+\frac{1}{2}} e^{-2n+O(\frac{1}{n})}} \cdot \frac{1}{\sqrt{n}}$$

$$= \frac{C}{\sqrt{2}}.$$

This gives $C = \sqrt{2\pi}$ and proves the theorem. Observe that the proof gives the first term of an error estimate. ∎

We next state a more general result and deduce some interesting consequences. A proof is given in Appendix D. For this we need a definition. The Bernoulli polynomials $B_n(x)$ are defined by

$$\frac{te^{xt}}{e^t - 1} = \sum_{n=0}^{\infty} B_n(x) \frac{t^n}{n!}. \qquad (1.4.4)$$

The Bernoulli numbers are given by $B_n(0) = B_n$ for $n \geq 1$.

Theorem 1.4.2 *For a complex number x not equal to zero or a negative real number,*

$$\log \Gamma(x) = \frac{1}{2} \log 2\pi + \left(x - \frac{1}{2}\right) \log x - x + \sum_{j=1}^{m} \frac{B_{2j}}{(2j-1)2j} \frac{1}{x^{2j-1}}$$

$$- \frac{1}{2m} \int_0^\infty \frac{B_{2m}(t - [t])}{(x+t)^{2m}} dt. \qquad (1.4.5)$$

The value of $\log x$ is the branch with $\log x$ real when x is real and positive.

The expansion of $\log \Gamma(x)$ in (1.4.5) is an asymptotic series since the integral is easily seen to be $O(x^{-2m+1})$ for $|\arg x| \leq \pi - \delta, \delta > 0$.

From this theorem the following corollary is immediately obtained.

Corollary 1.4.3 *For $\delta > 0$ and $|\arg x| \leq \pi - \delta$,*

$$\Gamma(x) \sim \sqrt{2\pi}\, x^{x-1/2} e^{-x} \quad as \quad |x| \to \infty.$$

Corollary 1.4.4 *When $x = a + ib$, $a_1 \leq a \leq a_2$ and $|b| \to \infty$, then*

$$|\Gamma(a + ib)| = \sqrt{2\pi}\,|b|^{a-1/2} e^{-\pi|b|/2}[1 + O(1/|b|)],$$

where the constant implied by O depends only on a_1 and a_2.

Proof. Take $|b| > 1, a > 0$. It is easy to check that the Bernoulli polynomial $B_2 - B_2(t) = t - t^2$. Thus $\frac{1}{2}|B_2 - B_2(t)| \leq \frac{1}{2}|t(1 - t)| \leq \frac{1}{8}$ for $0 \leq t \leq 1$. So (1.4.5) with $m = 1$ is

$$\log \Gamma(a + ib) = \left(a + ib - \frac{1}{2}\right)\log(a + ib) - (a + ib) + \frac{1}{2}\log 2\pi + R(x),$$

and

$$|R(x)| \leq \frac{1}{8}\int_0^\infty \frac{dt}{|t + x|^2} = \frac{1}{8}\int_0^\infty \frac{dt}{(a + t)^2 + b^2} = \frac{1}{8|b|}\tan^{-1}\frac{|b|}{a}, \quad b \neq 0.$$

Now

$$\mathrm{Re}\left[\left(a + ib - \frac{1}{2}\right)\log(a + ib)\right] = \left(a - \frac{1}{2}\right)\log(a^2 + b^2)^{1/2} - b \arctan\frac{b}{a}.$$

Also,

$$\log(a^2 + b^2)^{1/2} = \frac{1}{2}\log b^2 + \frac{1}{2}\log\left(1 + \frac{a^2}{b^2}\right) = \log|b| + O\left(\frac{1}{b^2}\right).$$

Moreover,

$$\arctan\frac{b}{a} + \arctan\frac{a}{b} = \begin{cases} \frac{\pi}{2}, & \text{if } b > 0, \\ -\frac{\pi}{2}, & \text{if } b < 0. \end{cases}$$

This gives

$$-b\arctan\frac{b}{a} = -b\left[\pm\frac{\pi}{2} - \frac{a}{b} + O\left(\frac{1}{b^2}\right)\right]$$

$$= -\frac{\pi}{2}|b| + a + O\left(\frac{1}{b^2}\right).$$

Putting all this together gives

$$\log|\Gamma(a + ib)| = \left(a - \frac{1}{2}\right)\log|b| - \frac{\pi}{2}|b| + \frac{1}{2}\log 2\pi + O\left(\frac{1}{|b|}\right).$$

The condition $a > 0$ is removed by a finite number of uses of the functional equation (1.1.6) and the corollary follows. Observe that the proof only uses $a = o(|b|)$ rather than a bounded.   ∎

Corollary 1.4.5   *For* $|\arg x| \leq \pi - \delta, \delta > 0$,

$$\psi(x) = \log x - \frac{1}{2x} - \sum_{j=1}^{m} \frac{B_{2j}}{(2j)} \frac{1}{x^{2j}} + O\left(\frac{1}{x^{2m}}\right).$$

Corollary 1.4.4 shows that $\Gamma(a + ib)$ decays exponentially in the imaginary direction. This can be anticipated from the reflection formula, for

$$\Gamma\left(\frac{1}{2} + ib\right)\Gamma\left(\frac{1}{2} - ib\right) = \frac{\pi}{\cosh \pi b},$$

or

$$\left|\Gamma\left(\frac{1}{2} + ib\right)\right|^2 = \frac{2\pi}{e^{\pi b} + e^{-\pi b}} \sim 2\pi e^{-\pi|b|} \quad \text{as} \quad b \to \pm\infty,$$

or

$$\left|\Gamma\left(\frac{1}{2} + ib\right)\right| \sim \sqrt{2\pi} e^{-\pi|b|/2} \quad \text{as} \quad b \to \pm\infty.$$

Similarly,

$$\Gamma(ib)\Gamma(-ib) = \frac{\pi}{-ib \sin \pi bi} = \frac{2\pi}{b(e^{\pi b} - e^{-\pi b})}$$

and

$$|\Gamma(ib)| \sim \sqrt{2\pi} |b|^{-1/2} e^{-\pi|b|/2} \quad \text{as} \quad b \to \pm\infty.$$

Since $\Gamma(x)$ increases rapidly on the positive real axis and decreases rapidly in the imaginary direction, there should be curves going to infinity on which a normalized version of $\Gamma(x)$ has a nondegenerate limit. Indeed, there are. See Exercise 18.

1.5 Gauss's Multiplication Formula for $\Gamma(mx)$

The factorization

$$(a)_{2n} = 2^{2n}\left(\frac{a}{2}\right)_n \left(\frac{a+1}{2}\right)_n$$

together with the definition of the gamma function leads immediately to Legendre's duplication formula contained in the next theorem.

Theorem 1.5.1

$$\Gamma(2a)\Gamma\left(\frac{1}{2}\right) = 2^{2a-1}\Gamma(a)\Gamma\left(a + \frac{1}{2}\right). \qquad (1.5.1)$$

This proof suggests that one should consider the more general case: the factorization of $(a)_{mn}$, where m is a positive integer. This gives Guass's formula.

Theorem 1.5.2

$$\Gamma(ma)(2\pi)^{(m-1)/2} = m^{ma-1/2}\Gamma(a)\Gamma\left(a + \frac{1}{m}\right)\cdots\Gamma\left(a + \frac{m-1}{m}\right). \quad (1.5.2)$$

Proof. The same argument almost gives (1.5.2). What it gives is (1.5.2) but with

$$(2\pi)^{\frac{m-1}{2}} m^{-\frac{1}{2}} \quad \text{replaced by} \quad \Gamma\left(\frac{1}{m}\right)\cdots\Gamma\left(\frac{m-1}{m}\right) =: P. \quad (1.5.3)$$

To show that (1.5.3) is true, we show that

$$P^2 = \frac{(2\pi)^{m-1}}{m}.$$

By the reflection formula

$$\Gamma\left(\frac{k}{m}\right)\Gamma\left(1 - \frac{k}{m}\right) = \frac{\pi}{\sin\frac{\pi k}{m}}.$$

So it is enough to prove

$$2^{m-1}\sin\frac{\pi}{m}\sin\frac{2\pi}{m}\cdots\sin\frac{(m-1)\pi}{m} = m.$$

Start with the factorization

$$\frac{x^m - 1}{x - 1} = \prod_{k=1}^{m-1}(x - \exp(2k\pi i/m)).$$

Let $x \to 1$ to obtain

$$m = \prod_{k=1}^{m-1}(1 - \exp(2k\pi i/m))$$

$$= 2^{m-1}\sin\frac{\pi}{m}\sin\frac{2\pi}{m}\cdots\sin\frac{(m-1)\pi}{m}.$$

This proves (1.5.3). ■

Remark 1.5.1 A different proof of (1.5.1) or (1.5.2) that uses the asymptotic formula for $\Gamma(x)$ and the elementary property $\Gamma(x + 1) = x\Gamma(x)$ is also possible. In fact it is easily verifed that

$$g(x) = 2^{2x-1}\frac{\Gamma(x)\Gamma(x + 1/2)}{\Gamma(1/2)\Gamma(2x)}$$

satisfies the relation $g(x + 1) = g(x)$. Stirling's formula implies that $g(x) \sim 1$ as $x \to \infty$ so that $\lim_{n\to\infty} g(x+n) = 1$ when n is an integer. Since $g(x+n) = g(x)$ we can conclude that $g(x) = 1$. A similar proof may be given for Gauss's formula. This is left to the reader.

An elegant proof of the multiplication formula using the integral definition of the gamma function is due to Liouville [1855]. We reproduce it here.

The product of the gamma functions on the right side of (1.5.2) is

$$\int_0^\infty e^{-x_1}x_1^{a-1}dx_1 \int_0^\infty e^{-x_2}x_2^{a+(1/m)-1}dx_2 \cdots \int_0^\infty e^{-x_m}x_m^{a+((m-1)/m)-1}dx_m$$

$$= \int_0^\infty \int_0^\infty \cdots \int_0^\infty e^{-(x_1+x_2+\cdots+x_m)}x_1^{a-1}x_2^{a+(1/m)-1}$$

$$\cdots x_m^{a+((m-1)/m)-1}dx_1dx_2\cdots dx_m.$$

Introduce a change of variables:

$$x_1 = \frac{z^m}{x_2\cdots x_m}, \quad x_2 = x_2, \ldots, x_m = x_m.$$

The Jacobian is easily seen to be

$$\frac{mz^{m-1}}{x_2x_3\cdots x_m}$$

and the integral can be written

$$\int_0^\infty \cdots \int_0^\infty \exp\left[-\left(x_2 + x_3 + \cdots + x_m + \frac{z^m}{x_2x_3\cdots x_m}\right)\right]$$

$$\times \left(\frac{z^m}{x_2\cdots x_m}\right)^{a-1} x_2^{a+(1/m)-1} \cdots x_m^{a+((m-1)/m)-1}\frac{mz^{m-1}}{x_2x_3\cdots x_m}dzdx_2\cdots dx_m.$$

Set $t = x_2 + x_3 + \cdots + x_m + z^m/(x_2x_3\cdots x_m)$, and rewrite the integral as

$$m\int_0^\infty \int_0^\infty \cdots \int_0^\infty e^{-t}z^{ma-1}x_2^{(1/m)-1}x_3^{(2/m)-1} \cdots x_m^{((m-1)/m)-1}dzdx_2\cdots dx_m. \tag{1.5.4}$$

First compute

$$I = \int_0^\infty \cdots \int_0^\infty e^{-t}\prod_{j=1}^{m-1} x_{j+1}^{(j/m)-1}dx_2dx_3\cdots dx_m.$$

Clearly,

$$\frac{dI}{dz} = -mz^{m-1}\int_0^\infty \cdots \int_0^\infty e^{-t}\prod_{j=1}^{m-1} x_{j+1}^{(j/m)-1}\frac{dx_2\cdots dx_m}{x_2\cdots x_m}.$$

Now introduce a change of variables,

$$x_2 = z^m/(x_1x_3\cdots x_m), \quad x_3 = x_3, \ldots, x_m = x_m,$$

and

$$t_1 = x_3 + x_4 + \cdots + x_m + x_1 + z^m/(x_3 \cdots x_m x_1).$$

The Jacobian is

$$J = \frac{-z^m}{x_1^2 x_3 \cdots x_{m-1}},$$

and $\frac{dI}{dz}$ is given by

$$\frac{dI}{dz} = mz^{m-1} \int_0^\infty \cdots \int_0^\infty e^{-t_1}|J|\left(\frac{z^m}{x_1 x_3 \cdots x_m}\right)^{(1/m)-1}$$

$$\cdot \prod_{j=2}^{m-1} x_{j+1}^{(j/m)-1} \frac{dx_1 dx_3 \cdots dx_m}{z^m/x_1}$$

$$= -m \int_0^\infty \cdots \int_0^\infty e^{-t_1} \prod_{j=2}^{m-1} x_{j+1}^{(j/m)-1} x_1^{((m-1)/m)-1} dx_3 \cdots dx_m dx_1$$

$$= -mI.$$

Therefore,

$$I = Ce^{-mz}.$$

To find C, set $z = 0$ in the integral for I as well as in the above equation and equate to get

$$\Gamma\left(\frac{1}{m}\right)\Gamma\left(\frac{2}{m}\right) \cdots \Gamma\left(\frac{m-1}{m}\right) = C.$$

By (1.5.3), $C = (2\pi)^{(m-1)/2} m^{-1/2}$ and $I = (2\pi)^{(m-1)/2} m^{-1/2} e^{-mz}$. Substitution in (1.5.4) gives

$$\Gamma(a)\Gamma(a + 1/m) \cdots \Gamma(a + (m-1)/m) = m^{1/2}(2\pi)^{(m-1)/2} \int_0^\infty e^{-mz} z^{ma-1} dz$$

$$= m^{1/2-ma}(2\pi)^{\frac{m-1}{2}} \Gamma(ma),$$

which is Gauss's formula.

Remark 1.5.2 We pointed out earlier that $1/\Gamma(x)$ is a half of $\sin \pi x$. In this sense the duplication formula is the analog of the double angle formula

$$\sin 2\pi x = 2 \sin \pi x \sin \pi \left(x + \frac{1}{2}\right).$$

This is usually written as $\sin 2\pi x = 2 \sin \pi x \cos \pi x$ and so is thought of as a special case of the addition for $\sin(x + y)$. The gamma function does not have an addition formula.

1.6 Integral Representations for Log $\Gamma(x)$ and $\psi(x)$

In (1.2.13), we obtained

$$\psi(x) - \psi(1) = \sum_{k=0}^{\infty} \left(\frac{1}{k+1} - \frac{1}{x+k} \right) = \sum_{k=0}^{\infty} \frac{x-1}{(k+1)(x+k)}$$

from the product for $1/\Gamma(x)$. We start this section by rederiving it from the beta integral. Note that, for $x > 1$,

$$-(x-1) \int_0^{1-\epsilon} t^{x-2} \log(1-t)dt = \sum_{k=0}^{\infty} \frac{(x-1)(1-\epsilon)^{x+k}}{(k+1)(x+k)}$$

by term-by-term integration, which is valid because of uniform convergence in $[0, 1 - \epsilon]$. Now let $\epsilon \to 0$. By Abel's continuity theorem for power series,

$$-(x-1) \int_0^1 t^{x-2} \log(1-t)dt = \sum_{k=0}^{\infty} \frac{(x-1)}{(k+1)(x+k)}.$$

We can introduce $\log(1-t)$ in the beta integral $\int_0^1 t^{x-2}(1-t)^y dt$ by taking the derivative with respect to y. In that case,

$$(x-1)\frac{\partial}{\partial y} \int_0^1 t^{x-2}(1-t)^y dt = \frac{\partial}{\partial y} \frac{\Gamma(x)\Gamma(y+1)}{\Gamma(x+y)},$$

or

$$(x-1) \int_0^1 t^{x-2}(1-t)^y \log(1-t)dt$$

$$= \frac{\Gamma(x)\Gamma'(y+1)\Gamma(x+y) - \Gamma(x)\Gamma(y+1)\Gamma'(x+y)}{\Gamma(x+y)^2}.$$

The case $y = 0$ gives the necessary result. The differentiation is justified since the integrands involved are continuous. Some care should also be taken of the fact that the integrals are improper. The details are easy and left to the reader. The next theorem gives the integral representations of $\psi(x)$ due to Dirichlet and Gauss.

Theorem 1.6.1   *For* $\operatorname{Re} x > 0$,

(i)
$$\psi(x) = \int_0^{\infty} \frac{1}{z} \left(e^{-z} - \frac{1}{(1+z)^x} \right) dz \qquad \text{(Dirichlet)},$$

(ii)
$$\psi(x) = \int_0^{\infty} \left(\frac{e^{-z}}{z} - \frac{e^{-xz}}{1 - e^{-z}} \right) dz \qquad \text{(Gauss)}.$$

Proof. (i) Evaluate the integral $\int_0^\infty \int_1^s e^{-tz} dt\, dz$ in two different ways by changing the order of integration to get the formula

$$\int_0^\infty \frac{e^{-z} - e^{-sz}}{z} dz = \log s. \tag{1.6.1}$$

Similarly, the double integral

$$\int_0^\infty \int_0^\infty s^{x-1} \frac{e^{-s-z} - e^{-s(1+z)}}{z} ds\, dz$$

when first integrated with respect to z yields (by (1.6.1))

$$\int_0^\infty e^{-s} s^{x-1} \log s\, ds = \frac{d}{dx} \int_0^\infty e^{-s} s^{x-1} ds = \Gamma'(x).$$

If we integrate the double integral with respect to s we get

$$\Gamma(x) \int_0^\infty \frac{1}{z} \left(e^{-z} - \frac{1}{(1+z)^x} \right) dz.$$

Equate the last two expressions to get Dirichlet's formula.

(ii) Gauss's formula is obtained from Dirichlet's by a change of variables:

$$\psi(x) = \lim_{\delta \to 0^+} \left(\int_\delta^\infty \frac{e^{-z}}{z} dz - \int_\delta^\infty \frac{dz}{z(1+z)^x} \right)$$

$$= \lim_{\delta \to 0^+} \left(\int_\delta^\infty \frac{e^{-z}}{z} dz - \int_{\log(1+\delta)}^\infty \frac{e^{-tx}}{1 - e^{-t}} dt \right)$$

$$= \lim_{\delta \to 0^+} \left\{ \int_\delta^{\log(1+\delta)} \frac{e^{-z}}{z} dz + \int_{\log(1+\delta)}^\infty \left(\frac{e^{-t}}{t} - \frac{e^{-tx}}{1 - e^{-t}} \right) dt \right\}$$

$$= \int_0^\infty \left(\frac{e^{-z}}{z} - \frac{e^{-xz}}{1 - e^{-z}} \right) dz,$$

since

$$\left| \int_\delta^{\log(1+\delta)} \frac{e^{-z}}{z} dz \right| < \int_{\log(1+\delta)}^\delta \frac{1}{z} dz = \log \frac{\delta}{\log(1+\delta)} \to 0 \quad \text{as} \quad \delta \to 0.$$

This proves (ii). ∎

The integrated form of the last theorem is given in the next result.

Theorem 1.6.2 *For* $\operatorname{Re} x > 0$,

(i) $$\log \Gamma(x) = \int_0^\infty \left((x - 1)e^{-t} - \frac{(1+t)^{-1} - (1+t)^{-x}}{\log(1+t)} \right) \frac{dt}{t}$$

and

(ii) $$\log \Gamma(x) = \int_0^\infty \left((x-1)e^{-t} - \frac{e^{-t} - e^{-xt}}{1 - e^{-t}} \right) \frac{dt}{t}.$$

Proof. The integrals in Theorem 1.6.1 are uniformly convergent for $\operatorname{Re} x \geq \delta > 0$, so we can integrate from 1 to x under the sign of integration. The integrals in Theorem 1.6.2 are the corresponding integrated forms.

A change of variables $u = e^{-t}$ in (ii) gives

$$\log \Gamma(x) = \int_0^1 \left(\frac{1 - u^{x-1}}{1 - u} - x + 1 \right) \frac{du}{\log u}. \tag{1.6.2}$$

There are two other integrals for $\log \Gamma(x)$ due to Binet that are of interest. These are given in the next theorem. A proof of one of them is sketched and the other is left as an exercise. See Exercise 43. ∎

Theorem 1.6.3   *For* $\operatorname{Re} x > 0$,

(i) $$\log \Gamma(x) = \left(x - \frac{1}{2} \right) \log x - x + \frac{1}{2} \log 2\pi + \int_0^\infty \left(\frac{1}{2} - \frac{1}{t} + \frac{1}{e^t - 1} \right) \frac{e^{-tx}}{t} dt$$

and

(ii) $$\log \Gamma(x) = \left(x - \frac{1}{2} \right) \log x - x + \frac{1}{2} \log 2\pi + 2 \int_0^\infty \frac{\arctan(t/x)}{e^{2\pi t} - 1} dt.$$

Proof.   Gauss's formula in Theorem 1.6.1 together with Equation (1.6.1) give

$$\psi(x+1) = \frac{d}{dx} \log \Gamma(x+1) = \frac{1}{2x} + \log x - \int_0^\infty \left(\frac{1}{2} - \frac{1}{t} + \frac{1}{e^t - 1} \right) e^{-tx} dt.$$

Integrate from 1 to x, changing the order of integration to get

$$\log \Gamma(x+1) = \left(x + \frac{1}{2} \right) \log x - x + 1 + \int_0^\infty \left(\frac{1}{2} - \frac{1}{t} + \frac{1}{e^t - 1} \right) \frac{e^{-tx} - e^{-t}}{t} dt.$$

Use $\log \Gamma(x+1) = \log \Gamma(x) + \log x$ to rewrite the above formula as

$$\log \Gamma(x) = \left(x - \frac{1}{2} \right) \log x - x + 1 + \int_0^\infty \left(\frac{1}{2} - \frac{1}{t} + \frac{1}{e^t - 1} \right) \frac{e^{-tx}}{t} dt - I,$$

where

$$I = \int_0^\infty \left(\frac{1}{2} - \frac{1}{t} + \frac{1}{e^t - 1} \right) \frac{e^{-t}}{t} dt. \tag{1.6.3}$$

Stirling's formula applied above gives $I = 1 - (1/2) \log 2\pi$.

The second Binet formula can be used to derive the asymptotic expansion for $\log \Gamma(x)$ contained in Corollary 1.4.5.

Expand $1/(e^{2\pi t} - 1)$ by the geometric series and integrate term by term to see that

$$\int_0^\infty \frac{t^{2k-1}}{e^{2\pi t} - 1} dt = \frac{\Gamma(2k)\zeta(2k)}{(2\pi)^{2k}} = (-1)^{k-1} \frac{B_{2k}}{4k}. \tag{1.6.4}$$

The last equality comes from Theorem 1.2.4. Now,

$$\frac{1}{1+z^2} = 1 - z^2 + z^4 - \cdots + (-1)^{n-1} z^{2n-2} + (-1)^n \frac{z^{2n}}{1+z^2}$$

gives, after integration,

$$\arctan(t/x) = \frac{t}{x} - \frac{1}{3}\frac{t^3}{x^3} + \frac{1}{5}\frac{t^5}{x^5} \cdots + \frac{(-1)^{n-1}}{2n-1}\frac{t^{2n-1}}{x^{2n-1}} + \frac{(-1)^n}{x^{2n-1}} \int_0^t \frac{z^{2n} dz}{x^2 + z^2}.$$

Substitute this in Binet's formula (ii) and use (1.6.4) to arrive at

$$\log \Gamma(x) = \left(x - \frac{1}{2}\right) \log x - x + \frac{1}{2} \log 2\pi + \sum_{j=1}^n \frac{B_{2j}}{2j(2j-1)x^{2j-1}}$$

$$+ \frac{2(-1)^n}{x^{2n-1}} \int_0^\infty \left(\int_0^t \frac{z^{2n}}{x^2 + z^2} dz\right) \frac{dt}{e^{2\pi t} - 1}.$$

For $|\arg x| \le \frac{\pi}{2} - \epsilon, \epsilon > 0$, it can be seen that $|\frac{x^2}{x^2+z^2}| \le \csc z\epsilon$ for all $z \ge 0$. This implies that the last term involving the integral is $O(\frac{1}{|x|^{2n+1}})$. So we have the asymptotic series but only for $|\arg x| \le \frac{\pi}{2} - \epsilon$ instead of $|\arg x| \le \pi - \epsilon$. Whittaker and Watson [1940, §13.6] show how to extend the range of validity. It is also possible to derive an asymptotic formula for $\log \Gamma(x)$ from Binet's first formula. See Wang and Guo [1989, §3.12]. For references to the works of Gauss, Dirichlet, Binet, and others, see Whittaker and Watson [1940, pp. 235–259]. ■

1.7 Kummer's Fourier Expansion of Log Γ(x)

Kummer [1847] discovered the following theorem:

Theorem 1.7.1 *For $0 < x < 1$*

$$\log \frac{\Gamma(x)}{\sqrt{2\pi}} = -\frac{1}{2} \log(2 \sin \pi x) + \frac{1}{2}(\gamma + \log 2\pi)(1 - 2x) + \frac{1}{\pi} \sum_{k=1}^\infty \frac{\log k}{k} \sin 2\pi kx,$$

where γ is Euler's constant.

Proof. Start with the identity

$$-\log(1 - e^{2\pi ix}) = e^{2\pi ix} + \frac{e^{4\pi ix}}{2} + \frac{e^{6\pi ix}}{3} + \cdots, \quad 0 < x < 1.$$

The real and imaginary parts are

$$- \log(2 \sin \pi x) = \sum_{k=1}^{\infty} \frac{\cos 2k\pi x}{k} \tag{1.7.1}$$

and

$$\frac{\pi}{2}(1 - 2x) = \sum_{k=1}^{\infty} \frac{\sin 2k\pi x}{k}. \tag{1.7.2}$$

Since $\log \Gamma(x)$ is differentiable in $0 < x < 1$, it has a Fourier expansion

$$\log \Gamma(x) = C_0 + 2 \sum_{k=1}^{\infty} C_k \cos 2k\pi x + 2 \sum_{k=1}^{\infty} D_k \sin 2k\pi x,$$

where

$$C_k = \int_0^1 \log \Gamma(x) \cos 2k\pi x \, dx \quad \text{and} \quad D_k = \int_0^1 \log \Gamma(x) \sin 2k\pi x \, dx. \tag{1.7.3}$$

We use Kummer's method to compute C_k and D_k. The C_k are easy to find. Take the logarithm of Euler's reflection formula (1.2.1):

$$\log \Gamma(x) + \log \Gamma(1 - x) = \log 2\pi - \log(2 \sin \pi x)$$
$$= \log 2\pi + \cos 2\pi x + \frac{1}{2} \cos 4\pi x + \cdots .$$

The Fourier series of $\log \Gamma(x)$ gives

$$\log \Gamma(x) + \log \Gamma(1 - x) = 2C_0 + 4C_1 \cos 2\pi x + 4C_2 \cos 4\pi x + \cdots .$$

Equating the last two relations gives

$$C_0 = \frac{1}{2} \log 2\pi \quad \text{and} \quad C_k = \frac{1}{4k} \quad \text{for } k \geq 1.$$

Now use integral (1.6.2) for $\log \Gamma(x)$ in (1.7.3) so that

$$D_k = \int_0^1 \int_0^1 \left(\frac{1 - u^{x-1}}{1 - u} - x + 1 \right) \frac{\sin 2k\pi x \, du \, dx}{\log u}.$$

But

$$\int_0^1 \sin 2k\pi x \, dx = 0, \qquad \int_0^1 x \sin 2k\pi x \, dx = -\frac{1}{2k\pi},$$

and

$$\int_0^1 u^{x-1} \sin 2k\pi x \, dx = \frac{(1 - u)2k\pi}{u((\log u)^2 + 4k^2\pi^2)}.$$

The first two integrals are easy to solve and the third is the imaginary part of

$$\frac{1}{u} \int_0^1 e^{x(\log u + 2k\pi i)} dx = \frac{1}{u} \cdot \frac{u - 1}{\log u + 2k\pi i}.$$

Therefore,

$$D_k = \int_0^1 \left(\frac{-2k\pi}{u((\log u)^2 + 4k^2\pi^2)} + \frac{1}{2k\pi} \right) \frac{du}{\log u},$$

or, with $u = e^{-2k\pi t}$,

$$D_k = \frac{1}{2k\pi} \int_0^\infty \left(\frac{1}{1+t^2} - e^{-2k\pi t} \right) \frac{dt}{t}.$$

Take $k = 1$ and we have

$$D_1 = \frac{1}{2\pi} \int_0^\infty \left(\frac{1}{1+t^2} - e^{-2\pi t} \right) \frac{dt}{t}.$$

Moreover, $x = 1$ in Dirichlet's formula (Theorem 1.6.1) gives

$$-\frac{\gamma}{2\pi} = \frac{1}{2\pi} \int_0^\infty \left(e^{-t} - \frac{1}{1+t} \right) \frac{dt}{t},$$

where γ is Euler's constant. Therefore,

$$D_1 - \frac{\gamma}{2\pi} = \frac{1}{2\pi} \int_0^\infty \frac{e^{-t} - e^{-2\pi t}}{t} dt + \frac{1}{2\pi} \int_0^\infty \left(\frac{1}{1+t^2} - \frac{1}{1+t} \right) \frac{dt}{t}.$$

By (1.6.1), the first integral is $\log 2\pi$ and a change of variables from t to $1/t$ shows that the second integral is 0. Thus

$$D_1 = \frac{\gamma}{2\pi} + \frac{1}{2\pi} \log 2\pi.$$

To find D_k, observe that

$$kD_k - D_1 = \frac{1}{2\pi} \int_0^\infty \frac{e^{-2\pi t} - e^{-2k\pi t}}{t} dt = \frac{1}{2\pi} \log k,$$

where the integral is once again evaluated by (1.6.1). Thus

$$D_k = \frac{1}{2k\pi} (\gamma + \log 2k\pi), \quad k = 1, 2, 3, \ldots.$$

The Fourier expansion is then

$$\log \Gamma(x) = \frac{1}{2} \log 2\pi + \sum_{k=1}^\infty \frac{\cos 2\pi kx}{2k} + \frac{1}{\pi}(\gamma + \log 2\pi) \sum_{k=1}^\infty \frac{\sin 2k\pi x}{2k}$$

$$+ \frac{1}{\pi} \sum_{k=1}^\infty \frac{\log k}{k} \sin 2k\pi x.$$

Apply (1.7.1) and (1.7.2) to get the result. ∎

Kummer's expansion for $\log (\Gamma(x)/\sqrt{2\pi})$ and Theorem 1.3.4 have applications in number theory. Usually they give different ways of deriving the same result. This suggests that the Hurwitz zeta function itself has a Fourier expansion from which Kummer's result can be obtained. Such a result exists and is simply the functional equation for the Hurwitz function:

$$\zeta(x, s) = \frac{2\Gamma(1-s)}{(2\pi)^{1-s}} \left\{ \sin \frac{1}{2}\pi s \sum_{m=1}^{\infty} \frac{\cos 2m\pi x}{m^{1-s}} + \cos \frac{1}{2}\pi s \sum_{m=1}^{\infty} \frac{\sin 2m\pi x}{m^{1-s}} \right\}. \quad (1.7.4)$$

The functional equation for the Riemann zeta function is a particular case of this when $x = 1$. See Exercises 24 and 25 for a proof of (1.7.4) and another derivation of Kummer's formula.

1.8 Integrals of Dirichlet and Volumes of Ellipsoids

Dirichlet found a multidimensional extension of the beta integral which is useful in computing volumes. We follow Liouville's exposition of Dirichlet's work. Liouville's [1839] presentation was inspired by the double integral evaluation of the beta function by Jacobi and Poisson.

Theorem 1.8.1 *If V is a region defined by $x_i \geq 0$, $i = 1, 2, \ldots, n$, and $\sum x_i \leq 1$, then for* $\operatorname{Re} \alpha_i > 0$,

$$\int \cdots \int_V x_1^{\alpha_1-1} x_2^{\alpha_2-1} \cdots x_n^{\alpha_n-1} dx_1 \cdots dx_n = \frac{\prod_{i=1}^{n} \Gamma(\alpha_i)}{\Gamma(1+\sum \alpha_i)}.$$

Proof. The proof is by induction. The formula is clearly true for $n = 1$. Assume it is true for $n = k$. Then for a $(k + 1)$-dimensional V

$$\int \cdots \int_V x_1^{\alpha_1-1} x_2^{\alpha_2-1} \cdots x_{k+1}^{\alpha_{k+1}-1} dx_1 dx_2 \cdots dx_{k+1}$$

$$= \int_0^1 \int_0^{1-x_1} \cdots \int_0^{1-x_1-x_2-\cdots-x_k} x_1^{\alpha_1-1} x_2^{\alpha_2-1} \cdots x_{k+1}^{\alpha_{k+1}-1} dx_{k+1} \cdots dx_1$$

$$= \frac{1}{\alpha_{k+1}} \int_0^1 \cdots \int_0^{1-x_1-\cdots-x_{k-1}} x_1^{\alpha_1-1} \cdots x_k^{\alpha_k-1}$$

$$\cdot (1 - x_1 - \cdots - x_k)^{\alpha_{k+1}} dx_k dx_{k-1} \cdots dx_1 .$$

Now set $x_k = (1 - x_1 - \cdots - x_{k-1})t$ to get

$$\frac{1}{\alpha_{k+1}} \int_0^1 \int_0^{1-x_1} \cdots \int_0^{1-x_1-\cdots-x_{k-2}} \int_0^1 x_1^{\alpha_1-1} \cdots x_{k-1}^{\alpha_{k-1}-1}$$

$$\cdot (1 - x_1 - \cdots - x_{k-1})^{\alpha_k+\alpha_{k+1}} t^{\alpha_k-1} (1 - t)^{\alpha_{k+1}} dt\, dx_{k-1} \cdots dx_1$$

$$= \frac{\Gamma(\alpha_k)\Gamma(\alpha_{k+1}+1)}{\alpha_{k+1}\Gamma(\alpha_k+\alpha_{k+1}+1)} \int_0^1 \int_0^{1-x_1} \cdots \int_0^{1-x_1-\cdots-x_{k-2}}$$

$$\cdot x_1^{\alpha_1-1} \cdots x_{k-1}^{\alpha_{k-1}-1} (1 - x_1 - \cdots - x_{k-1})^{\alpha_k+\alpha_{k+1}} dx_{k-1} \cdots dx_1.$$

Compare this with the integral to which the change of variables was applied and use induction to get

$$\frac{\Gamma(\alpha_k)\Gamma(\alpha_{k+1}+1)}{\alpha_{k+1}\Gamma(\alpha_k+\alpha_{k+1}+1)} \cdot \frac{(\alpha_k+\alpha_{k+1})\left(\prod_{i=1}^{k-1}\Gamma(\alpha_i)\right)\Gamma(\alpha_k+\alpha_{k+1})}{\Gamma\left(1+\sum_{i=1}^{k+1}\alpha_i\right)}.$$

This reduces to the expression in the theorem. ∎

Corollary 1.8.2 *If V is the region enclosed by $x_i \geq 0$ and $\sum(x_i/a_i)^{p_i} \leq 1$, then*

$$\int\cdots\int_V x_1^{\alpha_1-1}x_2^{\alpha_2-1}\cdots x_n^{\alpha_n-1}dx_1\cdots dx_n = \frac{\prod\left(a_i^{\alpha_i}/p_i\right)\Gamma((\alpha_i/p_i))}{\Gamma\left(1+\sum(\alpha_i/p_i)\right)}.$$

Proof. Apply the change of variables, $y_i = (x_i/a_i)^{p_i}$, $i = 1, \ldots, n$. Then

$$\frac{\partial x_i}{\partial y_i} = \frac{1}{p_i}\frac{x_i}{y_i}$$

and the Jacobian is

$$\frac{1}{p_1 p_2 \cdots p_n} \cdot \frac{x_1 x_2 \cdots x_n}{y_1 y_2 \cdots y_n}.$$

The integral becomes

$$\frac{a_1^{\alpha_1} a_2^{\alpha_2} \cdots a_n^{\alpha_n}}{p_1 p_2 \cdots p_n} \int\cdots\int_{\tilde{V}} y_1^{(\alpha_1/p_1)-1} \cdots y_n^{(\alpha_n/p_n)-1} dy_1 dy_2 \cdots dy_n,$$

where $\tilde{V}$ is defined by $y_i \geq 0$ and $\sum y_i \leq 1$. The corollary now follows from the theorem. ∎

Corollary 1.8.3 *The volume enclosed by $\sum(x_i/a_i)^{p_i} \leq 1$, $x_i \geq 0$ is $\frac{\prod_{i=1}^{n} a_i \Gamma(1+1/p_i)}{\Gamma(1+\sum 1/p_i)}$.*
In particular the volume of the n-dimensional ellipsoid $\sum(x_i/a_i)^2 \leq 1$ is

$$\frac{\pi^{n/2} a_1 a_2 \cdots a_n}{\Gamma(1+n/2)}.$$

Proof. For the first part of the corollary take $\alpha_i = 1$. For the particular case take $p_i = 2$ and use the fact that $\Gamma(\frac{3}{2}) = \frac{1}{2}\sqrt{\pi}$. ∎

Corollary 1.8.4 *If V is given by $x_i \geq 0$ and $\sum(\frac{x_i}{a_i})^{p_i} \leq \lambda$ in Dirichlet's integral, then its value is*

$$\lambda^{\sum(\alpha_i/p_i)} \frac{\prod\left(a_i^{\alpha_i}/p_i\right)\Gamma(\alpha_i/p_i)}{\Gamma\left(1+\sum(\alpha_i/p_i)\right)}.$$

Liouville also gave the following extension of Dirichlet's result, which can be proven in the same way.

Theorem 1.8.5   *If V consists of $x_i \geq 0$, $t_1 \leq \sum(x_i/a_i)^{p_i} \leq t_2$ and f is a continuous function on (t_1, t_2), then*

$$\int \cdots \int_V x_1^{\alpha_1-1} \cdots x_n^{\alpha_n-1} f\{(x_1/a_1)^{p_1} + \cdots + (x_n/a_n)^{p_n}\} dx_1 \cdots dx_n$$

$$= \frac{\Pi a_i^{\alpha_i} \Gamma(\alpha_i/p_i)/p_i}{\Gamma(\Sigma \alpha_i/p_i)} \int_{t_1}^{t_2} u^{\Sigma(\alpha_i/p_i)-1} f(u) du.$$

A related integral is given next.

Theorem 1.8.6   *If V is the set $x_i \geq 0$, $\sum_{i=1}^{n} x_i = 1$, then*

$$\int \cdots \int_V x_1^{\alpha_1-1} \cdots x_n^{\alpha_n-1} dx_1 \cdots dx_{n-1} = \frac{\Pi \Gamma(\alpha_i)}{\Gamma(\Sigma \alpha_i)}.$$

This is a surface integral rather than a volume integral, but it can be evaluated directly by induction or from Corollary 1.8.2. It is also a special case of Theorem 1.8.5 when $f(u)$ is taken to be the delta function at $u = 1$. This function is not continuous, but it can be approximated by continuous functions.

1.9 The Bohr–Mollerup Theorem

The problem posed by Euler was to find a continuous function of $x > 0$ that equaled $n!$ at $x = n$, an integer. Clearly, the gamma function is not the unique solution to this problem. The condition of convexity (defined below) is not enough, but the fact that the gamma function occurs so frequently gives some indication that it must be unique in some sense. The correct conditions for uniqueness were found by Bohr and Mollerup [1922]. In fact, the notion of logarithmic convexity was extracted from their work by Artin [1964] (the original German edition appeared in 1931) whose treatment we follow here.

Definition 1.9.1   *A real valued function f on (a, b) is convex if*

$$f(\lambda x + (1 - \lambda)y) \leq \lambda f(x) + (1 - \lambda) f(y)$$

for $x, y \in (a, b)$ and $0 < \lambda < 1$.

Definition 1.9.2   *A positive function f on (a, b) is logarithmically convex if $\log f$ is convex on (a, b).*

It is easy to verify that if f is convex in (a, b) and $a < x < y < z < b$, then

$$\frac{f(y) - f(x)}{y - x} \leq \frac{f(z) - f(x)}{z - x} \leq \frac{f(z) - f(y)}{z - y}. \tag{1.9.1}$$

With these definitions we can state the Bohr–Mollerup theorem:

Theorem 1.9.3 *If f is a positive function on $x > 0$ and (i) $f(1) = 1$, (ii) $f(x + 1) = xf(x)$, and (iii) f is logarithmically convex, then $f(x) = \Gamma(x)$ for $x > 0$.*

Proof. Suppose n is a positive integer and $0 < x < 1$. By conditions (i) and (ii) it is sufficient to prove the theorem for such x. Consider the intervals $[n, n + 1]$, $[n + 1, n + 1 + x]$, and $[n + 1, n + 2]$. Apply (1.9.1) to see that the difference quotient of $\log f(x)$ on these intervals is increasing. Thus

$$\log \frac{f(n + 1)}{f(n)} \leq \frac{1}{x} \log \frac{f(n + 1 + x)}{f(n + 1)} \leq \log \frac{f(n + 2)}{f(n + 1)}.$$

Simplify this by conditions (i) and (ii) to get

$$x \log n \leq \log \left[\frac{(x + n)(x + n - 1) \cdots xf(x)}{n!} \right] \leq x \log(n + 1).$$

Rearrange the inequalities as follows:

$$0 \leq \log \frac{x(x + 1) \cdots (x + n)}{n! n^x} + \log f(x) \leq x \log \left(1 + \frac{1}{n} \right).$$

Therefore,

$$f(x) = \lim_{n \to \infty} \frac{n! n^x}{x(x + 1) \cdots (x + n)} = \Gamma(x),$$

and the theorem is proved. ∎

This theorem can be made the basis for the development of the theory of the gamma and beta functions. As examples, we show how to derive the formulas

$$\Gamma(x) = \int_0^\infty e^{-t} t^{x-1} dt, \quad x > 0,$$

and

$$\int_0^1 t^{x-1}(1 - t)^{y-1} dt = \frac{\Gamma(x)\Gamma(y)}{\Gamma(x + y)}, \quad x > 0 \quad \text{and} \quad y > 0. \tag{1.9.2}$$

We require Hölder's inequality, a proof of which is sketched in Exercise 6. We state the inequality here for the reader's convenience. If f and g are measurable nonnegative functions on (a, b), so that the integrals on the right in (1.9.3) are finite, and p and q positive real numbers such that $1/p + 1/q = 1$, then

$$\int_a^b fg\, dx \leq \left(\int_a^b f^p dx \right)^{1/p} \left(\int_a^b g^q dx \right)^{1/q}. \tag{1.9.3}$$

It is clear that we need to check only condition (iii) for $\log \Gamma(x)$. This condition can be written as

$$\Gamma(\alpha x + \beta y) \leq \Gamma(x)^\alpha \Gamma(y)^\beta, \quad \alpha > 0, \beta > 0 \quad \text{and} \quad \alpha + \beta = 1. \tag{1.9.4}$$

Now observe that

$$\Gamma(\alpha x + \beta y) = \int_0^\infty (e^{-t}t^{x-1})^\alpha (e^{-t}t^{y-1})^\beta dt$$

and apply Hölder's inequality with $\alpha = 1/p$ and $\beta = 1/q$ to get (1.9.4).

To prove (1.9.2) consider the function

$$f(x) = \frac{\Gamma(x+y)B(x, y)}{\Gamma(y)}.$$

Once again we require the functional relation (1.1.14) for $B(x, y)$. This is needed to prove that $f(x+1) = xf(x)$. It is evident that $f(1) = 1$ and we need only check the convexity of $\log f(x)$. The proof again uses Hölder's inequality in exactly the same way as for the gamma function.

We state another uniqueness theorem, the proof of which is left to the reader.

Theorem 1.9.4 *If $f(x)$ is defined for $x > 0$ and satisfies (i) $f(1) = 1$, (ii) $f(x + 1) = xf(x)$, and (iii) $\lim_{n\to\infty} f(x+n)/[n^x f(n)] = 1$, then $f(x) = \Gamma(x)$.*

For other uniqueness theorems the reader may consult Artin [1964] or Anastassiadis [1964]. See Exercises 26–30 at the end of the chapter. Finally, we note that Ahern and Rudin [1996] have shown that $\log |\Gamma(x + iy)|$ is a convex function of x in $\operatorname{Re} x \geq 1/2$. See Exercise 55.

1.10 Gauss and Jacobi Sums

The integral representation of the gamma function is

$$\frac{\Gamma(x)}{c^x} = \int_0^\infty e^{-ct} t^x \frac{dt}{t}.$$

Here dt/t should be regarded as the invariant measure on the multiplicative group $(0, \infty)$, since

$$\frac{d(ct)}{ct} = \frac{dt}{t}.$$

To find the finite field analog one should, therefore, look at the integrand $e^{-ct}t^x$. The functions e^{-ct} and t^x can be viewed as solutions of certain functional relations. This point of view suggests the following analogs.

Theorem 1.10.1 *Suppose f is a homomorphism from the additive group of real numbers R to the multiplicative group of nonzero complex numbers $\mathbb{C}^*$, that is,*

$$f : R \to \mathbb{C}^*$$

and

$$f(x + y) = f(x)f(y). \tag{1.10.1}$$

If f is differentiable with $f'(0) = c \neq 0$, then $f(x) = e^{cx}$.

Remark 1.10.1   We have assumed that $f(x) \neq 0$ for any x but, in fact, the relation $g(x + y) = g(x)g(y)$, where $g : R \to \mathbb{C}$, implies that if g is zero at one point it vanishes everywhere.

Proof.   First observe that $f(0+0) = f(0)^2$ by (1.10.1). So $f(0) = 1$, since $f(0)$ cannot be 0. Now, by the definition of the derivative,

$$f'(x) = \lim_{t \to 0} \frac{f(x + t) - f(x)}{t} = \lim_{t \to 0} \frac{f(x)f(t) - f(x)}{t}$$

$$= f(x) \lim_{t \to 0} \frac{f(t) - f(0)}{t}$$

$$= cf(x).$$

So $f(x) = e^{cx}$.   ∎

Remark 1.10.2   In the above theorem it is enough to assume that f is continuous or just integrable. To see this, choose a $y \in R$ such that $\int_0^y f(t)dt \neq 0$. Then $f(x) \int_0^y f(t)dt = \int_0^y f(x + t)dt = \int_x^{x+y} f(t)dt$. So

$$f(x) = \frac{\int_x^{x+y} f(t)dt}{\int_0^y f(t)dt}.$$

This equation implies that if f is integrable, then it must be continuous and hence differentiable.

Corollary 1.10.2   *Suppose g is a homomorphism from the multiplicative group of positive reals R^+ to $\mathbb{C}^*$, that is,*

$$g(xy) = g(x)g(y). \tag{1.10.2}$$

Then $g(x) = x^c$ for some c.

Proof.   Consider the map $f = g \circ \exp : R \to \mathbb{C}^*$, where $\exp(x) = e^x$. Then f satisfies (1.10.1) and $g(e^x) = e^{cx}$. This implies the result.   ∎

A finite field has p^n elements, where p is prime and n is a positive integer. For simplicity we take $n = 1$, so the field is isomorphic to $\mathbb{Z}(p)$, the integers modulo p. The analog of f in (1.10.1) is a homomorphism

$$\psi : \mathbb{Z}(p) \to \mathbb{C}^*.$$

Since $\mathbb{Z}(p)$ is a cyclic group of order p generated by 1 we need only specify $\psi(1)$. Also, $\psi(1)^p = \psi(0) = 1$ and we can choose any of the pth roots of unity as the value of $\psi(1)$. We therefore have p different homomorphisms

$$\psi_j(x) = e^{2\pi i j x/p}, \quad j = 0, 1, \ldots, p-1. \tag{1.10.3}$$

These are called the additive characters of the field. In a similar way the multiplicative characters are the $p-1$ characters defined by the homomorphisms from $\mathbb{Z}(p)^*$ to $\mathbb{C}^*$. Here $\mathbb{Z}(p)^* = \mathbb{Z}(p) - \{0\}$. Since $\mathbb{Z}(p)^*$ is a cyclic group of order $p-1$, we have an isomorphism $\mathbb{Z}(p)^* \cong \mathbb{Z}(p-1)$. The $p-1$ characters on $\mathbb{Z}(p)^*$ can be defined by means of this isomorphism and (1.10.3). We denote a multiplicative character by either χ or η, unless otherwise stated.

It is now clear how to define the "gamma" function for a finite field.

Definition 1.10.3   *For an additive character ψ_i and multiplicative character χ_i we define the Gauss sums $g_j(\chi_i)$, $j = 0, 1, \ldots, p-1$ by the formula*

$$g_j(\chi_i) = \sum_{x=0}^{p-1} \chi_i(x)\psi_j(x), \tag{1.10.4}$$

where we extend the domain of χ_i by setting $\chi_i(0) = 0$.

It is sufficient to consider $g(\chi) := g_1(\chi)$, for when $j \neq 0$,

$$g_j(\chi) = \sum_x \chi(x)\psi_j(x) = \sum_x \chi(x)e^{2\pi i j x/p}$$

$$= \overline{\chi(j)}\chi(j) \sum_x \chi(x)e^{2\pi i j x/p}$$

$$= \overline{\chi(j)} \sum_x \chi(jx)e^{2\pi i j x/p}$$

$$= \overline{\chi(j)}g(\chi). \tag{1.10.5}$$

This formula corresponds to $\int_0^\infty e^{-jx}x^{s-1}dx = \Gamma(s)/j^s$, where j is a nonzero complex number with positive real part. When $j = 0$ in (1.10.4) the sum is $\sum_x \chi(x)$, which can be shown to be zero when $\chi(x) \neq 1$ for at least one value of x.

Theorem 1.10.4   *For a character χ,*

$$\sum_x \chi(x) = \begin{cases} 0 & \text{if } \chi \neq id, \\ p-1 & \text{if } \chi = id. \end{cases} \tag{1.10.6}$$

Remark 1.10.3   The identity character is the one that takes the value 1 at each point in $\mathbb{Z}(p)^*$.

Proof.   The result is obvious for $\chi = id$. If $\chi \neq id$, there is a $y \in \mathbb{Z}(p)^*$ such that $\chi(y) \neq 1$. Then

$$\chi(y)\sum_x \chi(x) = \sum_x \chi(xy) = \sum_x \chi(x),$$

which implies the theorem. There is a dual to (1.10.6) given by the following theorem: ∎

Theorem 1.10.5   *For the sum over all characters we have*

$$\sum_{\chi} \chi(x) = \begin{cases} 0 & \text{if } x \neq 1, \\ p-1 & \text{if } x = 1. \end{cases} \tag{1.10.7}$$

Proof.   It is sufficient to observe that if $x \neq 1$, then there is a character χ such that $\chi(x) \neq 1$. The theorem may now be proved as before. ∎

We now define the analog of the beta function.

Definition 1.10.6   *For two multiplicative characters χ and η the Jacobi sum is defined by*

$$J(\chi, \eta) = \sum_{x+y=1} \chi(x)\eta(y). \tag{1.10.8}$$

The following theorem gives some elementary properties of the Jacobi sum. We denote the trivial or identity character by e. The reader should notice that the last result is the analog of the formula $B(x, y) = \Gamma(x)\Gamma(y)/[\Gamma(x + y)]$.

Theorem 1.10.7   *For nontrivial characters χ and η, the following properties hold:*

$$J(e, \chi) = 0. \tag{1.10.9}$$

$$J(e, e) = p - 2. \tag{1.10.10}$$

$$J(\chi, \chi^{-1}) = -\chi(-1). \tag{1.10.11}$$

$$\text{If } \chi\eta \neq e, \quad \text{then} \quad J(\chi, \eta) = \frac{g(\chi)g(\eta)}{g(\chi\eta)}. \tag{1.10.12}$$

Remark 1.10.4   From the definition of characters it is clear that the product of two characters is itself a character and so the set of characters forms a group. The additive characters form a cyclic group of order p and the multiplicative characters a cyclic group of order $p - 1$. Also, $\chi^{-1}(x) = \chi(x^{-1}) = 1/\chi(x)$ and since $|\chi(x)| = 1$ it follows that $\chi^{-1}(x) = \bar{\chi}(x)$.

Proof.   The first part of the theorem is a restatement of Theorem 1.10.3 and the second part is obvious. To prove (1.10.11), begin with the definition

$$J(\chi, \chi^{-1}) = \sum_{x} \chi(x)\chi^{-1}(1 - x) = \sum_{x \neq 0,1} \chi(x(1 - x)^{-1}).$$

Now note that as x runs through $2, \ldots, p - 1$, then $x(1 - x)$ runs through $1, \ldots, p-2$. The value $y = p-1 \equiv -1 \pmod{p}$ is not assumed because $x = y(1+y)^{-1}$. Therefore,

$$J(\chi, \chi^{-1}) = \sum_{y \neq -1} \chi(y) = -\chi(-1),$$

by Theorem 1.10.4. This proves the third part. The proof of the fourth part is very similar to Poisson's or Jacobi's proofs of the analogous formula for the beta function. Here one multiplies two Gauss sums and by a change of variables arrives at a product of a Jacobi sum and a Gauss sum. Thus, for $\chi\eta \neq e$,

$$g(\chi)g(\eta) = \sum_x \chi(x)e^{2\pi ix/p} \sum_y \eta(y)e^{2\pi iy/p}$$

$$= \sum_{x,y} \chi(x)\eta(y)e^{2\pi i(x+y)/p}$$

$$= \sum_{x+y=0} \chi(x)\eta(y) + \sum_{t=x+y\neq 0} \chi(x)\eta(t-x)e^{2\pi it/p}.$$

The first sum is $\sum_x \chi(x)\eta(-x) = \eta(-1)\sum_x \chi\eta(x) = 0$ since $\chi\eta \neq id$. The second sum with $x = st$ is

$$\sum_{t\neq 0,s} \chi(s)\chi(t)\eta(t)\eta(1-s)e^{2\pi it/p} = \sum_t \chi\eta(t)e^{2\pi it/p} \sum_s \chi(s)\eta(1-s)$$

$$= g(\chi\eta)J(\chi,\eta).$$

This proves the fourth part of the theorem. ∎

We were able to evaluate $\Gamma(s)$ in a nice form for positive integer values and half-integer values of s. Evaluations of special cases of Gauss sums are also possible and important, but in any case the magnitude of the Gauss sum can always be found.

Theorem 1.10.8   *For nontrivial multiplicative and additive characters χ and ψ,*

$$\left| \sum_x \chi(x)\psi(x) \right| = \sqrt{p}.$$

Proof.   By (1.10.5) it is enough to prove that $|g_1(\chi)|^2 \equiv |g(\chi)|^2 = p$:

$$|g(\chi)|^2 = \sum_x \chi(x)e^{2\pi ix/p} \sum_y \bar{\chi}(y)e^{-2\pi iy/p}$$

$$= \sum_{xy\neq 0} \chi(xy^{-1})e^{2\pi i(x-y)/p}.$$

Set $x = ty$. Then

$$|g(\chi)|^2 = \sum_{ty\neq 0} \chi(t)e^{2\pi iy(t-1)/p}$$

$$= \sum_1^{p-1} \chi(1) + \sum_{t\neq 0 \text{ or } 1} \chi(t) \sum_{y\neq 0} e^{2\pi iy(t-1)/p}.$$

The first sum is $p - 1$ and the inner sum in the second term is -1. Thus

$$|g(\chi)|^2 = p - 1 - \sum_{t \neq 1} \chi(t) = p - 1 + 1 = p,$$

and the result is proved. ∎

Corollary 1.10.9 *If χ, η and $\chi\eta$ are nontrivial characters, then*

$$|J(\chi, \eta)| = \sqrt{p}. \tag{1.10.13}$$

Proof. This follows from Theorems 1.10.7 and 1.10.8. ∎

As an interesting consequence we have:

Corollary 1.10.10 *If $p = 4n + 1$ is a prime, then there exist integers a and b such that $p = a^2 + b^2$.*

Proof. The group $\mathbb{Z}(p)^*$ is of order $p - 1 = 4n$, which is also isomorphic to the group of multiplicative characters on $\mathbb{Z}(p)^*$. Since the latter group is cyclic there exists a character χ of order 4 that takes the value $\pm 1, \pm i$. It follows that $J(\chi, \chi) = a + bi$ for integers a and b. Since $\chi^2 \neq id$, apply Corollary 1.10.9 to obtain the desired result. ∎

Corollary 1.10.10 is a theorem of Fermat, though Euler was the first to publish a proof. See Weil [1983, pp. 66–69]. Later we shall prove a more refined result that gives the number of representations of a positive integer as a sum of two squares. This will come from a formula that involves yet another analog of the beta integral.

We have seen that characters can be defined for cyclic groups. Since any abelian group is a direct product of cyclic groups, it is not difficult to find all the characters of an abelian group and their structure. The following observation may be sufficient here:

If χ_1 is a character of a abelian group G_1, and χ_2 of G_2, then we can define a character $\chi : G_1 \times G_2 \to \mathbb{C}^*$ by $\chi(x, y) = \chi_1(x)\chi_2(y)$.

We thus obtain n additive characters of $\mathbb{Z}(n)$ and $\phi(n)$ multiplicative characters of $\mathbb{Z}(n)^*$. The Gauss and Jacobi sums for these more general characters can be defined in the same way as before. Gauss [1808] found one derivation of the law of quadratic reciprocity by evaluating the Gauss sum arising from the quadratic character. (A character $\chi \neq id$ is a quadratic character when $\chi^2 = id$.) Details of this connection are in Exercise 37 at the end of the chapter. One problem that arises here, and which Gauss dealt with, is evaluating the sum $G = \sum_{x=0}^{N-1} e^{2\pi i x^2/N}$. As in Theorem 1.10.8 one can show that $G^2 = \pm N$ depending on whether $N \equiv 1$ (4) or 3 (4). The problem is to determine the appropriate square root for obtaining G. According to Gauss, it took him four years to settle this question. Dirichlet's evaluation of $\sum_{x=0}^{N=1} e^{2\pi i x^2/N}$ by means of Fourier series is given in Exercise 32.

Jacobi and Eisenstein also considered the more general Jacobi sum

$$J(\chi_1, \chi_2, \ldots, \chi_\ell) = \sum_{t_1 + \cdots + t_\ell = 1} \chi_1(t_1)\chi_2(t_2) \cdots \chi_\ell(t_\ell). \qquad (1.10.14)$$

This is the analog of the general beta integral in Theorem 1.8.6. Eisenstein's result, corresponding to the formula in Theorem 1.8.6, follows.

Theorem 1.10.11   *If* $\chi_1, \chi_2, \ldots, \chi_\ell$ *are nontrivial characters and* $\chi_1\chi_2\cdots\chi_\ell$ *is nontrivial, then*

$$J(\chi_1, \chi_2, \ldots, \chi_\ell) = \frac{g(\chi_1)g(\chi_2)\cdots g(\chi_\ell)}{g(\chi_1\chi_2\cdots\chi_\ell)}. \qquad (1.10.15)$$

The proof of this is similar to that of Theorem 1.10.7, and the reader should fill in the details.

In Section 1.8 the volume of n-dimensional objects of the form $a_1 x_1^{s_1} + a_2 x_2^{s_2} + \cdots + a_k x_k^{s_k} \le b$ was determined by means of the gamma function. In the same way, for finite fields, the number of points satisfying $a_1 x_1^{s_1} + a_2 x_2^{s_2} + \cdots + a_k x_k^{s_k} = b$ can be found in terms of Gauss sums. Gauss himself first found the number of points on such (but simpler) hypersurfaces and used this to evaluate some specific Gauss sums. Weil [1949] observed that it is easier to reverse the process and obtain the number of points in terms of Gauss sums. For an account of this the reader should see Weil [1974]. It may be mentioned that Weil's famous conjectures concerning the zeta function of algebraic varieties over finite fields are contained in his 1949 paper. It also contains the references to Gauss's works. One may also consult Ireland and Rosen [1991] for more on Jacobi and Gauss sums and for references to the papers of Jacobi and Eisenstein.

The form of the Gauss sums also suggests that they are connected with Fourier transforms. Let $\mathcal{F}$ denote the vector space of all complex valued functions on $\mathbb{Z}(N)$, the integers modulo N. Let F be the Fourier transform on $\mathcal{F}$ defined by

$$(Ff)(n) = \frac{1}{\sqrt{N}} \sum_{x=0}^{N-1} f(x)e^{2\pi inx/N}. \qquad (1.10.16)$$

It can be shown that the trace of this Fourier transform with respect to the basis $\{\delta_0, \delta_1, \ldots, \delta_N\}$, where

$$\delta_x(y) = \begin{cases} 0, & x \ne y, \\ 1, & x = y, \end{cases}$$

is the quadratic Gauss sum $\sum_{x=0}^{N-1} e^{2\pi ix^2/N}$. Schur [1921] gave another evaluation of this sum from this fact. The details are given in Exercise 47. One first proves that the fourth power of F is the identity so that the eigenvalues are ± 1, $\pm i$ and the essential problem is to find the multiplicity of these eigenvalues.

Discrete or finite Fourier analysis was not applied extensively before 1965 because of the difficulty of numerical computation. This changed when Cooley and Tukey [1965] introduced an algorithm they called the Fast Fourier Transform (FFT) to reduce the computation by several orders of magnitude. The reader may wish to consult the paper of Auslander and Tolimieri [1979] for an introduction to FFT, which emphasizes the connection with group theory. Some of the earlier instances of an FFT algorithm are mentioned here. Computational aspects are also interesting. See de Boor [1980] and Van Loan [1992, §1.3].

1.11 A Probabilistic Evaluation of the Beta Function

When α and β are positive integers,

$$\int_0^1 x^{\alpha-1}(1-x)^{\beta-1}dx = \frac{(\alpha-1)!(\beta-1)!}{(\alpha+\beta-1)!}.$$

It seems that it should be possible to arrive at this result by a combinatorial argument. But working with only a finite number of objects could not give an integral. Here is a combinatorial-cum-probabilistic argument that evaluates the integral. Choose points at random from the unit interval $[0, 1]$. Assume that the probability that a point lies in a subinterval (a, b) is $b - a$. Fix an integer n and let $P(x_k < t)$ denote the probability that, of n points chosen at random, exactly k of them have values less than t. The probability density function for $P(x_k < t)$ is

$$\rho(t) = \lim_{\Delta t \to 0} \frac{P(x_k < t + \Delta t) - P(x_k < t)}{\Delta t}.$$

Now

$P(x_k < t + \Delta t) - P(x_k < t)$

  = the probability that one point lies in $(t, t + \Delta t)$,

     $k - 1$ points less than t and $n - k$ points greater than $t + \Delta t$,

     + the probability that two points lie in $(t, t + \Delta t)$,

     $k - 2$ points less than t and $n - k$ points greater than $t + \Delta t$,

     $+ \cdots$.

Since there are n points, the number of ways that one point is in $(t, t + \Delta t)$, $k - 1$ points are less than t, and $n - k$, are greater than $t + \Delta t$ is

$$\binom{n}{1}\binom{n-1}{k-1}\binom{n-k}{n-k} = n\binom{n-1}{k-1}.$$

The probability of each such event is $\Delta t t^{k-1}(1-t-\Delta t)^{n-k}$, and since the events are mutually exclusive, we get

$$P(x_k < t + \Delta t) - P(x_k < t)$$
$$= n \binom{n-1}{k-1} \Delta t \, t^{k-1}(1-t-\Delta t)^{n-k}$$
$$+ \frac{n(n-1)}{2} \binom{n-2}{k-2} (\Delta t)^2 t^{k-2}(1-t-\Delta t)^{n-k}$$
$$+ \cdots$$
$$= n \binom{n-1}{k-1} t^{k-1}(1-t-\Delta t)^{n-k} \Delta t + O((\Delta t)^2).$$

Therefore,

$$\rho(t) = n \binom{n-1}{k-1} t^{k-1}(1-t)^{n-k}.$$

Since

$$\int_0^1 \rho(t) dt = 1,$$

we obtain

$$\int_0^1 t^{k-1}(1-t)^{n-k} dt = \frac{(k-1)!(n-k)!}{n!} = \frac{\Gamma(k)\Gamma(n-k+1)}{\Gamma(n+1)}.$$

We made use of probability theory here to indicate its relationship with the beta function. Though we do not use it elsewhere, probability theory can be used to derive formulas involving some extensions of the beta function.

1.12 The p-adic Gamma Function

In number theory there are completions of the rationals other than the reals that are of great importance. These are the p-adic completions of the rationals. There is an analog of the gamma function defined on the p-adic numbers that is useful. The following is a very brief account of the p-adic gamma function. The interested reader should consult the references given later.

Suppose a is an integer and p a prime. Define $\mathrm{ord}_p a$ to be the highest power of p that divides a. Let Q be the set of rational numbers. For $x = a/b \in Q$, where a and b are integers, define $\mathrm{ord}_p x = \mathrm{ord}_p a - \mathrm{ord}_p b$. The p-adic norm $|\;|_p$ on Q is defined by

$$|x|_p = \begin{cases} 1/p^{\mathrm{ord}_p x}, & x \neq 0, \\ 0, & x = 0. \end{cases}$$

Thus in the *p*-adic norm, p^n gets small as n gets large. In contrast, for negative values of n, p^n becomes big. So it is reasonable to write numbers in powers of p. An integer would have an expansion of the form

$$a_0 + a_1 p + a_2 p^2 + \cdots + a_n p^n,$$

where $a_i \in \{0, 1, 2, \ldots, p - 1\}$. For rational numbers negative powers of p will also be involved. The *p*-adic norm is non-Archimedean, that is,

$$\|x + y\|_p \leq \max(\|x\|_p, \|y\|_p),$$

and so the triangle inequality holds. This gives a metric on Q.

We can then obtain a completion of Q with this metric in the same way as we get the real numbers by taking the ordinary metric on Q. This involves taking the Cauchy sequence of rationals. The *p*-adic completion is denoted by Q_p. The *p*-adic numbers can be represented by the series

$$\frac{b_{-m}}{p^m} + \frac{b_{-m+1}}{p^{m-1}} + \cdots + \frac{b_{-1}}{p} + b_0 + b_1 p + b_2 p^2 + \cdots.$$

The subset of Q_p which contains all numbers with nonnegative powers of p forms a ring denoted by Z_p. This is the ring of *p*-adic integers. The positive integers $\mathbb{Z}^+$ form a dense subset of Z_p. This makes sense because a member of Z_p can be represented as an infinite series

$$a_0 + a_1 p + a_2 p^2 + \cdots, \qquad a_i \in \{0, 1, \ldots p - 1\},$$

and the partial sums are integers that converge to the *p*-adic number. So, if there is a function f defined on the positive integers and the values of f at two integers that are *p*-adically close are close to each other, then f has a unique continuous extension to Z_p.

Define a function f on the positive integers n by the formula

$$f(n) = (-1)^n \prod_{\substack{k=1 \\ p \nmid k}}^{n} k.$$

It is not difficult to show that $f(n + p^m \ell) \equiv f(n) \bmod p^m$, where n, m, and ℓ are positive integers. Now $n + p^n \ell$ and n are *p*-adically close to each other and the values of f at these points are also *p*-adically close. Consequently, f has an extension to Z_p. This extension gives the *p*-adic gamma function due to Morita [1975]. The *p*-adic gamma function is defined by $\Gamma_p(x) = -f(x - 1)$. This function also has a functional relation and other useful properties. There is a formula of Gross and Koblitz [1979] that gives the Gauss sum as a product of values of the *p*-adic gamma function.

A good treatment of the p-adic numbers and functions is given in Koblitz [1977]. An account of the p-adic gamma function and the Gross–Koblitz formula is available in Lang [1980], including a reference to a paper by "Boyarsky" [1980]. In fact, p-adic extensions of the beta function, and more generally, the Mellin transform, are also available.

Exercises

1. Use the change of variables $s = ut$ to show that

$$\Gamma(x)\Gamma(y) = \int_0^\infty \int_0^\infty t^{x-1}s^{y-1}e^{-(s+t)}dt\,ds$$

   is $\Gamma(x + y)B(x, y)$.   (Poisson)

2. Let $I = \int_0^\infty e^{-x^2}dx$. Observe that $I^2 = \int_0^\infty \int_0^\infty e^{-(x^2+y^2)}dx\,dy$. Evaluate this double integral by converting to polar coordinates and show that $I = \sqrt{\pi}/2$.

3. A proof of Wallis's formula is sketched below:
   (a) Show that

   $$\int_0^\infty \frac{dx}{x^2 + a} = \frac{\pi}{2\sqrt{a}}.$$

   (b) Take the derivative of both sides n times with respect to the parameter a to conclude that

   $$\int_0^\infty \frac{dx}{(x^2 + a)^{n+1}} = \frac{1 \cdot 3 \cdot 5 \cdots 2n - 1}{2 \cdot 4 \cdot 6 \cdots 2n} \frac{\pi}{2} \frac{1}{a^{n+(1/2)}}.$$

   (c) Set $x = y/\sqrt{n}$, $a = 1$, let $n \to \infty$, and use Exercise 2 to obtain Wallis's formula.

4. Evaluate $\int_{-1}^1 (1 - t^2)^{x-1}dt$ in two different ways to prove the duplication formula given in Theorem 1.5.1. To get another proof evaluate

   $$\int_0^{\pi/2} \sin^{2x-1} 2\theta\, d\theta$$

   in two ways.

5. Suppose that f is twice differentiable. Show that $f'' \geq 0$ is equivalent to $f(\alpha x + \beta y) \leq \alpha f(x) + \beta f(y)$ for α and β nonnegative and $\alpha + \beta = 1$.

6. Convexity can be used to prove some important inequalities, for example, Hölder's inequality:

   $$\left| \int_a^b fg\,dx \right| \leq \left\{ \int_a^b |f|^p dx \right\}^{1/p} \left\{ \int_a^b |g|^q dx \right\}^{1/q},$$

   where f and g are integrable functions and $\frac{1}{p} + \frac{1}{q} = 1$. We sketch a proof here.

(a) Note that e^x is a convex function; use this and the result of Exercise 5 to show that if u and v are nonnegative real numbers then

$$uv \le \frac{u^p}{p} + \frac{v^q}{q}.$$

Equality holds if and only if $u^p = v^q$.

(b) Deduce Hölder's inequality from (a).

It might be appropriate to call this the Rogers–Hölder inequality since Rogers [1888] had the result before Hölder [1889]. Other important results of L. J. Rogers are discussed later in the book.

7. Here is another proof of the functional relation

$$B(x, y) = \frac{x + y}{y} B(x, y + 1).$$

Write

$$B(x, y + 1) = \int_0^1 t^{x+y-1} \left(\frac{1 - t}{t} \right)^y dt$$

and perform an integration by parts to show that

$$B(x, y + 1) = \frac{y}{x + y} B(x, y).$$

8. Show that

$$\frac{B(x, y)}{c^y} = \int_0^\infty \frac{t^{x-1} dt}{(c + t)^{x+y}}.$$

Take the derivative with respect to c and derive the functional equation

$$B(x, y) = \frac{x + y}{y} B(x, y + 1).$$

Give a similar argument using

$$\int_0^c t^{x-1}(c - t)^y dt.$$

9. Write Gauss's formula as

$$\Gamma(x) = \frac{\prod_{k=0}^{n-1} \Gamma((x + k)/n)}{(2\pi)^{(n-1)/2} n^{(1/2)-x}}.$$

Show that the right side satisfies all the conditions of the Bohr–Mollerup theorem. This proves the formula.

10. Give a proof of Gauss's formula by using the definition of $\Gamma(x)$.

11. Prove Gauss's formula by the method given in the remark after Theorem 1.5.2.

12. It is clear from $\Gamma(x + 1) = x\Gamma(x)$ that $\int_x^{x+1} \log \Gamma(t) dt = x \log x - x + C$. Show that $C = \frac{1}{2} \log 2\pi$. Stirling's formula will work, but there is a more elegant argument using Gauss's multiplication formula first.

13. There is another beta integral due to Cauchy defined by

$$C(x, y) = \int_{-\infty}^{\infty} \frac{dt}{(1+it)^x(1-it)^y} = \frac{\pi 2^{2-x-y}\Gamma(x+y-1)}{\Gamma(x)\Gamma(y)}, \quad \mathrm{Re}\,(x+y) > 1.$$

(a) To prove this, show:

    (i) Integration by parts gives $C(x, y + 1) - \frac{x}{y}C(x + 1, y)$.

    (ii) Write

$$C(x, y) = \int_{-\infty}^{\infty} \frac{(-1-it)+2}{(1+it)^x(1-it)^{y+1}}dt$$

$$= 2C(x, y+1) - C(x-1, y+1).$$

This together with (i) gives

$$C(x, y) = \frac{2y}{x+y-1}C(x, y+1).$$

(iii) Iteration gives

$$C(x, y) = \frac{2^{2n}(x)_n(y)_n}{(x+y-1)_{2n}}C(x+n, y+n).$$

$$C(x+n, y+n) = \int_{-\infty}^{\infty} \frac{dt}{(1+it)^{x+n}(1-it)^{y+n}},$$

$$= \int_{-\infty}^{\infty} \frac{dt}{(1+t^2)^n(1+it)^x(1-it)^y}.$$

Set $t \to t/\sqrt{n}$ in the second integral and let $n \to \infty$.

(b) The substitution $t = \tan\theta$ leads to an important integral. Find it.

14. Use the method for obtaining Stirling's formula to show that

$$\frac{1}{\sqrt{1}} + \frac{1}{\sqrt{2}} + \cdots + \frac{1}{\sqrt{n}} = 2\sqrt{n} + C + \frac{1}{2\sqrt{n}} + O\left(\frac{1}{n^{3/2}}\right),$$

where

$$C = -(1+\sqrt{2})\left(1 - \frac{1}{\sqrt{2}} + \frac{1}{\sqrt{3}} - \frac{1}{\sqrt{4}} + \cdots\right).$$

Sum

$$c_n = \sum_{k=1}^{n}(c_k - c_{k-1}) \quad \text{with} \quad c_n = \sum_{k=1}^{n} \frac{1}{\sqrt{k}} - 2\sqrt{n},$$

and use some algebra to change $c_n - c_{n-1}$ to an expression that goes to zero like $n^{-3/2}$ to show that

$$\sum_{n=1}^{\infty} \frac{1}{\sqrt{n}[\sqrt{n} + \sqrt{n-1}]^2} = (\sqrt{2}+1)\sum_{n=1}^{\infty} \frac{(-1)^{n+1}}{\sqrt{n}}.$$

See Ramanujan [1927, papers 9 and 13] for further results of this type.

15. Here is an outline of a real variable proof of (1.2.1). Let

$$g(x) = \frac{\pi}{\tan \pi x} - \lim_{N \to \infty} \sum_{-N}^{N} \frac{1}{n+x}.$$

(a) $$g'(x) = -\pi^2 / \sin^2 \pi x + \sum_{-\infty}^{\infty} 1/(n+x)^2.$$

(b) $g'(x)$ is continuous for $0 \le x \le 1$ if $g'(0) = g'(1) = 0$.

(c) $$g'(x/2) + g'((x+1)/2) = 4g'(x).$$

(d) Let $M = \max_{0 \le x \le 1} |g'(x)|$. Then $M \le M/2$ so $M = 0$.

(e) $$g(x/2) - g((x+1)/2) = 2\pi/(\sin \pi x) - 2\sum_{-\infty}^{\infty}(-1)^n/(n+x).$$

(f) $g(x+1) = g(x)$.

(g) $g(x) = $ constant.

(h) $$\int_0^\infty t^{x-1}/(1+t)dt = \sum_{n=0}^\infty (-1)^n/(n+x) + \sum_{n=0}^\infty (-1)^n/(n+1-x)$$

$$= \sum_{-\infty}^\infty (-1)^n/(n+x),$$

so (1.2.1) holds. This proof, due to Herglotz, was published by Carathéodory [1954, pp. 269–270]. Bochner's [1979] review of the collected works of Herglotz also includes this proof.

16. The following is Dedekind's [1853] proof of $\Gamma(x)\Gamma(1-x) = \pi/\sin \pi x$. Set

$$\phi(x) = \int_0^\infty \frac{t^{x-1}}{1+t}dt.$$

(a) Show that

$$\int_0^\infty \frac{t^{x-1}}{st+1}dt = \phi(x)s^{-x}$$

and

$$\int_0^\infty \frac{t^{x-1}}{t+s}dt = \phi(x)s^{x-1}.$$

(b) Deduce that

$$\phi(x)\frac{(s^{x-1}-s^{-x})}{s-1} = \int_0^\infty \frac{t^{x-1}(t-1)}{(st+1)(t+s)}dt.$$

(c) Use the second formula in (a) to get

$$[\phi(x)]^2 = \int_0^\infty \frac{1}{s+1}\left(\int_0^\infty \frac{t^{x-1}}{t+s}dt\right)ds;$$

then change the order of integration to obtain

$$[\phi(x)]^2 = \int_0^\infty \frac{t^{x-1}\log t}{t-1}\,dt.$$

(d) Deduce

$$\int_{1-y}^y [\phi(x)]^2 dx = \int_0^\infty \frac{t^{y-1}-t^{-y}}{t-1}\,dt.$$

(e) Integrate (b) with respect to s over $(0, \infty)$ and use (d) to derive

$$\phi(x)\int_{1-x}^x [\phi(t)]^2 dt = 2\int_0^\infty \frac{t^{x-1}\log t}{1+t}\,dt = 2\phi'(x).$$

(f) Show that $\phi(x) = \phi(1-x)$ implies $\phi'(\tfrac{1}{2}) = 0$ and

$$\int_{1-x}^x [\phi(t)]^2 dt = 2\int_{1/2}^x [\phi(t)]^2 dt.$$

(g) Deduce that

$$\phi(x)\int_{1/2}^x [\phi(t)]^2 dt = \phi'(x).$$

(h) Show that ϕ satisfies the differential equation $\phi\phi'' - (\phi')^2 = \phi^4$.
(i) Solve the differential equation with initial condition $\phi(\tfrac{1}{2}) = \pi$ and $\phi'(\tfrac{1}{2}) = 0$ to get $\phi(x) = \pi \csc \pi x$.

17. Show that

$$\int_0^1 \frac{t^{x-1}(1-t)^{y-1}dt}{[at+b(1-t)]^{x+y}} = \frac{\Gamma(x)\Gamma(y)}{a^x b^y \Gamma(x+y)}, \quad \mathrm{Re}\,x > 0,\ \mathrm{Re}\,y > 0,\ a > 0.$$

18. Show that

$$\lim_{t\to\infty} \frac{\Gamma(t+ix\sqrt{t})\Gamma(t-ix\sqrt{t})}{\Gamma(t)\Gamma(t)} = e^{-x^2}.$$

19. Prove that for $a > 0$,

$$\int_0^\infty \frac{\sin ax}{x^b}\,dx = \frac{1}{2}\pi a^{b-1}\frac{\csc(\pi b/2)}{\Gamma(b)}, \quad 0 < \mathrm{Re}\,b < 2,$$

and

$$\int_0^\infty \frac{\cos ax}{x^b}\,dx = \frac{1}{2}\pi a^{b-1}\frac{\sec(\pi b/2)}{\Gamma(b)}, \quad 0 < \mathrm{Re}\,b < 1.$$

20. For $\lambda > 0$, $x > 0$ and $-\pi/2 < \alpha < \pi/2$, prove that

$$\int_0^\infty t^{x-1}e^{-\lambda t\cos\alpha}\cos(\lambda t\sin\alpha)\,dt = \lambda^{-x}\Gamma(x)\cos\alpha x$$

and

$$\int_0^\infty t^{x-1} e^{-\lambda t \cos\alpha} \sin(\lambda t \sin\alpha) dt = \lambda^{-x} \Gamma(x) \sin\alpha x.$$

21. Prove that $\pi^{-s/2}\Gamma(s/2)\zeta(s) = \pi^{-(1-s)/2}\Gamma((1-s)/2)\zeta(1-s)$ as follows:
    (a) Observe that

$$\sum_{n=1}^\infty \frac{\sin(2n+1)x}{2n+1} = (-1)^m \pi/4 \quad \text{for } m\pi < x < (m+1)\pi, m=0, 1, \dots.$$

    (b) Multiply the equation by $x^{s-1}(0 < s < 1)$ and integrate over $(0, \infty)$. Show that the left side is $\Gamma(s)\sin(s\pi/2)(1 - 2^{-s-1})\zeta(s+1)$ and that the right represents an analytic function for $\operatorname{Re} s < 1$ and is equal to $2(1 - 2^{s+1})\zeta(1-s)$ for $\operatorname{Re} s < 0$.
    (c) Deduce the functional equation for the zeta function.   (Hardy)
22. Let C be a contour that starts at infinity on the negative real axis, encircles the origin once in the positive direction, and returns to negative infinity. Prove that

$$\frac{1}{\Gamma(s)} = \frac{1}{2\pi i} \int_C e^t t^{-s} dt.$$

This formula holds for all complex s.
    (a) Note that the integral represents an analytic function of s.
    (b) C may be taken to be a line from $-\infty$ to $-\delta$, then a circle of radius δ in the positive direction, and finally a line from $-\delta$ to $-\infty$. Show that

$$\int_C e^t t^{-s} dt = 2i \sin\pi s \int_\delta^\infty e^{-u} u^{-s} du + I,$$

    where I is the integral on the circle $|t| = \delta$.
This representation of the gamma function is due to Hankel; see Whittaker and Watson [1940, p. 244].
23. Prove that

$$\zeta(x, s) = \frac{e^{-i\pi s}\Gamma(1 - s)}{2\pi i} \int_C \frac{t^{s-1} e^{-xt}}{1 - e^{-t}} dt,$$

where C starts at infinity on the positive real axis, encircles the origin once in the positive direction, excluding the points $\pm 2n\pi i$, $n \geq 1$ an integer, and returns to positive infinity.
    Hint: First prove that

$$\zeta(x, s) = \frac{1}{\Gamma(s)} \int_0^\infty \frac{t^{s-1} e^{-xt}}{1 - e^{-t}} dt$$

and then apply the ideas of the previous exercise. Note also that $\zeta(x, s)$ is now defined as a meromorphic function by the contour integral with a simple pole at $s = 1$.

24. Prove the functional equation

$$\zeta(x, s) = \frac{2\Gamma(1-s)}{(2\pi)^{1-s}} \left\{ \sin(\pi s/2) \sum_{m=1}^{\infty} \frac{\cos 2m\pi x}{m^{1-s}} + \cos(s\pi/2) \sum_{m=1}^{\infty} \frac{\sin 2m\pi x}{m^{1-s}} \right\}.$$

Hint: Let C_n denote the line along the positive real axis from ∞ to $(2n+1)\pi$, then a square with corners $(2n+1)\pi(\pm 1 \pm i)$, and then the line from $(2n+1)$ to ∞. Show that

$$\int_C \frac{t^{s-1}e^{-xt}}{1-e^{-t}} dt = \int_{C_n} \frac{t^{s-1}e^{-xt}}{1-e^{-t}} dt$$

$$- \text{the sum of the residues at} \pm 2m\pi i, m = 1, \ldots, n,$$

where C is the curve in the previous exercise.

Note that the sum of the residues at $\pm 2m\pi i$ is

$$-2(2m\pi)^{s-1}e^{i\pi s} \sin(2m\pi x + \pi s/2).$$

Now let $n \to \infty$ and show that $\int_{C_n} \to 0$.

25. Show that the functional equation for $\zeta(x, s)$ easily implies
   (a) the functional equation for $\zeta(s)$,
   (b) Kummer's Fourier expansion for $\log \Gamma(x)/\sqrt{2\pi}$.
   The next five problems are taken from Artin [1964].

26. For $0 < x < \infty$, let $\phi(x)$ be positive and continuously twice differentiable satisfying (a) $\phi(x + 1) = \phi(x)$, (b) $\phi(\frac{x}{2})\phi(\frac{x+1}{2}) = d\phi(x)$, where d is a constant. Prove that ϕ is a constant.
   Hint: Let $g(x) = \frac{d^2}{dx^2} \log \phi(x)$. Observe that $g(x+1) = g(x)$ and $\frac{1}{4}(g(\frac{x}{2})+g(\frac{x+1}{2})) = g(x)$.

27. Show that $\phi(x) = \Gamma(x)\Gamma(1-x) \sin \pi x$ satisfies the conditions of the previous problem. Deduce Euler's reflection formula.

28. Prove that a twice continuously differentiable function f that is positive in $0 < x < \infty$ and satisfies (a) $f(x+1) = xf(x)$ and (b) $2^{2x-1} f(x)f(x+\frac{1}{2}) = \sqrt{\pi} f(2x)$ is identical to $\Gamma(x)$.

29. It is enough to assume that f is continuously differentiable in the previous problem. This is implied by the following: If g is continuously differentiable, $g(x+1) = g(x)$, and $g(\frac{x}{2}) + g(\frac{x+1}{2}) = g(x)$, then $g \equiv 0$.
   Hint: Observe that

$$\frac{1}{2^n} \sum_{k=0}^{2^n-1} g'((x+k)/2^n) = g'(x).$$

The left side tends to $\int_0^1 g'(x)dx = g(1) - g(0) = 0$ as $n \to \infty$.

30. Prove that the example $g(x) = \sum_{n=1}^{\infty} \frac{1}{2^n} \sin(2^n \pi x)$ shows that just continuity is insufficient in the previous problem.

31. Suppose f and g are differentiable functions such that $f(x+y) = f(x)f(y) - g(x)g(y)$ and $g(x+y) = f(x)g(y) + g(x)f(y)$. Prove that $f(x) = e^{ax} \cos bx$ and $g(x) = e^{ax} \sin bx$, unless $f(x) = g(x) \equiv 0$.

32. Prove that $\sum_{x=0}^{N-1} e^{2\pi i x^2/N} = \frac{1+i^{-N}}{1-i}\sqrt{N}$, where $i = \sqrt{-1}$.

(a) Set $f(t) = \sum_{x=0}^{N-1} e^{2\pi i (x+t)^2/N}$, $0 \le t \le 1$. Note that $f(0) = f(1)$ and extend $f(t)$ as a periodic function to the whole real line.

(b) Note that $f(t) = \sum_{-\infty}^{\infty} a_n e^{2\pi i n t}$, where $a_n = \int_0^1 f(t) e^{-2\pi i n t} dt$. Conclude that $f(0) = \sum_{x=0}^{N-1} e^{2\pi i x^2/N} = \sum_{-\infty}^{\infty} a_n$.

(c) Show that $a_n = e^{-2\pi i N n^2/4} \int_{-Nn/2}^{N(1-n/2)} e^{2\pi i y^2/N} dy$.

(d) Show that

$$\sum_{-\infty}^{\infty} a_n = \left(\sum_{n\,\text{even}} \int_{-Nn/2}^{N-nN/2} + i^{-N} \sum_{n\,\text{odd}} \int_{-Nn/2}^{N-nN/2} \right) e^{2\pi i y^2/N} dy$$

$$= (1 + i^{-N}) \int_{-\infty}^{\infty} e^{2\pi i y^2/N} dy.$$

(e) Use Exercise 19 to evaulate the integral. Another way is to take $N = 1$ in (d).   (Dirichlet)

33. If p is an odd prime, then there is exactly one character χ_2 that maps $\mathbb{Z}(p)^*$ onto $\{\pm 1\}$. Recall that $\mathbb{Z}(p)^*$ is the integers modulo p without 0. Prove that $\chi_2(a) = 1$ if and only if $x^2 = a \bmod p$ is solvable, that is, a is a square in $\mathbb{Z}(p)^*$. Usually one writes $\chi_2(a) = (\frac{a}{p})$, which is called the Legendre symbol.

34. Prove that if a is a positive integer prime to p, then $a^{p-1/2} \equiv (\frac{a}{p}) \pmod{p}$. Here p is an odd prime. (Use the fact that $\mathbb{Z}(p)^*$ is a cyclic group.)

35. For p an odd prime, use the previous problem to prove that $(\frac{-1}{p}) = (-1)^{(p-1)/2}$ and $(\frac{2}{p}) = (-1)^{(p^2-1)/8}$. (Use $2^{p/2} = (e^{\pi i/4} + e^{-\pi i/4})^p \equiv (e^{p\pi i/4} + e^{-p\pi i/4}) \pmod{p}$. Consider the two cases $p \equiv \pm 1 \pmod{8}$ and $p \equiv \pm 3 \pmod{8}$ separately.)

36. Prove the law of quadratic reciprocity: For odd primes p and q, $(\frac{p}{q})(\frac{q}{p}) = (-1)^{\frac{p-1}{2} \cdot \frac{q-1}{2}}$.

(a) For $S = \sum_{x=1}^{p-1} (\frac{x}{p}) e^{2\pi i x/p}$, show that $S^2 = (-\frac{1}{p})p$. (The proof is similar to that of Theorem 1.10.8.)

(b) Use (a) and Exercise 34 to prove that $S^{q-1} \equiv (-1)^{\frac{p-1}{2} \cdot \frac{q-1}{2}} (\frac{p}{q}) \pmod{q}$.

(c) Show that $S^q \equiv \sum_{x=1}^{p-1} (\frac{x}{p}) e^{2\pi i q x/p} \equiv (\frac{q}{p}) S \pmod{q}$.

(d) Deduce the reciprocity theorem from (b) and (c).   (Gauss)

37. For integers a and N with $N > 0$, define $G(a, N) = \sum_{x=0}^{N-1} e^{2\pi i a x^2/N}$.

(a) For p prime, show that $G(1, p) = \sum_{x=1}^{p-1} (\frac{x}{p}) e^{2\pi i x/p}$.

(b) For p prime show that $G(a, p) = (\frac{a}{p}) G(1, p)$.

(c) Prove that $G(q, p)G(p, q) = G(1, pq)$ when p and q are odd primes.

(d) Now use the result of Exercise 32 to deduce the reciprocity law. (Gauss) For a discussion of Exercises 32–37 and for references, see Scharlau and Opolka [1985, Chapters 6 and 8].

38. Prove Theorems 1.8.5 and 1.8.6.

39. Prove Theorem 1.9.4.

40. Weierstrass's approximation theorem: Suppose f is a continuous function on a closed and bounded interval, which we can choose to be $[0, 1]$ without any loss of generality. The following exercise shows that f can be uniformly approximated by polynomials on $[0, 1]$.

(a) Show that it is enough to prove the result for $f(0) = f(1) = 0$. Now extend f continuously to the whole real line by taking $f \equiv 0$ on $x < 0$ and $x > 1$.

(b) Observe that

$$Q_n(t) = \frac{(2n + 1)!}{2^{2n+1}(n!)^2}(1 - t^2)^n$$

is a polynomial such that

$$\int_{-1}^{1} Q_n(t)dt = 1.$$

Show that $P_n(x) = \int_{-1}^{1} f(x + t)Q_n(t)dt$ is a polynomial in x for $x \in [0, 1]$.

(c) Use Stirling's formula to show that for $\delta > 0$ and $\delta < |t| < 1$, $Q_n(t) \to 0$ uniformly as $n \to \infty$.

(d) Note that for $0 \le x \le 1$, $P_n(x) - f(x) = \int_{-1}^{1}[f(x+t) - f(x)]Q_n(t)dt$. To show that $P_n(x) \to f(x)$ uniformly on $[0, 1]$, break up the integral into three parts, $\int_{-1}^{-\delta} + \int_{-\delta}^{\delta} + \int_{\delta}^{1}$, and use (c).

41. Prove Plana's formula (see Whittaker and Watson [1940, p. 145] for references to Plana): For positive integers m and n

$$\sum_{k=m}^{n} \phi(k) = \frac{\phi(m) + \phi(n)}{2} + \int_{m}^{n} \phi(x)dx - i \int_{0}^{\infty}$$

$$\times \frac{\phi(n + iy) - \phi(m + iy) - \phi(n - iy) + \phi(m - iy)}{e^{2\pi y} - 1}dy,$$

where $\phi(x + iy)$ is a bounded analytic function in $m \le x \le n$.

Hint:

(a) Consider the integral $\int_C \phi(z)/(e^{-2\pi iz} - 1)dz$ where C is a suitable indented rectangle with vertices $k, k + 1, k + 1 + Li$, and $k + Li$. Then let $L \to \infty$.

(b) Now replace i with $-i$ in the contour C and repeat the process in (a).

(c) Add the results in (a) and (b) and sum over k.

42. (i) In Plana's formula let $m = 0$, $n \to \infty$, and suppose that $\phi(n) \to 0$, $\phi(n \pm iy) \to 0$, to get

$$\sum_{k=0}^{\infty} \phi(k) = \frac{1}{2}\phi(0) + \int_0^{\infty} \phi(x)dx + i \int_0^{\infty} \frac{\phi(iy) - \phi(-iy)}{e^{2\pi y} - 1} dy.$$

(ii) Deduce Hermite's formula (for reference, see Whittaker and Watson, [1940, p. 269])

$$\zeta(x, s) = \frac{x^{-s}}{2} + \frac{x^{1-s}}{s-1} + 2\int_0^{\infty} \frac{(x^2 + t^2)^{-s/2} \sin(s \arctan t/x)}{e^{2\pi t} - 1} dt.$$

(iii) Conclude that $\zeta(x, 2) = \dfrac{1}{2x^2} + \dfrac{1}{x} + \displaystyle\int_0^{\infty} \dfrac{4xt \, dt}{(x^2 + t^2)^2(e^{2\pi t} - 1)}.$

43. (a) For $\psi(x) = \Gamma'(x)/\Gamma(x)$, note that $\psi'(x) = \zeta(x, 2)$.

(b) Deduce that

$$\psi(x) = \ln x - \frac{1}{2x} - \int_0^{\infty} \frac{2t \, dt}{(x^2 + t^2)(e^{2\pi t} - 1)}.$$

(Use part (iii) of the previous exercise.)

(c) Deduce Binet's second formula

$$\ln \Gamma(x) = \left(x - \frac{1}{2}\right) \ln x - x + \frac{1}{2}\ln(2\pi) + 2\int_0^{\infty} \frac{\arctan(t/x)}{e^{2\pi t} - 1} dt,$$

where x is complex and $\operatorname{Re} x > 0$.

(d) Use Hermite's formula in the previous problem to obtain Lerch's formula (1.3.7) for $(\frac{\partial}{\partial s}\zeta(x, s))_{s=0}$.

44. Prove the following properties of Bernoulli polynomials:

(a) $B_q(x + 1) - B_q(x) = qx^{q-1}$.

(b) $$\sum_{n=M}^{N-1} n^q = \frac{1}{q+1}\{B_{q+1}(N) - B_{q+1}(M)\}.$$

(c) $$B_n(x) = \sum_{k=0}^{n} \binom{n}{k} B_k x^{n-k}.$$

(d) $B_n(1 - x) = (-1)^n B_n(x)$.

(e) $$B_n(\ell x) = \ell^{n-1} \sum_{k=0}^{\ell-1} B_n\left(x + \frac{k}{\ell}\right).$$

45. Prove that

(a) $$B_{2q-1}(x - [x]) = 2(-1)^q (2q - 1)! \sum_{n=1}^{\infty} \frac{\sin 2\pi nx}{(2\pi n)^{2q-1}}, \quad q \geq 1,$$

and

$$B_{2q}(x - [x]) = 2(-1)^{q-1}(2q)! \sum_{n=1}^{\infty} \frac{\cos 2\pi nx}{(2\pi n)^{2q}}.$$

(b) Deduce

$$\zeta(2q) = (-1)^{q-1} \frac{(2\pi)^{2q}}{(2q)!} \frac{B_{2q}}{2}, \quad q \geq 1,$$

and

$$\sum_{n=0}^{\infty} \frac{(-1)^n}{(2n+1)^{2q-1}} = (-1)^q \frac{(2\pi)^{2q-1}}{2(2q-1)!} B_{2q-1}(1/4), \quad q \geq 1.$$

46. Prove that

$$B_{2n} = G_{2n} - \sum_{(p-1)|2n} \frac{1}{p},$$

where G_{2n} is some integer and p is a prime such that $p - 1$ divides $2n$.

(Clausen–von Staudt)

Hint: Define $\sum_{n=0}^{\infty} \frac{a_n}{n!} x^n \equiv \sum_{n=0}^{\infty} \frac{b_n}{n!} x^n \pmod{k}$ if k divides $a_n - b_n$ for all $n \geq 0$. Show that

(a) $(e^z - 1)^3 \equiv 2\left(\frac{z^3}{3!} + \frac{z^5}{5!} + \frac{z^7}{7!} + \cdots\right) \pmod 4$.

(b) For prime p,

$$(e^z - 1)^{p-1} \equiv -\left(\frac{z^{p-1}}{(p-1)!} + \frac{z^{2(p-1)}}{(2p-2)!} + \frac{z^{3(p-1)}}{(3p-3)!} + \cdots\right) \pmod p.$$

(c) For composite $m > 4$

$$(e^z - 1)^{m-1} \equiv 0 \pmod m.$$

(d) $\qquad \dfrac{z}{e^z - 1} = 1 - \dfrac{e^z - 1}{2} + \dfrac{(e^z - 1)^2}{3} - \dfrac{(e^z - 1)^3}{4} + \cdots.$

Deduce the result on Bernoulli numbers. (See Pólya and Szegö [1972, Vol. II, p. 339].

47. Let $C(\mathbb{Z}(n))$, where $\mathbb{Z}(n)$ is the integers modulo n, be the set of all complex functions on $\mathbb{Z}(n)$, where n is an odd positive integer. Define $F : C(\mathbb{Z}(n)) \rightarrow C(\mathbb{Z}(n))$ by

$$(Ff)(x) = \frac{1}{\sqrt{n}} \sum_{k=0}^{n-1} f(k)e^{2\pi ikx/n} \quad \text{for } x \in \mathbb{Z}(n).$$

(a) Show that Trace $F = \frac{1}{\sqrt{n}} \sum_{k=0}^{n-1} e^{2\pi ik^2/n}$.

Hint: Use the functions $\delta_x, x \in \mathbb{Z}(n)$, where $\delta_x(y) = 0, x \neq y$, and $\delta_x(x) = 1$, as a basis for $\mathbb{Z}(n)$.

(b) Prove that $(F^2 f)(x) = f(-x)$. Conclude that $F^4 = id$ and hence that $\pm 1, \pm i$ are the eigenvalues of F. Let m_1, m_2, m_3, m_4 be the multiplicities of $1, i, -1$, and $-i$ respectively. Thus $m_1 + m_2 + m_3 + m_4 = 1$.

(c) Show that Trace $F^2 = 1$ and conclude that $m_1 - m_2 + m_3 - m_4 = 1$.

(d) Show that $|\frac{1}{\sqrt{n}} \sum_{k=0}^{n-1} e^{2\pi i k^2/n}|^2 = 1$. Use (a) to get $(m_1 - m_2)^2 + (m_3 - m_4)^2 = 1$.

(e) Prove that

$$\det F = i^{(m_2 - m_4) + (m_1 - m_3) - (n+1)/2} = \begin{cases} (m_1 - m_3)i^{(1-n)/2}, & n \equiv 1(4), \\ (m_2 - m_4)i^{(1-n)/2}, & n \equiv 3(4), \end{cases}$$

and also

$$\det F = \det \left(\frac{1}{\sqrt{n}} e^{2\pi ixy/n} \right)_{0 \le x, y \le n-1} = K i^{(1-n)/2},$$

where K is a positive number.

(f) Show that $m_1 = a + 1$, and $m_2 = m_3 = m_4 = a$ when $n = 4a + 1$ and $m_1 = m_2 = m_3 = a$ and $m_4 = a - 1$ when $n = 4a - 1$.

(g) Obtain the value of $\frac{1}{\sqrt{n}} \sum_{k=0}^{n-1} e^{2\pi i k^2/n}$ for n odd.   (Schur)

Let m be a positive integer and let χ be a character on the group $\mathbb{Z}(m)^*$. The function χ can then be defined on all the integers by setting $\chi(k) = 0$ when $\gcd(k, m) > 1$. Clearly χ has period m. We call χ primitive if it does not have a smaller period. Also, χ is even if $\chi(-1) = 1$ and odd if $\chi(-1) = -1$.

Also, define

$$g_k(\chi) = \sum_{n=0}^{m-1} \chi(n)e^{2\pi ikn/m} \quad \text{and} \quad g_1(\chi) = g(\chi).$$

48. For $a \in \mathbb{Z}(p)$, let $N(x^n = a)$ denote the number of solutions of the equation $x^n = a$. If $n \mid p - 1$, then prove that

$$N(x^n = a) = 1 + \sum_{\substack{\chi^n = id \\ \chi \ne id}} \chi(a),$$

where the sum is over all nontrivial characters of order dividing n.

Let a be a nonzero integer. Consider the elliptic curve E defined by $x_0 x_2^2 - x_1^3 - a x_0^3 = 0$, which in affine coordinates is $y^2 = x^3 + a$. Suppose $p \ne 2$ or 3 is a prime that does not divide a. Then $y^2 = x^3 + a$ is an elliptic curve over $\mathbb{Z}(p)$ with a point at infinity. If N_p denotes the number of $\mathbb{Z}(p)$ points on the curve, then $N_p = 1 + N(y^2 = x^3 + a)$.

(a) Show that if $p \equiv 2 \pmod 3$, then $N_p = p + 1$.

(b) Let $p \equiv 1 \pmod 3$ and let χ_3 and χ_2 denote the cubic and quadratic characters of $\mathbb{Z}(p)^*$. Note that $N(y^2 = x^3 + a) = \sum_{u+v=a} N(y^2 = u)$

$N(x^3 = -v)$. Deduce that

$$N_p = p + 1 + \chi_2\chi_3(a)J(\chi_2, \chi_3) + \overline{\chi_2\chi_3}(a)\ \overline{J(\chi_2, \chi_3)}.$$

(c) Show that if $N_p = p + 1 - a_p$ then $|a_p| \le 2\sqrt{p}$.

49. By the method used in the previous problem, show that

$$|N(x^3 + y^3 = 1) - p + 2| \le 2\sqrt{p}.$$

For Exercises 48 and 49, see Ireland and Rosen [1991, Chapters 8 and 18].

50. Prove that if χ is primitive, then

$$g_k(\chi) = \begin{cases} \bar{\chi}(k)g(\chi) & \text{when} \quad \gcd(k, m) = 1, \\ 0 & \text{when} \quad \gcd(k, m) > 1. \end{cases}$$

Define the Dirichlet L-function by

$$L(\chi, s) := \sum_{n=1}^{\infty} \frac{\chi(n)}{n^s}.$$

The series converges for $\mathrm{Re}\, s > 0$, when χ is a nontrivial character, that is, $\chi(n) \ne 1$ for at least one $n \in \mathbb{Z}(m)^*$.

51. (a) Prove that when χ is nontrivial

$$L(\chi, 1) = \frac{-1}{m} \sum_{k=1}^{m-1} g_k(\chi) \log(1 - e^{-2\pi ik/m}).$$

(b) Show that if χ is primitive

$$L(\chi, 1) = -\frac{\chi(-1)g(\chi)}{m} \sum_{k \in \mathbb{Z}(m)^*} \bar{\chi}(k) \log(1 - e^{-2\pi ik/m})$$

$$= -\frac{\chi(-1)g(\chi)}{m} \sum_{k \in \mathbb{Z}(m)^*} \bar{\chi}(k) \left(\log \sin \frac{k\pi}{m} + \frac{k\pi i}{m} \right).$$

(c) Prove that when χ is even, $\sum \bar{\chi}(k)k = 0$, and also, when χ is odd, $\sum \bar{\chi}(k) \log \sin \frac{k\pi}{m} = 0$.

(d) Prove that

$$L(\chi, 1) = \begin{cases} -\frac{2g(\chi)}{m} \sum_{\substack{k \in \mathbb{Z}(m)^* \\ k < m/2}} \bar{\chi}(k) \log \sin \frac{k\pi}{m}, & \text{when } \chi \text{ is even,} \\ \frac{\pi i g(\chi)}{m^2} \sum_{k \in \mathbb{Z}(m)^*} \bar{\chi}(k)k, & \text{when } \chi \text{ is odd.} \end{cases}$$

52. Prove that

(a) $1 - \frac{1}{3} + \frac{1}{5} - \frac{1}{7} + \cdots = \frac{\pi}{4}$ (Madhava–Leibniz)

(b) $1 + \frac{1}{3} - \frac{1}{5} - \frac{1}{7} + \frac{1}{9} + \frac{1}{11} - \cdots = \frac{\pi}{2\sqrt{2}}$ (Newton)

(c) $1 - \frac{1}{2} + \frac{1}{4} - \frac{1}{5} + \frac{1}{7} - \frac{1}{8} + \cdots = \frac{\pi}{3\sqrt{3}}$ (Euler)

(d) $1 + \frac{1}{2} - \frac{1}{3} + \frac{1}{4} - \frac{1}{5} - \frac{1}{6} + \frac{1}{8} + \cdots = \frac{\pi}{\sqrt{7}}$ (Euler)

(e) $1 - \frac{1}{2} - \frac{1}{3} + \frac{1}{4} + \frac{1}{6} - \frac{1}{7} - \frac{1}{8} + \frac{1}{9} + \frac{1}{11} - \cdots = \frac{2}{\sqrt{5}} \log \frac{1+\sqrt{5}}{2}$

The series for $\pi/4$, usually called Leibniz's formula, was known to Madhava in the fourteenth century. See Roy [1990]. Newton [1960, p. 156] produced his series in response to Leibniz's formula by evaluating the integral

$$\int_0^1 \frac{1+x^2}{1+x^4} dx$$

in two different ways. Series (c) and (d) are attributed to Euler by Scharlau and Opolka [1985, pp. 30 and 83].

Define the generalized Bernoulli numbers by the formula

$$\sum_{a=1}^{m} \frac{\chi(a) x e^{ax}}{e^{mx} - 1} = \sum_{n=0}^{\infty} B_{n,\chi} \frac{x^n}{n!}.$$

53. (a) Prove the following functional equation for $L(\chi, s)$, χ primitive:

$$L(\chi, s) = \frac{g(\chi)}{2i^\delta} \left(\frac{2\pi}{m} \right)^s \frac{L(\bar{\chi}, 1-s)}{\Gamma(s) \cos \frac{\pi(s-\delta)}{2}},$$

where $\delta = 0$ or 1 according as χ is even or odd.

Hint: Consider the integral

$$\int_C \frac{t^{s-1} \sum_{a=1}^{m} \chi(a) e^{at}}{e^{mt} - 1} dt,$$

where C is as in problems 23 and 24. Follow the procedure given in those problems.

(b) For any integer $n \geq 1$, show that

$$L(\chi, 1-n) = -\frac{B_{n,\chi}}{n}.$$

(c) For $n \geq 1$ and $n \equiv \delta \pmod{2}$ (δ as defined in (a)), prove that

$$L(\chi, n) = (-1)^{\frac{n-\delta}{2}+1} \frac{g(\chi)}{2i^\delta} \left(\frac{2\pi}{m} \right)^n \frac{B_{n,\bar{\chi}}}{n!}.$$

54. Let P be any point between 0 and 1. Show that

$$\int_P^{(1+,0+,1-,0-)} t^{\alpha-1}(1-t)^{\beta-1} dt = \frac{-4\pi^2 e^{\pi i (\alpha+\beta)}}{\Gamma(1-\alpha)\Gamma(1-\beta)\Gamma(\alpha+\beta)}.$$

The notation implies that the integration is over a contour that starts at P, encircles the point 1 in the positive (counterclockwise) direction, returns to P, then encircles the origin in the positive direction, and returns to P. The $1-, 0-$ indicates that now the path of integration is in the clockwise direction, first around 1 and then 0. See Whittaker and Watson [1940, pp. 256–257].

55. Let $G(z) = \log \Gamma(z)$. Show that

(a) If $x \geq 1/2$, then Re $G''(x + iy) > 0$ for all real y.

(b) If $x \leq 1/2$, then Re $G''(x + iy) < 0$ for all sufficiently large y.

(c) If $1/2 \leq a < b$, then

$$\arg \frac{\Gamma(b + iy)}{\Gamma(a + iy)}$$

is an increasing function of y on $(-\infty, \infty)$.

(d) The conclusion in (c) also holds if $0 < a < 1/2$ and $b > 1 - a$.

(Ahern and Rudin)

56. Show that

$$2. \sum_{k=1}^{\infty} \frac{(-1)^{k+1} k^2}{k^3 + 1} = \qquad \frac{1}{3} - \frac{1}{3}\log 2 + \frac{\pi}{3} \text{ sech } (\pi\sqrt{3}/2).$$

This problem was given without the value by Amend [1996]. FOXTROT

2

The Hypergeometric Functions

Almost all of the elementary functions of mathematics are either hypergeometric or ratios of hypergeometric functions. A series Σc_n is hypergeometric if the ratio c_{n+1}/c_n is a rational function of n. Many of the nonelementary functions that arise in mathematics and physics also have representations as hypergeometric series.

In this chapter, we introduce three important approaches to hypergeometric functions. First, Euler's fractional integral representation leads easily to the derivation of essential identities and transformations of hypergeometric functions. A second-order linear differential equation satisfied by a hypergeometric function provides a second method. This equation was also found by Euler and then studied by Gauss. Still later, Riemann observed that a characterization of second-order equations with three regular singularities gives a powerful technique, involving minimal calculation, for obtaining formulas for hypergeometric functions. Third, Barnes expressed a hypergeometric function as a contour integral, which can be seen as a Mellin inversion formula. Some integrals that arise here are really extensions of beta integrals. They also appear in the orthogonality relations for some special orthogonal polynomials.

Perceiving their significance, Gauss gave a complete list of contiguous relations for $_2F_1$ functions. These have numerous applications. We show how they imply some continued fraction expansions for hypergeometric functions and also contain three-term recurrence relations for hypergeometric orthogonal polynomials. We discuss one case of the latter in this chapter, namely, Jacobi polynomials.

2.1 The Hypergeometric Series

A hypergeometric series is a series $\sum c_n$ such that c_{n+1}/c_n is a rational function of n. On factorizing the polynomials in n, we obtain

$$\frac{c_{n+1}}{c_n} = \frac{(n + a_1)(n + a_2) \cdots (n + a_p)x}{(n + b_1)(n + b_2) \cdots (n + b_q)(n + 1)}. \tag{2.1.1}$$

The x occurs because the polynomial may not be monic. The factor $(n + 1)$ may result from the factorization, or it may not. If not, add it along with the compensating factor $(n + 1)$ in the numerator. At present, a reason for inserting this factor is to introduce $n!$ in the hypergeometric series $\sum c_n$. This is a convenient factor to have in a hypergeometric series, since it often occurs naturally for many cases that are significant enough to have been given names. Later in this chapter we shall give a more intrinsic reason.

From (2.1.1) we have

$$\sum_{n=0}^{\infty} c_n = c_0 \sum_{n=0}^{\infty} \frac{(a_1)_n \cdots (a_p)_n}{(b_1)_n \cdots (b_q)_n} \frac{x^n}{n!} =: c_0 \, {}_pF_q \left(\begin{matrix} a_1, \ldots, a_p \\ b_1, \ldots, b_q \end{matrix} ; x \right). \quad (2.1.2)$$

Here the b_i are not negative integers or zero, as that would make the denominator zero. For typographical reasons, we shall sometimes denote the sum on the right side of (2.1.2) by $_pF_q(a_1, \ldots, a_p; b_1, \ldots, b_q; x)$ or by $_pF_q$. It is natural to apply the ratio test to determine the convergence of the series (2.1.2). Thus,

$$\left| \frac{c_{n+1}}{c_n} \right| \leq \frac{|x| n^{p-q-1} (1 + |a_1|/n) \cdots (1 + |a_p|/n)}{|(1 + 1/n)(1 + b_1/n) \cdots (1 + b_q/n)|}.$$

An immediate consequence of this is the following:

Theorem 2.1.1 *The series $_pF_q(a_1, \ldots, a_p; b_1, \ldots, b_q; x)$ converges absolutely for all x if $p \leq q$ and for $|x| < 1$ if $p = q + 1$, and it diverges for all $x \neq 0$ if $p > q + 1$ and the series does not terminate.*

Proof. It is clear that $|c_{n+1}/c_n| \to 0$ as $n \to \infty$ if $p < q$. For $p = q + 1$, $\lim_{n \to \infty} |c_{n+1}/c_n| = |x|$, and for $p > q + 1$, $|c_{n+1}/c_n| \to \infty$ as $n \to \infty$. This proves the theorem. ∎

The case $|x| = 1$ when $p = q + 1$ is of great interest. The next result gives the conditions for convergence in this case.

Theorem 2.1.2 *The series $_{q+1}F_q(a_1, \ldots, a_{q+1}; b_1, \ldots, b_q; x)$ with $|x| = 1$ converges absolutely if $\mathrm{Re}(\sum b_i - \sum a_i) > 0$. The series converges conditionally if $x = e^{i\theta} \neq 1$ and $0 \geq \mathrm{Re}(\sum b_i - \sum a_i) > -1$ and the series diverges if $\mathrm{Re}(\sum b_i - \sum a_i) \leq -1$.*

Proof. The coefficient of nth term in $_{q+1}F_q$ is

$$\frac{(a_1)_n \cdots (a_{q+1})_n}{(b_1)_n \cdots (b_q)_n n!},$$

and the definition of the gamma function implies that this term is

$$\sim \frac{\Pi \Gamma(b_i)}{\Pi \Gamma(a_i)} n^{\Sigma a - \Sigma b - 1}$$

as $n \to \infty$. Usually one invokes Stirling's formula to obtain this, but that is not necessary. See Formula (1.4.3). The statements about absolute convergence and divergence follow immediately. The part of the theorem concerning conditional convergence can be proved by summation by parts. ∎

This chapter will focus on a study of the special case $_2F_1(a, b; c; x)$, though more general series will be considered in a few places. The $_2F_1$ series was studied extensively by Euler, Pfaff, Gauss, Kummer, and Riemann and most of the present chapter and the next one is a discussion of their fundamental ideas.

We saw that $_2F_1(a, b; c; x)$ diverges in general for $x = 1$ and $\text{Re}(c - a - b) \leq 0$. The next theorem due to Gauss describes the behavior of the series as $x \to 1^-$. A proof is given later in the text, where it arises naturally.

Theorem 2.1.3   *If* $\text{Re}(c - a - b) < 0$, *then*

$$\lim_{x \to 1^-} \frac{_2F_1(a, b; c; x)}{(1 - x)^{c-a-b}} = \frac{\Gamma(c)\Gamma(a + b - c)}{\Gamma(a)\Gamma(b)};$$

and for $c = a + b$,

$$\lim_{x \to 1^-} \frac{_2F_1(a, b; a + b; x)}{\log(1/(1 - x))} = \frac{\Gamma(a + b)}{\Gamma(a)\Gamma(b)}.$$

The next result about partial sums of $_2F_1(a, b; c; 1)$ is due to Hill [1908]. It can be stated more generally for $_{p+1}F_p$. The proof is left as an exercise.

Theorem 2.1.4   *Let* s_n *denote the* n*th partial sum of* $_2F_1(a, b; c; 1)$. *For* $\text{Re}(c - a - b) < 0$,

$$s_n \sim \frac{\Gamma(c)n^{a+b-c}}{\Gamma(a)\Gamma(b)(a + b - c)},$$

and for $c = a + b$,

$$s_n \sim \frac{\Gamma(c)\log n}{\Gamma(a)\Gamma(b)}.$$

The theorem is easily believable when we note that the nth term is

$$\sim \frac{\Gamma(c)}{\Gamma(a)\Gamma(b)} n^{a+b-c-1}.$$

The necessary result would now follow if we replace the sum with an integral.

Many of the elementary functions have representations as hypergeometric series. Here are some examples:

$$\log(1 + x) = x \, {}_2F_1\left(\begin{matrix} 1, 1 \\ 2 \end{matrix}; -x\right);$$

(2.1.3)

$$\tan^{-1} x = x \, {}_2F_1\left(\begin{matrix} 1/2, 1 \\ 3/2 \end{matrix}; -x^2\right);$$

(2.1.4)

$$\sin^{-1} x = x \, {}_2F_1\left(\begin{matrix} 1/2, 1/2 \\ 3/2 \end{matrix}; x^2\right);$$

(2.1.5)

$$(1 - x)^{-a} = {}_1F_0\left(\begin{matrix} a \\ - \end{matrix}; x\right).$$

(2.1.6)

This last relation is merely the binomial theorem. We also have

$$\sin x = x \, {}_0F_1\left(\begin{matrix} - \\ 3/2 \end{matrix}; -x^2/4\right);$$

(2.1.7)

$$\cos x = {}_0F_1\left(\begin{matrix} - \\ 1/2 \end{matrix}; \frac{-x^2}{4}\right);$$

(2.1.8)

$$e^x = {}_0F_0\left(\begin{matrix} - \\ - \end{matrix}; x\right).$$

(2.1.9)

The next set of examples uses limits:

$$e^x = \lim_{b \to \infty} {}_2F_1\left(\begin{matrix} 1, b \\ 1 \end{matrix}; \frac{x}{b}\right);$$

(2.1.10)

$$\cosh x = \lim_{a,b \to \infty} {}_2F_1\left(\begin{matrix} a, b \\ 1/2 \end{matrix}; \frac{x^2}{4ab}\right);$$

(2.1.11)

$${}_1F_1\left(\begin{matrix} a \\ c \end{matrix}; x\right) = \lim_{b \to \infty} {}_2F_1\left(\begin{matrix} a, b \\ c \end{matrix}; \frac{x}{b}\right);$$

(2.1.12)

$${}_0F_1\left(\begin{matrix} - \\ c \end{matrix}; x\right) = \lim_{a,b \to \infty} {}_2F_1\left(\begin{matrix} a, b \\ c \end{matrix}; \frac{x}{ab}\right).$$

(2.1.13)

The example of $\log(1 - x) = -x \, {}_2F_1(1, 1; 2; x)$ shows that though the series converges for $|x| < 1$, it has a continuation as a single-valued function in the complex plane from which a line joining 1 to ∞ is deleted. This describes the general situation; a ${}_2F_1$ function has a continuation to the complex plane with branch points at 1 and ∞.

Definition 2.1.5   *The hypergeometric function $_2F_1(a, b; c; x)$ is defined by the series*

$$\sum_{n=0}^{\infty} \frac{(a)_n (b)_n}{(c)_n n!} x^n$$

for $|x| < 1$, and by continuation elsewhere.

When the words "hypergeometric function" are used, they usually refer to the function $_2F_1(a, b; c; x)$. We will usually follow this tradition, but when referring to a hypergeometric series it will not necessarily mean just $_2F_1$. Hypergeometric series will be the series defined in (2.1.2).

2.2 Euler's Integral Representation

Contained in the following theorem is an important integral representation of the $_2F_1$ function due to Euler [1769, Vol. 12, pp. 221–230]. This integral also has an interpretation as a fractional integral as discussed in Section 2.9.

Theorem 2.2.1 *If* $\operatorname{Re} c > \operatorname{Re} b > 0$, *then*

$$_2F_1\left(\begin{matrix} a, b \\ c \end{matrix}; x\right) = \frac{\Gamma(c)}{\Gamma(b)\Gamma(c-b)} \int_0^1 t^{b-1}(1-t)^{c-b-1}(1-xt)^{-a} dt$$

in the x plane cut along the real axis from 1 to ∞. Here it is understood that $\arg t = \arg(1-t) = 0$ *and* $(1-xt)^{-a}$ *has its principal value.*

Proof. Suppose at first that $|x| < 1$. Expand $(1-xt)^{-a}$ by the binomial theorem given in (2.1.6) so that the right side of the formula becomes

$$\frac{\Gamma(c)}{\Gamma(b)\Gamma(c-b)} \sum_{n=0}^{\infty} \frac{(a)_n}{n!} x^n \int_0^1 t^{n+b-1}(1-t)^{c-b-1} dt.$$

This is a beta integral, which in terms of the gamma function is

$$\frac{\Gamma(n+b)\Gamma(c-b)}{\Gamma(n+c)}.$$

Substitute this in the last expression to get

$$\frac{\Gamma(c)}{\Gamma(b)} \sum_{n=0}^{\infty} \frac{(a)_n \Gamma(n+b)}{n! \Gamma(n+c)} x^n = {_2F_1}\left(\begin{matrix} a, b \\ c \end{matrix}; x\right).$$

This proves the result for $|x| < 1$. Since the integral is analytic in the cut plane, the theorem holds for x in this region as well. ∎

The integral in Theorem 2.2.1 may be viewed as the analytic continuation of the $_2F_1$ series, but only when $\operatorname{Re} c > \operatorname{Re} b > 0$. The function $(1-xt)^{-a}$ in the integrand is in general multivalued and one may study the multivalued nature of $_2F_1(a, b; c; x)$ using this integral. To discuss analytic continuation more deeply would require some ideas from the theory of Riemann surfaces, which goes beyond the scope of this book. See Klein [1894].

It is also important to note that we view $_2F_1(a, b; c; x)$ as a function of four complex variables $a, b, c,$ and x instead of just x. It is easy to see that $\frac{1}{\Gamma(c)} {_2F_1}(a, b; c; x)$ is an entire function of a, b, c if x is fixed and $|x| < 1$, for in this case the series converges uniformly in every compact domain of the a, b, c space. Analytic continuation may be applied to the parameters a, b, c. The results may at first be obtained under some restrictions, and then extended. For example:

Theorem 2.2.2 (Gauss [1812])   *For* $\text{Re}(c - a - b) > 0$, *we have*

$$\sum_{n=0}^{\infty} \frac{(a)_n (b)_n}{n!(c)_n} = {}_2F_1\left(\begin{matrix} a, b \\ c \end{matrix}; 1\right) = \frac{\Gamma(c)\Gamma(c - a - b)}{\Gamma(c - a)\Gamma(c - b)}.$$

Proof.   Let $x \to 1^-$ in Euler's integral for ${}_2F_1$. The result is, by Abel's continuity theorem,

$$\begin{aligned}
{}_2F_1\left(\begin{matrix} a, b \\ c \end{matrix}; 1\right) &= \frac{\Gamma(c)}{\Gamma(b)\Gamma(c - b)} \int_0^1 t^{b-1}(1 - t)^{c-a-b-1} dt \\
&= \frac{\Gamma(c)\Gamma(c - a - b)}{\Gamma(c - a)\Gamma(c - b)},
\end{aligned}$$

when $\text{Re}\, c > \text{Re}\, b > 0$ and $\text{Re}(c - a - b) > 0$. The condition $\text{Re}\, c > \text{Re}\, b > 0$ may be removed by continuation. It is, however, instructive to give a proof that does not appeal to the principle of analytic continuation.

Our first goal is to prove the relationship

$$\qquad {}_2F_1\left(\begin{matrix} a, b \\ c \end{matrix}; 1\right) = \frac{(c - a)(c - b)}{c(c - a - b)} {}_2F_1\left(\begin{matrix} a, b \\ c + 1 \end{matrix}; 1\right). \qquad (2.2.1)$$

If

$$A_n = \frac{(a)_n (b)_n}{n!(c)_n} \quad \text{and} \quad B_n = \frac{(a)_n (b)_n}{n!(c + 1)_n},$$

then

$$c(c - a - b)A_n - (c - a)(c - b)B_n = \frac{(a)_n (b)_n}{n!(c + 1)_{n-1}}\left[c - a - b - \frac{(c - a)(c - b)}{c + n}\right]$$

and

$$c(nA_n - (n + 1)A_{n+1}) = \frac{(a)_n (b)_n}{n!(c + 1)_{n-1}}\left[n - \frac{(a + n)(b + n)}{c + n}\right].$$

So, since the right sides in the last two expresions are equal,

$$c(c - a - b)A_n = (c - a)(c - b)B_n + cnA_n - c(n + 1)A_{n+1}$$

and

$$c(c - a - b)\sum_0^N A_n = (c - a)(c - b)\sum_0^N B_n - c(N + 1)A_{N+1}.$$

Now let $N \to \infty$ and observe that $(N + 1)A_{N+1} \sim 1/N^{c-a-b} \to 0$, because $\text{Re}(c - a - b) > 0$. This proves (2.2.1). Iterate this relation n times to get

$$\frac{\Gamma(c-a)\Gamma(c-b)}{\Gamma(c)\Gamma(c-a-b)} {}_2F_1\left(\begin{matrix} a, b \\ c \end{matrix}; 1\right) = \frac{\Gamma(c+n-a)\Gamma(c+n-b)}{\Gamma(c+n)\Gamma(c+n-a-b)} {}_2F_1\left(\begin{matrix} a, b \\ c+n \end{matrix}; 1\right).$$

It is an easy verification that the right side $\to 1$ as $n \to \infty$. This proves the theorem for $\text{Re}(c - a - b) > 0$. The theorem is called Gauss's summation formula. ∎

The case where one of the upper parameters is a negative integer, thereby making the ${}_2F_1$ a finite sum, is worthy of note. This result was essentially known to the thirteenth century Chinese mathematician Chu and rediscovered later. See Askey [1975, Chapter 7].

Corollary 2.2.3 (Chu–Vandermonde)

$$_2F_1\left(\begin{matrix} -n, a \\ c \end{matrix}; 1\right) = \frac{(c-a)_n}{(c)_n}.$$

Euler's integral for ${}_2F_1$ can be generalized to ${}_pF_q$. Rewrite it as

$$_2F_1\left(\begin{matrix} a, b \\ c \end{matrix}; x\right) = \frac{\Gamma(c)}{\Gamma(b)\Gamma(c-b)} \int_0^1 t^{b-1}(1-t)^{c-b-1} {}_1F_0(a; xt)\,dt.$$

Thus, integrating a ${}_1F_0$ with respect to the beta distribution $t^{b-1}(1-t)^{c-b-1}$ gives a ${}_2F_1$, that is, a parameter b is added in the numerator and c in the denominator of the original ${}_1F_0(a; t)$.

More generally, we have

$$_{p+1}F_{q+1}\left(\begin{matrix} a_1, \ldots, a_p, a_{p+1} \\ b_1, \ldots, b_q, b_{q+1} \end{matrix}; x\right) = \frac{\Gamma(b_{q+1})}{\Gamma(a_{p+1})\Gamma(b_{q+1}-a_{p+1})} \int_0^1 t^{a_{p+1}-1}$$

$$\cdot (1-t)^{b_{q+1}-a_{p+1}-1} {}_pF_q\left(\begin{matrix} a_1, \ldots, a_p \\ b_1, \ldots, b_q \end{matrix}; xt\right) dt \tag{2.2.2}$$

when $\text{Re}\, b_{q+1} > \text{Re}\, a_{p+1} > 0$. This condition is needed for the convergence of the integral. By a change of variables the expression on the right of (2.2.2) also equals

$$\frac{\Gamma(b_{q+1})x^{1-b_{q+1}}}{\Gamma(a_{p+1})\Gamma(b_{q+1}-a_{p+1})} \int_0^x t^{a_{p+1}-1}(x-t)^{b_{q+1}-a_{p+1}-1} {}_pF_q\left(\begin{matrix} a_1, \ldots, a_p \\ b_1; \ldots, b_q \end{matrix}; t\right) dt.$$

$$\tag{2.2.3}$$

Note also that (2.2.2) can be used to change the value of a denominator or numerator parameter in ${}_pF_q(a_1, \ldots, a_p; b_1, \ldots, b_q; x)$. For example, take $a_{p+1} = b_q$ in

(2.2.2) to get

$$
{}_pF_q\left(\begin{array}{c} a_1,\ldots,a_p \\ b_1,\ldots,b_{q-1},b_{q+1} \end{array}; x\right) = \frac{\Gamma(b_{q+1})}{\Gamma(b_q)\Gamma(b_{q+1}-b_q)}
$$

$$
\cdot \int_0^1 t^{b_q-1}(1-t)^{b_{q+1}-b_q-1}\,{}_pF_q\left(\begin{array}{c} a_1,\ldots,a_p \\ b_1,\ldots,b_q \end{array}; xt\right)dt. \qquad (2.2.4)
$$

It should be remarked that when x is a complex variable in (2.2.2) to (2.2.4), then the ${}_pF_q$ is in general a multivalued function. Thus, the variable x has to be restricted to a domain where the ${}_pF_q$ in the integrand is single valued. One must take care to state the conditions for single-valuedness. We note a special case of (2.2.4).

Theorem 2.2.4   *For* $\mathrm{Re}\,c > \mathrm{Re}\,d > 0$, $x \neq 1$, *and* $|\arg(1-x)| < \pi$,

$$
{}_2F_1\left(\begin{array}{c} a,b \\ c \end{array}; x\right) = \frac{\Gamma(c)}{\Gamma(d)\Gamma(c-d)}\int_0^1 t^{d-1}(1-t)^{c-d-1}\,{}_2F_1\left(\begin{array}{c} a,b \\ d \end{array}; xt\right)dt.
$$

One pecularity of Euler's integral for ${}_2F_1$ is that the ${}_2F_1$ is obviously symmetric in the upper parameters a and b, whereas it is not evident that the integral remains the same when a and b are interchanged. Erdélyi [1937] has presented a double integral from which the two representations can be obtained:

$$
\frac{[\Gamma(c)]^2}{\Gamma(a)\Gamma(b)\Gamma(c-a)\Gamma(c-b)}\int_0^1\int_0^1 t^{b-1}s^{a-1}(1-t)^{c-b-1}
$$

$$
\cdot (1-s)^{c-a-1}(1-tsx)^{-c}dtds. \qquad (2.2.5)
$$

The next theorem gives an important application of Euler's integral to the derivation of two transformation formulas of hypergeometric functions.

Theorem 2.2.5

$$
{}_2F_1\left(\begin{array}{c} a,b \\ c \end{array}; x\right) = (1-x)^{-a}\,{}_2F_1\left(\begin{array}{c} a,c-b \\ c \end{array}; \frac{x}{x-1}\right) \quad \text{(Pfaff)}, \qquad (2.2.6)
$$

$$
{}_2F_1\left(\begin{array}{c} a,b \\ c \end{array}; x\right) = (1-x)^{c-a-b}\,{}_2F_1\left(\begin{array}{c} c-a,c-b \\ c \end{array}; x\right) \quad \text{(Euler)}. \qquad (2.2.7)
$$

Proof.   Replace t with $1-s$ in Euler's integral (Theorem 2.2.1) to obtain

$$
{}_2F_1\left(\begin{array}{c} a,b \\ c \end{array}; x\right) = \frac{\Gamma(c)}{\Gamma(b)\Gamma(c-b)}\int_0^1 (1-x+xs)^{-a}(1-s)^{b-1}s^{c-b-1}ds
$$

$$
= \frac{(1-x)^{-a}\Gamma(c)}{\Gamma(b)\Gamma(c-b)}\int_0^1\left(1-\frac{xs}{x-1}\right)^{-a}s^{c-b-1}(1-s)^{b-1}ds.
$$

This proves Pfaff's [1797] transformation for $\operatorname{Re} c > \operatorname{Re} b > 0$. The complete result follows by continuation of c and b.

The hypergeometric function is symmetric in the parameters a and b, so we apply Pfaff's transformation to itself:

$$
{}_2F_1\left(\begin{matrix} a, b \\ c \end{matrix}; x\right) = (1-x)^{-a}\left(1 - \frac{x}{x-1}\right)^{-c+b} {}_2F_1\left(\begin{matrix} c-a, c-b \\ c \end{matrix}; x\right).
$$

This is Euler's [1794] formula and the theorem is proved. ∎

The right-hand series in Pfaff's transformation converges for $|x/(x-1)| < 1$. This condition is implied by $\operatorname{Re} x < 1/2$; so we have a continuation of the series ${}_2F_1(a, b; c; x)$ to this region by Pfaff's formula.

The following two examples give an indication of the power of the transformation formulas. By Pfaff's transformation,

$$
\tan^{-1} x = x \; {}_2F_1\left(\begin{matrix} 1/2, 1 \\ 3/2 \end{matrix}; -x^2\right) = \frac{x}{\sqrt{1+x^2}} \; {}_2F_1\left(\begin{matrix} 1/2, 1/2 \\ 3/2 \end{matrix}; \frac{x^2}{1+x^2}\right)
$$

$$
= \sin^{-1}\frac{x}{\sqrt{1+x^2}}.
$$

In Chapter 1 we showed how the gamma function could be used to develop some aspects of trigonometric functions, starting with the series definitions of the sine and cosine functions. The above relation can now be the basis for the connection between the trigonometric functions and a right triangle.

For the second example, write Euler's transformation as

$$
(1-x)^{a+b-c} {}_2F_1\left(\begin{matrix} a, b \\ c \end{matrix}; x\right) = {}_2F_1\left(\begin{matrix} c-a, c-b \\ c \end{matrix}; x\right).
$$

Equate the coefficient of x^n on both sides to get

$$
\sum_{j=0}^{n} \frac{(a)_j (b)_j (c-a-b)_{n-j}}{j!(c)_j (n-j)!} = \frac{(c-a)_n (c-b)_n}{n!(c)_n}.
$$

Rewrite this as:

Theorem 2.2.6 (Pfaff–Saalschütz)

$$
{}_3F_2\left(\begin{matrix} -n, a, b \\ c, 1+a+b-c-n \end{matrix}; 1\right) = \frac{(c-a)_n (c-b)_n}{(c)_n (c-a-b)_n}. \tag{2.2.8}
$$

Gauss's ${}_2F_1$ sum (Theorem 2.2.2) follows from this by letting $n \to \infty$. The limiting procedure may be justified by Tannery's theorem, which is a discrete form of Lebesgue's dominated convergence theorem. Theorem 2.2.6 was first discoverd by Pfaff [1797a] and rediscovered by Saalschütz [1890]. It is often called Saalschütz's theorem but this nomenclature does not give due credit to

Pfaff. Surprisingly, a special case seems to have been found by Chu. See Takács
[1973].

Remark 2.2.1   The Chu–Vandermonde identity (Corollary 2.2.3) gives the sum
of a terminating $_2F_1$. The Pfaff–Saalschütz identity involves a special type of
terminating $_3F_2$. The sum of denominator parameters is one more than the sum
of the numerator parameters. Such a series is called balanced. This identity
was obtained by a factorization of a $_2F_1$ and it is worth noting that the Chu–
Vandermonde identity can be derived from a factorization of a $_1F_0$. Thus, one may
equate the coefficients of x^n in

$$(1 - x)^{-a}(1 - x)^{-b} = (1 - x)^{-(a+b)}$$

to get an equivalent identity:

$$\sum_{k=0}^{n} \frac{(a)_k (b)_{n-k}}{k!(n - k)!} = \frac{(a + b)_n}{n!}.$$

The right side of Pfaff's transformation formula when expanded as a series
equals

$$\sum_{k=0}^{\infty} \frac{(a)_k (c - b)_k}{(c)_k k!} (-x)^k (1 - x)^{-k-a}$$

$$= \sum_{k=0}^{\infty} \sum_{j=0}^{\infty} \frac{(a)_k (c - b)_k}{(c)_k k!} (-x)^k \frac{(a + k)_j}{j!} x^j.$$

Note that $(a)_k (a + k)_j = (a)_{j+k}$; then write $j + k = n$ to see that the sum is

$$\sum_{n=0}^{\infty} \frac{(a)_n}{n!} \sum_{k=0}^{n} \frac{(-n)_k (c - b)_k}{(c)_k k!} x^n = \sum_{n=0}^{\infty} \frac{(a)_n (b)_n}{n!(c)_n} x^n,$$

where the inner sum was evaluated by the Chu–Vandermonde identity. This gives
another proof of Pfaff's transformation.

The following definition is suggested by the Pfaff–Saalschütz formula.

Definition 2.2.7   A series

$$_{p+1}F_p \left(\begin{array}{c} a_1, \ldots, a_{p+1} \\ b_1, \ldots, b_p \end{array} ; x \right)$$

is called balanced if $x = 1$, one of the numerator parameters is a negative integer,
and $a_1 + \cdots + a_{p+1} + 1 = b_1 + \cdots + b_p$.

Remark 2.2.2   The Pfaff–Saalschütz identity can be written as

$$(c)_n (c + a + b)_n {}_3F_2 \left(\begin{array}{c} -n, -a, -b \\ c, 1 - a - b - n - c \end{array} ; 1 \right) = (c + a)_n (c + b)_n.$$

This is a polynomial identity in a, b, c. Dougall [1907] took the view that both sides of this equation are polynomials of degree n in a. Therefore, the identity is true if both sides are equal for $n + 1$ distinct values of a. Clearly the result is true when $n = 0$. Assume the result true for $n = 0, 1, \ldots, k - 1$. Now set $n = k$. By symmetry in a and n, it follows that the identity is true for $a = 0, 1, \ldots, k - 1$. These are k values, so if we can find one more value of a for which the identity holds, then it is proved. Note that

$$\frac{(c + a + b)_n}{(1 - a - b - n - c)_j} = (-1)^j (c + a + b)_{n-j}.$$

So, if $a = -b - c$ then both sides of the identity are equal to $(-a)_n(-b)_n$. This proves the identity. Dougall showed that a more general identity could be proved by this method. The identity is

$$_7F_6 \left(\begin{array}{c} a, 1 + \tfrac{1}{2}a, -b, -c, -d, -e, -n \\ \tfrac{1}{2}a, 1 + a + b, 1 + a + c, 1 + a + d, 1 + d + e, 1 + a + n \end{array} ; 1 \right)$$

$$= \frac{(1 + a)_n (1 + a + b + c)_n (1 + a + b + d)_n (1 + a + c + d)_n}{(1 + a + b)_n (1 + a + c)_n (1 + a + d)_n (1 + a + b + c + d)_n}, \quad (2.2.9)$$

where $1 + 2a + b + c + d + e + n = 0$ and n is a positive integer. This condition means that the series terminates and the sum of the denominator parameters is 2 more than the sum of the numerator parameter. Such a series is called 2-balanced. (A 1-balanced series is balanced as in Pfaff–Saalschütz.) Note that the sum of the parameters in a column in this $_7F_6$ add up to the same quantity. Thus, $1 + a = 1 + \tfrac{1}{2}a + \tfrac{1}{2}a = 1 + a + b - b = 1 + a + c - c$ and so on. This type of series is called well poised. Dougall's identity thus gives the sum of a class of well-poised 2-balanced $_7F_6$. These series are called very well poised because the series contains the factor

$$\frac{\left(\tfrac{a}{2} + 1 \right)_k}{\left(\tfrac{a}{2} \right)_k} = \frac{a + 2k}{a}.$$

This identity was also discovered by Ramanujan at around the same time as Dougall's discovery. See Hardy [1940, p. 102]. A proof of the sum (2.2.9) is given later. To obtain another important identity from Dougall, let $n \to \infty$ to get

$$_5F_4 \left(\begin{array}{c} a, a/2 + 1, -b, -c, -d \\ a/2, a + b + 1, a + c + 1, a + d + 1 \end{array} ; 1 \right)$$

$$= \frac{\Gamma(a + b + 1)\Gamma(a + c + 1)\Gamma(a + d + 1)\Gamma(a + b + c + d + 1)}{\Gamma(a + 1)\Gamma(a + b + c + 1)\Gamma(a + b + d + 1)\Gamma(a + c + d + 1)} \quad (2.2.10)$$

when $\mathrm{Re}(a + b + c + d + 1) > 0$. Then take $d = -a/2$ to get

$$
{}_3F_2\left(\begin{array}{c} a, -b, -c \\ a + b + 1, a + c + 1 \end{array}; 1\right)
$$

$$
= \frac{\Gamma\left(\frac{a}{2} + 1\right)\Gamma(a + b + 1)\Gamma(a + c + 1)\Gamma\left(\frac{a}{2} + b + c + 1\right)}{\Gamma(a + 1)\Gamma\left(\frac{a}{2} + b + 1\right)\Gamma\left(\frac{a}{2} + c + 1\right)\Gamma(a + b + c + 1)}. \qquad (2.2.11)
$$

This gives the sum of a general well-poised ${}_3F_2$ series. It is due to Dixon [1903]. The limiting procedure above may be justified by Tannery's theorem. A more general result than (2.2.10) had been found by Rogers [1895]. This will be given later.

Remark 2.2.3   We have seen that the Chu–Vandermonde identity can be obtained from the Euler's integral in Theorem 2.2.1, which is an immediate consequence of the value of the beta function. A type of converse holds. The Chu–Vandermonde identity is a discrete form of the beta integral formula,

$$
\int_0^1 t^{a-1}(1 - t)^{b-1}dt = \frac{\Gamma(a)\Gamma(b)}{\Gamma(a + b)}.
$$

By Remark 2.2.1, we have

$$
\frac{n!}{(a + b)_n}\sum_{k=0}^{n}\frac{(a)_k(b)_{n-k}}{k!(n - k)!} = 1.
$$

We briefly sketch the argument showing that a limiting form of this identity is the beta integral formula. Rewrite the identity as

$$
\frac{(n + 1)!}{(a + b)_n} \cdot \frac{1}{n + 1}\sum_{k=0}^{n}\frac{(a)_k(b)_{n-k}}{k!(n - k)!} = 1
$$

or

$$
\lim_{n \to \infty}\frac{1}{n + 1}\sum_{k=0}^{n}\frac{[(k + 1)^{1-a}(a)_k/k!][(n + 1 - k)^{1-b}(b)_{n-k}/(n - k)!]}{(n + 1)^{1-a-b}(a + b)_n/n!}
$$

$$
\cdot \left(\frac{k + 1}{n + 1}\right)^{a-1}\left(1 - \frac{k}{n + 1}\right)^{b-1} = 1.
$$

Recall that by definition

$$
\lim_{k \to \infty}\frac{(a)_k(k + 1)^{1-a}}{k!} = \frac{1}{\Gamma(a)}.
$$

If we break up the sum as

$$\sum_{k=0}^{\log n} + \sum_{\log n}^{n-\log n} + \sum_{n-\log n}^{n},$$

the first and third sums go to zero and the second tends to

$$\frac{\Gamma(a+b)}{\Gamma(a)\Gamma(b)} \int_0^1 t^{a-1}(1-t)^{b-1} dt \quad \text{as} \quad n \to \infty.$$

This expression equals 1 and we have the result. The reader should try to find the beta integral that corresponds to Gauss's formula for $_2F_1$ at $x = 1$.

2.3 The Hypergeometric Equation

The hypergeometric function satisfies a second-order differential equation with three regular singular points. This equation was found by Euler [1769] and was extensively studied by Gauss [1812] and Kummer [1836]. Riemann [1857] introduced a more abstract approach, which is very important. Our treatment will basically follow Riemann, in a more explicit form given by Papperitz [1889]. The reader who has never seen series solutions of differential equations with regular singular points might find it helpful to read Appendix F first.

Let $p(x)$ and $q(x)$ be meromorphic functions. Suppose that the equation

$$\frac{d^2y}{dx^2} + p(x)\frac{dy}{dx} + q(x)y = 0 \tag{2.3.1}$$

has regular singularities at the finite points α, β, γ and that the indicial equations at these points have solutions $a_1, a_2; b_1, b_2;$ and c_1, c_2 respectively. Assume that $a_1 - a_2, b_1 - b_2,$ and $c_1 - c_2$ are not integers. Set $x = 1/t$ so that the differential equation is transformed to

$$\frac{d^2y}{dt^2} + \left(\frac{2}{t} - \frac{1}{t^2}p(1/t)\right)\frac{dy}{dt} + \frac{1}{t^4}q(1/t)y = 0. \tag{2.3.2}$$

Since ∞ is an ordinary point, $2x - x^2 p(x)$ and $x^4 q(x)$ are analytic at ∞. Moreover, since α, β, γ are regular singular points,

$$p(x) = \frac{A}{x-\alpha} + \frac{B}{x-\beta} + \frac{C}{x-\gamma} + u_1(x)$$

and

$$(x-\alpha)(x-\beta)(x-\gamma)q(x) = \frac{D}{x-\alpha} + \frac{E}{x-\beta} + \frac{F}{x-\gamma} + u_2(x),$$

where $u_1(x)$ and $u_2(x)$ are analytic functions.

The last two relations together with the analyticity of $2x - x^2 p(x)$ and $x^4 q(x)$ at infinity imply that $A + B + C = 2$ and $u_1(x) = u_2(x) \equiv 0$. Suppose a solution has the form $\sum_{n=0}^{\infty} a_n (x - \alpha)^{n+\lambda}$, where the exponent λ satisfies the indicial equation

$$\lambda(\lambda - 1) + \lambda A + \frac{D}{(\alpha - \beta)(\alpha - \gamma)} = 0.$$

Since a_1 and a_2 are roots of this equation,

$$a_1 + a_2 = 1 - A$$

and

$$a_1 a_2 = \frac{D}{(\alpha - \beta)(\alpha - \gamma)}.$$

Therefore,

$$A = 1 - a_1 - a_2 \quad \text{and} \quad D = (\alpha - \beta)(\alpha - \gamma) a_1 a_2.$$

Similarly,

$$B = 1 - b_1 - b_2 \quad \text{and} \quad E = (\beta - \alpha)(\beta - \gamma) b_1 b_2,$$

and

$$C = 1 - c_1 - c_2 \quad \text{and} \quad F = (\gamma - \alpha)(\gamma - \beta) c_1 c_2.$$

Since $A + B + C = 2$, it follows that the exponents of the differential equation are related by the equation

$$a_1 + a_2 + b_1 + b_2 + c_1 + c_2 = 1. \tag{2.3.3}$$

We summarize the results in the following theorem due to Papperitz [1889].

Theorem 2.3.1   *A differential equation with three singular points α, β, γ and exponents $a_1, a_2; b_1, b_2;$ and c_1, c_2 respectively has the form*

$$\frac{d^2 y}{dx^2} + \left\{ \frac{1 - a_1 - a_2}{x - \alpha} + \frac{1 - b_1 - b_2}{x - \beta} + \frac{1 - c_1 - c_2}{x - \gamma} \right\} \frac{dy}{dx}$$

$$+ \frac{y}{(x - \alpha)(x - \beta)(x - \gamma)} \left\{ \frac{(\alpha - \beta)(\alpha - \gamma) a_1 a_2}{x - \alpha} \right.$$

$$+ \frac{(\beta - \alpha)(\beta - \gamma) b_1 b_2}{x - \beta} + \left. \frac{(\gamma - \alpha)(\gamma - \beta) c_1 c_2}{x - \gamma} \right\} = 0,$$

and the exponents satisfy (2.3.3).

It is customary to take the regular singularities at 0, 1, and ∞. So let $\alpha = 0$, $\beta = 1$, and $\gamma \to \infty$ in the above differential equation to obtain

$$x^2(x-1)^2\frac{d^2y}{dx^2} + \{(1-a_1-a_2)x(x-1)^2 + (1-b_1-b_2)x^2(x-1)\}\frac{dy}{dx}$$
$$+ \{a_1a_2(1-x) + b_1b_2x + c_1c_2x(x-1)\}y = 0. \tag{2.3.4}$$

The hypergeometric equation is obtained from this one by another simplification. Write this equation in the form (2.3.1). If y satisfies (2.3.1) and $y = x^\lambda f$, then f satisfies

$$\frac{d^2f}{dx^2} + \left(p(x) + \frac{2\lambda}{x}\right)\frac{df}{dx} + \left(q(x) + \frac{\lambda p(x)}{x} + \frac{\lambda(\lambda-1)}{x^2}\right)f = 0.$$

This equation also has $0, 1, \infty$ as singular points, but the exponents are different. Equation (2.3.4) has exponents a_1 and a_2 at 0; the new equation has exponents $a_1 - \lambda$ and $a_2 - \lambda$ at zero. The exponents at ∞, however, are $c_1 + \lambda$ and $c_2 + \lambda$. By this procedure we can arrange that one exponent at 0 and one exponent at 1 be equal to 0. (At $x = 1$, we set $y = (1-x)^\lambda f(x)$.) Thus the new equation has exponents $0, a_2 - a_1$; $0, b_2 - b_1$; $c_1 + a_1 + b_1$, $c_2 + a_1 + b_1$. This brings considerable simplification in (2.3.4) since the terms a_1a_2 and b_1b_2 vanish. It is traditional to write $a = c_1 + a_1 + b_1$, $b = c_2 + a_1 + b_1$, and $c = 1 + a_1 - a_2$. After simplification, the equation becomes

$$x(1-x)\frac{d^2y}{dx^2} + [c - (a+b+1)x]\frac{dy}{dx} - aby = 0. \tag{2.3.5}$$

This is Euler's hypergeometric differential equation. It has regular singularities at 0, 1, and ∞ with exponents $0, 1-c$; $0, c-a-b$; and a, b respectively. Unless specifically stated, we assume that c, $a - b$, and $c - a - b$ are not integers.

Riemann [1857] denoted the set of all solutions of the equation in Theorem 2.3.1 by

$$P\begin{Bmatrix} \alpha & \beta & \gamma & \\ a_1 & b_1 & c_1 & x \\ a_2 & b_2 & c_2 & \end{Bmatrix}. \tag{2.3.6}$$

In particular, the set of solutions of (2.3.5) is denoted by

$$P\begin{Bmatrix} 0 & \infty & 1 & \\ 0 & a & 0 & x \\ 1-c & b & c-a-b & \end{Bmatrix}.$$

Our earlier discussion implies that

$$x^\lambda (1-x)^\mu P \left\{ \begin{matrix} 0 & \infty & 1 \\ a_1 & c_1 & b_1 & x \\ a_2 & c_2 & b_2 \end{matrix} \right\}$$

$$= P \left\{ \begin{matrix} 0 & \infty & 1 \\ a_1 + \lambda & c_1 - \lambda - \mu & b_1 + \mu & x \\ a_2 + \lambda & c_2 - \lambda - \mu & b_2 + \mu \end{matrix} \right\}. \tag{2.3.7}$$

Every conformal mapping of the Riemann sphere $\mathbb{C} \cup \{\infty\}$ is of the form

$$t = \frac{\lambda x + \mu}{\delta x + \nu},$$

where $\lambda\nu - \mu\delta = 1$. Such a mapping takes any set of three distinct points $\{\alpha, \beta, \gamma\}$ to another set of three distinct points $\{\alpha_1, \beta_1, \gamma_1\}$. In this case

$$P \left\{ \begin{matrix} \alpha & \beta & \gamma \\ a_1 & b_1 & c_1 & x \\ a_2 & b_2 & c_2 \end{matrix} \right\} = P \left\{ \begin{matrix} \alpha_1 & \beta_1 & \gamma_1 \\ a_1 & b_1 & c_1 & t \\ a_2 & b_2 & c_2 \end{matrix} \right\}. \tag{2.3.8}$$

This is easily checked. Moreover, there are six linear fractional transformations that will map a set of three points to a permutation of the three points. For example, the set $\{0, 1, \infty\}$ will be mapped to itself by the mappings

$$x \to x, 1-x, \frac{1}{x}, \frac{1}{1-x}, 1 - \frac{1}{x} = \frac{x-1}{x}, \frac{1}{1-1/x} = \frac{x}{x-1}. \tag{2.3.9}$$

We note a few particular cases of (2.3.7)–(2.3.9):

$$P \left\{ \begin{matrix} 0 & \infty & 1 \\ 0 & a & 0 & x \\ 1-c & b & c-a-b \end{matrix} \right\} = P \left\{ \begin{matrix} 0 & \infty & 1 \\ 0 & a & 0 & 1-x \\ c-a-b & b & 1-c \end{matrix} \right\}$$

$$\tag{2.3.10}$$

$$= P \left\{ \begin{matrix} 0 & \infty & 1 \\ 0 & 0 & a & \frac{x}{x-1} \\ 1-c & c-a-b & b \end{matrix} \right\} \tag{A}$$

$$= (1-x)^{-a} P \left\{ \begin{matrix} 0 & \infty & 1 \\ 0 & a & 0 & \frac{x}{x-1} \\ 1-c & c-b & b-a \end{matrix} \right\} \tag{B}$$

(Note: $\left(1 - \frac{x}{x-1}\right)^a = (1-x)^{-a}$.)

$$= P \left\{ \begin{matrix} 0 & \infty & 1 & \\ a & 0 & 0 & \frac{1}{x} \\ b & 1-c & c-a-b & \end{matrix} \right\} \tag{C}$$

$$= x^{-a} P \left\{ \begin{matrix} 0 & \infty & 1 & \\ 0 & a & 0 & \frac{1}{x} \\ b-a & 1-c+a & c-a-b & \end{matrix} \right\} \tag{D}$$

$$= x^{-b} P \left\{ \begin{matrix} 0 & \infty & 1 & \\ 0 & b & 0 & \frac{1}{x} \\ a-b & 1-c+b & c-a-b & \end{matrix} \right\} \tag{E}$$

$$= (1-x)^{c-a-b} P \left\{ \begin{matrix} 0 & \infty & 1 & \\ 0 & c-a & 0 & x \\ 1-c & c-b & a+b-c & \end{matrix} \right\} \tag{F}$$

$$= x^{1-c} P \left\{ \begin{matrix} 0 & \infty & 1 & \\ 0 & 1+a-c & 0 & x \\ c-1 & 1+b-c & c-a-b & \end{matrix} \right\}. \tag{G}$$

Now from the full set of solutions $P\{\ \}$ of (2.3.5) we choose two solutions about $x = 0$ which form a basis. For a solution of the form $x^\lambda \sum_0^\infty a_n x^n$, λ is either 0 or $1 - c$. When $\lambda = 0$, the coefficients a_n satisfy

$$a_n = \frac{(a)_n (b)_n}{(1)_n (c)_n}.$$

So one solution is $_2F_1(a, b; c; x)$. If c is not an integer, then the hypergeometric equation has only one independent solution analytic at $x = 0$. In particular, $_2F_1(a, b; c; x)$ is the only solution analytic at $x = 0$ and with value 1 at $x = 0$. The other solution is of the form $W = x^{1-c} g$, where g is analytic at $x = 0$. It follows from (2.3.10G) that $g = k_2 F_1(a + 1 - c, b + 1 - c; 2 - c; x)$, where k is a constant. Therefore, the independent solutions are $_2F_1(a, b; c; x)$ and $x^{1-c} {}_2F_1(a + 1 - c, b + 1 - c; 2 - c; x)$. In a similar way, (2.3.10) immediately implies that two independent solutions at $x = 1$ are

$$_2F_1(a, b; a + b + 1 - c; 1 - x)$$

and

$$(1-x)^{c-a-b} {}_2F_1(c - a, c - b; c + 1 - a - b; 1 - x)$$

and at ∞ are

$$(-x)^{-a} {}_2F_1(a, a + 1 - c; a + 1 - b; 1/x)$$

and

$$(-x)^{-b} {}_2F_1(b, b + 1 - c; b + 1 - a; 1/x).$$

The powers of -1 have been introduced for convenience in expressing some later formulas.

It is important to note that some of the hypergeometric transformation formulas can also be obtained from (2.3.10). Pfaff's transformation $_2F_1(a, b; c; x) = (1 - x)^{-a}{}_2F_1(a, c - b; c; x/(x - 1))$ is an immediate consequence of (2.3.10B), combined with the fact that there is only one solution analytic and equal to 1 at $x = 0$. Euler's transformation $_2F_1(a, b; c; x) = (1 - x)^{c-a-b}{}_2F_1(c - a, c - b; c; x)$ follows from (2.3.10F).

A number of relations involving hypergeometric functions arise from the fact that the hypergeometric equation has two independent solutions, so that any three solutions must be linearly related.

Theorem 2.3.2

$$_2F_1\left(\begin{matrix} a, b \\ a + b + 1 - c \end{matrix}; 1 - x\right)$$
$$= A_2F_1\left(\begin{matrix} a, b \\ c \end{matrix}; x\right) + Bx^{1-c}{}_2F_1\left(\begin{matrix} 1 + a - c, 1 + b - c \\ 2 - c \end{matrix}; x\right), \quad (2.3.11)$$

where

$$A = \frac{\Gamma(a + b + 1 - c)\Gamma(1 - c)}{\Gamma(a + 1 - c)\Gamma(b + 1 - c)} \quad and \quad B = \frac{\Gamma(c - 1)\Gamma(a + b + 1 - c)}{\Gamma(a)\Gamma(b)}.$$

$$_2F_1\left(\begin{matrix} a, b \\ c \end{matrix}; x\right) = C(-x)^{-a}{}_2F_1\left(\begin{matrix} a, a - c + 1 \\ a - b + 1 \end{matrix}; \frac{1}{x}\right)$$
$$+ D(-x)^{-b}{}_2F_1\left(\begin{matrix} b, b - c + 1 \\ b - a + 1 \end{matrix}; \frac{1}{x}\right) \quad (2.3.12)$$

where

$$C = \frac{\Gamma(c)\Gamma(b - a)}{\Gamma(c - a)\Gamma(b)} \quad and \quad D = \frac{\Gamma(c)\Gamma(a - b)}{\Gamma(c - b)\Gamma(a)}.$$

Proof.   When $x = 0$ and $\mathrm{Re}\, c < 1$, Gauss's summation formula gives

$$A = \frac{\Gamma(a + b + 1 - c)\Gamma(1 - c)}{\Gamma(a + 1 - c)\Gamma(b + 1 - c)}.$$

Observe that $\mathrm{Re}\, c < 1$ was used twice. First it was used to make the second term on the right vanish; it is also the condition for the series on the left to converge at $x = 0$. Then $x = 1$ gives (for $\mathrm{Re}(c - a - b) > 0$)

$$1 = A\frac{\Gamma(c)\Gamma(c - a - b)}{\Gamma(c - a)\Gamma(c - b)} + B\frac{\Gamma(2 - c)\Gamma(c - a - b)}{\Gamma(1 - a)\Gamma(1 - b)}.$$

After some tedious trigonometric calculation, which comes in after applying Euler's reflection formula to the second term on the right, we arrive at the value of B required by the theorem. This proves (2.3.11).

Suppose $\operatorname{Re} b > \operatorname{Re} a$. The right side of (2.3.12) as $x \to \infty$ is $\sim C(-x)^{-a}$. To see the behavior of the left side, apply Pfaff's transformation. Then

$$\,_2F_1\left(\begin{matrix} a, b \\ c \end{matrix}; x\right) = (1-x)^{-a}\,_2F_1\left(\begin{matrix} a, c-b \\ c \end{matrix}; \frac{x}{x-1}\right)$$

$$\sim (-x)^{-a}\,_2F_1\left(\begin{matrix} a, c-b \\ c \end{matrix}; 1\right)$$

$$= (-x)^{-a}\frac{\Gamma(c)\Gamma(b-a)}{\Gamma(c-a)\Gamma(b)}.$$

The assumption that $\operatorname{Re} b > \operatorname{Re} a$ was used in the last step to evaluate the $\,_2F_1$ by Gauss's formula. It follows that

$$C = \frac{\Gamma(c)\Gamma(b-a)}{\Gamma(c-a)\Gamma(b)}.$$

The value of D follows from the symmetry in a and b. ∎

Corollary 2.3.3

$$\,_2F_1\left(\begin{matrix} a, b \\ c \end{matrix}; x\right) = \frac{\Gamma(c)\Gamma(c-a-b)}{\Gamma(c-a)\Gamma(c-b)}\,_2F_1\left(\begin{matrix} a, b \\ a+b+1-c \end{matrix}; 1-x\right)$$

$$+ \frac{\Gamma(c)\Gamma(a+b-c)}{\Gamma(a)\Gamma(b)}(1-x)^{c-a-b}\,_2F_1\left(\begin{matrix} c-a, c-b \\ 1+c-a-b \end{matrix}; 1-x\right).$$

$$(2.3.13)$$

$$\,_2F_1\left(\begin{matrix} -n, b \\ c \end{matrix}; x\right) = \frac{(c-b)_n}{(c)_n}\,_2F_1\left(\begin{matrix} -n, b \\ b+1-n-c \end{matrix}; 1-x\right) \quad \text{(Pfaff)} \quad (2.3.14)$$

Proof. In (2.3.11), replace x by $1-x$ and c by $a+b+1-c$. Then (2.3.14) follows from (2.3.13). Just take $a = -n$ and recall that $\frac{1}{\Gamma(-n)} = 0$ when n is a nonnegative integer. ∎

The first part of Theorem 2.1.3 also follows from (2.3.13) above. It should also be noted that, since Pfaff's formula

$$\,_2F_1(a, b; c; x) = (1-x)^{-a}\,_2F_1(a, c-b; c; x/(x-1))$$

gives a continuation of $\,_2F_1$ from $|x| < 1$ to $\operatorname{Re} x < \frac{1}{2}$, then (2.3.13) gives the continuation to $\operatorname{Re} x > 1/2$ cut along the real axis from $x = 1$ to $x = \infty$. The cut comes from the branch points of $(1-x)^{c-a-b}$, and once this function is defined on a Riemann surface, $\,_2F_1(a, b; c; x)$ is also defined there.

Now consider the function

$$S(x) = \int_0^\infty \frac{dt}{(1+t)^{1/3}(x+t)}.$$

We show how Theorem 2.3.2 can be employed to find the asymptotic expansion of the function given above. Wong [1989, p. 18] used this function to demonstrate that a certain amount of care should be taken when finding the asymptotic expansion of a function. In this instance, when the method of integration by parts is applied, one gets the expansion

$$S(x) \sim -\sum_{n=1}^\infty \frac{3^n(n-1)!}{2 \cdot 5 \cdots (3n-1)} x^{-n} \quad \text{as} \quad x \to +\infty, \qquad (2.3.15)$$

which is obviously incorrect, since the integral is positive and every term of the expansion is negative. However, for $t > 1$ we have

$$(1+t)^{-1/3} = \sum_{n=0}^\infty \frac{(1/3)_n(-1)^n}{n!} t^{-n-1/3}.$$

If this series is substituted in the integral, then term-by-term integration produces the divergent integrals

$$\int_0^\infty \frac{t^{-n-1/3}}{x+t} dt.$$

If these are interpreted in a "distributional" sense the value of the above integral can be set equal to

$$\frac{2\pi}{\sqrt{3}} \frac{(-1)^n}{x^{n+1/3}}.$$

With this interpretation, $S(x)$ has the expansion (after termwise integration)

$$S(x) \sim \frac{2\pi}{\sqrt{3}} \sum_{n=0}^\infty \frac{\left(\frac{1}{3}\right)_n}{n!} x^{-n-1/3} \quad \text{as} \quad x \to \infty. \qquad (2.3.16)$$

The correct result is, however, the sum of the two expansions in (2.3.15) and (2.3.16). We obtain this from Theorem 2.3.2. First note that for $\mathrm{Re}(a+1-c) > 0$ and $\mathrm{Re}\, b > 0$,

$$\int_0^\infty t^{b-1}(1+t)^{c-b-1}(1+xt)^{-a} dt$$
$$= \frac{\Gamma(a+1-c)\Gamma(b)}{\Gamma(a+b+1-c)} {}_2F_1\left(\begin{matrix} a, b \\ a+b+1-c \end{matrix}; 1-x\right). \qquad (2.3.17)$$

This follows from Euler's integral representation of a ${}_2F_1$ given in Theorem 2.2.1. To reduce this integral to Euler's form, set $t = u/(1-u)$. From (2.3.17) it follows that

$$S(x) = \frac{1}{x} \int_0^\infty (1+t)^{-1/3}(1+t/x)^{-1}dt$$

$$= \frac{\Gamma(1)\Gamma\left(\frac{1}{3}\right)}{x\Gamma\left(\frac{4}{3}\right)} {}_2F_1\left(\begin{matrix}1, 1\\4/3\end{matrix}; 1 - \frac{1}{x}\right)$$

$$= \frac{3}{x} {}_2F_1\left(\begin{matrix}1, 1\\4/3\end{matrix}; 1 - \frac{1}{x}\right).$$

Apply (2.3.11) to get

$$S(x) = \frac{3}{x}\left\{ \frac{\Gamma(4/3)\Gamma(-2/3)}{\Gamma(1/3)\Gamma(1/3)} {}_2F_1\left(\begin{matrix}1, 1\\5/3\end{matrix}; \frac{1}{x}\right) \right.$$

$$\left. + \frac{\Gamma(2/3)\Gamma(4/3)}{\Gamma(1)\Gamma(1)} x^{2/3} {}_2F_1\left(\begin{matrix}1/3, 1/3\\1/3\end{matrix}; \frac{1}{x}\right) \right\}$$

$$= \frac{3}{x}\left\{ -\frac{1}{2^2}F_1\left(\begin{matrix}1, 1\\5/3\end{matrix}; \frac{1}{x}\right) + \frac{1}{3}\frac{\pi}{\sin \pi/3} x^{2/3} {}_2F_1\left(\begin{matrix}1/3, 1/3\\1/3\end{matrix}; \frac{1}{x}\right) \right\}$$

$$= -\frac{3}{2x^2}F_1\left(\begin{matrix}1, 1\\5/3\end{matrix}; \frac{1}{x}\right) + \frac{2\pi}{\sqrt{3}}\frac{1}{x^{1/3}}{}_2F_1\left(\begin{matrix}1/3, 1/3\\1/3\end{matrix}; \frac{1}{x}\right).$$

This is equivalent to the sum of the series in (2.3.15) and (2.3.16).

Remark 2.3.1 We have seen that $_2F_1(a, b; c; x)$ is one solution of the hypergeometric equation. This fact was used to show that another independent solution is

$$x^{1-c}{}_2F_1(a + 1 - c, b + 1 - c; 2 - c; x).$$

Here we show how the other solutions can be obtained formally from the series for $_2F_1$. We should write the hypergeometric series as a bilateral series

$$\sum_{n=-\infty}^{\infty} \frac{(a)_n(b)_n}{(c)_n\Gamma(n + 1)}x^n.$$

Since $\Gamma(1 + x)$ has poles at $x = -1, -2, \ldots$, the series has no negative powers of x. It is clear that a change of variables $n \to n + m$, where m is an integer, does not change the bilateral series. Consider the transformation $n \to n + \alpha$, where α is a noninteger. Now the series takes the form

$$\sum_{n=-\infty}^{\infty} \frac{(a)_{n+\alpha}(b)_{n+\alpha}x^{n+\alpha}}{(c)_{n+\alpha}(1)_{n+\alpha}} = \frac{(a)_\alpha(b)_\alpha}{(c)_\alpha(1)_\alpha}x^\alpha \sum_{n=-\infty}^{\infty} \frac{(a + \alpha)_n(b + \alpha)}{(c + \alpha)_n(1 + \alpha)_n}x^n. \quad (A)$$

In the latter expression, the terms with negative values of n vanish if we set $c + \alpha = 1$ or $1 + \alpha = 1$. The last condition gives back the original $_2F_1$ series. The first case, where $\alpha = 1 - c$, gives

$$x^{1-c} \sum_{n=0}^{\infty} \frac{(a + 1 - c)_n (b + 1 - c)_n}{(1)_n (2 - c)_n} = x^{1-c} {}_2F_1 \left(\begin{array}{c} a + 1 - c, b + 1 - c \\ 2 - c \end{array} ; x \right),$$

which is the second independent solution. The solutions at ∞ are obtained in a similar manner by changing n to $-n$. In this case (A) becomes

$$k x^{\alpha} \sum_{n=-\infty}^{\infty} \frac{(a + \alpha)_{-n} (b + \alpha)_{-n}}{(c + \alpha)_{-n} (1 + \alpha)_{-n}} x^{-n},$$

where k is a constant. Since $(a + \alpha)_{-n} = (-1)^n / (1 - \alpha - a)_n$, write the last series as

$$k x^{\alpha} \sum_{n=-\infty}^{\infty} \frac{(1 - c - \alpha)_n x^{-n} (-\alpha)_n}{(1 - a - \alpha)_n (1 - b - \alpha)_n} x^{-n}.$$

Once again, to eliminate that portion of the sum involving negative values of n, take either $\alpha = -a$ or $\alpha = -b$. In the first case we get $cx^{-a} {}_2F_1(a + 1 - c, a; a + 1 - b; 1/x)$ and in the second case $cx^{-b} {}_2F_1(b + 1 - c, b; b + 1 - a; 1/x)$.

Remark 2.3.2   In Section 2.1 we proved that the series $_2F_1(a, b; c; x)$ has radius of convergence 1. We now reverse the point of view taken in Remark 2.3.1 and see how to obtain convergence from the theory of differential equations. Since the singularities of the equation are at 0, 1, and ∞, the radius of convergence is at least 1. If it is more than 1, then the series is an entire function. Moreover, by (2.3.12) it is a linear combination of $x^{-a} f_1(x)$ and $x^{-b} f_2(x)$, which are solutions at ∞. This is possible only if either a or b is an integer. Otherwise, both solutions at ∞ are multivalued. (Both a and b cannot be integers since $a - b$ is not an integer.) Liouville's theorem shows us that the integer must be negative and that the $_2F_1$ is a polynomial. Hence, if the $_2F_1$ is an infinite series then the radius of convergence must be 1.

We now consider the case where c is an integer. Suppose c is a positive integer; then

$$F_1 := \frac{\Gamma(a)\Gamma(b)}{\Gamma(c)} {}_2F_1 \left(\begin{array}{c} a, b \\ c \end{array} ; x \right) = \sum_{k=0}^{\infty} \frac{\Gamma(a + k)\Gamma(b + k)}{k!\Gamma(c + k)} x^k$$

and

$$F_2 := \frac{\Gamma(a + 1 - c)\Gamma(b + 1 - c)}{\Gamma(2 - c)} x^{1-c} {}_2F_1 \left(\begin{array}{c} a + c - 1, b + 1 - c \\ 2 - c \end{array} ; x \right)$$

$$= x^{1-c} \sum_{k=0}^{\infty} \frac{\Gamma(a + 1 - c + k)\Gamma(b + 1 - c + k)}{k!\Gamma(2 - c + k)} x^k$$

are equal. To find the second solution in this case, suppose a and b are not negative integers. Consider the limit

$$\lim_{c \to n} \frac{F_1 - F_2}{c - n} = \frac{\partial}{\partial c}(F_1 - F_2)\,|_{c=n}$$

$$= -\sum_{k=0}^{\infty} \frac{\Gamma(a+k)\Gamma(b+k)\Gamma'(n+k)}{k!\Gamma(n+k)\Gamma(n+k)}x^k$$

$$+ x^{1-n} \log x \sum_{k=0}^{\infty} \frac{\Gamma(a+1-n+k)\Gamma(b+1-n+k)}{k!\Gamma(2-n+k)}x^k$$

$$+ x^{1-n} \sum_{k=0}^{\infty} \frac{\Gamma(a+1-n+k)\Gamma(b+1-n+k)}{k!\Gamma(2-n+k)}$$

$$\cdot \left[\frac{\Gamma'(a+1-n+k)}{\Gamma(a+1-n+k)} + \frac{\Gamma'(b+1-n+k)}{\Gamma(b+1-n+k)}\right]x^k$$

$$- \lim_{c \to n} x^{1-n} \sum_{k=0}^{\infty} \frac{\Gamma(a+1-c+k)\Gamma(b+1-c+k)}{k!\Gamma(2-c+k)}$$

$$\cdot \frac{\Gamma'(2-c+k)}{\Gamma(2-c+k)}x^k.$$

The second series is

$$\frac{\Gamma(a)\Gamma(b)}{\Gamma(n)} \log x \, {}_2F_1\left(\begin{matrix} a, b \\ n \end{matrix}; x\right)$$

and the first $n-1$ terms in the third series are zero because $\frac{1}{\Gamma(2-n+k)} = 0$ for $k = 0, 1, \ldots, n-2$. The same is not true in the fourth series, because $\Gamma'(2-c+k)/\Gamma(2-c+k)$ has poles at these points. By Euler's reflection formula,

$$\frac{\Gamma'(1-x)}{[\Gamma(1-x)]^2} = \frac{\Gamma'(x)}{\Gamma(x)\Gamma(1-x)} + \cos \pi x \Gamma(x).$$

Set $x = n - k - 1$ to get

$$\lim_{c \to n} \frac{\Gamma'(2-c+k)}{[\Gamma(2-c+k)]^2} = (-1)^{n-k-1}\Gamma(n-k-1).$$

The fourth series can now be written as

$$\sum_{k=1}^{n-1}(-1)^{k-1}\frac{(k-1)!\Gamma(a-k)\Gamma(b-k)}{(n-k-1)!}x^{-k}$$

$$- \sum_{k=0}^{\infty} \frac{\Gamma(a+1-n+k)\Gamma(b+1-n+k)}{k!\Gamma(n+k)}x^k.$$

So, when a and b are not negative integers, the second solution is

$$_2F_1\left(\begin{matrix} a, b \\ n \end{matrix}; x\right) \log x$$

$$+ \sum_{k=0}^{\infty} \frac{(a)_k (b)_k}{k!(n)_k} \{\psi(a+k) + \psi(b+k) - \psi(1+k) - \psi(n+k)\} x^k$$

$$+ \frac{(n-1)!}{\Gamma(a)\Gamma(b)} \sum_{k=1}^{n-1} (-1)^{k-1} \frac{(k-1)!\Gamma(a-k)\Gamma(b-k)}{(n-k-1)!} x^{-k}, \qquad (2.3.18)$$

where $\psi(x) = \Gamma'(x)/\Gamma(x)$.

If a is a negative integer, say $-m$, then $\psi(a+k)$ is undefined for some values of k. Consequently, the solution given above does not work. To resolve this difficulty observe that

$$\lim_{a \to -m} \{\psi(a+k) - \psi(a)\} = \psi(1+m-k) - \psi(1+m) \quad \text{for } k \le m. \quad (2.3.19)$$

Now, if $\psi(a)\, _2F_1(a, b; c; x)$ is subtracted from the second term in (2.3.18), then the resulting series is again a solution of the hypergeometric equation except that now we may let a tend to the negative integer $-m$. The reader should verify (2.3.19) and also that in this case the second solution is

$$_2F_1\left(\begin{matrix} -m, b \\ n \end{matrix}; x\right) \log x$$

$$+ \sum_{k=0}^{m} \frac{(-m)_k (b)_k}{k!(n)_k} \{\psi(1+m-k) + \psi(b+k) - \psi(n+k) - \psi(1+k)\} x^k$$

$$- \frac{(n-1)!}{\Gamma(b)} \sum_{k=1}^{n-1} \frac{(k-1)!\Gamma(b-k)}{(n-k-1)!(m+1)_k} x^{-k}. \qquad (2.3.20)$$

The case where both a and b are negative integers may be treated in the same way. When $c = 0, -1, -2, \ldots$, then the indicial equation shows that the first solution is

$$x^{1-c}\,_2F_1\left(\begin{matrix} a-c+1, b-c+1 \\ 2-c \end{matrix}; x\right).$$

The second solution in this case can be obtained from (2.3.18) by replacing a and b with $a-c+1$ and $b-c+1$ respectively.

Theorem 2.3.2 and its corollary must be modified when c, $a-b$, or $c-a-b$ is an integer. The reader should work out the necessary changes.

An interesting history of the hypergeometric equation is contained in Gray [1986].

2.4 The Barnes Integral for the Hypergeometric Function

In a sequence of papers published in the period 1904–1910, Barnes developed an alternative method of treating the hypergeometric function $_2F_1$. A cornerstone of this structure is a contour integral representation of $_2F_1(a, b; c; x)$. An understanding of this representation can be obtained through the concept of a Mellin transform. We begin with a simple and familiar example: $\Gamma(s) = \int_0^\infty x^{s-1} e^{-x} dx$. It turns out that it is possible to recover the integrated function e^{-x} in terms of a complex integral involving $\Gamma(s)$. This inversion formula is given by

$$e^{-x} = \frac{1}{2\pi i} \int_{c-i\infty}^{c+i\infty} x^{-s} \Gamma(s) ds, \quad c > 0. \tag{2.4.1}$$

This can be proved by Cauchy's residue theorem. Take a rectangular contour L with vertices $c \pm iR, c - (N + \frac{1}{2}) \pm iR$, where N is a positive integer. The poles of $\Gamma(s)$ inside this contour are at $0, -1, \ldots, -N$ and the residues are $(-1)^j/j!$ at $j = 0, 1, \ldots, N$. Cauchy's theorem gives $\frac{1}{2\pi i} \int_L x^{-s} \Gamma(s) ds = \sum_{j=0}^{N} (-1)^j x^j/j!$. Now let R and N tend to infinity and use Theorem 1.4.1 and Corollary 1.4.4 to show that the integral on L minus the line joining $c - iR$ to $c + iR$ tends to zero. This proves (2.4.1).

The Mellin transform of a function $f(x)$ is defined by the integral $F(s) = \int_0^\infty x^{s-1} f(x) dx$. We have studied other examples of Mellin transforms in Chapter 1. The integral $\int_0^1 x^{s-1}(1-x)^{t-1} dx$ is the transform of

$$(1-x)_+^{t-1} = \begin{cases} (1-x)^{t-1}, & 0 < x < 1, \\ 0, & x \geq 1, \end{cases}$$

and

$$\int_0^\infty \frac{x^{s-1}}{(1+x)^t} dx$$

is the transform of $f(x) = 1/(1+x)^t$. Once again, one can prove that

$$(1-x)_+^{t-1} = \frac{\Gamma(t)}{2\pi i} \int_{c-i\infty}^{c+i\infty} x^{-s} \frac{\Gamma(s)}{\Gamma(s+t)} ds, \quad \text{Re } t > 0 \quad \text{and} \quad c > 0, \tag{2.4.2}$$

and

$$\frac{1}{(1+x)^t} = \frac{1}{2\pi i \Gamma(t)} \int_{c-i\infty}^{c+i\infty} x^{-s} \Gamma(s) \Gamma(t-s) ds, \quad 0 < c < \text{Re } t. \tag{2.4.3}$$

The phenomenon exhibited by (2.4.1) through (2.4.3) continues to hold for a fairly large class of functions $f(x)$. Thus, if $F(x) = \int_0^\infty x^{s-1} f(x) dx$ then $f(x) = \frac{1}{2\pi i} \int_{c-i\infty}^{c+i\infty} x^{-s} F(s) ds$ is true for a class of functions. We do not develop this theory as a whole but prove a few interesting cases. The Mellin transform is further discussed in Chapter 10 with a different motivation.

The above discussion shows that if we want a complex integral representation for the hypergeometric function we should find its Mellin transform. Now,

$$\int_0^\infty x^{s-1}\,{}_2F_1\!\left(\begin{matrix} a,\,b \\ c \end{matrix};\,-x\right)dx$$

$$= \int_0^\infty x^{s-1}\frac{\Gamma(c)}{\Gamma(b)\Gamma(c-b)}\int_0^1 t^{b-1}(1-t)^{c-b-1}(1+xt)^{-a}dtdx$$

$$= \frac{\Gamma(c)}{\Gamma(b)(c-b)}\int_0^1 t^{b-1}(1-t)^{c-b-1}\int_0^\infty \frac{x^{s-1}}{(1+xt)^a}dxdt$$

$$= \frac{\Gamma(s)\Gamma(a-s)\Gamma(c)}{\Gamma(a)\Gamma(b)\Gamma(c-b)}\int_0^1 t^{b-s-1}(1-t)^{c-b-1}dt$$

$$= \frac{\Gamma(s)\Gamma(a-s)\Gamma(c)}{\Gamma(a)\Gamma(b)\Gamma(c-b)}\frac{\Gamma(b-s)\Gamma(c-b)}{\Gamma(c-s)}$$

$$= \frac{\Gamma(c)}{\Gamma(a)\Gamma(b)}\frac{\Gamma(s)\Gamma(a-s)\Gamma(b-s)}{\Gamma(c-s)}. \tag{2.4.4}$$

These formal steps can be justified by assuming $\min(\operatorname{Re} a, \operatorname{Re} b) > \operatorname{Re} s > 0$. The auxiliary condition $\operatorname{Re} c > \operatorname{Re} b$ can be removed by analytic continuation or via contiguous relations, which are treated in Section 2.5. Note that we integrated $_2F_1$ at $-x$ because, in general, $_2F_1(a, b; c; x)$ has branch points at $x = 1$ and $x = \infty$ and in (2.4.4) the integral is over the positive real axis. There is another proof of (2.4.4) in Exercise 35.

We expect, by inversion, that

$$\frac{\Gamma(a)\Gamma(b)}{\Gamma(c)}\,{}_2F_1\!\left(\begin{matrix} a,\,b \\ c \end{matrix};\,x\right) = \frac{1}{2\pi i}\int_{k-i\infty}^{k+i\infty}\frac{\Gamma(s)\Gamma(a-s)\Gamma(b-s)}{\Gamma(c-s)}(-x)^{-s}ds,$$

$$\tag{2.4.5}$$

where $\min(\operatorname{Re} a, \operatorname{Re} b) > k > 0$ and $c \neq 0, -1, -2, \ldots$. This is Barnes's formula and it is the basis for an alternative development of the theory of hypergeometric functions. It should be clear that we can represent a $_pF_q$ by a similar integral. The precise form of Barnes's [1908] theorem is given next.

Theorem 2.4.1

$$\frac{\Gamma(a)\Gamma(b)}{\Gamma(c)}\,{}_2F_1\!\left(\begin{matrix} a,\,b \\ c \end{matrix};\,x\right) = \frac{1}{2\pi i}\int_{-i\infty}^{i\infty}\frac{\Gamma(a+s)\Gamma(b+s)\Gamma(-s)}{\Gamma(c+s)}(-x)^s ds,$$

$|\arg(-x)| < \pi$. *The path of integration is curved, if necessary, to separate the poles* $s = -a - n$, $s = -b - n$, *from the poles* $s = n$, *where* n *is an integer* ≥ 0. *(Such a contour can always be drawn if* a *and* b *are not negative integers.)*

Proof. Let L be the closed contour formed by a part of the curve used in the theorem from $-(N+\frac{1}{2})i$ to $(N+\frac{1}{2})i$ together with the semicircle of radius $N+\frac{1}{2}$

drawn to the right with 0 as center. We first show that the above integral defines an analytic function in $|\arg(-x)| \leq \pi - \delta$, $\delta > 0$. By Euler's reflection formula, write the integrand as

$$-\frac{\Gamma(a+s)\Gamma(b+s)(-x)^s \pi}{\Gamma(c+s)\Gamma(1+s)\sin s\pi}.$$

By Corollary 1.4.3, this expression is asymptotic to

$$-s^{a+b-c-1}\frac{\pi(-x)^s}{\sin s\pi}.$$

Set $s = it$ to get

$$-(it)^{a+b-c-1}2\pi i \frac{e^{it(\log|x|+i\arg(-x))}}{e^{-\pi t}-e^{\pi t}} = O(|t|^{a+b-c-1}e^{-|t|\delta})$$

for $|\arg(-x)| \leq \pi - \delta$. This estimate shows that the integral represents an analytic function in $|\arg(-x)| \leq \pi - \delta$ for every $\delta > 0$, and hence it is analytic in $|\arg(-x)| < \pi$. We now show that the integral represents the series

$$\sum_{n=0}^{\infty} \frac{\Gamma(a+n)\Gamma(b+n)}{n!\Gamma(c+n)}x^n \quad \text{for } |x| < 1.$$

This will prove the theorem by continuation. (Note that, if we start with the series for $_2F_1(a, b; c; x)$, then Barnes's integral gives the continuation to the cut region $|\arg(-x)| < \pi$.)

On the semicircular part of the contour L the integrand is

$$O(N^{a+b-c-1})\frac{(-x)^s}{\sin s\pi}$$

for large N. For $s = (N + \frac{1}{2})e^{i\theta}$ and $|x| < 1$,

$$\frac{(-x)^s}{\sin s\pi} = O\left[e^{(N+\frac{1}{2})(\cos\theta \log|x| - \sin\theta \arg(-x) - \pi|\sin\theta|)}\right].$$

Since $-\pi + \delta \leq \arg(-x) \leq \pi - \delta$, the last expression is

$$O\left[e^{(N+\frac{1}{2})(\cos\theta \log|x| - \delta|\sin\theta|)}\right].$$

In $0 \leq |\theta| \leq \frac{\pi}{4}$, $\cos\theta \geq \frac{1}{\sqrt{2}}$ and in $\frac{\pi}{4} \leq |\theta| \leq \frac{\pi}{2}$, $|\sin\theta| \geq \frac{1}{\sqrt{2}}$. So, since $\log|x| < 0$, the integrand is $O(N^{a+b-c-1}e^{\frac{1}{\sqrt{2}}(N+\frac{1}{2})\log|x|})$ for $0 \leq |\theta| \leq \frac{\pi}{4}$ and $O(N^{a+b-c-1}e^{-\frac{1}{\sqrt{2}}\delta(N+\frac{1}{2})})$ for $\frac{\pi}{4} \leq |\theta| \leq \frac{\pi}{2}$. This implies that the integral on the semicircle $\to 0$ as $N \to \infty$. Since the pole $s = n$ of the integrand has residue

$$\frac{\Gamma(a+n)\Gamma(b+n)}{n!\Gamma(c+n)}x^n,$$

the theorem is proved. ∎

We can recover the asymptotic expansion contained in (2.3.12) from Barnes's integral quite easily. Suppose $a - b$ is not an integer. Move the line of integration to the left by m units and collect the residues at $s = -a - n$ and $s = -b - n$. The residue at $s = -a - n$ is

$$(-1)^n \frac{\Gamma(b - a - n)\Gamma(a + n)(-x)^{-a-n}}{n!\Gamma(c - a - n)}$$

$$= (-x)^{-a} \frac{\Gamma(a)\Gamma(b - a)}{\Gamma(c - a)} \cdot \frac{(a)_n(1 + a - c)_n}{n!(1 + a - b)_n}(x)^{-n}$$

after a little simplification. Thus,

$$\frac{\Gamma(a)\Gamma(b)}{\Gamma(c)} {}_2F_1\left(\begin{matrix} a, b \\ c \end{matrix}; x\right) = \frac{1}{2\pi i} \int_{-m-i\infty}^{-m+i\infty} \frac{\Gamma(a + s)\Gamma(b + s)\Gamma(-s)(-x)^s}{\Gamma(c + s)} ds$$

$$+ (-x)^{-a} \frac{\Gamma(a)\Gamma(b - a)}{\Gamma(c - a)} \sum_0^{m(a)} \frac{(a)_n(1 + a - c)_n}{n!(1 + a - b)_n} x^{-n}$$

$$+ (-x)^{-b} \frac{\Gamma(b)\Gamma(a - b)}{\Gamma(c - b)} \sum_0^{m(b)} \frac{(b)_n(1 + b - c)_n}{n!(1 + b - a)_n} x^{-n},$$

$$(2.4.6)$$

where $m(a)$ is the largest integer n such that $a + n \leq m$. We define $m(b)$ similarly. The integral is equal to

$$-\frac{1}{2\pi i} x^{-m} \int_{-i\infty}^{i\infty} \frac{\Gamma(a - m + s)\Gamma(b - m + s)\pi}{\Gamma(c - m + s)\Gamma(1 - m + s)\sin \pi s}(-x)^s ds.$$

For $|\arg(-x)| \leq \pi - \delta, \delta > 0$, the last integral is a bounded function of m and x. This implies that the expression is $O(1/x^m)$, so that we have an asymptotic expansion for ${}_2F_1(a, b; c; x)$ in (2.4.6). If $a - b$ is an integer, then some of the poles of $\Gamma(a + s)\Gamma(b + s)$ are double poles and logarithmic terms are involved. The reader should work out this case as an exercise.

To gain insight into the next result of Barnes [1910], suppose $F(s)$ and $G(s)$ are the Mellin transforms of $f(x)$ and $g(x)$ respectively. The problem is to determine how the Mellin transform of $f(x)g(x)$ is related to $F(s)$ and $G(s)$. Formally, it is easily seen that

$$\int_0^\infty x^{s-1} f(x)g(x)dx = \frac{1}{2\pi i} \int_0^\infty x^{s-1} g(x) \int_{c-i\infty}^{c+i\infty} F(t)x^{-t}dtdx$$

$$= \frac{1}{2\pi i} \int_{c-i\infty}^{c+i\infty} F(t) \int_0^\infty x^{s-t-1} g(x)dxdt$$

$$= \frac{1}{2\pi i} \int_{c-i\infty}^{c+i\infty} F(t)G(s - t)dt. \qquad (2.4.7)$$

The case of interest to us is where $s = 1$. Then

$$\int_0^\infty f(x)g(x)dx = \frac{1}{2\pi i} \int_{c-i\infty}^{c+i\infty} F(t)G(1-t)dt. \qquad (2.4.8)$$

Apply this to the Mellin pairs

$$f(x) = \frac{x^b}{(1+x)^a}, \qquad F(s) = \frac{\Gamma(b+s)\Gamma(a-b-s)}{\Gamma(a)}$$

and

$$g(x) = \frac{x^d}{(1+x)^c}, \qquad G(s) = \frac{\Gamma(d+s)\Gamma(c-d-s)}{\Gamma(c)}$$

to obtain

$$\frac{1}{2\pi i} \int_{k-i\infty}^{k+i\infty} \frac{\Gamma(b+s)\Gamma(a-b-s)\Gamma(d+1-s)\Gamma(c-d-1+s)}{\Gamma(a)\Gamma(c)} ds$$

$$= \int_0^\infty \frac{x^{b+d}}{(1+x)^{a+c}} dx = \frac{\Gamma(b+d+1)\Gamma(a+c-b-d-1)}{\Gamma(a+c)}$$

for a suitable k. By renaming the parameters, this can be written as

$$\frac{1}{2\pi i} \int_{-i\infty}^{i\infty} \Gamma(a+s)\Gamma(b+s)\Gamma(c-s)\Gamma(d-s)ds$$

$$= \frac{\Gamma(a+c)\Gamma(a+d)\Gamma(b+c)\Gamma(b+d)}{\Gamma(a+b+c+d)}. \qquad (2.4.9)$$

This formula is due to Barnes and the above proof is due to Titchmarsh [1937]. It is correct when $\mathrm{Re}(a, b, c, d) > 0$. We give another proof, because we have not developed the general theory of Mellin transforms in a rigorous way here. But first note that, if we take $f(x) = x_+^a(1-x)_+^{b-a-1}$ and $g(x) = x_+^{c-1}(1-x)_+^{d-c-1}$ in (2.4.8), we get

$$\frac{1}{2\pi i} \int_{k-i\infty}^{k+i\infty} \frac{\Gamma(a+s)\Gamma(c-s)}{\Gamma(b+s)\Gamma(d-s)} ds = \frac{\Gamma(a+c)\Gamma(b+d-a-c-1)}{\Gamma(b-a)\Gamma(d-c)\Gamma(b+d-1)},$$
$$\max(-a, -b) < k < \min(c, d). \qquad (2.4.10)$$

Theorem 2.4.2 *If the path of integration is curved to separate the poles of Γ $(a+s)\Gamma(b+s)$ from the poles of $\Gamma(c-s)\Gamma(d-s)$, then*

$$I := \frac{1}{2\pi i} \int_{-i\infty}^{i\infty} \Gamma(a+s)\Gamma(b+s)\Gamma(c-s)\Gamma(d-s)ds$$

$$= \frac{\Gamma(a+c)\Gamma(a+d)\Gamma(b+c)\Gamma(b+d)}{\Gamma(a+b+c+d)}.$$

Note that $a+c, a+d, b+c, b+d$ cannot be 0 or a negative integer.

Proof. As in the proof of the previous theorem, use Euler's reflection formula to write the integrand as

$$\frac{\Gamma(a+s)\Gamma(b+s)}{\Gamma(1-c+s)(1-d+s)} \cdot \frac{\pi^2}{\sin\pi(c-s)\sin\pi(d-s)}.$$

Also, let L be a closed contour formed by a part of the curve in the theorem together with a semicircle of radius R to the right of the imaginary axis. By Stirling's theorem (Corollary 1.4.3), the integrand is $O(s^{a+b+c+d-2}e^{-2\pi|\mathrm{Im}\, s|})$ as $|s| \to \infty$ on L. So the integral in Theorem 2.4.2 converges, but for $\int_L$ we see that Im s can be arbitrarily small when $|s|$ is large. Thus, we have to assume that $\mathrm{Re}(a+b+c+d-1) < 0$ to ensure that $\int_L$ on the semicircle tends to 0 as $R \to \infty$. By Cauchy's residue theorem,

$$I = \sum_{n=0}^{\infty} \frac{\Gamma(a+c+n)\Gamma(b+c+n)\Gamma(d-c-n)(-1)^n}{n!}$$

$$+ \sum_{n=0}^{\infty} \frac{\Gamma(a+d+n)\Gamma(b+d+n)\Gamma(c-d-n)(-1)^n}{n!}$$

$$= \Gamma(a+c)\Gamma(b+c)\Gamma(d-c){}_2F_1\left(\begin{matrix} a+c, b+c \\ 1+c-d \end{matrix}; 1\right)$$

$$+ \Gamma(a+d)\Gamma(b+d)\Gamma(c-d){}_2F_1\left(\begin{matrix} a+d, b+d \\ 1+d-c \end{matrix}; 1\right).$$

The ${}_2F_1$s can be summed by Gauss's formula. After some simplification using Euler's reflection formula and trigonometry, the right side of the theorem is obtained. This is under the condition $\mathrm{Re}(a+b+c+d-1) < 0$. The complete result follows by analytic continuation of the parameters a, b, c, d. ∎

Theorem 2.4.2 is the integral analog of Gauss's summation of the ${}_2F_1$ at $x = 1$. Moreover, if we let $b = e - it$, $d = f - it$, and $s = itx$ in the theorem and let $t \to \infty$, we get, after some reduction employing Stirling's formula,

$$\int_0^1 x^{a+c-1}(1-x)^{e+f-1}dx = \frac{\Gamma(a+c)\Gamma(e+f)}{\Gamma(a+c+e+f)}.$$

Thus, Barnes's integral formula is an extension of the beta integral on $(0, 1)$ and so will be called Barnes's beta integral. It is also called Barnes's first lemma.

Theorem 2.4.2 can also be used to prove that

$${}_2F_1\left(\begin{matrix} a, b \\ c \end{matrix}; x\right) = \frac{\Gamma(c)\Gamma(c-a-b)}{\Gamma(c-a)\Gamma(c-b)}{}_2F_1\left(\begin{matrix} a, b \\ a+b-c+1 \end{matrix}; 1-x\right)$$

$$+ \frac{\Gamma(c)\Gamma(a+b-c)}{\Gamma(a)\Gamma(b)}(1-x)^{c-a-b}{}_2F_1\left(\begin{matrix} c-a, c-b \\ c-a-b+1 \end{matrix}; 1-x\right),$$

$c - a - b \neq$ integer. The proof is an exercise. This result was derived from the hypergeometric differential equation in the previous section.

The next theorem gives an integral analog of the Pfaff–Saalschütz identity.

Theorem 2.4.3   *For a suitably curved line of integration, so that the decreasing sequences of poles lie to the left and the increasing sequence of poles lies to the right of the contour:*

$$\frac{1}{2\pi i} \int_{-i\infty}^{i\infty} \frac{\Gamma(a+s)\Gamma(b+s)\Gamma(c+s)\Gamma(1-d-s)\Gamma(-s)}{\Gamma(e+s)} ds$$

$$= \frac{\Gamma(a)\Gamma(b)\Gamma(c)\Gamma(1-d+a)\Gamma(1-d+b)\Gamma(1-d+c)}{\Gamma(e-a)\Gamma(e-b)\Gamma(e-c)}$$

where $d + e = a + b + c + 1$.

Proof.   Start with the following special case of Theorem 2.4.2:

$$\frac{\Gamma(c-a)\Gamma(c-b)\Gamma(a+n)\Gamma(b+n)}{\Gamma(c+n)}$$

$$= \frac{1}{2\pi i} \int_{-i\infty}^{i\infty} \Gamma(a+s)\Gamma(b+s)\Gamma(n-s)\Gamma(c-a-b-s) ds.$$

Multiply both sides by $(d)_n / [n!(e)_n]$ and sum with respect to n. The result is

$$\frac{\Gamma(c-a)\Gamma(c-b)\Gamma(a)\Gamma(b)}{\Gamma(c)} {}_3F_2\left(\begin{array}{c} a, b, d \\ c, e \end{array} ; 1\right)$$

$$= \frac{1}{2\pi i} \int_{-i\infty}^{i\infty} \sum_{n=0}^{\infty} \frac{(d)_n}{n!(e)_n} \Gamma(a+s)\Gamma(b+s)\Gamma(n-s)\Gamma(c-a-b-s) ds$$

$$= \frac{1}{2\pi i} \int_{-i\infty}^{i\infty} \Gamma(a+s)\Gamma(b+s)\Gamma(c-a-b-s)\Gamma(-s){}_2F_1\left(\begin{array}{c} -s, d \\ e \end{array} ; 1\right) ds$$

$$= \frac{1}{2\pi i} \int_{-i\infty}^{i\infty} \frac{\Gamma(e)}{\Gamma(e-d)} \frac{\Gamma(a+s)\Gamma(b+s)\Gamma(e-d+s)\Gamma(c-a-b-s)\Gamma(-s)}{\Gamma(e+s)} ds$$

$$(2.4.11)$$

Take $c = d$, so that the ${}_3F_2$ on the left becomes ${}_2F_1$. Thus we have

$$\frac{\Gamma(a)\Gamma(b)\Gamma(e)\Gamma(e-a-b)\Gamma(c-a)\Gamma(c-b)}{\Gamma(c)\Gamma(e-a)\Gamma(e-b)}$$

$$= \frac{\Gamma(e)}{\Gamma(e-c)} \frac{1}{2\pi i} \int_{-i\infty}^{i\infty} \frac{\Gamma(a+s)\Gamma(b+s)\Gamma(e-c+s)\Gamma(c-a-b-s)\Gamma(-s)}{\Gamma(e+s)} ds.$$

This is the required result after renaming the parameters. Note that some of the operations carried out in the proof can be done only after appropriate restrictions on the parameters. These restrictions can be removed later by analytic continuation. The reader can check the details. ∎

A corollary of the proof of the previous theorem is the following interesting formula:

Theorem 2.4.4

$$
{}_3F_2\left(\begin{matrix} a, b, c \\ d, e \end{matrix}; 1\right) = \frac{\Gamma(d)\Gamma(d-a-b)}{\Gamma(d-a)\Gamma(d-b)}\,{}_3F_2\left(\begin{matrix} a, b, e-c \\ e, 1+a+b-d \end{matrix}; 1\right)
$$

$$
+ \frac{\Gamma(d)\Gamma(e)\Gamma(d+e-a-b-c)\Gamma(a+b-d)}{\Gamma(a)\Gamma(b)\Gamma(d+e-a-b)\Gamma(e-c)}
$$

$$
\cdot\, {}_3F_2\left(\begin{matrix} d-a, d-b, d+e-a-b-c \\ d+e-a-b, d+1-a-b \end{matrix}; 1\right).
$$

Proof.   As a consequence of Cauchy's theorem, (2.4.11) is equal to

$$
\frac{\Gamma(e)}{\Gamma(e-d)}\left[\sum_{n=0}^{\infty}\frac{\Gamma(a+n)\Gamma(b+n)\Gamma(e-d+n)\Gamma(c-a-b-n)(-1)^n}{n!\,\Gamma(e+n)} + \sum_{n=0}^{\infty}\right.
$$

$$
\cdot\,\frac{\Gamma(c-b+n)\Gamma(c-a+n)\Gamma(c+e-a-b-d+n)\Gamma(a+b-c-n)(-1)^n}{n!\,\Gamma(c+e-a-b+n)}\Bigg].
$$

Set this equal to the ${}_3F_2$ on the left of (2.4.11) and the theorem is obtained after reduction. ∎

Corollary 2.4.5   *If $d+e=a+b+c+1$, then*

$$
{}_3F_2\left(\begin{matrix} a, b, c \\ d, e \end{matrix}; 1\right) = \frac{\Gamma(d)\Gamma(e)\Gamma(d-a-b)\Gamma(e-a-b)}{\Gamma(d-a)\Gamma(d-b)\Gamma(e-a)\Gamma(e-b)}
$$

$$
+ \frac{1}{a+b-d}\,\frac{\Gamma(d)\Gamma(e)}{\Gamma(a)\Gamma(b)\Gamma(d+e-a-b)}
$$

$$
\times {}_3F_2\left(\begin{matrix} d-a, d-b, 1 \\ d+e-a-b, d+1-a-b \end{matrix}; 1\right).
$$

Proof.   When $d+e=a+b+c+1$, the first ${}_3F_2$ on the right in Theorem 2.4.4 becomes ${}_2F_1$. Evaluate the ${}_2F_1$ by Gauss's formula and get the result. ∎

Note that when a or b is a negative integer the second expression on the right vanishes because of the factor $1/[\Gamma(a)\Gamma(b)]$ and we recover the Pfaff–Saalschütz formula. Thus this result is the nonterminating form of the Pfaff–Saalschütz identity.

The second term on the right in Theorem 2.4.4 vanishes if we take $c = e + n - 1$, where $n \geq 1$ is an integer. The formula obtained is

$$_3F_2\left(\begin{array}{c} a, b, e + n - 1 \\ d, e \end{array}; 1\right) = \frac{\Gamma(d)\Gamma(d - a - b)}{\Gamma(d - a)\Gamma(d - b)} {_3F_2}\left(\begin{array}{c} a, b, 1 - n \\ a + b - d + 1, e \end{array}; 1\right).$$

$$(2.4.12)$$

This leads to an interesting result about the partial sums of $_2F_1(a, b; e; 1)$. Set $d = a + b + n + \epsilon$ and let $\epsilon \to 0$ to get

$$_2F_1\left(\begin{array}{c} a, b \\ e \end{array}; 1\right) [\text{to } n \text{ terms}] = \frac{\Gamma(a + n)\Gamma(b + n)}{\Gamma(a + b + n)\Gamma(n)} {_3F_2}\left(\begin{array}{c} a, b, e + n - 1 \\ e, a + b + n \end{array}; 1\right).$$

$$(2.4.13)$$

A particular case, where $a = b = \frac{1}{2}$ and $e = 1$, was given by Ramanujan in the following striking form:

$$\frac{1}{n} + \left(\frac{1}{2}\right)^2 \frac{1}{n+1} + \left(\frac{1.3}{2.4}\right)^2 \frac{1}{n+2} + \cdots$$

$$= \left\{\frac{\Gamma(n)}{\Gamma(n + \frac{1}{2})}\right\}^2 \left\{1 + \left(\frac{1}{2}\right)^2 + \left(\frac{1.3}{2.4}\right)^2 + \cdots \quad \text{to} \quad n \quad \text{terms}\right\}.$$

Bailey [1931, 1932] also proved the next, more general, theorem.

Theorem 2.4.6

$$\frac{\Gamma(x + m)\Gamma(y + m)}{\Gamma(m)\Gamma(x + y + m)} {_3F_2}\left(\begin{array}{c} x, y, u + m - 1 \\ u, x + y + m \end{array}; 1\right) \quad \text{to} \quad n \quad \text{terms}$$

$$= \frac{\Gamma(x + n)\Gamma(y + n)}{\Gamma(n)\Gamma(x + y + n)} {_3F_2}\left(\begin{array}{c} x, y, u + n - 1 \\ u, x + y + n \end{array}; 1\right) \quad \text{to} \quad m \quad \text{terms}.$$

There is a simple proof from Theorem 3.3.3.

Remark The Mellin transform can be seen as the Fourier transform carried over to the multiplicative group $(0, \infty)$ by means of the exponential function. In $F(s) = \int_0^\infty x^{s-1} f(x)dx$, write $s = \sigma + it$ and $x = e^{2\pi u}$ to get $F(\sigma + it) = 2\pi \int_{-\infty}^\infty (f(e^{2\pi u}) \cdot e^{2\pi\sigma u})e^{2\pi itu}du =: 2\pi \int_{-\infty}^\infty g(u)e^{2\pi itu}du$. A new feature in the theory of Mellin transforms is that $F(s)$ is analytic in a vertical strip.

Just as the gamma function, a special case of a Mellin transform, has a finite field analog, so does the more general case. Let F_q denote a finite field with q elements and F_q^* its multiplicative part. Let f be a complex-valued function on F_q^*. Its Mellin transform is defined on the group of characters of F_q^* as $F(\chi) = \sum_{a \in F_q^*} \chi(a)f(a)$. The reader may verify that this has an inversion given by $f(s) = \frac{1}{q-1}\sum_\chi \bar{\chi}(a)F(\chi)$. There is also an analog of Barnes's formula (Theorem 2.4.2) due to Helversen–Pasotto [1978]. A proof based on Mellin transforms is given in Helversen–Pasotto and Solé [1993].

2.5 Contiguous Relations

Gauss defined two hypergeometric functions to be contiguous if they have the same power-series variable, if two of the parameters are pairwise equal, and if the third pair differ by 1. We use $F(a\pm)$ to denote $_2F_1(a \pm 1, b; c; x)$ respectively. $F(b\pm)$ and $F(c\pm)$ are defined similarly. Gauss [1812] showed that a hypergeometric function and any two others contiguous to it are linearly related. Since there are six functions contiguous to a given $_2F_1$, we get $\binom{6}{2} = 15$ relations. In fact, there are only nine different relations, if the symmetry in a and b is taken into account. These relations can be iterated, so any three hypergeometric functions whose parameters differ by integers are linearly related. These relations are called contiguous relations. In this section we show how Gauss's fifteen relations are derived. Then we briefly point out connections with continued fractions and orthogonal polynomials.

Contiguous relations can be iterated and we use the word contiguous in the more general sense when the parameters differ by integers. It is easily verified that

$$\frac{d}{dx} {}_2F_1\left(\begin{matrix} a, b \\ c \end{matrix}; x\right) = \frac{ab}{c} {}_2F_1\left(\begin{matrix} a+1, b+1 \\ c+1 \end{matrix}; x\right). \tag{2.5.1}$$

Since this $_2F_1$ satisfies the equation

$$x(1-x)y'' + [c - (a+b+1)x]y' - aby = 0,$$

we get the contiguous relation

$$x(1-x)\frac{(a+1)(b+1)}{c(c+1)} {}_2F_1\left(\begin{matrix} a+2, b+2 \\ c+2 \end{matrix}; x\right)$$

$$+ \frac{(c - (a+b+1)x)}{c} {}_2F_1\left(\begin{matrix} a+1, b+1 \\ c+1 \end{matrix}; x\right) - {}_2F_1\left(\begin{matrix} a, b \\ c \end{matrix}; x\right) = 0. \tag{2.5.2}$$

By means of transformation formulas, this can be changed into other contiguous relations.

Apply Pfaff's transformation

$${}_2F_1\left(\begin{matrix} a, b \\ c \end{matrix}; x\right) = (1-x)^{-a} {}_2F_1\left(\begin{matrix} a, c - b \\ c \end{matrix}; \frac{x}{x-1}\right)$$

to each term in the above equation. After a little simplification where we set $u = x/(x-1)$ and replace $c - b$ by b, the result is

$${}_2F_1\left(\begin{matrix} a, b \\ c \end{matrix}; u\right) = \frac{c + (a - b + 1)u}{c} {}_2F_1\left(\begin{matrix} a+1, b \\ c+1 \end{matrix}; u\right)$$

$$- \frac{(a+1)(c - b + 1)u}{c(c+1)} {}_2F_1\left(\begin{matrix} a+2, b \\ c+2 \end{matrix}; u\right). \tag{2.5.3}$$

This is a contiguous relation due to Euler, who derived it in a different way. If we apply Euler's transformation

$$
{}_2F_1\left(\begin{matrix} a,b \\ c \end{matrix}; x\right) = (1-x)^{c-a-b} \cdot {}_2F_1\left(\begin{matrix} c-a, c-b \\ c \end{matrix}; x\right)
$$

to (2.5.2), then we get another contiguous relation:

$$
(1-x){}_2F_1\left(\begin{matrix} a,b \\ c \end{matrix}; x\right) = \frac{c - (2c-a-b+1)x}{c}{}_2F_1\left(\begin{matrix} a,b \\ c+1 \end{matrix}; x\right)
$$

$$
+ \frac{(c-a+1)(c-b+1)}{c(c+1)}{}_2F_1\left(\begin{matrix} a,b \\ c+2 \end{matrix}; x\right). \qquad (2.5.4)
$$

This is one of the relations Gauss obtained. Euler's method for obtaining (2.5.3) was to use his integral representation of ${}_2F_1$. By direct integration he found a formula of which the following is a particular case:

$$
a\int_0^1 t^{a-1}(1-t)^{c-a-1}(1-tx)^{-b}dt
$$

$$
= (c+(a+1-b)x)\int_0^1 t^a(1-t)^{c-a-1}(1-tx)^{-b}dt
$$

$$
- (c-b+1)x\int_0^1 t^{a+1}(1-t)^{c-a-1}(1-tx)^{-b}dt. \qquad (2.5.3')
$$

This is identical with (2.5.3). See Exercise 23 for a simple way of proving this identity. As another example of how the integral can be used, observe that

$$
(1-xt)^{-a} = (1-xt)^{-a-1}(1-xt) = (1-xt)^{-a-1}[1-x+(1-t)x].
$$

Substitute the right side in

$$
{}_2F_1\left(\begin{matrix} a,b \\ c \end{matrix}; x\right) = \frac{\Gamma(c)}{\Gamma(b)\Gamma(c-b)}\int_0^1 (1-xt)^{-a}t^{b-1}(1-t)^{c-b-1}dt
$$

to obtain

$$
{}_2F_1\left(\begin{matrix} a,b \\ c \end{matrix}; x\right) = (1-x){}_2F_1\left(\begin{matrix} a+1,b \\ c \end{matrix}; x\right) + \frac{(c-b)x}{c}{}_2F_1\left(\begin{matrix} a+1,b \\ c+1 \end{matrix}; x\right).
$$

The above examples show how contiguous relations arise. Now we give a derivation of Gauss's basic contiguous relations. It is enough to obtain a set of six relations from which Gauss's fifteen are obtained by equating the $\binom{6}{2}$ pairs of them. The first three in this set are

$$x\frac{dF}{dx} = a(F(a+) - F),\qquad(2.5.5)$$

$$x\frac{dF}{dx} = b(F(b+) - F),\qquad(2.5.6)$$

$$x\frac{dF}{dx} = (c - 1)(F(c-) - F),\qquad(2.5.7)$$

where

$$F := {}_2F_1\left(\begin{matrix} a, b \\ c \end{matrix}; x\right) \quad \text{and} \quad F(a+) := {}_2F_1\left(\begin{matrix} a+1, b \\ c \end{matrix}; x\right)$$

and so on. The nth term of $F(a+) - F$ is

$$\left[\frac{(a+1)_n(b)_n}{n!(c)_n} - \frac{(a)_n(b)_n}{n!(c)_n}\right]x^n = \frac{(a+1)_{n-1}(b)_n}{n!(c)_n}(a+n-a)x^n$$

$$= \frac{n}{a}\frac{(a)_n(b)_n}{n!(c)_n}x^n,$$

and

$$x\frac{dF}{dx} = \sum_{n=0}^{\infty}\frac{n(a)_n(b)_n}{n!(c)_n}x^n.$$

This proves (2.5.5). Formula (2.5.6) follows by symmetry in a and b, and (2.5.7) is proved similarly. To obtain the other three equations, set $\delta := x\frac{d}{dx}$ and verify that the hypergeometric equation can be written as

$$[\delta(\delta + c - 1) - x(\delta + a)(\delta + b)]y = 0.$$

So,

$$[\delta(\delta + c - 1) - x(\delta + a - 1)(\delta + b)]F(a-) = 0.$$

Now

$$\delta(\delta + c - 1) = (\delta + a - 1)(\delta + c - a) - (a - 1)(c - a)$$

so that

$$[(\delta + c - a) - x(\delta + b)](\delta + a - 1)F(a-) = (c - a)(a - 1)F(a-).$$

Apply (2.5.5) in the form $(\delta + a - 1)F(a-) = (a - 1)F$ to the above equation to get

$$[(\delta + c - a) - x(\delta + b)]F = (c - a)F(a-),$$

or

$$x(1 - x)\frac{dF}{dx} = (c - a)F(a-) + (a - c + bx)F.\qquad(2.5.8)$$

The remaining two relations are

$$x(1-x)\frac{dF}{dx} = (c-b)F(b-) + (b-c+ax)F, \qquad (2.5.9)$$

$$c(1-x)\frac{dF}{dx} = (c-a)(c-b)F(c+) + c(a+b-c)F. \qquad (2.5.10)$$

Formula (2.5.9) is obtained from (2.5.8) by symmetry in a and b, and (2.5.10) is proved in a manner similar to (2.5.8). Gauss's fifteen contiguous relations are obtained by equating two values of $x\frac{dF}{dx}$ in (2.5.5) to (2.5.7) and in (2.5.8) to (2.5.10). For example,

$$[c - 2a - (b-a)x]F + a(1-x)F(a+) - (c-a)F(a-) = 0$$

follows from (2.5.5) and (2.5.8), whereas (2.5.5) and (2.5.10) give

$$c[a - (c-b)x]F - ac(1-x)F(a+) + (c-a)(c-b)xF(c+) = 0.$$

We have seen a special case of the last contiguous relation. Set $x = 1$ to obtain

$$c(c-a-b)F\left(\begin{matrix} a, b \\ c \end{matrix}; 1\right) = (c-a)(c-b)F\left(\begin{matrix} a, b \\ c+1 \end{matrix}; 1\right).$$

We used this relation in Section 2.2 to derive Gauss's formula for $_2F_1(a, b; c; 1)$. The reader should derive a few more contiguous relations as an exercise. It must also be clear that once a contiguous relation is given, it is very easy to verify by considering the coefficient of x^n in each term. For example,

$$_2F_1\left(\begin{matrix} a, b \\ c \end{matrix}; x\right) = {_2F_1}\left(\begin{matrix} a, b+1 \\ c+1 \end{matrix}; x\right) - \frac{a(c-b)}{c(c+1)}x\,{_2F_1}\left(\begin{matrix} a+1, b+1 \\ c+2 \end{matrix}; x\right)$$

$$\qquad (2.5.11)$$

is true because

$$\frac{(a)_n(b+1)_n}{n!(c+1)_n} - \frac{a(c-b)}{c(c+1)}\frac{(a+1)_{n-1}(b+1)_{n-1}}{(n-1)!(c+2)_{n-1}} = \frac{(a)_n(b)_n}{n!(c)_n}.$$

Gauss used (2.5.11) to obtain an interesting continued fraction for the ratio of two associated hypergeometric series. Rewrite (2.5.11) as

$$\frac{_2F_1(a, b; c; x)}{_2F_1(a, b+1; c+1; x)} = 1 - \frac{a(c-b)}{c(c+1)}x \cdot \frac{1}{\frac{_2F_1(a, b+1; c+1; x)}{_2F_1(a+1, b+1; c+2; x)}}$$

$$= 1 - \cfrac{u_1 x}{1 - \cfrac{v_1 x}{1 - \cfrac{u_2 x}{1 - \cfrac{v_2 x}{\cdots,}}}}$$

where

$$u_n = \frac{(a+n-1)(c-b+n-1)}{(c+2n-2)(c+2n-1)} \quad \text{and} \quad v_n = \frac{(b+n)(c-a+n)}{(c+2n-1)(c+2n)}.$$

Surprisingly, Euler had earlier found a different continued fraction for $_2F_1(a, b;$ $c; x)/_2F_1(a, b+1; c+1; x)$. This comes from (2.5.3). Interchange a and b in (2.5.3) and rewrite it as

$$c\frac{_2F_1(a, b; c; x)}{_2F_1(a, b+1; c+1; x)} = c + (1+b-a)x - \frac{(b+1)(c-a+1)x}{(c+1)\frac{_2F_1(a, b+1; c+1; x)}{_2F_1(a, b+2; c+2; x)}}.$$

We now have the continued fraction

$$c + (1+b-a)x - \frac{(b+1)(c-a+1)x}{(c+1+(2+b-a)x)_-} \frac{(b+2)(c-a+2)x}{(c+2+(3+b-a)x)_-} \cdots.$$

It is clear that numerous examples of continued fractions for appropriate ratios of hypergeometric functions can be obtained in this way. Note that (2.5.2), which is just the differential equation for the hypergeometric equation, also gives a continued fraction for $_2F_1(a+1, b+1; c+1; x)/_2F_1(a, b; c; x)$. It involves quadratic terms in x and is given in Exercise 26. We have not given conditions for convergence of these infinite continued fractions. The reader should see Lorentzen and Waadeland [1992] for a discussion of convergence of continued fractions. Also see Berndt [1985, pp. 136–137] for the reference to Euler's continued fraction and for the work of Ramanujan on this topic.

One way of arriving at a connection between hypergeometric functions and orthogonal polynomials (defined below) is through a formula of Jacobi, which we now derive. Multiply the hypergeometric equation by $x^{c-1}(1-x)^{a+b-c}$ and write it as

$$\frac{d}{dx}[x(1-x)x^{c-1}(1-x)^{a+b-c}y'] = abx^{c-1}(1-x)^{a+b-c}y,$$

where $y = {_2F_1}(a, b; c; x)$.

From (2.5.1) it follows that the derivative of a hypergeometric function is again hypergeometric, so that it satisfies the differential equation with a, b, c changed to $a+1, b+1, c+1$. By induction, this implies that

$$\frac{d}{dx}[x^k(1-x)^k My^{(k)}] = (a+k-1)(b+k-1)x^{k-1}(1-x)^{k-1}My^{(k-1)},$$

where $M = x^{c-1}(1-x)^{a+b-c}$. Consequently, we have the recurrence relation

$$\frac{d^k}{dx^k}[x^k(1-x)^k My^{(k)}] = (a+k-1)(b+k-1)\frac{d^{k-1}}{dx^{k-1}}$$
$$\cdot [x^{k-1}(1-x)^{k-1}My^{(k-1)}]$$
$$= (a)_k(b)_k My.$$

Substitute

$$y^{(k)} = \frac{(a)_k (b)_k}{(c)_k} {}_2F_1 \left(\begin{matrix} a+k, b+k \\ c+k \end{matrix} ; x \right)$$

in the above equation to get

$$\frac{d^k}{dx^k} \left[x^k (1-x)^k M_2 F_1 \left(\begin{matrix} a+k, b+k \\ c+k \end{matrix} ; x \right) \right] = (c)_k M_2 F_1 \left(\begin{matrix} a, b \\ c \end{matrix} ; x \right).$$

If b is a negative integer $-n$, then for $k = n$ we have Jacobi's formula

$$_2F_1 \left(\begin{matrix} -n, a \\ c \end{matrix} ; x \right) = \frac{x^{1-c}(1-x)^{c+n-a}}{(c)_n} \frac{d^n}{dx^n} [x^{c+n-1}(1-x)^{a-c}]. \qquad (2.5.12)$$

For a more symmetrical expression in (2.5.12) set $x = (1-y)/2$, $c = \alpha + 1$, and $a = n + \alpha + \beta + 1$ to get

$$_2F_1 \left(\begin{matrix} -n, n+\alpha+\beta+1 \\ \alpha+1 \end{matrix} ; \frac{1-y}{2} \right)$$

$$= \frac{(1-y)^{-\alpha}(1+y)^{-\beta}}{(\alpha+1)_n 2^n} (-1)^n \frac{d^n}{dx^n} [(1-y)^{n+\alpha}(1+y)^{n+\beta}]. \qquad (2.5.13)$$

Definition 2.5.1 *The Jacobi polynomial of degree n is defined by*

$$P_n^{(\alpha,\beta)}(x) := \frac{(\alpha+1)_n}{n!} {}_2F_1 \left(\begin{matrix} -n, n+\alpha+\beta+1 \\ \alpha+1 \end{matrix} ; \frac{1-x}{2} \right).$$

One of its fundamental properties is that

$$\int_{-1}^{+1} P_n^{(\alpha,\beta)}(x) P_m^{(\alpha,\beta)}(x)(1-x)^\alpha (1+x)^\beta dx$$

$$= \frac{2^{\alpha+\beta+1}\Gamma(n+\alpha+1)\Gamma(n+\beta+1)}{(2n+\alpha+\beta+1)\Gamma(n+\alpha+\beta+1)n!} \delta_{mn}. \qquad (2.5.14)$$

This is easy to prove. Use (2.5.13) and integration by parts.

Remark 2.5.1 Formula (5.13), which can be written as

$$(1-x)^\alpha (1+x)^\beta P_n^{(\alpha,\beta)}(x) = \frac{(-1)^n}{2^n n!} \frac{d^n}{dx^n} [(1-x)^{n+\alpha}(1+x)^{n+\beta}], \qquad (2.5.13')$$

is often called the Rodrigues formula for Jacobi polynomials. The particular case for Legendre polynomials, where $\alpha = \beta = 0$, was published by O. Rodrigues in 1816 in an Ecole Polytechnique journal. Unfortunately, Rodrigues's paper did not receive much attention. The formula was rediscovered independently by J. Ivory in 1822 and Jacobi in 1827. It is amusing to note that Jacobi later suggested to Ivory that they write a joint paper on this important formula and publish it in France as it was not known there. Their paper appeared in Liouville's journal in 1837.

Interestingly, Rodrigues's teacher, Laplace, in the course of his work in probability (1810–11), found a similar formula [see (6.1.3)] for Hermite polynomials. For references, see Roy [1993].

A set of polynomials $\{p_n(x)\}$ is called orthogonal if there is a positive measure $d\mu(x)$ with finite moments of all orders so that

$$\int_{-\infty}^{\infty} p_n(x) p_m(x) d\mu(x) = 0, \quad m \neq n.$$

Thus, the Jacobi polynomials are orthogonal with respect to the measure

$$d\mu(x) = \begin{cases} (1-x)^\alpha (1+x)^\beta dx, & -1 < x < 1, \\ 0, & \text{otherwise}. \end{cases}$$

We shall see in Chapter 5 that any set of orthogonal polynomials satisfies a three-term recurrence relation:

$$xp_n(x) = A_n p_{n+1}(x) + B_n p_n(x) + C_n p_{n-1}(x),$$
$$p_0(x) = 1, \, p_{-1}(x) = 0,$$

$$A_n, B_n, C_{n+1} \text{ real}, \qquad A_n C_{n+1} > 0, \quad n = 0, 1, \ldots .$$

Conversely, any set of polynomials that satisfies this recurrence relation is orthogonal with respect to a positive measure, which may not be unique.

The three-term recurrence relation for Jacobi polynomials comes from the contiguous relation

$$2b(c-a)(b-a-1)_2F_1\left(\begin{matrix} a-1, b+1 \\ c \end{matrix} ; x \right)$$

$$- [(1-2x)(b-a-1)_3 + (b-a)(b+a-1)(2c-b-a-1)]_2F_1\left(\begin{matrix} a, b \\ c \end{matrix} ; x \right)$$

$$- 2a(b-c)(b-a+1)_2F_1\left(\begin{matrix} a+1, b-1 \\ c \end{matrix} ; x \right) = 0, \qquad (2.5.15)$$

after proper identification. In particular we require that $a = -n$, where n is a positive integer. But (2.5.15) continues to hold when the series does not terminate. Another contiguous relation that gives a set of orthogonal polynomials is

$$a(1-x)_2F_1\left(\begin{matrix} a+1, b \\ c \end{matrix} ; x \right) + [c - 2a - (b-a)x]_2F_1\left(\begin{matrix} a, b \\ c \end{matrix} ; x \right)$$

$$- (c-a)_2F_1\left(\begin{matrix} a-1, b \\ c \end{matrix} ; x \right) = 0. \qquad (2.5.16)$$

We shall study orthogonal polynomials in detail in later chapters. That will provide the natural setting for some of the contiguous relations in the sense implied by the above remarks.

Kummer [1836] considered the problem of extending the contiguous relations to $_pF_q$, but he stopped with the remark that for $_3F_2(a, b, c; d, e; x)$ the formulas are more complicated. In particular, the linear contiguous relations require four functions, a $_3F_2$ and three contiguous to it. Kummer also noted that only when $x = 1$ did the formulas simplify. That is the key to three-term relations for some higher p and q. These will be discussed in the next chapter.

Remark 2.5.2 It should be noted that the continued fraction from Gauss's relation (2.5.11) contains a continued fraction for $\arctan x$ as a special case. The Taylor series for $\arctan x$ converges very slowly when $x = 1$, but the convergence of the continued fraction is extremely rapid and was once useful in computing approximations of π. See Exercise 25.

Remark 2.5.3 The orthogonality relation (2.5.14) is a generalization of the well-known fact from trigonometry:

$$\int_0^\pi \cos m\theta \cos n\theta d\theta = 0 \quad \text{for } m \neq n. \tag{2.5.17}$$

This becomes clear by setting $x = \cos\theta$ and $\cos n\theta = T_n(x)$. Then $T_n(x)$ is a polynomial of degree n and (2.5.17) becomes

$$\int_{-1}^1 T_m(x)T_n(x)(1-x)^{-1/2}(1+x)^{-1/2}dx = 0 \quad \text{for } m \neq n.$$

It is not difficult to show that $T_n(x) = C P_n^{(-1/2,-1/2)}(x)$, where $C = (2n)!/[2^{2n}(n!)^2]$. Another set of polynomials is

$$U_n(\cos\theta) = \frac{\sin(n+1)\theta}{\sin\theta}.$$

$T_n(x)$ and $U_n(x)$ are called Chebyshev polynomials of the first and second kind respectively. The three-term recurrence relation for $T_n(x)$ is

$$xT_n(x) = \frac{1}{2}T_{n+1}(x) + \frac{1}{2}T_{n-1}(x).$$

This is the trigonometric identity

$$2\cos\theta\cos n\theta = \cos(n+1)\theta + \cos(n-1)\theta.$$

Several properties of the Chebyshev polynomials translate to elementary trigonometric identities and they form the starting point for generalizations to Jacobi and other sets of orthogonal polynomials. For this reason the reader should keep them in mind when studying the "classical" orthogonal polynomials. A number of exercises at the end of this chapter deal with Chebyshev polynomials. We also refer to

$$V_n(\cos\theta) = \frac{\sin\{(2n+1)\theta/2\}}{\sin\theta/2} \quad and \quad W_n(\cos\theta) = \frac{\cos\{(2n+1)\theta/2\}}{\cos\theta/2}$$

as Chebyshev polynomials of the third and fourth kinds respectively. These polynomials are of lesser importance.

2.6 Dilogarithms

All the examples given in Section 2.1 of special functions expressible as hypergeometric functions were either $_2F_1$ or of lower level. The dilogarithm function
is an example of a $_3F_2$. This function was first discussed by Euler and later by
many other mathematicians including Abel and Kummer. But it is only in the
past two decades that it has begun to appear in several different mathematical
contexts. Its growing importance is reflected in the two books devoted to it and
its generalization, the polylogarithm. See Lewin [1981, 1981a]. Here we give
a few elementary properties of dilogarithms. The reader may also wish to see
Kirillov [1994] and Zagier [1989]. The latter paper gives a number of interesting
applications in number theory and geometry.

The dilogarithm is defined by the series

$$\mathrm{Li}_2(x) := \sum_{n=1}^{\infty} \frac{x^n}{n^2}, \quad \text{for } |x| \le 1. \tag{2.6.1}$$

From the Taylor expansion of $\log(1-t)$ it follows that

$$\mathrm{Li}_2(x) = -\int_0^x \frac{\log(1-t)}{t} dt. \tag{2.6.2}$$

The integral is defined as a single-valued function in the cut plane $\mathbb{C} - [1, \infty)$; so
we have an analytic continuation of $\mathrm{Li}_2(x)$ to this region. The multivaluedness of
$\mathrm{Li}_2(x)$ can also be studied easily. There are branch points at 1 and ∞. If $\mathrm{Li}_2(x)$
is continued along a loop that winds around $x = 1$ once, then the value of Li_2
changes to $\mathrm{Li}_2(x) - 2\pi i \log x$. This is easily seen from the integral definition.

We now obtain the hypergeometric representation from the integral

$$\mathrm{Li}_2(x) = \int_0^x {}_2F_1\left(\begin{matrix}1,1\\2\end{matrix};t\right)dt = x\int_0^1 {}_2F_1\left(\begin{matrix}1,1\\2\end{matrix};xu\right)du = x\,{}_3F_2\left(\begin{matrix}1,1,1\\2,2\end{matrix};x\right)$$

by formula (2.2.2). Though it is possible to develop the properties of the dilogarithm without any reference to the theory of hypergeometric series, we note one
example where Pfaff's transformation is applicable.

Theorem 2.6.1 $\text{Li}_2(x) + \text{Li}_2(x/(x-1)) = -\frac{1}{2}[\log(1-x)]^2$ *(Landen's transformation).*

Proof. By Pfaff's transformation (Theorem 2.2.5)

$$\text{Li}_2(x) = \int_0^x {}_2F_1\left(\begin{matrix}1,1\\2\end{matrix};t\right)dt = \int_0^x {}_2F_1\left(\begin{matrix}1,1\\2\end{matrix};\frac{t}{t-1}\right)\frac{dt}{1-t}.$$

Set $u = t/(t-1)$ in the last integral to get

$$-\int_0^{x/(x-1)} {}_2F_1\left(\begin{matrix}1,1\\2\end{matrix};u\right)\frac{du}{1-u}$$

$$= -\int_0^{x/(x-1)} {}_2F_1\left(\begin{matrix}1,1\\2\end{matrix};u\right)du - \int_0^{x/(x-1)}\frac{u}{u-1}{}_2F_1\left(\begin{matrix}1,1\\2\end{matrix};u\right)du.$$

The first integral is $\text{Li}_2(x/(x-1))$ and the second integral is

$$-\int_0^{x/(x-1)}\frac{\log(1-u)}{1-u}du = -\frac{1}{2}[\log(1-x)]^2.$$

This proves the result. ∎

We give another proof since it involves a different expression for the dilogarithm as a hypergeometric function. This expression is given by

$$\text{Li}_2(x) = \lim_{\epsilon \to 0}\frac{1}{\epsilon^2}\left\{{}_2F_1\left(\begin{matrix}\epsilon,\epsilon\\1+\epsilon\end{matrix};x\right) - 1\right\}. \tag{2.6.3}$$

Let x be in the region $\{x\,||x| \le \delta < 1\} \cap \{x\,||x/(x-1)| \le \delta < 1\} = S_\delta$, where $\delta > 0$. Apply Pfaff's transformation to (2.6.3) to get

$$\text{Li}_2(x) = \lim_{\epsilon \to 0}\frac{1}{\epsilon^2}\left\{(1-x)^{-\epsilon}{}_2F_1\left(\begin{matrix}\epsilon,1\\1+\epsilon\end{matrix};\frac{x}{x-1}\right) - 1\right\}$$

$$= \lim_{\epsilon \to 0}\frac{1}{\epsilon^2}\left\{\left(1 - \epsilon\log(1-x) + \frac{\epsilon^2}{2}[\log(1-x)]^2 + O(\epsilon^3)\right)\right.$$

$$\left. \cdot \left(1 + \epsilon\sum_{n=1}^{\infty}\frac{1}{n+\epsilon}\left(\frac{x}{x-1}\right)^n\right) - 1\right\}.$$

Now,

$$\sum_{n=1}^{\infty}\frac{1}{n+\epsilon}\left(\frac{x}{x-1}\right)^n = \sum_{n=1}^{\infty}\frac{1}{n}\left(\frac{x}{x-1}\right)^n + \sum_{n=1}^{\infty}\left(\frac{1}{n+\epsilon} - \frac{1}{n}\right)\left(\frac{x}{x-1}\right)^n$$

$$= -\log\left(1 - \frac{x}{x-1}\right) - \epsilon\sum_{n=1}^{\infty}\frac{1}{n(n+\epsilon)}\left(\frac{x}{x-1}\right)^n.$$

Thus,

$$
\begin{aligned}
\mathrm{Li}_2(x) &= \lim_{\epsilon \to 0} \frac{1}{\epsilon^2} \left\{ \left(1 - \epsilon \log(1-x) + \frac{\epsilon^2}{2} [\log(1-x)]^2 + O(\epsilon^3) \right) \right. \\
&\quad \cdot \left. \left(1 + \epsilon \log(1-x) - \epsilon^2 \sum_{n=1}^{\infty} \frac{1}{n(n+\epsilon)} \left(\frac{x}{x-1} \right)^n + O(\epsilon^3) \right) - 1 \right\} \\
&= \lim_{\epsilon \to 0} \frac{1}{\epsilon^2} \left\{ -\frac{\epsilon^2}{2} [\log(1-x)]^2 - \epsilon^2 \sum_{n=1}^{\infty} \frac{1}{n(n+\epsilon)} \left(\frac{x}{x-1} \right)^n + O(\epsilon^3) \right\} \\
&= -\frac{1}{2} [\log(1-x)]^2 - \mathrm{Li}_2 \left(\frac{x}{x-1} \right).
\end{aligned}
$$

The limit operation may be justified by the fact that for $x \in S_\delta$ and $|\epsilon| < 1/2$, the relevant series represent analytic functions of x and ϵ. This proves the theorem again. There is another proof in Exercise 38.

Theorem 2.6.2

$$\text{If } \omega^n = 1, \quad \text{then} \quad \frac{1}{n} \mathrm{Li}_2(x^n) = \sum_{k=0}^{n-1} \mathrm{Li}_2(\omega^k x). \quad (2.6.4)$$

$$\frac{1}{2} \mathrm{Li}_2(x^2) = \mathrm{Li}_2(x) + \mathrm{Li}_2(-x). \quad (2.6.5)$$

$$\mathrm{Li}_2(x) + \mathrm{Li}_2(1-x) = \frac{\pi^2}{6} - \log x \log(1-x). \quad (2.6.6)$$

Proof. To prove (2.6.4) start with the factorization $(1 - t^n) = (1-t)(1-\omega t) \cdots$ $(1 - \omega^{n-1} t)$. Take the logarithm and integrate to get

$$
\begin{aligned}
-\int_0^x \frac{\log(1-t^n)}{t} dt &= -\int_0^x \frac{\log(1-t)}{t} dt \\
&\quad - \int_0^x \frac{\log(1-\omega t)}{t} dt \cdots - \int_0^x \frac{\log(1-\omega^{n-1}t)}{t} dt.
\end{aligned}
$$

A change of variables shows that the integral on the left is $\frac{1}{n} \mathrm{Li}_2(x^n)$. This proves (2.6.4), and (2.6.5) follows by taking $n = 2$.

To derive (2.6.6) integrate by parts:

$$\mathrm{Li}_2(x) = -\int_0^x \frac{\log(1-t)}{t} dt = -\log x \log(1-x) - \int_0^x \frac{\log t}{1-t} dt.$$

The last integral, after a change of variables $u = 1 - t$, is

$$-\int_1^{1-x} \frac{\log(1-u)}{u} du = \mathrm{Li}_2(1-x) - \mathrm{Li}_2(1).$$

But

$$\text{Li}_2(1) = \sum_1^\infty \frac{1}{n^2} = \zeta(2) = \frac{\pi^2}{6}$$

by Theorem 1.2.4. The proof of the theorem is complete. ∎

Apparently, the only values x at which $\text{Li}_2(x)$ can be computed in terms of more elementary functions are the eight values $x = 0, \pm 1, \frac{1}{2}, \frac{-1\pm\sqrt{5}}{2}, \frac{1-\sqrt{5}}{2}, \frac{3-\sqrt{5}}{2}$.

Theorem 2.6.3

$$\text{Li}_2(0) = 0, \tag{2.6.7}$$

$$\text{Li}_2(1) = \frac{\pi^2}{6}, \tag{2.6.8}$$

$$\text{Li}_2(-1) = -\frac{\pi^2}{12}, \tag{2.6.9}$$

$$\text{Li}_2\left(\frac{1}{2}\right) = \frac{\pi^2}{12} - \frac{1}{2}[\log 2]^2, \tag{2.6.10}$$

$$\text{Li}_2\left(\frac{3-\sqrt{5}}{2}\right) = \frac{\pi^2}{15} - \frac{1}{4}\left[\log\left(\frac{3-\sqrt{5}}{2}\right)\right]^2, \tag{2.6.11}$$

$$\text{Li}_2\left(\frac{\sqrt{5}-1}{2}\right) = \frac{\pi^2}{10} - \left[\log\left(\frac{\sqrt{5}-1}{2}\right)\right]^2, \tag{2.6.12}$$

$$\text{Li}_2\left(\frac{1-\sqrt{5}}{2}\right) = -\frac{\pi^2}{15} + \frac{1}{2}\left[\log\left(\frac{\sqrt{5}-1}{2}\right)\right]^2, \tag{2.6.13}$$

$$\text{Li}_2\left(-\frac{1+\sqrt{5}}{2}\right) = \frac{-\pi^2}{10} + \frac{1}{2}\left[\log\left(\frac{\sqrt{5}+1}{2}\right)\right]^2. \tag{2.6.14}$$

Proof. Relation (2.6.7) is obvious and (2.6.8) was done in the proof of Theorem 2.6.2.

For (2.6.9), observe that

$$\text{Li}_2(-1) = \sum_1^\infty \frac{(-1)^n}{n^2} = -\left(1 - \frac{1}{2^2} + \frac{1}{3^2} - \frac{1}{4^2} + \cdots\right)$$

$$= -\left[1 + \frac{1}{2^2} + \frac{1}{3^2} + \frac{1}{4^2} + \cdots - 2\left(\frac{1}{2^2} + \frac{1}{4^2} + \cdots\right)\right]$$

$$= -\left[\frac{\pi^2}{6} - \frac{2}{2^2} \cdot \frac{\pi^2}{6}\right] = \frac{-\pi^2}{12}.$$

Set $x = \frac{1}{2}$ in (2.6.6) to get (2.6.10).

The identities (2.6.11) and (2.6.12) can be derived as follows: Landen's transformation and (2.6.5) combine to give

$$\text{Li}_2\left(\frac{x}{x-1}\right) + \frac{1}{2}\text{Li}_2(x^2) - \text{Li}_2(-x) = -\frac{1}{2}[\log(1-x)]^2. \qquad (2.6.15)$$

Set the variables in the first two dilogarithmic functions equal to each other. Then $x/(x-1) = x^2$ and $x^2 - x - 1 = 0$. A solution of this is $x = (1 - \sqrt{5})/2$. Substitute this x in (2.6.15) to obtain

$$\frac{3}{2}\text{Li}_2\left(\frac{3-\sqrt{5}}{2}\right) - \text{Li}_2\left(\frac{\sqrt{5}-1}{2}\right) = -\frac{1}{2}\left[\log\left(\frac{\sqrt{5}+1}{2}\right)\right]^2.$$

To find another equation involving $\text{Li}_2((3-\sqrt{5})/2)$ and $\text{Li}_2((1-\sqrt{5})/2)$ take $x = (3-\sqrt{5})/2$ in (2.6.6) to arrive at

$$\text{Li}_2\left(\frac{3-\sqrt{5}}{2}\right) + \text{Li}_2\left(\frac{\sqrt{5}-1}{2}\right) = \frac{\pi^2}{6} - \log\left(\frac{3-\sqrt{5}}{2}\right)\log\left(\frac{\sqrt{5}-1}{2}\right).$$

Now solve these equations to obtain the necessary result. The proofs of the formulas (2.6.13) and (2.6.14) are left as exercises. ∎

There are also two variable equations for the dilogarithm. The following is usually attributed to Abel though it was published earlier by Spence. See Lewin [1981a] for references. The formula is

$$\text{Li}_2\left[\frac{x}{1-x} \cdot \frac{y}{1-y}\right] = \text{Li}_2\left[\frac{x}{1-y}\right] + \text{Li}_2\left[\frac{y}{1-x}\right] - \text{Li}_2(x) - \text{Li}_2(y)$$
$$- \log(1-x)\log(1-y). \qquad (2.6.16)$$

This is easily verified by partial differentiation with respect to x or y and is left to the reader.

More generally we can define the polylogarithm by the series

$$\text{Li}_m x := \sum_{n=1}^{\infty} \frac{x^n}{n^m} \quad \text{for } |x| \leq 1, m = 2, 3, \ldots. \qquad (2.6.17)$$

The relation

$$\frac{d}{dx}\text{Li}_m(x) = \frac{1}{x}\text{Li}_{m-1}(x)$$

is easy to show and one can use this to define the analytic continuation of $\text{Li}_m(x)$. The polylogarithm can be expressed as a hypergeometric function as well. The formula is

$$\text{Li}_m(x) = x_{m+1}F_m\left(\begin{array}{c} 1, 1, \ldots, 1 \\ 2, \ldots, 2 \end{array}; x\right).$$

We do not go into the properties of this function any further but instead refer the reader to Zagier's [1989] article and the books mentioned earlier.

2.7 Binomial Sums

One area where hypergeometric identities are very useful is in the evaluation of single sums of products of binomial coefficients. The essential character of such sums is revealed by writing them as hypergeometric series. Sums of binomial coefficients that appear to be very different from one another turn out to be examples of the same hypergeometric series. One reason for this is that binomial coefficients can be taken apart and then rearranged to take many different forms. A few examples given below will explain these points.

Consider the sum

$$S = \sum_{j=0}^{n} (-1)^j \frac{\binom{k}{j}\binom{k-1-j}{n-j}}{j+1} = \sum_{j=0}^{n} c_j.$$

To write this as a hypergeometric series, look at the ratio c_{j+1}/c_j as we did when we defined these series. A simple calculation shows that

$$\frac{c_{j+1}}{c_j} = \frac{(j-k)(j-n)}{(j-k+1)(j+2)}.$$

So

$$S = c_0 \sum_{j=0}^{n} \frac{(-k)_j(-n)_j}{(-k+1)_j(2)_j} \quad \text{and} \quad c_0 = \binom{k-1}{n}.$$

Now, as explained in Section 2.1, we could introduce $j! = (1)_j$ in the numerator and denominator to get

$$S = \binom{k-1}{n} \, {}_3F_2\left(\begin{matrix} -n, -k, 1 \\ -k+1, 2 \end{matrix}; 1\right).$$

We have learned to sum two $_3F_2$ series: the balanced and the well-poised series (see Section 2.2). This is neither though it is actually "nearly" poised. However, we have not yet considered this type of series. But there is another way out. Note that the denominator has $(2)_j$, which can be written as $(1)_{j+1}$. Then

$$S = \binom{k-1}{n} \frac{(-k)}{(n+1)(k+1)} \sum_{j=0}^{n} \frac{(-k-1)_{j+1}(-n-1)_{j+1}}{(1)_{j+1}(-k)_{j+1}}$$

$$= \binom{k-1}{n} \frac{(-k)}{(n+1)(k+1)} \sum_{\ell=1}^{n+1} \frac{(-k-1)_\ell(-n-1)_\ell}{(1)_\ell(-k)_\ell}$$

$$= \binom{k-1}{n} \frac{(-k)}{(n+1)(k+1)} \left[{}_2F_1\left(\begin{matrix} -n-1, -k-1 \\ -k \end{matrix}; 1\right) - 1\right].$$

The $_2F_1$ can be evaluated by the Chu–Vandermonde formula and after simplification we get

$$S = \frac{1}{k+1}\left[\binom{k}{n+1} + (-1)^n\right].$$

As another example take the sum

$$S = \sum_{k\geq 0}\binom{n+k}{m+2k}\binom{2k}{k}\frac{(-1)^k}{k+1} = \sum_{k\geq 0} c_k.$$

Here

$$\frac{c_{k+1}}{c_k} = \frac{(k+n+1)\left(k+\frac{1}{2}\right)(k-m+n)}{\left(k+\frac{m}{2}+1\right)\left(k+\frac{m+1}{2}\right)(k+2)},$$

after simplification. Thus

$$S = \binom{n}{m}\left[{}_3F_2\left(\begin{array}{c}-1-n+m, n, -\frac{1}{2}\\ \frac{m-1}{2}, \frac{m}{2}\end{array}; 1\right) - 1\right].$$

The $_3F_2$ is balanced and we can apply the Pfaff–Saalschütz identity. We get

$$S = \binom{n}{m}\left[\frac{\left(\frac{m-1}{2}-n\right)_{n+1-m}\left(\frac{m}{2}\right)_{n+1-m}}{\left(\frac{m-1}{2}\right)_{n+1-m}\left(\frac{m}{2}-n\right)_{n+1-m}} - 1\right],$$

which simplifies to $\binom{n-1}{m-1}$. The reader may verify that

$$\sum_{k\geq 0}\binom{m-r+s}{k}\binom{n+r-s}{n-k}\binom{r+k}{m+n}$$

also reduces to an example of a Pfaff–Saalschütz series.

As the final example take the series

$$\sum_{k=-\ell}^{\ell}(-1)^k\binom{2\ell}{\ell+k}\binom{2m}{m+k}\binom{2n}{n+k} = S,$$

where we are assuming that $\ell = \min(\ell, m, n)$. This reduces to the series

$$\frac{(-1)^\ell (2m)!(2n)!}{(m-\ell)!(m+\ell)!(n-\ell)!(n+\ell)!}\,{}_3F_2\left(\begin{array}{c}-2\ell, -m-\ell, -n-\ell\\ m-\ell+1, n-\ell+1\end{array}; 1\right).$$

This is a well-poised series (see Section 2.2) that can be summed by Dixon's formula. However, this result cannot be applied directly because we get a term

$\Gamma(1 - \ell)/\Gamma(1 - 2\ell)$ that is undefined. A way around this is to use the following case of Dixon's formula:

$$_3F_2\left(\begin{matrix} -2\ell - 2\epsilon, \, -m - \ell - \epsilon, \, -n - \ell - \epsilon \\ m - \ell - \epsilon + 1, \, n - \ell - \epsilon + 1 \end{matrix}; 1\right)$$

$$= \frac{\Gamma(1 - \ell - \epsilon)\Gamma(1 + m - \ell - \epsilon)\Gamma(1 + m + n + \ell + \epsilon)}{\Gamma(1 - 2\ell - 2\epsilon)\Gamma(1 + m)\Gamma(1 + n)\Gamma(1 + m + n)}.$$

Apply Euler's reflection formula to the right side to get

$$\frac{\sin \pi(2\ell + 2\epsilon)}{\sin \pi(\ell + \epsilon)} \frac{\Gamma(2\ell + 2\epsilon)}{\Gamma(\ell + \epsilon)}$$
$$\cdot \frac{\Gamma(1 + m - \ell - \epsilon)\Gamma(1 + n - \ell - \epsilon)\Gamma(1 + m + n + \ell + \epsilon)}{\Gamma(1 + m)\Gamma(1 + n)\Gamma(1 + m + n)}.$$

In the limit as $\epsilon \to 0$, this expression is

$$2(-1)^\ell \frac{(2\ell - 1)!}{(\ell - 1)!} \frac{(m - \ell)!(n - \ell)!(m + n + \ell)!}{m!n!(m + n)!}.$$

Thus

$$S = \frac{(\ell + m + n)!(2\ell)!(2m)!(2n)!}{(\ell + m)!(\ell + n)!(m + n)!\ell!m!n!}.$$

Another example that gives a well-poised $_3F_2$ is $\sum_{k=1}^{n} 2k\binom{2p}{k+p}\binom{2n}{k+n}$. Examples can be multiplied but the discussion above is sufficient to explain how hypergeometric identities apply to the evaluation of binomial sums. See Exercise 29 in Chapter 3.

2.8 Dougall's Bilateral Sum

The bilateral series

$$\sum_{-\infty}^{\infty} \frac{\Gamma(a + n)\Gamma(b + n)}{\Gamma(c + n)\Gamma(d + n)}$$

is the subject of this section. In fact, the hypergeometric series $_2F_2(a, b; c; 1)$ should be regarded as a special case of the bilateral series where $d = 1$, since $1/\Gamma(n) = 0$ for nonpositive integers n. This explains why we introduced $1/n!$ in the nth term of the hypergeometric series.

The above remark gives us a way of evaluating the above bilateral series. For $d = 1, 2, \ldots$, the sum reduces to a series that can be evaluated by Gauss's summation of $_2F_1(a, b; c; 1)$. The following theorem of Carlson allows one to evaluate the bilateral series from its values at $d = 1, 2, \ldots$.

Theorem 2.8.1 *If $f(z)$ is analytic and bounded for Re $z \geq 0$ and if $f(z) = 0$ for $z = 0, 1, 2, \ldots$, then $f(z)$ is identically zero.*

Remark The boundedness condition can be relaxed. We need only assume that $f(z) = O(e^{k|z|})$, where $k < \pi$. The simple proof given below of the particular case is due to Selberg [1944].

Proof. As a consequence of Cauchy's residue theorem,

$$f(a) = \frac{(a-1)(a-2)\cdots(a-n)}{2\pi i} \int_{-i\infty}^{i\infty} \frac{f(z)}{(z-a)(z-1)\cdots(z-n)} dz$$

for $n > a > 0$. Then, for $a \geq 1$,

$$|f(a)| \leq \frac{[a]!(n-[a])!}{2\pi} \int_{-\infty}^{\infty} \frac{|f(it)|dt}{\sqrt{(a^2+t^2)(1+t^2)\cdots(n^2+t^2)}}$$

$$\leq \frac{[a]!(n-[a])!}{2\pi n!} \int_{-\infty}^{\infty} \frac{|f(it)|}{1+t^2} dt.$$

Let $n \to \infty$ to see that $f(a) = 0$ for all real $a \geq 1$. This implies the theorem. ∎

Theorem 2.8.2 (Dougall) *For $1 + \mathrm{Re}(a+b) < \mathrm{Re}(c+d)$,*

$$\sum_{n=-\infty}^{\infty} \frac{\Gamma(a+n)\Gamma(b+n)}{\Gamma(c+n)\Gamma(d+n)} = \frac{\pi^2}{\sin \pi a \sin \pi b} \frac{\Gamma(c+d-a-b-1)}{\Gamma(c-a)\Gamma(d-a)\Gamma(c-b)\Gamma(d-b)}.$$

Proof. For Re $d > \mathrm{Re}(a+b-c) + 1$, the functions on both sides are bounded analytic functions of d. Let m be an integer in this half plane. For $d = m$ the series on the left is

$$\sum_{n=-m+1}^{\infty} \frac{\Gamma(a+n)\Gamma(b+n)}{\Gamma(c+n)\Gamma(m+n)}$$

$$= \frac{\Gamma(a-m+1)\Gamma(b-m+1)}{\Gamma(c-m+1)} \sum_{\ell=0}^{\infty} \frac{(a-m+1)_\ell(b-m+1)_\ell}{(c-m+1)_\ell \ell!}.$$

This series can be summed by Gauss's $_2F_1$ formula. Thus Dougall's result can be verified for d equal to an integer in the half plane. Carlson's theorem now implies Theorem 2.8.2. ∎

Gauss's $_2F_1$ sum is itself a consequence of Theorem 2.8.1. We have to prove that

$$_2F_1\left(\begin{matrix} -b, a \\ c \end{matrix}; 1\right) = \frac{\Gamma(c)\Gamma(c-a+b)}{\Gamma(c-a)\Gamma(c+b)}.$$

This relation is true for $b = 1, 2, \ldots$ by the Chu–Vandermonde identity, and then by the above argument the general case follows.

For a different example where Carlson's theorem applies, consider the formula

$$\int_0^1 x^{\alpha-1}(1-x)^{\beta-1}ds = \frac{\Gamma(\alpha)\Gamma(\beta)}{\Gamma(\alpha+\beta)}.$$

It is easy to prove this by induction when α is a positive integer. We saw this in Chapter 1. The integral is bounded and analytic in $\operatorname{Re}\alpha \geq \delta > 0$ and so is the right side of the formula. This proves the result.

2.9 Fractional Integration by Parts and Hypergeometric Integrals

Theorem 2.2.1 gives Euler's integral representation for a hypergeometric function. One drawback of this representation is that the symmetry in the parameters a and b of the function is not obvious in the integral. We observed that Erdélyi's double integral (2.2.5) gives the two different representations by changing the order of integration. In this section, we show how fractional integration by parts can be used to transform the integral for $_2F_1(a, b; c; x)$ to another integral in which a and b have been interchanged. In fact, fractional integration by parts is a powerful tool and we also use it to prove a formula of Erdélyi. This contains some of the integral formulas considered in this chapter as special cases. It also has implications in the theory of orthogonal polynomials, which will be discussed in Chapter 6.

Let

$$(If)(x) := \int_a^x f(t)dt$$

and

$$(I_2 f)(x) := \int_a^x \int_a^t f(t_1)dt_1 dt = \int_a^x (x-t)f(t)dt.$$

Inductively, it follows that, for a positive integer n,

$$(I_n f)(x) := \int_a^x \int_a^t \cdots \int_a^{t_{n-1}} f(t_{n-1})dt_{n-1}\cdots dt_1 dt$$

$$= \frac{1}{(n-1)!}\int_a^x (x-t)^{n-1}f(t)dt.$$

A fractional integral I_α, for $\operatorname{Re}\alpha > 0$, is then defined by

$$(I_\alpha f)(x) := \frac{1}{\Gamma(\alpha)}\int_a^x (x-t)^{\alpha-1}f(t)dt. \tag{2.9.1}$$

The restriction $\operatorname{Re}\alpha > 0$ can be removed by using contour integrals. An interpretation of Euler's integral for $_2F_1(a, b; c; x)$ as a fractional integral is now evident.

The fractional derivatives can also be defined by the formal relation

$$\frac{d^v w^\mu}{dw^v} = \frac{\Gamma(\mu + 1)}{\Gamma(\mu - v + 1)} w^{\mu - v}, \tag{2.9.2}$$

when the right side is meaningful.

To state the formula for fractional integration by parts, suppose u and v are functions defined by

$$u = \sum_{r=0}^{\infty} A_r (x - a)^{\rho + r - 1}, \qquad v = \sum_{s=0}^{\infty} B_s (b - x)^{\sigma + s - 1}.$$

Then

$$\int_a^b u \frac{d^v v}{d(b - x)^v} dx = \int_a^b v \frac{d^v u}{d(x - a)^v} dx, \tag{2.9.3}$$

provided that the integrals exist. This can be verified directly by substituting the series for u and v and their derivatives, which are obtained by applying (2.9.2) term by term. The two are seen to be identical after integration term by term. It is noteworthy that if $\operatorname{Re} v < 0$ in (2.9.3), then (2.9.3) is equivalent to the identity

$$\int_a^b u(x) \left[\int_x^b (y - x)^{\alpha - 1} v(y) dy \right] dx = \int_a^b v(x) \left[\int_a^x (x - y)^{\alpha - 1} u(y) dy \right] dx,$$

where $v = -\alpha$. This formula holds because both sides equal the double integral

$$\int\int u(x) v(y) (y - x)^{\alpha - 1} dx dy.$$

We now show how to transform one integral representation of $_2F_1(a, b; c; x)$ to the other. It is clear from (2.9.2) that

$$\frac{\Gamma(c)}{\Gamma(b)\Gamma(c - b)} \int_0^1 t^{b - 1} (1 - t)^{c - b - 1} (1 - xt)^{-a} dt$$

$$= \frac{\Gamma(c)}{\Gamma(b)\Gamma(c - a)} \int_0^1 t^{b - 1} (1 - xt)^{-a} \frac{d^{b - a} (1 - t)^{c - a - 1}}{d(1 - t)^{b - a}} dt. \tag{2.9.4}$$

By the integration by parts formula (2.9.3), we get

$$\frac{\Gamma(c)}{\Gamma(b)\Gamma(c - a)} \int_0^1 (1 - t)^{c - a - 1} \frac{d^{b - a} t^{b - 1} (1 - xt)^{-a}}{dt^{b - a}} dt. \tag{2.9.5}$$

The binomial theorem and (2.9.2) give

$$\frac{d^{b-a}t^{b-1}(1-xt)^{-a}}{dt^{b-a}} = \frac{d^{b-a}}{dt^{b-a}} \sum_{r=0}^{\infty} \frac{\Gamma(a+r)t^{b+r-1}x^r}{\Gamma(a)r!}$$

$$= \sum_{r=0}^{\infty} \frac{\Gamma(b+r)t^{a+r-1}x^r}{\Gamma(a)r!}$$

$$= \frac{\Gamma(b)}{\Gamma(a)}t^{a-1}(1-xt)^{-b}.$$

Substitute this in (2.9.5) to obtain

$$\frac{\Gamma(c)}{\Gamma(a)\Gamma(c-a)} \int_0^1 t^{a-1}(1-t)^{c-a-1}(1-xt)^{-b}dt.$$

This is the expression in (2.9.4) with a and b interchanged; our claim is proved.

As another example of the use of fractional integration by parts, we re-prove the following formula contained in Theorem 2.2.4:

$$_2F_1\left(\begin{matrix} a,b \\ c \end{matrix}; x\right) = \frac{\Gamma(c)}{\Gamma(d)\Gamma(c-d)} \int_0^1 t^{d-1}(1-t)^{c-d-1} {}_2F_1\left(\begin{matrix} a,b \\ d \end{matrix}; xt\right)dt, \quad (2.9.6)$$

when $\operatorname{Re} c > \operatorname{Re} d > 0$, $x \neq 1$, $|\arg(1-x)| < \pi$. Use (2.9.2) to see that (2.9.4) is equal to

$$\frac{\Gamma(c)}{\Gamma(b)\Gamma(c-d)} \int_0^1 t^{b-1}(1-xt)^{-a} \frac{d^{b-d}(1-t)^{c-d-1}}{d(1-t)^{b-d}}dt$$

$$= \frac{\Gamma(c)}{\Gamma(b)\Gamma(c-d)} \int_0^1 (1-t)^{c-d-1} \frac{d^{b-d}t^{b-1}(1-xt)^{-a}}{dt^{b-d}}dt.$$

Also,

$$\frac{d^{b-d}t^{b-1}(1-xt)^{-a}}{dt^{b-d}} = \frac{d^{b-d}}{dt^{b-d}} \sum_{r=0}^{\infty} \frac{\Gamma(a+r)t^{b+r-1}x^r}{\Gamma(a)r!}$$

$$= \sum_{r=0}^{\infty} \frac{\Gamma(a+r)\Gamma(b+r)t^{d+r-1}x^r}{\Gamma(a)\Gamma(d+r)r!}$$

$$= \frac{\Gamma(b)}{\Gamma(d)}t^{d-1} {}_2F_1\left(\begin{matrix} a,b \\ d \end{matrix}; xt\right).$$

Substitute this in the last integral to complete the proof of (2.9.6).

We now state and prove the formula of Erdélyi [1939] mentioned earlier.

Theorem 2.9.1   *For* $\operatorname{Re} c > \operatorname{Re} \mu > 0$, $x \neq 1$, $|\arg(1-x)| < \pi$, *we have*

$$_2F_1\left(\begin{matrix} a,b \\ c \end{matrix}; x\right) = \frac{\Gamma(c)}{\Gamma(\mu)\Gamma(c-\mu)} \int_0^1 t^{\mu-1}(1-t)^{c-\mu-1}(1-xt)^{\lambda-a-b}$$

$$\cdot {}_2F_1\left(\begin{matrix} \lambda-a, \lambda-b \\ \mu \end{matrix}; xt\right) {}_2F_1\left(\begin{matrix} a+b-\lambda, \lambda-\mu \\ c-\mu \end{matrix}; \frac{(1-t)x}{1-xt}\right)dt.$$

Proof. Apply Euler's transformation in Theorem 2.2.5 to the $_2F_1$ inside the integral in (2.9.6). The result is

$$_2F_1\left({a,b \atop c};x\right) = \frac{\Gamma(c)}{\Gamma(\lambda)\Gamma(c-\lambda)} \int_0^1 t^{\lambda-1}(1-t)^{c-\lambda-1}$$
$$\cdot (1-xt)^{\lambda-a-b}\,_2F_1\left({\lambda-a,\lambda-b \atop \lambda};xt\right)dt.$$

By (2.9.2) and the series representation of $_2F_1$, we see that

$$\frac{t^{\lambda-1}}{\Gamma(\lambda)}\,_2F_1\left({\lambda-a,\lambda-b \atop \lambda};xt\right) = \frac{d^{\mu-\lambda}}{dt^{\mu-\lambda}}\left\{\frac{t^{\mu-1}}{\Gamma(\mu)}\,_2F_1\left({\lambda-a,\lambda-b \atop \mu};xt\right)\right\}.$$

Substitute this in the last integral and use the fractional-integration-by-parts formula (2.9.3) to get

$$_2F_1\left({a,b \atop c};x\right) = \frac{\Gamma(c)}{\Gamma(\mu)} \int_0^1 t^{\mu-1}\,_2F_1\left({\lambda-a,\lambda-b \atop \mu};xt\right)$$
$$\cdot \frac{d^{\mu-\lambda}}{d(1-t)^{\mu-\lambda}}\left\{\frac{(1-t)^{c-\lambda-1}}{\Gamma(c-\lambda)}(1-xt)^{\lambda-a-b}\right\}dt.$$

Write the expression in curly braces as

$$(1-x)^{\lambda-a-b}\frac{(1-t)^{c-\lambda-1}}{\Gamma(c-\lambda)}\left(1+\frac{1-t}{1-x}x\right)^{\lambda-a-b}$$
$$= (1-x)^{\lambda-a-b}\sum\frac{(a+b-\lambda)_r}{r!\Gamma(c-\lambda)}\left(\frac{x}{x-1}\right)^r(1-t)^{c-\lambda+r-1}.$$

Take the $(\mu-\lambda)$th derivative of this expression to obtain

$$(1-x)^{\lambda-a-b}\frac{(1-t)^{c-\mu-1}}{\Gamma(c-\mu)}\,_2F_1\left({a+b-\lambda,c-\lambda \atop c-\mu};\frac{(1-t)x}{x-1}\right).$$

By Pfaff's transformation (Theorem 2.2.5), this is equal to

$$(1-xt)^{\lambda-a-b}\frac{(1-t)^{c-\mu-1}}{\Gamma(c-\mu)}\,_2F_1\left({a+b-\lambda,\lambda-\mu \atop c-\mu};\frac{(1-t)x}{1-xt}\right).$$

Substitute this in the last integral. The result is proved.   ∎

Exercises

1. Complete the proof of the second part of Theorem 2.1.2 concerning conditional convergence.

2. Suppose that $\left|\sum_{k=1}^{n} a_k\right|$ and $\sum_{k=1}^{n} |a_k|$ tend to infinity in the same way, that is,

$$\sum_{k=1}^{n} |a_k| < K \left|\sum_{k=1}^{n} a_k\right|$$

for all n and K is independent of n. Prove that

$$\lim_{n \to \infty} \frac{\sum_{k=1}^{n} b_k}{\sum_{k=1}^{n} a_k} = \lim_{n \to \infty} \frac{b_n}{a_n},$$

provided that the right-hand limit exists.

3. Use the result in Exercise 2 to prove Theorem 2.1.4.

4. (a) Show that $\frac{1}{2}((1 + x)^n + (1 - x)^n) = {}_2F_1(-n/2, -(n + 1)/2; 1/2; x^2)$. Find a similar expression for $\frac{1}{2}((1 + x)^n - (1 - x)^n)$.

(b) Show that $(1 + x)^n = 1 + nx {}_2F_1(1 - n, 1; 2; -x)$.

5. Derive the Chu–Vandermonde identity by equating the coefficient of x^n on each side of $(1 - x)^{-a}(1 - x)^{-b} = (1 - x)^{-(a+b)}$.

6. Suppose that $\log(1 + x)$ is defined by the series (2.1.3). Use Pfaff's transformation (Theorem 2.2.5) to show that $\log(1 + x) = -\log(1 + x)^{-1}$.

7. Show that Pfaff's transformation is equivalent to

$${}_2F_1\left(\begin{matrix} -n, c - b \\ c \end{matrix}; 1\right) = \frac{(b)_n}{(c)_n}, \quad n = 0, 1, \ldots.$$

This is one way of removing the restriction $\operatorname{Re} c > \operatorname{Re} b > 0$ used in the proof of Theorem 2.2.5.

8. Prove the identities (2.1.10) to (2.1.13).

9. Show that ${}_1F_1(a; c; x) = e^x {}_1F_1(c - a; c; -x)$.

10. Suppose x is a complex number not equal to zero or a negative integer. Show that

$$\Gamma(x) = \frac{1}{x} {}_1F_1\left(\begin{matrix} x \\ x + 1 \end{matrix}; -1\right) + \int_1^\infty t^{x-1} e^{-t} dt$$

$$= \sum_{n=0}^{\infty} \frac{(-1)^n}{n!(n + x)} + \int_1^\infty t^{x-1} e^{-t} dt.$$

Note that the ${}_1F_1$ series exhibits the poles and residues of $\Gamma(x)$.

11. Show that

$$\int_0^\infty e^{-st} t^{\alpha-1} {}_pF_q\left(\begin{matrix} a_1, \ldots, a_p \\ b_1, \ldots, b_q \end{matrix}; xt\right) dt$$

$$= \frac{\Gamma(\alpha)}{s^\alpha} {}_{p+1}F_q\left(\begin{matrix} a_1, \ldots, a_p, \alpha \\ b_1, \ldots, b_q \end{matrix}; \frac{x}{s}\right),$$

when $p \le q$, $\operatorname{Re} s > 0$, $\operatorname{Re} \alpha > 0$, and term-by-term integration is permitted.

12. Show that

$$\cos mx = {}_2F_1\left(\begin{array}{c} \frac{m}{2}, -\frac{m}{2} \\ \frac{1}{2} \end{array}; \sin^2 x\right),$$

$$\sin mx = m \sin x\, {}_2F_1\left(\begin{array}{c} \frac{1+m}{2}, \frac{1-m}{2} \\ \frac{3}{2} \end{array}; \sin^2 x\right).$$

13. Prove that the functions

$$y_1 = \left\{ {}_2F_1\left(\begin{array}{c} a, b \\ a + b + \frac{1}{2} \end{array}; x\right)\right\}^2$$

and

$$y_2 = {}_3F_2\left(\begin{array}{c} 2a, 2b, a + b \\ 2a + 2b, a + b + \frac{1}{2} \end{array}; x\right)$$

both satisfy the differential equation

$$x^2(x - 1)y''' - 3x\left(a + b + \frac{1}{2} - (a + b + 1)x\right)y''$$
$$+ \{[2(a^2 + b^2 + 4ab) + 3(a + b) + 1]x - (a + b)(2a + 2b + 1)\}y'$$
$$+ 4ab(a + b)y = 0.$$

Thus prove Clausen's identity (see Clausen [1828]),

$$\left\{ {}_2F_1\left(\begin{array}{c} a, b \\ a + b + \frac{1}{2} \end{array}; x\right)\right\}^2 = {}_3F_2\left(\begin{array}{c} 2a, 2b, a + b \\ 2a + 2b, a + b + \frac{1}{2} \end{array}; x\right).$$

Also prove that

$${}_2F_1\left(\begin{array}{c} a, b \\ a + b + \frac{1}{2} \end{array}; x\right) {}_2F_1\left(\begin{array}{c} \frac{1}{2} - a, \frac{1}{2} - b \\ \frac{3}{2} - a - b \end{array}; x\right)$$
$$= {}_3F_2\left(\begin{array}{c} a - b + \frac{1}{2}, b - a + \frac{1}{2}, \frac{1}{2} \\ a + b + \frac{1}{2}, \frac{3}{2} - a - b \end{array}; x\right).$$

14. Show that the Pfaff–Saalschütz identity (2.2.8) can be written in a completely symmetric form as

$${}_3F_2\left(\begin{array}{c} a, b, c \\ d, e \end{array}; 1\right) = \frac{\pi^2 \Gamma(d)\Gamma(e)[\cos \pi d \cos \pi e + \cos \pi a \cos \pi b \cos \pi c]}{\Gamma(d - a)\Gamma(d - b)\Gamma(d - c)\Gamma(e - a)\Gamma(e - b)\Gamma(e - c)}$$

when the series terminates naturally and $d + e = a + b + c + 1$. This observation was made by R. William Gosper.

15. Prove that

$$T_n(x) = \sum_{k=0}^{[n/2]} \binom{n}{2k} x^{n-2k}(x^2 - 1)^k,$$

where $T_n(x)$ is the Chebyshev polynomial of the first kind. Find a similar expression for $U_n(x)$, the Chebyshev polynomial of the second kind.

16. Prove the following analog of Fermat's little theorem: If x is a positive integer and p an odd prime then $T_p(x) \equiv T_1(x) \pmod{p}$.

17. Pell's equation is $x^2 - Dy^2 = 1$, where D is a square free positive integer. Let S be the set of all positive solutions (x, y) of Pell's equation. Let (x_1, y_1) be the solution with least x in S. Show that $(T_n(x_1), y_1 U_{n-1}(x_1)), n \geq 1$, is a solution and thus that Pell's equation has infinitely many solutions, if it has one.

18. Let $S_n(x) = U_{n-1}(x)$ with $U_{-1}(x) = 0$. Prove that $\gcd(S_n(x), S_m(x)) = S_{\gcd(m,n)}(x)$, where x, m, and n are positive integers.

19. (a) Show that

$$\int_{-1}^{1} P_n^{(\alpha,-1/2)}(2x^2 - 1)p(x)(1 - x^2)^\alpha dx = 0,$$

where $p(x)$ is any polynomial of degree $\leq 2n - 1$. Deduce that

$$P_{2n}^{(\alpha,\alpha)}(x) = \frac{\Gamma(2n + \alpha + 1)n!}{\Gamma(n + \alpha + 1)(2n)!} P_n^{(\alpha,-1/2)}(2x^2 - 1).$$

(b) Show that

$$P_{2n+1}^{(\alpha,\alpha)}(x) = \frac{\Gamma(2n + \alpha + 2)n!}{\Gamma(n + \alpha + 1)(2n + 1)!} x P_n^{(\alpha,1/2)}(2x^2 - 1).$$

(c) What do (a) and (b) mean for the Chebyshev polynomials of the first and second kind respectively?

20. Prove that

$$P_n^{(\alpha,\beta)}(x) = (-1)^n \frac{(\beta + 1)_n}{n!} {}_2F_1\left(\begin{array}{c} -n, n + \alpha + \beta + 1 \\ \beta + 1 \end{array}; \frac{1 + x}{2}\right)$$

$$= (-1)^n P_n^{(\beta,\alpha)}(-x)$$

$$= \frac{(\alpha + 1)_n}{n!}\left(\frac{1 + x}{2}\right)^n {}_2F_1\left(\begin{array}{c} -n, -n - \beta \\ \alpha + 1 \end{array}; \frac{x - 1}{x + 1}\right)$$

$$= \frac{(n + \alpha + \beta + 1)_n}{n!}\left(\frac{x - 1}{2}\right)^n {}_2F_1\left(\begin{array}{c} -n, -n - \alpha \\ -\alpha - \beta - 2n \end{array}; \frac{2}{1 - x}\right).$$

21. Let $x = \cos\theta$. Prove that

$$\frac{d^{n-1}\sin^{2n-1}\theta}{dx^{n-1}} = \frac{(-1)^{n-1}}{n} \cdot \frac{(2n)!}{2^n n!} \sin n\theta. \qquad \text{(Jacobi)}$$

22. Prove that the Jacobi polynomials $P_n^{(\alpha,\beta)}(x)$ satisfy the three-term recurrence relation

$$2(n+1)(n+\alpha+\beta+1)(2n+\alpha+\beta)P_{n+1}^{(\alpha,\beta)}(x)$$
$$= (2n+\alpha+\beta+1)\{(2n+\alpha+\beta+2)(2n+\alpha+\beta)x + \alpha^2 - \beta^2\}$$
$$\times P_n^{(\alpha,\beta)}(x) - 2(n+\alpha)(n+\beta)$$
$$\times (2n+\alpha+\beta+2)P_{n-1}^{(\alpha,\beta)}(x) = 0, \quad n = 1, 2, 3, \ldots.$$

(Compare this with (2.5.15).)

23. Prove Euler's contiguous relation expressed as the integral formula (2.5.3′) by observing that

$$0 = \int_0^1 \frac{d}{dt}(t^a(1-t)^{c-a}(1-tx)^{1-b})dt, \quad \operatorname{Re} c > \operatorname{Re} a > 0.$$

24. Prove the following contiguous relations:
(a) $(c - 2a - (b-a)x)F + a(1-x)F(a+) - (c-a)F(a-) = 0.$
(b) $(c - a - 1)F + aF(a+) - (c-1)F(c-) = 0.$
(c) $(b-a)(1-x)F - (c-a)F(a-) + (c-b)F(b-) = 0.$
(d) $c(b - (c-a)x)F - bc(1-x)F(b+) - (c-b)F(b-) = 0.$

25. Prove

(a) $$_2F_1\left(\begin{matrix}a, 1\\ c\end{matrix}; x\right) = \frac{1}{1-}\ \frac{\frac{a}{c}x}{1-}\ \frac{\frac{(c-a)x}{c(c+1)}}{1-}\cdots.$$

(b) $$\log(1+x) = \frac{x}{1+}\ \frac{1^2 x}{2+}\ \frac{1^2 x}{3+}\ \frac{2^2 x}{4+}\ \frac{2^2 x}{5+}\cdots.$$

(c) $$\arctan x = \frac{x}{1+}\ \frac{1^2 x^2}{3+}\ \frac{2^2 x^2}{5+}\ \frac{3^2 x^2}{7+}\cdots.$$

(d) $$\frac{\pi}{4} = \frac{1}{1+}\ \frac{1^2}{3+}\ \frac{2^2}{5+}\ \frac{3^2}{7+}\cdots.$$

The 9th approximant gives $\pi/4$ correctly up to seven decimal places.

(e) $$\log\frac{x+1}{x-1} = \frac{2}{x-}\ \frac{\frac{1}{2}}{\frac{3}{2}x-}\ \frac{\frac{2}{3}}{\frac{5}{3}x-}\ \frac{\frac{3}{4}}{\frac{7}{4}x-}\cdots.$$

(f) $$\frac{\arcsin x}{\sqrt{1-x^2}} = \frac{x}{1-}\ \frac{1\cdot 2x^2}{3-}\ \frac{1\cdot 2x^2}{5-}\ \frac{3\cdot 4x^2}{7-}\ \frac{3\cdot 4x^2}{9-}\cdots.$$

26. Show that

$$_2F_1\left(\begin{matrix} a, b \\ c \end{matrix}; x\right) \Big/ {}_2F_1\left(\begin{matrix} a+1, b+1 \\ c+1 \end{matrix}; x\right)$$

$$= \frac{c - (a+b+1)x}{c} + \frac{x(1-x)\frac{(a+1)(b+1)}{c(c+1)}}{c+1-\frac{(a+b+3)x}{c+1}} + \frac{x(1-x)\frac{(a+2)(b+2)}{(c+1)(c+2)}}{c+2-\frac{(a+b+5)x}{c+2}} + \cdots.$$

27. Use Barnes's integral representation of $_2F_1$ and Barnes's beta integral (Theorem 2.4.2) to prove formula (2.3.13). Also consider the case where $c - a - b$ is an integer.

28. Multiply the equation in (2.3.13) by $x^{d-1}(1-x)^{e-d-1}$ and integrate over $(0, 1)$ to obtain another proof of Theorem 2.4.4.

29. Prove the following formulas of Ramanujan [1927, paper 11]

(a) $$\int_0^\infty \frac{1 + (x/(b+1))^2}{1 + (x/a)^2} \cdot \frac{1 + (x/(b+2))^2}{1 + (x/(a+1))^2} \cdots dx$$

$$= \frac{\sqrt{\pi}}{2} \frac{\Gamma\left(a + \frac{1}{2}\right)\Gamma(b+1)\Gamma\left(b - a + \frac{1}{2}\right)}{\Gamma(a)\Gamma\left(b + \frac{1}{2}\right)\Gamma(b - a + 1)} \quad \text{for } 0 < a < b - \frac{1}{2}.$$

(b) $$\int_0^\infty \frac{dx}{(1 + (x/a)^2)(1 + (x/(a+1))^2)\cdots(1 + (x/b)^2)(1 + (x/(b+1))^2)\cdots}$$

$$= \frac{\sqrt{\pi}}{2} \frac{\Gamma(a)\Gamma\left(a + \frac{1}{2}\right)\Gamma(b)\Gamma\left(b + \frac{1}{2}\right)\Gamma(a+b)}{\Gamma\left(a + b + \frac{1}{2}\right)}, \quad a > 0, b > 0.$$

30. Prove that for Re $t > 0$

$$\sum_{n=-\infty}^{\infty} e^{-(n+a)^2\pi t} = \frac{1}{\sqrt{t}} \sum_{n=-\infty}^{\infty} e^{-n^2\pi/t} e^{2\pi ina}.$$

Hint: Denote the left side by $f(a)$ and note that f has period one. Expand f as a Fourier series

$$\sum_{n=-\infty}^{\infty} A_n e^{2\pi ina}$$

and observe that

$$A_n = \int_{-\infty}^{\infty} e^{-\pi t y^2} e^{-2\pi iny} dy.$$

31. (a) Let χ be a nontrivial even primitive character (mod N). Prove that

$$\sum_{n=-\infty}^{\infty} \chi(n)e^{-n^2\pi t} = \frac{g(\chi)}{\sqrt{N^2 t}} \sum_{n=-\infty}^{\infty} \bar{\chi}(n)e^{-n^2\pi/(N^2 t)},$$

where Re $t > 0$ and

$$g(\chi) = \sum_{a=1}^{N} \chi(a)e^{2\pi ia/N}.$$

Hint: First observe that

$$\sum_{n=-\infty}^{\infty} \chi(n)e^{-n^2\pi t} = \sum_{a=1}^{N} \chi(a) \sum_{n=-\infty}^{\infty} e^{-\pi t(Nn+a)^2}.$$

Then apply the result of the previous exercise.

(b) Let χ be a nontrivial odd primitive character (mod N). Prove that

$$\sum_{n=-\infty}^{\infty} n\chi(n)e^{-n^2\pi t} = \frac{-ig(\chi)}{N^2 t^{3/2}} \sum_{n=-\infty}^{\infty} n\bar{\chi}(n)e^{-n^2\pi/(N^2 t)}.$$

32. (a) Show that $\pi^{-s}\Gamma(s)\zeta(2s)$ is the Mellin transform of

$$\sum_{n=1}^{\infty} e^{-n^2\pi t} \quad \text{for Re } s > 1.$$

(b) Use Exercise 30 to show that

$$\pi^{-s/2}\Gamma\left(\frac{s}{2}\right)\zeta(s) = \int_1^{\infty} \left(t^{(s/2)-1} + t^{((1-s)/2)-1}\right) \sum_{n=1}^{\infty} e^{-n^2\pi t}\, dt$$

$$-\frac{1}{s} - \frac{1}{1-s}.$$

(c) Observe that the expression on the right in (b) does not change under $s \to 1 - s$. Deduce the analytic continuation and functional equations of the zeta function.

33. Obtain the functional equation of $L(\chi, s)$, χ primitive (mod N), using Exercise 31 and the idea of Exercise 32. (By Exercise 1.53 the functional equation is

$$L(\chi, s) = \frac{g(\chi)}{2i^\delta}\left(\frac{2\pi}{N}\right)^s \frac{L(\bar{\chi}, 1-s)}{\Gamma(s)\cos\pi(s-\delta)/2},$$

where $\delta = 0$ or 1 depending on whether χ is even or odd.)

34. Assuming the functional equation of the zeta function, apply Mellin inversion to prove that

$$\sum_{n=-\infty}^{\infty} e^{-n^2\pi t} = \frac{1}{\sqrt{t}} \sum_{n=-\infty}^{\infty} e^{-n^2\pi/t}.$$

35. Evaluate the Mellin transform of $_2F_1(a, b; c; -x)$ as follows:

$$\int_0^{\infty} x^{s-1}{}_2F_1\left(\begin{matrix} a, b \\ c \end{matrix}; -x\right) dx$$

$$= \int_0^{\infty} x^{s-1}(1+x)^{-a}{}_2F_1\left(\begin{matrix} a, c-b \\ c \end{matrix}; \frac{x}{x+1}\right) dx.$$

Set $u = x/(x+1)$ and integrate the $_2F_1$ series term by term.

36. Let $F_i(s)$ be the Mellin transform of $f_i(x)$. Show that

$$\frac{1}{2\pi i} \int_{k-i\infty}^{k+i\infty} F_1(s)F_2(s)F_3(s)\,ds$$

$$= \int_0^\infty \int_0^\infty f_1(u)f_2(u)f_3\left(\frac{1}{uv}\right)\frac{du\,dv}{uv}.$$

Deduce Barnes's continuous extension (Theorem 2.4.3) of the Pfaff–Saalschütz identity.

37. (a) Prove that

$$\sum_{n=1}^\infty n^{s-1}e^{2\pi i n a} = \Gamma(s)(2\pi)^{-s}e^{is\pi/2}\sum_{n=-\infty}^\infty (a+n)^{-s}, \quad \text{Im } a > 0, \text{Re } s > 1,$$

by using Carlson's theorem. (The formula is due to Lipschitz.)

(b) Expand $\sum_{-\infty}^\infty (a+n)^{-s}$, Im $a > 0$, as a Fourier series $\Sigma A_m e^{2\pi i m a}$. Express A_m as an integral over $(-\infty, \infty)$. Use the result in (a) to deduce Hankel's formula for $1/\Gamma(s)$. See Exercise 1.22.

38. (a) Verify Theorem 2.6.1 by differentiating both sides of the equation.

(b) Verify the Abel–Spence identity (2.6.16).

(c) Prove the identities (2.6.13) and (2.6.14).

39. (a) Prove that

$$\text{Li}_2(x) - \text{Li}_2(y) + \text{Li}_2\left(\frac{y}{x}\right) + \text{Li}_2\left[\frac{1-x}{1-y}\right] - \text{Li}_2\left[\frac{y(1-x)}{x(1-y)}\right]$$

$$= \frac{\pi^2}{6} - \log x \log\left[\frac{1-x}{1-y}\right].$$

(b) Prove that

$$\text{Prove that Li}_2\left[\frac{x(1-y)^2}{y(1-x)^2}\right] = \text{Li}_2\left[-x\cdot\frac{1-y}{1-x}\right] + \text{Li}_2\left[-\frac{1}{y}\frac{1-y}{1-x}\right]$$

$$+ \frac{x(1-y)}{y(1-x)} + \text{Li}_2\left[\frac{1-y}{1-x}\right] + \frac{1}{2}\log 2y. \qquad\text{(Kummer)}$$

See Lewin [1981a].

(c) Prove that

$$\text{Li}_2(x) + \text{Li}_2(y) - \text{Li}_2(xy)$$

$$= \text{Li}_2\left(\frac{x(1-y)}{1-xy}\right) + \text{Li}_2\left(\frac{y(1-x)}{1-xy}\right)$$

$$+ \log\left(\frac{1-x}{1-xy}\right)\log\left(\frac{1-y}{1-xy}\right). \qquad\text{(Rogers)}$$

(d) Show that if $0 < x < 1$ and $f(x) \in C^2((0,1))$ and satisfies (2.6.6) and the functional relation in part (c), then $f(x) = \text{Li}_2(x)$. See Rogers [1907].

40. Suppose χ is a primitive Dirichlet character mod N. Show that

$$L(\chi, 2) = \sum_{n=1}^{\infty} \frac{\chi(n)}{n^2} = \frac{g(\chi)}{N} \sum_{n=1}^{N} \bar{\chi}(n) \mathrm{Li}_2(e^{-2\pi i n/N}),$$

where $g(\chi)$ is the Gauss sum defined in Exercise 31.

41. Suppose n is a positive integer. Define

$$\Lambda(n) = \begin{cases} \log p, & n = p^k, \quad \text{a power of a prime,} \\ 0, & \text{otherwise.} \end{cases}$$

(a) Show $\zeta(s) = \prod_p (1 - p^{-s})^{-1}$ for $\mathrm{Re}\, s > 1$.

(b) Show that

$$-\frac{\zeta'(s)}{\zeta(s)} = \sum_{n=1}^{\infty} \frac{\Lambda(n)}{n^s} \quad \text{for } \mathrm{Re}\, s > 1.$$

(c) Let $\psi(x) = \sum_{n \leq x} \Lambda(n)$. Show that $\psi(x) = 0(x)$ and

$$\int_0^{\infty} x^{-s-1} \psi(x) dx = -\frac{1}{s} \frac{\zeta'(s)}{\zeta(s)}, \quad \mathrm{Re}\, s > 1.$$

(d) Prove the inversion formula

$$\psi(x) = -\frac{1}{2\pi i} \int_{c-i\infty}^{c+i\infty} \frac{x^s}{s} \frac{\zeta'(s)}{\zeta(s)} ds, \quad c > 1 \quad \text{and } x \text{ not an integer.}$$

(e) Let

$$\psi_1(x) = \frac{1}{x} \int_0^x \psi(t) dt.$$

Show that

$$\int_0^{\infty} x^{-s-1} \psi_1(x) dx = -\frac{1}{s(s+1)} \frac{\zeta'(s)}{\zeta(s)}, \quad \mathrm{Re}\, s > 1.$$

(f) Prove the inversion of (e),

$$\psi_1(x) = -\frac{1}{2\pi i} \int_{c-i\infty}^{c+i\infty} \frac{x^s}{s(s+1)} \frac{\zeta'(s)}{\zeta(s)} ds, \quad c > 1 \quad \text{and } x \text{ not an integer.}$$

(g) Show that the Mellin transform of $\sum_{n=1}^{\infty} \Lambda(n) e^{-nx}$ is $-\Gamma(s)\zeta'(s)/\zeta(s)$.

(h) Prove the Mellin inversion formula of (g), that is,

$$-\frac{1}{2\pi i} \int_{c-i\infty}^{c+i\infty} x^{-s} \Gamma(s) \frac{\zeta'(s)}{\zeta(s)} ds = \sum_{1}^{\infty} \Lambda(n) e^{-nx}, \quad c > 1, \; x > 0.$$

42. Let a, b, c, d be complex numbers.

(a) Suppose $c \neq$ an integer and Re $(a + b) > -1$. Prove that

$$\sum_{k=-\infty}^{\infty} \frac{(-1)^k}{(c+k)\Gamma(a+1-k)\Gamma(b+1+k)}$$

$$= \frac{\pi}{\sin c\pi \, \Gamma(a+c+1)\Gamma(b-c+1)}.$$

(b) Suppose Re $(a + b + c + d) > -1$. Prove that

$$\sum_{k=-\infty}^{\infty} \frac{1}{\Gamma(a-k+1)\Gamma(b-k+1)\Gamma(c+k+1)\Gamma(d+k+1)}$$

$$= \frac{\Gamma(a+b+c+d+1)}{\Gamma(a+c+1)\Gamma(b+c+1)\Gamma(a+d+1)\Gamma(b+d+1)}.$$

(c) Which beta integral is extended to the sum in part (b)?

43. This problem gives a quick way of obtaining the differential equation for $y = {}_2F_1(a, b; c; x)$. From the series for y and Euler's transformation, we have

(a)
$$y' = \frac{ab}{c} {}_2F_1(a+1, b+1; c+1; x)$$

$$= \frac{ab}{c}(1-x)^{c-a-b-1}{}_2F_1(c-a, c-b; c+1; x),$$

and

(b)
$$\frac{d}{dx}\left[x^{c-1}{}_2F_1(a, b; c; x)\right] = (c-1)x^{c-2}{}_2F_1(a, b; c-1; x).$$

Use (a) and (b) to see that

$$\frac{d}{dx}[x^c(1-x)^{a+b+1-c}y'] = abx^{c-1}{}_2F_1(c-a, c-b; c; x)$$

$$= abx^{c-1}(1-x)^{a+b-c}y.$$

The differential equation is obtained after computing the derivative on the left side.

44. Evaluate

$$\int_{-\pi/2}^{\pi/2} (1 + e^{2i\theta})^\alpha (1 + e^{-2i\theta})^\beta \, d\theta$$

by use of the binomial theorem and Theorem 2.2.1, and so show that

$$\int_0^{\pi/2} (\cos\theta)^{\alpha+\beta} \cos(\alpha-\beta)\theta \, d\theta = \frac{\pi\Gamma(\alpha+\beta+1)}{2^{\alpha+\beta+1}\Gamma(\alpha+1)\Gamma(\beta+1)}.$$

3
Hypergeometric Transformations and Identities

Gauss's work on the hypergeometric equation contains a discussion of the monodromy question for the solutions of this equation. Gauss found and analyzed a quadratic transformation of hypergeometric functions; this apparently led him to the problem of monodromy. Unlike the linear (fractional) transformations of these functions, of which Pfaff's formula in Theorem 2.2.5 is an example, quadratic transformations exist only under certain conditions on the parameters. Nevertheless, they are important and useful. We have given some applications of these transformations after deriving a few basic formulas. An interesting application deals with the problem of proving Gauss's arithmetic-geometric mean to be expressible as an elliptic integral.

This chapter also contains a discussion of some methods for the summation of certain types of hypergeometric series. We use a quadratic transformation to obtain Dixon's identity for a well-poised $_3F_2$ at $x = 1$. We then apply a method of Bailey to derive identities for special types of $_{p+1}F_p$ with $2 \leq p \leq 6$, including Dougall's identity, which was mentioned in Remark 2.2.2 in the previous chapter. An important transformation formula due to Whipple is obtained by the same method. Just as Barnes's integral on the product of gamma functions was an analog of Gauss's $_2F_1$ sum, these identities also have integral analogs and we discuss them. The hypergeometric identities provide a systematic approach to the evaluation of single sums of binomial coefficients.

Contiguous relations for hypergeometric series contain an enormous amount of hidden information. There are three-term relations for balanced $_4F_3$ functions as noted by Wilson [1977] and independently by Raynal [1979] and for $_3F_2$ functions at $x = 1$, a fact pointed out much earlier by Kummer. We describe Wilson's simple technique for deriving the contiguous relations that contain the three-term recurrence for Wilson polynomials. These $_4F_3$ polynomials contain a whole range of classical orthogonal polynomials as special or limiting cases. We devote a section of this chapter to the definition and orthogonality of Wilson polynomials. They are orthogonal with respect to a weight function that occurs as the integrand in an integral analog of a $_5F_4$ identity.

Gosper, Zeilberger, and Wilf have done significant work toward devising computer algorithms for finding and proving hypergeometric identities. We discuss the Wilf–Zeilberger method and compare it with that of Pfaff; both are applications of contiguous relations.

3.1 Quadratic Transformations

Exercise 2.19 asks the reader to prove the following relation satisfied by Jacobi polynomials:

$$P_{2n}^{(\alpha,\alpha)}(x) = \frac{n!(\alpha+1)_{2n}}{(2n)!(\alpha+1)_n} P_n^{(\alpha,-1/2)}(2x^2-1). \tag{3.1.1}$$

This important formula can be stated in hypergeometric form as

$$_2F_1\left(\begin{matrix} -2n, 2n+2\alpha+1 \\ \alpha+1 \end{matrix}; (1-x)/2\right) = {}_2F_1\left(\begin{matrix} -n, n+\alpha+\frac{1}{2} \\ \alpha+1 \end{matrix}; 1-x^2\right). \tag{3.1.2}$$

Note that the $_2F_1$ on the left is linear in x and the one on the right is quadratic in x. This is an example of a quadratic transformation. In this section we give the fundamental results on quadratic transformations with two free parameters.

It is natural to suspect that (3.1.2) continues to hold when the two series do not terminate. This can be shown directly by rewriting (3.1.2) as

$$_2F_1\left(\begin{matrix} 2a, 2b \\ a+b+\frac{1}{2} \end{matrix}; x\right) = {}_2F_1\left(\begin{matrix} a, b \\ a+b+\frac{1}{2} \end{matrix}; 4x(1-x)\right) \tag{3.1.3}$$

and expanding the right-hand side as a power series in x. The coefficient of x^n is a balanced $_3F_2$, which can be summed by the Pfaff–Saalschütz identity (Theorem 2.2.6). However, two cases have to be considered: n even and n odd. An equivalent identity (3.1.4) does not have this problem.

Apply Pfaff's transformation (Theorem 2.2.5) to the left-hand side and re-label the parameters and variable. The result we want to prove is

Theorem 3.1.1 *For all x where the two series converge,*

$$_2F_1\left(\begin{matrix} a, b \\ a-b+1 \end{matrix}; x\right) = (1-x)^{-a} {}_2F_1\left(\begin{matrix} a/2, (1+a)/2-b \\ a-b+1 \end{matrix}; \frac{-4x}{(1-x)^2}\right). \tag{3.1.4}$$

Proof. Write the series on the right as

$$\sum_{k=0}^{\infty} \frac{(a/2)_k(-b+(a+1)/2)_k}{k!(a-b+1)_k}(-4x)^k(1-x)^{-a+2k}$$

$$= \sum_{k=0}^{\infty} \frac{(a/2)_k(-b+(a+1)/2)_k}{k!(a-b+1)_k}(-4x)^k \sum_{j=0}^{\infty} \frac{(a+2k)_j}{j!}x^j.$$

It is easy to see that the coefficient of x^n in the last expression is

$$\sum_{k=0}^{n} \frac{(a/2)_k(-b+(a+1)/2)_k(-4)^k(a+2k)_{n-k}}{(a-b+1)_k k!(n-k)!}. \tag{3.1.5}$$

Now observe that

$$(a+2k)_{n-k} = \frac{(a)_{n+k}}{(a)_{2_k}} = \frac{(a)_n(a+n)_k}{2^{2k}(a/2)_k((a+1)/2)_k}.$$

So (3.1.5) is the same as

$$\frac{(a)_n}{n!}\sum_{k=0}^{n} \frac{(-b+(a+1)/2)_k(a+n)_k(-n)_k}{(a-b+1)_k((a+1)/2)_k k!}.$$

Application of the Pfaff–Saalschütz identity shows that this balanced $_3F_2$ is equal to

$$\frac{(a)_n(b)_n}{n!(a-b+1)_n}.$$

Clearly, this is also the coefficient of x^n on the left side of (3.1.4). This proves the theorem. ■

Corollary 3.1.2 (Kummer [1836])

$$_2F_1\left(\begin{matrix} a, b \\ a-b+1 \end{matrix}; -1\right) = \frac{\Gamma(a-b+1)\Gamma((a/2)+1)}{\Gamma(a+1)\Gamma((a/2)-b+1)}.$$

Proof. Let $x \to -1$ in (3.1.4) and conclude by Abel's continuity theorem that

$$_2F_1\left(\begin{matrix} a, b \\ a-b+1 \end{matrix}; -1\right) = 2^{-a}{}_2F_1\left(\begin{matrix} a/2, \frac{1}{2}(a+1)-b \\ a-b+1 \end{matrix}; 1\right).$$

Now sum the $_2F_1$ on the right by Gauss's summation formula (Theorem 2.2.2). The corollary follows after an application of Legendre's duplication formula (Theorem 1.5.1). ■

The quadratic transformation in Theorem 3.1.1 holds when the two series converge. This is no longer true for (3.1.3). Both sides of (3.1.3) converge for $\frac{1}{2} < x < 1$. Letting $x \to 1$ on both sides gives

$$_2F_1\left(\begin{matrix} 2a, 2b \\ a+b+\frac{1}{2} \end{matrix}; 1\right) = 1.$$

This implies, by Gauss's summation, that

$$\frac{\Gamma\left(a+b+\frac{1}{2}\right)\Gamma\left(\frac{1}{2}-a-b\right)}{\Gamma\left(b-a+\frac{1}{2}\right)\Gamma\left(a-b+\frac{1}{2}\right)} = \frac{\cos\pi(a-b)}{\cos\pi(a+b)} = 1. \tag{3.1.6}$$

This identity is not true in general, although it is true when a or b is an integer. When a or b is a negative integer, (3.1.3) holds for all x. Otherwise it holds in the connected component of $x = 0$ when both $|x| < 1$ and $|4x(1-x)| < 1$. Gauss was the first to remark on this, although his paper remained unpublished until after analytic continuation was discovered. Gauss understood that a hypergeometric function is many valued, citing $\sin^{-1} x$ as a similar case; thus, he saw that (3.1.3) does not hold for all x where the two sides converge. See Exercise 6 for the correct identity when $\frac{1}{2} < x < 1$.

The set $\frac{1}{2} < x < 1$ in (3.1.3) maps to $x < -1$ in (3.1.4). The function on the right-hand side is evaluated at a point inside the unit circle, but on the left an analytic continuation is needed to make sense of the function.

Our first evaluation of $_2F_1(a, b; c; x)$ at $x = 1$ depended on Euler's integral representation

$$_2F_1\left(\begin{matrix} a, b \\ c \end{matrix}; x\right) = \frac{\Gamma(c)}{\Gamma(a)\Gamma(c - a)} \int_0^1 t^{a-1}(1 - t)^{c-a-1}(1 - xt)^{-b}dt.$$

It is possible to derive Kummer's identity in Corollary 3.1.2 by taking $x = -1$ in the integral. In this case, $c = a - b + 1$, which makes the powers of $1 - t$ and $1 - xt$ equal. The other fundamental quadratic transformation comes from taking the powers of t and $1 - t$ to be the same, that is, $c = 2a$.

Theorem 3.1.3 *For all x where the series converge*

$$_2F_1\left(\begin{matrix} a, b \\ 2a \end{matrix}; x\right) = \left(1 - \frac{x}{2}\right)^{-b} {}_2F_1\left(\begin{matrix} b/2, (b + 1)/2 \\ a + \frac{1}{2} \end{matrix}; \left(\frac{x}{2 - x}\right)^2\right). \tag{3.1.7}$$

Proof. The left side equals

$$\frac{\Gamma(2a)}{\Gamma(a)\Gamma(a)} \int_0^1 (1 - xt)^{-b}\left[\frac{1}{4} - \left(\frac{1}{2} - t\right)\right]^{a-1} dt.$$

Substitute $s = 1 - 2t$ or $t = (1 - s)/2$ and simplify to get

$$\frac{\Gamma(2a)(1 - (x/2))^{-b}}{2^{2a-1}\Gamma(a)^2} \int_{-1}^1 \left(1 - \frac{sx}{x - 2}\right)^{-b}(1 - s^2)^{a-1}ds$$

$$= \frac{\Gamma(2a)(1 - (x/2))^{-b}}{2^{2a-1}\Gamma(a)^2} \sum_{n=0}^{\infty} \frac{(b)_n}{n!}\left(\frac{x}{x - 2}\right)^n \int_{-1}^1 s^n(1 - s^2)^{a-1}ds.$$

When n is odd, the last integral is zero. Otherwise it is

$$\int_0^1 u^{m-\frac{1}{2}}(1 - u)^{a-1}du = \frac{\Gamma\left(m + \frac{1}{2}\right)\Gamma(a)}{\Gamma\left(m + a + \frac{1}{2}\right)},$$

where $n = 2m$. So

$$_2F_1\left(\begin{matrix} a, b \\ 2a \end{matrix}; x\right) = \frac{\Gamma(2a)(1 - (x/2))^{-b}}{\Gamma(a)^2 2^{2a-1}} \sum_{m=0}^{\infty} \frac{(b)_{2m}\Gamma\left(m + \frac{1}{2}\right)\Gamma(a)}{(1)_{2m}\Gamma\left(m + a + \frac{1}{2}\right)}\left(\frac{x}{2 - x}\right)^{2m}$$

$$= \frac{\Gamma(2a)\Gamma\left(\frac{1}{2}\right)(1 - (x/2))^{-b}}{\Gamma(a)\Gamma\left(a + \frac{1}{2}\right)2^{2a-1}} {}_2F_1\left(\begin{matrix} b/2, (b + 1)/2 \\ a + \frac{1}{2} \end{matrix}; \left(\frac{x}{2 - x}\right)^2\right).$$

$$\tag{3.1.8}$$

An application of Legendre's duplication formula proves the theorem. ∎

Remark It is worth observing that Legendre's duplication formula follows from (3.1.8) by taking $x = 0$.

Theorems 3.1.1 and 3.1.3 contain the basic quadratic transformations. Others can be derived from these two by using the fractional linear transformations or the three-term relations connecting different solutions of the hypergeometric differential equation.

Apply Pfaff's transformation to the right side of (3.1.4) to obtain

$$_2F_1\left(\begin{array}{c} a, b \\ a - b + 1 \end{array}; x\right) = (1 + x)^{-a}\,_2F_1\left(\begin{array}{c} a/2, (a + 1)/2 \\ a - b + 1 \end{array}; \frac{4x}{(1 + x)^2}\right). \qquad (3.1.9)$$

Replace $4x/(1 + x)^2$ with x to derive the equivalent formula

$$_2F_1\left(\begin{array}{c} a/2, (a + 1)/2 \\ a - b + 1 \end{array}; x\right) = 2^a(1 + \sqrt{1 - x})^{-a}\,_2F_1\left(\begin{array}{c} a, b \\ a - b + 1 \end{array}; \frac{1 - \sqrt{1 - x}}{1 + \sqrt{1 - x}}\right). \qquad (3.1.10)$$

A combination of (3.1.10) and (3.1.7) is another useful transformation, namely,

$$_2F_1\left(\begin{array}{c} a, b \\ 2b \end{array}; \frac{4x}{(1 + x)^2}\right) = (1 + x)^{2a}\,_2F_1\left(\begin{array}{c} a, a + \frac{1}{2} - b \\ b + \frac{1}{2} \end{array}; x^2\right). \qquad (3.1.11)$$

To prove this, replace x with $4x/(1 + x)^2$ in (3.1.7) and interchange a and b to get

$$_2F_1\left(\begin{array}{c} a, b \\ 2b \end{array}; \frac{4x}{(1 + x)^2}\right) = (1 + x)^{2a}(1 + x^2)^{-a}\,_2F_1\left(\begin{array}{c} a/2, (a + 1)/2 \\ b + \frac{1}{2} \end{array}; \left(\frac{2x}{1 + x^2}\right)^2\right).$$

By (3.1.10) it follows that the right side of the last formula is equal to the right side of (3.1.11). This proves (3.1.11).

To see how three-term relations can be used to derive more quadratic transformations, recall formula (2.3.13),

$$_2F_1\left(\begin{array}{c} a, b \\ c \end{array}; x\right) = \frac{\Gamma(c)\Gamma(c - a - b)}{\Gamma(c - a)\Gamma(c - b)}\,_2F_1\left(\begin{array}{c} a, b \\ a + b - c + 1 \end{array}; 1 - x\right)$$

$$+ \frac{\Gamma(c)\Gamma(a + b - c)}{\Gamma(a)\Gamma(b)}(1 - x)^{c - a - b}\,_2F_1\left(\begin{array}{c} c - a, c - b \\ c - a - b + 1 \end{array}; 1 - x\right).$$

Apply this to (3.1.3). The result is

$$_2F_1\left(\begin{array}{c} 2a, 2b \\ a + b + \frac{1}{2} \end{array}; \frac{x + 1}{2}\right) = \frac{\Gamma(a + b + \frac{1}{2})\Gamma(\frac{1}{2})}{\Gamma(a + \frac{1}{2})\Gamma(b + \frac{1}{2})}\,_2F_1\left(\begin{array}{c} a, b \\ 1/2 \end{array}; x^2\right)$$

$$- x\frac{\Gamma(a + b + \frac{1}{2})\Gamma(-\frac{1}{2})}{\Gamma(a)\Gamma(b)}\,_2F_1\left(\begin{array}{c} a + \frac{1}{2}, b + \frac{1}{2} \\ 3/2 \end{array}; x^2\right). \qquad (3.1.12)$$

Some general remarks about quadratic transformations are now in order. There are two sides to the standard quadratic transformations: the linear side and the quadratic side. On the linear side, the variable is linear or the image of linear under the linear fractional transformation $x(x-1)^{-1}$. The parameters on the linear side are restricted by one condition, which comes from writing the $_2F_1$ as an integral and equating two of the exponents of the functions in the integrand. This gives $c = 2a$, $c = a - b + 1$, and $a + b = 1$. To obtain the complete list, symmetry in the parameters a and b and the Pfaff transformation are used. The complete list of conditions is

$$c = 2a, c = 2b, c = a - b + 1, c = b - a + 1, a + b = 1, c = \frac{a + b + 1}{2}.$$

$$(3.1.13)$$

On the quadratic side, the variable appears in a quadratic form; the parameters are restricted by requiring that those in the numerator differ by $\frac{1}{2}$, or the denominator parameters differ by $\frac{1}{2}$ (that is, $c = \frac{1}{2}$ or $c = \frac{3}{2}$), or the Pfaff transformation is used. The conditions are

$$a = b + \frac{1}{2}, b = a + \frac{1}{2}, c = a + b + \frac{1}{2}, c = a + b - \frac{1}{2}, c = \frac{1}{2}, c = \frac{3}{2}.$$

$$(3.1.14)$$

Note that we use a, b, c to denote generic parameters and that they may differ in a specific formula.

The two sides of the quadratic transformation are split into two groups; one has $c = 2a$ or $c = 2b$ on the linear side and any of the other conditions (3.1.13) in the other group. On the quadratic side, one set has $c = \frac{1}{2}$ or $c = \frac{3}{2}$, and the other set has the remaining conditions in (3.1.14). Notice that (3.1.7) connects $c = 2a$ with $b = a + \frac{1}{2}$ and that (3.1.4) connects $c = a - b + 1$ with $c = (a + b + 1)/2$. The transformations that go to the group $c = \frac{1}{2}$ and $c = \frac{3}{2}$ come from three-term relations. An example of this is (3.1.12).

Notice a simple way to distinguish those quadratic transformations that come from (3.1.4) from those that come from (3.1.7) via linear transformations. In (3.1.4) the denominator parameters are equal; in (3.1.7) they differ except for one value. Observe that (3.1.11) is not one of these quadratic transformations. Both sides have conditions in (3.1.13), and they were shown to be equal by showing that each is equal to a function with one of the conditions in (3.1.14) satisfied.

Let us return to the proof of Theorem 3.1.1 and note that the balanced $_3F_2$ that occurs in its proof has one specialization not needed to sum it. There is likely to be a more general quadratic transformation with one more degree of freedom. It would be nice if this were a transformation of a general $_2F_1$, but our discussion of the connection between the quadratic transformation and Euler's integral makes this unlikely. The extra freedom comes at the $_3F_2$ level. Whipple found the following quadratic transformation:

$$_3F_2\left(\begin{matrix} a, b, c \\ a - b + 1, a - c + 1 \end{matrix} ; x\right)$$

$$= (1 - x)^{-a} {}_3F_2\left(\begin{matrix} a - b - c + 1, a/2, (a + 1)/2 \\ a - b + 1, a - c + 1 \end{matrix} ; \frac{-4x}{(1 - x)^2}\right). \qquad (3.1.15)$$

The proof of this formula is very similar to that of Theorem 3.1.1. The coefficient of x^n in the expression on the right of (3.1.15) is again a balanced $_3F_2$. The details are left to the reader. There exist examples of cubic transformations, though these are not as well understood as quadratic transformations. Two examples are given in Exercise 38.

Kummer [1836] gave some quadratic transformations of $_2F_1$s with one free parameter. The parameter was chosen so that more than one transformation could be applied to the $_2F_1$ function. Consider the following two fundamental transformations:

$$_2F_1\left(\begin{matrix} a, b \\ a - b + 1 \end{matrix} ; x\right) = (1 + x)^{-a} {}_2F_1\left(\begin{matrix} a/2, (a + 1)/2 \\ a - b + 1 \end{matrix} ; 4x/(1 + x)^2\right)$$

and

$$_2F_1\left(\begin{matrix} a, b \\ 2b \end{matrix} ; x\right) = \left(1 - \frac{x}{2}\right)^{-a} {}_2F_1\left(\begin{matrix} a/2, (a + 1)/2 \\ b + (1/2) \end{matrix} ; \left(\frac{x}{2 - x}\right)^2\right).$$

The $_2F_1$s on the left become identical when we make the denominator parameters equal. This means $2b = a - b + 1$ or $b = (a + 1)/3$. In this case

$$\left(1 - \frac{x}{2}\right)^{-a} {}_2F_1\left(\begin{matrix} a/2, (a + 1)/2 \\ b + (1/2) \end{matrix} ; \left(\frac{x}{2 - x}\right)^2\right)$$

$$= {}_2F_1\left(\begin{matrix} a, (a + 1)/3 \\ (2a + 2)/3 \end{matrix} ; x\right)$$

$$= (1 + x)^{-a} {}_2F_1\left(\begin{matrix} a/2, (a + 1)/2 \\ (2a + 2)/3 \end{matrix} ; \frac{4x}{(1 + x)^2}\right)$$

$$= (1 - x)^{-a} {}_2F_1\left(\begin{matrix} a/2, (a + 1)/6 \\ (2a + 2)/3 \end{matrix} ; -\frac{4x}{(1 - x)^2}\right). \qquad (3.1.16)$$

The last equation was obtained by an application of the Pfaff transformation. Let

$x \to -1$ in the first and last expressions of (3.1.16) to get

$$\,_2F_1\left(\begin{matrix} a/2, (a+1)/2 \\ (2a+5)/6 \end{matrix}; \frac{1}{9}\right) = \left(\frac{4}{3}\right)^{-a} \,_2F_1\left(\begin{matrix} a/2, (a+1)/6 \\ 2(a+1)/3 \end{matrix}; 1\right)$$

$$= \left(\frac{3}{4}\right)^a \frac{\sqrt{\pi}\,\Gamma((2a+2)/3)}{\Gamma((a+4)/6)\Gamma((a+1)/2)}. \qquad (3.1.17)$$

Now use Gauss's quadratic transformation (3.1.3) and then Pfaff's transformation (2.2.6) to obtain

$$\,_2F_1\left(\begin{matrix} a, (a+1)/3 \\ 2(a+1)/3 \end{matrix}; x\right)$$

$$= \,_2F_1\left(\begin{matrix} a/2, (a+1)/6 \\ 2(a+1)/3 \end{matrix}; 4x(1-x)\right)$$

$$= (1 - 4x + 4x^2)^{-a/2} \,_2F_1\left(\begin{matrix} a/2, (a+1)/2 \\ 2(a+1)/3 \end{matrix}; \frac{4x^2 - 4x}{4x^2 - 4x + 1}\right). \qquad (3.1.18)$$

Recall that Gauss's formula holds in the connected component of the region $|4x(1-x)| < 1$ that contains the origin. Thus $\sqrt{1 - 4x + 4x^2} = 1 - 2x$. (A similar argument applies in the derivation of the last equation in (3.1.16).) Combine (3.1.16) and (3.1.17) to get

$$(1+x)^{-a} \,_2F_1\left(\begin{matrix} a/2, (a+1)/2 \\ 2(a+1)/3 \end{matrix}; \frac{4x}{(1+x)^2}\right)$$

$$= \,_2F_1\left(\begin{matrix} a/2, (a+1)/6 \\ 2(a+1)/3 \end{matrix}; 4x(1-x)\right)$$

$$= (1 - 2x)^{-a} \,_2F_1\left(\begin{matrix} a/2, (a+1)/2 \\ 2(a+1)/3 \end{matrix}; \frac{4x^2 - 4x}{4x^2 - 4x + 1}\right). \qquad (3.1.19)$$

Let $x \to \frac{1}{2}$ in the first equation to arrive at

$$\,_2F_1\left(\begin{matrix} a/2, (a+1)/2 \\ 2(a+1)/3 \end{matrix}; \frac{8}{9}\right) = \left(\frac{2}{3}\right)^{-a} \,_2F_1\left(\begin{matrix} a/2, (a+1)/6 \\ 2(a+1)/3 \end{matrix}; 1\right)$$

$$= \left(\frac{3}{2}\right)^a \frac{\sqrt{\pi}\,\Gamma((2a+2)/3)}{\Gamma((a+4)/6)\Gamma((a+1)/2)}. \qquad (3.1.20)$$

The result in (2.3.13) connects the following three $_2F_1$s:

$$\,_2F_1\left(\begin{matrix} a/2, (a+1)/2 \\ 2(a+1)/3 \end{matrix}; 8/9\right), \quad \,_2F_1\left(\begin{matrix} a/2, (a+1)/2 \\ (2a+5)/6 \end{matrix}; 1/9\right),$$

and

$$\,_2F_1\left(\begin{matrix} (a+4)/6, (a+1)/6 \\ (7-2a)/6 \end{matrix}; 1/9\right).$$

Thus the third $_2F_1$ can also be computed. In Exercise 2 we give a few more results of Kummer on quadratic transformations with one free parameter.

3.2 The Arithmetic-Geometric Mean and Elliptic Integrals

Definition 3.2.1   *Let $0 < k < 1$. Following Legendre, the integral*

$$\int_0^x \frac{dt}{\sqrt{(1-t^2)(1-k^2t^2)}}, \quad \text{for } x \in [-1, 1] \tag{3.2.1}$$

is called an elliptic integral of the first kind. If $x = 1$, this definite integral is called a complete elliptic integral and denoted by K. Thus

$$K := K(k) := \int_0^1 \frac{dt}{\sqrt{(1-t^2)(1-k^2t^2)}} = \int_0^{\pi/2} \frac{d\theta}{\sqrt{1-k^2\sin^2\theta}}. \tag{3.2.2}$$

To fully understand the integral (3.2.1), one has to study its inverse, which is a Jacobi elliptic function. This is similar to looking at the integral $\int_0^x (1-t^2)^{-1/2}dt$ as the inverse of the sine function. We shall take a very brief look at the Jacobi elliptic functions in Chapter 10. Here we consider how some theory of hypergeometric functions can be used to obtain interesting results about integral (3.2.2). First note that the binomial expansion of the integrand $(1 - k^2 \sin^2 \theta)^{-1/2}$, integrated term by term, gives

$$K(k) = \frac{\pi}{2} \, {}_2F_1\!\left(\begin{matrix} 1/2, 1/2 \\ 1 \end{matrix}; k^2\right). \tag{3.2.3}$$

Now replace x with $(1 - \sqrt{1-x^2})/(1+\sqrt{1-x^2})$ in the quadratic transformation (3.1.11) to obtain

$$_2F_1\!\left(\begin{matrix} a, b \\ 2b \end{matrix}; x^2\right) = \left(\frac{1+\sqrt{1-x^2}}{2}\right)^{-2a} {}_2F_1\!\left(\begin{matrix} a, a-b+1/2 \\ b+1/2 \end{matrix}; \left(\frac{1-\sqrt{1-x^2}}{1+\sqrt{1-x^2}}\right)^2\right).$$

Apply this to the hypergeometric function in (3.2.3). The result is

$$K(k) = \frac{2}{1+k'} K\!\left(\frac{1-k'}{1+k'}\right), \tag{3.2.4}$$

where $k'^2 = 1 - k^2$. To iterate this result, we introduce the following notation:

$$k_0 := k, \ k'_m := \sqrt{1-k_m^2}, \quad \text{and} \quad k_{m+1} := \frac{1-k'_m}{1+k'_m}, \quad m = 0, 1, 2, \ldots. \tag{3.2.5}$$

It follows from (3.2.4) and (3.2.5) that

$$K(k) = \prod_{m=0}^n \frac{2}{1+k'_m} K(k_{m+1}).$$

Observe that

$$k_{m+1} = \frac{1 - \sqrt{1 - k_m^2}}{1 + \sqrt{1 - k_m^2}} < 1 - \sqrt{1 - k_m^2} < 1 - \left(1 - k_m^2\right) < k_m^2.$$

So

$$k_m < k^{2^m} \quad \text{and} \quad k_m \to 0 \quad \text{as} \quad m \to \infty.$$

Moreover,

$$\prod_{m=0}^{\infty} \frac{1 + k_m'}{2} = \prod_{m=0}^{\infty} \left(1 - \frac{1 - k_m'}{2}\right)$$

converges, since $\sum_{m=0}^{\infty}(1 - \sqrt{1 - k_m^2}) < \sum_{m=0}^{\infty} k^{2^m}$ converges. Thus

$$K(k) = \prod_{m=0}^{\infty} \frac{2}{1 + k_m'} K(0) = \frac{\pi}{2} \prod_{m=0}^{\infty} \frac{2}{1 + k_m'}.$$

The k_m's can be obtained successively from

$$k_{m+1}' = \frac{2\sqrt{k_m'}}{1 + k_m'}. \tag{3.2.6}$$

The right side is the ratio of the geometric mean and the arithmetic mean of 1 and k_m'.

Now suppose there are two sequences $\{a_n\}$ and $\{b_n\}$ with $a_0 = 1$ and $b_0 = k'$ such that

$$\frac{b_n}{a_n} = k_n' \quad \text{and} \quad \frac{a_{n+1}}{a_n} = \frac{1 + k_n'}{2}. \tag{3.2.7}$$

Then

$$a_n = \frac{a_n}{a_{n-1}} \cdot \frac{a_{n-1}}{a_{n-2}} \cdots \frac{a_1}{a_0} = \prod_{m=0}^{n-1} \frac{1 + k_m'}{2}$$

and

$$\lim_{n \to \infty} a_n = \prod_{m=0}^{\infty} \frac{1 + k_m'}{2}.$$

Moreover, $k_m' \to 1$ implies that

$$\lim_{n \to \infty} b_n = \lim_{n \to \infty} a_n.$$

From (3.2.6) and (3.2.7) it follows that

$$a_{n+1} = \frac{a_n + b_n}{2} \quad \text{and} \quad b_{n+1} = \sqrt{a_n b_n}. \tag{3.2.8}$$

Conversely, it is easy to see that if two sequences $\{a_n\}$ and $\{b_n\}$ satisfy (3.2.8) with $a_0 = 1$ and $b_0 = k'$, then (3.2.7) also holds. The common limit of the two sequences is called the arithmethic-geometric mean of the sequences and it is equal to

$$\left[\frac{2}{\pi} \int_0^1 \frac{dt}{\sqrt{(1 - t^2)(1 - k^2 t^2)}} \right]^{-1}.$$

This result is due to Lagrange and Gauss independently. See Cox [1984]. We state it as a theorem after formally defining the arithmetic-geometric mean.

Definition 3.2.2 *Suppose $\{a_n\}$ and $\{b_n\}$ are two sequences such that $a =: a_0$ and $b =: b_0$ are real with $a \geq b > 0$ and $a_{n+1} = (a_n + b_n)/2$, $b_{n+1} = \sqrt{a_n b_n}$. Then the two sequences converge to a common limit $M(a, b)$, called the arithmetic-geometric mean.*

Theorem 3.2.3

$$\frac{1}{M(a, b)} = \frac{2}{\pi} \int_0^{\pi/2} \frac{d\theta}{(a^2 \cos^2 \theta + b^2 \sin^2 \theta)^{1/2}}.$$

Proof. It is clear from the definition that $M(\lambda a, \lambda b) = \lambda M(a, b)$. So $M(a, b) = a M(1, b/a)$. The theorem follows from the result in the previous paragraph by taking $k' = b/a$. ∎

The fact that the two sequences in Definition 3.2.2 have a common limit can be obtained directly. It is easy to see that

$$a = a_0 \geq a_1 \geq a_2 \geq \cdots \geq a_n \geq \cdots \geq b_n \geq b_{n-1} \geq \cdots \geq b_0 = b.$$

Also,

$$a_{n+1} - b_{n+1} \leq a_{n+1} - b_n = \frac{a_n + b_n}{2} - b_n = \frac{a_n - b_n}{2}.$$

Thus

$$a_n - b_n \leq \frac{a - b}{2^n},$$

and the two sequences converge to the same limit. From the equation

$$a_{n+1} - b_{n+1} = \frac{(a_n - b_n)^2}{4(a_{n+1} + b_{n+1})}$$

it follows that the convergence is quadratic. Gauss saw this rapid convergence by computing the numerical example with $a = \sqrt{2}$ and $b = 1$. The first few values of a_n and b_n are

n	a_n	b_n
0	1.41421356237	1.00000000000
1	1.20710678118	1.18920711500
2	1.19815694809	1.19812352149
3	1.19814023479	1.19814023467
4	1.19814023473	1.19814023473

Gauss calculated up to twenty-one decimal places but the above table is sufficent to illustrate our point. Somewhat later he calculated the ratio $\pi/\tilde{\omega}$ where

$$\tilde{\omega} = 2 \int_0^1 (1 - x^4)^{-1/2} dx.$$

On May 30, 1799, he noted in his diary that $M(\sqrt{2}, 1)$ and $\pi/\tilde{\omega}$ agreed to eleven decimal places and he conjectured that they were equal. Later he proved the more general result contained in Theorem 3.2.3. Once one has made this conjecture, however, it is not too difficult to prove the result. If $I(a, b)$ denotes the integral in the theorem, then it is enough to prove

$$I(a, b) = I(a_1, b_1), \tag{3.2.9}$$

for then

$$I(a, b) = I(a_1, b_1) = \cdots = I(a_m, b_m) = \cdots = \lim_{m \to \infty} I(a_m, b_m).$$

To prove (3.2.9), Gauss defined a new variable θ_1 by

$$\sin \theta = \frac{2a \sin \theta_1}{a + b + (a - b) \sin^2 \theta_1}$$

and asserted that

$$(a^2 \cos^2 \theta + b^2 \sin^2 \theta)^{-1/2} d\theta = \left(a_1^2 \cos^2 \theta_1 + b_1^2 \sin^2 \theta_1\right)^{-1/2} d\theta_1.$$

This requires some computation, and (3.2.9) follows.

We started with an elliptic integral and then introduced the arithmetic-geometric mean. Clearly Gauss was approaching the problem from the opposite direction, as presented above. Let us now consider another way of arriving at the elliptic integral from the arithmetic-geometric mean. Gauss also studied the function

$$f(x) = \frac{1}{M(1 + x, 1 - x)} = \frac{1}{M(1, \sqrt{1 - x^2})}.$$

Now

$$f\left(\frac{2t}{1 + t^2}\right) = \frac{1}{M\left(1 + \frac{2t}{1+t^2}, 1 - \frac{2t}{1+t^2}\right)} = \frac{1}{M\left(1, \sqrt{1 - \frac{4t^2}{(1+t^2)^2}}\right)}$$

$$= \frac{1 + t^2}{M(1 + t^2, 1 - t^2)} = (1 + t^2) f(t^2). \tag{3.2.10}$$

Assume that f is analytic about $t = 0$. We want the analytic function that satisfies $f(0) = 1$ and the functional relation in (3.2.10). Since f is clearly even, $f(x) = g(x^2)$ for some function g and

$$g\left(\frac{4t^2}{(1+t^2)^2}\right) = (1+t^2)g(t^4).$$

Replace t^2 with x to get

$$g\left(\frac{4x}{(1+x)^2}\right) = (1+x)g(x^2).$$

Write $g(x) = \sum_{n=0}^{\infty} a_n x^n$, and use this functional equation to get

$$a_1 = a_0/4, \quad a_2 = \frac{1^2 \cdot 3^2}{2^2 \cdot 4^2} a_0, \quad a_3 = \frac{1^2 \cdot 3^2 \cdot 5^2}{2^2 \cdot 4^2 \cdot 6^2} a_0.$$

This suggests that

$$g(x) = {}_2F_1\left(\begin{matrix} 1/2, 1/2 \\ 1 \end{matrix}; x^2\right) = \frac{2}{\pi} K.$$

Formula (3.1.11),

$$_2F_1\left(\begin{matrix} a, b \\ 2b \end{matrix}; 4x/(1+x)^2\right) = (1+x)^{2a} {}_2F_1\left(\begin{matrix} a, a + \frac{1}{2} - b \\ b + \frac{1}{2} \end{matrix}; x^2\right),$$

suggests the same identity. It is possible to show directly that $f(x) = g(x^2)$ is analytic. See the first few pages of Borwein and Borwein [1987]. However, it is easier to do the argument in the other direction, as done above.

Definition 3.2.4 *The complete elliptic integral of the second kind is defined as*

$$E := E(k) := \int_0^{\pi/2} (1 - k^2 \sin^2 \theta)^{1/2} d\theta. \tag{3.2.11}$$

A theorem of Legendre connects the complete elliptic integral of the first kind with that of the second kind. Before proving it we state a lemma about the Wronskian of an hypergeometric equation. (For the reference to Legendre's book, where the result appears, see Whittaker and Watson [1940, p. 520].)

Definition 3.2.5 *If y_1 and y_2 are two solutions of a second-order differential equation, then their Wronskian is $W(y_1, y_2) := y_1 y_2' - y_2 y_1'$.*

Lemma 3.2.6 *If y_1 and y_2 are two independent solutions of the hypergeometric equation $y'' + (c - (a + b + 1)x)y' - aby = 0$, then*

$$W(y_1, y_2) = \frac{A}{x^c(1-x)^{a+b-c+1}},$$

where A is a constant.

Proof. Multiply the equation

$$x(1-x)y_2'' + (c - (a+b+1)x)y_2' - aby_2 = 0$$

by y_1 and subtract from it the equation obtained by interchanging y_1 and y_2. The result is

$$x(1-x)\left(y_1 y_2'' - y_2 y_1''\right) + (c - (a+b+1)x)\left(y_1 y_2' - y_2 y_1'\right) = 0$$

or

$$x(1-x)W'(y_1, y_2) + (c - (a+b+1)x)W = 0.$$

Now solve this equation to verify the result in the lemma. ∎

We shall need particular cases of the following two independent solutions of the general hypergeometric equation:

$$y_1 = {}_2F_1\left(\begin{matrix} a, b \\ c \end{matrix}; x\right), \tag{3.2.12}$$

$$y_2 = x^{1-c}(1-x)^{c-a-b}{}_2F_1\left(\begin{matrix} 1-a, 1-b \\ 1-a-b+c \end{matrix}; 1-x\right). \tag{3.2.13}$$

Observe that from (3.2.11)

$$E(k) = \frac{\pi}{2}{}_2F_1\left(\begin{matrix} \frac{1}{2}, -\frac{1}{2} \\ 1 \end{matrix}; k^2\right). \tag{3.2.14}$$

Theorem 3.2.7 $EK' + E'K - KK' = \frac{\pi}{2}$, where $K' := K(k')$, $E' := E(k')$, and $k'^2 = 1 - k^2$.

Proof. Set $x = k^2$ so that $1 - x = k'^2$. The contiguous relation (2.5.9) gives us

$$x(1-x)\frac{dK}{dx} = \frac{1}{2}E - \frac{1}{2}(1-x)K. \tag{3.2.15}$$

Similarly,

$$-x(1-x)\frac{dK'}{dx} = \frac{1}{2}E' - \frac{x}{2}K'. \tag{3.2.16}$$

Multiply (3.2.15) by K' and (3.2.16) by K and add to get

$$EK' + E'K - KK' = 2x(1-x)W(K', K).$$

With $a = b = \frac{1}{2}$ and $c = 1$, Lemma 3.2.6 gives $W(K', K) = A/x(1-x)$. So $EK' + E'K - KK'$ is a constant. An examination of the asymptotic behavior of K as $x \to 1$ shows that the constant must be $(\pi/2)$. This is left to the reader. ∎

It is possible to prove the following more general result of Elliot [1904] in exactly the same way. The proof is left as an exercise. The formula in Theorem 3.2.7 is called Legendre's relation.

Theorem 3.2.8

$$
{}_2F_1\left(\begin{array}{c}\frac{1}{2}+a, -\frac{1}{2}-c\\ a+b+1\end{array}; x\right){}_2F_1\left(\begin{array}{c}\frac{1}{2}-a, c+\frac{1}{2}\\ b+c+1\end{array}; 1-x\right)
$$

$$
+{}_2F_1\left(\begin{array}{c}a+\frac{1}{2}, \frac{1}{2}-c\\ a+b+1\end{array}; x\right){}_2F_1\left(\begin{array}{c}-(a+\frac{1}{2}), c+\frac{1}{2}\\ b+c+1\end{array}; 1-x\right)
$$

$$
-{}_2F_1\left(\begin{array}{c}a+\frac{1}{2}, \frac{1}{2}-c\\ a+b+1\end{array}; x\right){}_2F_1\left(\begin{array}{c}\frac{1}{2}-a, c+\frac{1}{2}\\ b+c+1\end{array}; 1-x\right)
$$

$$
= \frac{\Gamma(a+b+1)\Gamma(b+c+1)}{\Gamma(a+b+c+\frac{3}{2})\Gamma(b+\frac{1}{2})}.
$$

Salamin [1976] and Brent [1976] independently combined Legendre's relation with the arithmetic-geometric mean to find an algorithm for approximating π. We conclude this section with a brief sketch of this application. Some of the details are left for the reader to work out.

Lemma 3.2.9 *If $\{a_n\}$ and $\{b_n\}$ are sequences in Definition 3.2.2, then*

$$
2J_{n+1} - J_n = a_n b_n I_n,
$$

where

$$
J_n := \int_0^{\pi/2} \left(a_n^2 \cos^2 \theta + b_n^2 \sin^2 \theta\right)^{1/2} d\theta,
$$

and I_n has $-\frac{1}{2}$ as the power in the integrand.

Lemma 3.2.10

$$
E(k) = \left(1 - \sum_{n=0}^{\infty} 2^{n-1} c_n^2\right) K(k),
$$

where $c_n^2 = a_n^2 - b_n^2$.

Proof. From (3.2.9) we know that $I_n = I(a, b) =: I$. By Lemma 3.2.9,

$$
2\left(J_{n+1} - a_{n+1}^2 I\right) - \left(J_n - a_n^2 I\right)
$$
$$
= \left(a_n b_n - 2a_{n+1}^2 + a_n^2\right) I
$$
$$
= \frac{1}{2} c_n^2 I.
$$

Rewrite this equation as

$$2^{n+1}\left(J_{n+1} - a_{n+1}^2 I\right) - 2^n\left(J_n - a_n^2 I\right) = 2^{n-1} c_n^2 I$$

and sum it from $n = 0$ to $n = m$ to get

$$J - 2^{m+1}\left(J_{m+1} - a_{m+1}^2 I\right) = \left(a^2 - \sum_{n=0}^m 2^{n-1} c_n^2\right) I, \qquad (3.2.17)$$

where $J := J_0$. Now,

$$2^{m+1}\left(J_{m+1} - a_{m+1}^2 I_{m+1}\right) = 2^{m+1} c_{m+1}^2 \int_0^{\pi/2} \frac{-\sin^2\theta\, d\theta}{\sqrt{a_{m+1}^2 \cos^2\theta + b_{m+1}^2 \sin^2\theta}}.$$

Since c_{m+1}^2 tends to zero quadratically, the last term tends to zero. Let $m \to \infty$ in (3.2.17) and then take $a = 1$ and $b = k'$. The lemma is proved. ∎

Theorem 3.2.11

$$\pi = \frac{M^2(\sqrt{2}, 1)}{1 - \sum_{n=0}^\infty 2^n c_n^2},$$

where $c_n^2 = a_n^2 - b_n^2$ with $a_0 = 1$ and $b_0 = \frac{1}{\sqrt{2}}$.

Proof. Take $k = \frac{1}{\sqrt{2}}$. Then $k' = \frac{1}{\sqrt{2}}$ and Legendre's relation becomes

$$\frac{\pi}{2} = (2E - K)K, \quad \text{where } E = E\left(\frac{1}{\sqrt{2}}\right) \quad \text{and} \quad K = K\left(\frac{1}{\sqrt{2}}\right).$$

This implies

$$\left[1 - \sum_{n=0}^\infty 2^n c_n^2\right] K^2 = \frac{\pi}{2}.$$

Since $K = \pi/(2M(1, \frac{1}{\sqrt{2}}))$, the result follows. ∎

An algorithm based on this theorem has been used to compute millions of digits of π. Define

$$\pi_m := \frac{2 a_{m+1}^2}{1 - \sum_{n=0}^m 2^n c_n^2}.$$

Then π_m increases monotonically to π. Note that $c_n = \sqrt{a_n^2 - b_n^2} = c_{n-1}^2/4a_n$. The a_n and b_n are computed by the arithmetic-geometric mean algorithm. For more information on the computation of π, see Berggren, Borwein, and Borwein [1997].

3.3 Transformations of Balanced Series

In the previous chapter we saw how a general $_2F_1$ transforms under a fractional linear transformation and how to evaluate the sum of the series when $x = 1$. In the case of quadratic transformations, there were restrictions on the parameters. For higher $_{p+1}F_p$, transformations and summation formulas do not exist in general. There are, however, two classes of hypergeometric series for which some results can be obtained.

Definition 3.3.1   *A hypergeometric series*

$$_{p+1}F_p\left(\begin{matrix} a_0, \ldots, a_p \\ b_1, \ldots, b_p \end{matrix}; x\right)$$

is called k-balanced where k is a positive integer, if $x = 1$, if one of the a_is is a negative integer, and if

$$k + \sum_{i=0}^{p} a_i = \sum_{i=1}^{p} b_i.$$

The condition that the series terminates may seem artificial, but without it many results do not hold. The case $k = 1$ is very important, and then the series is called balanced or Saalschützian.

Definition 3.3.2   *If the parameters in the hypergeometric series satisfy the relations*

$$a_0 + 1 = a_1 + b_1 = \cdots = a_p + b_p$$

the series is called well poised. It is nearly poised if all but one of the pairs of parameters have the same sum.

In Section 3.4 we shall give a connection between the two kinds of series considered in Definitions 3.3.1 and 3.3.2. We begin with a study of balanced series. The main theorem of this section is the following result of Whipple, which transforms a balanced $_4F_3$ to another balanced $_4F_3$.

Theorem 3.3.3

$$_4F_3\left(\begin{matrix} -n, a, b, c \\ d, e, f \end{matrix}; 1\right) = \frac{(e-a)_n(f-a)_n}{(e)_n(f)_n}$$

$$\times \,_4F_3\left(\begin{matrix} -n, a, d-b, d-c \\ d, a+1-n-e, a+1-n-f \end{matrix}; 1\right),$$

where

$$a + b + c - n + 1 = d + e + f.$$

Proof.   Start with Euler's transformation:

$$_2F_1\left(\begin{matrix} a, b \\ c \end{matrix}; x\right) = (1 - x)^{c-a-b} {}_2F_1\left(\begin{matrix} c - a, c - b \\ c \end{matrix}; x\right).$$

Rewrite this with different parameters:

$$(1 - x)^{f-d-e} {}_2F_1\left(\begin{matrix} f - d, f - e \\ f \end{matrix}; x\right) = {}_2F_1\left(\begin{matrix} d, e \\ f \end{matrix}; x\right).$$

Suppose $c - a - b = f - d - e$ and multiply the two identities to get

$$_2F_1\left(\begin{matrix} a, b \\ c \end{matrix}; x\right) {}_2F_1\left(\begin{matrix} f - d, f - e \\ f \end{matrix}; x\right) = {}_2F_1\left(\begin{matrix} c - a, c - b \\ c \end{matrix}; x\right) {}_2F_1\left(\begin{matrix} d, e \\ f \end{matrix}; x\right).$$

The coefficient of x^n on the left side is

$$\sum_{k=0}^{n} \frac{(a)_k (b)_k (f - d)_{n-k} (f - e)_{n-k}}{(c)_k k! (f)_{n-k} (n - k)!}.$$

This expression can be rewritten as

$$\frac{(f - d)_n (f - e)_n}{n! (f)_n} {}_4F_3\left(\begin{matrix} a, b, 1 - f - n, -n \\ c, d - f - n + 1, e - f - n + 1 \end{matrix}; 1\right).$$

Equating this to the coefficient of x^n on the right side, we obtain

$$_4F_3\left(\begin{matrix} -n, a, b, -f - n + 1 \\ c, d - f - n + 1, e - f - n + 1 \end{matrix}; 1\right)$$
$$= \frac{(d)_n (e)_n}{(f - d)_n (f - e)_n} {}_4F_3\left(\begin{matrix} -n, c - a, c - b, 1 - f - n \\ c, 1 - d - n, 1 - e - n \end{matrix}; 1\right).$$

This is equivalent to the statement of the theorem. This result is due to Whipple [1926]. For a different proof see Remark 3.4.2.   ∎

The next result was given by Sheppard [1912].

Corollary 3.3.4

$$_3F_2\left(\begin{matrix} -n, a, b \\ d, e \end{matrix}; 1\right) = \frac{(d - a)_n (e - a)_n}{(d)_n (e)_n} {}_3F_2\left(\begin{matrix} -n, a, a + b - n - d - e + 1 \\ a - n - d + 1, a - n - e + 1 \end{matrix}; 1\right).$$

Proof.   Let $f \to \infty$ but keep $f - c$ fixed in Theorem 3.3.3 so that

$$_4F_3\left(\begin{matrix} -n, a, b, c \\ d, e, f \end{matrix}; 1\right) \to {}_3F_2\left(\begin{matrix} -n, a, b \\ d, e \end{matrix}; 1\right).$$

A similar change takes place on the right side and the end result is

$$_3F_2\left(\begin{matrix} -n, a, b \\ d, e \end{matrix}; 1\right) = \frac{(e-a)_n}{(e)_n} {}_3F_2\left(\begin{matrix} -n, a, d-b \\ d, a+1-n-e \end{matrix}; 1\right).$$

Sheppard's transformation is obtained by applying this transformation to itself. The corollary is proved. ■

The formula in Corollary 3.3.4 has some interesting special cases. For example, suppose the left side is k-balanced, that is,

$$d + e = k - n + a + b.$$

Then the right side is a sum of k terms. In particular, $k = 1$ gives back the Pfaff–Saalschütz identity.

Corollary 3.3.5

$$_3F_2\left(\begin{matrix} a, b, c \\ d, e \end{matrix}; 1\right) = \frac{\Gamma(e)\Gamma(d+e-a-b-c)}{\Gamma(e-a)\Gamma(d+e-b-c)} {}_3F_2\left(\begin{matrix} a, d-b, d-c \\ d, d+e-b-c \end{matrix}; 1\right)$$

when the two series converge.

Proof. Let $n \to \infty$ and keep $f + n$ fixed. Since the number of terms of the series tends to infinity, Tannery's theorem may be used to justify the calculation. The left side in Theorem 3.3.3 becomes

$$_3F_2\left(\begin{matrix} a, b, c \\ d, e \end{matrix}; 1\right).$$

To find the right side, write

$$\frac{(e-a)_n(f-a)_n}{(e)_n(f)_n} = \frac{\Gamma(e-a+n)\Gamma(f-a+n)}{\Gamma(e-a)\Gamma(f-a)} \cdot \frac{\Gamma(e)\Gamma(f)}{\Gamma(e+n)\Gamma(f+n)}$$

$$= \frac{\Gamma(e)}{\Gamma(e-a)} \frac{\Gamma(a-f+1)\Gamma(-n-f+1)}{\Gamma(a-f-n+1)\Gamma(-f+1)} \cdot \frac{\Gamma(e-a+n)}{\Gamma(e+n)},$$

where Euler's reflection formula was used to derive the second equality. Recall that $1 - f - n = d + e - a - b - c$. So

$$\frac{(e-a)_n(f-a)_n}{(e)_n(f)_n} = \frac{\Gamma(e)\Gamma(d+e-a-b-c)}{\Gamma(e-a)\Gamma(d+e-b-c)} \cdot \left(\frac{\Gamma(a-f+1)\Gamma(e-a+n)}{\Gamma(-f+1)\Gamma(e+n)}\right).$$

As $n \to \infty$ and $n + f$ is fixed, we have $-f \to \infty$ and the expression in parentheses equals 1 in the limit. The corollary follows. ■

Corollary 3.3.5 was given by Kummer [1836]. If we apply Kummer's transformation to itself we get a theorem of Thomae [1879]:

Corollary 3.3.6

$$_3F_2\left(\begin{matrix} a, b, c \\ d, e \end{matrix}; 1\right) = \frac{\Gamma(d)\Gamma(e)\Gamma(s)}{\Gamma(a)\Gamma(s+b)\Gamma(s+c)}{}_3F_2\left(\begin{matrix} d-a, e-a, s \\ s+b, s+c \end{matrix}; 1\right),$$

where $s = d + e - a - b - c$.

3.4 Whipple's Transformation

The main result of this section is an important formula of Whipple [1926] that connects a terminating well-poised $_7F_6$ with a balanced $_4F_3$. We prove it by a method of Bailey, which requires that we know the value of a general well-poised $_3F_2$ at $x = 1$. In Chapter 2, we showed that the latter result, known as Dixon's theorem, is a consequence of Dougall's theorem. See Remark 2.2.2 in Chapter 2. Because Dougall's theorem is itself a corollary of Whipple's transformation, it would be nice if we had a direct proof of Dixon's formula, that is, one that does not use Dougall's formula. Several such proofs are known. We give one that follows from a quadratic transformation given in Section 3.1.

Theorem 3.4.1

$$_3F_2\left(\begin{matrix} a, -b, -c \\ 1+a+b, 1+a+c \end{matrix}; 1\right)$$

$$= \frac{\Gamma((a/2)+1)\Gamma(a+b+1)\Gamma(a+c+1)\Gamma((a/2)+b+c+1)}{\Gamma(a+1)\Gamma((a/2)+b+1)\Gamma((a/2)+c+1)\Gamma(a+b+c+1)}.$$

Proof. If $a = -n$, a negative integer, in the quadratic transformation (3.1.15), then both sides of the equation are polynomials in x. Take $x = 1$. If a is an even negative integer, then we get

$$_3F_2\left(\begin{matrix} -2n, b, c \\ 1-2n-b, 1-2n-c \end{matrix}; 1\right) = \frac{(2n)!(b+c+n)_n}{(b+n)_n(c+n)_n n!}.$$

If a is an odd negative integer, then

$$_3F_2\left(\begin{matrix} -2n-1, b, c \\ -2n-b, -2n-c \end{matrix}; 1\right) = 0.$$

Thus Theorem 3.4.1 is verified when a is a negative integer. Now suppose that c is a positive integer and a is arbitrary. In this case both sides are rational functions of a and the identity is true for an infinite number of values of a. Thus, we have shown that the identity holds if c is an integer and a and b are arbitrary. In the general case, for Re $c > $ Re$(-a/2 - b - 1)$, both sides of the identity are bounded analytic functions of c and equal for $c = 1, 2, 3, \dots$. By Carlson's theorem the result is proved. ∎

Kummer's identity, which gives the value of a well-poised $_2F_1$ at $x = -1$, is a corollary of Dixon's theorem. To see this, let $c \to \infty$.

Remark 3.4.1   We noted earlier that the balanced identities in their simplest form come from the factorization

$$(1-x)^{-a}(1-x)^{-b} = (1-x)^{-a-b}.$$

In a similar sense, the well-poised series comes from

$$(1-x)^{-b}(1+x)^{-b} = (1-x^2)^{-b}.$$

Equate the coefficient of x^{2n} from both sides to get

$$\sum_{k=0}^{2n}(-1)^k \frac{(b)_k(b)_{2n-k}}{k!(2n-k)!} = \frac{(b)_n}{n!}$$

or

$$_2F_1\left(\begin{array}{c} -2n, b \\ 1-2n-b \end{array}; -1\right) = \frac{(b)_n(2n)!}{n!(b)_{2n}}.$$

This is Kummer's identity for

$$_2F_1\left(\begin{array}{c} a, b; -1 \\ a-b+1 \end{array}\right)$$

when a is a negative even integer. As in the proof of Theorem 3.4.1, we can now obtain the general result of Kummer. The $_2F_1$ result was so special that Kummer failed to realize that series of a similar nature could be studied at the $_3F_2$ and higher levels. This is not surprising for well-poised series are not on the surface.

We also need the following lemma to prove Whipple's theorem. It is proved by Bailey's [1935] method mentioned at the beginning of this section.

Lemma 3.4.2

$$_5F_4\left(\begin{array}{c} a, b, c, d, -m \\ a-b+1, a-c+1, a-d+1, a+m+1 \end{array}; 1\right)$$

$$= \frac{(a+1)_m((a/2)-d+1)_m}{((a/2)+1)_m(a-d+1)_m} {}_4F_3\left(\begin{array}{c} (a/2), a-b-c+1, d, -m \\ a-b+1, a-c+1, d-m-a/2 \end{array}; 1\right).$$

Proof.   By the Pfaff–Saalschütz identity for a balanced $_3F_2$ we have

$$\sum_{r=0}^{n} \frac{(-n)_r(a-b-c+1)_r(a+n)_r}{r!(a-b+1)_r(a-c+1)_r} = \frac{(b)_n(c)_n}{(a-b+1)_n(a-c+1)_n}.$$

So

$$
{}_5F_4\left(\begin{array}{c} a, b, c, d, -m \\ a-b+1, a-c+1, a-d+1, a+m+1 \end{array}; 1\right)
$$

$$
= \sum_{n=0}^{m} \frac{(a)_n (d)_n (-m)_n}{n!(a-d+1)_n (a+m+1)_n} \sum_{r=0}^{n} \frac{(-n)_r (a-b-c+1)_r (a+n)_r}{r!(a-b+1)_r (a-c+1)_r}
$$

$$
= \sum_{r=0}^{m} \sum_{n=r}^{m} \frac{(-1)^r (a)_{n+r} (d)_n (-m)_n (a-b-c+1)_r}{(n-r)! r!(a-b+1)_r (a-c+1)_r (a-d+1)_n (a+m+1)_n}
$$

$$
\text{(set } t = n - r),
$$

$$
= \sum_{r=0}^{m} \sum_{t=0}^{m-r} \frac{(a)_{t+2r} (d)_{t+r} (-m)_{t+r} (a-b-c+1)_r (-1)^r}{t! r!(a-b+1)_r (a-c+1)_r (a-d+1)_{t+r} (a+m+1)_{t+r}}
$$

$$
= \sum_{r=0}^{m} \frac{(a)_{2r} (d)_r (-m)_r (a-b-c+1)_r (-1)^r}{r!(a-b+1)_r (a-c+1)_r (a-d+1)_r (a+m+1)_r}
$$

$$
\cdot \sum_{t=0}^{m-r} \frac{(a+2r)_t (d+r)_t (-m+r)_t}{t!(a-d+r+1)_t (a+m+r+1)_t}.
$$

The inner sum can be computed by Dixon's identity (Theorem 3.4.1). An easy calculation then gives the required relation. ∎

Lemma 3.4.2 transforms a terminating well-poised ${}_5F_4$ to a balanced ${}_4F_3$.

Corollary 3.4.3

$$
{}_5F_4\left(\begin{array}{c} a, (a/2)+1, c, d, -m \\ a/2, a-c+1, a-d+1, a+m+1 \end{array}; 1\right) = \frac{(a+1)_m (a-c-d+1)_m}{(a-c+1)_m (a-d+1)_m}.
$$

Proof. Take $b = (a/2) + 1$ in Lemma 3.4.2. The ${}_4F_3$ reduces to a balanced ${}_3F_2$. ∎

Theorem 3.4.4

$$
{}_7F_6\left(\begin{array}{c} a, (a/2)+1, b, c, d, e, -m \\ a/2, a-b+1, a-c+1, a-d+1, a-e+1, a+m+1 \end{array}; 1\right)
$$

$$
= \frac{(a+1)_m (a-d-e+1)_m}{(a-d+1)_m (a-e+1)_m} \, {}_4F_3\left(\begin{array}{c} a-b-c+1, d, e, -m \\ a-b+1, a-c+1, d+e-a-m \end{array}; 1\right).
$$

Proof. The proof of this theorem is exactly the same as that of Lemma 3.4.2 except that one uses Corollary 3.4.3 instead of Dixon's theorem. Thus

$$
{}_7F_6\left(\begin{matrix} a, (a/2)+1, b, c, d, e, -m \\ a/2, a-b+1, a-c+1, a-d+1, a-e+1, a+m+1 \end{matrix}; 1\right)
$$

$$
= \sum_{n=0}^{m} \frac{(a)_n((a/2)+1)_n(d)_n(e)_n(-m)_n}{n!((a/2))_n(a-d+1)_n(a-e+1)_n(a+m+1)_n}
$$

$$
\cdot \sum_{r=0}^{n} \frac{(-n)_r(a-b-c+1)_r(a+n)_r}{r!(a-b+1)_r(a-c+1)_r}.
$$

After a calculation similar to the one in Lemma 3.4.2, this sum equals

$$
\sum_{r=0}^{m} \frac{(a)_{2r}((a/2)+1)_r(d)_r(e)_r(-m)_r}{r!(a-b+1)_r(a-c+1)_r(a-d+1)_r(a/2)_r(a-e+1)_r(a+m+1)_r}
$$

$$
\cdot \sum_{t=0}^{m-r} \frac{(a+2r)_t((a/2)+r+1)_t(d+r)_t(e+r)_t(-m+r)_t}{t!((a/2)+r)_t(a-d+r+1)_t(a-e+r+1)_t(a+m+r+1)_t}.
$$

The inner sum can be evaluated by Corollary 3.4.3 and the result follows after a straightforward calculation. ∎

Since the two elementary identities in Remark 3.4.1 are not related, we see the very surprising nature of Whipple's identity. In a later chapter we give a more natural proof of Whipple's theorem as a consequence of some properties of Jacobi polynomials. We refer to the above ${}_7F_6$ as a very well poised ${}_7F_6$. The word "very" refers to the factor

$$
\frac{((a/2)+1)_k}{(a/2)_k} = \frac{a+2k}{a}.
$$

The ${}_5F_4$ in Corollary 3.4.3 is also very well poised.

Remark 3.4.2 Theorem 3.3.2 is a particular case of Theorem 3.4.4 because of the symmetry in the parameters b, c, d, e in the ${}_7F_6$.

Whipple also stated a more general form of Theorem 3.4.4.

Theorem 3.4.5

$$
{}_7F_6\left(\begin{matrix} a, (a/2)+1, b, c, d, e, f \\ a/2, a-b+1, a-c+1, a-d+1, a-e+1, a-f+1 \end{matrix}; 1\right)
$$

$$
= \frac{\Gamma(a-d+1)\Gamma(a-e+1)\Gamma(a-f+1)\Gamma(a-d-e-f+1)}{\Gamma(a+1)\Gamma(a-e-f+1)\Gamma(a-d-e+1)\Gamma(a-d-f+1)}
$$

$$
\cdot {}_4F_3\left(\begin{matrix} a-b-c+1, d, e, f \\ a-b+1, a-c+1, d+e+f-a \end{matrix}; 1\right),
$$

provided the series on the right side terminates and the one on the left converges.

Proof. This is a consequence of Carlson's theorem and Theorem 3.4.4. The reader should work out the details or see Bailey [1935, p. 40]. ∎

Theorem 3.4.6

$$
{}_6F_5\left(\begin{matrix} a, (a/2)+1, b, c, d, e \\ a/2, a-b+1, a-c+1, a-d+1, a-e+1 \end{matrix}; -1\right)
$$
$$
= \frac{\Gamma(a-d+1)\Gamma(a-e+1)}{\Gamma(a+1)\Gamma(a-d-e+1)} {}_3F_2\left(\begin{matrix} a-b-c+1, d, e \\ a-b+1, a-c+1 \end{matrix}; 1\right).
$$

Proof. Let $m \to \infty$ in Theorem 3.4.4 to prove the result. ∎

Observe that Theorem 3.4.6 connects a general ${}_3F_2$ at $x = 1$ with a very well poised ${}_6F_5$ at $x = -1$.

3.5 Dougall's Formula and Hypergeometric Identities

Set $2a + 1 = b + c + d + e - m$ in Theorem 3.4.4. The ${}_4F_3$ reduces to a balanced ${}_3F_2$, which can be summed. The result is Dougall's formula.

Theorem 3.5.1

$$
{}_7F_6\left(\begin{matrix} a, (a/2)+1, b, c, d, e, -m \\ a/2, a-b+1, a-c+1, a-d+1, a-e+1, a+m+1 \end{matrix}; 1\right)
$$
$$
= \frac{(a+1)_m (a-b-c+1)_m (a-b-d+1)_m (a-c-d+1)_m}{(a-b+1)_m (a-c+1)_m (a-d+1)_m (a-b-c-d+1)_m},
$$

when $2a + 1 = b + c + d + e - m$. This formula sums a 2-balanced very well poised ${}_7F_6$.

In the following identities, convergence conditions need to be imposed. They are not explicitly stated, since they are easy to work out in each case.

Corollary 3.5.2

$$
{}_5F_4\left(\begin{matrix} a, (a/2)+1, c, d, e \\ a/2, a-c+1, a-d+1, a-e+1 \end{matrix}; 1\right)
$$
$$
= \frac{\Gamma(a-c+1)\Gamma(a-d+1)\Gamma(a-e+1)\Gamma(a-c-d-e+1)}{\Gamma(a+1)\Gamma(a-d-e+1)\Gamma(a-c-e+1)\Gamma(a-c-d+1)}.
$$

Proof. Substitute $b = 2a - c - d - e + m + 1$ in Theorem 3.5.1 and let $m \to \infty$. This procedure may be justified by Tannery's theorem. The corollary follows. ∎

One may also derive this corollary from Corollary 3.4.3 of the previous section by an application of Carlson's theorem. Dixon's formula follows from Corollary 3.5.2 by taking $e = a/2$. The next corollary gives the value of a very well poised $_4F_3$ at $x = -1$.

Corollary 3.5.3

$$_4F_3\left(\begin{matrix} a, (a/2) + 1, c, d \\ a/2, a - c + 1, a - d + 1 \end{matrix}; -1\right) = \frac{\Gamma(a - c + 1)\Gamma(a - d + 1)}{\Gamma(a + 1)\Gamma(a - c - d + 1)}.$$

Proof. Let $e \to -\infty$ in Corollary 3.5.2 and the result follows; or else take $b + c = a + 1$ in Theorem 3.4.6. ∎

Here are a few more summation formulas.

Theorem 3.5.4

(i) $$_2F_1\left(\begin{matrix} a, b \\ (a + b + 1)/2 \end{matrix}; 1/2\right) = \frac{\Gamma(1/2)\Gamma((a + b + 1)/2)}{\Gamma((a + 1)/2)\Gamma((b + 1)/2)}.$$

(ii) $$_2F_1\left(\begin{matrix} a, 1 - a \\ c \end{matrix}; 1/2\right) = \frac{\Gamma(c/2)\Gamma((c + 1)/2)}{\Gamma((c + a)/2)\Gamma((c - a + 1)/2)}.$$

Proof. Let $x \to -1$ in Pfaff's transformation (Theorem 2.2.5),

$$_2F_1\left(\begin{matrix} a, c - b \\ c \end{matrix}; x\right) = (1 - x)^{-a}{}_2F_1\left(\begin{matrix} a, b; \\ c \end{matrix}; x/(x - 1)\right),$$

to get

$$_2F_1\left(\begin{matrix} a, c - b \\ c \end{matrix}; -1\right) = 2^{-a}{}_2F_1\left(\begin{matrix} a, b \\ c \end{matrix}; 1/2\right).$$

There are two ways in which the series on the left becomes well poised so that it can be summed by Kummer's identity (Corollary 3.1.2). These cases are (i) $2c - b = a + 1$ and (ii) $a + c = c - b + 1$ or $a + b = 1$. The two parts of the theorem follow immediately. ∎

Theorem 3.5.5

(i) $$_3F_2\left(\begin{matrix} a, b, c \\ (a + b + 1)/2, 2c \end{matrix}; 1\right)$$

$$= \frac{\Gamma(1/2)\Gamma(c + (1/2))\Gamma((a + b + 1)/2)\Gamma(c - (a + b - 1)/2)}{\Gamma((a + 1)/2)\Gamma((b + 1)/2)\Gamma(c - (a - 1)/2)\Gamma(c - (b - 1)/2)}.$$

(ii) $3F_2\left(\begin{matrix} a, b, c \\ e, f \end{matrix}; 1\right)$

$$= \frac{\pi\Gamma(e)\Gamma(f)}{2^{2c-1}\Gamma((a+e)/2)\Gamma((a+f)/2)\Gamma((b+e)/2)\Gamma((b+f)/2)},$$

when $a + b = 1$ and $e + f = 2c + 1$.

Proof. These results follow from Thomae's formula (Corollary 3.3.6) by choosing parameters appropriately. To obtain (i), choose the parameters so that the right side becomes well poised. Thus $d = (a + b + 1)/2$ and $e = 2c$ and Thomae's formula gives

$$3F_2\left(\begin{matrix} a, b, c \\ (a+b+1)/2, 2c \end{matrix}; 1\right) = \frac{\Gamma((a+b+1)/2)\Gamma(2c)\Gamma(c - (a+b-1)/2)}{\Gamma(a)\Gamma(c - (a-b-1)/2)\Gamma(2c - (a+b-1)/2)}$$

$$\cdot 3F_2\left(\begin{matrix} 2c - a, (b-a+1)/2, c - (a+b-1)/2 \\ 2c - (a+b-1)/2, c - (a-b-1)/2 \end{matrix}; 1\right).$$

Now apply Dixon's identity to get (i). Note that we require $c > (a + b - 1)/2$. The $3F_2$ also exists without this condition when $c = -n$, a negative integer, if the series is taken to terminate with $n + 1$ terms. The value of this series can be found, but it is not what one gets by letting c tend to $-n$.

(ii) To prove this identity choose the parameters so that the right side is the $3F_2$ given by (i). Thus take $a + b = 1$ and $e + f = 2c + 1$ to get

$$3F_2\left(\begin{matrix} a, b, c \\ e, f \end{matrix}; 1\right) = \frac{\Gamma(e)\Gamma(f)\Gamma(c)}{\Gamma(a)\Gamma(b+c)\Gamma(2c)} 3F_2\left(\begin{matrix} e - a, f - a, c \\ b + c, 2c \end{matrix}; 1\right).$$

An application of (i) at this point gives (ii). ∎

Note that Theorem 3.5.4(i) and (ii) are limiting cases of Theorem 3.5.5(i) and (ii) respectively. Let $c \to \infty$ to see this. Part (i) of the last theorem is due to Watson [1925], who proved it for the case where $a = -n$, a negative integer. Watson's theorem can also be obtained by equating the coefficient of x^n on each side of the quadratic transformation (3.1.11). The general case then follows by Carlson's theorem. Another way is to multiply Equation (3.1.3) by $(x - x^2)^{c-1}$ and integrate over $(0, 1)$. This only works in the terminating case.

Remark We end this section with the following comments on well-poised series. Let

$$f(x) = {}_{q+1}F_q\left(\begin{matrix} -n, a_1, a_2, \ldots, a_q \\ 1 - n - a_1, \ldots, 1 - n - a_q \end{matrix}; -x\right).$$

Then the polynomial $f(x)$ satisfies the relation

$$f(x) = (-1)^{qn} x^n f(1/x).$$

A polynomial $g(x)$ that satisfies $g(x) = x^n g(1/x)$ is called a reciprocal polynomial of degree n. These polynomials have the form

$$g(x) = a_0 + a_1 x + a_2 x^2 + \cdots + a_2 x^{n-2} + a_1 x^{n-1} + a_0 x^n.$$

Note that $f(x)$ is reciprocal if either q or n is even. It is easy to check that if g is the reciprocal polynomial given above then

$$g(x) = \bar{a}_0(1+x)^n + \bar{a}_1 x(1+x)^{n-2} + \cdots + \bar{a}_\nu x^\nu (1+x)^{n-2\nu} \quad (3.5.1)$$

for some $\bar{a}_0, \bar{a}_1, \ldots, \bar{a}_\nu$ which can be defined in terms of $a_0, a_1, \ldots$ Here $\nu = [n/2]$. It can be shown that

$$a_j = \sum_{0 \le 2r \le n} \binom{n-2r}{j-r} \bar{a}_r \quad (3.5.2)$$

and

$$\bar{a}_j = \sum_{0 \le i \le j} (-1)^{j+i} \binom{n-j-i}{j-i} \frac{n-2i}{n-j-i} a_i. \quad (3.5.3)$$

Note that (3.5.1) can also be written as

$$g(x) = (1+x)^n \sum_{k=0}^{\nu} \bar{a}_k \frac{x^k}{(1+x)^{2k}}. \quad (3.5.4)$$

Observe the connection of (3.5.4) with the quadratic transformations (3.1.4) and (3.1.15).

3.6 Integral Analogs of Hypergeometric Sums

In Chapter 2 we saw two integrals of Barnes that were continuous analogs of Gauss's $_2F_1$ identity and the Pfaff–Saalschütz $_3F_2$ identity. There are other Barnes-type integrals that are analogs of the higher $_pF_q$ sums we have considered in this chapter. The path of integration in the integrals will be parallel to the imaginary axis but suitably deformed so that the increasing sequence of poles of the integrand is separated by the contour from the decreasing sequence of poles.

The following theorem of Bailey is an analog of Corollary 3.5.2, which sums a very well poised $_5F_4$ at $x = 1$. See Bailey [1935, p. 47].

Theorem 3.6.1

$$\frac{1}{2\pi i} \int \frac{\Gamma(a+s)\Gamma((a/2)+1+s)\Gamma(b+s)\Gamma(c+s)\Gamma(d+s)\Gamma(b-a-s)\Gamma(-s)ds}{\Gamma((a/2)+s)\Gamma(a-c+1+s)\Gamma(a-d+1+s)}$$

$$= \frac{\Gamma(b)\Gamma(c)\Gamma(d)\Gamma(b+c-a)\Gamma(b+d-a)}{2\Gamma(a-c-d+1)\Gamma(b+c+d-a)}.$$

Proof. The proof is similar to that of Theorem 2.4.2. The reader should fill in the details. The residues at the poles of $\Gamma(b-a-s)\Gamma(-s)$ to the right of contour give the integral as the sum of two very well poised $_5F_4$. These can be summed by Corollary 3.5.2. The result follows. ∎

A different and useful form of Theorem 3.6.1 was given by Wilson [1978]. We note it here.

Theorem 3.6.2

$$\frac{1}{2\pi i} \int \frac{\Gamma(a+s)\Gamma(a-s)\Gamma(b+s)\Gamma(b-s)\Gamma(c+s)\Gamma(c-s)\Gamma(d+s)\Gamma(d-s)}{\Gamma(2s)\Gamma(-2s)} ds$$

$$= \frac{2\Gamma(a+b)\Gamma(a+c)\Gamma(a+d)\Gamma(b+c)\Gamma(b+d)\Gamma(c+d)}{\Gamma(a+b+c+d)}.$$

Here the contour is along the imaginary axis but suitably deformed. As always, we are assuming that a, b, c, d are such that this can be done.

Proof. In Theorem 3.6.1 replace a with $2a$; b, c, d with $b+a, c+a$, and $d+a$ respectively; and s with $s-a$. We get

$$\frac{1}{2\pi i} \int \frac{\Gamma(a+s)\Gamma(s+1)\Gamma(b+s)\Gamma(c+s)\Gamma(d+s)\Gamma(b-s)\Gamma(a-s)}{\Gamma(s)\Gamma(1-c+s)\Gamma(1-d+s)} ds$$

$$= \frac{\Gamma(a+b)\Gamma(a+c)\Gamma(a+d)\Gamma(b+c)\Gamma(b+d)}{2\Gamma(1-c-d)\Gamma(a+b+c+d)}.$$

Use Euler's reflection formula to rewrite this as

$$\frac{1}{2\pi i} \int \frac{\Gamma(a+s)\Gamma(b+s)\Gamma(c+s)\Gamma(d+s)\Gamma(a-s)\Gamma(b-s)\Gamma(c-s)\Gamma(d-s)}{\Gamma(2s)\Gamma(-2s)}$$

$$\cdot \frac{\sin(c-s)\pi \sin(d-s)\pi}{-\sin s\pi \cos s\pi \sin(c+d)\pi} ds$$

$$= \frac{2\Gamma(a+b)\Gamma(a+c)\Gamma(a+d)\Gamma(b+c)\Gamma(b+d)\Gamma(c+d)}{\Gamma(a+b+c+d)}.$$

The trigonometric expression in the integrand is

$$1 - \frac{\sin c\pi \sin d\pi \cos^2 s\pi + \cos c\pi \cos d\pi \sin^2 s\pi}{\sin(c+d)\pi \sin s\pi \cos s\pi}.$$

Now observe that the second term involving the trigonometric functions changes sign when s is changed to $-s$. Hence that part of the integral vanishes and the theorem is proved. ∎

If a, b, c, d are positive or $a = \bar{b}$ and/or $c = \bar{d}$ and the real parts are positive, then we can write Wilson's formula as

$$\frac{1}{2\pi} \int_0^\infty \left| \frac{\Gamma(a+ix)\Gamma(b+ix)\Gamma(c+ix)\Gamma(d+ix)}{\Gamma(2ix)} \right|^2 dx$$

$$= \frac{\Gamma(a+b)\Gamma(a+c)\Gamma(a+d)\Gamma(b+c)\Gamma(b+d)\Gamma(c+d)}{\Gamma(a+b+c+d)}.$$

The result in (3.6.1) was given also by de Branges [1972, 1972a]. Formula (3.6.1) continues to hold when one of the parameters is zero. The following corollary is the result of letting $d \to \infty$:

$$\frac{1}{2\pi} \int_0^\infty \left| \frac{\Gamma(a+ix)\Gamma(b+ix)\Gamma(c+ix)}{\Gamma(2ix)} \right|^2 dx = \Gamma(a+b)\Gamma(a+c)\Gamma(b+c).$$
$$(3.6.1)$$

There is also an analog of the Dougall $_7F_6$ formula. This is the next theorem, also given by Bailey.

Theorem 3.6.3

$$\frac{1}{2\pi i} \int \frac{\Gamma(a+s)\Gamma((a/2)+1+s)\Gamma(b+s)\Gamma(c+s)\Gamma(d+s)}{\Gamma((a/2)+s)\Gamma(a-c+1+s)\Gamma(a-d+1+s)}$$

$$\cdot \frac{\Gamma(e+s)\Gamma(f+s)\Gamma(b-a-s)\Gamma(-s)}{\Gamma(a-e+1+s)\Gamma(a-f+1+s)} ds$$

$$= \frac{\Gamma(b)\Gamma(c)\Gamma(d)\Gamma(e)\Gamma(f)\Gamma(b+c-a)}{2\Gamma(a-d-e+1)\Gamma(a-c-e+1)\Gamma(a-c-d+1)}$$

$$\cdot \frac{\Gamma(b+d-a)\Gamma(b+e-a)\Gamma(b+f-a)}{\Gamma(a-c-f+1)\Gamma(a-d-f+1)\Gamma(a-e-f+1)}, \quad (3.6.2)$$

when $2a + 1 = b + c + d + e + f$.

Proof.   It is not possible to evaluate this integral as in Theorem 3.6.1. The two $_7F_6$ series cannot be summed by Dougall's formula unless they terminate. However, if they terminate, a contour separating the increasing and decreasing sequences of poles cannot be constructed. It is possible, nonetheless, to give a proof that is the integral analog of the proof of Theorem 3.4.4. Start with the formula in

Theorem 2.4.3 in the following form:

$$\frac{\Gamma(d+s)\Gamma(e+s)\Gamma(f+s)}{\Gamma(a-d+1+s)\Gamma(a-e+1+s)\Gamma(a-f+1+s)}$$

$$= \frac{1}{\Gamma(a-d-e+1)\Gamma(a-d-f+1)\Gamma(a-e-f+1)}$$

$$\cdot \frac{1}{2\pi i} \int \frac{\Gamma(d+t)\Gamma(e+t)\Gamma(f+t)\Gamma(a-d-e-f+1-t)\Gamma(s-t)}{\Gamma(a+1+s+t)} dt.$$

$$(3.6.3)$$

The left side of (3.6.4) is a part of the integrand in (3.6.3). Substitute this in (3.6.3) to get

$$\frac{1}{2\pi i} \int \frac{\Gamma(d+t)\Gamma(e+t)\Gamma(f+t)\Gamma(a-d-e-f+1-t)}{\Gamma(a-d-e+1)\Gamma(a-d-f+1)\Gamma(a-e-f+1)}$$

$$\cdot \frac{1}{2\pi i} \int \frac{\Gamma(a+s)\Gamma((a/2)+1+s)\Gamma(b+s)\Gamma(c+s)}{\Gamma((a/2)+s)\Gamma(a-c+1+s)}$$

$$\cdot \frac{\Gamma(b-a-s)\Gamma(s-t)\Gamma(-s)}{\Gamma(a+1+s+t)} ds dt.$$

Evaluate the inner integral by Theorem 3.6.1. The resulting integral then reduces to 1, which can be computed by (3.6.4) when $2a+1 = b+c+d+e+f$. The theorem follows. ∎

An analog of Dixon's well-poised $_3F_2$ can also be derived. Take $d = a/2$ in Theorem 3.6.1 to get the required result:

$$\frac{1}{2\pi i} \int \frac{\Gamma(a+s)\Gamma(b+s)\Gamma(c+s)\Gamma(b-a-s)\Gamma(-s)}{\Gamma(a-c+1+s)} ds$$

$$= \frac{\Gamma(b)\Gamma(c)\Gamma(a/2)\Gamma(b+c-a)\Gamma(b-(a/2))}{2\Gamma((a/2)-c+1)\Gamma(b+c-(a/2))}$$

$$(3.6.4)$$

To obtain a more symmetrical form, replace a with $2a$ and b, c with $b+a, c+a$ respectively and s with $s-a$ to get

$$\frac{1}{2\pi i} \int \frac{\Gamma(a+s)\Gamma(b+s)\Gamma(c+s)\Gamma(b-s)\Gamma(a-s)}{\Gamma(1-c+s)} ds$$

$$= \frac{\Gamma(a)\Gamma(b)\Gamma(a+b)\Gamma(a+c)\Gamma(b+c)}{2\Gamma(1-c)\Gamma(a+b+c)}.$$

$$(3.6.5)$$

Now apply the reflection formula and do the simplification (as employed in the reduction from Theorem 3.6.1 to Wilson's integral) to get

$$\frac{1}{2\pi i} \int \Gamma(a+s)\Gamma(b+s)\Gamma(c+s)\Gamma(a-s)\Gamma(b-s)\Gamma(c-s)\cos s\pi ds$$

$$= \frac{\Gamma(a)\Gamma(b)\Gamma(c)\Gamma(a+b)\Gamma(a+c)\Gamma(b+c)}{2\Gamma(a+b+c)}. \tag{3.6.6}$$

If a, b, and c are positive, we can write the formula as

$$\frac{2}{\pi} \int_0^\infty |\Gamma(a+ix)\Gamma(b+ix)\Gamma(c+ix)|^2 \cosh \pi x dx$$

$$= \frac{\Gamma(a)\Gamma(b)\Gamma(c)\Gamma(a+b)\Gamma(a+c)\Gamma(b+c)}{\Gamma(a+b+c)}. \tag{3.6.7}$$

3.7 Contiguous Relations

In the previous chapter we gave the three-term contiguous relations of Gauss for the $_2F_1$ functions. More generally, there are $(q+2)$-term relations for $_pF_q$s with $p \leq q+2$. Under certain conditions, these relations become three-term relations. Kummer observed that it was possible to obtain such relations for the $_3F_2$s when $x=1$. Bailey [1954] gave a procedure using the differential equation satisfied by the $_3F_2$ to produce these relations. A simpler method was given by Wilson [1978]. This applies to more general $_pF_q$s. In this section we use Wilson's method to derive his results on the three-term contiguous relations for balanced $_4F_3$s. These contain the three-term recurrence relations for a set of orthogonal polynomials due to Wilson, which contain the "classical" orthogonal polynomials as special cases.

Before describing Wilson's method, we note that it is possible to obtain three-term relations for the $_3F_2$s from the contiguous relations for the $_2F_1$s by integration. For example, multiply the equation

$$(b-a)F + aF(a+) - bF(b+) = 0$$

by $x^{d-1}(1-x)^{e-d-1}$ and integrate over $(0, 1)$ to get

$$(b-a)_3F_2\left(\begin{matrix} a, b, d \\ c, e \end{matrix}; 1\right) + a_3F_2\left(\begin{matrix} a+1, b, d \\ c, e \end{matrix}; 1\right) - b_3F_2\left(\begin{matrix} a, b+1, d \\ c, e \end{matrix}; 1\right) = 0.$$

As another example, apply this procedure to

$$_2F_1\left(\begin{matrix} a, b+1 \\ c+1 \end{matrix}; 1\right) - _2F_1\left(\begin{matrix} a, b \\ c \end{matrix}; x\right) = \frac{(c-b)ax}{c(c+1)} {_2F_1}\left(\begin{matrix} a+1, b+1 \\ c+2 \end{matrix}; x\right)$$

to arrive at

$$_3F_2\left(\begin{matrix} a, b+1, d \\ c+1, e \end{matrix}; 1\right) - _3F_2\left(\begin{matrix} a, b, d \\ c, e \end{matrix}; 1\right) = \frac{a(c-b)d}{c(c+1)e} {_3F_2}\left(\begin{matrix} a+1, b+1, d+1 \\ c+2, e+1 \end{matrix}; 1\right).$$

Let us now turn to Wilson's procedure for systematically deriving all the contiguous relations of a balanced $_4F_3$. Note that if only one of the parameters in a balanced $_4F_3$ is altered, the new $_4F_3$ is not balanced.

Definition 3.7.1 *Given a balanced $_4F_3$, a contiguous $_4F_3$ is obtained by altering two parameters by ± 1 in such a way that the new series is also balanced. As before, a relation among contiguous functions is called a contiguous relation.*

Denote the balanced $_4F_3(a, b, c, d; e, f, g; 1)$ by F. By definition, one of the numerator parameters is $-n$ and the sum of the denominator parameters is one more than the sum of the numerator parameters. There are $2 \times \binom{7}{2} = 42$ $_4F_3$s contiguous to F. Consider the difference $F(a-, b+) - F$. Since

$$\frac{(a-1)_k(b+1)_k(c)_k(d)_k}{k!(e)_k(f)_k(g)_k} - \frac{(a)_k(b)_k(c)_k(d)_k}{k!(e)_k(f)_k(g)_k}$$

$$= \frac{(a)_{k-1}(b+1)_{k-1}(c)_k(d)_k}{k!(e)_k(f)_k(g)_k}[(a-1)(b+k) - (a+k-1)b]$$

$$= \frac{(a)_{k-1}(b+1)_{k-1}(c)_k(d)_k}{(k-1)!(e)_k(f)_k(g)_k}(a-b-1),$$

we have

$$F(a-, b+) - F = \frac{(a-b-1)cd}{efg}F_+(a-), \tag{3.7.1}$$

where F_+ is obtained from F by increasing every parameter by 1. Similarly,

$$F(a-, e-) - F = \frac{(a-e)bcd}{(e-1)efg}F_+(a-), \tag{3.7.2}$$

$$F(a+, e+) - F = \frac{(e-a)bcd}{e(e+1)fg}F_+(e+), \tag{3.7.3}$$

$$F(e+, f-) - F = \frac{(e-f+1)abcd}{e(e+1)fg}F_+(e+). \tag{3.7.4}$$

By symmetry in the parameters, we now have expressions for all the differences between F and a function contiguous to it. Some of the contiguous relations are immediate corollaries of these relations. By (3.7.1),

$$F(a-, c+) - F = \frac{(a-c-1)bd}{efg}F_+(a-).$$

Therefore, equating the two expressions for $F_+(a-)$, we get

$$b(a-c-1)(F(a-, b+) - F) = c(a-b-1)(F(a-, c+) - F).$$

The other contiguous relations would follow once we find an equation connecting F with $F_+(a-)$ and $F_+(e+)$. Of course, this equation would imply the other necessary relations by symmetry. To derive the required relation, take $a = -n$ and apply the transformation in Theorem 3.3.1 to obtain

$$\frac{(f)_n(g)_n}{(f-b)_n(g-b)_n} F = {}_4F_3\left(\begin{array}{c} a,b,e-c,e-d \\ e,e+f-c-d,e+g-c-d \end{array};1\right) =: \tilde{F},$$

$$\frac{(f+1)_n(g+1)_n}{(f-b)_n(g-b)_n} F_+(a-) = {}_4F_3\left(\begin{array}{c} a,b+1,e-c,e-d \\ e+1,e+f-c-d,e+g-c-d \end{array};1\right)$$
$$= \tilde{F}(b+,e+),$$

and

$$\frac{(f+1)_{n-1}(g+1)_{n-1}}{(f-b)_{n-1}(g-b)_{n-1}} F_+(e+)$$
$$= {}_4F_3\left(\begin{array}{c} a+1,b+1,e-c+1,e-d+1 \\ e+2,e+f-c-d+1,e+g-c-d+1 \end{array};1\right) = \tilde{F}_+(e+).$$

Now the connection between $\tilde{F}$, $\tilde{F}(b+,e+)$, and $\tilde{F}_+(e+)$ is given by (3.7.3). This implies the relation among F, $F_+(a-)$, and $F_+(e+)$, which is given by

$$fgF - (f-a)(g-a)F_+(a-) + \frac{a(e-b)(e-c)(e-d)}{e(e+1)} F_+(e+) = 0. \quad (3.7.5)$$

In this derivation we also used the fact that $a+b+c+d+1 = e+f+g$. We note that for (3.7.5), it does not matter which numerator parameter is a negative integer. This is true because F is a rational function of the five free parameters. Thus we also have

$$fgF - (f-b)(g-b)F_+(b-) + \frac{b(e-a)(e-c)(e-d)}{e(e+1)} F_+(e+) = 0.$$

Eliminate $F_+(e+)$ from the last two equations to get

$$b(e-a)(f-a)(g-a)F_+(a-) - a(e-b)(f-b)(g-b)F_+(b-)$$
$$+ (a-b)efgF = 0. \quad (3.7.6)$$

The final relation we need is similarly obtained from (3.7.5). It is

$$\frac{(e-a)(e-b)(e-c)(e-d)}{e(e+1)} F_+(e+)$$
$$- \frac{(f-a)(f-b)(f-c)(f-d)}{f(f+1)} F_+(f+) + g(e-f)F = 0. \quad (3.7.7)$$

All the contiguous relations can now be obtained from (3.7.1) to (3.7.7). As one more example, substitute the values of $F_+(a-)$ and $F_+(b-)$ from (3.7.1) into

(3.7.6) to get

$$\frac{b(e-a)(f-a)(g-a)}{a-b-1}(F(a-,b+)-F)$$

$$-\frac{a(e-b)(f-b)(g-b)}{b-a-1}(F(a+,b-)-F)+cd(a-b)F=0. \qquad (3.7.8)$$

The contiguous relations for the $_3F_2$s are found by letting $n \to \infty$ in the relations for the $_4F_3$s. One may also write down the fundamental relations corresponding to (3.7.1) to (3.7.7) and derive the others from these. We give a few examples. In (3.7.1), let $a = -n$ and $n \to \infty$ to get, after renaming the parameters,

$$F(a+)-F = \frac{bc}{de}F_+, \qquad (3.7.9)$$

where F stands for a general $_3F_2$ at $x = 1$. In (3.7.2), let $d = -n$ and $e = -n+a+b+c-f-g-1$ and $n \to \infty$. After renaming parameters,

$$F(a-)-F = -\frac{bc}{de}F_+(a-). \qquad (3.7.10)$$

It is possible to get (3.7.10) from (3.7.1) too. Similarly, we have

$$F(d-)-F = \frac{abc}{(d-1)de}F_+ \qquad (3.7.11)$$

and

$$F(d+)-F = -\frac{abc}{d(d+1)e}F_+(d+). \qquad (3.7.12)$$

By taking limits in different ways in (3.7.5) we get two more relations:

$$deF - a(d+e-a-b-c-1)F_+ - (d-a)(e-a)F_+(a-) = 0 \quad (3.7.13)$$

and

$$eF - (e-a)F_+(a-) - \frac{a(d-b)(d-c)}{d(d+1)}F_+(d+) = 0. \qquad (3.7.14)$$

The rest of the relations can be found in a similar manner. In fact, all the $_3F_2$ contiguous relations follow from (3.7.9) to (3.7.14) if symmetry in the parameters is also used.

3.8 The Wilson Polynomials

Consider the polynomial p_n of degree $n \geq 0$ defined by the relation

$$\tilde{p}_n(x^2) = {}_4F_3\left(\begin{array}{c}-n, n+a+b+c+d-1, a-ix, a+ix \\ a+b, a+c, a+d\end{array}; 1\right) \qquad (3.8.1)$$

where a, b, c, d are real and positive. From the contiguous relation (3.7.8) we find that p_n satisfies the three-term recurrence relation

$$A_n(\tilde{p}_{n+1}(x) - \tilde{p}_n(x)) + C_n(\tilde{p}_{n-1}(x) - \tilde{p}_n(x)) + (a^2 + x)\tilde{p}_n(x) = 0, \quad (3.8.2)$$

where

$$A_n = \frac{(n+a+b+c+d-1)(n+a+b)(n+a+c)(n+a+d)}{(2n+a+b+c+d-1)(2n+a+b+c+d)}$$

and

$$C_n = \frac{n(n+c+d-1)(n+b+d-1)(n+b+c-1)}{(2n+a+b+c+d-2)(2n+a+b+c+d-1)}.$$

As we have remarked before, since $A_n C_{n+1} > 0$ for $n \geq 0$, the polynomials $\tilde{p}_n$ are orthogonal with respect to some positive weight function. In fact, it will be shown that

$$\frac{1}{2\pi} \int_0^\infty \left| \frac{\Gamma(a+ix)\Gamma(b+ix)\Gamma(c+ix)\Gamma(d+ix)}{\Gamma(2ix)} \right|^2 p_n(x^2) p_m(x^2) dx$$
$$= \delta_{m,n} n! (n+a+b+c+d-1)_n$$
$$\times \frac{\Gamma(a+b+n)\Gamma(a+c+n)\cdots\Gamma(c+d+n)}{\Gamma(a+b+c+d+2n)}, \qquad (3.8.3)$$

where

$$p_n(x^2) = (a+b)_n(a+c)_n(a+d)_n \tilde{p}_n(x^2).$$

The relations given above continue to hold when $a = \bar{b}$ and/or $c = \bar{d}$ and the real parts of these parameters are positive.

Definition 3.8.1   *The Wilson polynomials $p_n(x)$ are defined by*

$$p_n(x^2) = (a+b)_n(a+c)_n(a+d)_n$$
$$\cdot {}_4F_3\left(\begin{matrix} -n, n+a+b+c+d-1, a-ix, a+ix \\ a+b, a+c, a+d \end{matrix}; 1\right)$$

where a, b, c, d are complex parameters.

It is evident from the definition that $p_n(x)$ is symmetric in b, c, and d. An application of Theorem 3.3.3 shows that symmetry in all four parameters a, b, c, and d exists. The Wilson polynomials are orthogonal with respect to the integrand of Theorem 3.6.2 as a weight function. We denote the integrand by $f(s)$.

Theorem 3.8.2 *With the contour and the parameters a, b, c, and d as in Theorem 3.6.2,*

$$\frac{1}{2\pi i}\int f(s)p_n(-s^2)p_m(-s^2)ds = \delta_{m,n}2n!(n+a+b+c+d-1)_n$$

$$\cdot \frac{\Gamma(a+b+n)\Gamma(a+c+n)\Gamma(a+d+n)\Gamma(b+c+n)\Gamma(b+d+n)\Gamma(c+d+n)}{\Gamma(a+b+c+d+2n)}.$$

Proof. First observe that we can write

$$p_m(-s^2) = \sum_{k=0}^{m} A_k(b-s)_k(b+s)_k,$$

where A_k are suitable constants. We compute

$$\frac{1}{2\pi i}\int f(s)p_n(-s^2)(b-s)_k(b+s)_k ds$$

$$= (a+b)_n(a+c)_n(b+c)_n \sum_{j=0}^{n} \frac{(-n)_j(n+a+b+c+d-1)_j}{(a+b)_j(a+c)_j(b+c)_j j!}$$

$$\cdot \frac{1}{2\pi i}\int_C f(s)(a-s)_j(a+s)_j(b-s)_k(b+s)_k ds. \tag{3.8.4}$$

The integral in the sum can be rewritten as

$$\frac{1}{2\pi i}\int_C \frac{\Gamma(a+j+s)\Gamma(a+j-s)\Gamma(b+k+s)\Gamma(b+k-s)\Gamma(c+s)\Gamma(c-s)\Gamma(d+s)\Gamma(d-s)}{\Gamma(2s)\Gamma(-2s)}ds.$$

This integral can be evaluated by Theorem 3.6.2. After a little simplification we see that (3.8.4) is equal to

$$\frac{2\Gamma(a+b+k)\Gamma(a+c+n)\Gamma(a+d+n)\Gamma(b+c+k)\Gamma(b+d+k)\Gamma(c+d)(a+b)_n}{\Gamma(a+b+c+d+k)}$$

$$\cdot {}_3F_2\left(\begin{array}{c}-n,n+a+b+c+d-1,a+b+k\\a+b,a+b+c+d+k\end{array};1\right).$$

This ${}_3F_2$ is balanced and so can be summed by the Pfaff–Saalschütz identity. We then get

$$\frac{1}{2\pi i}\int f(s)p_n(-s^2)(b-s)_k(b+s)_k ds = 2(-k)_n$$

$$\cdot \frac{\Gamma(a+b+k)\Gamma(a+c+n)\Gamma(a+d+n)\Gamma(b+c+k)\Gamma(b+d+k)\Gamma(c+d+n)}{\Gamma(a+b+c+d+n+k)}.$$

The factor $(-k)_n$ is zero for $k < n$. By symmetry in a and b we know that

$$p_n(-s^2) = \frac{(-n)_n}{n!}(n+a+b+c+d-1)_n(b-s)_n(b+s)_n$$

$$+ \sum_{k=0}^{n-1} A_k(b-s)_k(b+s)_k.$$

This completes the proof of the theorem. ∎

The result in (3.8.3) follows from Theorem 3.8.2 when a, b, c, and d are positive or when $a = \bar{b}, c = \bar{d}$, and the parameters have positive real parts. The Wilson polynomials contain many sets of orthogonal polynomials as limiting or special cases. Here we show how the Jacobi polynomials (introduced in the previous chapter) are derived.

Set $a = b = (\alpha + 1)/2, c = \bar{d} = (\beta + 1)/2 + i\omega$, and $x = \omega\sqrt{(1 - t)/2}$ in the $_4F_3$ in Definition 3.8.1. Let $\omega \to \infty$. We get, except for a constant factor, the Jacobi polynomial

$$_2F_1\left(\begin{array}{c} -n, n + \alpha + \beta + 1 \\ \alpha + 1 \end{array}; \frac{1 - t}{2}\right).$$

3.9 Quadratic Transformations – Riemann's View

Riemann exploited to the fullest degree the idea that a function is determined to a large extent by its singularities. An example of this was given in the previous chapter in a discussion of the hypergeometric differential equation. Here we show how the ideas developed there give Riemann's basic result on quadratic transformations.

The hypergeometric equation has regular singularities at $0, 1$, and ∞ but no other singularities. Suppose the exponents at 0 are 0 and $1/2$. By a change of variables $x = t^2$, the solutions at 0 become analytic. But now singularities are introduced at $t = \pm 1$ and we get another hypergeometric equation. To see the details, note that (2.3.4) of Chapter 2 shows that

$$P\left\{\begin{array}{ccc} 0 & \infty & 1 \\ 0 & c_1 & b_1 \quad x \\ 1/2 & c_2 & b_2 \end{array}\right\} \tag{3.9.1}$$

is the set of solutions of the equation

$$\frac{d^2y}{dx^2} + \left(\frac{1}{2x} + \frac{1 - b_1 - b_2}{x - 1}\right)\frac{dy}{dx} + \left(\frac{b_1b_2}{x - 1} + c_1c_2\right)\frac{y}{x(x - 1)} = 0. \tag{3.9.2}$$

The change of variables $x = t^2$ implies

$$\frac{dy}{dx} = \frac{1}{2t}\frac{dy}{dt} \quad \text{and} \quad \frac{d^2y}{dx^2} = \frac{1}{2t}\left[\frac{-1}{2t^2}\frac{dy}{dt} + \frac{1}{2t}\frac{d^2y}{dt^2}\right].$$

Substitute this in (3.9.2) to obtain an equation that can be written as

$$\frac{d^2y}{dt^2} + \left(\frac{1 - b_1 - b_2}{t - 1} + \frac{1 - b_1 - b_2}{t + 1}\right)\frac{dy}{dt}$$

$$+ \left(\frac{2b_1b_2}{t - 1} + \frac{2b_1b_2}{t + 1} + 4c_1c_2\right)\frac{y}{t^2 - 1} = 0. \tag{3.9.3}$$

Apply Theorem 2.3.1 to conclude that the set of solutions of (3.9.3) is

$$
P\left\{\begin{matrix} -1 & \infty & 1 & \\ b_1 & 2c_1 & b_1 & t \\ b_2 & 2c_2 & b_2 & \end{matrix}\right\}.
$$

It follows that

Theorem 3.9.1

$$
P\left\{\begin{matrix} 0 & \infty & 1 & \\ 0 & c_1 & b_1 & t^2 \\ 1/2 & c_2 & b_2 & \end{matrix}\right\} = P\left\{\begin{matrix} -1 & \infty & 1 & \\ b_1 & 2c_1 & b_1 & t \\ b_2 & 2c_2 & b_2 & \end{matrix}\right\} \tag{3.9.4}
$$

$$
= P\left\{\begin{matrix} 0 & \infty & 1 & \\ b_1 & 2c_1 & b_1 & \dfrac{1+t}{2} \\ b_2 & 2c_2 & b_2 & \end{matrix}\right\}.
$$

This is Riemann's theorem on quadratic transformations. Let us derive the two basic quadratic transformations for the $_2F_1$ contained in Theorems 3.1.1 and 3.1.3. Write (3.9.4) as

$$
P\left\{\begin{matrix} 0 & \infty & 1 & \\ c & c & 2a & \dfrac{x+1}{x-1} \\ d & d & 2b & \end{matrix}\right\} = P\left\{\begin{matrix} 0 & \infty & 1 & \\ c & a & 0 & 1-x^2 \\ d & b & 1/2 & \end{matrix}\right\}.
$$

Set $c = 0$, and replace $2b$ with $1 - 2b + a$, $2a$ with a, and d with $b - a$ to get

$$
P\left\{\begin{matrix} 0 & \infty & 1 & \\ 0 & 0 & a & x \\ b-a & b-a & a-2b+1 & \end{matrix}\right\}
$$

$$
= P\left\{\begin{matrix} 0 & \infty & 1 & \\ 0 & a/2 & 0 & -\dfrac{4x}{(1-x)^2} \\ b-a & \frac{a+1}{2}-b & 1/2 & \end{matrix}\right\}
$$

or

$$
(1-x)^a P\left\{\begin{matrix} 0 & \infty & 1 & \\ 0 & a & 0 & x \\ b-a & b & 1-2b & \end{matrix}\right\}
$$

$$
= P\left\{\begin{matrix} 0 & \infty & 1 & \\ 0 & a/2 & 0 & -\dfrac{4x}{(1-x)^2} \\ b-a & \frac{a+1}{2}-b & 1/2 & \end{matrix}\right\}.
$$

Since there is only one solution analytic at 0 with value 1 at that point, we get

$$(1-x)^a{}_2F_1\left(\begin{matrix} a,b \\ a-b+1 \end{matrix}; x\right) = {}_2F_1\left(\begin{matrix} a/2, (a+1)/2 - b \\ a-b+1 \end{matrix}; -\frac{4x}{(1-x)^2}\right).$$

This is the result of Theorem 3.1.1. Similarly, we have

$$P\left\{\begin{matrix} 0 & \infty & 1 \\ a & 0 & c \\ b & 1/2 & d \end{matrix} \ \frac{1}{x^2}\right\} = P\left\{\begin{matrix} 0 & \infty & 1 \\ 2a & c & c \\ 2b & d & d \end{matrix} \ \frac{2}{1+x}\right\}$$

or

$$P\left\{\begin{matrix} 0 & \infty & 1 \\ 2a & c & c \\ 2b & d & d \end{matrix} \ x\right\} = P\left\{\begin{matrix} 0 & \infty & 1 \\ a & 0 & c \\ b & 1/2 & d \end{matrix} \ \left(\frac{x}{2-x}\right)^2\right\}.$$

With appropriate changes in the parameters we can write this as

$$(1-x)^{a/2}P\left\{\begin{matrix} 0 & \infty & 1 \\ 0 & a & 0 \\ 1-2b & b & b-a \end{matrix} \ x\right\}$$

$$= \left(1 - \left(\frac{x}{2-x}\right)^2\right)^{a/2} P\left\{\begin{matrix} 0 & \infty & 1 \\ 0 & a/2 & 0 \\ \frac{1}{2}-b & (a+1)/2 & b-a \end{matrix} \ \left(\frac{x}{2-x}\right)^2\right\}$$

or

$${}_2F_1\left(\begin{matrix} a,b \\ 2b \end{matrix}; x\right) = \left(1 - \frac{x}{2}\right)^{-a} {}_2F_1\left(\begin{matrix} a/2, (a+1)/2 \\ b+\frac{1}{2} \end{matrix}; \left(\frac{x}{2-x}\right)^2\right).$$

This proves Theorem 3.1.3 once again.

The differential equation with regular singularities at -1, 1, and ∞ but no other regular singularities when written as

$$(1-x^2)y'' - 2xy' + \left[\nu(\nu+1) - \frac{\mu^2}{1-x^2}\right]y = 0$$

is called Legendre's differential equation. It is clear that the set of solutions is

$$P\left\{\begin{matrix} -1 & \infty & 1 \\ \mu/2 & \nu+1 & \mu/2 \\ -\mu/2 & -\nu & -\mu/2 \end{matrix} \ x\right\}. \tag{3.9.5}$$

A comparison of (3.9.5) with (3.9.4) shows that quadratic transformations apply to the solutions of the Legendre equation.

3.10 Indefinite Hypergeometric Summation

We have derived a number of hypergeometric identities in this chapter. In this section we consider the problem of evaluating partial sums of hypergeometric series. Gosper [1978] has given an algorithm that yields the value of such a sum provided it is a hypergeometric term. To make this more precise, suppose that $\sum_{k=1}^{n} c_k$ is a partial sum of a hypergeometric series. The problem is to find a function S_n such that

$$c_k = S_k - S_{k-1}, \tag{3.10.1}$$

when it is assumed that S_k/S_{k-1} is a rational function of k. We refer to such an S_k as a hypergeometric term. From (3.10.1) it is clear that

$$\sum_{k=1}^{n} c_k = S_n - S_0.$$

Write

$$\frac{c_k}{c_{k-1}} = \frac{p_k}{p_{k-1}} \cdot \frac{q_k}{r_k}, \tag{3.10.2}$$

where p_k, q_k, and r_k are polynomials in k satisfying

$$\gcd(q_k, r_{k+j}) = 1, \tag{3.10.3}$$

for all integers $j \geq 0$. The necessity of this condition will become evident later, but q_k and r_k can be chosen so that (3.10.3) is true. Condition (3.10.3) implies that if $k + \alpha$ and $k + \beta$ are factors of q_k and r_k respectively, then $\alpha - \beta$ cannot be a nonnegative integer. Suppose that, initially in the decomposition (3.10.2), $\gcd(q_k, r_{k+j}) = g_k$. Then replace q_k, r_k, and p_k with

$$q'_k = \frac{q_k}{g_k}, \quad r'_k = \frac{r_k}{g_{k-j}}, \quad \text{and } p'_k = p_k g_k g_{k-1} \cdots g_{k-j+1}.$$

It is easy to check that

$$\frac{c_k}{c_{k-1}} = \frac{p'_k}{p'_{k-1}} \cdot \frac{q'_k}{r'_k} \quad \text{and} \quad \gcd(q'_k, r'_{k+j}) = 1.$$

It is now clear that after a finite number of repetitions of this process, condition (3.10.3) will hold for all $j \geq 0$. The next step is to write

$$S_k = \frac{q_{k+1}}{p_k} f_k c_k. \tag{3.10.4}$$

Substitute in (3.10.1) and use (3.10.2) to find a relation for f_k, which is the only unknown in (3.10.4). Then

$$p_k = q_{k+1} f_k - r_k f_{k-1}. \tag{3.10.5}$$

By the condition on S_k, we see that f_k is a rational function of k. We show that it is a polynomial. Let

$$f_k = \ell_k / m_k,$$

where ℓ_k and m_k are polynomials in k with no common factors. Suppose m_k is not independent of k. Let j be the largest nonnegative integer such that $k + \lambda$ and $k + \lambda + j$ are both factors of m_k. Substitute the expression for f_k in (3.10.5) to get

$$p_k m_k m_{k-1} = q_{k+1} \ell_k m_{k-1} - r_k \ell_{k-1} m_k. \tag{3.10.6}$$

Since $k + \lambda - 1 \mid m_{k-1}$, the last equation implies that $k + \lambda - 1 \mid r_k \ell_{k-1} m_k$. But $\gcd(m_{k-1}, \ell_{k-1}) = 1$ and $k + \lambda - 1$ does not divide m_k by the maximality of j. So

$$k + \lambda - 1 \mid r_k. \tag{3.10.7}$$

Similarly, the fact that $k + \lambda + j \mid q_{k+1} \ell_k m_{k-1}$ implies that

$$k + \lambda + j \mid q_{k+1} \quad \text{or} \quad k + \lambda + j - 1 \mid q_k. \tag{3.10.8}$$

By (3.10.7) and (3.10.8), $\gcd(q_k, r_{k+j}) \neq 1$. This contradicts (3.10.3) and we can conclude that f is a polynomial of degree d, say, given by

$$f_k = a_0 k^d + a_1 k^{d-1} + \cdots + a_d. \tag{3.10.9}$$

Substitute this expression for f_k in (3.10.5) to obtain a system of linear equations satisfied by $a_0, a_1, \ldots, a_d$. If this is a consistent system the values $a_0, a_1, \ldots, a_d$ are obtained by solving the equations. From f_k we get S_k by (3.10.4).

To obtain the possible degrees of the polynomial f_k, write (3.10.5) as

$$p_k = (q_{k+1} - r_k)\frac{(f_k + f_{k-1})}{2} + (q_{k+1} + r_k)\frac{(f_k - f_{k-1})}{2}. \tag{3.10.10}$$

There are two cases. First suppose that

$$\deg(q_{k+1} + r_k) \le \deg(q_{k+1} - r_k) = d'.$$

Since $\deg(f_k - f_{k-1})/2 < d$, it follows that

$$d = \deg p_k - d'.$$

Now suppose that

$$(q_{k+1} + r_k)/2 = bk^{d'} + \cdots, \quad b \neq 0$$

and

$$(q_{k+1} - r_k)/2 = ck^{d'-1} + \cdots.$$

Use these expressions in (3.10.10) to obtain

$$p_k = (a_0c + a_0bd/2)k^{d+d'-1} + \cdots .$$

If $a_0c + a_0bd/2 \neq 0$, then

$$d = \deg p_k - d' + 1.$$

Otherwise,

$$d = -2c/b \quad \text{and} \quad d > \deg p_k - d' + 1.$$

The last value of d is used only if it is an integer greater than $\deg p_k - d' + 1$. This completes Gosper's algorithm. It decides whether a partial sum of a hypergeometric series can be expressed as a hypergeometric term and gives its value if it does.

Zeilberger [1982] extended the scope of this algorithm by taking c_k as a function of two variables n and k rather than just k. We discuss the Wilf–Zeilberger method here and in the next section. This method is very powerful in proving hypergeometric identities.

Suppose the identity to be proved can be written as

$$\sum_k T(n, k) = A(n),$$

where $A(n) \neq 0$ and $n \geq 0$. Divide both sides by $A(n)$ and write the identity as

$$\sum_k F(n, k) = 1. \tag{3.10.11}$$

This implies that

$$\sum_k (F(n+1, k) - F(n, k)) = 0.$$

Earlier we were trying to express $F(n, k)$, or rather $T(n, k)$, as the difference $S_{k+1} - S_k$, but this is often not possible. As an example, consider the sum

$$\sum_{k=0}^{j} \frac{(-1)^k(-n)_k}{k!}$$

when $j < n$. By going through the steps in Gosper's algorithm, it can be seen that this sum is not expressible as a hypergeometric term.

In Zeilberger's method one tries to write the difference $F(n+1, k) - F(n, k)$ as $S_{k+1} - S_k$. This improves the situation. Suppose there is a function $G(n, k)$ such that

$$F(n+1, k) - F(n, k) = G(n, k+1) - G(n, k). \tag{3.10.12}$$

This function G can be determined by Gosper's algorithm. Then

$$\sum_{k=-L}^{K} (F(n+1,k) - F(n,k)) = G(n, K+1) - G(n, -L). \qquad (3.10.13)$$

If we assume that G satisfies the property

$$\lim_{k \to \pm\infty} G(n,k) = 0, \qquad (3.10.14)$$

then it follows that

$$\sum_{k} F(n,k) = \text{constant}.$$

It is then sufficient to verify the identity for one value of n, say $n = 0$.

Thus to prove (3.10.11), find a G that satisfies (3.10.12) and (3.10.14). If this G exists, then (3.10.11) is known to be true after it is verified for $n = 0$. This method works for a very large class of identities. The reader may consult Petkovšek, Wilf, and Zeilberger [1996] and Nemes, Petkovšek, Wilf, and Zeilberger [1997] for examples and further results.

As an example, consider the identity

$$\sum_{k} \frac{(-1)^k (-n)_k}{k! 2^n} = 1.$$

Here

$$F(n,k) = \frac{(-1)^k (-n)_k}{k! 2^n},$$

and one can show that

$$G(n,k) = \frac{(-1)^k (-n)_{k-1}}{(k-1)! 2^{n+1}}$$

satisfies (3.10.12) and (3.10.14). The identity is therefore verified since it is true for $n = 0$.

The next section contains further illustrations of the Wilf–Zeilberger method and a comparison with a related method.

3.11 The W–Z Method

In a series of papers, Zeilberger (sometimes jointly with Wilf) developed a technique he called creative telescoping. The method is often referred to as the W–Z method in that an important component of this work was presented in the Wilf–Zeilberger paper. The method is surprisingly easy to describe in full, but for simplicity we shall apply it to some elementary identities and then compare this method with that of Pfaff.

Suppose that there is a linear, homogeneous recurrence relation that we wish to prove for a particular sum. In other words, suppose

$$S(n) = \sum_{k=-\infty}^{\infty} F(n, k),$$

where for each n, $F(n, k)$ is zero for all but finitely many k. Suppose we expect that

$$\alpha(n)S(n) = \beta(n)S(n-1)$$

(so that in fact $S(n) = S(0) \prod_{j=1}^{n} (\beta(j)/\alpha(j))$). Then the W–Z method constructs a function $G(n, k)$ (which again for each n is 0 for all but finitely many k) so that

$$\alpha(n)F(n, k) - \beta(n)F(n-1, k) = G(n, k) - G(n, k-1).$$

Then the desired recurrence follows immediately. Therefore,

$$\alpha(n)S(n) - \beta(n)S(n-1) = \sum_{k} (\alpha(n)F(n, k) - \beta(n)F(n-1, k))$$
$$= \sum_{k} (G(n, k) - G(n, k-1))$$
$$= 0$$

because the final sum telescopes. This example illustrates the appropriateness of the label "creative telescoping."

The best way to appreciate this is through some examples. Consider the Chu–Vandermonde summation. We wish to prove

$$S_v(n) = \frac{(b-a)_n}{(b)_n}, \quad \text{where } S_v(n) = \sum_{k=0}^{n} F_v(n, k)$$

and

$$F_v(n, k) = \frac{(-n)_k (a)_k}{k!(b)_k}.$$

In other words, we wish to find $G(n, k)$ so that

$$(b+n-1)F_v(n, k) - (b-a+n-1)F_v(n-1, k) = G(n, k) - G(n, k-1).$$

$$(3.11.1)$$

Zeilberger has fully implemented on the computer an algorithm for finding $G(n, k)$. For problems involving hypergeometric series such as $S_v(n)$, $G(n, k)$ has been shown by Zeilberger to be of the following form:

$$G(n, k) = R(n, k)F(n-1, k).$$

However, we can easily work out the value of $G(n, k)$ by inspection. In (3.11.1) set $k = 0$. Then, assuming $G(n, -1) = 0$, we see

$$G(n, 0) = (b + n - 1)F_v(n, 0) - (b - a + n - 1)F_v(n - 1, 0)$$
$$= (b + n - 1) - (b - a + n - 1)$$
$$= a.$$

Set $k = 1$ in (3.11.1). Then

$$G(n, 1) = G(n, 0) + (b + n - 1)F_v(n, 1) - (b - a + n - 1)F_v(n - 1, 1)$$
$$= a + (b + n - 1)\left(-\frac{na}{b}\right) - (b - a + n - 1)\left(-\frac{(n - 1)a}{b}\right)$$
$$= \frac{(-n + 1)a}{b}(a + 1)$$
$$= (a + 1)F_v(n - 1, 1).$$

In this manner we can work out (either by hand or with the use of a computer algebra system) the conjecture that in fact

$$G(n, k) = (a + k)F_v(n - 1, k). \tag{3.11.2}$$

Once conjectured, the proof of (3.11.2) is pure algebra. We have

$$(b + n - 1)F_v(n, k) - (b - a + n - 1)F_v(n - 1, k)$$
$$= \frac{(b + n - 1)(-n)_k(a)_k}{k!(b)_k} - \frac{(b - a + n - 1)(-n + 1)_k(a)_k}{k!(b)_k}$$
$$= \frac{(-n + 1)_{k-1}(a)_k}{k!(b)_k}[(b + n - 1)(-n) - (b - a + n - 1)(-n + k)]$$
$$= \frac{(-n + 1)_{k-1}(a)_k}{k!(b)_k}(-an - k(b - a + n - 1)). \tag{3.11.3}$$

However,

$$G(n, k) - G(n, k - 1)$$
$$= \frac{(a + k)(1 - n)_k(a)_k}{k!(b)_k} - \frac{(a + k - 1)(1 - n)_{k-1}(a)_{k-1}}{(k - 1)!(b)_{k-1}}$$
$$= \frac{(-n + 1)_{k-1}(a)_k}{k!(b)_k}[-an - k(b - a + n - 1)]$$
$$= (b + n - 1)F_v(n, k) - (b - a + n - 1)F_v(n - 1, k),$$

(by (3.11.3)). Hence creative telescoping shows us that

$$(b + n - 1)S_v(n) = (b - a + n - 1)S_v(n - 1),$$

and so by iteration $S_v(n) = (b - a)_n/(b)_n$, as desired.

Now turn to the Pfaff–Saalschütz summation. This summation was initially treated in Chapter 2. It may be stated as follows:

$$S_p(n) = \frac{(c-a)_n(c-b)_n}{(c)_n(c-a-b)_n},$$

where $S_p(n) = \sum_{k=0}^{n} F_p(n,k)$ with (3.11.4)

$$F_p(n,k) = \frac{(-n)_k(a)_k(b)_k}{k!(c)_k(1-n+a+b-c)_k}.$$

Note that (3.11.4) is clearly equivalent to

$$(c+n-1)(c-a-b+n-1)S_p(n)$$
$$= (c-a+n-1)(c-b+n-1)S_p(n-1).$$

We proceed as before. We let

$$G_p(n,k) = R_p(n,k)F_p(n-1,k)$$

and we wish to construct the rational function $R_p(n,k)$ so that

$$(c+n-1)(c-a-b+n-1)F_p(n,k)$$
$$-(c-a+n-1)(c-b+n-1)F_p(n-1,k) = G_p(n,k) - G_p(n,k-1).$$
(3.11.5)

Consequently, with $k=0$ in (3.11.5) we get

$$G_p(n,0) = (c+n-1)(c-a-b+n-1) - (c-a+n-1)(c-b+n-1)$$
$$= (c+n-1)(-a) + a(c-b+n-1)$$
$$= -ab.$$

Now by (3.11.5) with $k=1$,

$$G_p(n,1) = G_p(n,0) + (c+n-1)(c-a-b+n-1)\frac{ab(1-n)}{c(2-n+a+b-c)}$$
$$- (c-a+n-1)(c-b+n-1)\frac{(2-n)ab}{c(3-n+a+b-c)}$$
$$= -(a+1)(b+1)F_p(n-1,1).$$

Thus as in (3.11.2) we may conjecture

$$G_p(n,k) = -(a+k)(b+k)F_p(n-1,k). \tag{3.11.6}$$

The proof of this conjecture is again merely an algebraic exercise. In each case it turns out that

$$(c + n - 1)(c - a - b + n - 1) - (c - a + n - 1)(c - b + n - 1)F_p(n - 1, k)$$
$$= G_p(n, k) - G_p(n, k - 1)$$
$$= \left(-\frac{(a + k)(b + k)(1 - n)_k (a)_k (b)_k}{k!(c)_k (2 - n + a + b - c)_k} \right.$$
$$\left. + \frac{(a + k - 1)(b + k - 1)(1 - n)_{k-1}(a)_{k-1}(b)_{k-1}}{(k - 1)!(c)_{k-1}(2 - n + a + b - c)_{k-1}} \right),$$

which simplifies to our desired result. So (3.11.4) has been proved by the W–Z method.

We now turn to Bailey's $_4F_3$ summation, which is somewhat more difficult. Our object here is to prove

$$_4F_3\left(\begin{matrix} a/2, (a + 1)/2, b + n, -n \\ b/2, (b + 1)/2, a + 1 \end{matrix} ; 1 \right) = \frac{(b - a)_n}{(b)_n}. \qquad (3.11.7)$$

Note that this is a balanced $_4F_3$. In notation suitable for the W–Z method, we wish to prove

$$S_B(n) = \frac{(b - a)_n}{(b)_n},$$

where $S_B(n) = \sum_{k=0}^n F_B(n, k)$ with

$$F_B(n, k) = \frac{(a)_{2k}(b + n)_k(-n)_k}{k!(b)_{2k}(a + 1)_k}.$$

In this instance, our first expectation is that we can prove

$$(b + n - 1)S_B(n) - (b - a + n - 1)S_B(n - 1) = 0.$$

However, the method utilized in the previous two cases fails initially. It is at this moment that we realize how useful computer algebra is in such matters. It turns out that we can use the W–Z method to obtain a second-order recurrence, namely

$$(n + 1)(-n - b + a)F_B(n, k)$$
$$+ (-a^2 + ba - a + 2nb + 3b + 2 + 4n + 2n^2)F_B(n + 1, k)$$
$$- (b + n + 1)(a + n + 2)F_B(n + 2, k)$$
$$= G_B(n, k) - G_B(n, k - 1),$$

where

$$G_B(n, k) = -\frac{(a + 1 + 2k)(a + 2k)(b + n + k)(n + 1)}{(b + n)(n - k + 1)} F_B(n, k).$$

This implies that

$$(n + 1)(-n - b + a)S_B(n) + (-a^2 + ba - a + 2nb + 3b + 2$$
$$+ 4n + 2n^2)S_B(n + 1) - (b + n + 1)(a + n + 2)S_B(n + 2) = 0. \quad (3.11.8)$$

Also, since $S_B(0) = 1$, $S_B(1) = (b - a)/b$, and since $(b - a)_n/(b)_n$ satisfies the above recurrence, we see that (3.11.7) is proved.

A couple of observations should be made here. First, an identity like (3.11.7) often arises in practice as a conjecture. In other words, if (3.11.7) is true, then something useful follows (c.f. Andrews and Burge [1993]). Consequently, one usually knows the form of the desired summation identity before looking for a proof. Second, suppose that we are dealing with a more complicated identity where we perhaps had not determined exactly what the summation should be like. Here Marco Petkovšek has produced an auxiliary algorithm for the W–Z method. It finds the minimal recurrence satisfied by the sum in question. Thus in this instance Petkovšek's algorithm applied to (3.11.7) would reveal (3.11.8).

There is a somewhat different summation method due to Pfaff. This method is less algorithmic than the W–Z method. However, it spreads out the algebraic complications to systems of recurrences. Consequently, it may provide new summations in addition to the one we wish to prove and it may allow the required algebra to be considerably simpler than that required by the W–Z method. Pfaff's method rather resembles the W–Z method; however, it allows the various additional parameters in the summation to play an important role. Pfaff's method begins very simply. We merely subtract term by term the sum at $n - 1$ from the sum at n.

Consider the Chu–Vandermonde summation once again. We phrase the problem slightly differently. Let

$$S_v(n, a, b) = \sum_{j=0}^{n} \frac{(-n)_j(a)_j}{j!(b)_j}.$$

Now we note

$$S_v(n, a, b) - S_v(n - 1, a, b) = \sum_{j=0}^{n} \left(\frac{(-n)_j(a)_j}{j!(b)_j} - \frac{(1 - n)_j(a)_j}{j!(b)_j} \right)$$

$$= \sum_{j=0}^{n} \frac{(a)_j(1 - n)_{j-1}}{j!(b)_j}((-n) - (-n + j))$$

$$= -\sum_{j=1}^{n} \frac{(a)_j(1 - n)_{j-1}}{(j - 1)!(b)_j}$$

$$= -\frac{a}{b} \sum_{j=1}^{n-1} \frac{(a + 1)_j(1 - n)_j}{j!(b + 1)_j}$$

$$= -\frac{a}{b} S_v(n - 1, a + 1, b + 1). \quad (3.11.9)$$

Note that (3.11.9) together with $S_v(0, a, b) = 1$ uniquely defines $S_v(n, a, b)$. But if

$$\sigma_v(n, a, b) = \frac{(b - a)_n}{(b)_n},$$

then

$$\sigma_v(n, a, b) - \sigma_v(n - 1, a, b) = \frac{(b - a)_n}{(b)_n} - \frac{(b - a)_{n-1}}{(b)_{n-1}}$$

$$= \frac{(b - a)_{n-1}}{(b)_n}(-a)$$

$$= -\frac{a}{b} \frac{(b - a)_{n-1}}{(b + 1)_{n-1}} = -\frac{a}{b} S_v(n - 1, a + 1, b + 1).$$

Hence

$$S_v(n, a, b) = \sigma_v(n, a, b) = (b - a)_n / (b)_n$$

as desired.

Pfaff–Saalschütz summation by Pfaff's method is as follows: Let

$$S_p(n, a, b, c) = \sum_{j=0}^{n} \frac{(-n)_j (a)_j (b)_j}{j!(c)_j (1 - n + a + b - c)_j}.$$

Now we note

$$S_p(n, a, b, c) - S_p(n - 1, a, b, c)$$

$$= \sum_{j=0}^{n} \frac{(-n)_j (a)_j (b)_j}{j!(c)_j (1 - n + a + b - c)_j} - \frac{(1 - n)_j (a)_j (b)_j}{j!(c)_j (2 - n + a + b - c)_j}$$

$$= \sum_{j=0}^{n} \frac{(a)_j (b)_j (1 - n)_{j-1}}{j!(c)_j (1 - n + a + b - c)_{j-1}} ((-n)(1 - n + a + b - c + j)$$

$$- (n + j)(1 - n + a + b - c))$$

$$= n(a + b + 1 - c) \sum_{j=1}^{n} \cdot \frac{(a)_j (b)_j (1 - n)_{j-1}}{(j - 1)!(c)_j (1 - n + a + b - c)_{j-1}}$$

$$= \frac{n(a + b + 1 - c)ab}{c(1 - n + a + b - c)(2 - n + a + b - c)}$$

$$\cdot \sum_{j=0}^{n-1} \frac{(1 - n)_j (a + 1)_j (b + 1)_j}{j!(c + 1)_j (3 - n + a + b - c)_j}$$

$$= \frac{-(a + b + 1 - c)ab}{c(1 - n + a + b - c)(2 - n + a + b - c)} S_p(n - 1, a + 1, b + 1, c + 1).$$

$$(3.11.10)$$

The rest follows by showing that $(c-a)_n(c-b)_n/[(c)_n(c-a-b)_n]$ satisfies the same recurrence and is equal to 1 when n equals 0.

This is precisely the proof given by Pfaff for this formula in 1797. We finally look at Bailey's $_4F_3$ summation by Pfaff's method.

Just as the W–Z method did not work as expected, so too does Pfaff's method require a new twist. Here we wish to prove that

$$S_B(n, a, b) = \frac{(b-a)_n}{(b)_n}, \tag{3.11.11}$$

where

$$S_B(n, a, b) = \sum_{j=0}^{n} \frac{(a)_{2j}(b+n)_j(-n)_j}{j!(b)_{2j}(a+1)_j}.$$

Subtracting term by term we find

$$S_B(n, a, b) - S_B(n-1, a, b)$$
$$= \frac{a(1-b-2n)}{b(b+1)} T_B(n-1, a+2, b+2), \tag{3.11.12}$$

where

$$T_B(n, a, b) = \sum_{j=0}^{n} \frac{(a)_{2j}(b+n-1)_j(-n)_j}{j!(b)_{2j}(a)_j}. \tag{3.11.13}$$

We have thus introduced a new sum. Calculation of this sum for $n = 1, 2$, and 3 suggests that

$$T_B(n, a, b) = \frac{(b-a)_n}{(b+2n-1)(b)_{n-1}}. \tag{3.11.14}$$

We now try term-by-term comparison of $T_B(n, a, b)$ with $S_B(n, a, b)$ and $S_B(n-1, a, b)$. The second comparison yields

$$T_B(n, a, b) - S_B(n-1, a, b) = -\frac{(b+n-1)(a+n)}{b(b+1)} T_B(n-1, a+2, b+2). \tag{3.11.15}$$

The two recurrences (3.11.12) and (3.11.15) together with the initial values $S_B(0, a, b) = T_B(0, a, b) = 1$ completely define $S_B(n, a, b)$ and $T_B(n, a, b)$. It is now again an easy algebraic exercise to see that $(b-a)_n/(b)_n$ and $(b-a)_n/((b+2n-1)(b)_{n-1})$ satisfy the same recurrences and initial conditions. Consequently, we have not only proved (3.11.11) but we have also proved (3.11.14).

3.12 Contiguous Relations and Summation Methods

From (2.5.6) and (2.5.8) we get one of Gauss's contiguous relations, namely

$$(c - a - b)F = (c - a)F(a-) - b(1 - x)F(b+).$$

Set $a = -n + 1$ to find

$$(c + n - 1)_2F_1(-n, b; c; x) - (c - b + n - 1)_2F_1(-n + 1, b; c; x)$$
$$= b(1 - x) \sum_{j \geq 0} \frac{(-n + 1)_j(b + 1)_j}{j!(c)_j} x^j$$
$$= \sum_{j \geq 0} \left(\frac{(-n + 1)_j(b)_{j+1}}{j!(c)_j} - \frac{(-n + 1)_{j-1}(b)_j}{(j - 1)!(c)_{j-1}} \right) x^j.$$

Now put $x = 1$ to see that precisely the creative telescoping of Zeilberger reduces the right-hand side to 0, and we have produced the W–Z proof of the Chu–Vandermonde sum. Similarly,

$$(c + n - 1)(c - a - b + n - 1)_3F_2(-n, a, b; c, 1 - n + a + b - c; x)$$
$$- (c - a + n - 1)(c - b + n - 1)_3F_2(-n + 1, a, b; c, 2 - n + a + b - c; x)$$
$$= -(1 - x) \sum_{j \geq 0} \frac{(1 - n)_j(a)_{j+1}(b)_{j+1}x^j}{j!(c)_j(2 - n + a + b - c)_j}$$
$$= \sum_{j \geq 0} \left(-\frac{(1 - n)_{j-1}(a)_j(b)_j}{(j - 1)!(c)_{j-1}(2 - n + a + b - c)_{j-1}} \right.$$
$$\left. + \frac{(1 - n)_j(a)_{j+1}(b)_{j+1}}{j!(c)_j(2 - n + a + b - c)_j} \right) x^j$$
$$= -ab(1 - x)_3F_2(1 - n, a + 1, b + 1; c, 2 - n + a + b - c; x),$$

and if we set $x = 1$ we get the W–Z proof of the Pfaff–Saalschütz summation. Thus we see how the W–Z method is an effective algorithm for discovering useful instances of contiguous relations. In the case of Bailey's $_4F_3$, the W–Z method fails to find a first-order recurrence because there is no three-term contiguous relation relating

$$(b + n - 1)_4F_3\left(\begin{matrix} a/2, (a + 1)/2, b + n, -n \\ b/2, (b + 1)/2, a + 1 \end{matrix}; x \right)$$

$$- (b - a + n - 1)_4F_3\left(\begin{matrix} a/2, (a + 1)/2, b + n - 1, 1 - n \\ b/2, (b + 1)/2, a + 1 \end{matrix}; x \right)$$

to a third $_4F_3$ multiplied by a factor including $(1 - x)$. However, when one moves to four terms, such a relation holds and the W–Z proof of Bailey's summation follows by setting $x = 1$.

Pfaff's is even more obviously a method of contiguous relations. In his method x is set equal to 1 before we begin. The full Pfaff proof of Chu–Vandermonde is given by directly establishing the contiguous relation

$$
{}_2F_1\left({-n, a \atop b}; 1\right) - {}_2F_1\left({-n+1, a \atop b}; 1\right) = -\frac{a}{b^2}{}_2F_1\left({-n+1, a+1 \atop b+1}; 1\right).
$$

The Pfaff proof of Pfaff–Saalschütz is just

$$
{}_3F_2\left({-n, a, b \atop c, 1-n+a+b-c}; 1\right) - {}_3F_2\left({1-n, a, b \atop c, 2-n+a+b-c}; 1\right)
$$
$$
= \frac{n(a+b+1-c)ab}{c(1-n+a+b-c)(2-n+a+b-c)}{}_3F_2\left({1-n, a+1, b+1 \atop c+1, 3-n+a+b-c}; 1\right).
$$

The proof of Bailey's formula relies entirely on two contiguous relations:

$$
{}_4F_3\left({a/2, (a+1)/2, b+n, -n \atop b/2, (b+1)/2, a+1}; 1\right) - {}_4F_3\left({a/2, (a+1)/2, b+n-1, 1-n \atop b/2, (b+1)/2, a+1}; 1\right)
$$
$$
= \frac{a(1-b-2n)}{b(b+1)}{}_4F_3\left({(a/2)+1, (a+3)/2, b+n, 1-n \atop b/2+1, (b+3)/2, a+2}; 1\right)
$$

and

$$
{}_4F_3\left({a/2, (a+1)/2, b+n-1, -n \atop b/2, (b+1)/2, a}; 1\right)
$$
$$
- {}_4F_3\left({a/2, (a+1)/2, b+n-1, 1-n \atop b/2, (b+1)/2, a+1}; 1\right)
$$
$$
= -\frac{(b+n-1)(a+n)}{b(b+1)}{}_4F_3\left({(a/2)+1, (a+3)/2, b+n, 1-n \atop (b/2)+1, (b+3)/2, a+2}; 1\right).
$$

The discoveries of Wilf and Zeilberger truly revolutionized the study of summations of terminating hypergeometric series. An important offshoot of this work is the MAPLE implementations of these algorithms also prepared by Zeilberger. Peripheral to this accomplishment has been a philosophical debate by one of us (Andrews [1994]) with Zeilberger [1994] about the implication of these discoveries for artificial intelligence.

Currently the internal constraints in MAPLE have prevented a W–Z proof of

$$
{}_5F_4\left({-2n, x+2n+1, x-z+\frac{1}{2}, x+n-1, z+n+1 \atop (x/2)+1, (x+1)/2, 2z+2n+1, 2x-2z}; 1\right)
$$
$$
= \frac{(1/2)_n(2z-x)_n(2z-x+n+2)_n}{(x+1)_{n-3}(x+2n-2)_3(x-z)_n(z+n+\frac{1}{2})_n}. \tag{3.12.1}
$$

The Pfaff-method proof involves a gigantic simultaneous treatment of twenty identities established in the manner discussed earlier. Undoubtedly, the improvements
of software and hardware will eventually yield a W–Z proof of (3.12.1).

It is to be hoped that this contest between methods will serve to make clear that
progress occurs when human thought aided by machines applies itself to any given
problem. What should not get lost in the shuffle, however, is the observation that
the Pfaff and W–Z methods are valuable applications of the classical theory of
contiguous relations to summation problems. It would be hard to believe that these
are the only such methods buried in contiguous relations, and further investigations
are clearly merited.

Exercises

1. Verify the following quadratic transformation formulas:

(a) $\displaystyle {}_2F_1\left(\begin{matrix} 2a, 2b \\ a+b+\frac{1}{2} \end{matrix}; x\right) = (1-2x)^{-2a}\, {}_2F_1\left(\begin{matrix} a, a+\frac{1}{2} \\ a+b+\frac{1}{2} \end{matrix}; \frac{4x(x-1)}{(2x-1)^2}\right).$

(b) $\displaystyle {}_2F_1\left(\begin{matrix} 2a, b \\ 2b \end{matrix}; x\right) = (1-x)^{-a}\, {}_2F_1\left(\begin{matrix} a, b-a \\ b+\frac{1}{2} \end{matrix}; \frac{x^2}{4(x-1)}\right).$

(c) $\displaystyle {}_2F_1\left(\begin{matrix} a, b \\ 2b \end{matrix}; x\right) = (1-x)^{-a/2}\, {}_2F_1\left(\begin{matrix} a, 2b-a \\ b+\frac{1}{2} \end{matrix}; -\frac{(1-\sqrt{1-x})^2}{4\sqrt{1-x}}\right).$

(d) $\displaystyle {}_2F_1\left(\begin{matrix} a, b \\ a-b+1 \end{matrix}; x\right) = (1+\sqrt{x})^{-2a}\, {}_2F_1\left(\begin{matrix} a, a-b+\frac{1}{2} \\ 2a-2b+1 \end{matrix}; \frac{4\sqrt{x}}{(1+\sqrt{x})^2}\right).$

(e) $\displaystyle \frac{2\sqrt{\pi}\,\Gamma(a-b+1)}{\Gamma\left(a+\frac{1}{2}\right)\Gamma(-b+1)}(1+x)^a\, {}_2F_1\left(\begin{matrix} a, b \\ 1/2 \end{matrix}; -x\right)$

$\displaystyle = {}_2F_1\left(\begin{matrix} 2a, -2b+1 \\ a-b+1 \end{matrix}; \frac{1}{2}\sqrt{\frac{x}{1+x}}+\frac{1}{2}\right)$

$\displaystyle + {}_2F_1\left(\begin{matrix} 2a, -2b+1 \\ a-b+1 \end{matrix}; -\frac{1}{2}\sqrt{\frac{x}{1+x}}+\frac{1}{2}\right).$

2. Verify the following one-parameter transformations of Kummer [1836]:

(a) $\displaystyle \left(\frac{1+\sqrt{(1-x)}}{2}\right)^{-2a}\, {}_2F_1\left(\begin{matrix} a, (4a+1)/6 \\ (2a+5)/6 \end{matrix}; \left(\frac{1-\sqrt{1-x}}{1+\sqrt{1-x}}\right)^2\right)$

$\displaystyle = (1+x)^{-a}\, {}_2F_1\left(\begin{matrix} (a/2), (a+1)/2 \\ (2(a+1))/3 \end{matrix}; \frac{4x}{(1+x)^2}\right).$

(b)
$$(1 + \sqrt{x})^{-2a} {}_2F_1\left(\begin{matrix} a, (4a+1)/6 \\ (4a+1)/3 \end{matrix} ; \frac{4\sqrt{x}}{(1+\sqrt{x})^2} \right)$$

$$= \left(1 - \frac{x}{2}\right)^{-a} {}_2F_1\left(\begin{matrix} a/2, (a+1)/2 \\ (2a+5)/6 \end{matrix} ; \left(\frac{x}{2-x}\right)^2 \right).$$

(c)
$$\left(\frac{1 + \sqrt{(1-x)}}{2} \right)^{-2a} {}_2F_1\left(\begin{matrix} 2a, a + \frac{1}{4} \\ a + \frac{3}{4} \end{matrix} ; \frac{\sqrt{(1-x)} - 1}{\sqrt{1-x} + 1} \right)$$

$$= (1+x)^{-a} {}_2F_1\left(\begin{matrix} a/2, (a+1)/2 \\ a + (3/4) \end{matrix} ; \frac{4x}{(1+x)^2} \right).$$

3. Show that

(a)
$$ {}_2F_1\left(\begin{matrix} a/2, (2-a)/6 \\ (2a+5)/6 \end{matrix} ; -1/8 \right) = 2^{-a/2} {}_2F_1\left(\begin{matrix} a/2, (a+1)/6 \\ 2(a+1)/3 \end{matrix} ; 1 \right)$$

$$= 2^{-a/2} \frac{\Gamma((2a+2)/3)\Gamma(1/2)}{\Gamma((a+1)/2)\Gamma((a+4)/6)}.$$

(b)
$$ {}_2F_1\left(\begin{matrix} 2a, a + \frac{1}{4} \\ a + \frac{3}{4} \end{matrix} ; \frac{\sqrt{2} - 1}{\sqrt{2} + 1} \right)$$

$$= (4 - 2\sqrt{2})^{-2a} \frac{\Gamma\left(a + \frac{3}{4}\right)\Gamma(1/4)}{\Gamma((2a+3)/4)\Gamma((2a+1)/4)}.$$

4. Deduce Kummer's identity in Corollary 3.1.2 from Euler's integral for ${}_2F_1$.

5. Note that (3.1.2) is equivalent to

(a)
$$ {}_2F_1\left(\begin{matrix} -2n, 2b \\ -n+b+\frac{1}{2} \end{matrix} ; x \right) = {}_2F_1\left(\begin{matrix} -n, b \\ -n+b+\frac{1}{2} \end{matrix} ; 4x(1-x) \right).$$

Here n is a positive integer.
Multiply by $x^{c-1}(1-x)^{d-c-1}$ and integrate over $(0, 1)$ to get

(b)
$$ {}_3F_2\left(\begin{matrix} -2n, 2b, c \\ -n+b+\frac{1}{2}, d \end{matrix} ; 1 \right) = {}_4F_3\left(\begin{matrix} -n, b, c, d-c \\ -n+b+\frac{1}{2}, d/2, (d+1)/2 \end{matrix} ; 1 \right).$$

Deduce that

(c)
$$ {}_3F_2\left(\begin{matrix} 2a, 2b, -k \\ a+b+\frac{1}{2}, d \end{matrix} ; 1 \right) = {}_4F_3\left(\begin{matrix} a, b, -k, d+k \\ a+b+\frac{1}{2}, d/2, (d+1)/2 \end{matrix} ; 1 \right),$$

where k is a positive integer.

(d) Let $d = -\frac{k}{x} + \epsilon$ and $k \to \infty$ to see that (a) holds without restriction on n.

6. Show that for $0 < x < \frac{1}{2}$,

$$
{}_2F_1\left(\begin{matrix} 2a, 2b \\ a+b+\frac{1}{2} \end{matrix}; 1-x\right)
$$

$$
= \frac{\Gamma(a+b+\frac{1}{2})\Gamma(\frac{1}{2}-a-b)}{\Gamma(a-b+\frac{1}{2})\Gamma(b-a+\frac{1}{2})}{}_2F_1\left(\begin{matrix} a, b \\ a+b+\frac{1}{2} \end{matrix}; 4x(1-x)\right)
$$

$$
+ \frac{\Gamma(a+b+\frac{1}{2})\Gamma(a+b-\frac{1}{2})}{\Gamma(2a)\Gamma(2b)} x^{\frac{1}{2}-a-b}(1-x)^{\frac{1}{2}-a-b}
$$

$$
\cdot {}_2F_1\left(\begin{matrix} \frac{1}{2}-a, \frac{1}{2}-b \\ \frac{3}{2}-a-b \end{matrix}; 4x(1-x)\right).
$$

7. Prove formulas (3.2.3) and (3.2.14) for the elliptic integrals E and K.

8. Prove Elliot's result contained in Theorem 3.2.8.

9. Prove Lemma 3.2.9.

10. Let $\theta_2(q) := \sum_{-\infty}^{\infty} q^{(n+\frac{1}{2})^2}$, $\theta_3(q) := \sum_{-\infty}^{\infty} q^{n^2}$, $\theta_4(q) := \sum_{-\infty}^{\infty} (-1)^n q^{n^2}$ for $|q| < 1$.

    (a) Prove that $(\theta_3^2(q) + \theta_4^2(q))/2 = \theta_3^2(q^2)$ and $\sqrt{\theta_3^2(q)\theta_4^2(q)} = \theta_4^2(q^2)$.

    (b) Deduce that the arithmetic-geometric mean of $\theta_3^2(q)$ and $\theta_4^2(q)$ is 1 (i.e., $M(\theta_3^2(q), \theta_4^2(q)) = 1$).

    (c) Prove that $\theta_3^2(q) - \theta_3^2(q^2) = \theta_2^2(q^2)$.

    (d) Deduce from (c) and (a) that $\theta_3^2(q^2) - \theta_2^2(q^2) = \theta_4^2(q)$ and $\theta_3^4(q) = \theta_4^4(q) + \theta_2^4(q)$.

    (e) For $0 < q < 1$, let $k := k(q) := \theta_2^2(q)/\theta_3^2(q)$. Prove that $0 < k < 1$ and $M(1, k') = \theta_3^{-2}(q)$, where $k'^2 = 1 - k^2$. Also prove that

$$
K(k) = \frac{\pi}{2}\theta_3^2(q).
$$

11. (a) Use Exercise 2.30 to show that $\sqrt{x}\, \theta_3(e^{-\pi x}) = \theta_3(e^{-\pi/x})$ and $\sqrt{x}\, \theta_2(e^{-\pi x}) = \theta_4(e^{-\pi/x})$.

    (b) With $k(q)$ as in Exercise 10, show that $k(e^{-\pi x}) = k'(e^{-\pi/x})$.

    (c) Prove that $\frac{M(1,k')}{M(1,k)} = x$ or $\frac{K'(k)}{K(k)} = x$.

    (d) Show that the unique solution of $\theta_2^2(q)/\theta_3^2(q) = k$ for $0 < k < 1$ is $q = e^{-\pi K'/K}$.

12. Prove formula (3.2.9), that $I(a, b) = I((a+b)/2, \sqrt{ab})$.

13. Let $a < b < c$. Prove that

$$
\int_a^b \frac{dx}{\sqrt{(x-a)(x-b)(x-c)}} = \frac{\pi}{M(\sqrt{c-a}, \sqrt{c-b})},
$$

$$
\int_b^c \frac{dx}{\sqrt{(x-a)(x-b)(x-c)}} = \frac{i\pi}{M(\sqrt{c-a}, \sqrt{b-a})}.
$$

14. Multiply Euler's transformation

$$
_2F_1\left(\begin{matrix} a, b \\ c \end{matrix}; x\right) = (1-x)^{c-a-b}\,_2F_1\left(\begin{matrix} c-a, c-b \\ c \end{matrix}; x\right)
$$

by $x^{d-1}(1-x)^{e-d-1}$ and integrate over $(0,1)$ to obtain Corollary 3.3.5.

15. Prove that

$$
_3F_2\left(\begin{matrix} -n, -a, -b \\ c, 2-n-a-b-c \end{matrix}; 1\right) = \frac{(b+c-1)_n(a+c)_n}{(a+b+c-1)_n(c)_n}
$$

$$
\cdot\left[1 + \frac{an}{(b+c-1)(a+c+n-1)}\right].
$$

16. Prove Watson's identity in Theorem 3.5.5(i) by multiplying Equation (3.1.2),

$$
_2F_1\left(\begin{matrix} -2n, 2b \\ b-n+\frac{1}{2} \end{matrix}; x\right) = \,_2F_1\left(\begin{matrix} -n, b \\ b-n+\frac{1}{2} \end{matrix}; 4x(1-x)\right),
$$

by $x^{c-1}(1-x)^{c-1}$ and then integrating over $(0, 1)$. (Note that n is a nonnegative integer in the above formula. Otherwise the formula does not hold over $(0, 1)$.)

17. (a) Show that

$$
_5F_4\left(\begin{matrix} a, b, c, d, -n \\ 1+a-b, 1+a-c, 1+a-d, 1+a+n \end{matrix}; 1\right)
$$

$$
= \frac{(1+a)_n(1+a-b-c)_n\left(\frac{a}{2}-b+1\right)_n\left(\frac{a}{2}-c+1\right)_n}{(1+a-b)_n(1+a-c)_n\left(\frac{a}{2}+1\right)_n\left(\frac{a}{2}-b-c+1\right)_n},
$$

when $3a+2 = 2(b+c+d-n)$.

(b) Prove that

$$
_4F_3\left(\begin{matrix} a, b, c, -n \\ 1+a-b, 1+a-c, 1+a+n \end{matrix}; 1\right)
$$

$$
= \frac{(a+1)_n(a+1-2b)_{2n}\left(\frac{1}{2}\right)_n}{(a+1-b)_n(a+1)_{2n}\left(\frac{1}{2}-b\right)_n}
$$

$$
= \frac{\Gamma(a-2b+2n+1)\Gamma\left(n+\frac{1}{2}\right)\Gamma(a-b+1)\Gamma(a+n+1)\Gamma\left(\frac{1}{2}-b\right)}{\Gamma(a-2b+1)\Gamma\left(\frac{1}{2}\right)\Gamma(a-b+n+1)\Gamma(a+2n+1)\Gamma\left(\frac{1}{2}-b+n\right)},
$$

when $1+2a = 2b+2c-2n$.

(c) Deduce that

$$
_4F_3\left(\begin{matrix} -k, a, b, c \\ 1-a-k, 1-b-k, 1-c-k \end{matrix}; 1\right) = \frac{(2a)_k(2b)_k(a+b)_k}{(a)_k(b)_k(2a+2b)_k},
$$

when $1-2c = 2a+2b+2k$.

(d) Deduce Clausen's identity:

$$\left[{}_2F_1\left(\begin{matrix} a, b \\ a + b + \frac{1}{2} \end{matrix}; x\right)\right]^2 = {}_3F_2\left(\begin{matrix} 2a, 2b, a + b \\ a + b + \frac{1}{2}, 2a + 2b \end{matrix}; x\right).$$

(*Hint*: Equate the coefficient of x^n on both sides.) Note that a different proof of this identity was given in Exercise 2.13.

18. Prove that

(a) $\dfrac{(a - d - e + 1)_n}{(a - d + 1)_n(a - e + 1)_n}$

$$\cdot {}_4F_3\left(\begin{matrix} a - b - c + 1, d, e, -n \\ a - b + 1, a - c + 1, d + e - a - n \end{matrix}; 1\right)$$

$$= \dfrac{(a - b - c + 1)_n}{(a - b + 1)_n(a - c + 1)_n}$$

$$\cdot {}_4F_3\left(\begin{matrix} a - d - e + 1, b, c, -n \\ a - d + 1, a - e + 1, b + c - a - n \end{matrix}; 1\right).$$

(b) ${}_4F_3\left(\begin{matrix} a, (a/2) + 1, b, c \\ a/2, a - b + 1, a - c + 1 \end{matrix}; -1\right) = \dfrac{\Gamma(a - b + 1)\Gamma(a - c + 1)}{\Gamma(a + 1)\Gamma(a - b - c + 1)}.$

19. (a) Use the method of Lemma 3.4.2 to prove that

$${}_4F_3\left(\begin{matrix} a, b, c, -n \\ a - b + 1, a - c + 1, d \end{matrix}; 1\right) = \dfrac{(d - a)_n}{(d)_n}$$

$$\cdot {}_5F_4\left(\begin{matrix} a - d + 1, a/2, (a + 1)/2, a - b - c + 1, -n \\ a - b + 1, a - c + 1, a - d - n + 1)/2, (a - d - n)/2 + 1 \end{matrix}; 1\right).$$

This transforms a nearly poised ${}_4F_3$ into a balanced ${}_5F_4$. Deduce that

(b) ${}_3F_2\left(\begin{matrix} a, (a/2) + 1, -n \\ a/2, d \end{matrix}; 1\right) = \dfrac{(d - a - n - 1)(d - a)_{n-1}}{(d)_n}.$

(c) ${}_3F_2\left(\begin{matrix} a, b, -n \\ a - b + 1, 2b - n + 1 \end{matrix}; 1\right) = \dfrac{(a - 2b)_n((a/2) + 1 - b)_n(-b)_n}{(a - b + 1)_n((a/2) - b)_n(-2b)_n}.$

(d) ${}_4F_3\left(\begin{matrix} a, (a/2) + 1, b, -n \\ a/2, a - b + 1, 2b - n + 1 \end{matrix}; 1\right) = \dfrac{(a - 2b)_n(-b)_n}{(a - b + 1)_n(-2b)_n}.$

(e) ${}_4F_3\left(\begin{matrix} a, (a/2) + 1, b, -n \\ a/2, a - b + 1, 2b + 2 - n \end{matrix}; 1\right)$

$$= \dfrac{(a - 2b - 1)_n((a + 1)/2 - b)_n(-b - 1)_n}{(a - b + 1)_n((a/2) - b - (1/2))_n(-2b - 1)_n}.$$

(f) $_4F_3\left(\begin{array}{c}-n, b, c, e \\ 1-n-b, 1-n-c, d\end{array}; 1\right) = \dfrac{(d-e)_n}{(d)_n}$.

$\cdot {}_5F_4\left(\begin{array}{c}e, 1-n-b-c, -n/2, (1-n)/2, 1-n-d \\ 1-n-b, 1-n-c, (1+e-d-n)/2, (e-d-n)/2+1\end{array}; 1\right)$.

(Whipple)

See Bailey [1935, §§ 4.5 and 4.7] for the reference to Whipple.

20. Prove that

$$\left[{}_2F_1\left(\begin{array}{c}a, b \\ c\end{array}; x\right)\right]^2 = \sum_{n=0}^{\infty} \frac{(2a)_n(2b)_n\left(c-\frac{1}{2}\right)_n}{(c)_n(2c-1)_n n!}$$

$$\cdot {}_4F_3\left(\begin{array}{c}-n/2, (1-n)/2, \frac{1}{2}, a+b+\frac{1}{2}-c \\ a+\frac{1}{2}, b+\frac{1}{2}, \frac{3}{2}-n-c\end{array}; 1\right) x^n.$$

(*Hint*: Apply Exercise 19(f) to the $_4F_3$ that appears after squaring the $_2F_1$. Then apply Theorem 3.3.3.)

21. Obtain the transformation in Exercise 19(a) by multiplying Whipple's quadratic transformation for $_3F_2$ in (3.1.15) by x^{d+n-1} and equating the coefficients of x^n.

22. Derive the formula

$$_5F_4\left(\begin{array}{c}a, a/2+1, b, c, -n \\ a/2, a-b+1, a-c+1, d\end{array}; 1\right) = \frac{(d-a-n-1)(d-a)_{n-1}}{(d)_n}$$

$$\cdot {}_5F_4\left(\begin{array}{c}a/2+1, (a+1)/2, a-b-c+1, a-d+1, -n \\ (a-d-n+3)/2, (a-d-n+2)/2, a-b+1, a-c+1\end{array}; 1\right)$$

by using the formula in Exercise 19(b) instead of Dixon's theorem in the proof of Lemma 3.4.2. The above formula transforms a nearly poised $_5F_4$ into a balanced $_5F_4$. See Bailey [1935, §4.5].

23. (a) By letting $a \to 0$ in Corollary 3.5.2, evaluate the sum

$$\sum_{-\infty}^{\infty} \frac{(b)_n(c)_n(d)_n}{(1-b)_n(1-c)_n(1-d)_n}.$$

(b) More generally, show that

$$\sum_{-\infty}^{\infty} \frac{(a+b)_n(a+c)_n(a+d)_n}{(a-b+1)_n(a-c+1)_n(a-d+1)_n}$$

$$= \frac{\Gamma(a-b+1)\Gamma(a-c+1)\Gamma(a-d+1)\Gamma(1-b-c-d)\Gamma(b)\Gamma(c)\Gamma(d)}{\Gamma(a+b)\Gamma(a+c)\Gamma(a+d)\Gamma(1-b-c)\Gamma(1-b-d)\Gamma(1-c-d)}.$$

24. (a) Observe that Dougall's formula can be written as

$$
{}_7F_6\left(\begin{matrix} k, k/2 + 1, k + b - a, k + c - a, k + d - a, a + n, -n \\ k/2, a - b + 1, a - c + 1, a - d + 1, k - a - n + 1, k + n + 1 \end{matrix}; 1\right)
$$

$$
= \frac{(k+1)_n (b)_n (c)_n (d)_n}{(a-k)_n (a-b+1)_n (a-c+1)_n (a-d+1)_n},
$$

when $k = 2a - b - c - d + 1$.

(b) In the proof of Lemma 3.4.2 use (a) instead of the Pfaff–Saalschütz's identity to get

$$
{}_9F_8\left(\begin{matrix} a, a/2 + 1, b, c, d, e, f, g, -n \\ a/2, a - b + 1, a - c + 1, a - d + 1, a - e + 1, a - f + 1, a - g + 1, a + n + 1 \end{matrix}; 1\right)
$$

$$
= \frac{(a+1)_n (k - e + 1)_n (k - f + 1)_n (k - g + 1)_n}{(k+1)_n (a - e + 1)_n (a - f + 1)_n (a - g + 1)_n}
$$

$$
\cdot {}_9F_8\left(\begin{matrix} k, k/2 + 1, k + b - a, k + c - a, k + d - a, e, f, g, -n \\ k/2, a - b + 1, a - c + 1, a - d + 1, k - e + 1, k - f + 1, k - g + 1, k + n + 1 \end{matrix}; 1\right),
$$

when $k = 2a - b - c - d + 1$ and $b + c + d + e + f + g - n = 3a + 2$.

(c) Deduce Theorem 3.4.5 from (b).

See Bailey [1935, §4.3].

25. (a) Show that

$$
{}_4F_3\left(\begin{matrix} a, a/2 + 1, b, c \\ a/2, a - b + 1, a - c + 1 \end{matrix}; 1\right)
$$

$$
= \frac{\Gamma(a - b + 1)\Gamma(a - c + 1)\Gamma\big((a+1)/2\big)\Gamma\big((a+1)/2 - b - c\big)}{\Gamma(a+1)\Gamma(a - b - c + 1)\Gamma\big((a+1)/2 - b\big)\Gamma\big((a+1)/2 - c\big)}.
$$

(b) Add this identity to the one in 18(b) to obtain a formula for

$$
{}_7F_6\left(\begin{matrix} \dfrac{a}{2}, \dfrac{a+1}{2}, \dfrac{a}{4} + 1, \dfrac{b}{2}, \dfrac{b+1}{2}, \dfrac{c}{2}, \dfrac{d+1}{2} \\ \dfrac{a}{4}, \dfrac{a-b}{2} + 1, \dfrac{a-b+1}{2}, \dfrac{1}{2}, \dfrac{a-c}{2}, \dfrac{a-c+1}{2} + 1 \end{matrix}; 1\right).
$$

26. Show that

(a) $$ 1 - 5\left(\frac{1}{2}\right)^3 + 9\left(\frac{1 \cdot 3}{2 \cdot 4}\right)^3 - 13\left(\frac{1 \cdot 3 \cdot 5}{2 \cdot 4 \cdot 6}\right)^3 + \cdots = \frac{2}{\pi}. $$

(b) $$ 1 + 9\left(\frac{1}{4}\right)^4 + 17\left(\frac{1 \cdot 5}{4 \cdot 8}\right)^4 + 25\left(\frac{1 \cdot 5 \cdot 9}{4 \cdot 8 \cdot 12}\right)^4 + \cdots = \frac{2^{3/2}}{\sqrt{\pi}\{\Gamma(3/4)\}^2}. $$

(c) $$ 1 - 5\left(\frac{1}{2}\right)^5 + 9\left(\frac{1 \cdot 3}{2 \cdot 4}\right)^5 - 13\left(\frac{1 \cdot 3 \cdot 5}{2 \cdot 4 \cdot 6}\right)^5 + \cdots = \frac{2}{\{\Gamma(3/4)\}^4}. $$

(d) $$ 1 - \left(\frac{1}{2}\right)^3 + \left(\frac{1 \cdot 3}{2 \cdot 4}\right)^3 - \cdots = \left\{\frac{\Gamma(9/8)}{\Gamma(5/4)\Gamma(7/8)}\right\}^2. \qquad \text{(Ramanujan)} $$

27. Show that

(a)
$$s + (s+2)\left(\frac{s}{1}\right)^3 + (s+4)\left\{\frac{s(s+1)}{1 \cdot 2}\right\}^3 + \cdots$$

$$= \frac{\sin s\pi}{\pi} \cdot \frac{\Gamma((s+1)/2)\Gamma((1-3s)/2)}{[\Gamma((1-s)/2)]^2}.$$

(b) $s - (s+2)\left(\frac{s}{1}\right)^3 + (s+4)\left\{\frac{s(s+1)}{1 \cdot 2}\right\}^3 - \cdots = \frac{\sin s\pi}{\pi}.$ (Dougall)

For Exercises 26 and 27, see Hardy [1940, pp. 105–6] or Bailey [1935, p. 96].

28. Prove that

$$\frac{\pi^2}{4} = 1 + \sum_{n=1}^{\infty} \frac{2^{2n+1}(n!)^2}{(2n+2)!}.$$ (Takebe Kenko)

See Roy [1990] for reference.

29. Evaluate the sums

(a)
$$\sum_{k\geq 0} \binom{n+k}{m+2k}\binom{2k}{k}\frac{(-1)^k}{k+1},$$

(b)
$$\sum_{k=0}^{n} \binom{n}{k}\binom{r}{k}\binom{x+n+r-k}{n+r},$$

(c)
$$\sum_{k\geq 0} \binom{n}{k}\binom{2k}{k}(-1/2)^k,$$

(d)
$$\sum_{k\geq 0} \binom{2n+1}{2p+2k+1}\binom{p+k}{k},$$

(e)
$$\sum_{k\geq 0} \binom{2n}{2p+2k}\binom{p+k}{k},$$

(f)
$$\sum_{k\geq 0} \binom{m}{k}\binom{n}{k-j}\binom{p+k}{m+n},$$

(g)
$$\sum_{k=0}^{2n} (-1)^k \binom{2n}{k}\binom{2a}{a-n+k}\binom{2b}{b-n+k},$$

(h)
$$\sum_{k\geq 0} \binom{m-r+s}{k}\binom{n+r-s}{n-k}\binom{r+k}{m+n},$$

(i)
$$\sum_{k=0}^{n} \binom{n}{k}^2 \binom{m+2n-k}{2n},$$

(j)
$$\sum_{k=0}^{n} \frac{2n}{n+k} \binom{n+k}{2k} \binom{2k}{k} \frac{(-1)^k}{k+p},$$

(k)
$$\sum_{k=0}^{n} (-1)^k \binom{n}{k} \binom{2n+m}{k} \binom{2n}{k}^{-1}.$$

30. Prove that

$$_7F_6\left(\begin{matrix} a, 1+a/2, d/2, (d+1)/2, a-d, 1+2a-d+m, -m \\ a/2, 1+a-d/2, a+(1-d)/2, 1+d, d-a-m, 1+a+m \end{matrix} ; 1 \right)$$

$$= \frac{(1+a)_m (1+2a-2d)_m}{1+a-d)_m (1+2a-d)_m}.$$

See Bailey [1935, p. 98].

31. Prove that

(a)
$$_2F_1\left(\begin{matrix} a, b \\ a+b-\frac{1}{2} \end{matrix} ; x \right) {}_2F_1\left(\begin{matrix} a, b \\ a+b+\frac{1}{2} \end{matrix} ; x \right)$$

$$= {}_3F_2\left(\begin{matrix} 2a, 2b, a+b \\ 2a+2b-1, a+b+\frac{1}{2} \end{matrix} ; x \right),$$

(b)
$$_2F_1\left(\begin{matrix} a, b \\ a+b-\frac{1}{2} \end{matrix} ; x \right) {}_2F_1\left(\begin{matrix} a, b-1 \\ a+b-\frac{1}{2} \end{matrix} ; x \right)$$

$$= {}_3F_2\left(\begin{matrix} 2a, 2b-1, a+b-1 \\ 2a+2b-2, a+b-\frac{1}{2} \end{matrix} ; x \right). \qquad \text{(Orr)}$$

See Bailey [1935, p. 86].

32. Prove Theorem 3.6.1.

33. Prove that

$$\frac{\Gamma(x+m)\Gamma(y+m)}{\Gamma(m)\Gamma(x+y+m)} {}_3F_2\left(\begin{matrix} x, y, v+m-1 \\ v, x+y+m \end{matrix} ; 1 \right) \quad \text{to } n \text{ terms}$$

$$= \frac{\Gamma(x+n)\Gamma(y+n)}{\Gamma(n)\Gamma(x+y+n)} {}_3F_2\left(\begin{matrix} x, y, v+n-1 \\ v, x+y+n \end{matrix} ; 1 \right) \quad \text{to } m \text{ terms}.$$

34. Prove formula (3.8.2).

35. Prove that Wilson's polynomial in Definition 8.1 is symmetric in a, b, c, and d.

36. Show that

$$_7F_6\left(\begin{matrix} a, (a+2)/2, b, c, d, e, -n \\ (a/2), a+1-b, a+1-c, a+1-d, a+1-e, a+1+n \end{matrix}; 1\right)$$

$$= \frac{(a+1)_n(a-b-c)_n(a-b-d)_n(a-c-d)_n}{(a+1-b)_n(a+1-c)_n(a+1-d)_n(a-b-c-d)_n}$$

$$\cdot \left[1 + \frac{n(n+2a-b-c-d)(a-b-c-d)}{(a-b-c)(a-b-d)(a-c-d)}\right],$$

when $e = 2a + n - b - c - d$. The $_7F_6$ is 4-balanced and very well poised.

37. Prove formulas (3.5.2), (3.5.3), and (3.5.4) connected with reciprocal polynomials.

38. Prove Bailey's cubic transformations:

(a)
$$_3F_2\left(\begin{matrix} a, 2b-a-1, a+2-2b \\ b, a+\frac{3}{2}-b \end{matrix}; \frac{x}{4}\right)$$

$$= (1-x)^{-a}{}_3F_2\left(\begin{matrix} \frac{a}{3}, \frac{a+1}{3}, \frac{a+2}{3} \\ b, a-b+\frac{3}{2} \end{matrix}; \frac{-27x}{4(1-x)^3}\right).$$

(b)
$$_3F_2\left(\begin{matrix} a, b-\frac{1}{2}, a+1-b \\ 2b, 2a+2-2b \end{matrix}; x\right)$$

$$= \left(1-\frac{x}{4}\right)^{-a}{}_3F_2\left(\begin{matrix} \frac{a}{3}, \frac{a+1}{3}, \frac{a+2}{3} \\ b, a+\frac{3}{2}-b \end{matrix}; \frac{27x^2}{(4-x)^3}\right).$$

For comments on these cubic transformations and for the reference to Bailey, see Askey [1994].

39. Show that

$$_2F_1\left(\begin{matrix} a, a+1/2 \\ 2a+1 \end{matrix}; x\right) = \left(\frac{2}{1+\sqrt{1-x}}\right)^{2a}$$

and

$$_2F_1\left(\begin{matrix} a, a+1/2 \\ 2a \end{matrix}; x\right) = \frac{1}{\sqrt{1-x}}\left(\frac{2}{1+\sqrt{1-x}}\right)^{2a-1}.$$

40. Define

$$\phi(x; n) = \prod_{j=0}^{n-1}(a_j + xb_j),$$

$$\phi(x; 0) = 1.$$

If

$$f(n) = \sum_{k=0}^{n} (-1)^k \binom{n}{k} \phi(k; n) g(k),$$

show that

$$g(n) = \sum_{k=0}^{n} (-1)^k \binom{n}{k} \frac{q_k + k b_k}{\phi(n; k+1)} f(k).$$

See Gould and Hsu [1973].

41. By appropriate choices of $g(k)$ and $\phi(k; n)$, show that

$$_3F_2\left(\begin{array}{c} -n, n+a, 1+a-b-c \\ 1+a-b, 1+a-c \end{array}; 1\right) = \frac{(b)_n (c)_n}{(1+a-b)_n (1+a-c)_n}$$

gives the sum of the terminating very well poised $_5F_4$.

4

Bessel Functions and Confluent Hypergeometric Functions

In this chapter, we discuss the confluent hypergeometric equation and the related Bessel and Whittaker equations. The Bessel equation is important in mathematical physics because it arises from the Laplace equation when there is cylindrical symmetry. The confluent hypergeometric equation is obtained when we start with a second-order differential equation whose only singularities are regular singularities at 0, b, and ∞; we let $b \to \infty$. The resulting equation has ∞ as an irregular singular point obtained from a confluence of two regular singularities. Thus, the confluent equation can be derived from the hypergeometric equation by changing the independent variable x to x/b and letting $b \to \infty$. The solutions are $_1F_1$ functions, and some properties of these functions are limits of properties of $_2F_1$ functions. However, it is often easier to derive the results directly than to justify the limiting procedures.

Whittaker transformed the confluent equation to one in which the coefficient of the first derivative is zero. Solutions of this equation are called Whittaker functions. We find their series and integral representations and their asymptotic behavior and then give some important examples such as the error function and the parabolic cylinder function.

The Bessel equation can be derived from a particular Whittaker equation and can be solved to obtain the Bessel functions to which we devote a good portion of this chapter. These functions are also important for their role in Fourier transforms in several variables. We present some integral representations of Bessel functions due to Poisson, Gegenbauer, and others. Later, we discuss some interesting finite and infinite integrals involving Bessel functions as integrands. Some of these are really limits of generating functions for Jacobi polynomials.

The sine and cosine functions are particular cases of Bessel functions. Thus, it is useful to look for generalizations of formulas for these trigonometric functions to Bessel functions. Nicholson found a remarkable extension of $\sin^2 x + \cos^2 x = 1$ to Bessel functions; he expressed it as an integral formula. We present Nicholson's formula and later show how Lorch and Szego used it to derive results about zeros of Bessel functions.

We end this chapter with a discussion of some work of Saff and Varga on zero-free regions for the sequence of polynomials that are partial sums of the exponential function and, more generally, of the $_1F_1$ functions.

4.1 The Confluent Hypergeometric Equation

It is easily seen that the hypergeometric series

$$y = {}_pF_q\left(\begin{matrix} a_1, \ldots, a_p \\ b_1, \ldots, b_q \end{matrix}; x\right)$$ (4.1.1)

is a formal solution of the differential equation

$$\{\delta(\delta + b_1 - 1) \cdots (\delta + b_q - 1) - x(\delta + a_1) \cdots (\delta + a_p)\}y = 0, \quad (4.1.2)$$

where

$$\delta = x\frac{d}{dx}.$$

When $p > 2$ or $q > 1$, this equation is of order $\max(p, q + 1) > 2$, and the resulting equation is not as useful as the hypergeometric equation. When $q = 1$ and $p = 0$ or 1, the equation is still of second order with a regular singular point at $x = 0$, but the other singular point is at $x = \infty$ and is an irregular singular point. Although irregular singular points cause serious problems, it is still possible to say something about the solutions near them.

We consider the case where $p = q = 1$. Then the equation is

$$\left\{x\frac{d}{dx}\left(x\frac{d}{dx} + c - 1\right) - x\left(x\frac{d}{dx} + a\right)\right\}y = 0,$$

or

$$xy'' + (c - x)y' - ay = 0. \quad (4.1.3)$$

This equation can be obtained from the hypergeometric equation

$$x(1 - x)y'' + \{c - (a + b + 1)x\}y' - aby = 0$$

by the following process. Replace x with x/b, so that the new equation has singular points at $0, b$, and ∞. Now let $b \to \infty$ so that infinity is a confluence of two singularities. The resulting Equation (4.1.3) is called the confluent hypergeometric equation.

When c is not an integer, two independent solutions of the hypergeometric equation around $x = 0$ are

$$_2F_1\left(\begin{matrix} a, b \\ c \end{matrix}; x\right) \quad \text{and} \quad x^{1-c}{}_2F_1\left(\begin{matrix} a+1-c, b+1-c \\ 2-c \end{matrix}; x\right).$$

Replace x with x/b in these expressions and let $b \to \infty$ to get

$$_1F_1\left(\begin{matrix} a \\ c \end{matrix}; x\right) \quad \text{and} \quad x^{1-c}{}_1F_1\left(\begin{matrix} a+1-c \\ 2-c \end{matrix}; x\right). \tag{4.1.4}$$

These are two independent solutions of (4.1.3) around $x = 0$. They are valid over the whole complex plane, since $_1F_1$ is an entire function. One has to be somewhat more careful to find the solutions around infinity. A solution of the hypergeometric equation about infinity is given by

$$x^{-a}{}_2F_1\left(\begin{matrix} a, a+1-c \\ a+1-b \end{matrix}; \frac{1}{x}\right).$$

When x is changed to x/b and $b \to \infty$, this expression tends termwise to

$$x^{-a}{}_2F_0\left(\begin{matrix} a, a+1-c \\ - \end{matrix}; -\frac{1}{x}\right).$$

This series diverges, so it does not directly give a solution of Equation (4.1.3). However, it is possible to find an integral representation of a solution of (4.1.3) that has this series as an asymptotic expansion. To find this integral representation, start with Euler's hypergeometric integral

$$x^{-a}{}_2F_1\left(\begin{matrix} a+1-c, a \\ a+1-b \end{matrix}; \frac{b}{x}\right)$$
$$= x^{-a}\frac{\Gamma(a+1-b)(-b)^{-a}}{\Gamma(a)\Gamma(1-b)}\int_0^{-b}\left(1+\frac{t}{x}\right)^{c-a-1}t^{a-1}\left(1+\frac{t}{b}\right)^{-b}dt. \tag{4.1.5}$$

It is possible to let $b \to -\infty$ in the integral, though (4.1.5) no longer makes sense in the limit. The right side of (4.1.5) tends to

$$\frac{x^{-a}}{\Gamma(a)}\int_0^\infty e^{-t}t^{a-1}\left(1+\frac{t}{x}\right)^{c-a-1}dt = \frac{1}{\Gamma(a)}\int_0^\infty e^{-xt}t^{a-1}(1+t)^{c-a-1}dt. \tag{4.1.6}$$

This integral converges for Re $a > 0$ and Re $x > 0$. It is easy to verify that (4.1.6) is a solution of the confluent equation (4.1.3).

Remark 4.1.1 Here is another way of arriving at (4.1.6). Let

$$y(x) = \int_0^\infty e^{-xt}f(t)dt.$$

Then

$$xy'' + (c - x)y' - ay$$

$$= \int_0^\infty (xt^2 - (c - x)t - a)e^{-xt} f(t)dt$$

$$= \int_0^\infty \left[\left(-\frac{\partial}{\partial t} e^{-xt} \right) t^2 - \left(\frac{\partial}{\partial t} e^{-xt} \right) t - (a + ct)e^{-xt} \right] f(t)dt$$

$$= \int_0^\infty e^{-xt} \{ [t^2 f(t)]' + [tf(t)]' - (a + ct) f(t) \} dt = 0.$$

The last equation holds when

$$\frac{f'(t)}{f(t)} = \frac{a - 1 + (c - 2)t}{t(t + 1)} = \frac{a - 1}{t} + \frac{c - a - 1}{t + 1}$$

or

$$f(t) = t^{a-1}(1 + t)^{c-a-1},$$

and we get integral (4.1.6) once again.

Suppose $x \geq 1$ in (4.1.6). By Taylor's theorem

$$\left(1 + \frac{t}{x} \right)^{c-a-1} = 1 + \sum_{k=1}^{n-1} \frac{(-1)^k (a + 1 - c)_k}{k!} \frac{t^k}{x^k}$$

$$+ \frac{(-1)^n (a + 1 - c)_n}{n!} \frac{t^n}{x^n} \left(1 + \frac{\theta t}{x} \right)^{c-a-n-1}, \qquad (4.1.7)$$

where $0 < \theta < 1$. So

$$\frac{1}{\Gamma(a)} \int_0^\infty e^{-t} t^{a-1} \left(1 + \frac{t}{x} \right)^{c-a-1} dt = \sum_{k=0}^{n-1} \frac{(a + 1 - c)_k (a)_k}{k!} \left(-\frac{1}{x} \right)^k + R_n(x),$$

where

$$R_n(x) = \frac{(-1)^n (a + 1 - c)_n}{n! \Gamma(a) x^n} \int_0^\infty e^{-t} t^{a+n-1} \left(1 + \frac{\theta t}{x} \right)^{c-a-n-1} dt.$$

The integral converges and

$$R_n(x) = 0 \left(\frac{1}{x^n} \right).$$

Thus we see that, except for a constant factor,

$$(-x)^{-a} {}_2F_0 \left(\begin{matrix} a, a + 1 - c \\ - \end{matrix} ; -\frac{1}{x} \right)$$

gives an asymptotic expansion of a solution of the confluent hypergeometric equation when $x > 1$ is large. In fact, we need not restrict ourselves to positive x if instead of (4.1.7) we use

$$\left(1 + \frac{t}{x}\right)^{-m} = \sum_{k=0}^{n-1} \frac{(m)_k}{k!} \left(-\frac{t}{x}\right)^k + \frac{(m)_n}{n!} \left(1 + \frac{t}{x}\right)^{-m} \int_0^{t/x} u^n (1+u)^{m-1} du.$$

(4.1.8)

This holds as long as $1 + \frac{t}{x}$ is not a negative real number. To remove the restriction $\mathrm{Re}\, x > 0$, which is necessary for convergence in (4.1.6), consider the integral

$$\int_\infty^{(0+)} e^{-xt} t^{a-1} (1+t)^{c-a-1} dt$$

(4.1.9)

or

$$x^{-a} \int_\infty^{(0+)} e^{-t} t^{a-1} \left(1 + \frac{t}{x}\right)^{c-a-1} dt.$$

(4.1.10)

These integrals are also solutions of the confluent equation, but without the restrictions on x and a needed in (4.1.6). From (4.1.10) and (4.1.8), we can once again obtain the $_2F_0$ asymptotic expansion for large $|x|$, when $|\arg x| \leq \pi - \delta < \pi$.

Relations among solutions of the hypergeometric equation suggest corresponding relations among solutions of the confluent equation. These can then be proved rigorously. Similarly, transformations of hypergeometric functions imply transformations of the $_1F_1$ function. The following are a few examples.

In Pfaff's transformation,

$$_2F_1\left(\begin{matrix} a, b \\ c \end{matrix}; x\right) = (1-x)^{-b} {}_2F_1\left(\begin{matrix} b, c-a \\ c \end{matrix}; \frac{x}{x-1}\right),$$

(2.2.6)

change x to x/b and let $b \to \infty$ to get Kummer's first transformation,

$$_1F_1\left(\begin{matrix} a \\ c \end{matrix}; x\right) = e^x {}_1F_1\left(\begin{matrix} c-a \\ a \end{matrix}; -x\right).$$

(4.1.11)

A similar procedure applied to the quadratic transformation,

$$_2F_1\left(\begin{matrix} a, b \\ 2a \end{matrix}; \frac{4x}{(1+x)^2}\right) = (1+x)^{2a} {}_2F_1\left(\begin{matrix} a, a + \frac{1}{2} - b \\ b + \frac{1}{2} \end{matrix}; x^2\right),$$

(3.1.11)

leads to Kummer's second transformation,

$$_1F_1\left(\begin{matrix} a \\ 2a \end{matrix}; 4x\right) = e^{2x} {}_0F_1\left(\begin{matrix} - \\ a + 1/2 \end{matrix}; x^2\right).$$

(4.1.12)

Finally, the three-term relation

$$(-x)^{-a} {}_2F_1\left(\begin{matrix} a, a+1-c \\ a+1-b \end{matrix}; \frac{1}{x}\right)$$

$$= \frac{\Gamma(1-c)\Gamma(a+1-b)}{\Gamma(a+1-c)\Gamma(1-b)} {}_2F_1\left(\begin{matrix} a, b \\ c \end{matrix}; x\right)$$

$$+ \frac{\Gamma(c-1)\Gamma(a+1-b)}{\Gamma(a)\Gamma(c-b)} (-x)^{1-c} {}_2F_1\left(\begin{matrix} a+1-c, b+1-c \\ 2-c \end{matrix}; x\right)$$

suggests that

$$\frac{\Gamma(1-c)}{\Gamma(a+1-c)}{}_1F_1\left(\begin{matrix}a\\c\end{matrix};x\right) + \frac{\Gamma(c-1)}{\Gamma(a)}x^{1-c}{}_1F_1\left(\begin{matrix}a+1-c\\2-c\end{matrix};x\right)$$

$$\sim x^{-a}{}_2F_0\left(\begin{matrix}a,a+1-c\\-\end{matrix};-\frac{1}{x}\right). \tag{4.1.13}$$

Formulas (4.1.11) and (4.1.12) can be proved directly. Thus, the coefficient of x^n on the right side of (4.1.11) is

$$\sum_{k=0}^{n}\frac{(c-a)_k(-1)^k}{(c)_k k!(n-k)!} = \frac{1}{n!}{}_2F_1\left(\begin{matrix}-n,c-a\\c\end{matrix};1\right)$$

$$= \frac{(a)_n}{n!(c)_n},$$

which is the coefficient of x^n on the left side. There is a similar proof of (4.1.12). We give a proof of (4.1.13) in the next section where we approach this topic from a different point of view.

4.2 Barnes's Integral for $_1F_1$

We can find the contour integral representation for $_1F_1(a;c;x)$ by computing its Mellin transform. This is similar to finding such a representation for the hypergeometric function. Let

$$I = \int_0^\infty x^{s-1}{}_1F_1\left(\begin{matrix}a\\c\end{matrix};-x\right)dx.$$

By Kummer's first transformation (4.1.11),

$$I = \int_0^\infty x^{s-1}e^{-x}{}_1F_1\left(\begin{matrix}c-a\\c\end{matrix};x\right)dx$$

$$= \int_0^\infty \sum_{n=0}^{\infty}\frac{(c-a)_n}{(c)_n n!}e^{-x}x^{s+n-1}dx$$

$$= \Gamma(s)\,{}_2F_1\left(\begin{matrix}c-a,s\\c\end{matrix};1\right) = \frac{\Gamma(c)}{\Gamma(a)}\frac{\Gamma(s)\Gamma(a-s)}{\Gamma(c-s)}.$$

By Mellin inversion, we should have

$$\Gamma(a)\,{}_1F_1\left(\begin{matrix}a\\c\end{matrix};-x\right) = \frac{\Gamma(c)}{2\pi i}\int_{-i\infty}^{i\infty}\frac{\Gamma(a-s)\Gamma(s)}{\Gamma(c-s)}x^{-s}ds \tag{4.2.1}$$

or

$$\Gamma(a)\,{}_1F_1\left(\begin{matrix}a\\c\end{matrix};x\right) = \frac{\Gamma(c)}{2\pi i}\int_{-i\infty}^{i\infty}\frac{\Gamma(a+s)\Gamma(-s)}{\Gamma(c+s)}(-x)^s ds. \tag{4.2.2}$$

Of course, once we have seen Barnes's integral for a $_2F_1$, this can be written by analogy. In (4.2.2) we have $-x > 0$, but this can be extended. The next theorem gives the extension and is due to Barnes.

Theorem 4.2.1 For $|\arg(-x)| < \pi/2$ and a not a negative integer or zero,

$$\Gamma(a)\,_1F_1(a; c; x) = \frac{\Gamma(c)}{2\pi i} \int_{-i\infty}^{i\infty} \frac{\Gamma(a+s)}{\Gamma(c+s)}\Gamma(-s)(-x)^s ds,$$

where the path of integration is curved, if necessary, to separate the negative poles from the positive ones.

The proof follows the same lines as that of Theorem 2.4.1. The reader should work out the details.

Again as in Chapter 2, this representation of $_1F_1$ can be used to obtain an asymptotic expansion by moving the line of integration to the left. The residues come from the poles of $\Gamma(a+s)$ at $s = -a - n$. The result is contained in the next theorem.

Theorem 4.2.2 For $\operatorname{Re} x < 0$,

$$_1F_1(a; c; x) \sim \frac{\Gamma(c)}{\Gamma(c-a)}(-x)^{-a}\,_2F_0\left(\begin{matrix} a, a+1-c \\ - \end{matrix} ; -\frac{1}{x}\right).$$

Corollary 4.2.3 For $\operatorname{Re} x > 0$,

$$_1F_1(a; c; x) \sim \frac{\Gamma(c)e^x}{\Gamma(a)x^{c-a}}\,_2F_0\left(\begin{matrix} c-a, 1-a \\ - \end{matrix} ; \frac{1}{x}\right).$$

Proof. This follows from Theorem 4.2.2 after an application of (4.1.11). ∎

Now note that the $_2F_0$ in Theorem 4.2.2 suggests the integral

$$J = \frac{1}{2\pi i} \int_{-i\infty}^{i\infty} \Gamma(-s)\Gamma(1 - c - s)\Gamma(a+s)x^s ds. \tag{4.2.3}$$

Again the line of integration is suitably curved. By moving the line of integration to the left and picking up the residues at $s = -k - a$, where $k \geq 0$ is an integer, we get

$$J = \Gamma(a)\Gamma(1 + a - c)x^{-a} \sum_{k=0}^{n} \frac{(a)_n(1 + a - c)_n}{n!}\left(-\frac{1}{x}\right)^n$$

$$+ \frac{1}{2\pi i} \int_{-a-n-i\infty}^{-a-n+i\infty} \Gamma(-s)\Gamma(1 - c - s)\Gamma(a+s)x^s ds. \tag{4.2.4}$$

To ensure the validity of this formula, we need an estimate of the integrand on $s = \sigma + iT$, where T is large and $-a - n \leq \sigma \leq 0$. By Stirling's formula (see

Corollary 1.4.4),

$$|\Gamma(-s)\Gamma(1-c-s)\Gamma(a+s)x^s|$$

$$\sim (2\pi)^{3/2} T^{\text{Re}(a-c-1/2)} e^{-T(\arg x + 3\pi/2)} e^{-\frac{\pi}{2}|\text{Im}(a+c)|} \left| \frac{x}{M} \right|^\sigma .$$

The expression on the right-hand side dies out exponentially when $|\arg x| \le 3\pi/2 - \delta < 3\pi/2$. We assume this condition and (4.2.4) is then true. The last integral in (4.2.4) is equal to

$$\frac{x^{-a-n}}{2\pi i} \int_{-i\infty}^{i\infty} \Gamma(a+n-s)\Gamma(1+a-c+n-s)\Gamma(s-n)x^s ds = 0(x^{-a-n})$$

when $|x|$ is large. Thus

$$J \sim \Gamma(a)\Gamma(1+a-c)x^{-a} {}_2F_0\left(\begin{matrix} a, 1+a-c \\ - \end{matrix} ; -\frac{1}{x} \right), \qquad (4.2.5)$$

the asymptotic expansion being valid for $|\arg x| < 3\pi/2$.

However, when the line of integration is moved to the right, it can be seen that

$$J = \Gamma(a)\Gamma(1-c) \, {}_1F_1\left(\begin{matrix} a \\ c \end{matrix} ; x \right) + \Gamma(a+1-c)\Gamma(c-1)x^{1-c} {}_1F_1\left(\begin{matrix} a+1-c \\ 2-c \end{matrix} ; x \right)$$

$$(4.2.6)$$

when $|\arg x| < 3\pi/2$. This proves the next theorem.

Theorem 4.2.4 *For* $|\arg x| < 3\pi/2$, *(4.2.5) and (4.2.6) hold and*

$$\Gamma(a)\Gamma(1-c) {}_1F_1\left(\begin{matrix} a \\ c \end{matrix} ; x \right) + \Gamma(a+1-c)\Gamma(c-1)x^{1-c} {}_1F_1\left(\begin{matrix} a+1-c \\ 2-c \end{matrix} ; x \right)$$

$$\sim \Gamma(a)\Gamma(a+1-c)x^{-a} {}_2F_0\left(\begin{matrix} a, a+1-c \\ - \end{matrix} ; \frac{-1}{x} \right).$$

Observe that this is the same as relation (4.1.13).

This theorem gives the linear combination of the two independent ${}_1F_1$ that produce the recessive solution of the confluent equation. This is of special interest in numerical work. To clarify this point, consider the simpler equation $y'' - y = 0$, which has independent solutions $\sinh x$ and $\cosh x$ as well as e^x and e^{-x}. This equation has an essential singularity at ∞ and, in the neighborhood of this point for $\text{Re } x > 0$, e^{-x} is the recessive solution. Any other solution independent of e^{-x} is a dominant solution. Thus the combination $Ae^x + Be^{-x}$ can be computed very accurately from values of e^x and e^{-x}. However, $A \cosh x - B \sinh x$ creates problems, especially when $A \approx B$ and x has a large positive real part.

4.3 Whittaker Functions

Whittaker [1904] gave another important form of the confluent equation. This is obtained from Kummer's equation (4.1.3) by a transformation that eliminates the first derivative from the equation. Set $y = e^{x/2}x^{-c/2}\omega(x)$ in (4.1.3). The equation satisfied by ω is

$$\omega'' + \left[-\frac{1}{4} + \left(\frac{c}{2} - a\right)\frac{1}{x} + \frac{c}{2}\left(1 - \frac{c}{2}\right)\frac{1}{x^2}\right]\omega = 0.$$

Two independent solutions of this equation can have a more symmetric form if we set

$$c = 1 + 2m, \qquad \frac{c}{2} - a = k$$

or

$$m = \frac{c-1}{2}, \qquad a = \frac{1}{2} + m - k.$$

The result is Whittaker's equation:

$$W'' + \left\{-\frac{1}{4} + \frac{k}{x} + \frac{\frac{1}{4} - m^2}{x^2}\right\}W = 0. \tag{4.3.1}$$

From the solutions (4.1.4) of (4.1.3), it is clear that when $2m$ is not an integer, two independent solutions of (4.3.1) are

$$M_{k,m}(x) = e^{-x/2}x^{\frac{1}{2}+m}{}_1F_1\left(\begin{matrix}\frac{1}{2} + m - k\\1 + 2m\end{matrix}; x\right) \tag{4.3.2}$$

and

$$M_{k,-m}(x) = e^{-x/2}x^{\frac{1}{2}-m}{}_1F_1\left(\begin{matrix}\frac{1}{2} - m - k\\1 - 2m\end{matrix}; x\right). \tag{4.3.3}$$

The solutions $M_{k,\pm m}(x)$ are called Whittaker functions. Because of the factors $x^{\frac{1}{2}\pm m}$, the functions are not single valued in the complex plane. Usually one restricts x to $|\arg x| < \pi$.

Formulas for ${}_1F_1$ obviously carry over to the Whittaker functions. Kummer's first formula, for example, takes the form

$$x^{-\frac{1}{2}-m}M_{k,m}(x) = (-x)^{-\frac{1}{2}-m}M_{-k,m}(-x). \tag{4.3.4}$$

A drawback of the functions $M_{k,\pm m}(x)$ is that one of them is not defined when $2m$ is an integer. Moreover, the asymptotic behavior of the solution of Whittaker's equation is not easily obtained from these functions. So we use the integral in

(4.1.10) to derive another Whittaker function, $W_{k,m}(x)$. This is defined by

$$
W_{k,m}(x) := -\frac{1}{2\pi i} \Gamma\left(k + \frac{1}{2} - m\right) e^{-x/2} x^k
$$
$$
\cdot \int_\infty^{(0+)} (-t)^{-k-\frac{1}{2}+m} \left(1 + \frac{t}{x}\right)^{k-\frac{1}{2}+m} e^{-t} dt, \qquad (4.3.5)
$$

where $\arg x$ takes its principal value and the contour does not contain the point $t = -x$. Moreover, $|\arg(-t)| \le \pi$, and when t approaches 0 along the contour, $\arg(1 + t/x) \to 0$. This makes the integrand single valued. It is easily verified that $W_{k,m}(x)$ is also a solution of (4.3.1). Note that Whittaker's equation (4.3.1) is unchanged when x and k change sign. Thus $W_{-k,m}(-x)$ is also a solution and is independent of $W_{k,m}(x)$. This is clear when one considers the asymptotic expansion for $W_{k,m}(x)$. The reader should verify that the remarks after (4.1.10) imply that

$$
W_{k,m}(x) \sim e^{-x/2} x^k \,_2F_0\left(\frac{1}{2} - k + m, \frac{1}{2} - k - m \atop - \;;\; -\frac{1}{x}\right), \qquad |x| \to \infty, \quad (4.3.6)
$$

when $|\arg x| \le \pi - \delta < \pi$. Consequently,

$$
W_{\pm k,m}(\pm x) = e^{\pm x/2} (\pm x)^{\pm k} \left\{1 + 0\left(\frac{1}{x}\right)\right\}.
$$

This shows that $W_{k,m}(x)$ and $W_{-k,m}(-x)$ are linearly independent.

4.4 Examples of $_1F_1$ and Whittaker Functions

This section contains some important examples of $_1F_1$ and Whittaker functions that occur frequently enough in mathematics, statistics, and physics to be given names.

(a) The simplest example is given by

$$
e^x = \,_1F_1(a; a; x). \qquad (4.4.1)
$$

(b) The error function is defined by

$$
\operatorname{erf} x = \frac{2}{\sqrt{\pi}} \int_0^x e^{-t^2} dt = 1 - \operatorname{erfc} x \quad (x \text{ real}), \qquad (4.4.2)
$$

where

$$
\operatorname{erfc} x = \frac{2}{\sqrt{\pi}} \int_x^\infty e^{-t^2} dt.
$$

It is easy to see that $\operatorname{erf} x = \frac{2x}{\sqrt{\pi}} \,_1F_1(1/2; 3/2; x)$.

To express the error function in terms of $W_{k,m}(x)$ we need to write (4.3.5) as an integral over $(0, \infty)$. Assume that Re $(k - \frac{1}{2} - m) < 0$; then (4.3.5) can be written as

$$W_{k,m}(x) = e^{-x/2}x^k\Gamma\left(k + \frac{1}{2} - m\right)\frac{\sin\pi\left(k + \frac{1}{2} - m\right)}{\pi}$$

$$\cdot \int_0^\infty e^{-t}t^{-k-\frac{1}{2}+m}(1+t/x)^{k-\frac{1}{2}+m}dt$$

$$= \frac{e^{-x/2}x^k}{\Gamma\left(\frac{1}{2} - k + m\right)}\int_0^\infty e^{-t}t^{-k-\frac{1}{2}+m}(1+t/x)^{k-\frac{1}{2}+m}dt. \qquad (4.4.3)$$

The integral converges for Re $(k - \frac{1}{2} - m) < 0$ and Re $x > 0$. Note the relation of (4.4.3) with the integral in (4.1.6). Now set $t = u^2 - s^2$, where u is the new variable. Then

$$W_{k,m}(x) = \frac{e^{-x/2}x^k2e^{s^2}}{\Gamma\left(\frac{1}{2} - k + m\right)}\int_s^\infty (u^2-s^2)^{-k-\frac{1}{2}+m}\left(\frac{x+u^2-s^2}{x}\right)^{k-\frac{1}{2}+m}e^{-u^2}u\,du.$$

Set $k = -1/4$, $m = 1/4$, and $x = s^2$ to get

$$W_{-1/4,1/4}(s^2) = 2e^{s^2/2}\sqrt{s}\int_s^\infty e^{-u^2}du.$$

Thus

$$\text{erf } x = 1 - \frac{1}{\sqrt{\pi}}e^{-x^2/2}x^{-1/2}W_{-1/4,1/4}(x^2). \qquad (4.4.4)$$

An asymptotic expansion for erf x can be derived from this formula and (4.3.6).

(c) The incomplete gamma function is defined by

$$\gamma(a, x) = \int_0^x e^{-t}t^{a-1}dt = \Gamma(a) - \int_x^\infty e^{-t}t^{a-1}dt = \Gamma(a) - \Gamma(a, x). \qquad (4.4.5)$$

After expanding e^{-t} as a series in t and term-by-term integration, it is clear that

$$\gamma(a, x) = \frac{x^a}{a}\,_1F_1(a; a + 1; x). \qquad (4.4.6)$$

The reader may also verify that

$$\Gamma(a, x) = e^{-x/2}x^{\frac{a-1}{2}}W_{\frac{a-1}{2}, \frac{a}{2}}(x). \qquad (4.4.7)$$

(d) The logarithmic integral li(x) is defined by

$$\text{li}(x) = \int_0^x \frac{dt}{\log t}.$$

Check that

$$\mathrm{li}(x) = -(-\log x)^{-1/2}x^{1/2}W_{-1/2,0}(-\log x).$$

If x is complex, take $|\arg(-\log x)| < \pi$.

Additional examples of Whittaker functions such as the integral sine and cosine and Fresnel integrals are given in Exercise 4.

(e) The parabolic cylinder functions are also particular cases of the Whittaker functions. To see how these functions arise, consider the Laplace equation

$$\frac{\partial^2 u}{\partial x^2} + \frac{\partial^2 u}{\partial y^2} + \frac{\partial^2 u}{\partial z^2} = 0. \tag{4.4.8}$$

The coordinates of the parabolic cylinder ξ, η, z are defined by

$$x = \frac{1}{2}(\xi^2 - \eta^2), \quad y = \xi\eta, \quad z = z. \tag{4.4.9}$$

Apply the change of variables (4.4.9) to the Laplace equation. The result after some calculation is

$$\frac{1}{\xi^2 + \eta^2}\left(\frac{\partial^2 u}{\partial \xi^2} + \frac{\partial^2 u}{\partial \eta^2}\right) + \frac{\partial^2 u}{\partial z^2} = 0.$$

This equation has particular solutions of the form $U(\xi)V(\eta)W(z)$, which can be obtained by separation of variables. The equation satisfied by U, for example, has the form

$$\frac{d^2 U}{d\xi^2} + (\sigma\xi^2 + \lambda)U = 0,$$

where σ and λ are constants. After a slight change in variables, this equation can be written as

$$\frac{d^2 y}{dx^2} + \left(n + \frac{1}{2} - \frac{1}{4}x^2\right)y = 0. \tag{4.4.10}$$

Equation (4.4.10) is called Weber's equation. It can be checked that

$$D_n(x) = 2^{\frac{n}{2}+\frac{1}{4}}x^{-\frac{1}{2}}W_{\frac{n}{2}+\frac{1}{4},-\frac{1}{4}}(x^2/2) \qquad (|\arg x| < 3\pi/4) \tag{4.4.11}$$

is a solution of (4.4.10). The constant factor is chosen to make the coefficient of the first term in the asymptotic expansion of $D_n(x)$ equal to one. $D_n(x)$ is called the parabolic cylinder function. When n is a positive integer, $D_n(x)$ is $e^{-\frac{1}{4}x^2}$ times a polynomial, which, except for a constant factor, is $H_n(x/\sqrt{2})$, where $H_n(x)$ is the Hermite polynomial of degree n. These polynomials will be studied in Chapter 6.

(f) In the study of scattering of charged particles by spherically symmetric potentials, we can take (see Schiff [1947, Chapter V]) the solution of the Schrödinger equation

$$-\frac{\hbar^2}{2\mu}\nabla^2 u + Vu = Eu$$

to be of the form

$$u(r, \theta) = \sum_{\ell=0}^{\infty} \frac{y_\ell(r)}{r} P_\ell(\cos\theta),$$

where P_ℓ is the Legendre polynomial of degree ℓ and y_ℓ satisfies the equation

$$\frac{d^2 y}{dr^2} + \left[k^2 - U(r) - \frac{\ell(\ell+1)}{r^2} \right] y = 0,$$

$$k^2 = \frac{2\mu E}{\hbar^2}, \qquad U(r) = \frac{2\mu V(r)}{\hbar^2}.$$

By a change of variables we can take $k = 1$. The Coulomb potential is given by $U(r) = 2\eta/r$, so the equation for y is

$$\frac{d^2 y}{dr^2} + \left[1 - \frac{2\eta}{r} - \frac{\ell(\ell+1)}{r^2} \right] y = 0.$$

Comparison of this equation with Whittaker's equation (4.3.1) shows that

$$y_\ell = r^{\ell+1} e^{-ir} {}_1F_1(\ell + 1 - i\eta; 2\ell + 2; 2ir).$$

The function

$$\Phi_\ell(\eta, r) := e^{-ir} {}_1F_1(\ell + 1 - i\eta; 2\ell + 2; 2ir)$$

is called the Coulomb wave function.

4.5 Bessel's Equation and Bessel Functions

Bessel functions are important in mathematical physics because they are solutions of the Bessel equation, which is obtained from Laplace's equation when there is cylindrical symmetry. The rest of this chapter gives an account of some elementary properties of Bessel functions.

When $k = 0$ and $m = \alpha$ in Whittaker's equation (4.3.1), we get

$$\frac{d^2 W}{d\xi^2} + \left[-\frac{1}{4} + \frac{1/4 - \alpha^2}{\xi^2} \right] W = 0.$$

If we set $y(x) = \sqrt{x}\, W(2ix)$, then y satisfies the equation

$$\frac{d^2 y}{dx^2} + \frac{1}{x}\frac{dy}{dx} + (1 - \alpha^2/x^2)y = 0. \tag{4.5.1}$$

This equation is called Bessel's equation of order α. It is easily verified that

$$J_\alpha(x) := \frac{(x/2)^\alpha}{\Gamma(\alpha+1)} {}_0F_1\left(\begin{array}{c} - \\ \alpha+1 \end{array}; -\left(\frac{x}{2}\right)^2\right) \tag{4.5.2}$$

is a solution of (4.5.1). $J_\alpha(x)$ is the Bessel function of the first kind of order α. From (4.1.12), we have another representation of $J_\alpha(x)$. This is

$$J_\alpha(x) = \frac{(x/2)^\alpha}{\Gamma(\alpha+1)} e^{-ix} {}_1F_1\left(\begin{array}{c} \alpha+1/2 \\ 2\alpha+1 \end{array}; 2ix\right). \tag{4.5.3}$$

Equation (4.5.1) is unchanged when α is replaced by $-\alpha$. This means that $J_{-\alpha}(x)$ is also a solution of (4.5.1). One can check directly that when α is not an integer, $J_\alpha(x)$ and $J_{-\alpha}(x)$ are linearly independent solutions. When α is an integer, say $\alpha = n$, then

$$J_{-n}(x) = (-1)^n J_n(x). \tag{4.5.4}$$

Therefore $J_{-n}(x)$ is linearly dependent on $J_n(x)$. A second linearly independent solution can be found as follows. Since $(-1)^n = \cos n\pi$, we see that $J_\alpha(x)\cos\pi\alpha - J_{-\alpha}(x)$ is a solution of (4.5.1), which vanishes when α is an integer. Define

$$Y_\alpha(x) := \frac{J_\alpha(x)\cos\pi\alpha - J_{-\alpha}(x)}{\sin\pi\alpha}. \tag{4.5.5}$$

When $\alpha = n$ is an integer, $Y_\alpha(x)$ is defined as a limit. By L'Hopital's rule,

$$Y_n(x) = \lim_{\alpha\to n} Y_\alpha(x) = \frac{1}{\pi}\left\{\frac{\partial J_\alpha}{\partial \alpha} - (-1)^n \frac{\partial J_{-\alpha}}{\partial \alpha}\right\}\Bigg|_{\alpha=n}. \tag{4.5.6}$$

Note that

$$J_\alpha(x) = \sum_{k=0}^\infty \frac{(-1)^k (x/2)^{2k+\alpha}}{k!\,\Gamma(k+\alpha+1)}.$$

This implies that $J_\alpha(x)$ is an entire function in α. Thus the functions $\frac{\partial J_\alpha}{\partial \alpha}$ and $\frac{\partial J_{-\alpha}}{\partial \alpha}$ in (4.5.6) are meaningful. Moreover, as functions of x, $J_\alpha(x)$ are analytic functions of x in a cut plane. Thus we can verify that $Y_n(x)$ is a solution of Bessel's equation (4.5.1) when $\alpha = n$ is an integer. We can conclude that (4.5.5) is a solution of (4.5.1) in all cases. $Y_\alpha(x)$ is called a Bessel function of the second kind.

Substitution of the series for $J_\alpha(x)$ in (4.5.6) gives, after simplification,

$$Y_n(x) = \frac{2}{\pi} J_n(x) \ln\frac{x}{2} - \frac{1}{\pi}\sum_{k=0}^{n-1}\frac{(n-k-1)!}{k!}(x/2)^{2k-n}$$

$$- \frac{1}{\pi}\sum_{k=0}^\infty \frac{(-1)^k}{k!(n+k)!}[\psi(n+k+1)+\psi(k+1)](x/2)^{2k+n}. \tag{4.5.7}$$

Here n is a nonnegative integer, $|\arg x| < \pi$, and $\psi(x) = \Gamma'(x)/\Gamma(x)$.

Note that Bessel's equation can be written

$$\frac{d}{dx}\left(x\frac{dy}{dx}\right) + \left(x - \frac{\alpha^2}{x}\right)y = 0. \tag{4.5.8}$$

Suppose α is not an integer. It is easy to deduce from (4.5.8) that

$$J_{-\alpha}(x)\frac{d}{dx}\left(x\frac{dJ_\alpha(x)}{dx}\right) - J_\alpha(x)\frac{d}{dx}\left(x\frac{dJ_{-\alpha}(x)}{dx}\right) = 0$$

or

$$x\left[J_{-\alpha}(x)J_\alpha'(x) - J_\alpha(x)J_{-\alpha}'(x)\right] = C = \text{constant}.$$

To find C, let $x \to 0$ and use the series (4.5.2) and Euler's reflection formula. The result is

$$C = 2\sin\alpha\pi/\pi.$$

Thus the Wronskian $W(J_\alpha(x), J_{-\alpha}(x)) = J_\alpha(x)J_{-\alpha}'(x) - J_{-\alpha}(x)J_\alpha'(x)$ is given by

$$W(J_\alpha(x), J_{-\alpha}(x)) = -2\sin\alpha\pi/\pi x,$$

for $\alpha \neq$ integer, and

$$W(J_\alpha(x), Y_\alpha(x)) = 2/\pi x$$

not only when $\alpha \neq$ integer, but also for $\alpha = n$ by continuity.

Many differential equations can be reduced to the Bessel equation (4.5.1). For example,

$$u = x^a J_\alpha(bx^c)$$

satisfies

$$u'' + \frac{(1 - 2a)}{x}u' + \left[(bcx^{c-1})^2 + \frac{a^2 - \alpha^2 c^2}{x^2}\right]u = 0. \tag{4.5.9}$$

When $x = 1/2$, $b = 2/3$, $c = 3/2$, and $\alpha^2 = 1/a$, this equation reduces to

$$u'' + xu = 0. \tag{4.5.10}$$

This is the Airy equation and it has a turning point at $x = 0$, so solutions oscillate for $x > 0$ and are eventually monotonic when $x < 0$. As such, solutions of the Airy equation can be used to approximate solutions to many other more complicated differential equations that have a turning point. For example, the differential equation after (6.1.12) has $x = \sqrt{2n+1}$ as a turning point. Airy functions can be used to uniformly approximate Hermite polynomials in a two-sided neighborhood of the turning point. See Erdélyi [1960].

4.6 Recurrence Relations

There are two important differentiation formulas for Bessel functions:

$$\frac{d}{dx} x^\alpha J_\alpha(x) = \sum_{n=0}^\infty \frac{(-1)^n (2n + 2\alpha) x^{2n+2\alpha-1}}{\Gamma(n + \alpha + 1) n! 2^{2n+\alpha}}$$

$$= \sum_{n=0}^\infty \frac{(-1)^n x^{2n+2\alpha-1}}{\Gamma(n + \alpha) n! 2^{2n+\alpha-1}} = x^\alpha J_{\alpha-1}(x) \qquad (4.6.1)$$

and, similarly,

$$\frac{d}{dx} x^{-\alpha} J_\alpha(x) = -x^{-\alpha} J_{\alpha+1}(x). \qquad (4.6.2)$$

From the series for cosine and sine,

$$J_{1/2}(x) = \sqrt{\frac{2}{\pi x}} \sin x \qquad (4.6.3)$$

and

$$J_{-1/2}(x) = \sqrt{\frac{2}{\pi x}} \cos x. \qquad (4.6.4)$$

Rewrite (4.6.1) and (4.6.2) as

$$\alpha J_\alpha(x) + x J_\alpha'(x) = x J_{\alpha-1}(x)$$

and

$$-\alpha J_\alpha(x) + x J_\alpha'(x) = -x J_{\alpha+1}(x).$$

Elimination of the derivative J_α' gives

$$J_{\alpha-1}(x) + J_{\alpha+1}(x) = \frac{2\alpha}{x} J_\alpha(x). \qquad (4.6.5)$$

Elimination of $J_\alpha(x)$ gives

$$J_{\alpha-1}(x) - J_{\alpha+1}(x) = 2 J_\alpha'(x). \qquad (4.6.6)$$

It follows from (4.6.1) and (4.6.2) that

$$\left(\frac{1}{x} \frac{d}{dx}\right)^n (x^\alpha J_\alpha(x)) = x^{\alpha-n} J_{\alpha-n}(x) \qquad (4.6.7)$$

and

$$\left(\frac{1}{x} \frac{d}{dx}\right)^n (x^{-\alpha} J_\alpha(x)) = (-1)^n x^{-\alpha-n} J_{\alpha+n}(x). \qquad (4.6.8)$$

When these are applied to (4.6.3) and (4.6.4), we obtain

$$J_{n+1/2}(x) = (-1)^n \sqrt{\frac{2}{\pi x}} x^{n+1} \left(\frac{1}{x} \frac{d}{dx}\right)^n \left(\frac{\sin x}{x}\right) \qquad (4.6.9)$$

and

$$J_{-n-1/2}(x) = \sqrt{\frac{2}{\pi x}} x^{n+1} \left(\frac{1}{x}\frac{d}{dx}\right)^n \left(\frac{\cos x}{x}\right). \tag{4.6.10}$$

The following two formulas can now be proved by induction (the details are left to the reader):

$$J_{n+1/2}(x) = \sqrt{\frac{2}{\pi x}} \Bigg\{ \sin(x - n\pi/2) \sum_{k=0}^{[n/2]} \frac{(-1)^k (n+2k)!}{(2k)!(n-2k)!(2x)^{2k}}$$

$$+ \cos(x - n\pi/2) \sum_{k=0}^{[(n-1)/2]} \frac{(-1)^k (n+2k+1)!}{(2k+1)!(n-2k-1)!(2x)^{2k+1}} \Bigg\},$$

$$\tag{4.6.11}$$

$$J_{-n-1/2}(x) = \sqrt{\frac{2}{\pi x}} \Bigg\{ \cos(x + n\pi/2) \sum_{k=0}^{[n/2]} \frac{(-1)^k (n+2k)!}{(2k)!(n-2k)!(2x)^{2k}}$$

$$- \sin(x + n\pi/2) \sum_{k=0}^{[(n-1)/2]} \frac{(-1)^k (n+2k+1)!}{(2k+1)!(n-2k-1)!(2x)^{2k+1}} \Bigg\}.$$

$$\tag{4.6.12}$$

4.7 Integral Representations of Bessel Functions

Set $y = x^\alpha u$ in Bessel's equation

$$y'' + \frac{1}{x}y' + (1 - \alpha^2/x^2)y = 0.$$

Then u satisfies the equation

$$xu'' + (2\alpha + 1)u' + xu = 0. \tag{4.7.1}$$

Since equations with linear coefficients have Laplace integrals as solutions, let

$$u = A \int_C e^{xt} f(t)dt,$$

where A is a constant. Substitute this in (4.7.1). Then

$$0 = \int_C f(t)(xt^2 + (2\alpha + 1)t + x)e^{xt}dt$$

$$= \int_C f(t)\left[(t^2 + 1)\frac{\partial}{\partial t} + (2\alpha + 1)t\right]e^{xt}dt$$

$$= [e^{xt}(t^2 + 1)f(t)]_C + \int_C e^{xt}\left\{-\frac{\partial}{\partial t}[(t^2 + 1)f(t)] + (2\alpha + 1)tf(t)\right\}dt$$

after integration by parts. This equation is satisfied when

$$[e^{xt}(t^2 + 1)f(t)]_C = 0 \tag{4.7.2}$$

and

$$\frac{\partial}{\partial t}[(t^2 + 1)f(t)] = (2\alpha + 1)f(t). \tag{4.7.3}$$

Equation (4.7.3) holds when $f(t) = (t^2 + 1)^{\alpha - 1/2}$. Replace t with $\sqrt{-1}t$ so that (4.7.2) holds when C is the line joining -1 and 1 and $\mathrm{Re}\,\alpha > -1/2$. The above calculations imply that

$$y = Ax^\alpha \int_{-1}^{1} e^{ixt}(1 - t^2)^{\alpha - 1/2}dt \tag{4.7.4}$$

is a solution of Bessel's equation. We assume that $\arg(1 - t^2) = 0$. To see that (4.7.4) gives an integral representation of $J_\alpha(x)$ for $\mathrm{Re}\,\alpha > -1/2$, expand the exponential function in the integrand as a series and integrate. The result, after an easy calculation involving beta integrals, is

$$y(x) = Ax^\alpha \Gamma(\alpha + 1/2) \sum_{k=0}^{\infty} \frac{(-1)^k x^{2k} \Gamma(k + 1/2)}{(2k)!\Gamma(\alpha + k + 1)}.$$

An application of Legendre's duplication formula (Theorem 1.5.1),

$$2^{2k}\Gamma(k + 1)\Gamma(k + 1/2) = \sqrt{\pi}(2k)!,$$

gives

$$y(x) = A\sqrt{\pi}\Gamma(\alpha + 1/2)2^\alpha J_\alpha(x).$$

Therefore,

$$J_\alpha(x) = \frac{1}{\sqrt{\pi}\Gamma(\alpha + 1/2)}(x/2)^\alpha \int_{-1}^{1} e^{ixt}(1 - t^2)^{\alpha - 1/2} \tag{4.7.5}$$

when $\mathrm{Re}\,\alpha > -1/2$. Put $t = \cos\theta$ to get the Poisson integral representation

$$J_\alpha(x) = \frac{1}{\sqrt{\pi}\Gamma(\alpha + 1/2)}(x/2)^\alpha \int_0^\pi e^{ix\cos\theta} \sin^{2\alpha}\theta d\theta$$

$$= \frac{1}{\sqrt{\pi}\Gamma(\alpha + 1/2)}(x/2)^\alpha \int_0^\pi \cos(x\cos\theta)\sin^{2\alpha}\theta d\theta, \tag{4.7.6}$$

for $\mathrm{Re}\,\alpha > -1/2$. An important consequence of (4.7.5) is Gegenbauer's formula, which gives a Bessel function as an integral of an ultraspherical polynomial. The formula is

$$J_{\nu+n}(x) = \frac{(-i)^n \Gamma(2\nu)n!(x/2)^\nu}{\Gamma(\nu + 1/2)\Gamma(1/2)\Gamma(2\nu + n)} \int_0^\pi e^{ix\cos\theta} \sin^{2\nu}\theta C_n^\nu(\cos\theta)d\theta \tag{4.7.7}$$

for $\mathrm{Re}\,\nu > -1/2$. When $\nu \to 0$, we get Bessel's integral (4.9.11) for $J_n(x)$.

To prove this, take $\alpha = \nu + n$ in (4.7.5) and integrate by parts n times to get

$$J_{\nu+n}(x) = \frac{(i)^n (x/2)^\nu}{2^n \Gamma(\nu+n+1/2)\Gamma(1/2)} \int_{-1}^1 e^{ixt} \frac{d^n}{dt^n}(1-t^2)^{\nu+n-1/2} dt.$$

By Rodrigues's formula (2.5.13),

$$\frac{d^n(1-t^2)^{\nu+n-1/2}}{dt^n} = \frac{(-2)^n n! \Gamma(\nu+n+1/2)\Gamma(2\nu)}{\Gamma(\nu+1/2)\Gamma(2\nu+n)}(1-t^2)^{\nu-1/2}C_n^\nu(t).$$

Use of this in the previous integral gives Gegenbauer's formula (4.7.7).

Condition (4.7.2) is also valid when C is a closed contour on which $e^{ixt}(t^2 - 1)^{\alpha+1/2}$ returns to its initial value after t moves around the curve once. We take C as shown in Figure 4.1, and write an integral on C as

$$\int_A^{(1+,-1-)} f(t)dt.$$

Here $1+$ means that 1 is circled in the positive direction, and $-1-$ means -1 is circled in the negative direction. We are interested in the integral

$$y(x) = x^\alpha \int_A^{(1+,-1-)} e^{ixt}(t^2-1)^{\alpha-1/2}dt,$$

which is defined for all α, since C does not pass through any singularity of the integrand. When $\operatorname{Re}\alpha > -1/2$, we can deform C into a pair of lines from -1 to 1 and back. We choose $\arg(t^2 - 1) = 0$ at A; then

$$y(x) = x^\alpha \left[\int_{-1}^1 e^{ixt}[(1-t^2)e^{-\pi i}]^{\alpha-1/2}dt + \int_1^{-1} e^{ixt}[(1-t^2)e^{\pi i}]^{\alpha-1/2}dt \right]$$

$$= x^\alpha 2i \sin\left(\frac{1}{2} - \alpha\right)\pi \int_{-1}^1 e^{ixt}(1-t^2)^{\alpha-1/2}dt$$

$$= \frac{2\pi i \sqrt{\pi}}{\Gamma(\frac{1}{2} - \alpha)} J_\alpha(x).$$

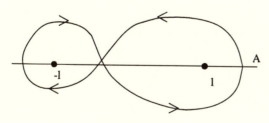

Figure 4.1

This gives Hankel's formula:

$$J_\alpha(x) = \frac{\Gamma\left(\frac{1}{2} - \alpha\right)}{\sqrt{\pi}} (x/2)^\alpha \frac{1}{2\pi i} \int_A^{(1+,-1-)} e^{ixt}(t^2 - 1)^{\alpha - 1/2} dt, \qquad (4.7.8)$$

when $\alpha \neq \frac{2n+1}{2}$, $n = 0, 1, 2, \ldots$, and $\arg(t^2 - 1) = 0$ at A.

We now prove another formula of Hankel,

$$J_{-\alpha}(x) = \frac{\Gamma(1/2 - \alpha)e^{\pi i \alpha}(x/2)^\alpha}{\sqrt{\pi}} \frac{1}{2\pi i} \int_{i\infty}^{(-1+,1+)} e^{ixt}(t^2 - 1)^{\alpha - 1/2} dt, \quad (4.7.9)$$

when $\mathrm{Re}\, x > 0$, $-3\pi < \arg(t^2 - 1) < \pi$, and $\alpha + \frac{1}{2} \neq 1, 2, \ldots$. The contour in (4.7.9) is shown in Figure 4.2. It is assumed that the contour lies outside the unit circle.

To prove (4.7.9) we need the following formula of Hankel (see Exercise 1.22):

$$\frac{1}{\Gamma(z)} = \frac{1}{2\pi i} \int_{-\infty}^{(0+)} e^t t^{-z} dt = \frac{1}{2\pi i} \int_{-\infty e^{i\theta}}^{(0+)} e^t t^{-z} dt, \qquad (4.7.10)$$

for $|\theta| < \pi/2$.

The expansion of $(t^2 - 1)^{\alpha - 1/2}$ in powers of $1/t$ converges uniformly on the contour. We have

$$(t^2 - 1)^{\alpha - 1/2} = t^{2\alpha - 1} \sum_{k=0}^{\infty} \frac{\left(\frac{1}{2} - \alpha\right)_k}{k!} t^{-2k},$$

and $-\frac{3\pi}{2} < \arg t < \frac{\pi}{2}$. So

$$x^\alpha \int_{i\infty}^{(-1+,1+)} e^{ixt}(t^2 - 1)^{\alpha - 1/2} dt = \sum_{k=0}^{\infty} \frac{x^\alpha \left(\frac{1}{2} - \alpha\right)_k}{k!} \int_{i\infty}^{(-1+,1+)} t^{2\alpha - 1 - 2k} e^{ixt} dt.$$

$$(4.7.11)$$

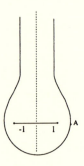

Figure 4.2

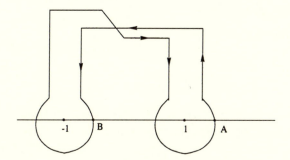

Figure 4.3

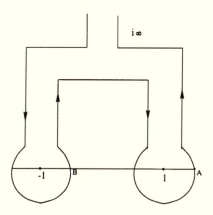

Figure 4.4

Since $\operatorname{Re} x > 0$, we have $|\arg x| = |\theta| < \frac{1}{2}\pi$. Set $u = e^{i\pi/2}xt$. Then

$$\int_{i\infty}^{(-1+,1+)} t^{2\alpha-1-2k} e^{ixt} dt = (-1)^k e^{-\alpha\pi i} x^{2k-2\alpha} \int_{-\infty e^{i\theta}}^{(0+)} e^u u^{2\alpha-1-2k} du$$

$$= (-1)^k e^{-\alpha\pi i} x^{2k-2\alpha} \frac{2\pi i}{\Gamma(1-2\alpha+2k)}.$$

Substitute this in (4.7.11) and apply Legendre's duplication formula. This proves (4.7.9).

Now modify the paths in (4.7.7) and (4.7.8) to those shown in Figures 4.3 and 4.4 respectively. Take $\operatorname{Re} x > 0$ in (4.7.7) and (4.7.8). When the horizontal parts of these curves are made to go to infinity we get

$$J_\alpha(x) = \frac{\Gamma\left(\frac{1}{2}-\alpha\right)(x/2)^\alpha}{\sqrt{\pi}} \frac{1}{2\pi i} \left[\int_{1+i\infty}^{(1+)} + \int_{-1+i\infty}^{(-1-)} e^{ixt}(t^2-1)^{\alpha-1/2} dt\right]$$

$$(4.7.12)$$

and

$$J_{-\alpha}(x) = \frac{\Gamma\left(\frac{1}{2}-\alpha\right)(x/2)^\alpha}{\sqrt{\pi}} \frac{e^{\alpha\pi i}}{2\pi i} \left[\int_{1+i\infty}^{(1+)} + \int_{-1+i\infty}^{(-1+)} e^{ixt}(t^2-1)^{\alpha-1/2}dt \right].$$

(4.7.13)

In (4.7.12) $\arg(t^2-1)$ is 0 at A and π at B, whereas in (4.7.13) $\arg(t^2-1)$ is 0 at A and $-\pi$ at B. To make $\arg(t^2-1) = \pi$ at B in (4.7.13), we multiply $(t^2-1)^{\alpha-1/2}$ in the second integral by the factor $e^{-2(\alpha-1/2)\pi i}$. Formula (4.7.13) may now be written (after reversing the direction of the contour in the second integral) as

$$J_{-\alpha}(x) = \frac{\Gamma\left(\frac{1}{2}-\alpha\right)(x/2)^\alpha}{\sqrt{\pi}\,2\pi i} \cdot \left[e^{\pi\alpha i} \int_{1+i\infty}^{(1+)} e^{ixt}(t^2-1)^{\alpha-1/2}dt \right.$$

$$\left. + e^{-\pi\alpha i} \int_{-1+i\infty}^{(-1-)} e^{ixt}(t^2-1)^{\alpha-\frac{1}{2}}dt \right].$$

(4.7.14)

The form of (4.7.12) and (4.7.14) suggests the following two functions, which are called Bessel functions of the third kind or Hankel functions:

$$H_\alpha^{(1)}(x) := \frac{i}{\sin\alpha\pi}[e^{-\alpha\pi i}J_\alpha(x) - J_{-\alpha}(x)]$$

(4.7.15)

and

$$H_\alpha^{(2)}(x) := \frac{-i}{\sin\alpha\pi}[e^{\alpha\pi i}J_\alpha(x) - J_{-\alpha}(x)].$$

(4.7.16)

These are more simply written in terms of $J_\alpha(x)$ and of $Y_\alpha(x)$, the Bessel function of the second kind. Thus,

$$H_\alpha^{(1)}(x) = J_\alpha(x) + iY_\alpha(x),$$

(4.7.17)

$$H_\alpha^{(2)}(x) = J_\alpha(x) - iY_\alpha(x),$$

(4.7.18)

$$H_\alpha^{(1)}(x) = \frac{\Gamma\left(\frac{1}{2}-\alpha\right)}{\sqrt{\pi}\,\pi i}(x/2)^\alpha \int_{1+i\infty}^{(1+)} e^{ixt}(t^2-1)^{\alpha-1/2}dt,$$

(4.7.19)

and

$$H_\alpha^{(2)}(x) = \frac{\Gamma\left(\frac{1}{2}-\alpha\right)}{\sqrt{\pi}\,\pi i}(x/2)^\alpha \int_{-1+i\infty}^{(-1-)} e^{ixt}(t^2-1)^{\alpha-1/2}dt.$$

(4.7.20)

The integral formulas for $H_\alpha^{(1)}(x)$ and $H_\alpha^{(2)}(x)$ hold when $\mathrm{Re}\,x > 0$, $\alpha + \frac{1}{2} \neq 1, 2, \ldots$. Moreover, $\arg(t^2-1) = -\pi$ at $1+i\infty$ and $\arg(t^2-1) = \pi$ at $-1+i\infty$.
    Note also that

$$H_{-1/2}^{(1)}(x) = \sqrt{\frac{2}{\pi x}}(\cos x + i\sin x) = \sqrt{\frac{2}{\pi x}}e^{ix} = H_{1/2}^{(2)}(x) \quad (4.7.21)$$

and

$$H_{1/2}^{(1)}(x) = \sqrt{\frac{2}{\pi x}} e^{-ix} = H_{-1/2}^{(2)}(x). \tag{4.7.22}$$

References to the work of Bessel, Poisson, Gegenbauer, and Hankel can be found in Watson [1944, Chapters 2, 3, and 6].

4.8 Asymptotic Expansions

Set $t = 1 + iu/x$ in (4.7.19). The integral becomes

$$-e^{ix}(e^{\pi i/2}x)^{-\alpha-1/2}2^{\alpha-1/2} \int_{\infty}^{(0+)} e^{-u}(-u)^{\alpha-1/2} \left(1 + \frac{iu}{2x}\right)^{\alpha-1/2} du. \tag{4.8.1}$$

Compare this with (4.3.5) and (4.3.6) to obtain the asymptotic expansion

$$H_\alpha^{(1)}(x) \sim \sqrt{\frac{2}{\pi x}} e^{i\left(x - \frac{\alpha\pi}{2} - \frac{\pi}{4}\right)} {}_2F_0 \left(\begin{matrix} \frac{1}{2} + \alpha, \frac{1}{2} - \alpha \\ - \end{matrix}; \frac{1}{2ix}\right). \tag{4.8.2}$$

Hankel introduced the notation

$$(\alpha, k) := (-1)^k \frac{\left(\frac{1}{2} - \alpha\right)_k \left(\frac{1}{2} + \alpha\right)_k}{k!}$$

$$= \frac{(4\alpha^2 - 1^2)(4\alpha^2 - 3^2)\cdots(4\alpha^2 - (2k-1)^2)}{2^{2k}k!}.$$

Then (4.8.2) can be written as

$$H_\alpha^{(1)}(x) = \sqrt{\frac{2}{\pi x}} e^{i\left(x - \frac{\alpha\pi}{2} - \frac{\pi}{4}\right)} \left[\sum_{j=0}^{k-1} \frac{(-1)^j(\alpha, j)}{(2ix)^j} + O(x^{-k})\right]. \tag{4.8.3}$$

A similar argument gives

$$H_\alpha^{(2)}(x) = \sqrt{\frac{2}{\pi x}} e^{-i\left(x - \frac{\alpha\pi}{2} - \frac{\pi}{4}\right)} \left[\sum_{j=0}^{k-1} \frac{(\alpha, j)}{(2ix)^j} + O(x^{-k})\right]. \tag{4.8.4}$$

Since

$$J_\alpha(x) = \frac{H_\alpha^{(1)}(x) + H_\alpha^{(2)}(x)}{2} \quad \text{and} \quad Y_\alpha(x) = \frac{H_\alpha^{(1)}(x) - H_\alpha^{(2)}(x)}{2i},$$

we have from (4.8.3) and (4.8.4)

$$J_\alpha(x) \sim \sqrt{\frac{2}{\pi x}} \left[\cos\left(x - \frac{\alpha\pi}{2} - \frac{\pi}{4}\right) \sum_{j=0}^{\infty} \frac{(-1)^j(\alpha, 2j)}{(2x)^{2j}} \right.$$

$$\left. - \sin\left(x - \frac{\alpha\pi}{2} - \frac{\pi}{4}\right) \sum_{j=0}^{\infty} \frac{(-1)^j(\alpha, 2j+1)}{(2x)^{2j+1}}\right] \tag{4.8.5}$$

and

$$Y_\alpha(x) \sim \sqrt{\frac{2}{\pi x}} \left[\sin\left(x - \frac{\alpha\pi}{2} - \frac{\pi}{4}\right) \sum_{j=0}^{\infty} (-1)^j \frac{(\alpha, 2j)}{(2x)^{2j}} \right.$$
$$\left. + \cos\left(x - \frac{\alpha\pi}{2} - \frac{\pi}{4}\right) \sum_{j=0}^{\infty} (-1)^j \frac{(\alpha, 2j+1)}{(2x)^{2j+1}} \right] \qquad (4.8.6)$$

when $|\arg x| < \pi$. Note that (4.6.11) and (4.6.12) are special cases of (4.8.5).

4.9 Fourier Transforms and Bessel Functions

Many special functions arise in the study of Fourier transforms. In Chapter 6, we shall see a connection between Hermite polynomials, which were mentioned in Section 4.4, and Fourier transforms in one variable. Here we consider a connection with Bessel functions and as a byproduct obtain a generating function for $J_n(x)$. Start with the Fourier transform in two dimensions. We have

$$F(u, v) = \frac{1}{2\pi} \int_{-\infty}^{\infty} \int_{-\infty}^{\infty} f(x, y) e^{i(xu+yv)} dx dy. \qquad (4.9.1)$$

Introduce polar coordinates in both (x, y) and (u, v) by

$$x = r \cos\theta, \quad y = r \sin\theta; \quad u = R \cos\phi, \quad v = R \sin\phi.$$

Then

$$F(u, v) = \frac{1}{2\pi} \int_0^{\infty} \int_0^{2\pi} f(r \cos\theta, r \sin\theta) e^{irR \cos(\theta-\phi)} r d\theta dr.$$

Expand f as a Fourier series in θ,

$$f(r \cos\theta, r \sin\theta) = \sum_{n=-\infty}^{\infty} f_n(r) e^{in\theta},$$

to get

$$F(u, v) = \sum_{n=-\infty}^{\infty} \int_0^{\infty} f_n(r) r \left[\frac{1}{2\pi} \int_0^{2\pi} e^{in\theta} e^{irR \cos(\theta-\phi)} d\theta \right] dr. \qquad (4.9.2)$$

The relation with Bessel functions comes from the inner integral. Since the integrand is periodic (of period 2π) it is sufficient to consider the integral

$$F_n(x) = \frac{1}{2\pi} \int_0^{2\pi} e^{ix \cos\theta} e^{in\theta} d\theta. \qquad (4.9.3)$$

Expand the exponential and integrate term by term to get

$$F_n(x) = \frac{1}{2\pi} \sum_{k=0}^{\infty} \frac{i^k x^k}{k!} \int_0^{2\pi} \cos^k \theta \, e^{in\theta} \, d\theta.$$ (4.9.4)

Now

$$2^k \cos^k \theta = (e^{i\theta} + e^{-i\theta})^k = e^{ik\theta} + \binom{k}{1} e^{i(k-2)\theta} + \cdots + \binom{k}{1} e^{-i(k-2)\theta} + e^{-ik\theta}.$$

So, writing $k = n + 2m$ we have

$$F_n(x) = \sum_{m=0}^{\infty} \frac{i^{n+2m} x^{n+2m}}{(n+2m)! 2^{n+2m}} \binom{n+2m}{m}$$

$$= i^n J_n(x)$$ (4.9.5)

This relation is interesting as it gives the Fourier expansion of $e^{ix \cos \theta}$:

$$e^{ix \cos \theta} = \sum_{n=-\infty}^{\infty} i^n J_n(x) e^{in\theta}$$

$$= J_0(x) + 2 \sum_{n=1}^{\infty} i^n J_n(x) \cos n\theta.$$ (4.9.6)

The last equation follows from $J_{-n}(x) = (-1)^n J_n(x)$. Equating the real and imaginary parts gives

$$\cos(x \cos \theta) = J_0(x) + 2 \sum_{n=1}^{\infty} (-1)^n J_{2n}(x) \cos 2n\theta$$ (4.9.7)

and

$$\sin(x \cos \theta) = 2 \sum_{n=0}^{\infty} (-1)^n J_{2n+1}(x) \cos(2n+1)\theta.$$ (4.9.8)

For an interesting special case, take $\theta = \pi/2$ in (4.9.7) to get

$$1 = J_0(x) + 2 \sum_{n=1}^{\infty} J_{2n}(x).$$ (4.9.9)

It is worth mentioning Miller's algorithm at this point. The series (4.9.9) shows that for any given $x = x_0$ and sufficiently large n, $J_{2n}(x_0)$ is small. So take $J_{2n}(x_0)$ to be 0 and $J_{2(n-1)}(x_0) = c$, which is to be determined. Use the recurrence relation (4.6.5) to compute $J_{2(n-2)}(x_0)$ and so on down to $J_2(x_0)$ and $J_0(x_0)$ as multiples of c. Since (4.9.9) can be approximated by

$$J_0(x_0) + 2 \sum_{k=1}^{n} J_{2k}(x_0) \approx 1,$$

we obtain an approximate value of c and hence also of the values of the Bessel functions $J_{2k}(x_0)$. This is an example of Miller's algorithm. See Gautschi [1967, p. 46].

There is another way of looking at (4.9.6). Put $t = ie^{i\theta}$. Then

$$\exp(x(t - 1/t)/2) = \sum_{n=-\infty}^{\infty} J_n(x)t^n. \qquad (4.9.10)$$

Thus, $\exp(x(t - 1/t)/2)$ is the generating function for Bessel functions of integer order.

Replace θ with $\frac{\pi}{2} - \theta$ in (4.9.3); then use (4.9.5) and the periodicity of the integrand to get Bessel's formula

$$
\begin{aligned}
J_n(x) &= \frac{1}{2\pi} \int_{-\pi}^{\pi} \exp(-in\theta + ix \sin\theta)d\theta \\
&= \frac{1}{2\pi} \int_0^{\pi} \exp(-in\theta + ix \sin\theta)d\theta + \frac{1}{2\pi} \int_0^{\pi} \exp(in\theta - ix \sin\theta)d\theta \\
&= \frac{1}{\pi} \int_0^{\pi} \cos(n\theta - x \sin\theta)d\theta. \qquad (4.9.11)
\end{aligned}
$$

In Section 4.7, we obtained this from Poisson's integral formula. We can go in the opposite direction and derive

$$J_n(x) = \frac{(x/2)^n}{\pi(1/2)_n} \int_0^{\pi} \cos(x \cos\theta) \sin^{2n}\theta d\theta \qquad (4.9.12)$$

from (4.9.11) by using Jacobi's formula given in Exercise 2.21,

$$\frac{d^{n-1} \sin^{2n-1}\theta}{dy^{n-1}} = \frac{(-1)^{n-1}}{n} 2^n (1/2)_n \sin n\theta, \quad y = \cos\theta.$$

In (4.9.12), n is a nonnegative integer. But this restriction can be removed. First multiply both sides of (4.9.12) by $(2/x)^n \Gamma(n + 1)$. Then both sides are bounded analytic functions of n for Re $n > -1/2$. By Carlson's theorem, we can conclude that (4.9.12) is true for these values of n.

We end this section with a proof of the inequalities:

For x real, $|J_0(x)| \le 1$,   and $|J_m(x)| \le 1/\sqrt{2}$ for $m = 1, 2, 3, \ldots$.
$$(4.9.13)$$

The first inequality follows from (4.9.12), but we have another proof which verifies them all at once. Change t to $-t$ in the generating function (4.9.10) to get

$$\exp(-x(t - 1/t)/2) = \sum_{n=-\infty}^{\infty} (-1)^n J_n(x)t^n. \qquad (4.9.14)$$

Multiply (4.9.10) by (4.9.14) to get

$$1 = \sum_{n=-\infty}^{\infty} J_m(x)t^m \sum_{n=-\infty}^{\infty} (-1)^n J_n(x)t^n$$

$$= \sum_{n=-\infty}^{\infty} t^n \sum_{m=-\infty}^{\infty} (-1)^m J_m(x) J_{n-m}(x).$$

Equate the coefficients of the powers of t and use $J_{-n}(x) = (-1)^n J_n(x)$ to obtain

$$\sum_{n=-\infty}^{\infty} J_n^2(x) = J_0^2(x) + 2\sum_{n=1}^{\infty} J_n^2(x) = 1 \tag{4.9.15}$$

and

$$\sum_{m=-\infty}^{\infty} (-1)^m J_m(x) J_{n-m}(x) = 0 \quad \text{for } n \neq 0. \tag{4.9.16}$$

The inequalities in (4.9.13) follow from (4.9.15).

Remark 4.9.1 Observe that, in Bessel's formula (4.9.11), n has to be an integer, for when $n = \alpha$, a noninteger, the integral on the right side of (4.9.11) is no longer a solution of Bessel's equation (4.5.1). However, Poisson's integral formula (4.9.12) holds for all n as long as $\text{Re}\, n > -1/2$. We also remark that Jacobi obtained the direct transformation of (4.9.12) to (4.9.11) by the argument given here in reverse. For references and details of Jacobi's proof, see Watson [1944, §§2.3–2.32]. Note also that Jacobi's formula mentioned after (4.9.12) is really a consequence of Rodrigues's formula (2.5.13′) when applied to Chebyshev polynomials of the second kind.

4.10 Addition Theorems

In this section, we prove a useful addition theorem of Gegenbauer. First we show that

$$J_n(x + y) = \sum_{m=-\infty}^{\infty} J_m(x) J_{n-m}(y). \tag{4.10.1}$$

This follows immediately from the fact that

$$\exp((x + y)(t - 1/t)/2) = \exp(x(t - 1/t)/2) \exp(y(t - 1/t)/2),$$

for this implies

$$\sum_{n=-\infty}^{\infty} J_n(x + y)t^n = \sum_{m=-\infty}^{\infty} J_m(x)t^m \sum_{n=-\infty}^{\infty} J_n(x)t^n.$$

The result in (4.10.1) is obtained by equating the coefficients of t^n. Observe that (4.9.16) follows from this addition formula.

To state the second addition theorem, suppose a, b, and c are lengths of sides of a triangle and $c^2 = a^2 + b^2 - 2ab \cos \theta$. Then

$$J_0(c) = \sum_{m=-\infty}^{\infty} J_m(a) J_m(b) e^{im\theta}. \tag{4.10.2}$$

Set

$$d = ae^{i\theta} - b.$$

Then $c^2 = d\bar{d}$, so c and d have the same absolute value. Thus there is a real ψ such that

$$c = (ae^{i\theta} - b)e^{i\psi}.$$

A short calculation shows that the last relation implies

$$c \sin \phi = a \sin(\theta + \psi + \phi) - b \sin(\psi + \phi).$$

By (4.9.11),

$$J_0(c) = \frac{1}{2\pi} \int_0^{2\pi} e^{ic \sin \phi} d\phi$$

$$= \frac{1}{2\pi} \int_0^{2\pi} e^{i[a \sin(\theta + \psi + \phi) - b \sin(\psi + \phi)]} d\phi.$$

Since ψ is independent of ϕ and the integrand is periodic, by (4.9.10),

$$J_0(c) = \frac{1}{2\pi} \int_0^{2\pi} e^{i[a \sin(\theta + \phi) - b \sin \phi]} d\phi$$

$$= \sum_{m=-\infty}^{\infty} J_m(a) e^{im\theta} \frac{1}{2\pi} \int_0^{2\pi} e^{-ib \sin \phi} e^{im\phi} d\phi$$

$$= \sum_{m=-\infty}^{\infty} J_m(a) e^{im\theta} \frac{1}{2\pi} \int_0^{2\pi} e^{ib \sin \phi} e^{-im\phi} d\phi$$

$$= \sum_{m=-\infty}^{\infty} J_m(a) J_m(b) e^{im\theta}.$$

The last equation comes from (4.9.11).

Since $J_{-m}(x) = (-1)^m J_m(x)$, we can write the addition formula in the following form:

$$J_0(c) = J_0(a) J_0(b) + 2 \sum_{m=1}^{\infty} J_m(a) J_m(b) \cos m\theta. \tag{4.10.3}$$

Observe that

$$\frac{1}{c}\frac{d}{dc} = \frac{1}{ab\sin\theta}\frac{d}{d\theta}, \tag{4.10.4}$$

then apply this operator to (4.10.3), and use (4.6.2) to get

$$\frac{J_1(c)}{c} = 2\sum_{m=1}^{\infty} m J_m(a) J_m(b)\frac{\sin m\theta}{\sin\theta}.$$

Rewrite this as

$$\frac{J_1(c)}{c} = 2\sum_{m=0}^{\infty}(m+1)\frac{J_{1+m}(a)}{a}\frac{J_{1+m}(b)}{b}C_m^1(\cos\theta).$$

Apply (4.10.4) to the last formula; use (4.6.2) again to get

$$\frac{J_2(c)}{c^2} = 2^2\sum_{m=0}^{\infty}(m+2)\frac{J_{2+m}(a)}{a^2}\frac{J_{2+m}(b)}{b^2}C_m^2(\cos\theta).$$

In general, we have the following result for derivatives of ultraspherical polynomials:

$$\frac{d}{d\theta}C_n^\lambda(\cos\theta) = -2\lambda\sin\theta C_{n-1}^{\lambda+1}(\cos\theta).$$

Now apply induction to see that

$$\frac{J_\alpha(c)}{c^\alpha} = 2^\alpha\Gamma(\alpha)\sum_{m=0}^{\infty}(m+\alpha)\frac{J_{\alpha+m}(a)}{a^\alpha}\frac{J_{\alpha+m}(b)}{b^\alpha}C_m^\alpha(\cos\theta), \tag{4.10.5}$$

when $\alpha = 0, 1, 2, \ldots$. By (6.4.11), $C_m^\alpha(\cos\theta)$ is a polynomial in α; hence by the remarks we made after formula (4.9.12) the two sides of (4.10.5) are bounded analytic functions in a right half plane. Carlson's theorem now implies the truth of (4.10.5) for values of α in this half plane. By analytic continuation, (4.10.5) is then true for all α except $\alpha = 0, -1, -2, \ldots$. Equation (4.10.5) is called Gegenbauer's addition formula.

We state without proof the following result of Graf:

$$J_\alpha(c)\left(\frac{a - be^{-i\theta}}{a - be^{i\theta}}\right)^{\alpha/2} = \sum_{m=-\infty}^{\infty} J_{\alpha+m}(a)J_m(b)e^{im\theta} \tag{4.10.6}$$

when $b < a$. When a, b, and θ are complex, we require that $|be^{\pm i\theta}/a| < 1$ and $c \to a$ as $b \to 0$. Graf's formula contains (4.10.1) and (4.10.2) as special cases. See Watson [1944, §11.3]. Exercise 29 gives a proof of (4.10.6) when α is an integer.

4.11 Integrals of Bessel Functions

Expand the function $F(u, v)$ in (4.9.2) as a Fourier series:

$$F(u, v) = \sum_{n=-\infty}^{\infty} F_n(R)e^{in\phi}$$

$$= \int_0^\infty \frac{1}{2\pi} \int_0^{2\pi} e^{iRr\cos(\theta-\phi)} \sum_{n=-\infty}^{\infty} f_n(r)e^{in\theta} d\theta r\,dr$$

$$= \int_0^\infty \frac{1}{2\pi} \int_0^{2\pi} e^{iRr\cos\theta} \sum_{n=-\infty}^{\infty} f_n(r)e^{in(\theta+\phi)} d\theta r\,dr$$

$$= \int_0^\infty \frac{1}{2\pi} \int_0^{2\pi} \sum_{m=-\infty}^{\infty} i^m J_m(Rr)e^{in\theta} \sum_{n=-\infty}^{\infty} f_n(r)e^{in(\theta+\phi)} d\theta r\,dr$$

[by (4.9.6)]

$$= \sum_{n=-\infty}^{\infty} i^n \left[\int_0^\infty J_n(Rr) f_n(r)r\,dr \right] e^{in\phi}. \qquad (4.11.1)$$

Hence

$$(-i)^n F_n(R) = \int_0^\infty f_n(r) J_n(Rr)r\,dr. \qquad (4.11.2)$$

The inverse Fourier transform of (4.9.1) is

$$f(x, y) = \frac{1}{2\pi} \int_{-\infty}^\infty \int_{-\infty}^\infty F(u, v)e^{-i(xu+yv)} du\,dv.$$

If a calculation similar to (4.11.1) is performed, we obtain

$$f_n(r) = (-i)^n \int_0^\infty F_n(R) J_n(Rr)R\,dR. \qquad (4.11.3)$$

The integral in (4.11.2) is called the Hankel transform of order n of the function $f_n(r)$. Then (4.11.3) is called the inverse Hankel transform. For a function $f(x)$ that is smooth enough and vanishes sufficiently fast at infinity, we have more generally the Hankel pair of order α:

$$F(y) = \int_0^\infty f(x) J_\alpha(yx)x\,dx \qquad (4.11.4)$$

and

$$f(x) = \int_0^\infty F(y) J_\alpha(xy)y\,dy. \qquad (4.11.5)$$

To obtain an interesting integral, multiply Gegenbauer's formula (4.10.5) by $C_n^\alpha(\cos\theta)$ and use the orthogonality relation which follows from (2.5.14), namely

$$\int_{-1}^{1} C_m^\alpha(x)C_n^\alpha(x)(1-x^2)^{\alpha-\frac{1}{2}}dx = \frac{2^{1-2\alpha}\pi\Gamma(n+2\alpha)}{[\Gamma(\alpha)]^2(n+\alpha)n!}\delta_{mn}.$$

The result is

$$\int_0^\pi \frac{J_\alpha(c)}{c^\alpha}C_n^\alpha(\cos\theta)\sin^{2\alpha}\theta d\theta = \frac{2^{1-\alpha}\pi\Gamma(n+2\alpha)}{n!\Gamma(\alpha)}\frac{J_{\alpha+n}(a)J_{\alpha+n}(b)}{a^\alpha b^\alpha},$$

$$(4.11.6)$$

where a, b, and c are sides of a triangle, that is, $c^2 = a^2 + b^2 - 2ab\cos\theta$. Rescale a, b, and c and take $n = 0$ to arrive at

$$\int_0^\pi \frac{J_\alpha(cx)}{c^\alpha}\sin^{2\alpha}\theta d\theta = \frac{\pi\Gamma(2\alpha)}{2^{\alpha-1}\Gamma(\alpha)}\frac{J_\alpha(ax)J_\alpha(bx)}{x^\alpha(ab)^\alpha}.$$

Rewrite this as

$$\int_{|a-b|}^{a+b} \frac{[(a+b)^2 - c^2]^{\alpha-\frac{1}{2}}[c^2 - (a-b)^2]^{\alpha-\frac{1}{2}}}{c^\alpha}J_\alpha(cx)cdc$$

$$= \frac{2^{3\alpha-1}\sqrt{\pi}\Gamma(\alpha+1/2)J_\alpha(ax)J_\alpha(bx)(ab)^\alpha}{x^\alpha}. \qquad (4.11.7)$$

Then, by the Hankel inversion formula, for $\mathrm{Re}\,a > -1/2$,

$$\int_0^\infty J_\alpha(ax)J_\alpha(bx)J_\alpha(cx)x^{1-\alpha}dx = \frac{[c^2 - (a-b)^2]^{\alpha-\frac{1}{2}}[(a+b)^2 - c^2]^{\alpha-\frac{1}{2}}}{2^{3\alpha-1}\sqrt{\pi}\Gamma(\alpha+1/2)(abc)^\alpha}$$

$$(4.11.8)$$

for $|a-b| < c < a+b$. The value of the integral is 0 otherwise. If the formula for the area of a triangle (denoted by Δ) in terms of its sides is used, then the right side of (4.11.8) can be written

$$\frac{2^{\alpha-1}\Delta^{2\alpha-1}}{\sqrt{\pi}\Gamma(\alpha+\frac{1}{2})(abc)^\alpha}.$$

There are important generalizations of integral (4.7.6) due to Sonine. These are contained in the next theorem.

Theorem 4.11.1   *For* $\mathrm{Re}\,\mu > -1$ *and* $\mathrm{Re}\,\nu > -1$,

$$J_{\mu+\nu+1}(x) = \frac{x^{\nu+1}}{2^\nu\Gamma(\nu+1)}\int_0^{\pi/2} J_\mu(x\sin\theta)\sin^{\mu+1}\theta\cos^{2\nu+1}\theta d\theta \qquad (4.11.9)$$

and

$$\frac{x^\mu y^\nu J_{\mu+\nu+1}\{(x^2+y^2)^{1/2}\}}{(x^2+y^2)^{(\mu+\nu+1)/2}} = \int_0^{\pi/2} J_\mu(x\sin\theta)J_\nu(y\cos\theta)\sin^{\mu+1}\theta\cos^{\nu+1}\theta d\theta.$$

$$(4.11.10)$$

The integrals (4.11.9) and (4.11.10) are referred to as Sonine's first and second integrals.

Proof. (i) The proof is simple. Expand $J_\mu(x \sin \theta)$ as a power series and integrate term by term. Thus

$$\frac{x^{\nu+1}}{2^\nu \Gamma(\nu+1)} \int_0^{\pi/2} J_\mu(x \sin \theta) \sin^{\mu+1}\theta \cos^{2\nu+1}\theta \, d\theta$$

$$= \sum_{m=0}^\infty \frac{(-1)^m x^{\mu+\nu+2m+1}}{2^{\mu+\nu+2m} m! \Gamma(\mu+m+1)\Gamma(\nu+1)} \int_0^{\pi/2} \sin^{2\mu+2m+1}\theta \cos^{2\nu+1}\theta \, d\theta.$$

The last integral is a beta integral equal to

$$\frac{1}{2} \int_0^1 (1-t)^{\mu+m} t^\nu dt = \frac{\Gamma(\mu+m+1)\Gamma(\nu+1)}{2\Gamma(\mu+m+2)}.$$

Substitution of this in the above series gives the result.

(ii) In this case expand both $J_\mu(x \sin \theta)$ and $J_\nu(y \cos \theta)$ in power series and integrate term by term. The details are left to the reader.

Observe that (4.11.9) is a special case of (4.11.10). Divide both sides of (4.11.10) by y^ν and let $y \to 0$. The result is (4.11.9). ∎

Corollary 4.11.2 *For* $\operatorname{Re} \alpha > -1/2$,

$$J_\alpha(x) = \frac{(x/2)^\alpha}{\Gamma(\alpha+1/2)\sqrt{\pi}} \int_0^\pi \cos(x \cos \theta) \sin^{2\alpha}\theta \, d\theta. \qquad (4.7.6)$$

Proof. Take $\mu = -1/2$, $\nu + 1/2 = \alpha$ and recall that

$$J_{-1/2}(x) = \sqrt{\frac{2}{\pi x}} \cos x. \qquad ∎$$

Sonine's first integral (4.11.9) can also be written as

$$J_{\mu+\nu+1}(x) = \frac{x^{\nu+1}}{2^\nu \Gamma(\nu+1)} \int_0^\infty t^\mu (1-t^2)_+^\nu J_\mu(xt) t \, dt. \qquad (4.11.11)$$

By Hankel inversion,

$$2^\nu \Gamma(\nu+1) \int_0^\infty \frac{J_{\mu+\nu+1}(x)}{x^{\nu+1}} J_\mu(xt) x \, dx = t^\mu (1-t^2)_+^\nu \qquad (4.11.12)$$

for $\operatorname{Re} \mu > -1$, $\operatorname{Re} \nu > -1$.

We now turn to the computation of the Laplace transform of Bessel functions. Hankel evaluated the transform of $t^{\mu-1} J_\alpha(yt)$ in terms of a $_2F_1$ function. For special values of α and μ, the $_2F_1$ reduces to more elementary functions. We consider this class of integrals next. The simplest integral of this kind was found by Lipschitz. His result is the following: For Re $(x \pm iy) > 0$,

$$\int_0^\infty e^{-xt} J_0(yt)dt = \frac{1}{\sqrt{(x^2 + y^2)}}. \tag{4.11.13}$$

From the asymptotic expansion for Bessel functions (4.8.5), it is clear that Re$(x \pm iy) > 0$ is sufficient for the convergence of the integral. Use (4.9.11) to see that

$$\int_0^\infty e^{-xt} J_0(yt)dt = \frac{1}{\pi} \int_0^\infty e^{-xt} \int_0^\pi e^{iyt\cos\theta} d\theta dt$$

$$= \frac{1}{\pi} \int_0^\pi \frac{d\theta}{x - iy\cos\theta}$$

$$= \frac{1}{(x^2 + y^2)^{1/2}}.$$

The more general result, which gives the Laplace transform of $t^{\mu-1} J_\alpha(yt)$, is due to Hankel [1875].

Theorem 4.11.3   For Re $(\alpha + \mu) > 0$ and Re $(x \pm iy) > 0$,

$$\int_0^\infty e^{-xt} J_\alpha(yt)t^{\mu-1}dt = \frac{(y/2x)^\alpha \Gamma(\alpha + \mu)}{x^\mu \Gamma(\alpha + 1)}$$

$$\cdot {}_2F_1\left(\begin{array}{c} (\alpha + \mu)/2, (\alpha + \mu + 1)/2 \\ \alpha + 1 \end{array}; -\frac{y^2}{x^2}\right). \tag{4.11.14}$$

Proof.   First assume that $|y/x| < 1$. Substitute the series for $J_\nu(yt)$ in the integral to get

$$\sum_{m=0}^\infty \frac{(-1)^m (y/2)^{\alpha+2m}}{m!\Gamma(\alpha + m + 1)} \int_0^\infty t^{\mu+\alpha+2m-1} e^{-xt} dt$$

$$= \sum_{m=0}^\infty \frac{(-1)^m (y/2)^{\alpha+2m}}{m!\Gamma(\alpha + m + 1)} \frac{\Gamma(\alpha + \mu + 2m)}{x^{\alpha+\mu+2m}}.$$

Since $|y/x| < 1$, the final series is absolutely convergent. This justifies the term-by-term integration. So we have (4.11.14) under the restriction $|y/x| < 1$. The complete result follows upon analytic continuation, since both sides of (4.11.14) are analytic functions of y when Re $(x \pm iy) > 0$. This proves the theorem.   ∎

When $\mu = \alpha + 1$ or $\alpha + 2$, the ${}_2F_1$ in (4.11.14) reduces to a ${}_1F_0$, which can be summed by the binomial theorem for $|y/x| < 1$. The results are in the next corollary.

Corollary 4.11.4   *For* Re $(x \pm iy) > 0$,

$$\int_0^\infty e^{-xt} J_\alpha(yt) t^\alpha dt = \frac{(2y)^\alpha \Gamma(\alpha + 1/2)}{(x^2 + y^2)^{\alpha+1/2} \sqrt{\pi}}, \quad \text{when } \text{Re } \alpha > -1/2,$$

(4.11.15)

and

$$\int_0^\infty e^{-xt} J_\alpha(yt) t^{\alpha+1} dt = \frac{2y(2x)^\alpha \Gamma(\alpha + 3/2)}{(x^2 + y^2)^{\alpha+3/2} \sqrt{\pi}}, \quad \text{when } \text{Re } \alpha > -1.$$

(4.11.16)

Corollary 4.11.5   *For* Re $(x \pm iy) > 0$,

$$\int_0^\infty e^{-xt} J_\alpha(yt) t^{-1} dt = \frac{[(x^2 + y^2)^{1/2} - x]^\alpha}{\alpha y^\alpha}, \quad \text{when } \text{Re } \alpha > 0, \quad (4.11.17)$$

and

$$\int_0^\infty e^{-xt} J_\alpha(yt) dt = \frac{[(x^2 + y^2)^{1/2} - x]^\alpha}{y^\alpha (x^2 + y^2)^{1/2}}, \quad \text{when } \text{Re } \alpha > -1. \quad (4.11.18)$$

Proof.   Apply Exercise 3.39.   ∎

The formulas in the two corollaries are limits of some formulas for Jacobi polynomials introduced in Chapter 2. Recall that these are defined by

$$P_n^{(\alpha,\beta)}(x) = \frac{(\alpha+1)_n}{n!} {}_2F_1\left(\begin{matrix} -n, n+\alpha+\beta+1 \\ \alpha+1 \end{matrix}; \frac{1-x}{2}\right), \quad n = 0, 1, 2, \ldots.$$

Theorem 4.11.6   *For real α and β,*

$$\lim_{n\to\infty} n^{-\alpha} P_n^{(\alpha,\beta)}\left(\cos\frac{x}{n}\right) = \lim_{n\to\infty} n^{-\alpha} P_n^{(\alpha,\beta)}\left(1 - \frac{x^2}{2n^2}\right) = (x/2)^{-\alpha} J_\alpha(x).$$

Proof.   This follows easily from Tannery's theorem (or the Lebesgue dominated convergence theorem). Suppose α is not a negative integer. Termwise convergence is easily checked. Moreover, domination by a convergent series is seen from the fact that, for large n,

$$\frac{n^{-\alpha}(n+\alpha+\beta+1)_k(\alpha+1)_n}{k!(n-k)!(\alpha+1)_k 2^k n^{2k}} \le \frac{(2n+\alpha+\beta)^k(\alpha+1)_n n^{-\alpha}}{k!(\alpha+1)_k(2n)^k n!} \le \frac{C}{k!(\alpha+1)_k},$$

where C is a constant that holds for all k, $0 \le k \le n$.
When $\alpha = -\ell$ is a negative integer, use the fact that

$$\binom{n}{\ell} P_n^{(-\ell,\beta)}(x) = \binom{n+\beta}{\ell}\left(\frac{x-1}{2}\right)^\ell P_{n-\ell}^{(\ell,\beta)}(x) \quad \text{for } \ell \le n$$

to obtain the desired result.   ∎

The integral formulas (4.11.15) to (4.11.18) are limits of generating-function formulas for Jacobi polynomials, three of which will be proved in Chapter 6. The corresponding formulas are

$$\sum_{n=0}^{\infty} \frac{(\alpha + \beta + 1)_n P_n^{(\alpha,\beta)}(x) r^n}{(\alpha + 1)_n} = (1 - r)^{-\alpha - \beta - 1}{}_2F_1\left(\begin{array}{c} \frac{\alpha+\beta+1}{2}, \frac{\alpha+\beta+2}{2} \\ \alpha + 1 \end{array}; \frac{2r(x - 1)}{(1 - r)^2}\right)$$

(4.11.19)

$$\sum_{n=0}^{\infty} \frac{(2n + \alpha + \beta + 1)\Gamma(n + \alpha + \beta + 1) P_n^{(\alpha,\beta)}(x) r^n}{\Gamma(n + \beta + 1)}$$

$$= \frac{\Gamma(\alpha + \beta + 2)(1 - r)}{\Gamma(\beta + 1)(1 + r)^{\alpha+\beta+2}}{}_2F_1\left(\begin{array}{c} \frac{\alpha+\beta+2}{2}, \frac{\alpha+\beta+3}{2} \\ \beta + 1 \end{array}; \frac{2r(1 + x)}{(1 + r)^2}\right),$$

(4.11.20)

$$\sum_{n=0}^{\infty} \frac{\alpha}{n + \alpha} P_n^{(\alpha,-1)}(x) r^n = 2^{\alpha}(1 - r + R)^{-\alpha},$$

(4.11.21)

where $R = (1 - 2xr + r^2)^{1/2}$, and

$$\sum_{n=0}^{\infty} P_n^{(\alpha,\beta)}(x) r^n = 2^{\alpha+\beta} R^{-1}(1 - r + R)^{-\alpha}(1 + r + R)^{-\beta}.$$

(4.11.22)

A proof of a result more general than (4.11.21) is sketched in Exercise 7.31. The other generating-function formulas are proved in Chapter 6 (Section 6.4).

The next theorem gives another infinite integral of a Bessel function due to Hankel.

Theorem 4.11.7   *For* Re $(\mu + \nu) > 0$,

$$\int_0^{\infty} J_\nu(at)\ t^{\mu-1} e^{-p^2 t^2} dt = \frac{\Gamma\left(\frac{\mu+\nu}{2}\right)(a/2p)^\nu}{2p^\mu \Gamma(\nu + 1)}{}_1F_1\left(\begin{array}{c} (\mu + \nu)/2 \\ \nu + 1 \end{array}; -\frac{a^2}{4p^2}\right).$$

(4.11.23)

Proof.   The condition Re $(\mu + \nu) > 0$ is necessary for convergence at zero. The asymptotic behavior of $J_\nu(x)$ given by (4.8.5) shows that the integral converges absolutely. Thus the integral can be evaluated using term by term integration. This gives

$$\int_0^{\infty} J_\nu(at)\ e^{-p^2 t^2} t^{\mu-1} dt = \sum_{m=0}^{\infty} \frac{(-1)^m (a/2)^{\nu+2m}}{m! \Gamma(\nu + m + 1)} \int_0^{\infty} t^{\nu+\mu+2m-1} e^{-p^2 t^2} dt.$$

Since the integral on the right-hand side equals

$$\Gamma\left(\frac{\mu + \nu}{2} + m\right)\Big/2p^{\nu+\mu+2m},$$

the result follows.   ∎

Corollary 4.11.8   *For* Re $(\mu + \nu) > 0$,

$$\int_0^\infty J_\nu(at)\, t^{\mu-1} e^{-p^2 t^2}\, dt = \frac{\Gamma\left(\frac{\mu+\nu}{2}\right)(a/2p)^\nu e^{-a^2/4p^2}}{2p^\mu \Gamma(\nu+1)}\, {}_1F_1\left(\frac{\frac{\nu-\mu}{2}+1}{\nu+1}\,;\,\frac{a^2}{4p^2}\right).$$

(4.11.24)

Proof.   Apply Kummer's first ${}_1F_1$ transformation (4.1.11) to Hankel's formula in Theorem 4.11.7.   ∎

An important particular case we use later is

$$\int_0^\infty J_\nu(at)\, t^{\nu+1} e^{-p^2 t^2}\, dt = \frac{a^\nu}{(2p^2)^{\nu+1}} e^{-a^2/4p^2}, \qquad \text{Re } \nu > -1. \qquad (4.11.25)$$

See Watson [1944, Chapters 12 and 13] for references.

4.12 The Modified Bessel Functions

The differential equation

$$\frac{d^2 y}{dx^2} + \frac{1}{x}\frac{dy}{dx} - \left(1 + \frac{\alpha^2}{x^2}\right) y = 0, \qquad (4.12.1)$$

where x is real, arises frequently in mathematical physics. It is easily seen that $J_\alpha(ix)$ is a solution of this equation. Moreover, for x, real $e^{-\alpha\pi i/2} J_\alpha(x e^{\pi i/2})$ is a real function. We then define the modified Bessel function of the first kind as

$$I_\alpha(x) = e^{-\alpha\pi i/2} J_\alpha(x e^{\pi i/2}) \qquad (-\pi < \arg x \le \pi/2)$$

$$= e^{3\alpha\pi i/2} J_\alpha(x e^{-3\pi i/2}) \qquad \left(\frac{1}{2}\pi < \arg x \le \pi\right)$$

$$= (x/2)^\alpha \sum_{k=0}^\infty \frac{(x/2)^{2k}}{k!\,\Gamma(\alpha + k + 1)}. \qquad (4.12.2)$$

When α is not an integer, $I_\alpha(x)$ and $I_{-\alpha}(x)$ are two independent solutions of (4.12.1). When $\alpha = n$ is an integer, then

$$I_n(x) = I_{-n}(x).$$

To deal with this situation, define the modified Bessel function of the second kind:

$$K_\alpha(x) := \frac{\pi}{2\sin\alpha\pi}[I_{-\alpha}(x) - I_\alpha(x)]. \qquad (4.12.3)$$

It is immediately verified that

$$I_{1/2}(x) = \sqrt{\frac{2}{\pi x}}\sinh x \quad \text{and} \quad I_{-1/2}(x) = \sqrt{\frac{2}{\pi x}}\cosh x. \qquad (4.12.4)$$

Thus

$$K_{1/2}(x) = \sqrt{\frac{\pi}{2x}}e^{-x}. \qquad (4.12.5)$$

We see that $J_\alpha(x)$ corresponds to the sine and cosine functions whereas $I_\alpha(x)$ corresponds to the exponential function. Perhaps this is why the nineteenth century British mathematician, George Stokes, took $I_\alpha(x)$, rather than the Bessel function, as the fundamental function.

The asymptotic expansions for $I_\alpha(x)$ and $K_\alpha(x)$ can be obtained in the same way as those for $J_\alpha(x)$ and $Y_\alpha(x)$. Thus

$$K_\alpha(x) \sim \sqrt{\frac{\pi}{2x}}e^{-x}\left[1 + \sum_{n=1}^{\infty}\frac{(\alpha, n)}{(2x)^n}\right], \qquad (|\arg x| < 3\pi/2), \qquad (4.12.6)$$

$$I_\alpha(x) \sim \frac{e^x}{\sqrt{2\pi x}}\sum_{n=0}^{\infty}\frac{(-1)^n(\alpha, n)}{(2x)^n} + \frac{e^{-x+\left(\alpha+\frac{1}{2}\right)\pi i}}{\sqrt{2\pi x}}\sum_{n=0}^{\infty}\frac{(\alpha, n)}{(2x)^n},$$

$$(-\pi/2 < \arg x < 3\pi/2), \qquad (4.12.7)$$

and

$$I_\alpha(x) \sim \frac{e^x}{\sqrt{2\pi x}}\sum_{n=0}^{\infty}\frac{(-1)^n(\alpha, n)}{(2x)^n} + \frac{e^{-x-\left(\alpha+\frac{1}{2}\right)\pi i}}{\sqrt{2\pi x}}\sum_{n=0}^{\infty}\frac{(\alpha, n)}{(2x)^n},$$

$$(-3\pi/2 < \arg x < \pi/2). \qquad (4.12.8)$$

Here $(\alpha, n) = (-1)^n(\alpha + 1/2)_n(-\alpha + 1/2)_n/n!$.

4.13 Nicholson's Integral

Integral representations for modified Bessel functions can be obtained from those for Bessel functions. Similarly, there are formulas for integrals of modified Bessel functions. As one example, take $y = i$ and $\text{Re}\, x > 1$ in (4.11.18) to get

$$\int_0^\infty e^{-xt}I_\alpha(t)dt = \frac{[x - \sqrt{(x^2 - 1)}]^\alpha}{\sqrt{(x^2 - 1)}}. \qquad (4.13.1)$$

Set $x = \cosh \beta$; then (4.13.1) can be written as

$$\int_0^\infty e^{-t \cosh \beta} I_\alpha(t)\,dt = \frac{e^{-\alpha\beta}}{\sinh \beta}. \tag{4.13.2}$$

Now, since

$$K_\alpha(t) \sim \sqrt{\frac{\pi}{2t}} e^{-t} \quad \text{as } t \to \infty, \tag{4.13.3}$$

we see from (4.13.2), on replacing α with $-\alpha$, that

$$\int_0^\infty e^{-t \cosh \beta} K_\alpha(t)\,dt = \frac{\pi}{\sin \alpha\pi} \frac{\sinh \alpha\beta}{\sinh \beta}, \quad \text{when } \operatorname{Re}(\cosh \beta) > -1. \tag{4.13.4}$$

Let $\alpha \to 0$ to get

$$\int_0^\infty e^{-t \cosh \beta} K_0(t)\,dt = \frac{\beta}{\sinh \beta}. \tag{4.13.5}$$

When $\beta = i\pi/2$, we have

$$\int_0^\infty K_0(t)\,dt = \frac{\pi}{2}. \tag{4.13.6}$$

Nicholson's formula,

$$N(x) = \frac{8}{\pi^2} \int_0^\infty K_0(2x \sinh t) \cosh(2\alpha t)\,dt = J_\alpha^2(x) + Y_\alpha^2(x), \quad \operatorname{Re} x > 0, \tag{4.13.7}$$

generalizes the trigonometric identity $\sin^2 x + \cos^2 x = 1$ as can be seen by taking $\alpha = 1/2$ and applying (4.13.6). We present Wilkins's [1948] verification of (4.13.7). This is done by showing that both sides of (4.13.7) satisfy the same differential equation and then analyzing their asymptotic behavior.

We show first that

$$N(x) \sim \frac{2}{\pi x} \quad \text{as} \quad x \to \infty, \tag{4.13.8}$$

where $N(x)$ denotes the left side of (4.13.7). It is sufficient to prove that

$$\lim_{x\to\infty} x N(x) = \lim_{x\to\infty} \frac{8}{\pi^2} \int_0^\infty x K_0(2x \sinh t) \cosh t\,dt = \frac{2}{\pi}. \tag{4.13.9}$$

The second equation follows from (4.13.6). For the first equation, we show that

$$F(x, t) = x K_0(2x \sinh t)(\cosh 2\alpha t - \cosh t)$$

converges boundedly to 0. The dominated convergence theorem then implies
(4.13.9). From (4.13.3) we have

$$|F(x,t)| \le A_0(x \text{ csch } t)^{1/2} e^{-2x \sinh t} |\cosh 2\alpha t - \cosh t|$$
$$\le A_0(x/t)^{1/2} e^{-2x \sinh t} (2|\alpha| + 1) t (\sinh 2|\alpha| t + \sinh t).$$

The second inequality follows upon an application of the mean value theorem to
$\cosh 2\alpha t - \cosh t$, recalling the fact that $\text{csch } t \le 1/t$. Let $x \ge 1$. Since $\sinh t \ge t$
and $(xt)^{1/2} e^{-xt}$ is bounded, we have

$$|F(x,t)| \le A(\sinh 2|\alpha| t + \sinh t) e^{-x \sinh t}$$
$$\le A(\sinh 2|\alpha| t + \sinh t) e^{-\sinh t}.$$

This proves (4.13.9).

For the next step, check that the product of any two solutions of $y'' + py' + qy = 0$
satisfies the equation $y''' + 3py'' + (2p^2 + p' + 4q)y' + (4pq + 2q')y = 0$. Apply
this to Bessel's equation to see that $\{H_\nu^{(1)}(x)\}^2$, $\{H_\nu^{(2)}(x)\}^2$ and $J_\nu^2(x) + Y_\nu^2(x)$ are
independent solutions of

$$y''' + \frac{3}{x} y'' + \left(4 + \frac{1 - 4\alpha^2}{x^2} \right) y' + \frac{4}{x} y = 0.$$

Using differentiation under the integral sign, the reader should verify that $N(x)$
satisfies this differential equation. The differential equation

$$x K_0''(x) + K_0'(x) - x K_0(x) = 0$$

satisfied by $K_0(x)$ is also required in the calculation.

Thus we have

$$N(x) = A\{J_\alpha^2(x) + Y_\alpha^2(x)\} + B\{H_\alpha^{(1)}(x)\}^2 + C\{H_\alpha^{(2)}(x)\}^2.$$

Let $x \to \infty$ and use (4.8.3) to (4.8.6) to obtain

$$1 = A + e^{2i(x - \frac{1}{2}\alpha\pi - \frac{1}{4}\pi)} B + e^{-2i(x - \frac{1}{2}\alpha\pi - \frac{1}{4}\pi)} C + o(1).$$

Hence $B = C = 0$, $A = 1$, and Nicholson's formula is proved.

4.14 Zeros of Bessel Functions

It is easily seen that all nontrivial solutions of the Bessel equation (4.5.1) have
simple zeros except possibly at zero. The first derivatives of such solutions also
have simple zeros except possibly at zero and $\pm\alpha$.

From (4.8.5) we can conclude that for real α, $J_\alpha(x)$ changes sign infinitely often
as $x \to \infty$. This implies that $J_\alpha(x)$ and $J_\alpha'(x)$ have infinitely many positive zeros.
The conclusion for $J_\alpha'(x)$ follows from the mean value theorem.

Suppose that $j_{\alpha,1}, j_{\alpha,2}, \ldots$ are the positive zeros of $J_\alpha(x)$ in ascending order. Then, for $\alpha > -1$,

$$0 < j_{\alpha,1} < j_{\alpha+1,1} < j_{\alpha,2} < j_{\alpha+1,2} < j_{\alpha,3} < \cdots. \qquad (4.14.1)$$

From (4.6.1) and the mean value theorem it follows that between two zeros of $x^{\alpha+1} J_{\alpha+1}(x)$ there is a zero of $x^{\alpha+1} J_\alpha(x)$. Similarly, (4.6.2) implies that between two zeros of $x^{-\alpha} J_\alpha(x)$ there is a zero of $x^{-\alpha} J_{\alpha+1}(x)$. This proves (4.14.1).

When $\alpha \leq -1$, the zeros of $J_\alpha(x)$ and $J_{\alpha+1}(x)$ are still interlaced by the above argument, but the smallest zero of $J_{\alpha+1}(x)$ is closer to zero than that of $J_\alpha(x)$. It can also be proved that for $-2s < \alpha < -(2s+1)$, s a positive integer, $J_\alpha(x)$ has $4s$ complex zeros all with nonzero real parts. In contrast, when $-(2s+1) < \alpha < -2(s+1)$, s a nonnegative integer, $J_\alpha(x)$ has $4s+2$ complex zeros, two of which are purely imaginary. See Watson [1944, p. 483] for a proof.

Theorem 4.11.6 and Theorem 5.4.1 imply the following theorem about Bessel functions.

Theorem 4.14.1   *Let $x_{1n} > x_{2n} > \cdots$ be the zeros of $P_n^{(\alpha,\beta)}(x)$ in $[-1, 1]$ and let $x_{kn} = \cos\theta_{kn}$, $0 < \theta_{kn} < \pi$. Then for a fixed k,*

$$\lim_{n\to\infty} n\theta_{kn} = j_{\alpha,k}.$$

In particular, $J_\alpha(x)$ has an infinite number of positive zeros.

In the next chapter, we prove that all zeros of $P_n^{(\alpha,\beta)}(x)$ lie in $(-1, 1)$ when $\alpha, \beta > -1$. This, combined with Hurwitz's theorem, shows that $x^{-\alpha} J_\alpha(x)$ has only real zeros for $\alpha > -1$. For Hurwitz's theorem one may consult Hille [1962, p. 180].

Another method of obtaining the reality of the zeros of $J_\alpha(x)$ for $\alpha > -1$ is to establish the formula

$$(b^2 - a^2) \int_0^x t J_\alpha(at) J_\alpha(bt) dt = x \left[J_\alpha(bx) J_\alpha'(ax) - J_\alpha(ax) J_\alpha'(bx) \right]. \qquad (4.14.2)$$

To prove this, note that $J_\alpha(ax)$ satisfies the differential equation

$$\frac{1}{x} \frac{d}{dx}\left(x \frac{dy}{dx} \right) + \left(a^2 - \frac{\alpha^2}{x^2} \right) y = 0.$$

Multiply this equation by $J_\alpha(bx)$ and multiply the corresponding equation for $J_\alpha(bx)$ by $J_\alpha(ax)$; subtract to get

$$J_\alpha(bx) \frac{d}{dx}\left(x \frac{d J_\alpha(ax)}{dx} \right) - J_\alpha(ax) \frac{d}{dx}\left(x \frac{d J_\alpha(bx)}{dx} \right) = (b^2 - a^2) x J_\alpha(ax) J_\alpha(bx)$$

or

$$\frac{d}{dx}\left[x J_\alpha(bx) J_\alpha'(ax) - x J_\alpha(ax) J_\alpha'(bx)\right] = (b^2 - a^2) x J_\alpha(ax) J_\alpha(bx).$$

Formula (4.14.2) is simply an integrated form of this. Now if a is a complex zero of $J_\alpha(x)$, then so is $\bar{a}$. Take $x = 1$, $b = \bar{a}$ in (4.14.2) and note that the integrand $t J_\alpha(at) J_\alpha(\bar{a}t) > 0$. Hence the left side of (4.14.2) is nonzero but the right side is zero. This contradiction implies that $J_\alpha(x)$ has no complex zeros.

An argument using differential equations can also be given to show that $J_\alpha(x)$ has an infinity of positive real solutions for real α. This technique goes back to Sturm. The version of Sturm's comparison theorem given below is due to Watson [1944, p. 518].

Theorem 4.14.2   *Let $u_1(x)$ and $u_2(x)$ be the solutions of the equations*

$$\frac{d^2 u_1}{dx^2} + \phi_1(x) u_1 = 0, \qquad \frac{d^2 u_2}{dx^2} + \phi_2(x) u_2 = 0$$

such that when $x = a$

$$u_1(a) = u_2(a), \qquad u_1'(a) = u_2'(a).$$

Let $\phi_1(x)$ and $\phi_2(x)$ be continuous in the interval $a \leq x \leq b$, and $u_1'(x)$ and $u_2'(x)$ be continuous in the same interval. Then, if $\phi_1(x) \geq \phi_2(x)$ throughout the interval, $|u_2(x)|$ exceeds $|u_1(x)|$ as long as x lies between a and the first zero of $u_1(x)$ in the interval. Thus the first zero of $u_1(x)$ in the interval is on the left of the first zero of $u_2(x)$.

Proof.   Without loss of generality, we assume that $u_1(x)$ and $u_2(x)$ are both positive immediately to the right of $x = a$. Subtract u_2 times the first equation from u_1 times the second to get

$$u_1 \frac{d^2 u_2}{dx^2} - u_2 \frac{d^2 u_1}{dx^2} = (\phi_1(x) - \phi_2(x)) u_1 u_2 \geq 0.$$

Integration gives

$$\left[u_1 \frac{du_2}{dx} - u_2 \frac{du_1}{dx} \right]_a^x \geq 0.$$

Since the expression in the brackets vanishes at a, we have

$$u_1 \frac{du_2}{dx} - u_2 \frac{du_1}{dx} \geq 0.$$

Hence

$$\frac{d(u_2/u_1)}{dx} \geq 0$$

or

$$\left[\frac{u_2}{u_1}\right]_a^x \geq 0,$$

which implies that $u_2(x) \geq u_1(x)$. This proves the theorem. ∎

Suppose $|\alpha| > \frac{1}{2}$, α real, and take $\phi_1(x) = 1 - (\alpha^2 - 1/4)/x^2$ and $\phi_2(x) = 1 - (\alpha^2 - 1/4)/c^2$. Then for $x \geq c$, we have $\phi_1(x) \geq \phi_2(x)$. Note that $u_1 = x^{1/2} J_\alpha(x)$ is a solution of $u_1'' + \phi_1(x)u_1 = 0$. Denote the general solution by $x^{1/2} C_\alpha(x)$. It is clear that $u_2 = A \cos \omega x + B \sin \omega x$, where $\omega^2 = 1 - (\alpha^2 - 1/4)/c^2$. It follows from Theorem 4.14.2 that if c is any zero of $C_\alpha(x)$ then the next larger zero is at most $c + \pi/\omega$. When $|\alpha| \leq 1/2$, take $\phi_2(x) = \omega^2 < 1$. Thus for real α, $J_\alpha(x)$ has an infinite number of real zeros. Essentially, Sturm's theorem says that the greater the value of ϕ, the more rapid are the oscillations of the solutions of the equation as x increases.

Theorem 4.14.2 can also be used to prove that the forward differences of the positive zeros of $J_\alpha(x)$ are decreasing for $|\alpha| > 1/2$ and increasing for $|\alpha| < 1/2$. Suppose $|\alpha| > 1/2$ and let $j_{\alpha,n-1} < j_{\alpha n} < j_{\alpha,n+1}$ be three successive positive zeros of $J_\alpha(x)$. Now set $\phi_1(x) = 1 - (\alpha^2 - 1/4)/x^2$ and $\phi_2(x) = \phi_1(x - k)$, where $k = j_{\alpha n} - j_{\alpha,n-1}$. Now $\phi_1(x)$ is an increasing function, so $\phi_1(x) \geq \phi_2(x)$. Consider the interval $[j_{\alpha n}, j_{\alpha,n+1}]$. At $x = j_{\alpha n}$, $u_1 = J_\alpha(x) = 0$ and $u_2 = J_\alpha(x - k) = 0$. By Sturm's theorem, u_1 oscillates more rapidly and hence $j_{\alpha,n+1} - j_{\alpha,n} < j_{\alpha,n} < j_{\alpha,n-1}$. A similar argument applies to the case $|\alpha| < 1/2$. It should be clear that the same argument works for the general solution of Bessel's equation.

We end this section with an infinite product formula for $J_\alpha(x)$, α real. For large x, the asymptotic formula (4.8.5) for $J_\alpha(x)$ suggests that the asymptotic behavior of the zeros is given by

$$x \sim (m + (2\alpha + 1)/4)\pi. \tag{4.14.3}$$

Since the zeros of $J_\alpha(x)$ are real and simple, one expects that the number of zeros of $x^{-\alpha} J_\alpha(x)$ between the imaginary axis and the line $\text{Re } x = (m + (\alpha + 1)/4)\pi$, for large x, is m. This is true. See Watson [1944, §15.4]. It follows that the entire function $x^{-\alpha} J_\alpha(x)$ has the product formula

$$\Gamma(\alpha + 1)(x/2)^{-\alpha} J_\alpha(x)$$
$$= \prod_{n=1}^{\infty} \{(1 - x/j_{\alpha,n}) \exp(x/j_{\alpha,n})\} \prod_{n=1}^{\infty} \{(1 + x/j_{\alpha,n}) \exp(-x/j_{\alpha,n})\}. \tag{4.14.4}$$

This result continues to hold when α is not real. See Watson [1944, §15.41].

4.15 Monotonicity Properties of Bessel Functions

Sturm's comparison theorem for differential equations stated in the previous section gave information about the zeros of the solutions of the differential equation

$$\frac{d^2y}{dx^2} + \left(1 - \frac{\nu^2 - 1/4}{x^2}\right)y = 0. \tag{4.15.1}$$

Sturm also used his theorem to prove that the second forward differences of the positive zeros of any nontrivial solution of the above equation are all positive if $|\nu| < 1/2$ and all negative if $|\nu| > 1/2$. Lorch and Szego [1963] have greatly extended this result to higher-order differences. In this section we present one of their theorems. For further generalizations and other related results, the reader should see the references section.

Consider the differential equation

$$y'' + f(x)y = 0, \tag{4.15.2}$$

where $x \in I$, an open interval. Let $\lambda > -1$ and let $y(x)$ be an arbitrary solution of (4.15.2) with zeros at $x_1, x_2, \ldots$ in ascending order. Set

$$M_k = \int_{x_k}^{x_{k+1}} |y(x)|^\lambda dx, \quad k = 1, 2, \ldots. \tag{4.15.3}$$

Observe that when $\lambda = 0$, $M_k = \Delta x_k$, the difference of the successive zeros of $y(x)$. When $\lambda = 1$, then M_k gives the area under the arch formed by $y(x)$ from x_k to x_{k+1}.

We state the theorem of Lorch and Szego without proof. The reader may consult the original paper for a proof, which is somewhat lengthy. In the statement of the theorem, the notation $\Delta^n \mu_k$ is used to denote the nth forward difference of the sequence $\{\mu_k\}$. Thus,

$$\Delta^n \mu_k = \Delta^{n-1} \mu_{k+1} - \Delta^{n-1} \mu_k \quad \text{and} \quad \Delta^0 \mu_k = \mu_k.$$

Theorem 4.15.1 *Let y_1 and y_2 be two independent solutions of the differential equation (4.15.2) in a closed interval $\bar{I}$. Suppose that*

$$(-1)^n \frac{d^n}{dx^n}\{[y_1(x)]^2 + [y_2(x)]^2\} > 0 \quad \text{for } n = 0, 1, \ldots, N,$$

where the Nth derivative exists in the open interval I, and the lower-order derivatives are continuous in $\bar{I}$. Then

$$(-1)^n \Delta^n M_k > 0 \quad \text{for } n = 0, \ldots, N; \quad k = 1, 2, \ldots.$$

In particular

$$(-1)^{n-1} \Delta^n x_k > 0 \quad \text{for } n = 1, \ldots, N+1; \quad k = 1, 2, \ldots.$$

Moreover, if $\bar{y}(x)$ denotes another solution of (4.15.2) with zeros at $\bar{x}_1, \bar{x}_2, \ldots$ and if $x_1 > \bar{x}_1$, then

$$(-1)^n \Delta^n (x_k - \bar{x}_k) > 0 \quad for \ n = 0, \ldots, N; \quad k = 1, 2, \ldots.$$

This theorem yields results on Bessel functions when applied to equation (4.15.1). Two independent solutions of this equation are $\sqrt{x}J_\nu(x)$ and $\sqrt{x}Y_\nu(x)$. Let $\sqrt{x}C_\nu(x)$ denote the general solution. To apply Theorem 4.15.1, we have to study the expression

$$p(x) = x[J_\nu(x)]^2 + x[Y_\nu(x)]^2, \tag{4.15.4}$$

which can be represented by Nicholson's integral (4.13.7).

We need the following formula below:

$$K_0(x) = \int_0^\infty e^{-x \cosh t} dt. \tag{4.15.5}$$

For a proof, see Exercise 11. Now, by (4.13.7),

$$p'(x) = \frac{d}{dx} \left[x \{ J_\nu^2(x) + Y_\nu^2(x) \} \right]$$

$$= \frac{8}{\pi^2} \int_0^\infty \left[K_0(2x \sinh t) + 2x \sinh t \, K_0'(2x \sinh t) \right] \cosh 2\nu t \, dt.$$

Integrate the second term by parts to get

$$p'(x) = \frac{8}{\pi^2} [K_0(2x \sinh t) \tanh t \cosh 2\nu t]_0^\infty$$

$$+ \frac{8}{\pi^2} \int_0^\infty K_0(2x \sinh t) \left(\cosh 2\nu t - \frac{d}{dt} (\tanh t \cosh 2\nu t) \right) dt.$$

The first term on the right is zero, for by definition (4.12.3), it follows that $K_0(x)$ behaves like $\log x$ as $x \to 0$, while (4.12.6) gives the behavior of $K_0(x)$ as $x \to \infty$. Thus

$$p'(x) = \frac{8}{\pi^2} \int_0^\infty K_0(2x \sinh t) \tanh t \cosh 2\nu t [\tanh t - 2\nu \tanh 2\nu t] dt.$$

It is easy to check that the expression in brackets is negative for $|\nu| > 1/2$ and the rest of the integrand is positive. So $p'(x) < 0$. Similarly, it can be shown that

$$p^{(n)}(x) = \frac{8}{\pi^2} \int_0^\infty K_0^{(n-1)}(2x \sinh t)(2 \sinh t)^{n-1} \tanh t \cosh 2\nu t$$

$$\cdot \{\tanh t - 2\nu \tanh 2\nu t\} dt.$$

It is clear from (4.15.5) that $(-1)^n K_0^{(n)}(x) > 0$ for $x > 0, n = 0, 1, 2, \ldots$. Thus, the conditions of Theorem 4.15.1 hold for Equation (4.15.1) when $|\nu| > 1/2$. So we have the following corollary:

Corollary 4.15.2 Let $c_{vk}, \bar{c}_{vk}$ denote the kth positive zeros in ascending order of any pair of nontrivial solutions of Bessel's equation (4.15.1) with $|v| > 1/2$. Suppose $\lambda > -1$ and set

$$M_k = \int_{c_{vk}}^{c_{v,k+1}} x^{\lambda/2} |C_v(x)|^\lambda dx.$$

Then, for $k = 1, 2, \ldots,$

$$(-1)^n \Delta^n M_k > 0 \quad for \ n = 0, 1, \ldots,$$

$$(-1)^{n-1} \Delta^n c_{vk} > 0 \quad for \ n = 1, 2, \ldots,$$

$$(-1)^n \Delta^n (c_{v,m+k} - \bar{c}_{vk}) > 0 \quad for \ n = 0, 1, \ldots$$

with m a fixed nonnegative integer, provided that $c_{v,m+1} > \bar{c}_{v1}$.
   In particular,

$$(-1)^{n-1} \Delta^n j_{vk} > 0, \qquad (-1)^{n-1} \Delta^n y_{vk} > 0 \quad for \ n = 1, 2, \ldots$$

and

$$(-1)^n \Delta^n (j_{vk} - y_{vk}) > 0 \quad for \ n = 0, 1, \ldots,$$

where j_{vk}, y_{vk} denote the kth positive zeros of $J_v(x)$ and $Y_v(x)$ respectively.

Remark 4.15.1 Lorch, Muldoon, and Szego [1970] have extended Theorem 4.15.1 to a study of higher monotonicity properties of

$$M_k = \int_{x_k}^{x_{k+1}} W(x) |y(x)|^\lambda dx,$$

where $W(x)$ is a function subject to some restrictions. As an example, it is possible to take $W(x) = x^{-1/2}$ when y is a solution of (4.15.1). This implies the monotonicity of

$$(-1)^n \Delta^n \int_{c_{vk}}^{c_{vk+1}} |C_v(x)| dx,$$

where the integral is the area contained by an arch of a general Bessel function instead of $x^{1/2}$ times a Bessel function. For further extensions of the results see also Lorch, Muldoon, and Szego [1972].

4.16 Zero-Free Regions for $_1F_1$ Functions

We end this chapter with some results of Saff and Varga [1976] on zero-free regions for sequences of polynomials satisfying three-term recurrence relations. These polynomials can be partial sums of $_1F_1$ functions, so we may invoke a theorem of Hurwitz on zeros of an analytic function which is the limit of a sequence of analytic functions to obtain zero-free regions for $_1F_1$ functions.
   Saff and Varga's basic theorem is the following:

Theorem 4.16.1   *Let* $\{p_k(z)\}_0^n$ *be a finite sequence of polynomials satisfying*

$$p_k(z) = \left(\frac{z}{b_k} + 1\right) p_{k-1}(z) - \frac{z}{c_k} p_{k-2}(z), \quad k = 1, 2, \dots, n, \qquad (4.16.1)$$

where $p_{-1}(z) := 0$, $p_0(z) = p_0 \neq 0$, *and the* b_k *and* c_k *are positive real numbers.*
Let

$$\alpha := \min\{b_k(1 - b_{k-1}/c_k) : k = 1, 2, \dots, n\}, \quad b_0 = 0. \qquad (4.16.2)$$

Then, if $\alpha > 0$, *the parabolic region*

$$P_\alpha = \{z = x + iy : y^2 \leq 4\alpha(x + \alpha), x > -\alpha\} \qquad (4.16.3)$$

contains no zeros of $p_k(z)$, $\quad k = 1, 2, \dots, n.$

Proof.   Suppose $\tilde{z} \in P_\alpha$ is a fixed complex number that is not a zero of any $p_k(z)$, $k = 1, \dots, n$. Set

$$\mu_k := \mu_k(\tilde{z}) := \tilde{z} p_{k-1}(\tilde{z}) / b_k p_k(\tilde{z}) \quad \text{for } k = 1, \dots, n.$$

The proof depends on the following two facts:

1. The polynomials $p_k(z)$ and $p_{k-1}(z)$ have no zeros in common for each k, $k = 1, \dots, n$.
2. $\operatorname{Re} \mu_k \leq 1 \quad$ for $k = 1, \dots, n$.

Assume these results for now and suppose that for some k, $p_k(z)$ is zero at a point $z_0 \in P_\alpha$. Observe that $k \neq 1$, for, $p_1(z) = p_0(z + b_1)/b_1$ has a zero at $-b_1$, which by (4.16.2) is $\leq -\alpha$ and hence cannot be in P_α. Suppose $2 \leq k \leq n$. Since $p_k(z_0) = 0$, it follows from (4.16.1) that

$$(z_0/b_k + 1) p_{k-1}(z_0) = (z_0/c_k) p_{k-2}(z_0).$$

Fact 1 implies that $p_{k-1}(z_0) \neq 0$ because p_k and p_{k-1} have no common zeros. So we can divide by $p_{k-1}(z_0)$ to get

$$\frac{c_k}{b_{k-1} b_k}(z_0 + b_k) = \frac{z_0 p_{k-2}(z_0)}{b_{k-1} p_{k-1}(z_0)} = \mu_{k-1}(z_0). \qquad (4.16.4)$$

The second fact and the continuity of μ give $\operatorname{Re} \mu_{k-1}(z_0) \leq 1$. Then by (4.16.4)

$$\operatorname{Re} z_0 \leq -b_k(1 - b_{k-1}/c_k) \leq -\alpha,$$

which contradicts the assumption that $z_0 \in P_\alpha$. This means that $p_k(z)(k = 1, \dots, n)$ have no zeros in P_α. It remains only to prove our two assumptions. It is evident from (4.16.1) that none of the polynomials $p_k(z)$, $k = 0, 1, \dots, n$ vanish at 0. So suppose $p_k(z_0) = p_{k-1}(z_0) = 0$, where $z_0 \neq 0$, for some $k \geq 1$. By a repeated application of (4.16.1) we get $p_0(z_0) = 0$, which contradicts the assumption that $p_0 \neq 0$. Thus $p_k(z)$ and $p_{k-1}(z)$ have no common zeros.

We now prove that $\operatorname{Re} \mu_k(\tilde{z}) \leq 1$ by induction. Clearly,

$$\mu_1 = \frac{\tilde{z}}{\tilde{z} + b_1}.$$

Since $\tilde{z} \in P_\alpha$, $\operatorname{Re} \tilde{z} > -\alpha \geq -b_1$. Thus $\operatorname{Re} \mu_1 \leq 1$. Now, it follows easily from (4.16.1) that

$$\mu_k = \frac{\tilde{z}}{\tilde{z} + b_k - b_k c_k^{-1} b_{k-1} \mu_{k-1}},$$

or

$$\mu_k = T_k(\mu_{k-1}),$$

where $T_k(w)$ is a fractional linear transformation defined by

$$\xi = T_k(w) = \frac{\tilde{z}}{\tilde{z} + b_k - b_k c_k^{-1} b_{k-1} w}.$$

The function T_k has a pole at the point $w_k = (\tilde{z} + b_k)/(b_k c_k^{-1} b_{k-1})$ whose real part, by (4.16.2) and (4.16.3), is seen to be > 1. So T_k maps $\operatorname{Re} w \leq 1$ into a bounded disk with center $\xi_k = T_k(2 - \bar{w}_k)$, where $2 - \bar{w}_k$ is the point symmetric to the pole w_k with respect to the line $\operatorname{Re} w = 1$. By the definition of T_k

$$\xi_k = \frac{\tilde{z}}{2\operatorname{Re} \tilde{z} + 2b_k \left(1 - b_{k-1} c_k^{-1}\right)}.$$

Moreover, the point $0 = T_k(\infty)$ lies on the boundary of this disk so its radius is $|\xi_k|$. Thus the real part of any point in the disk does not exceed

$$\operatorname{Re} \xi_k + |\xi_k| = \frac{\operatorname{Re} \tilde{z} + |\tilde{z}|}{2\operatorname{Re} \tilde{z} + 2b_k \left(1 - b_{k-1} c_k^{-1}\right)} \leq \frac{\operatorname{Re} \tilde{z} + |\tilde{z}|}{2(\operatorname{Re} \tilde{z} + \alpha)} \leq 1,$$

where the first inequality follows from (4.16.2) and the second from (4.16.3). Now, by the induction hypothesis, $\operatorname{Re} \mu_{k-1} \leq 1$; thus $\operatorname{Re} \mu_k = \operatorname{Re} T_k(\mu_{k-1}) \leq 1$. This proves the theorem. ∎

Corollary 4.16.2 *For an infinite sequence of polynomials $\{p_k(z)\}_0^\infty$ satisfying the three-term relation (4.16.1), suppose that*

$$\alpha = \inf_{k \geq 1} \left\{ b_k \left(1 - b_{k-1} c_k^{-1}\right) \right\} > 0.$$

Then the region P_α defined in the theorem is zero-free for the polynomials $p_k(z)$. Moreover, if $p_k(z) \to f(z) \neq 0$ uniformly on compact subsets of P_α, then $f(z)$ is also zero-free in P_α.

Proof. The first part is obvious. For the second part use Hurwitz's theorem. ∎

The next corollary applies to the polynomial squence obtained from the partial sums of a power series.

Corollary 4.16.3 *Suppose* $s_k(z) := \sum_{j=0}^{k} a_j z^j$ *have strictly positive coefficients and*

$$\alpha := \min_{1 \le k \le n} \left\{ \left(\frac{a_{k-1}}{a_k} - \frac{a_{k-2}}{a_{k-1}} \right) \right\} > 0, \quad \text{where } a_{-1}/a_0 = 0.$$

Then the polynomials $s_k(z), k = 1, 2, \ldots, n$ *have no zeros in* P_α.

Proof. First observe that

$$s_k(z) = \left(\frac{a_k z}{a_{k-1}} + 1 \right) s_{k-1}(z) - \frac{a_k z}{a_{k-1}} s_{k-2}(z), \quad k = 1, 2, \ldots, n,$$

when $s_{-1} = 0$. Then apply Theorem 4.16.1 to obtain the required result. ∎

Note that a consequence of the above results is that the partial sums $s_n(z) = \sum_{0}^{n} z^n/n!$ of the exponential function have the parabolic region, $y^2 \le 4(x+1)$, $x > -1$, as a zero-free region. This region is sharp, both because of the zero at $z = -1$ for $s_1(z) = 1 + z$ and asymptotically as $n \to \infty$ in $s_n(z)$. The next corollary concerns the more general $_1F_1$ confluent hypergeometric function. The proof is left to the reader.

Corollary 4.16.4 *Suppose* $s_n(z)$ *is the nth partial sum of* $_1F_1(c; d; z)$. *Then* $s_n(z), n = 0, 1, 2, \ldots$ *have no zeros in the region*

(i) $P_{d/c}$, *if* $0 < d \le c$,
(ii) P_1, *if* $1 \le c \le d$,
(iii) $P_\alpha, \alpha = (2c - d + cd)/(c^2 + c)$, *if* $0 < c < 1$ *and* $c \le d < 2c/(1 - c)$.

Moreover, $_1F_1(c; d; z)$ *has no zeros in the corresponding interior region.*

For other applications of Theorem 4.16.1 see de Bruin, Saff, and Varga [1981].

Exercises

1. Show that

$$_1F_1(a; c; x) = \frac{\Gamma(c)}{\Gamma(a)\Gamma(c - a)} \int_0^1 e^{xt} t^{a-1} (1 - t)^{c-a-1} dt$$

when $\operatorname{Re} c > \operatorname{Re} a > 0$.

2. Let $_1F_1(a-) = {}_1F_1(a - 1; c; x)$ and define $_1F_1(a+)$ etc. in a similar way. Prove the contiguous relations:

(a) $(c - a) {}_1F_1(a-) + (2a - c + x) {}_1F_1 - a {}_1F_1(a+) = 0$,
(b) $c(c - 1) {}_1F_1(c-) - c(c - 1 + x) {}_1F_1 + (c - a)x {}_1F_1(c+) = 0$,
(c) $(a - c + 1) {}_1F_1 - a {}_1F_1(a+) + (c - 1) {}_1F_1(c-) = 0$,
(d) $c {}_1F_1 - c {}_1F_1(a-) - x {}_1F_1(c+) = 0$,
(e) $c(a + x) {}_1F_1 - (c - a)x {}_1F_1(c+) - ac {}_1F_1(a+) = 0$,
(f) $(a - 1 + x) {}_1F_1 + (c - a) {}_1F_1(a-) - (c - 1) {}_1F_1(c-) = 0$.

3. Prove the formulas

(a) $\quad \dfrac{d^n}{dx^n} {}_1F_1(a; c; x) = \dfrac{(a)_n}{(c)_n} {}_1F_1(a + n; c + n; x),$

(b) $\quad \dfrac{d^n}{dx^n}[e^{-x} {}_1F_1(a; c; x)] = (-1)^n \dfrac{(c - a)_n}{(c)_n} e^{-x} {}_1F_1(a; c + n; x).$

4. Express the following functions in terms of Whittaker functions:

(a) The sine integral $\mathrm{Si}(x) = \int_0^x t^{-1} \sin t\, dt.$

(b) The cosine integral $\mathrm{Ci}(x) = -\int_x^\infty t^{-1} \cos t\, dt.$

(c) The Fresnel integrals

$$C(x) = \int_0^x t^{-1/2} \cos t\, dt / \sqrt{2\pi},$$

$$S(x) = \int_0^x t^{-1/2} \sin t\, dt / \sqrt{2\pi}.$$

(d) The exponential integral

$$E_1(x) = \int_x^\infty \frac{e^{-t}}{t} dt.$$

5. Use (4.3.6) and (4.4.4) to derive an asymptotic expansion for erf x.

6. Prove that

$$\int_0^\infty e^{-sx} x^{c-1} {}_1F_1(a; c; x) {}_1F_1(a_1; c; \lambda x)dx$$

$$= \Gamma(c)(s - 1)^{-a}(s - \lambda)^{-a_1} s^{a+a_1-c} {}_2F_1(a, a_1; c; \lambda/[(s - 1)(s - \lambda)]).$$

7. Show that, for the parabolic cylinder function $D_n(x)$ given by (4.4.11), the following properties hold:

(a) $D_n(x) = \sqrt{\pi} 2^{n/2} e^{-x^2/2} {}_1F_1(-n/2; 1/2; x^2/2)/\Gamma((1 - n)/2)$
$\qquad - \sqrt{\pi} 2^{(n+1)/2} x e^{-x^2/4} {}_1F_1((1 - n)/2; 3/2; x^2/2)/\Gamma(-n/2).$

(b) $D_n(x) = (-1)^n e^{x^2/4} \dfrac{d^n}{dx^n}(e^{-x^2/2}).$

8. Prove the formulas (4.6.11) and (4.6.12) for $J_{n+1/2}(x)$ and $J_{-n-1/2}(x)$.

9. Show that for $\mathrm{Re}\,\alpha > -1/2$,

$$\Gamma(\alpha + 1/2)J_\alpha(x) = \frac{2}{\sqrt{\pi}}(x/2)^\alpha \int_0^{\pi/2} \cos(x \sin \theta)(\cos \theta)^{2\alpha} d\theta$$

and

$$\Gamma(\alpha + 1/2)I_\alpha(x) = \frac{1}{\sqrt{\pi}}(x/2)^\alpha \int_{-1}^1 e^{-xt}(1 - t^2)^{\alpha-1/2} dt.$$

Deduce that

$$|J_\alpha(x)| \le |x/2|^\alpha e^{|v|}/\Gamma(\alpha + 1), \quad \text{where} \quad x = u + iv.$$

10. Use (4.9.11) to obtain Neumann's formulas (Watson [1944, p. 32]):

$$J_n^2(x) = \frac{1}{\pi} \int_0^\pi J_{2n}(2x\sin\theta)d\theta = \frac{1}{\pi} \int_0^\pi J_0(2x\sin\theta)\cos 2n\theta d\theta.$$

11. Show that, for $|\arg x| < \pi/2$,

$$I_{-\alpha}(x) = \frac{\Gamma(1/2-\alpha)e^{2\pi i\alpha}(x/2)^\alpha}{2\pi i\Gamma(1/2)} \int_\infty^{(1+,-1+)} e^{-xt}(t^2-1)^{\alpha-1/2}dt.$$

Deduce that, when $\operatorname{Re}\alpha > -1/2$,

$$I_{-\alpha}(x) = \frac{\Gamma(1/2-\alpha)e^{2\pi i\alpha}(x/2)^\alpha}{2\pi i\Gamma(1/2)} \left[(1-e^{-4\pi i\alpha}) \int_1^\infty e^{-xt}(t^2-1)^{\alpha-1/2}dt \right.$$

$$\left. + i(e^{-\pi i\alpha} + e^{-3\pi i\alpha}) \int_{-1}^1 e^{-xt}(1-t^2)^{\alpha-1/2}dt \right].$$

Hence

$$K_\alpha(x) = \frac{\sqrt{\pi}(x/2)^\alpha}{\Gamma(\alpha+1/2)} \int_1^\infty e^{-xt}(t^2-1)^{\alpha-1/2}dt$$

$$= \frac{\sqrt{\pi}(x/2)^\alpha}{\Gamma(\alpha+1/2)} \int_0^\infty e^{-x\cosh\theta}\sinh^{2\alpha}\theta d\theta.$$

12. Prove that for $x > 0$ and $\alpha > -1/2$

$$K_\alpha(x) = \frac{2^\alpha\Gamma(\alpha+1/2)}{x^\alpha\sqrt{\pi}} \int_0^\infty \frac{\cos xt}{(1+t^2)^{\alpha+1/2}}dt.$$

13. Show that for $\alpha > -1$ and $c > 0$

$$J_\alpha(x) = \frac{(x/2)^\alpha}{2\pi i} \int_{c-i\infty}^{c+i\infty} t^{-\alpha-1}\exp\left(t - \frac{x^2}{4t}\right)dt.$$

14. Prove the following result of Sonine and Schafheitlin:

$$S := \int_0^\infty \frac{J_{\alpha-\beta}(at)J_{\gamma-1}(bt)}{t^{\gamma-\alpha-\beta}}dt$$

$$= \frac{b^{\gamma-1}\Gamma(\alpha)}{2^{\gamma-\alpha-\beta}a^{\alpha+\beta}\Gamma(\gamma)\Gamma(1-\beta)} {}_2F_1\left(\begin{matrix}\alpha,\beta\\\gamma\end{matrix}; \frac{b^2}{a^2}\right) \quad \text{for } 0 < b < a$$

and

$$S = \frac{a^{\alpha-\beta}\Gamma(\alpha)}{2^{\gamma-\alpha-\beta}b^{2\alpha-\gamma+1}\Gamma(\gamma-\alpha)\Gamma(\alpha-\beta+1)}$$

$$\cdot {}_2F_1\left(\begin{matrix}\alpha, \alpha-\gamma+1\\\alpha-\beta+1\end{matrix}; \frac{a^2}{b^2}\right) \quad \text{for } 0 < a < b,$$

provided the integral is convergent.

Consider the particular cases (a) $\beta = 0$, $\gamma - \alpha = 1$; (b) $\gamma = 3/2$, $\alpha + \beta = 1/2$; (c) $\gamma = 1/2$, $\alpha + \beta = -1/2$; (d) $\gamma = 3/2$, $\alpha + \beta = 3/2$; (e) $\gamma = 1/2$, $\alpha + \beta = 1/2$.

See Watson [1944, §13.4].

15. Show that when $a = b$ in Exercise 14, the result is

$$\int_0^\infty \frac{J_{\alpha-\beta}(at)J_{\gamma-1}(at)dt}{t^{\gamma-\alpha-\beta}} = \frac{(a/2)^{\gamma-\alpha-\beta-1}\Gamma(\gamma-\alpha-\beta)\Gamma(\alpha)}{2\Gamma(1-\beta)\Gamma(\gamma-\alpha)\Gamma(\gamma-\beta)},$$

provided that Re $\alpha > 0$ and Re $(\gamma - \alpha - \beta) > 0$.

16. Show that

$$J_\alpha(ax)J_\beta(bx) = \frac{(ax/2)^\alpha(bx/2)^\beta}{\Gamma(\beta+1)\Gamma(\alpha+1)}$$

$$\cdot \sum_{n=0}^\infty \frac{(-1)^n {}_2F_1(-n,-\alpha-n;\beta+1;b^2/a^2)(ax/2)^{2n}}{n!(\alpha+1)_n}.$$

Deduce that

$$J_\alpha(x)J_\beta(x) = \frac{(x/2)^{\alpha+\beta}}{\Gamma(\alpha+1)\Gamma(\beta+1)} \sum_{n=0}^\infty \frac{(-1)^n(\alpha+\beta+1)_{2n}(x/2)^{2n}}{(\alpha+1)_n(\beta+1)_n(\alpha+\beta+1)_n}.$$

17. Show that for $a, b > 0$ and $-1 < $ Re $\alpha < 2$Re $\beta + 3/2$

$$\int_0^\infty \frac{x^{\alpha+1}J_\alpha(bx)}{(x^2+a^2)^{\beta+1}}dx = \frac{a^{\alpha-\beta}b^\beta}{2^\alpha\Gamma(\alpha+1)}K_{\alpha-\beta}(ab).$$

(Note that $\int_0^\infty e^{-(x^2+a^2)t}t^\beta dt = \frac{\Gamma(\beta+1)}{(x^2+a^2)^{\beta+1}}$ for Re $\beta > -1$.)

18. Show that

$$K_\alpha(x) = \frac{(x/2)^\alpha}{2}\int_0^\infty e^{-t-x^2/4t}t^{-\alpha-1}dt, \qquad |\arg x| < \pi/4.$$

19. Prove that for $a > 0$, $b > 0$, $y > 0$, and Re $\beta > -1$

$$\int_0^\infty \frac{K_\alpha(a\sqrt{x^2+y^2})}{(x^2+y^2)^{\alpha/2}}J_\beta(bx)x^{\beta+1}dx$$

$$= \frac{b^\beta}{a^\alpha}\left(\sqrt{\frac{a^2+b^2}{y}}\right)^{\alpha-\beta-1}K_{\alpha-\beta-1}(y\sqrt{a^2+b^2}).$$

Consider the case $\alpha = 1/2$, $\beta = 0$.

20. Prove the following formula for Airy's integral:

$$\mathrm{Ai}(x) := \frac{1}{\pi}\int_0^\infty \cos(t^3+xt)dt = \frac{\sqrt{x}}{3\pi}K_{1/3}\left(\frac{2x\sqrt{x}}{3\sqrt{3}}\right).$$

See Watson [1944, §6.4].

21. Let $\phi(x)$ be a positive monotonic function in $C^1(a, b)$ and let $y(x)$ be any solution of the differential equation

$$y'' + \phi(x)y = 0.$$

Show that the relative maxima of $|y|$, as x increases from a to b, form an increasing or decreasing sequence accordingly as $\phi(x)$ decreases or increases.

   [*Hint*: For $f(x) = \{y(x)\}^2 + \{y'(x)\}^2/\phi(x)$ show that sgn $f'(x) = -$sgn $\phi'(x)$.]   (Sonine)

22. Suppose that $k(x)$ and $\phi(x)$ are positive and belong to $C'(a, b)$. If $y(x)$ is a solution of the equation

$$\{k(x)y'\} + \phi(x)y = 0,$$

then show that the relative maxima of $|y|$ form an increasing or decreasing sequence accordingly as $k(x)\phi(x)$ is decreasing or increasing.   (Butlewski)

23. Show that $u = x^a J_\alpha(bx^c)$ satisfies the differential equation

$$u'' + \frac{1 - 2a}{x}u' + \left[(bcx^{c-1})^2 + \frac{a^2 - \alpha^2 c^2}{x^2}\right]u = 0.$$

24. Take $\phi(x) = 1 + \frac{1/4 - \alpha^2}{x^2}$. Use (4.8.5) and Exercise 21 to prove that

$$\sup_{x \geq 0} \sqrt{x}|J_\alpha(x)| = \begin{cases} \sqrt{2/\pi} & \text{if } -1/2 \leq \alpha \leq 1/2, \\ \text{finite and } > \sqrt{2/\pi} & \text{if } \alpha > 1/2. \end{cases}$$

For Exercises 21, 22, and 24 and the references to Sonine and Butlewski, see Szegö [1975, pp. 166–167].

25. Let $\alpha = \lambda - 1/2, 0 < \lambda < 1$. Denote the positive zeros of $J_\alpha(x)$ by $j_1 < j_2 < j_3 < \cdots$ and the zeros of the ultraspherical polynomial $C_n^\lambda(\cos\theta)$ by $\theta_1 < \theta_2 < \cdots < \theta_n$. Use Theorem 4.14.2 to show that

$$\theta_k < j_k/(n + \lambda), \qquad k = 1, 2, \ldots, n.$$

[Note that $u = (\sin\theta)^\lambda C_n^\lambda(\cos\theta)$ satisfies the equation

$$\frac{d^2u}{d\theta^2} + \left\{(n + \lambda)^2 + \frac{\lambda(1 - \lambda)}{\sin^2\theta}\right\}u = 0,$$

and compare this with the equation satisfied by $\sqrt{\theta} J_\alpha\{(n + \lambda)\theta\}$.]

26. Suppose $-1/2 < \alpha \leq 1/2$ and $m\pi < x < (m + 1/2)\pi, m = 0, 1, 2, \ldots$. Show that $J_\nu(x)$ is positive for even m and negative for odd m. [Note that when $x = (m + \theta/2)\pi$ with $0 \leq \theta \leq 1$,

$$J_\alpha(x) = \frac{2(\pi/4)^\alpha}{\Gamma(\alpha + 1/2)\sqrt{\pi}(2m + \theta)^\alpha} \int_0^{2m+\theta} \frac{\cos(\pi u/2)}{\{(2m + \theta)^2 - u^2\}^{1/2-\alpha}} du$$

and show that

$$\operatorname{sgn} J_\alpha(m\pi + \theta\pi/2) = \operatorname{sgn}\left[(-1)^m\{v'_m + (v_m - v_{m-1})\right.$$

$$+ (v_{m-2} - v_{m-3}) + \cdots\}\right]$$

$$= \operatorname{sgn}(-1)^m,$$

where

$$(-1)^r v_r = \int_{2r-2}^{2r} \frac{\cos(\pi u/2)}{\{(2m+\theta)^2 - u^2\}^{1/2-\alpha}} du$$

and

$$(-1)^m v'_m = \int_{2m}^{2m+\theta} \frac{\cos(\pi u/2)du}{\{(2m+\theta)^2 - u^2\}^{1/2-\alpha}}\cdot\right]$$

27. Show that

$$\int_0^x t J_\alpha^2(t)dt = \frac{x}{2}\left[x\{J'_\alpha(x)\}^2 - J_\alpha(x)\frac{d}{dx}\{xJ'_\alpha(x)\}\right].$$

Deduce that $f(x) = AJ_\alpha(x) + BxJ'_\alpha(x) \not\equiv 0$ has no repeated zeros other than $x = 0$. (This result is due to Dixon. See Watson [1944, p. 480].)

28. Let $f(x) = AJ_\alpha(x) + BxJ'_\alpha(x)$ and $g(x) = CJ_\alpha(x) + DxJ'_\alpha(x)$ with $AD - BC \neq 0$. Prove that the positive zeros of $f(x)$ are interlaced with those of $g(x)$. (Show that $\phi(x) = f(x)/g(x)$ is monotonic.)

29. Prove Graf's formula (4.10.6) when α is an integer by using the identity

$$e^{a(t-1/t)/2}e^{-b(te^{-i\theta}-1/(te^{-i\theta}))/2} = e^{c(tu-1/(tu))/2},$$

where $u = (a - be^{-i\theta})/c$.

5

Orthogonal Polynomials

Although Murphy [1835] first defined orthogonal functions (which he called reciprocal functions), Chebyshev must be given credit for recognizing their importance. His work, done from 1855 on, was motivated by the analogy with Fourier series and by the theory of continued fractions and approximation theory. We start this chapter with a discussion of the Chebyshev polynomials of the first and second kinds. Some of their elementary properties suggest areas of study in the general situation. The rest of this chapter is devoted to the study of the properties of general orthogonal polynomials.

Orthogonal polynomials satisfy three-term recurrence relations; this illustrates their connection with continued fractions. We present some consequences of the three-term relations, such as the Christoffel–Darboux formula and its implications for the zeros of orthogonal polynomials. We also give Stieltjes's integral representation for continued fractions which arise from orthogonal polynomials.

In his theory on approximate quadrature, Gauss used polynomials that arise from the successive convergents of the continued fraction expansion of $\log(1 + x)/(1 - x)$. Later, Jacobi [1826] observed that these polynomials are Legendre polynomials and that their orthogonality played a fundamental role. We devote a section of this chapter to the Gauss quadrature formula and some of its consequences, especially for zeros of orthogonal polynomials. We also prove the Markov–Stieltjes inequalities for the constants that appear in Gauss's formula.

Finally, we employ a little elementary graph theory to find a continued fraction expansion for the moment-generating function. In the past two decades, combinatorial methods have been used quite successfully to study orthogonal polynomials.

5.1 Chebyshev Polynomials

We noted earlier that the example of the Chebyshev polynomials should be kept in mind when studying orthogonal polynomials. The Chebyshev polynomials of the first and second kinds, denoted respectively by $T_n(x)$ and $U_n(x)$, are defined by the formulas

$$P_n^{(-1/2,-1/2)}(x) = \frac{(2n)!}{2^{2n}(n!)^2} T_n(x) = \frac{(2n)!}{2^{2n}(n!)^2} \cos n\theta \qquad (5.1.1)$$

and

$$P_n^{(1/2,1/2)}(x) = \frac{(2n+2)!}{2^{2n+1}[(n+1)!]^2} U_n(x) = \frac{(2n+2)!}{2^{2n+1}[(n+1)!]^2} \frac{\sin(n+1)\theta}{\sin\theta}, \qquad (5.1.2)$$

where $x = \cos\theta$.

The orthogonality relation satisfied by $T_n(x)$ is given by

$$\int_{-1}^{+1} T_n(x)T_m(x)(1-x^2)^{-1/2}dx = 0, \quad \text{when} \quad m \neq n.$$

For $x = \cos\theta$, this is the elementary result:

$$\int_0^\pi \cos m\theta \cos n\theta \, d\theta = 0, \quad \text{when} \quad m \neq n.$$

Similarly, the orthogonality for (5.1.2) is contained in

$$\int_0^\pi \sin(n+1)\theta \sin(m+1)\theta \, d\theta = 0, \quad \text{when} \quad m \neq n.$$

To motivate our later discussion of orthogonal polynomials, we note a few results about Chebyshev polynomials. The three-term recurrence relation, for example, is given by

$$2x T_m(x) = T_{m+1}(x) + T_{m-1}(x), \qquad (5.1.3)$$

which is equivalent to

$$2\cos\theta \cos m\theta = \cos(m+1)\theta + \cos(m-1)\theta. \qquad (5.1.4)$$

The last relation is contained in the linearization formula

$$2\cos m\theta \cos n\theta = \cos(m+n)\theta + \cos(m-n)\theta \qquad (5.1.5)$$

or

$$T_m(x)T_n(x) = \frac{1}{2}(T_{m+n}(x) + T_{m-n}(x)). \qquad (5.1.6)$$

In a more general context, one is interested in the problem of determining the coefficients $a(k, m, n)$ in

$$p_m(x)p_n(x) = \sum_{k=0}^{m+n} a(k, m, n)p_k(x), \qquad (5.1.7)$$

where $\{p_n(x)\}$ is a sequence of polynomials with $p_n(x)$ of degree n exactly. A simple but important special case of this is

$$x^m x^n = x^{m+n}.$$

It is usually difficult to say very much about the coefficients $a(k, m, n)$. Later we shall see an important example which generalizes (5.1.5) and the next formula about Chebyshev polynomials of the second kind $U_n(x)$:

$$\frac{\sin(m+1)\theta}{\sin\theta} \frac{\sin(n+1)\theta}{\sin\theta} = \sum_{k=0}^{m\wedge n} \frac{\sin(m+n+1-2k)\theta}{\sin\theta}, \qquad (5.1.8)$$

where $m \wedge n = \min(m, n)$. This formula is easily verified by noting that $\sin(m+n+1-2k)\theta \sin\theta = \frac{1}{2}[\cos(m+n-2k)\theta - \cos(m+n-2k+2)\theta]$. The dual of (5.1.8) is given by

$$\sin(n+1)\theta \sin(n+1)\phi = \frac{n+1}{2} \int_{\theta-\phi}^{\theta+\phi} \sin(n+1)\psi \, d\psi, \qquad (5.1.9)$$

whereas the dual of (5.1.5) is essentially the same formula, that is,

$$\cos n\theta \cos n\phi = \frac{1}{2}(\cos n(\theta + \phi) + \cos n(\theta - \phi)).$$

In Fourier analysis, one represents a periodic function $f(x)$ by a series of sines and cosines. This involves analyzing partial sums of the form

$$\frac{1}{2}a_0 + \sum_{m=1}^{n}(a_m \cos m\theta + b_m \sin m\theta),$$

where

$$a_m = \frac{1}{\pi} \int_0^{2\pi} f(\phi) \cos m\phi \, d\phi \quad \text{and} \quad b_m = \frac{1}{\pi} \int_0^{2\pi} f(\phi) \sin m\phi \, d\phi.$$

Therefore,

$$\frac{1}{2}a_0 + \sum_{m=1}^{n}(a_m \cos m\theta + b_m \sin m\theta)$$

$$= \frac{1}{\pi} \int_0^{2\pi} \left[\frac{1}{2} + \sum_{m=1}^{n}(\cos m\theta \cos m\phi + \sin m\theta \sin m\phi)\right] f(\phi)d\phi$$

$$= \frac{1}{\pi} \int_0^{2\pi} \left[\frac{1}{2} + \sum_{m=1}^{n} \cos m(\phi - \theta)\right] f(\phi)d\phi.$$

By using the trigonometric identity

$$2\sin(\theta/2) \cos m\theta = \sin(m+1/2)\theta - \sin(m-1/2)\theta,$$

it can be verified that

$$\frac{1}{2} + \sum_{m=1}^{n} \cos m\theta = \frac{\sin\left(n + \frac{1}{2}\right)\theta}{2\sin(\theta/2)} =: D_n(\theta). \tag{5.1.10}$$

Thus, the sum inside the last integral is $D_n(\theta - \phi)$. The function $D_n(\theta)$ is called the Dirichlet kernel. Define the Chebyshev polynomials of the third kind by

$$V_n(x) = 2D_n(\theta), \quad \text{where} \quad x = \cos\theta. \tag{5.1.11}$$

It is easy to check that the sequence $V_n(x)$ is orthogonal with respect to $(1 - x)^{\frac{1}{2}}$ $(1 + x)^{-\frac{1}{2}}$ on $(-1, 1)$. Polynomials equivalent to $V_n(x)$ were studied by Viète. See Edwards [1987, p. 8].

A generalization of (5.1.10) is given by

$$1 + \sum_{m=1}^{n} 2\cos m\theta \cos m\phi = \frac{\cos(n+1)\theta \cos n\phi - \cos n\theta \cos(n+1)\phi}{\cos\theta - \cos\phi}. \tag{5.1.12}$$

This is in fact a particular case of the Christoffel–Darboux formula, which holds for general orthogonal polynomials. Note that the three-term recurrence (5.1.4) implies

$$2\cos m\theta \cos m\phi(\cos\theta - \cos\phi) = [\cos(m+1)\theta + \cos(m-1)\theta]\cos m\phi$$
$$- [\cos(m+1)\phi + \cos(m-1)\phi]\cos m\theta. \tag{5.1.13}$$

Adding this equation for the various values of m gives (5.1.12). To understand the reason for the factor 2 in the sum (5.1.12), observe that

$$\int_0^\pi \cos^2 m\theta \, d\theta = \begin{cases} \frac{\pi}{2} & \text{for } m \neq 0, \\ \pi & \text{for } m = 0. \end{cases} \tag{5.1.14}$$

The normalized function is, therefore, $\sqrt{\frac{2}{\pi}} \cos m\theta$ when $m \neq 0$ and $\frac{1}{\sqrt{\pi}}$ when $m = 0$.

The Poisson kernel for the Chebyshev polynomials defined by $\cos nx$ is given by the sum

$$1 + \sum_{m=1}^{\infty} (2\cos m\theta \cos m\phi)r^m =: P_r(\cos\theta, \cos\phi). \tag{5.1.15}$$

When $\phi = 0$, we have

$$1 + \sum_{m=1}^{\infty} 2\cos m\theta \, r^m = \frac{1 - r^2}{1 - 2r\cos\theta + r^2}, \tag{5.1.16}$$

which shows the positivity of the sum $P_r(\cos\theta, 1)$ for $|r| < 1$. This implies the positivity of $P_r(\cos\theta, \cos\phi)$ in $|r| < 1$, since

$$2\cos m\theta \cos m\phi = \cos m(\theta + \phi) + \cos m(\theta - \phi).$$

We end this section with some results concerning the zeros of $T_m(x)$, which can be given explicitly since we are dealing with $\cos m\theta$. Since $\cos m\theta = 0$ when $\theta = (2n + 1)\pi/2m$, it follows that $T_m(x)$ has m simple zeros in $(-1, 1)$, given by $\cos(2n + 1)\pi/2m$, for $n = 0, 1, 2, \ldots, m - 1$. Moreover, the zeros of $T_m(x)$ and $T_{m+1}(x)$ mutually separate each other. Also observe that between two successive zeros of $T_m(x)$, $\cos[(2k + 3)\pi/2m]$ and $\cos[(2k + 1)\pi/2m]$, there is a zero of $T_n(x)$ for $n > m$. This follows from the fact that we can always find a nonnegative integer $\ell \le n - 1$ such that

$$\frac{2k + 1}{2m} < \frac{2\ell + 1}{2n} < \frac{2k + 3}{2m}.$$

These properties of the zeros of $T_n(x)$ have generalizations to general orthogonal polynomials.

5.2 Recurrence

Let $\alpha(x)$ denote a nondecreasing function with an infinite number of points of increase in the interval $[a, b]$. The latter interval may be infinite. We assume that moments of all orders exist, that is, $\int_a^b x^n d\alpha(x)$ exists for $n = 0, 1, 2, \ldots$.

Definition 5.2.1   *We say that a sequence of polynomials $\{p_n(x)\}_0^\infty$, where $p_n(x)$ has exact degree n, is orthogonal with respect to the distribution $d\alpha(x)$ if*

$$\int_a^b p_n(x) p_m(x) d\alpha(x) = h_n \delta_{mn}. \tag{5.2.1}$$

In effect the next theorem says that $\{p_n(x)\}$ satisfies a three-term recurrence relation.

Theorem 5.2.2   *A sequence of orthogonal polynomials $\{p_n(x)\}$ satisfies*

$$p_{n+1}(x) = (A_n x + B_n) p_n(x) - C_n p_{n-1}(x) \quad \text{for} \quad n \ge 0, \tag{5.2.2}$$

where we set $p_{-1}(x) = 0$. Here A_n, B_n, and C_n are real constants, $n = 0, 1, 2, \ldots$, and $A_{n-1} A_n C_n > 0$, $n = 1, 2, \ldots$. If the highest coefficient of $p_n(x)$ is $k_n > 0$, then

$$A_n = \frac{k_{n+1}}{k_n}, \quad C_{n+1} = \frac{A_{n+1}}{A_n} \frac{h_{n+1}}{h_n},$$

where h_n is given by (5.2.1).

Proof. First determine A_n so that $p_{n+1}(x) - A_n x p_n(x)$ is a polynomial of degree n. Then

$$p_{n+1}(x) - A_n x p_n(x) = \sum_{k=0}^{n} b_k p_k(x) \qquad (5.2.3)$$

for some constants b_k. Note that, if $Q(x)$ is a polynomial of degree $m < n$, then by (5.2.1)

$$\int_a^b p_n(x) Q(x) d\alpha(x) = 0.$$

This implies that $b_k = 0$ for $k < n - 1$, as can be seen by multiplying both sides of (5.2.3) by $p_k(x)$ and integrating. This proves (5.2.2). It is clear that $A_n = k_{n+1}/k_n$. To derive the final result, multiply (5.2.2) by $p_{n-1}(x)$ and integrate to get

$$0 = A_n \int_a^b p_n(x) x p_{n-1}(x) d\alpha(x) - C_n \int_a^b p_{n-1}^2(x) d\alpha(x).$$

Since

$$x p_{n-1}(x) = \frac{k_{n-1}}{k_n} p_n(x) + \sum_{k=0}^{n-1} d_k p_k(x),$$

we get

$$\frac{A_n}{A_{n-1}} h_n - C_n h_{n-1} = 0.$$

This proves the theorem. ∎

Corollary 5.2.3 $h_n = (A_0/A_n) C_1 C_2 \cdots C_n h_0.$

This follows from the equation $h_n = A_{n-1} C_n h_{n-1}/A_n$ by iteration.

Corollary 5.2.3 shows that the L^2 norm of $p_n(x)$ can be computed from the recurrence relation. A converse of Theorem 5.2.2 also holds. If a sequence of polynomials $\{p_n(x)\}$ satisfies (5.2.2) then $\{p_n(x)\}$ is orthogonal with respect to a positive measure. This is usually called Favard's theorem. See Szegö [1975, §3.2] or Chihara [1978, p. 21].

Remark 5.2.1 The form of the recurrence relation given in (5.2.2) is the most useful when finding $p_{n+1}(x)$ from $p_n(x)$ and $p_{n-1}(x)$. However, there are other reasons for considering the three-term recurrence relation satisfied by orthogonal polynomials. Another useful form is

$$x p_n(x) = a_n p_{n+1}(x) + b_n p_n(x) + c_n p_{n-1}(x),$$

where a_n, b_n, and c_n are real. A similar calculation gives the relation

$$a_{n-1}h_n = c_n h_{n-1}.$$

This implies that $a_{n-1}c_n > 0$, $n = 1, 2, \ldots$. The L^2 norm of $p_n(x)$, that is, h_n, now has the form

$$h_n = h_0 \frac{c_1 c_2 \cdots c_n}{a_0 a_1 \cdots a_{n-1}}.$$

An important consequence of the recurrence relation in Theorem 5.2.2 is the following result, called the Christoffel–Darboux formula.

Theorem 5.2.4   *Suppose that the $p_n(x)$ are normalized so that*

$$h_n = \int_a^b p_n^2(x)d\alpha(x) = 1.$$

Then

$$\sum_{m=0}^{n} p_m(y)p_m(x) = \frac{k_n}{k_{n+1}} \frac{p_{n+1}(x)p_n(y) - p_{n+1}(y)p_n(x)}{x - y}, \tag{5.2.4}$$

where k_n is the highest coefficient of $p_n(x)$.

Proof.   The recurrence relation (5.2.2) implies that

$$p_n(y)p_{n+1}(x) = (A_n x + B_n)p_n(x)p_n(y) - C_n p_{n-1}(x)p_n(y)$$

and

$$p_n(x)p_{n+1}(y) = (A_n y + B_n)p_n(y)p_n(x) - C_n p_{n-1}(y)p_n(x).$$

Subtract and divide by $A_n(x - y)$ to get

$$\frac{1}{A_n} \frac{p_n(y)p_{n+1}(x) - p_n(x)p_{n+1}(y)}{x - y}$$
$$= p_n(x)p_n(y) + \frac{1}{A_{n-1}} \frac{p_{n-1}(y)p_n(x) - p_{n-1}(x)p_n(y)}{x - y}. \tag{5.2.5}$$

We have used the fact that $C_n = A_n/A_{n-1}$, since $h_n = 1$. Repeated application of (5.2.5) gives the required result when we observe that $A_n = k_{n+1}/k_n$.   ∎

Remark 5.2.2   If $h_n \neq 1$, then (5.2.4) takes the form

$$\sum_{m=0}^{n} \frac{p_m(y)p_m(x)}{h_m} = \frac{k_n}{k_{n+1}} \frac{p_{n+1}(x)p_n(y) - p_{n+1}(y)p_n(x)}{(x - y)h_n}.$$

The following theorem gives the confluent form of (5.2.4), that is, when $x = y$.

Theorem 5.2.5   *When $h_n = 1$, then*

$$\sum_{k=0}^{n} p_k^2(x) = \frac{k_n}{k_{n+1}} \left(p'_{n+1}(x) p_n(x) - p_{n+1}(x) p'_n(x) \right). \qquad (5.2.6)$$

Proof.   Write the right side of (5.2.4) as

$$\frac{k_n}{k_{n+1}} \frac{(p_{n+1}(x) - p_{n+1}(y)) p_n(y) - (p_n(x) - p_n(y)) p_{n+1}(y)}{(x - y)}$$

and let $y \to x$. The result follows.   ∎

Corollary 5.2.6 $p'_{n+1}(x) p_n(x) - p_{n+1}(x) p'_n(x) > 0$ *for all x.*

To conclude this section, we show how the three-term recurrence relation for Jacobi polynomials can be found. Earlier, we derived this formula from a contiguous relation, but this is hardly a practical idea. The methods given below can be extended to other hypergeometric orthogonal polynomials.

Consider the polynomials

$$p_n(x) = \frac{n!}{(\alpha + 1)_n} P_n^{(\alpha,\beta)}(x) = {}_2F_1 \left(\begin{matrix} -n, n + \alpha + \beta + 1 \\ \alpha + 1 \end{matrix} ; \frac{1 - x}{2} \right), \qquad (5.2.7)$$

$n = 0, 1, 2, \ldots$. Write the recurrence relation as

$$(1 - x) p_n(x) = A_n p_{n+1}(x) + B_n p_n(x) + C_n p_{n-1}(x), \quad n = 0, 1, \ldots,$$

where $p_{-1}(x) = 0$. To obtain A_n, equate the coefficients of $(1 - x)^{n+1}$. It remains to find B_n and C_n. Take $x = 1$ to get

$$0 = A_n + B_n + C_n$$

or

$$B_n = -(A_n + C_n).$$

From Remark 5.2.1, we see that

$$C_n = A_{n-1} h_n / h_{n-1},$$

where h_n is the L^2 norm of $p_n(x)$ and its value follows from the L^2 norm for Jacobi polynomials given in (2.5.14). Thus the recurrence relation is obtained.

The next method simultaneously yields the recurrence and the orthogonality of Jacobi polynomials. It is clear that

$$\left(\frac{1 - x}{2} \right) p_n(x) = A_{n+1} p_{n+1}(x) + A_n p_n(x) + A_{n-1} p_{n-1}(x) + \cdots + A_0 p_0(x).$$

$$(5.2.8)$$

Set $x = 1$ to get

$$A_n = -(A_{n+1} + A_{n-1} + A_{n-2} + \cdots + A_0). \tag{5.2.9}$$

This implies that

$$\left(\frac{1-x}{2}\right) p_n(x) = A_{n+1}(p_{n+1}(x) - p_n(x)) - A_{n-1}(p_n(x)$$
$$- p_{n-1}(x)) + \text{other terms.}$$

A short calculation shows that

$$p_{n+1}(x) - p_n(x) = -\frac{2n + \alpha + \beta + 2}{\alpha + 1}\left(\frac{1-x}{2}\right)$$
$$\cdot {}_2F_1\left(\begin{array}{c} -n, n + \alpha + \beta + 2 \\ \alpha + 2 \end{array}; \frac{1-x}{2}\right). \tag{5.2.10}$$

Therefore,

$$\left(\frac{1-x}{2}\right) p_n(x) = -\frac{1-x}{2}\frac{2n + \alpha + \beta + 2}{\alpha + 1} A_{n+1}\, {}_2F_1\left(\begin{array}{c} -n, n + \alpha + \beta + 2 \\ \alpha + 2 \end{array}; \frac{1-x}{2}\right)$$
$$+ \frac{1-x}{2}\frac{2n + \alpha + \beta}{\alpha + 1} A_{n-1}\, {}_2F_1\left(\begin{array}{c} -n+1, n + \alpha + \beta + 1 \\ \alpha + 2 \end{array}; \frac{1-x}{2}\right)$$
$$+ \text{other terms.}$$

Equating the highest power of $\frac{1-x}{2}$ gives

$$A_{n+1} = -\frac{(n + \alpha + 1)(n + \alpha + \beta + 1)}{(2n + \alpha + \beta + 1)(2n + \alpha + \beta + 2)}.$$

The next highest power gives

$$A_{n-1} = -\frac{n(n + \beta)}{(2n + \alpha + \beta)(2n + \alpha + \beta + 1)}.$$

Now one finds that with these values of A_n and A_{n-1} the sum of the first two terms on the right equals the polynomial on the left. Thus $A_{n-2} = A_{n-3} = \cdots = A_0 = 0$, and we have the three-term recurrence relation. Moreover, we have also proved the orthogonality of these polynomials, since by Favard's theorem such recurrence relations are satisfied only by polynomials orthogonal with respect to some positive measure.

5.3 Gauss Quadrature

The need to approximate an integral that cannot be evaluated exactly has existed since the start of calculus. Newton used the method of interpolating a function at n points and then integrating the interpolating function. He used polynomials to do the interpolation.

We use Lagrange interpolation polynomials here. Suppose $x_1 < x_2 < \cdots < x_n$ is a set of n numbers in an increasing sequence and $y_1, y_2, \ldots, y_n$ is an arbitrary set of numbers.

Definition 5.3.1 *The Lagrange interpolation polynomial is a polynomial of degree $n - 1$ that takes the value y_i at x_i for $i = 1, \ldots, n$. This polynomial is given by*

$$L_n(x) = \sum_{j=1}^{n} \frac{P(x)y_j}{P'(x_j)(x - x_j)}, \tag{5.3.1}$$

where

$$P(x) = (x - x_1) \cdots (x - x_n).$$

We write

$$\ell_j(x) := \frac{P(x)}{P'(x_j)(x - x_j)} \quad \text{for } j = 1, 2, \ldots, n. \tag{5.3.2}$$

It is clear that

$$\ell_j(x_k) = \delta_{jk}.$$

Thus if $f(x)$ is a continuous function whose values $f(x_i)$ are known at the points $x_i, i = 1, 2, \ldots, n$ in an interval $[a, b]$, then

$$L_n(x) = \sum_{j=1}^{n} \ell_j(x) f(x_j) \tag{5.3.3}$$

is a polynomial of degree $\leq n - 1$ which interpolates the function f in $[a, b]$.

Formula (5.3.3) can be applied to approximate integration. We have

$$\int_a^b f(x) d\alpha(x) \approx \sum_{j=1}^{n} f(x_j) \int_a^b \ell_j(x) d\alpha(x) = \sum_{j=1}^{n} \lambda_j f(x_j), \tag{5.3.4}$$

where

$$\lambda_j := \int_a^b \ell_j(x) d\alpha(x). \tag{5.3.5}$$

It is evident that (5.3.4) is exact if $f(x)$ is a polynomial of degree $\leq n - 1$. For, in this case $L_n(x) = f(x)$. There are a number of ways to measure how well the quadrature method approximates the integral. The most obvious is to see how large the difference is. There is another way that has been very fruitful: Require that the quadrature method be exact for as large a class of functions as possible.

For the interpolation method above, there are $2n$ parameters, λ_k and x_k. When the x_k are given in advance, it is easy to determine λ_k so that there is equality in

(5.3.4) for all functions $f(x)$ that are polynomials of degree at most $n-1$. One simply requires that the approximating polynomial agree with f at n points, so that the two are identically equal when f is a polynomial of degree at most $n-1$. This is as far as one can go by means of Lagrange interpolation at fixed points. However, if one does not require that the points x_k be fixed, there is a possibility of increasing the degree of the polynomial by one for each x_k, which is allowed to vary. The maximum degree should be $2n-1$ when all the x_ks are allowed to vary.

This appears to be a difficult nonlinear problem, for we seem to need to solve $2n$ equations that are linear in the λ_ks but nonlinear in the x_ks:

$$\sum_{k=1}^{n} \lambda_k x_k^j = \int_a^b x^j \, d\alpha(x), \quad j = 0, 1, \ldots, 2n-1.$$

The solution is contained in the next theorem, known as the Gauss quadrature formula. Before stating it we introduce some notation. Suppose $\{P_n(x)\}$ is a sequence of polynomials orthogonal with respect to the distribution $d\alpha(x)$, that is,

$$\int_a^b P_n(x) P_m(x) \, d\alpha(x) = 0 \quad \text{for } m \neq n. \tag{5.3.6}$$

Let $x_j = x_{jn} = x_{j,n}$, $j = 1, 2, \ldots, n$, denote the zeros of $P_n(x)$. We prove in the next section that these zeros are simple and lie in the interval $[a, b]$ used in (5.3.6). We saw an example of this in the Cheybshev polynomials of the first kind, $T_n(x)$, which had n simple zeros in $[-1, 1]$.

Theorem 5.3.2   *There are positive numbers $\lambda_1, \lambda_2, \ldots, \lambda_n$ such that for every polynomial $f(x)$ of degree at most $2n-1$*

$$\int_a^b f(x) \, d\alpha(x) = \sum_{j=1}^{n} \lambda_j f(x_j), \tag{5.3.7}$$

where x_j, $j = 1, \ldots, n$ are as defined after (5.3.6), and $\lambda_j = \lambda_{jn} = \lambda_{j,n}$.

Proof.   Let $f(x)$ be an arbitrary polynomial of any degree. Then by the division algorithm,

$$f(x) = P_n(x) Q(x) + R(x),$$

where $P_n(x)$ is as in (5.3.6) and deg $R \leq n-1$.

Since x_j are the zeros of $P_n(x)$, we have

$$f(x_j) = R(x_j) \quad \text{for } j = 1, 2, \ldots, n$$

and

$$\int_a^b f(x)d\alpha(x) = \int_a^b P_n(x)Q(x)d\alpha(x) + \sum_{j=1}^n \lambda_j f(x_j), \qquad (5.3.8)$$

where λ_j is defined in (5.3.5). Now (5.3.7) is exact if

$$\int_a^b P_n(x)Q(x)d\alpha(x) = 0. \qquad (5.3.9)$$

Since (5.3.8) is true for deg $Q(x) \le n - 1$, it follows that (5.3.7) is exact for polynomials $f(x)$ of degree $\le 2n - 1$. We have only to show the positivity of λ_j. For this, note that $\ell_j^2 - \ell_j$ is a polynomial of degree $2n - 2$ which vanishes at x_k, $k = 1, 2, \ldots, n$. So,

$$\ell_j^2 - \ell_j = P_n(x)Q(x),$$

where deg $Q \le n - 2$. Thus

$$\int_a^b (\ell_j^2 - \ell_j)d\alpha(x) = \int_a^b P_n(x)Q(x)d\alpha(x) = 0$$

and

$$\lambda_j = \int_a^b \ell_j(x)d\alpha(x) = \int_a^b \ell_j^2(x)d\alpha(x) > 0.$$

This proves the theorem. ■

Now, if $f(x)$ is not a polynomial of degree $\le 2n - 1$, then (5.3.7) is not exact, but we can use the right side as an approximation for the left side. Here the question of the error involved is of great importance. We do not go into this question in depth but merely prove that the right side of (5.3.7) tends to the left side as $n \to \infty$ if $f(x)$ is a continuous function.

Theorem 5.3.3   *If $f(x)$ is continuous on a finite interval $[a, b]$, then*

$$\lim_{n \to \infty} \sum_{j=1}^n \lambda_{jn} f(x_{jn}) = \int_a^b f(x)d\alpha(x),$$

where λ_{jn} and x_{jn} are as in Theorem 5.3.2.

Proof.   First note that by Weierstrass's approximation theorem (see Exercise 1.40), for every $\epsilon > 0$, there is a polynomial $p(x)$ such that

$$|f(x) - p(x)| < \epsilon/(2S) \quad \text{for all} \quad x \in [a, b].$$

Here

$$S = \sum_{k=1}^{n} \lambda_{nk} = \int_a^b d\alpha(x).$$

For notational convenience, denote

$$I(g) = \int_a^b g(x)d\alpha(x) \quad \text{and} \quad I_n(g) = \sum_{j=1}^{n} \lambda_{jn} g(x_{jn}),$$

where g is any continuous function in $[a, b]$. Then

$$|I(f) - I(p)| \le \int_a^b |f(x) - p(x)| d\alpha(x) < \epsilon/2$$

and

$$|I_n(f) - I_n(p)| \le \sum_{j=1}^{n} \lambda_{jn} |f(x_{jn}) - p(x_{jn})| < \epsilon/2,$$

so that

$$|I_n(f) - I(f)| \le |I_n(f) - I_n(p)| + |I_n(p) - I(p)| + |I(p) - I(f)|.$$

Take $2n - 1 \ge \deg p(x)$ so that $I_n(p) = I(p)$ and

$$|I_n(f) - I(f)| < \epsilon.$$

The conclusion of the theorem follows immediately from this inequality. ∎

Remark 5.3.1   Gauss considered the case where $d\alpha(x) = dx$ in Theorem 5.3.2. The orthogonal polynomials are then Legendre polynomials given by

$$P_n(x) = P_n^{(0,0)}(x) = \frac{(-1)^n}{2^n n!} \frac{d^n}{dx^n} (1 - x^2)^n$$

for the interval $[-1, 1]$.

Remark 5.3.2   When $d\alpha(x) = dx/\sqrt{1 - x^2}$, we get the Chebyshev polynomials of the first kind. In this case, one can prove that

$$\lambda_1 = \lambda_2 = \cdots = \lambda_n$$

and (5.3.7) reduces to

$$\int_{-1}^1 f(x) \frac{dx}{\sqrt{1 - x^2}} = \frac{\pi}{n} \sum_{j=1}^{n} f\left(\cos \frac{2j - 1}{2n} \pi\right),$$

when f is a polynomial of degree $\le 2n - 1$. A converse of this result is also true. See Natanson [1965] for this and for Exercises 3–10.

5.4 Zeros of Orthogonal Polynomials

We have seen that the Chebyshev polynomial $T_n(x)$ has n simple zeros in $[-1, 1]$. More generally, one can prove the same about Jacobi polynomials by using the representation $C(1-x)^{-\alpha}(1+x)^{-\beta}\frac{d^n}{dx^n}\{(1-x)^{n+\alpha}(1+x)^{n+\beta}\}$ and Rolle's theorem together with induction. The next theorem shows that a similar result is true for orthogonal polynomials in general.

Theorem 5.4.1 *Suppose that $\{P_n(x)\}$ is a sequence of orthogonal polynomials with respect to the distribution $d\alpha(x)$ on the interval $[a, b]$. Then $P_n(x)$ has n simple zeros in $[a, b]$.*

Proof. Suppose $P_n(x)$ has m distinct zeros $x_1, x_2, \ldots, x_m$ in $[a, b]$ that are of odd order. In that case

$$Q(x) = P_n(x)(x - x_1)(x - x_2) \cdots (x - x_m) \geq 0 \qquad (5.4.1)$$

for all x in $[a, b]$. If $m < n$, then by orthogonality

$$\int_a^b Q(x)dx = 0. \qquad (5.4.2)$$

However, the inequality in (5.4.1) implies that the integral in (5.4.2) should be strictly positive. This contradiction implies that $m = n$ and that the zeros are simple, yielding our result. ∎

For the next theorem we denote the zeros of $P_n(x)$ in increasing order by $x_{1n} < x_{2n} < \cdots < x_{nn}$.

Theorem 5.4.2 *The zeros of $P_n(x)$ and $P_{n+1}(x)$ separate each other.*

Proof. From Corollary 5.2.6,

$$P_{n+1}(x)P_n'(x) - P_n(x)P_{n+1}'(x) < 0.$$

Since $x_{k,n+1}$ is a zero of $P_{n+1}(x)$, we get

$$P_n(x_{k,n+1})P_{n+1}'(x_{k,n+1}) > 0.$$

The simplicity of the zeros implies that $P_{n+1}'(x_{k,n+1})$ and $P_{n+1}'(x_{k+1,n+1})$ have different signs. It follows that $P_n(x_{k,n+1})$ and $P_n(x_{k+1,n+1})$ have different signs. By the continuity of P_n, we know it has a zero between $x_{k,n+1}$ and $x_{k+1,n+1}$ for $k = 1, 2, \ldots, n$, and our result follows. ∎

We can obtain an extension of Theorem 5.4.2 by using the Gauss quadrature formula.

Theorem 5.4.3 *Let $m < n$. Between any two zeros of $P_m(x)$ there is a zero of $P_n(x)$.*

Proof.   Suppose there is no zero of $P_n(x)$ between x_{km} and $x_{k+1,m}$. Consider the polynomial

$$g(x) = \frac{P_m(x)}{(x - x_{km})(x - x_{k+1,m})}.$$

It is clear that $g(x)P_m(x) \geq 0$ for $x \notin (x_{km}, x_{k+1,m})$. By the Gauss quadrature formula

$$\int_a^b g(x) P_m(x) d\alpha(x) = \sum_{j=1}^n g(x_{jn}) P_m(x_{jn}).$$

Since $g(x_{jn}) P_m(x_{jn}) \geq 0$ and cannot vanish for all $j = 1, \ldots, n$, and $\lambda_j > 0$ for all j, we see that the sum is positive. The integral, however, is zero by orthogonality. This contradiction proves the result.   ∎

We conclude this section with the Markov–Stieltjes inequalities for the sums $\sum_{k=1}^j \lambda_k$, where $j \leq n$. Once again we let x_j, $j = 1, 2, \ldots, n$, denote the zeros of $P_n(x)$ in increasing order.

Theorem 5.4.4   *The Markov–Stieltjes inequalities*

$$\sum_{k=1}^{j-1} \lambda_k \leq \int_a^{x_j} d\alpha(x) \leq \sum_{k=1}^j \lambda_k$$

hold for $j = 1, 2, \ldots, n$.

Proof.   The Gauss quadrature formula is

$$\int_a^b f(x) d\alpha(x) = \sum_{k=1}^n \lambda_k f(x_k), \qquad (5.4.3)$$

where f is a polynomial of degree $\leq 2n - 1$. We have already noted that $f(x) \equiv 1$ gives

$$\sum_{k=1}^n \lambda_k = \int_a^b d\alpha(x).$$

If (5.4.3) were exact for the step function

$$f_j(x) = \begin{cases} 1, & x \leq x_j, \\ 0, & x > x_j, \end{cases}$$

we would have the value of $\sum_{k=1}^j \lambda_k$ as $\int_a^{x_j} d\alpha(x)$. But $f_j(x)$ is not a polynomial of degree $\leq 2n - 1$, so we use the following idea. Define polynomials $\phi_n(x, j)$

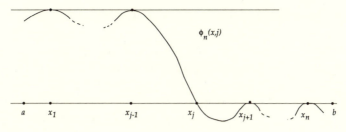

$$\text{Figure 5.1}$$

and $\Phi_n(x, j)$ (see Fig. 5.1) of degree $\leq 2n - 2$ such that

$$\phi_n(x_k, j) = \begin{cases} 1, & k = 1, \ldots, j - 1, \\ 0, & k = j, \ldots, n; \end{cases} \tag{5.4.4}$$

$$\Phi_n(x_k, j) = \begin{cases} 1, & k = 1, \ldots, j, \\ 0, & k = j + 1, \ldots, n; \end{cases} \tag{5.4.5}$$

and

$$\phi_n(x, j) \leq f_j(x) \leq \Phi_n(x, j). \tag{5.4.6}$$

We first assume the existence of these polynomials to derive the Markov–Stieltjes inequalities; we then prove that they exist. Use $\phi_n(x, j)$, a polynomial of degree $< 2n - 1$, in the Gauss quadrature formula (5.4.3), to get

$$\sum_{k=1}^{j-1} \lambda_k = \int_a^b \phi_n(x, j) d\alpha(x)$$
$$\leq \int_a^b f_j(x) d\alpha(x) = \int_a^{x_j} d\alpha(x).$$

Use of $\Phi_n(x, j)$ gives

$$\sum_{k=1}^{j} \lambda_k \geq \int_a^{x_j} d\alpha(x).$$

These inequalities may be written as

$$\sum_{k=1}^{j-1} \lambda_k \leq \int_a^{x_j} d\alpha(x) \leq \sum_{k=1}^{j} \lambda_k,$$

and we have the Markov–Stieltjes inequalities.

To show the existence of $\phi_n(x, j)$, note that the conditions (5.4.4) together with

$$\phi_n'(x_k, j) = 0, \quad k \neq j, \tag{5.4.7}$$

determine a polynomial of degree $\leq 2n - 2$. It is now sufficient to prove that $\phi_n(x, j)$ touches but does not cross the line $y = 1$ and crosses $y = 0$ at x_j and does not cross it again after that. Observe that, if $\phi_n(x, j)$ crosses $y = 1$ at v_1 points and $y = 0$ at $v_2 + 1$ points, then by (5.4.4) and (5.4.7), $\phi_n(x, j) - 1$ has at least $2(j - 1) + v_1$ zeros for $x \leq x_j$ and $\phi_n(x, j)$ has at least $2(n - j) + v_2 + 1$ zeros for $x > x_j$ counting multiplicity. Thus $\phi_n'(x, j)$ has at least $2(j - 1) + v_1 - 1 + 2(n - j) + v_2 = 2n - 3 + v_1 + v_2$ zeros. Since ϕ_n' is of degree $2n - 3$, v_1 and v_2 must be zero. This proves one part of the inequality (5.4.6). The other part is done in the same way and the theorem is proved. The second part is also a consequence of the first part; replace x_k with $b - x_k$ and ϕ with $1 - \phi$. ∎

5.5 Continued Fractions

Continued fractions of a certain type are closely connected with orthogonal polynomials. This connection has been extensively studied. We shall merely touch on this topic and prove an interesting result of Stieltjes.

Suppose $\{a_n\}_1^\infty$ and $\{b_n\}_0^\infty$ are sequences of complex numbers. One notation for an infinite continued fraction is

$$b_0 + \frac{a_1}{b_1+} \frac{a_2}{b_2+} \frac{a_3}{b_3+} \cdots . \tag{5.5.1}$$

We shall denote the nth convergent of this continued fraction by C_n. So

$$C_0 = b_0 =: \frac{A_0}{B_0}, \quad C_1 = b_0 + \frac{a_1}{b_1} = \frac{b_0 b_1 + a_1}{b_1} =: \frac{A_1}{B_1},$$

$$C_2 = b_0 + \frac{a_1}{b_1+} \frac{a_2}{b_2} = b_0 + \frac{a_1}{b_1 + a_2/b_2} = \frac{b_0(b_1 b_2 + a_2) + a_1 b_2}{b_1 b_2 + a_2} =: \frac{A_2}{B_2}.$$

Definition 5.5.1   *We say that the continued fraction (5.5.1) converges, if at most a finite number of C_n are undefined and $\lim_{n \to \infty} C_n$ exists. (In the above example, C_2 is undefined if $B_2 = b_1 b_2 + a_2 = 0$.)*

The sequences $\{A_n\}$ and $\{B_n\}$ defined above satisfy the three-term recurrence relations given in the next lemma.

Lemma 5.5.2   *For $n \geq 1$,*

$$A_n = b_n A_{n-1} + a_n A_{n-2}, \quad A_{-1} = 1 \tag{5.5.2}$$

and

$$B_n = b_n B_{n-1} + a_n B_{n-2}, \quad B_{-1} = 0. \tag{5.5.3}$$

Proof.   Since

$$C_n = b_0 + \frac{a_1}{b_1+} \frac{a_2}{b_2+} \cdots \frac{a_n}{b_n}$$

and

$$C_{n+1} = b_0 + \frac{a_1}{b_1 +} \frac{a_2}{b_2 +} \cdots \frac{a_n}{b_n +} \frac{a_{n+1}}{b_{n+1}},$$

we see that C_{n+1} is obtained from C_n by replacing a_n with $a_n b_{n+1}$ and b_n with $b_n b_{n+1} + a_{n+1}$. Suppose the result of the lemma is true up to n (it is clearly true for $n = 1$); we have

$$\begin{aligned}
A_{n+1} &= (b_n b_{n+1} + a_{n+1}) A_{n-1} + a_n b_{n+1} A_{n-2} \\
&= b_{n+1} (b_n A_{n-1} + a_n A_{n-2}) + a_{n+1} A_{n-1} \\
&= b_{n+1} A_n + a_{n+1} A_{n-1}.
\end{aligned}$$

This proves the result for A_n by induction; the proof for the sequence $\{B_n\}$ is similar. The lemma is proved. ∎

We prove the following lemma by a method similar to the proof of the Christoffel–Darboux identity.

Lemma 5.5.3

$$A_n B_{n-1} - B_n A_{n-1} = (-1)^{n+1} a_1 a_2 \cdots a_n, \qquad n \geq 1.$$

Proof. Multiply (5.5.2) by B_{n-1} and (5.5.3) by A_{n-1} and subtract to get

$$A_n B_{n-1} - B_n A_{n-1} = -a_n (A_{n-1} B_{n-2} - B_{n-1} A_{n-2}).$$

Now iterate to get the result. ∎

The recurrence relation (5.2.2) satisfied by a sequence of orthogonal polynomials $\{P_n(x)\}$ when compared with (5.5.3) suggests the consideration of the continued fraction

$$\frac{A_0}{A_0 x + B_0 -} \frac{C_1}{A_1 x + B_1 -} \frac{C_2}{A_2 x + B_2 -} \cdots . \tag{5.5.4}$$

In this case the nth convergent is a rational function whose denominator is $P_n(x)$. We denote the numerator by $P_n^*(x)$. The sequence $\{P_n^*(x)\}$ satisfies the same recursion, namely

$$P_{n+1}^*(x) = (A_n x + B_n) P_n^*(x) - C_n P_{n-1}^*(x), \quad n \geq 1, \tag{5.5.5}$$

but

$$P_0^*(x) = 0, \qquad P_1^*(x) = A_0.$$

Suppose that the sequence $\{P_n(x)\}$ is orthogonal with respect to the distribution $d\alpha(x)$ on $[a, b]$. The next result relates $P_n^*(x)$ to $P_n(x)$.

Theorem 5.5.4 *With $P_n(x)$ and $P_n^*(x)$ as defined above, we have*

$$P_n^*(x) = \delta \int_a^b \frac{P_n(x) - P_n(t)}{x - t} d\alpha(t), \quad n \geq 0, \tag{5.5.6}$$

where δ is a constant.

Proof. The result holds for $n = 0$ because in that case $P_0(x) = $ constant. For $n = 1$, $P_1(x) = k_1 x +$ constant. Thus the result holds in this case as well, if we adjust δ.

If $n \geq 2$, then denote the right side of (5.5.6) by $R_n(x)$ and observe that

$$R_n(x) - (A_{n-1}x + B_{n-1})R_{n-1}(x) + C_{n-1}R_{n-2}(x)$$

$$= \delta \int_a^b \frac{P_n(x) - P_n(t) - (A_{n-1}x + B_{n-1})(P_{n-1}(x) - P_{n-1}(t))}{x - t}$$

$$+ \frac{C_{n-1}(P_{n-2}(x) - P_{n-2}(t))}{x - t} d\alpha(t)$$

$$= \delta \int_a^b \frac{-(A_{n-1}t + B_{n-1})P_{n-1}(t) + (A_{n-1}x + B_{n-1})P_{n-1}(t)}{x - t} d\alpha(t)$$

$$= \delta A_{n-1} \int_a^b P_{n-1}(t) d\alpha(t) = 0.$$

This means that $R_n(x)$, which is the right side of (5.5.6), satisfies the same recurrence relation as P_n^* with identical initial values. This proves the theorem. ∎

The next result is an application of the Gauss quadrature formula. Let x_k, $k = 1$, $2, \ldots, n$, denote the zeros of $P_n(x)$.

Theorem 5.5.5 *Using the notation of Theorems 5.3.2 and 5.3.3, we have*

$$\frac{P_n^*(x)}{P_n(x)} = \delta \sum_{k=1}^n \frac{\lambda_k}{x - x_k}, \tag{5.5.7}$$

where δ is the constant appearing in (5.5.6).

Proof. The rational function $P_n^*(x)/P_n(x)$, expressed as a partial fraction, is

$$\frac{P_n^*(x)}{P_n(x)} = \sum_{k=1}^n \frac{P_n^*(x_k)}{P_n'(x_k)(x - x_k)}.$$

(Note that the degree of $P_n^*(x)$ is less than the degree of $P_n(x)$.) By Theorem 5.5.4, it follows that

$$\frac{P_n^*(x_k)}{P_n'(x_k)} = \delta \int_a^b \frac{P_n(t)}{P_n'(x_k)(t - x_k)} d\alpha(t) = \delta \lambda_k. \tag{5.5.8}$$

The last equality follows from the Gauss quadrature formula in Theorem 5.3.2. The result is proved. ∎

Theorem 5.5.5 is due to Stieltjes [1993, paper LXXXI]. The next theorem was presented by Markov [1895, p. 89].

Theorem 5.5.6 *Let $[a, b]$ be a finite interval. For any $x \notin [a, b]$*

$$\lim_{n \to \infty} \frac{P_n^*(x)}{P_n(x)} = \delta \int_a^b \frac{d\alpha(t)}{x - t}. \tag{5.5.9}$$

Proof. For any $x \notin [a, b]$, the function $\frac{1}{x-t}$ is a continuous function of t in $[a, b]$. This observation taken together with Theorems 5.3.3 and 5.5.5 imply the result. ∎

Remark 5.5.1 Since $\lambda_k > 0$, it is an immediate consequence of (5.5.8) that the zeros of P_n^* and P_n alternate.

Remark 5.5.2 In Theorem 5.5.6, we may take x to be a complex number that does not lie in $[a, b]$. If we denote the right side of (5.5.9) by $F(x)$, then the inversion formula of Stieltjes is given by

$$\alpha(c) - \alpha(d) = -\frac{1}{\pi} \lim_{v \to 0^+} \int_c^d \text{Im}\{F(u + iv)\} du.$$

Thus, the distribution can be recovered from F.

5.6 Kernel Polynomials

In Section 5.1 we saw that the partial sum of the Fourier series of a function when expressed as an integral gave us the Chebyshev polynomials of the third kind: $V_n(x) = \sin(n + 1/2)\theta / \sin(\theta/2)$, where $x = \cos\theta$. More generally, we get the kernel polynomials when we study partial sums involving orthogonal polynomials.

Let $\{p_n(x)\}$ be a sequence of polynomials orthogonal with respect to the distribution $d\alpha(t)$ on an interval $[a, b]$. As before, $-\infty \le a < b \le \infty$. Let f be a function such that $\int_a^b f(t)p_n(t)d\alpha(t)$ exists for all n.

The series corresponding to the Fourier series is given by

$$a_0 p_0(x) + a_1 p_1(x) + \cdots + a_n p_n(x) + \cdots, \tag{5.6.1}$$

where

$$a_n = \int_a^b f(t)p_n(t)d\alpha(t) \bigg/ \int_a^b \{p_n(t)\}^2 d\alpha(t). \tag{5.6.2}$$

In this section we assume that the denominator of a_n is one, that is, the sequence $\{p_n(x)\}$ is orthonormal. Then the nth partial sum $S_n(x)$ is given by

$$S_n(x) = \sum_{k=0}^{n} p_k(x) \int_a^b f(t) p_k(t) d\alpha(t)$$

$$= \int_a^b f(t) K_n(t, x) d\alpha(t), \tag{5.6.3}$$

where

$$K_n(t, x) = \sum_{k=0}^{n} p_k(t) p_k(x). \tag{5.6.4}$$

Definition 5.6.1   *For a sequence of orthonormal polynomials $\{p_n(x)\}$, the sequence $\{K_n(x_0, x)\}$, where*

$$K_n(x_0, x) = \sum_{k=0}^{n} p_k(x_0) p_k(x),$$

is called the kernel polynomial sequence.

Lemma 5.6.2   *If $Q(x)$ is a polynomial of degree $\leq n$, then*

$$Q(x) = \int_a^b K_n(t, x) Q(t) d\alpha(t).$$

Proof.   Clearly,

$$Q(x) = \sum_{k=0}^{n} a_k p_k(x)$$

for some constants a_k. Multiply both sides by $p_j(x)$ and integrate. Orthogonality gives us

$$\int_a^b Q(t) p_j(t) d\alpha(t) = a_j.$$

The lemma follows immediately.   ∎

Theorem 5.6.3   *Suppose $x_0 \leq a$ are both finite. The sequence $\{K_n(x_0, x)\}$ is orthogonal with respect to the distribution $(t - x_0) d\alpha(t)$.*

Proof.   In Lemma 5.6.2, let $Q(t) = (t - x_0) Q_{n-1}(t)$, where Q_{n-1} is an arbitrary polynomial of degree $n - 1$. The theorem follows.   ∎

Remark 5.6.1   A similar result is obtained when $b \leq x_0$ are both finite.

Remark 5.6.2   In the case of the Chebyshev polynomials $T_n(x)$ with $x_0 = a = -1$, we see that for $x = \cos\theta$,

$$K_n(-1, \cos\theta) = \frac{1}{2} - \cos\theta + \cos 2\theta - \cdots + (-1)^n \cos n\theta$$

$$= (-1)^n \frac{\cos(n + 1/2)\theta}{\cos(\theta/2)}.$$

The polynomials

$$W_n(x) = \frac{\cos\left(n + \frac{1}{2}\right)\theta}{\cos\frac{\theta}{2}}, \quad x = \cos\theta$$

are the Chebyshev polynomials of the fourth kind and Theorem 5.6.3 implies that the sequence $\{W_n(x)\}$ is orthogonal with respect to the weight function $\sqrt{\frac{1+x}{1-x}}$. If we choose $x_0 = b = 1$, then we get the Chebyshev polynomials of the third kind, $V_n(x)$, which are orthogonal with respect to the weight $[(1-x)/(1+x)]^{1/2}$.

Remark 5.6.3   Another straightforward consequence of Theorem 5.6.3 is that if $\{p_n(x)\}$ is orthogonal on $[a, b]$, then

$$\frac{p_n(x)}{p_n(a)} - \frac{p_{n+1}(x)}{p_{n+1}(a)} = \lambda_n q_n(x)(x - a),$$

where $\{q_n(x)\}$ is orthogonal with respect to $(t - a)d\alpha(t)$, for we have

$$K_n(a, x) = \sum_{k=0}^{n} \frac{p_k(x)p_k(a)}{h_k}$$

$$= \frac{A_n p_n(a)p_{n+1}(a)}{(x - a)h_n} \left[\frac{p_{n+1}(x)}{p_{n+1}(a)} - \frac{p_n(x)}{p_n(a)}\right]$$

$$= \mu_n q_n(x),$$

for a constant μ_n. The last equation also implies that

$$\mu_n q_n(x) - \mu_{n-1} q_{n-1}(x) = \frac{p_n(x)p_n(a)}{h_n}.$$

These results will be used in Chapter 6, Section 6.4.

The Christoffel–Darboux formula (Theorem 5.2.4) gives a compact expression for the kernel polynomials:

$$K_n(x_0, x) = \frac{k_n}{k_{n+1}} \frac{p_{n+1}(x)p_n(x_0) - p_{n+1}(x_0)p_n(x)}{x - x_0}. \tag{5.6.5}$$

If we choose x_0 to be the kth root of $p_n(x)$, that is, $x_0 = x_k$, then we can write

$$K_n(x_k, x) = -\frac{k_n}{k_{n+1}} \frac{p_{n+1}(x_k)p_n(x)}{x - x_k}. \tag{5.6.6}$$

This expression for K_n suggests a connection with the Gauss quadrature formula. In fact, we have the following theorem:

Theorem 5.6.4   *The numbers λ_k (or λ_{kn}) occurring in the Gauss quadrature formula are given by*

$$\lambda_k = -\frac{k_{n+1}}{k_n} \frac{1}{p_{n+1}(x_k)p'_n(x_k)};\qquad (5.6.7)$$

their reciprocal is

$$1/\lambda_k = K_n(x_k, x_k) = \sum_{k=0}^{n}(p_k(x_k))^2.\qquad (5.6.8)$$

Proof.   The expression for λ_k in Gauss's formula is

$$\lambda_k = \int_a^b \frac{p_n(t)d\alpha(t)}{p'_n(x_k)(t - x_k)}.$$

By (5.6.6) and Lemma 5.6.2,

$$\begin{aligned}\lambda_k &= -\frac{k_{n+1}}{k_n} \cdot \frac{1}{p_{n+1}(x_k)p'_n(x_k)} \int_a^b K_n(x_k, t)d\alpha(t)\\ &= -\frac{k_{n+1}}{k_n} \frac{1}{p_{n+1}(x_k)p'_n(x_k)}.\end{aligned}$$

This proves (5.6.7). To derive (5.6.8), let $x \to x_k$ in (5.6.6). Thus,

$$K(x_k, x_k) = -\frac{k_n}{k_{n+1}} p_{n+1}(x_k)p'_n(x_k).$$

This proves (5.6.8). ∎

The kernel polynomials also have a maximum property, as contained in the next theorem.

Theorem 5.6.5   *Let x_0 be any real number and $Q(x)$ an arbitrary polynomial of degree $\leq n$, normalized by the condition*

$$\int_a^b (Q(t))^2 d\alpha(t) = 1.$$

The maximum value of $(Q(x_0))^2$ is given by the polynomial

$$Q(x) = \pm K_n(x_0, x)/\sqrt{K_n(x_0, x_0)}$$

and the maximum itself is $K_n(x_0, x_0)$.

Proof.   Since $Q(x)$ is of degree $\leq n$, we have

$$Q(x) = a_0 p_0(x) + a_1 p_1(x) + \cdots + a_n p_n(x).$$

The normalization condition gives

$$a_0^2 + a_1^2 + \cdots + a_n^2 = 1.$$

By the Cauchy–Schwartz inequality

$$[Q(x_0)]^2 \le \sum a_k^2 \sum p_k^2 = \sum_{k=0}^{n} p_k^2(x_0) = K_n(x_0, x_0).$$

Equality holds when $a_k = A p_k(x_0)$, where A is determined by

$$A^2 \sum_{k=0}^{n} p_k^2(x_0) = 1.$$

This proves the theorem. ∎

5.7 Parseval's Formula

Let $L_\alpha^p(a, b)$ denote the class of functions f such that

$$\int_a^b |f|^p \, d\alpha(x) < \infty.$$

As always, we assume that $\int_a^b x^n \, d\alpha(x) < \infty$ for $n \ge 0$. In this section, we are interested in the space $L_\alpha^2(a, b)$. By the Cauchy–Schwartz inequality, we infer the existence of

$$\int_a^b f(x) x^n \, d\alpha(x),$$

for $n \ge 0$.

Theorem 5.7.1   *Suppose $f \in L_\alpha^2(a, b)$. Let $Q(x)$ be a polynomial of degree n, such that*

$$Q(x) = \sum_{k=0}^{n} a_k p_k(x),$$

where $\{p_n(x)\}$ is the orthonormal sequence of polynomials for $d\alpha$. The integral

$$\int_a^b [f(x) - Q(x)]^2 d\alpha(x) \tag{5.7.1}$$

becomes a minimum when

$$a_k = \int_a^b f(x) p_k(x) d\alpha(x). \tag{5.7.2}$$

Moreover, with a_k as in (5.7.2),

$$\sum_{k=0}^{n} a_k^2 \le \int_a^b [f(x)]^2 d\alpha(x). \tag{5.7.3}$$

Proof. Let $c_k = \int_a^b f(x) p_k(x) d\alpha(x)$. By the orthonormality of $\{p_n(x)\}$, we get

$$0 \le \int_a^b [f(x) - Q(x)]^2 d\alpha(x) = \int_a^b \left[f(x) - \sum_{k=0}^{n} a_k p_k(x) \right]^2 d\alpha(x)$$

$$= \int_a^b [f(x)]^2 d\alpha(x) - 2 \sum_{k=0}^{n} a_k c_k + \sum_{k=0}^{n} a_k^2$$

$$= \int_a^b [f(x)]^2 d\alpha(x) - \sum_{k=0}^{n} c_k^2 + \sum_{k=0}^{n} (a_k - c_k)^2.$$

The last expression assumes its least value when $a_k = c_k$. This proves both parts of the theorem. ■

Corollary 5.7.2   *For $f \in L_\alpha^2(a, b)$ and a_k as in (5.7.3), we have*

$$\sum_{n=0}^{\infty} a_n^2 \le \int_a^b [f(x)]^2 d\alpha(x). \tag{5.7.4}$$

Proof. The sequence of partial sums $s_n = \sum_{k=0}^{n} a_k^2$ is increasing and bounded. ■

The inequality (5.7.4) is called Bessel's inequality. We now seek the situation where equality holds. Assume that $[a, b]$ is a finite interval. We shall use the following result from the theory of integration.

Lemma 5.7.3   *For $f \in L_\alpha^2(a, b)$, and a given $\epsilon > 0$, there exists a continuous function g such that*

$$\int_a^b [f(x) - g(x)]^2 d\alpha(x) < \epsilon.$$

Theorem 5.7.4   *Let $[a, b]$ be a finite interval. With the notation of Theorem 5.7.1, we have Parseval's formula:*

$$\sum_{k=0}^{\infty} a_k^2 = \int_a^b [f(x)]^2 d\alpha(x). \tag{5.7.5}$$

Proof. Suppose g is as in Lemma 5.7.3. By Weierstrass's approximation theorem, for a given $\epsilon > 0$ there exists a polynomial $Q_n(x)$ such that

$$\int_a^b [g(x) - Q_n(x)]^2 d\alpha(x) < \epsilon.$$

This implies

$$\int_a^b [f(x) - Q_n(x)]^2 d\alpha(x) < 4\epsilon. \tag{5.7.6}$$

By Theorem 5.7.1, we may choose $Q_n(x) = \sum_{k=0}^n a_k p_k(x)$, where a_k is given by (5.7.2). As in the proof of Theorem 5.7.1, it follows from (5.7.6) that

$$\int_a^b [f(x)]^2 d\alpha(x) - \sum_{k=0}^n a_k^2 < 4\epsilon.$$

Since ϵ is arbitrary, we have proved the theorem. ∎

Corollary 5.7.5 *Suppose* $f \in L_\alpha^2(a, b)$, *where* $[a, b]$ *is a finite interval. If*

$$\int_a^b f(x) x^n d\alpha(x) = 0 \quad \text{for all integers} \quad n \geq 0, \tag{5.7.7}$$

then $f = 0$ *almost everywhere.*

Proof. Since $a_k = 0$ for all k, it follows from Parseval's formula (5.7.5) that

$$\int_a^b [f(x)]^2 d\alpha(x) = 0.$$

This implies the result. ∎

Exercises 24–28 give results similar to Corollary 5.7.5 for infinite intervals. Stone [1962] contains proofs of these results.

Remark 5.7.1 We could also argue as follows: With $Q_n(x)$ as in (5.7.6) it follows from (5.7.7) and the Cauchy–Schwartz inequality that

$$\left(\int_a^b [f(x)]^2 d\alpha(x) \right)^2 = \left(\int_a^b f(x)[f(x) - Q_n(x)] d\alpha(x) \right)^2$$
$$\leq \int_a^b [f(x)]^2 d\alpha(x) \int_a^b [f(x) - Q_n(x)]^2 d\alpha(x).$$

So

$$\int_a^b [f(x)]^2 d\alpha(x) \leq 4\epsilon$$

and the result follows.

Corollary 5.7.6 *With* f *as in Corollary 5.7.5 and* $s_n(x) = \sum_{k=0}^n a_k p_k(x)$, *where* a_k *is given by* (5.7.2), *it follows that*

$$\|s_n(x) - f(x)\|_2^2 = \int_a^b [s_n(x) - f(x)]^2 d\alpha(x) \to 0 \quad \text{as} \quad n \to \infty.$$

Proof.   This result is contained in the proof of Theorem 5.7.4.   ∎

The results of Theorem 5.7.4 and its corollaries are in general false when $[a, b]$ is not finite. As an example (see Exercise 1.20), one may take

$$d\alpha(x) = \exp(-x^\mu \cos \mu\pi)dx, \ f(x) = \sin(x^\mu \sin \mu\pi), \ \ 0 < \mu < 1/2.$$

There are, however, important examples of orthogonal polynomials over infinite intervals, such as the Laguerre polynomials on $(0, \infty)$ and Hermite polynomials on $(-\infty, \infty)$. We prove in the next chapter that Theorem 5.7.4 continues to hold in these cases.

5.8 The Moment-Generating Function

In this section we obtain a continued fraction expansion for the moment-generating function $\sum_{n\geq 0} \mu_n x^n$, where

$$\mu_n = (1, t^n) = \int_a^b t^n d\alpha(t). \tag{5.8.1}$$

The treatment here follows Godsil [1993] and the reader should consult this book for further information on the methods of algebraic combinatorics in the theory of orthogonal polynomials. We assume a minimal knowledge of graph theoretical terminology.

Let G be a graph with n vertices. The adjacency matrix $A = A(G)$ is the $n \times n$ matrix defined as follows: If the ith vertex is adjacent to the jth vertex, $A_{ij} = 1$; otherwise it is zero. An edge $\{i, j\}$ in G is considered to be composed of two arcs, (i, j) and (j, i). A walk in a graph is an alternating sequence of vertices and arcs where each arc joins the vertices before and after it in the sequence. If the first vertex is the same as the last one in the sequence, then it is a closed walk. The number of arcs in a walk is called the length of the walk. The following result is easily checked by induction: The number of walks in G from vertex i to vertex j of length m is given by $(A^m)_{ij}$, that is, the entry in the ith row and jth column of the matrix A^m.

We shall have to consider graphs with weighted arcs so that the entry $(A)_{ij}$ is the weight of the arc (i, j). We continue to denote this matrix by $A = A(G)$. Let

$$\phi(G, x) = \det(xI - A(G))$$

and

$$W_{ij}(G, x) = \sum_{n\geq 0} (A^n)_{ij} x^n.$$

Thus $W_{ij}(G, x)$ is the generating function for the set of all walks in G from vertex i to vertex j. Let $W(G, x)$ be the matrix whose entries are $W_{ij}(G, x)$. Then

$$W(G, x) = \sum_{n \geq 0} A^n x^n.$$

From the fact that $A \operatorname{adj}(A) = \det(A)I$, where $\operatorname{adj}(A)$ is the adjoint of A, it follows that

$$W(G, x) = x^{-1}\phi(G, x^{-1})^{-1}\operatorname{adj}(x^{-1}I - A). \tag{5.8.2}$$

This implies that

$$W_{ii}(G, x) = x^{-1}\phi(G\backslash i, x^{-1})/\phi(G, x^{-1}), \tag{5.8.3}$$

where $G\backslash i$ is the graph obtained from G by removing vertex i.

The connection of the above discussion with orthogonal polynomials is obtained as follows: Suppose $\{p_n(x)\}$ is an orthogonal polynomial sequence satisfying the three-term recurrence

$$p_{n+1}(x) = (x - a_n)p_n(x) - b_n p_{n-1}(x), \quad n \geq 1. \tag{5.8.4}$$

It is assumed that the polynomials are monic. Let A denote the matrix

$$\begin{pmatrix} a_0 & b_1 & & & \\ 1 & a_1 & b_2 & & \\ & 1 & a_2 & b_3 & \\ & & \ddots & \ddots & \end{pmatrix},$$

where the rows and columns of the matrix are indexed by the nonnegative integers. Let A_n denote the square matrix obtained from A but taking the first n rows and columns. Observe that when $\det(xI - A_n)$ is expanded about the last row, we get

$$\det(xI - A_n) = (x - a_{n-1})p_{n-1}(x) - b_{n-1}p_{n-2}(x) = p_n(x).$$

Thus $p_n(x)$ is the characteristic polynomial of A_n.

Observe that matrix A is the adjacency matrix of a particular weighted directed graph G whose vertex set is indexed by the nonnegative integers. If only the first n vertices of G are taken, then the adjacency matrix of the subgraph is A_n. Denote this subgraph by G_n.

We need the next lemma to derive the continued-fraction expression for the moment-generating function $\sum_{n \geq 0}(1, t^n)x^n$, which is understood in the sense of a formal power series. We assume that $\mu_0 = (1, 1) = 1$.

Lemma 5.8.1   *For nonnegative integers n,*

$$\mu_n = (1, x^n) = (A^n)_{00}.$$

Proof.   First note that $(A^k)_{00} = (A_n^k)_{00}$ for $k \leq 2n + 1$, because no closed walk starting at 0 and of length $\leq 2n+1$ can include a vertex beyond the nth vertex. This implies $(p_n(A))_{00} = (p_n(A_n))_{00}$. We have already noted that for $n \geq 1$, $p_n(x)$ is the characteristic polynomial of A_n. By the Cayley–Hamilton theorem, $p_n(A_n) = 0$, and hence

$$(1, p_n) = (p_n(A))_{00}$$

for $n \geq 1$ and for $n = 0$ by definition. Now x^n is a linear combination of $p_0, p_1, \ldots, p_n$, so the result follows.   ∎

Theorem 5.8.2   *With a_n and b_n as in (5.8.4),*

$$\sum_{n \geq 0}(1, t^n)x^n = \cfrac{1}{1 - xa_0-}\cfrac{x^2 b_1}{1 - xa_1-}\cfrac{x^2 b_2}{1 - xa_2-} \cdots .$$

Proof.   Let $A_{n,k}$ be the matrix obtained from A_n by removing the first k rows and columns. Set

$$q_{n-k}(x) = \det(I - x A_{n,k}).$$

Observe that

$$\phi(G_n, x) = \det(xI - A_n) = x^n \det(I - x^{-1}A_{n,0}) = x^n q_n(x^{-1})$$

and

$$\phi(G_n \backslash 0, x) = \det(xI - A_{n,1}) = x^{n-1} \det(I - x^{-1}A_{n,1}) = x^{n-1}q_{n-1}(x^{-1}).$$

By (5.8.3), we can conclude that

$$x^{-1}W_{00}(G_n, x^{-1}) = \frac{\phi(G_n \backslash 0, x)}{\phi(G_n, x)} = x^{-1}\frac{q_{n-1}(x^{-1})}{q_n(x^{-1})}. \tag{5.8.5}$$

Expansion of $\det(I - x A_n)$ about the first row gives

$$q_n(x) = (1 - xa_0)q_{n-1}(x) - x^2 b_1 q_{n-2}(x)$$

or

$$\frac{q_{n-1}(x)}{q_n(x)} = \frac{1}{1 - xa_0 - x^2 b_1 q_{n-2}(x)/q_{n-1}(x)}. \tag{5.8.6}$$

By Lemma 5.8.1,

$$\sum_{n\geq 0}(1,t^n)x^n = \sum_{n\geq 0}(A^n)_{00}x^n = W_{00}(G,x) = \lim_{n\to\infty}\frac{q_{n-1}(x)}{q_n(x)}.$$

This combined with (5.8.6) proves the theorem. ∎

Exercises

1. (a) Prove the positivity of the Poisson kernel for $T_n(x)$ in the interval $-1 < r < 1$ by showing that

$$1 + \sum_{m=1}^{\infty} 2\cos m\theta \; r^m = \frac{1-r^2}{1-2r\cos\theta + r^2}.$$

(b) Compute

$$\sum_{m=0}^{\infty} \frac{\sin(m+1)\theta}{\sin\theta}\frac{\sin(m+1)\phi}{\sin\phi}r^m,$$

which is the Poisson kernel for $U_n(x)$. Observe that it is positive in the interval $-1 < r < 1$.

(c) Show that the Poisson kernel for $\sin(n+1/2)\theta$ is

$$\sum_{n=0}^{\infty} r^n \sin(n+1/2)\theta \sin(n+1/2)\phi$$

$$= \frac{(1-r)\sin(\theta/2)\sin(\phi/2)[(1-r)^2+4r(1-\cos(\theta+\varphi))/2\cos(\theta-\phi)/2]}{[1-2r\cos(\theta+\phi)/2+r^2][1-2r\cos(\theta-\phi)/2+r^2]}.$$

2. Suppose f has continuous derivatives up to order n in $[a,b]$ and $x_1 < x_2 < \cdots < x_n$ are points in this interval. Prove the following Lagrange interpolation formula with remainder:

$$f(x) = L_n(x) + \frac{f^{(n)}(\xi)}{n!}(x-x_1)(x-x_2)\cdots(x-x_n),$$

where $a \leq \min(x,x_1,x_2,\ldots,x_n) < \xi < \max(x,x_1,\ldots,x_n) \leq b$. Here $L_n(x)$ is the Lagrange interpolation polynomial (defined by (5.3.1)) that takes the value $f(x_i)$ at x_i, $i = 1,2,\ldots,n$.

A discussion of the results in Exercises 3–10 can be found in Natanson [1965]. This book also contains the references to the works of Hermite and Fejér mentioned in the exercises.

3. With the notation of Exercise 2, suppose that

$$L_n(x) = \sum_{k=0}^{n} A_k(x-x_1)(x-x_2)\cdots(x-x_k).$$

Show that

$$A_{k-1} = \sum_{j=1}^{k} \frac{f(x_j)}{(x_j - x_1) \cdots (x_j - x_{j-1})(x_j - x_{j+1}) \cdots (x_j - x_k)}.$$

Now let $x_j = a + (j-1)h$ for $j = 1, 2, \ldots, n$. Show that

$$A_{k-1} = \frac{\Delta^{k-1} f(x_1)}{h^{k-1}(k-1)!},$$

where

$$\Delta f(x_j) = f(x_{j+1}) - f(x_j)$$

and

$$\Delta^\ell f(x_j) = \Delta(\Delta^{\ell-1} f(x_j)).$$

4. Let $\ell_j(x)$ be defined by (5.3.2) with $P(x) = (x - x_1)(x - x_2) \cdots (x - x_n)$. Check that $\ell'_j(x_j) = P''(x_j)/2P'(x_j)$. Now show that the function $H(x)$ defined by

$$H(x) = \sum_{j=1}^{n} y_j \left[1 - \frac{P''(x_j)}{P'(x_j)}(x - x_j) \right] \ell_j^2(x) + \sum_{j=1}^{n} y'_j (x - x_j)\ell_j^2(x)$$

(where $y_1, y_2, \ldots, y_n, y'_1, \ldots, y'_n$ is a given set of $2n$ real numbers) satisfies $H(x_j) = y_j$ and $H'(x_j) = y'_j$, where H' denotes the derivative of H. Prove also that if f is as in the previous problem with derivatives of order $2n$, then

$$f(x) = H(x) + \frac{f^{(2n)}(\xi)}{(2n)!} P^2(x),$$

where $y_j = f(x_j)$ and $y'_j = f'(x_j)$ in the definition of H. Again ξ lies in the same interval as before.

5. Apply Gauss quadrature to the formula for $f(x)$ in the previous problem to obtain

$$\int_a^b f(x)d\alpha(x) = \sum_{k=1}^{n} \lambda_k f(x_k) + \frac{f^{(2n)}(\eta)}{(2n)!} \int_a^b P_n^2(x)d\alpha(x), \quad a \le \eta \le b.$$

Here $\{P_n(x)\}$ is a sequence of polynomials orthogonal with respect to $d\alpha(x)$ on $[a, b]$ and the leading coefficient of $P_n(x)$ is one.

6. (a) Prove that, for the Legendre polynomials,

$$P_n(x) = \frac{(-1)^n}{2^n n!} \frac{d^n}{dx^n} (1 - x^2)^n,$$

$$\int_{-1}^{1} P_n^2(x)dx = \frac{2}{2n+1}.$$

(b) Use the Christoffel–Darboux formula to prove that

$$\int_{-1}^{1} \frac{P_n(t)P_{n-1}(x) - P_n(x)P_{n-1}(t)}{t - x}\,dt = \frac{2}{n}.$$

(c) Deduce from (b) that if x_k, $k = 1, \ldots, n$ are the zeros of $P_n(x)$, then

$$\int_{-1}^{1} \frac{P_n(t)}{t - x_k}\,dt = \frac{2}{n P_{n-1}(x_k)}.$$

(d) Use Exercise 5 and the above to obtain the formula

$$\int_{-1}^{1} f(x)\,dx = \sum_{k=1}^{n} \frac{2f(x_k)}{n P_{n-1}(x_k) P_n'(x_k)} + \frac{2^{2n+1}(n!)^4}{[(2n)!]^3} \frac{f^{(2n)}(\xi)}{2n + 1},$$

where $-1 \le \xi \le 1$.

7. Use Chebyshev polynomials to prove the following formula, which is similar to Exercise 6(d):

$$\int_{-1}^{1} \frac{f(x)}{\sqrt{1 - x^2}}\,dx = \frac{\pi}{n} \sum_{k=1}^{n} f\left(\cos \frac{(2k - 1)\pi}{2n}\right) + \frac{\pi}{(2n)!2^{2n-1}} f^{(2n)}(\xi)$$

with $-1 \le \xi \le 1$. Note that all the λ_k equal $\frac{\pi}{n}$ in this case. $\hspace{2em}$ (Hermite)

8. (a) Show that the roots of the Chebyshev polynomials $U_n(x)$ are $\cos \frac{k\pi}{n+1}$, $k = 1, 2, \ldots, n$ and those of $V_n(x)$ are $\cos \frac{2k\pi}{2n+1}$, $k = 1, 2, \ldots, n$. [Recall $T_n(\cos \theta) = \cos n\theta$ and $U_n(\cos \theta) = \sin n\theta / \sin \theta$].

(b) Prove the quadrature formulas

$$\int_{-1}^{1} \sqrt{1 - x^2} f(x)\,dx = \frac{\pi}{n + 1} \sum_{k=1}^{n} \sin^2 \frac{k\pi}{n + 1} f\left(\cos \frac{k\pi}{n + 1}\right)$$

$$+ \frac{\pi}{(2n)!2^{2n+1}} f^{(2n)}(\xi),$$

and

$$\int_{-1}^{1} \sqrt{\frac{1 - x}{1 + x}} f(x)\,dx = \frac{4\pi}{2n + 1} \sum_{k=1}^{n} \sin^2 \frac{k\pi}{2n + 1} f\left(\cos \frac{2k\pi}{2n + 1}\right)$$

$$+ \frac{\pi}{(2n)!2^{2n}} f^{(2n)}(\xi),$$

where $-1 \le \xi \le 1$. Obviously the various ξ are not necessarily the same.

9. Prove that

$$\int_{-1}^{1} \frac{T_n(x)dx}{T_n'(x_k)(x - x_k)} > 0,$$

where $x_k = \cos \frac{(2k-1)\pi}{2n}$.

Hint: Write the integral as

$$(-1)^{k-1} \frac{\sin \theta_k}{n} \int_0^\pi \frac{\cos n\theta}{\cos \theta - \cos \theta_k} \sin \theta \, d\theta,$$

where $\theta_k = \frac{(2k-1)\pi}{2n}$. Apply the Christoffel–Darboux formula and integrate term by term.   (Fejér)

10. Prove that

$$\int_{-1}^{1} \frac{U_n(x)}{U_n'(x_k)(x - x_k)} dx > 0, \qquad \text{(Fejér)}$$

where $U_n(x_k) = 0$.

11. Suppose $\{P_n(x)\}$ is an orthogonal polynomial sequence. Let x_k, $k = 1, 2, \ldots,$ n, denote the zeros of $P_n(x)$. Suppose that

$$\frac{P_{n-1}(x)}{P_n(x)} = \sum_{k=1}^{n} \frac{a_k}{x - x_k}$$

is a partial fraction decomposition of $P_{n-1}(x)/P_n(x)$. Prove that $a_k > 0$.

12. With the notation used in Lemma 9.5.2, show that

$$\frac{A_n}{B_n} = b_0 + \sum_{k=1}^{n} (-1)^{k+1} \frac{a_1 a_2 \cdots a_k}{B_{k-1} B_k}$$

provided $b_i \neq 0$, $B_i \neq 0$ $(1 \leq i \leq n)$.

13. Show that if $\{P_n(x)\}$ satisfies

$$P_n(x) = (A_{n-1}x + B_{n-1})P_{n-1}(x) - C_{n-1}P_{n-2}(x) \quad \text{for} \quad n \geq 1$$

and $P_{-1}(x) = 0$, then

$$P_n(x) = \begin{vmatrix} A_0 x + B_0 & 1 & 0 & 0 & 0 \\ C_1 & A_1 x + B_1 & 1 & 0 & 0 \\ 0 & C_2 & A_2 x + B_2 & 1 & 0 \\ \vdots & & \ddots & \ddots & & \ddots & \vdots \\ & & & A_{n-2}x + B_{n-2} & 1 \\ 0 & & & C_{n-1} & A_{n-1}x + B_{n-1} \end{vmatrix}.$$

14. Suppose $A_n = 1$ for all n in Exercise 13 and let $C_n = |d_n|^2 = d_n \bar{d}_n$. Then the zeros of $P_n(x)$ are the eigenvalues of the matrix:

$$
\begin{bmatrix}
-B_0 & d_1 & 0 & \cdots & 0 & 0 & 0 & 0 \\
d_1 & -B_1 & d_2 & \cdots & & 0 & 0 & 0 \\
0 & \bar{d}_2 & -B_2 & d_3 & & & & \vdots \\
\vdots & & \ddots & \ddots & & \ddots & & \\
& & & & & 0 & 0 & \\
0 & & & & \bar{d}_{n-2} & -B_{n-2} & d_{n-1} & \\
0 & & & & 0 & \bar{d}_{n-1} & -B_{n-1} &
\end{bmatrix}.
$$

15. Prove the following recurrence relations for Laguerre and Hermite polynomials respectively:

(a) $(n+1)L_{n+1}^\alpha(x) = (-x + 2n + \alpha + 1)L_n^\alpha(x) - (n+\alpha)L_{n-1}^\alpha(x),$

$$n = 0, 1, 2, 3, \ldots$$

(b) $H_{n+1}(x) = 2x H_n(x) - 2n H_{n-1}(x), \quad n = 0, 1, 2, \ldots,$

$H_0(x) = 1, \ H_{-1}(x) = 0.$

16. Let $\{p_n(x)\}_0^\infty$ be an orthonormal sequence of polynomials with respect to the distribution $d\alpha(x)$. Let

$$
\mu_n = \int_a^b x^n d\alpha(x), \ n = 0, 1, 2, \ldots.
$$

Show that

$$
p_n(x) = C_n
\begin{vmatrix}
\mu_0 & \mu_1 & \mu_2 & \cdots & \mu_n \\
\mu_1 & \mu_2 & \mu_3 & \cdots & \mu_{n+1} \\
\vdots & \vdots & & & \\
\mu_{n-1} & \mu_n & \mu_{n+1} & \cdots & \mu_{2n-1} \\
1 & x & x^2 & \cdots & x^n
\end{vmatrix},
$$

where C_n is a constant given by $C_n = (D_{n-1}D_n)^{-1/2}$, when D_n is the positive-valued determinant $[\mu_{k+m}]_{k,m=0,1,\ldots,n}$.

17. With the notation of Exercise 16, prove that

$$
p_n(x) = C_n
\begin{vmatrix}
\mu_0 x - \mu_1 & \mu_1 x - \mu_2 & \cdots & \mu_{n-1} x - \mu_n \\
\mu_1 x - \mu_2 & \mu_2 x - \mu_3 & \cdots & \mu_n x - \mu_{n+1} \\
\vdots & & \vdots & & \vdots \\
\mu_{n-1} x - \mu_n & \mu_n x - \mu_{n+1} & \cdots & \mu_{2n-2} x - \mu_{2n-1}
\end{vmatrix}.
$$

18. With the notation of Exercise 16, prove that

$$p_n(x) = \frac{C_n}{n!} \int_a^b \cdots \int_a^b \prod_{i=0}^{n-1} (x - x_i)$$
$$\cdot \prod_{0 \le i < j \le n-1} (x_i - x_j)^2 d\alpha(x_0) d\alpha(x_1) \cdots d\alpha(x_{n-1})$$

and

$$D_n = \frac{1}{(n+1)!} \int_a^b \cdots \int_a^b \prod_{0 \le i < j \le n} (x_i - x_j)^2 d\alpha(x_0) d\alpha(x_1) \cdots d\alpha(x_n).$$

(For the reference to Heine, see Szegö [1975, p. 27].)   (Heine)

19. Let $1 > x_1(\alpha, \beta) > x_2(\alpha, \beta) \cdots > x_n(\alpha, \beta) > -1$ be the roots of the Jacobi polynomial $P_n^{(\alpha,\beta)}(x)$. Show that, for $\alpha > -1$ and $\beta > -1$,

$$\frac{\partial x_k}{\partial \alpha} < 0, \quad \frac{\partial x_k}{\partial \beta} > 0, \quad k = 1, 2, \ldots, n.$$

Proceed as follows:

(a) Take $a = -1, b = 1$, and $d\alpha(x) = (1 - x)^\alpha (1 + x)^\beta$ in the Gauss quadrature formula (Theorem 5.3.2). Take the derivative with respect to α to get

$$\int_{-1}^1 f(x)(1 - x)^\alpha (1 + x)^\beta \log(1 - x) dx$$
$$= \sum_{j=1}^n \lambda_j f'(x_j) x_j'(\alpha) + \sum_{j=1}^n \lambda_j' f(x_j).$$

(b) Take $f(x) = \{P_n^{(\alpha,\beta)}(x)\}^2/(x - x_k)$ to show that

$$\int_{-1}^1 \{\log(1 - x) - \log(1 - x_k)\}(1 - x)^\alpha (1 + x)^\beta \frac{\left(P_n^{(\alpha,\beta)}(x)\right)^2}{x - x_k} dx$$
$$= \lambda_k(\alpha) \frac{\partial x_k}{\partial \alpha} \left(\frac{d P_n^{(\alpha,\beta)}}{dx}(x_k)\right)^2.$$

Now observe that the expression in curly braces and $x - x_k$ have opposite signs. Prove $\frac{\partial x_k}{\partial \beta} > 0$ in a similar way.   (Stieltjes)

20. This result generalizes Exercise 19. Let $\omega(x, \tau)$ be a weight function dependent on a parameter τ such that $\omega(x, \tau)$ is positive and continuous for $a < x < b, \tau_1 < \tau < \tau_2$. Assume that the continuity of the partial derivative $\frac{\partial \omega}{\partial \tau}$ for $a < x < b, \tau_1 < \tau < \tau_2$ and the convergence of the integrals

$$\int_a^b x^k \frac{\partial \omega(x, \tau)}{\partial \tau} dx, \quad k = 0, 1, \ldots, 2n - 1,$$

occur uniformly in every closed subinterval $\tau' \le \tau \le \tau''$ of (τ_1, τ_2). If the zeros of $P_n(x)=P_n(x, \tau)$ (the polynomials orthogonal with respect to $\omega(x, \tau)$) are $x_1(\tau)>x_2(\tau)> \cdots >x_n(\tau)$, then the kth zero $x_k(\tau)$ is an increasing function of τ provided that $\frac{\partial \omega}{\partial \tau}/\omega$ is an increasing function of x, $a<x<b$.

(Markov)

21. Let $\omega(x)$ and $W(x)$ be two positive, continuous weight functions on $[a, b]$. Let $W(x)/\omega(x)$ be increasing. Prove that if $\{x_k\}$ and $\{X_k\}$ denote the zeros of the corresponding orthogonal polynomials of degree n in decreasing order, then

$$x_k < X_k, \qquad k = 1, \ldots, n.$$

Hint: Take $\omega(x, \tau) = (1 - \tau)\omega(x) + \tau W(x)$ in Exercise 20.

See Szegö [1975, §§ 6.12 and 6.21] for Exercises 19–21 and for references.

22. Use the result of Exercise 20 to show that if the parameters α and β of the Jacobi polynomials lie in $[-1/2, 1/2]$, then the zeros of $P_n^{(\alpha,\beta)}(x)$ satisfy

$$\frac{2k - 1}{2n + 1}\pi \le x_k \le \frac{2k}{2n + 1}\pi, \qquad k = 1, 2, \ldots, n.$$

23. Suppose $\{p_n(x)\}_0^\infty$ is an orthonormal sequence of polynomials with respect to the distribution $d\alpha(x)$. Let

$$s_n(x, f) = s_n(x) = \sum_{k=0}^n c_k p_k(x), \quad \text{where} \quad c_k = \int_a^b f(x)p_k(x)d\alpha(x).$$

Prove that

$$f(x_0) - s_n(x_0) = \int_a^b [f(x_0) - f(x)]K_n(x, x_0)d\alpha(x).$$

24. Suppose $f : [0, \infty) \to R$ is continuous and $\lim_{x\to\infty} f(x) = 0$. Show that f can be uniformly approximated by functions of the form $e^{-\alpha x}p(x)$ where $p(x)$ is a polynomial, when α is a fixed positive number.

25. Prove that if $f \in L^p(0, \infty)$, $p \ge 1$ (or f is a bounded measurable function), and (for a given $\alpha > 0$)

$$\int_0^\infty f(x)e^{-\alpha x}x^n dx = 0 \quad \text{for} \quad n = 0, 1, 2, \ldots,$$

then $f(x) \equiv 0$ a.e.

26. Suppose $f : (-\infty, \infty) \to R$ is continuous and $\lim_{x\to\pm\infty} f(x) = 0$. Prove that f can be uniformly approximated by functions of the form $e^{-\alpha^2 x^2}p(x)$ where $p(x)$ is a polynomial.

27. Show that if $f \in L^p(-\infty, \infty)$, $p \ge 1$, and

$$\int_{-\infty}^\infty f(x)e^{-\alpha^2 x^2}x^n dx = 0, \quad n = 0, 1, 2, \ldots,$$

then $f(x) \equiv 0$ a.e.

28. Show that

$$K_n^{(\alpha,\beta)}(x, y) = \sum_{k=0}^{n} P_k^{(\alpha,\beta)}(x) P_k^{(\alpha,\beta)}(y) / h_k^{\alpha,\beta}$$

$$= \frac{2^{-\alpha-\beta}}{2n + \alpha + \beta + 2} \times \frac{\Gamma(n + 2)\Gamma(n + \alpha + \beta + 2)}{\Gamma(n + \alpha + 1)\Gamma(n + \beta + 1)}$$

$$\cdot \frac{P_{n+1}^{(\alpha,\beta)}(x) P_n^{(\alpha,\beta)}(y) - P_n^{(\alpha,\beta)}(x) P_{n+1}^{(\alpha,\beta)}(y)}{x - y}.$$

29. (a) Show that the Legendre polynomial $P_n(x)$ is a solution of

$$(1 - x^2)y'' - 2xy' + n(n + 1)y = 0.$$

(b) Show that $Q_n(x) = \frac{1}{2}\int_{-1}^{1} \frac{P_n(t)}{x-t}dt, x \notin [-1, 1]$ is another solution of the differential equation in (a).

(c) Show that $Q_n(x) = P_n(x)Q_0(x) - W_{n-1}(x)$, where $W_{n-1}(x)$ is a polynomial of degree $n - 1$ given by

$$W_{n-1}(x) = \frac{1}{2}\int_{-1}^{1} \frac{P_n(x) - P_n(t)}{x - t}dt.$$

For $-1 < x < 1$, define $Q_n(x) = P_n(x)Q_0(x) - W_{n-1}(x)$ with $Q_0(x) = \frac{1}{2}\log\frac{1+x}{1-x}$.

(d) Prove the following recurrence relations:

$$(2n + 1)x P_n(x) = (n + 1)P_{n+1}(x) + n P_{n-1}(x), \quad n = 0, 1, \ldots,$$

$$(2n + 1)x Q_n(x) = (n + 1)Q_{n+1}(x) + n Q_{n-1}(x), \quad n = 1, 2, \ldots.$$

(e) Prove that

$$\sum_{k=0}^{n}(2k + 1)Q_k(x)Q_k(y) = \frac{Q_0(y) - Q_0(x)}{x - y}$$

$$+ (n+1)\left[\frac{Q_{n+1}(x)Q_n(y) - Q_n(x)Q_{n+1}(y)}{x-y}\right]$$

and

$$\frac{1}{1 - x^2} + \sum_{k=0}^{n}(2k + 1)[Q_k(x)]^2$$

$$= (n + 1)\left[Q'_{n+1}(x)Q_n(x) - Q'_n(x)Q_{n+1}(x)\right].$$

(f) Show that $Q_n(x)$ has $n + 1$ zeros in $-1 < x < 1$.
See Frobenius [1871].

6

Special Orthogonal Polynomials

Special orthogonal polynomials began appearing in mathematics before the significance of such a concept became clear. Thus, Laplace used Hermite polynomials in his studies in probability while Legendre and Laplace utilized Legendre polynomials in celestial mechanics. We devote most of this chapter to Hermite, Laguerre, and Jacobi polynomials because these are the most extensively studied and have the longest history.

We reproduce Wilson's amazing derivation of the hypergeometric representation of Jacobi polynomials from the Gram determinant. This chapter also contains the derivation of the generating function of Jacobi polynomials, by two distinct methods. One method, due to Jacobi, uses Lagrange inversion. The other employs Hermite's beautiful idea on the form of the integral of the product of the generating function and a polynomial. This generating function is then used to obtain the behavior of the Jacobi polynomial $P_n^{(\alpha,\beta)}(x)$ for large n. We quote a theorem of Nevai to show how the asymptotic behavior of $P_n^{(\alpha,\beta)}(x)$ gives its weight function.

It is important to remember that the classical orthogonal polynomials are hypergeometric. We apply Bateman's fractional integral formula for hypergeometric functions, developed in Chapter 2, to derive integral representations of Jacobi polynomials. These are useful in proving positivity results about sums of Jacobi polynomials. We then use Whipple's transformation to obtain the linearization formula for the product of two ultraspherical polynomials. This clever idea is due to Bailey. One of the simplest examples of linearization is the formula

$$\cos m\theta \cos n\theta = \frac{1}{2}[\cos(m+n)\theta + \cos(m-n)\theta].$$

We observe that a linearization formula for a set of orthogonal polynomials is equivalent to the formula for an integral of the product of three of these polynomials.

We briefly discuss the connection between combinatorics and orthogonal polynomials. In recent years, this topic has been studied extensively by Viennot, Godsil, and many others. We content ourselves with two combinatorial evaluations of an integral of a product of three Hermite polynomials.

This chapter concludes with a brief introduction to q-ultraspherical polynomials. This discussion is motivated by a question raised and also answered by Feldheim and Lanzewizky: Suppose $f(z)$ is analytic and $|f(re^{i\theta})|^2$ is a generating function for a sequence of polynomials $p_n(\cos\theta)$. Do the $p_n(\cos\theta)$ produce an orthogonal polynomial sequence other than the ultraspherical polynomials? The answer involves an interesting nonlinear difference equation that can be neatly solved.

6.1 Hermite Polynomials

The normal integral $\int_{-\infty}^{\infty} e^{-x^2} dx$, which plays an important role in probability theory and other areas of mathematics, was computed in Chapter 1. The integrand e^{-x^2} has several interesting properties. For instance, it is essentially its own Fourier transform. In fact,

$$e^{-x^2} = \frac{1}{\sqrt{\pi}} \int_{-\infty}^{\infty} e^{-t^2} e^{2ixt} dt. \tag{6.1.1}$$

This can be proved in several ways. (See Exercise 1.) The integral is uniformly convergent in any disk $|x| \le r$ and is majorized in that region by the convergent integral

$$\frac{1}{\sqrt{\pi}} \int_{-\infty}^{\infty} e^{-t^2} e^{2rt} dt.$$

Thus the integral can be repeatedly differentiated with respect to x, and we have

$$\frac{d^n e^{-x^2}}{dx^n} = \frac{(2i)^n}{\sqrt{\pi}} \int_{-\infty}^{\infty} e^{-t^2} t^n e^{2ixt} dt. \tag{6.1.2}$$

The polynomials orthogonal with respect to the normal distribution e^{-x^2} are the Hermite polynomials. They can be defined by the formula

$$H_n(x) = (-1)^n e^{x^2} \frac{d^n e^{-x^2}}{dx^n}. \tag{6.1.3}$$

It is easy to check that $H_n(x)$ is a polynomial of degree n.

By (6.1.2), it is seen that

$$H_n(x) = \frac{(-2i)^n e^{x^2}}{\sqrt{\pi}} \int_{-\infty}^{\infty} e^{-t^2} t^n e^{2ixt} dt. \tag{6.1.4}$$

Let us first prove the orthogonality property of $H_n(x)$, namely

$$\int_{-\infty}^{\infty} e^{-x^2} H_n(x) H_m(x) dx = 2^n n! \sqrt{\pi}\, \delta_{mn}. \tag{6.1.5}$$

Consequent from the definition (6.1.3), we can write this integral as

$$(-1)^n \int_{-\infty}^{\infty} \frac{d^n e^{-x^2}}{dx^n} H_m(x)dx.$$

Suppose $n > m$ and integrate by parts n times. This shows that the integral is zero. The $m = n$ case will be considered a little later.

The Hermite polynomials have a simple generating function. Observe that the term t^n in the integrand suggests that we consider

$$\sum_{n=0}^{\infty} \frac{H_n(x)}{n!} r^n = \frac{e^{x^2}}{\sqrt{\pi}} \int_{-\infty}^{\infty} e^{-t^2} e^{2it(x-r)} dt. \tag{6.1.6}$$

The integral can be computed by (6.1.1). The result is the generating function for $H_n(x)$:

$$\sum_{n=0}^{\infty} \frac{H_n(x)}{n!} r^n = e^{2xr-r^2}. \tag{6.1.7}$$

This generating function is useful for deriving several properties of Hermite polynomials. For example, we have the following expression for these polynomials:

$$H_n(x) = \sum_{k=0}^{\lfloor n/2 \rfloor} \frac{(-1)^k n!}{k!(n-2k)!} (2x)^{n-2k}. \tag{6.1.8}$$

This can be obtained by writing

$$e^{2xr-r^2} = \sum_{p=0}^{\infty} \frac{(2x)^p}{p!} r^p \sum_{q=0}^{\infty} \frac{(-1)^q r^{2q}}{(2q)!}$$

and equating the coefficient of r^n on each side.

From (6.1.8), it follows that

$$\frac{d^n H_n(x)}{dx^n} = 2^n n!. \tag{6.1.9}$$

We can now complete the proof of (6.1.5). Integration by parts shows that

$$(-1)^n \int_{-\infty}^{\infty} \frac{d^n}{dx^n} e^{-x^2} H_n(x)dx = \int_{-\infty}^{\infty} e^{-x^2} \frac{d^n H_n(x)}{dx^n} dx = 2^n n! \sqrt{\pi},$$

where (6.1.9) is used in the final step.

We learned in the previous chapter that orthogonal polynomials satisfy three-term recurrence relations. To find this relation for Hermite polynomials, note that

$$F(x, r) = e^{2xr-r^2}$$

satisfies

$$\frac{\partial F}{\partial r} - (2x - 2r)F = 0.$$

Substitute the series in (6.1.7) for F to get

$$H_{n+1}(x) - 2x H_n(x) + 2n H_{n-1}(x) = 0, \quad n = 1, 2, \ldots. \tag{6.1.10}$$

This is the recurrence relation for Hermite polynomials. Another recurrence relation comes from

$$\frac{\partial F}{\partial x} - 2r F = 0.$$

This implies

$$H_n'(x) = 2n H_{n-1}(x), \quad n = 1, 2, \ldots \tag{6.1.11}$$

Eliminate $H_{n-1}(x)$ from (6.1.10) and (6.1.11) to obtain

$$H_{n+1}(x) - 2x H_n(x) + H_n'(x) = 0.$$

Differentiate this equation and use (6.1.11) again to get

$$H_n''(x) - 2x H_n'(x) + 2n H_n(x) = 0, \quad n = 0, 1, 2, \ldots.$$

Thus the Hermite polynomials $H_n(x)$ satisfy the second-order linear differential equation

$$u'' - 2xu' + 2nu = 0. \tag{6.1.12}$$

It is also worth noting here that the function

$$V(x) = e^{-x^2/2} H_n(x)$$

satisfies the differential equation

$$V'' + (2n + 1 - x^2)V = 0.$$

As another application of the integral representation (6.1.4) for $H_n(x)$, we derive a closed expression for the Poisson kernel for Hermite polynomials, namely

$$\sum_{n=0}^{\infty} \frac{H_n(x) H_n(y)}{2^n n!} r^n = (1 - r^2)^{-1/2} e^{[2xyr - (x^2 + y^2)r^2]/(1 - r^2)}. \tag{6.1.13}$$

By (6.1.4),

$$H_n(y) = \frac{(-2i)^n e^{y^2}}{\sqrt{\pi}} \int_{-\infty}^{\infty} e^{-s^2} s^n e^{2iys} \, ds.$$

So for $|r| < 1$, the left-hand side of (6.1.13) becomes

$$\frac{e^{x^2+y^2}}{\pi} \int_{-\infty}^{\infty} \int_{-\infty}^{\infty} e^{-s^2-t^2+2iys+2itx-2str} \, ds \, dt.$$

Now use the formula

$$\int_{-\infty}^{\infty} e^{-a^2 x^2 - 2bx} \, dx = \frac{\sqrt{\pi}}{a} e^{b^2/a^2}$$

twice in the double integral and (6.1.13) follows. The various formal processes can be justified by the absolute convergence of the integrals involved.

A different approach to the proof of (6.1.13) using the three-term recurrence is as follows: Denote the series on the left of (6.1.13) by $K(r, x, y)$. By (6.1.10) and (6.1.11),

$$\frac{\partial K}{\partial x} = \sum_{n=1}^{\infty} \frac{H_{n-1}(x) H_n(y)}{2^{n-1}(n-1)!} r^n = r \sum_{n=0}^{\infty} \frac{H_n(x) H_{n+1}(y)}{2^n n!} r^n$$

$$= 2ry \sum_{n=0}^{\infty} \frac{H_n(x) H_n(y)}{2^n n!} - r \sum_{n=1}^{\infty} \frac{H_n(x) H_{n-1}(y)}{2^{n-1}(n-1)!} r^n.$$

Thus

$$\frac{\partial K}{\partial x} = 2ryK - r \frac{\partial K}{\partial y}$$

and by the symmetry in x and y

$$\frac{\partial K}{\partial y} = 2rxK - r \frac{\partial K}{\partial x}.$$

The last two equations imply that

$$\frac{1}{K} \frac{\partial K}{\partial x} = \frac{2ry - 2r^2 x}{1 - r^2},$$

and so

$$\log K = \frac{2rxy - r^2 x^2}{1 - r^2} + g(y, r)$$

or

$$K = h(y, r) e^{(2rxy - r^2 x^2)/(1 - r^2)}.$$

Again by the symmetry in x and y, we can conclude that

$$K = c(r) e^{[2xyr - (x^2+y^2)r^2]/(1-r^2)}.$$

To find $c(r)$, set $x = y = 0$ to get

$$c(r) = \sum_{n=0}^{\infty} \frac{H_n^2(0)r^n}{2^n n!}.$$

From (6.1.8),

$$H_{2n}(0) = (-1)^n \frac{(2n)!}{n!} \quad \text{and} \quad H_{2n+1}(0) = 0. \qquad (6.1.14)$$

Therefore,

$$c(r) = \sum_{n=0}^{\infty} \frac{(2n)! r^{2n}}{2^{2n} n! n!} = \sum_{n=0}^{\infty} \frac{(1/2)_n}{n!} r^{2n} = (1 - r^2)^{-1/2}.$$

This proves (6.1.13) once again.

Remark 6.1.1 An interesting formula for Fourier transforms can be formally obtained from formula (6.1.13). Multiply both sides of (6.1.13) by $H_n(y)e^{-y^2}$ and integrate over $(-\infty, \infty)$ to get

$$\frac{1}{\sqrt{\pi}} \int_{-\infty}^{\infty} \frac{e^{-y^2 + [2xyr - (x^2+y^2)r^2]/(1-r^2)}}{\sqrt{1 - r^2}} H_n(y)dy = H_n(x)r^n.$$

The validity of this formula for $|r| < 1$ can be proven. Let $r \to i$ and we have, at least formally,

$$\frac{1}{\sqrt{2\pi}} \int_{-\infty}^{\infty} e^{ixy} e^{-y^2/2} H_n(y)dy = i^n e^{-x^2/2} H_n(x). \qquad (6.1.15)$$

This equation embodies the self-reciprocity of Hermite polynomials. It gives $e^{-x^2/2} H_n(x)$ as an eigenfunction of the Fourier transform with eigenvalue i^n. There are various ways of proving (6.1.15). See Exercise 11 for one method. Also see Exercise 12, which reproduces de Bruin's [1967] proof of Heisenberg's inequality using (6.1.15).

6.2 Laguerre Polynomials

The Laguerre polynomials are orthogonal with respect to the gamma distribution $e^{-x} x^\alpha dx$, where $\alpha > -1$. The definition of the Hermite polynomials and the proof of their orthogonality given by (6.1.5) suggest consideration of the polynomial

$$x^{-\alpha} e^x \frac{d^n}{dx^n} (e^{-x} x^{n+\alpha}).$$

An application of Leibniz's rule for derivatives shows that this expression is a polynomial of degree n. The Laguerre polynomial $L_n^\alpha(x)$ is defined by the formula

$$L_n^\alpha(x) := \frac{x^{-\alpha}e^x}{n!}\frac{d^n}{dx^n}(e^{-x}x^{n+\alpha}), \quad \text{for } n \geq 0. \tag{6.2.1}$$

It is easy to check that

$$L_n^\alpha(x) = \frac{(\alpha+1)_n}{n!}\sum_{k=0}^{n}\frac{(-n)_k x^k}{(\alpha+1)_k k!} = \frac{(\alpha+1)_n}{n!}\,{}_1F_1(-n;\alpha+1;x). \tag{6.2.2}$$

The orthogonality relation for the Laguerre polynomials is contained in

$$\int_0^\infty L_m^\alpha(x)L_n^\alpha(x)x^\alpha e^{-x}dx = \frac{\Gamma(\alpha+n+1)}{n!}\delta_{mn}, \quad \alpha > -1. \tag{6.2.3}$$

The integral on the left is

$$\frac{1}{n!}\int_0^\infty \frac{d^n}{dx^n}(e^{-x}x^{n+\alpha})L_m^\alpha(x)dx.$$

Suppose $n > m$ and integrate by parts n times to see that its value is 0. For $n = m$, first observe that, by (6.2.2),

$$\frac{d^n}{dx^n}L_n^\alpha(x) = (-1)^n,$$

so that n integration by parts gives (6.2.3) after an evaluation of the gamma integral at the last step.

The generating-function formula for $L_n^\alpha(x)$ is given by

$$\sum_{n=0}^\infty L_n^\alpha(x)r^n = \sum_{n=0}^\infty \frac{r^n(\alpha+1)_n}{n!}\sum_{k=0}^n \frac{(-n)_k x^k}{(\alpha+1)_k k!}$$

$$= \sum_{k=0}^\infty \frac{(-x)^k}{(\alpha+1)_k k!}\sum_{n=k}^\infty \frac{(\alpha+1)_n r^n}{(n-k)!}$$

$$= \sum_{k=0}^\infty \frac{(-x)^k r^k}{k!}\sum_{n=0}^\infty \frac{(\alpha+k+1)_n r^n}{n!}$$

$$= (1-r)^{-\alpha-1}\sum_{k=0}^\infty \frac{(-xr)^k}{k!(1-r)^k}$$

$$= (1-r)^{-\alpha-1}\exp(-xr/(1-r)). \tag{6.2.4}$$

Denote the generating function in (6.2.4) by $F(x,r)$. It is readily verified that

$$(1-r^2)\frac{\partial F}{\partial r} + [x - (1+\alpha)(1-r)]F = 0.$$

This gives the three-term recurrence relation

$$(n + 1)L_{n+1}^{\alpha}(x) + (x - \alpha - 2n - 1)L_n^{\alpha}(x) + (n + \alpha)L_{n-1}^{\alpha}(x) = 0, \quad (6.2.5)$$

where $n = 1, 2, 3, \ldots$.

Before deriving the differential equation for $L_n^{\alpha}(x)$, we obtain an interesting formula for the derivative of $L_n^{\alpha}(x)$. The formula is

$$x\frac{dL_n^{\alpha}(x)}{dx} = nL_n^{\alpha}(x) - (n + \alpha)L_{n-1}^{\alpha}(x), \quad \text{for } n \geq 1. \quad (6.2.6)$$

This arises from the identity

$$(1 - r)\frac{\partial F}{\partial x} + rF = 0,$$

which implies

$$\frac{dL_n^{\alpha}(x)}{dx} - \frac{dL_{n-1}^{\alpha}(x)}{dx} + L_{n-1}^{\alpha}(x) = 0, \quad \text{for } n \geq 1. \quad (6.2.7)$$

Eliminate $L_{n-1}^{\alpha}(x)$ from (6.2.5) and (6.2.7) to get

$$(x - n - 1)\frac{dL_n^{\alpha}(x)}{dx} + (n + 1)\frac{dL_{n+1}^{\alpha}(x)}{dx}$$
$$+ (2n + 2 + \alpha - x)L_n^{\alpha}(x) - (n + 1)L_{n+1}^{\alpha}(x) = 0,$$

for $n \geq 0$. Replace n by $n - 1$ in this equation and eliminate $(d/dx)L_{n-1}^{\alpha}(x)$ by means of (6.2.7) to get (6.2.6), the required result.

Now differentiate (6.2.6) and then apply (6.2.6) and (6.2.7) to arrive at

$$x\frac{d^2L_n^{\alpha}(x)}{dx^2} + (\alpha + 1 - x)\frac{dL_n^{\alpha}(x)}{dx} + nL_n^{\alpha}(x) = 0, \quad \text{for } n \geq 0, \quad (6.2.8)$$

Thus $u = L_n^{\alpha}(x)$ satisfies the second-order linear differential equation

$$xu'' + (\alpha + 1 - x)u' + nu = 0. \quad (6.2.9)$$

Because the normal integral is a particular case of the gamma integral, it should be possible to express Hermite polynomials in terms of Laguerre polynomials. Such a relationship exists and is given by

$$H_{2m}(x) = (-1)^m 2^{2m} m! L_m^{-1/2}(x^2) \quad (6.2.10)$$

and

$$H_{2m+1}(x) = (-1)^m 2^{2m+1} m! x L_m^{1/2}(x^2). \quad (6.2.11)$$

To prove that $H_{2m}(x) = C L_m^{-1/2}(x^2)$ for some constant C, it is sufficient to show that, for any polynomial $q(x)$ of degree $\leq 2m - 1$,

$$\int_{-\infty}^{\infty} L_m^{-1/2}(x^2)q(x)e^{-x^2}dx = 0.$$

A general polynomial is the sum of an even and an odd polynomial. When q is odd, the integral is obviously zero. When q is even, it can be written as $q(x) = r(x^2)$, where r is a polynomial of degree $\leq m - 1$. Then for $y = x^2$, the above integral becomes

$$\int_0^{\infty} L_m^{-1/2}(y)r(y)y^{-1/2}e^{-y}dy = 0,$$

by the orthogonality of $L_m^{-1/2}(y)$. The value of C can be found by setting $x = 0$. Relation (6.2.11) can be proved in the same way, or by differentiating (6.2.10).

There is another way in which the normal integral is related to the gamma integral. The normal integral is a limit of the gamma integral. This gives another connection between Laguerre and Hermite polynomials.

By Stirling's formula,

$$\int_0^{\infty} \left(\frac{x}{\alpha}\right)^{\alpha} e^{-(x-\alpha)} \frac{dx}{\sqrt{2\alpha}} \longrightarrow \sqrt{\pi} = \int_{-\infty}^{\infty} e^{-x^2}dx \quad \text{as } \alpha \to \infty.$$

A change of variables $x = \alpha + t/\sqrt{2\alpha}$ gives

$$\int_{-\sqrt{\alpha/2}}^{\infty} \left(1 + \sqrt{\frac{2}{\alpha}}u\right)^{\alpha} e^{-\sqrt{2\alpha}u}du \longrightarrow \int_{-\infty}^{\infty} e^{-x^2}dx \quad \text{as } \alpha \to \infty. \tag{6.2.12}$$

The orthogonality relation for $L_n^{\alpha}(x)$ implies

$$\int_0^{\infty} L_n^{\alpha}(x)^2 x^{\alpha} e^{-x}dx = \frac{\Gamma(n+\alpha+1)}{n!}.$$

Set $x = \alpha + \sqrt{2\alpha}u$. Then, by Stirling's formula,

$$\int_{-\sqrt{\alpha/2}}^{\infty} \left(\frac{2}{\alpha}\right)^n [L_n^{\alpha}(\alpha + \sqrt{2\alpha}u)]^2 \left(1 + \sqrt{\frac{2}{\alpha}}u\right)^{\alpha} e^{-\sqrt{2\alpha}u}du$$

$$\sim \sqrt{\pi} \frac{2^n (1 + n/\alpha)^{n+\alpha+1/2}e^{-n}}{n!} \quad \text{as } \alpha \to \infty. \tag{6.2.13}$$

A comparison of (6.2.12) and (6.2.13) suggests that

$$\lim_{\alpha \to \infty} \left(\frac{2}{\alpha}\right)^{n/2} L_n^{\alpha}(\alpha + \sqrt{2\alpha}x) = (-1)^n \frac{H_n(x)}{n!}. \tag{6.2.14}$$

This may be verified by using the generating functions for the Laguerre and Hermite polynomials and is left to the reader. Recurrence relations or the definitions (6.2.1) and (6.1.3) can also be used to give easy derivations.

We started the treatment of Hermite polynomials by expressing e^{-x^2} as an integral. A similar approach can be taken for Laguerre polynomials. It follows from (4.11.25) that

$$e^{-x}x^{n+\alpha} = \int_0^\infty (\sqrt{xt})^{n+\alpha} e^{-t} J_{n+\alpha}(2\sqrt{xt})dt.$$

By (4.6.1), this leads to the integral representation of $L_n^\alpha(x)$ given by

$$L_n^\alpha(x) = \frac{e^x x^{-\alpha/2}}{n!} \int_0^\infty t^{n+\alpha/2} J_\alpha(2\sqrt{xt})e^{-t}dt, \tag{6.2.15}$$

for $\alpha > -1$.

The reader should verify that the generating function for Laguerre polynomials follows easily from this. Also, since

$$J_{1/2}(x) = \sqrt{\frac{2}{\pi x}} \sin x \quad \text{and} \quad J_{-1/2}(x) = \sqrt{\frac{2}{\pi x}} \cos x,$$

one gets

$$L_n^{-1/2}(x) = \frac{2e^x}{n!\sqrt{\pi}} \int_0^\infty e^{-t^2} t^{2n} \cos(2\sqrt{xt})dt$$

and

$$L_n^{1/2}(x) = \frac{2e^x}{n!\sqrt{\pi x}} \int_0^\infty e^{-t^2} t^{2n+1} \sin(2\sqrt{xt})dt.$$

Compare these with (6.1.4) to get alternative proofs of (6.2.10) and (6.2.11).

We note a few elementary formulas that are useful:

$$\frac{d}{dx} L_n^\alpha(x) = -L_{n-1}^{\alpha+1}(x), \tag{6.2.16}$$

$$\frac{d}{dx}\left[x^\alpha L_n^\alpha(x)\right] = (n+\alpha)x^{\alpha-1} L_n^{\alpha-1}(x), \tag{6.2.17}$$

$$\frac{d}{dx}\left[e^{-x} L_n^\alpha(x)\right] = -e^{-x} L_n^{\alpha+1}(x), \tag{6.2.18}$$

$$\frac{d}{dx}\left[x^\alpha e^{-x} L_n^\alpha(x)\right] = (n+1)x^{\alpha-1} e^{-x} L_{n+1}^{\alpha-1}(x). \tag{6.2.19}$$

We now prove (6.2.19); the others can be proved in a similar way. By (6.2.2) and Kummer's transformation (4.1.11), we have

$$
\begin{aligned}
\frac{d}{dx}\left[x^\alpha e^{-x}L_n^\alpha(x)\right] &= \frac{(\alpha+1)_n}{n!}\frac{d}{dx}[x^\alpha e^{-x}{}_1F_1(-n;\alpha+1;x)] \\
&= \frac{(\alpha+1)_n}{n!}\frac{d}{dx}[x^\alpha\,{}_1F_1(n+\alpha+1;\alpha+1;-x)] \\
&= \frac{(\alpha+1)_n}{n!}\sum_{k=0}^\infty\frac{(n+\alpha+1)_k(-1)^k(k+\alpha)x^{k+\alpha-1}}{(\alpha+1)_k k!} \\
&= \frac{\alpha(\alpha+1)_n}{n!}x^{\alpha-1}\sum_{k=0}^\infty\frac{(n+\alpha+1)_k(-x)^k}{(\alpha)_k k!} \\
&= \frac{(\alpha)_{n+1}}{n!}x^{\alpha-1}e^{-x}{}_1F_1(-n-1;\alpha;x) \\
&= (n+1)x^{\alpha-1}e^{-x}L_{n+1}^{\alpha-1}(x).
\end{aligned}
$$

Formulas (6.2.17) and (6.2.18) can be written as integrals and then extended as fractional integrals. Formula (6.2.17) extends to

$$
x^{b+\mu-1}{}_1F_1(a;b+\mu;x) = \frac{\Gamma(b+\mu)}{\Gamma(b)\Gamma(\mu)}\int_0^x(x-t)^{\mu-1}t^{b-1}{}_1F_1(a;b;t)dt, \quad (6.2.20)
$$

for $\operatorname{Re}\mu > 1$. (See (2.2.4).) Write this as

$$
x^\beta L_n^\beta(x) = \frac{\Gamma(n+\beta+1)}{\Gamma(\beta-\alpha)\Gamma(n+\alpha+1)}\int_0^x(x-t)^{\beta-\alpha-1}t^\alpha L_n^\alpha(t)dt, \quad (6.2.21)
$$

when $\beta > \alpha$. Formula (6.2.18) extends in a slightly different way. The extension is given by

$$
e^{-x}L_n^\alpha(x) = \frac{1}{\Gamma(\beta-\alpha)}\int_x^\infty(t-x)^{\beta-\alpha-1}e^{-t}L_n^\beta(t)dt, \quad (6.2.22)
$$

when $\beta > \alpha$. We give a proof below that uses (6.2.21) and the orthogonality and the completeness of $L_n^\alpha(x)$. A proof of completeness is in Section 6.5. Observe that

$$
\begin{aligned}
&\int_0^\infty x^\beta L_n^\beta(x)L_m^\beta(x)e^{-x}dx \\
&= \frac{\Gamma(n+\beta+1)}{\Gamma(\beta-\alpha)\Gamma(n+\alpha+1)}\int_0^\infty L_m^\beta(x)e^{-x}dx\int_0^x(x-t)^{\beta-\alpha-1}t^\alpha L_n^\alpha(t)dt \\
&= \frac{\Gamma(n+\beta+1)}{\Gamma(\beta-\alpha)\Gamma(n+\alpha+1)}\int_0^\infty L_n^\alpha(t)t^\alpha\left[\int_t^\infty L_m^\beta(x)(x-t)^{\beta-\alpha-1}e^{-x}dx\right]dt.
\end{aligned}
$$

$$(6.2.23)$$

The orthogonality relation (6.2.3) applied to (6.2.23) implies

$$\int_0^\infty L_n^\alpha(t)t^\alpha \left[\int_t^\infty L_n^\beta(x)(x-t)^{\beta-\alpha-1}e^{-x}dx \right] dt$$

$$= \frac{\Gamma(\beta-\alpha)\Gamma(n+\alpha+1)}{\Gamma(n+1)}. \qquad (6.2.24)$$

By (6.2.3), (6.2.23), and (6.2.24)

$$\int_0^\infty L_n^\alpha(t)\left[\frac{1}{\Gamma(\beta-\alpha)} \int_t^\infty L_m^\beta(x)(x-t)^{\beta-\alpha-1}e^{-x}dx - L_m^\alpha(t)e^{-t} \right] dt = 0$$

for $n = 0, 1, \ldots$. Now the completeness of $L_n^\alpha(t)$ gives (6.2.22).

The formula for the Poisson kernel for Laguerre polynomials is given by

$$\sum_{n=0}^\infty \frac{n!L_n^\alpha(x)L_n^\alpha(y)r^n}{\Gamma(n+\alpha+1)} = (1-r)^{-1}e^{-(x+y)r/(1-r)}(xyr)^{-\alpha/2}I_\alpha\left(\frac{2\sqrt{xyr}}{1-r} \right),$$

$$(6.2.25)$$

when $|r| < 1$, $\alpha > -1$, and I_α is the modified Bessel function of order α. A simple proof of (6.2.25) is obtained by using the generating function (6.2.4) and the derivative formula (6.2.19). Write the left side of (6.2.25) as

$$\frac{1}{\Gamma(\alpha+1)} \sum_{n=0}^\infty L_n^\alpha(x)r^n \sum_{k=0}^n \frac{(-n)_k y^k}{(\alpha+1)_k k!}$$

$$= \sum_{k=0}^\infty \frac{(-yr)^k}{\Gamma(k+\alpha+1)k!} \sum_{n=0}^\infty (n+1)_k L_{n+k}^\alpha(x)r^n. \qquad (6.2.26)$$

To find a closed form for the inner sum, start with the generating-function formula

$$\sum_{n=0}^\infty L_n^{\alpha+k}(x)r^n = e^{-xr/(1-r)}/(1-r)^{\alpha+k+1}.$$

Multiply both sides by $x^{k+\alpha}e^{-x}$, take the kth derivative, and apply (6.2.19). The result is

$$x^\alpha e^{-x} \sum_{n=0}^\infty (n+1)_k L_{n+k}^\alpha(x)r^n$$

$$= \frac{1}{(1-r)^{\alpha+k+1}} \frac{d^k}{dx^k}\left[x^{k+\alpha}e^{-x/(1-r)} \right]$$

$$= \frac{1}{(1-r)^{\alpha+k+1}} \sum_{n=0}^\infty \frac{(-1)^n \frac{d^k}{dx^k}x^{n+k+\alpha}}{(1-r)^n n!}$$

$$= \frac{1}{(1-r)^{\alpha+k+1}} \sum_{n=0}^\infty \frac{(-x)^n(n+\alpha+1)_k x^\alpha}{(1-r)^n n!}$$

$$= \frac{(\alpha+1)_k x^\alpha}{(1-r)^{\alpha+k+1}} \,_1F_1(k+\alpha+1; \alpha+1; -x/(1-r)). \qquad (6.2.27)$$

Apply Kummer's transformation (4.1.11) to the ${}_1F_1$ to see that the expression (6.2.27) is equal to

$$\frac{x^\alpha}{(1-r)^{\alpha+k+1}}(\alpha+1)_k \, {}_1F_1(-k;\alpha+1;x/(1-r))e^{-x/(1-r)}$$

$$= \frac{x^\alpha}{(1-r)^{\alpha+k+1}}k!L_k^\alpha(x/(1-r))e^{-x/(1-r)}.$$

Use this for the inner sum in (6.2.26) to get

$$\sum_{n=0}^\infty \frac{n!L_n^\alpha(x)L_n^\alpha(y)r^n}{\Gamma(n+\alpha+1)}$$

$$= \frac{e^{-xr/(1-r)}}{(1-r)^{\alpha+1}}\sum_{k=0}^\infty \frac{L_k^\alpha(x/(1-r))}{\Gamma(k+\alpha+1)}\left(\frac{-yr}{1-r}\right)^k. \qquad (6.2.28)$$

The sum in the last expression can be written as

$$\sum_{k=0}^\infty \sum_{j=0}^k \frac{(-k)_j}{\Gamma(j+\alpha+1)j!k!}\left(\frac{x}{1-r}\right)^j\left(\frac{-yr}{1-r}\right)^k$$

$$= \sum_{j=0}^\infty \frac{(-x/(1-r))^j}{\Gamma(j+\alpha+1)j!}\sum_{k=0}^\infty \frac{(-yr/(1-r))^{k+j}(-k-j)_j}{(k+j)!}$$

$$= \sum_{j=0}^\infty \frac{(xyr/(1-r)^2)^j}{\Gamma(j+\alpha+1)j!}\sum_{k=0}^\infty \frac{(-yr/(1-r))^k}{k!}$$

$$= e^{-yr/(1-r)}\left(\frac{1-r}{\sqrt{xyr}}\right)^\alpha I_\alpha(2\sqrt{xyr}/(1-r)).$$

The last equation follows from the series expansion for the modified Bessel function $I_\alpha(x)$. This proves (6.2.25).

We can obtain the Hankel transform and its inverse from (6.2.25). The argument given here can be made rigorous. See Wiener [1933, pp. 64–70] where the Fourier inversion formula is derived from the Poisson kernel for Hermite polynomials.

Let

$$\psi_n(x) = \sqrt{\frac{n!}{\Gamma(n+\alpha+1)}}x^{\alpha/2}e^{-x/2}L_n^\alpha(x).$$

Then (6.2.25) can be written as

$$H(x,y,t) := \frac{e^{-(x+y)(1+t)/(1-t)}t^{-\alpha/2}I_\alpha(2\sqrt{xyt}/(1-t))}{1-t}$$

$$= \sum_{n=0}^\infty \psi_n(x)\psi_n(y)t^n. \qquad (6.2.29)$$

Let $f(x)$ be a sufficiently smooth function that dies away at infinity. Then $f(x)$ has the Fourier–Laguerre expansion

$$f(x) = \sum_{n=0}^{\infty} \left(\int_0^{\infty} f(y)\psi_n(y)dy \right) \psi_n(x).$$

Multiply (6.2.29) by $f(y)$ and integrate to get

$$\int_0^{\infty} f(y)H(x, y, t)dy = \sum_{n=0}^{\infty} \psi_n(x) \int_0^{\infty} f(y)\psi_n(y)dy \, t^n.$$

Let $t \to e^{-\pi i}$ to arrive at

$$\int_0^{\infty} f(y)e^{\alpha \pi i/2} I_\alpha(\sqrt{xy}e^{-\pi i/2})dy = \sum_{n=0}^{\infty}(-1)^n \int_0^{\infty} f(y)\psi_n(y)dy \, \psi_n(x)$$

$$=: g(x). \tag{6.2.30}$$

Thus,

$$g(x) = \int_0^{\infty} f(y)J_\alpha(\sqrt{xy})dy. \tag{6.2.31}$$

Now $g(x)$ has a Fourier–Laguerre expansion that, by the definition of $g(x)$, implies that

$$\int_0^{\infty} g(x)\psi_n(x)dx = (-1)^n \int_0^{\infty} f(x)\psi_n(x)dx.$$

By a derivation similar to that of (6.2.31), we have

$$\int_0^{\infty} g(x)J_\alpha(\sqrt{xy})dx = \sum_{n=0}^{\infty}(-1)^n \int_0^{\infty} g(y)\psi_n(y)dy \, \psi_n(x)$$

$$= \sum_{n=0}^{\infty} \int_0^{\infty} f(y)\psi_n(y)dy \, \psi_n(x)$$

$$= f(x). \tag{6.2.32}$$

We can write (6.2.31) and (6.2.32) as the Hankel pair

$$\int_0^{\infty} f(y^2)J_\alpha(xy)ydy = g(x^2),$$

$$\int_0^{\infty} g(x^2)J_\alpha(xy)xdx = f(y^2). \tag{6.2.33}$$

This may be the place to point out that (6.2.21) contains Sonine's first integral (4.11.11) as a limiting case. This follows from the fact that

$$\lim_{n\to\infty} n^{-\alpha}L_n^\alpha(x/n) = x^{-\alpha/2}J_\alpha(2\sqrt{x}). \tag{6.2.34}$$

If in the Hankel inverse of Sonine's first integral (4.11.12), we change x to x/t and, in the formula thus obtained, change t to $1/s$, we get

$$\int_0^\infty J_\mu(x) J_\lambda(xs) x^{\mu-\lambda} x \, dx = \frac{1}{\Gamma(\lambda - \mu) 2^{\lambda-\mu-1}} s^{-\lambda} (s^2 - 1)_+^{\lambda-\mu-1},$$

when $\lambda > \mu > -1$. The Hankel inverse of this is

$$x^{\mu-\lambda} J_\mu(x) = \frac{1}{\Gamma(\lambda - \mu) 2^{\lambda-\mu-1}} \int_1^\infty s^{-\lambda} (s^2 - 1)^{\lambda-\mu-1} J_\lambda(xs) s \, ds,$$

for $\mu < \lambda < 2\mu + 3/2$. Write it as

$$x^{-\mu} J_\mu(x) = \frac{2^{\mu-\lambda+1}}{\Gamma(\lambda - \mu)} \int_x^\infty t^{-\lambda+1} J_\lambda(t) (t^2 - x^2)^{\lambda-\mu-1} \, dt,$$

for $-1 < \mu < \lambda < 2\mu + 3/2$. This is the analog of (6.2.22). Two analogs of Sonine's second integral are

$$L_n^{(\alpha+\beta+1)}(x + y) = \sum_{k=0}^n L_k^\alpha(x) L_{n-k}^\beta(y) \tag{6.2.35}$$

and

$$\frac{L_{m+n}^\alpha(x)}{L_{m+n}^\alpha(0)} = \frac{\Gamma(\alpha + 1)\Gamma(\beta + 1)}{\Gamma(\alpha - \beta)}$$

$$\times \int_0^1 t^\beta (1 - t)^{\alpha-\beta-1} \frac{L_m^\beta(xt)}{L_m^\beta(0)} \frac{L_n^{\alpha-\beta-1}[x(1 - t)]}{L_n^{\alpha-\beta-1}(0)} \, dt. \tag{6.2.36}$$

Formula (6.2.36) is due to Feldheim [1943]. Formula (6.2.35) is an immediate consequence of the generating function (6.2.4), and (6.2.36) is proven by using the series representations of Laguerre polynomials, the value of the beta integral, and the Chu–Vandermonde sum. When $y = 0$, (6.2.35) is equivalent to

$$L_n^\beta(x) = \sum_{k=0}^n \frac{\Gamma(n - k + \beta - \alpha)}{\Gamma(n - k + 1)\Gamma(\beta - \alpha)} L_k^\alpha(x). \tag{6.2.37}$$

This is an easy consequence of the generating-function formula (6.2.4). The details are given in Section 7.1. This formula is equivalent to

$$\int_0^\infty L_n^\beta(x) L_k^\alpha(x) x^\alpha e^{-x} \, dx$$

$$= \frac{\Gamma(n - k + \beta - \alpha)\Gamma(k + \alpha + 1)}{\Gamma(n - k + 1)\Gamma(\beta - \alpha)\Gamma(k + 1)}. \tag{6.2.38}$$

This can be used to show that the Fourier–Laguerre expansion of $x^{\alpha-\beta}L_k^\alpha(x)$ in terms of $L_n^\beta(x)$ is given by the formula

$$x^\alpha e^{-x}L_k^\alpha(x)$$
$$= \sum_{n=k}^\infty \frac{\Gamma(n-k+\beta-\alpha)\Gamma(k+\alpha+1)\Gamma(n+1)}{\Gamma(n-k+1)\Gamma(\beta-\alpha)\Gamma(k+1)\Gamma(n+\beta+1)} L_n^\beta(x)x^\beta e^{-x},$$

$$(6.2.39)$$

for $\alpha > (\beta-1)/2$. To understand this condition on α and β, needed for convergence, see Theorem 6.5.3 and the remark after that. Note that (6.2.39) is the inversion of (6.2.37). Just as the Sonine integral and its inversion can be used to solve dual integral equations, (6.2.37) and (6.2.39) help to solve a dual sequence equation involving Laguerre polynomials.

Theorem 6.2.1   *Let α, λ, c be given, such that $c > (\lambda - 2\alpha - 1)/2$, $\alpha, \lambda > -1$. Then if a_n, b_n are given (and are small enough) and if*

$$a_n = \int_0^\infty x^c f(x)L_n^\alpha(x)x^\alpha e^{-x}dx, \quad n = 0, 1, \dots, N,$$

$$b_n = \int_0^\infty f(x)L_n^\lambda(x)x^\lambda e^{-x}dx, \quad n = N+1, N+2, \dots,$$

and if $\beta = \alpha + c$, then

$$f(x) = \sum_{n=0}^N \sum_{k=0}^n \frac{\Gamma(n-k+\beta-\alpha)}{\Gamma(n-k+1)\Gamma(\beta-\alpha)} a_k \frac{\Gamma(n+1)}{\Gamma(n+\beta+1)} L_n^\beta(x)$$

$$\cdot \sum_{n=N+1}^\infty \sum_{k=n}^\infty \frac{\Gamma(k-n+\lambda-\beta)\Gamma(k+1)}{\Gamma(k-n+1)\Gamma(\lambda-\beta)\Gamma(k+\alpha+1)} b_k L_n^\beta(x). \quad (6.2.40)$$

Proof. By (6.2.37), we have

$$\sum_{k=0}^n \frac{\Gamma(n-k+\beta-\alpha)}{\Gamma(n-k+1)\Gamma(\beta-\alpha)} a_k = \int_0^\infty f(x)L_n^\beta(x)x^\beta e^{-x}dx.$$

However, by (6.2.39) we have

$$\sum_{k=n}^\infty \frac{\Gamma(k-n+\lambda-\beta)\Gamma(n+\beta+1)\Gamma(k+1)}{\Gamma(k-n+1)\Gamma(\lambda-\beta)\Gamma(n+1)\Gamma(k+\lambda+1)} b_k$$

$$= \int_0^\infty f(x)L_n^\beta(x)x^\beta e^{-x}dx.$$

Now use the Fourier–Laguerre expansion of $f(x)$ to get (6.2.40). This makes more than merely formal sense if the b_ns are small enough. This proves the theorem. ∎

Similarly, (6.2.21) and (6.2.22) can be used to solve dual-series equations involving Laguerre polynomials.

Theorem 6.2.2 *Let $\alpha > \delta > -1$, and let $\alpha < \min(\delta + 1, 2\delta + 1)$. Suppose that*

$$f(x) = \sum_{n=0}^{\infty} a_n \frac{\Gamma(n + \delta + 1)}{\Gamma(n + \alpha + 1)} L_n^\alpha(x), \quad 0 \leq x < y,$$

and

$$g(x) = \sum_{n=0}^{\infty} a_n L_n^\alpha(x), \quad x < y < \infty.$$

Then

$$a_n = \frac{n!}{\Gamma(n + \delta + 1)} \left\{ \int_0^y \frac{1}{\Gamma(\delta - \alpha + 1)} \left[\frac{d}{dx} \int_0^x f(t) t^\alpha (x - t)^{\delta - \alpha} dt \right] L_n^\delta(x) e^{-x} dx \right.$$

$$\left. + \int_y^\infty \frac{1}{\Gamma(\alpha - \delta)} \left[\int_x^\infty g(t) e^{-t} (t - x)^{\alpha - \delta - 1} dt \right] L_m^\delta(x) x^\delta dx \right\}.$$

6.3 Jacobi Polynomials and Gram Determinants

If

$$c_n = \int_a^b x^n \, d\alpha(x)$$

are the moments with respect to a given distribution $d\alpha(x)$, then the polynomials

$$p_n(x) = \begin{vmatrix} c_0 & c_1 & c_2 & \cdots & c_n \\ c_1 & c_2 & c_3 & \cdots & c_{n+1} \\ \vdots & \vdots & & & \\ c_{n-1} & c_n & c_{n+1} & \cdots & c_{2n-1} \\ 1 & x & x^2 & & x^n \end{vmatrix} \qquad (6.3.1)$$

are orthogonal with respect to $d\alpha(x)$. It appears at first sight that it would be difficult to obtain a more useful representation of the polynomials from the determinant. Wilson [1978, 1991] showed that, in many interesting cases, it is possible and quite easy to get a hypergeometric representation of $p_n(x)$ from the determinant. In this section, we give his derivation of the hypergeometric form of the Jacobi polynomials.

We start with the following lemma, which generalizes the result contained in (6.3.1).

Lemma 6.3.1   *Suppose* $\{\phi_n\}_0^\infty$ *is a sequence of independent functions. The sequence of functions* $\{p_n(x)\}$ *given by*

$$p_n(x) = C_n \begin{vmatrix} \mu_{0,0} & \mu_{0,1} & \cdots & \mu_{0,n} \\ \mu_{1,0} & \mu_{1,1} & \cdots & \mu_{1,n} \\ \vdots & \vdots & & \vdots \\ \mu_{n-1,0} & \mu_{n-1,1} & \cdots & \mu_{n-1,n} \\ \phi_0 & \phi_1 & & \phi_n \end{vmatrix}, \qquad (6.3.2)$$

where

$$\mu_{i,j} = \int_a^b \phi_i(x)\phi_j(x)\,d\alpha(x),$$

and where C_n *is a constant, satisfies the relation*

$$\int_a^b p_n(x)\phi_m(x)\,d\alpha(x) = 0, \quad \text{for } m < n.$$

(Here $\alpha(x)$ *need not be a positive measure but all the integrals are assumed to exist.)*

Proof.   Expand the determinant (6.3.2) as

$$p_n(x) = C_n \sum_{k=0}^n A_k \phi_k(x).$$

Then

$$\int_a^b p_n(x)\phi_m(x)\,d\alpha(x) = C_n \sum_{k=0}^n A_k \mu_{m,k}. \qquad (6.3.3)$$

If $m \leq n - 1$, then the right-hand side of (6.3.3) represents a determinant in which two rows are identical; the value of such a determinant is zero. This proves the lemma. ■

Corollary 6.3.2   *Suppose* $\{\phi_n\}_0^\infty$ *and* $\{\psi_n\}_0^\infty$ *are two sequences. Define* $p_n(x)$ *as in (6.3.2) with*

$$\mu_{i,j} = \int_a^b \psi_i \phi_j \, d\alpha(x).$$

Then

$$\int_a^b p_n(x)\psi_m(x)\,d\alpha(x) = 0, \quad \text{for } m \leq n - 1.$$

We now specialize to the situation where the weight function is $\alpha'(x) = (1 - x)^\alpha$ $(1 + x)^\beta$ over the interval $(-1, 1)$. Our aim is to find $p_n(x)$. For this we choose

$\phi_k(x) = (1-x)^k$ and $\psi_k(x) = (1+x)^k$. Other choices of ϕ_k and ψ_k as polynomials of exact degree k would work, but these make the calculations simpler. In this case

$$
\begin{aligned}
\mu_{i,j} &= \int_{-1}^{1} (1-x)^{\alpha+j}(1+x)^{\beta+i}dx \\
&= 2^{\alpha+\beta+1+i+j}\frac{\Gamma(\alpha+1+j)\Gamma(\beta+1+i)}{\Gamma(\alpha+\beta+2+i+j)} \\
&= 2^{\alpha+\beta+1+i}\frac{\Gamma(\alpha+1)\Gamma(\beta+1+i)}{\Gamma(\alpha+\beta+2+i)} \cdot \frac{2^j(\alpha+1)_j}{(\alpha+\beta+2+i)_j}. \quad (6.3.4)
\end{aligned}
$$

Theorem 6.3.3 *The polynomial (6.3.2) with $\mu_{i,j}$ given by (6.3.4) is a constant multiple of the Jacobi polynomial*

$$
P_n^{(\alpha,\beta)}(x) = \frac{(\alpha+1)_n}{n!}\,_2F_1\left(\begin{matrix}-n, n+\alpha+\beta+1\\ \alpha+1\end{matrix}; \frac{1-x}{2}\right). \quad (6.3.5)
$$

Proof. From each row, $i = 0, \ldots, n-1$, pull out the first factor in (6.3.4) and absorb it in the constant C_n. We will continue to call it C_n, so it may differ from one line to the next. Then factor out $2^j(\alpha+1)_j$ from each of the columns, $j = 0, 1, \ldots, n$. The resulting determinant is

$$
C_n\begin{vmatrix}
\tilde{\mu}_{0,0} & \tilde{\mu}_{0,1} & \cdots & \tilde{\mu}_{0,n} \\
\tilde{\mu}_{1,0} & \tilde{\mu}_{1,1} & \cdots & \tilde{\mu}_{1,n} \\
\vdots & \vdots & & \vdots \\
\tilde{\mu}_{n-1,0} & \tilde{\mu}_{n-1,1} & \cdots & \tilde{\mu}_{n-1,n} \\
\tilde{\phi}_0 & \tilde{\phi}_1 & & \tilde{\phi}_n
\end{vmatrix},
$$

where

$$
\tilde{\mu}_{i,j} = \frac{1}{(\alpha+\beta+2+i)_j} \quad \text{and} \quad \tilde{\phi}_j(x) = \frac{(1-x)^j}{2^j(1+\alpha)_j}.
$$

Now expand about the last row to obtain the expression

$$
C_n\sum_{k=0}^{n}\frac{(-1)^k}{(\alpha+1)_k}\left(\frac{1-x}{2}\right)^k \det(\tilde{\mu}_{i,j})_{\substack{0\le i\le n-1\\ 0\le j\le n, j\ne k}}. \quad (6.3.6)
$$

Set $A = \alpha+\beta+2$. The problem now is to compute the determinant

$$
\Delta(A, n, k) = \det\left(\frac{1}{(A+i)_j}\right)_{\substack{0\le i\le n-1\\ 0\le j\le n, j\ne k}}, \qquad \Delta(A, 0, 0) = 1.
$$

Observe that the first column in this determinant consists of 1s when $k \ne 0$. We shall consider this case first. Subtract row $n-1$ from row n, then row $n-2$ from row $n-1$, and so on. This makes the first column zero except for the first entry, which is one. Expand the determinant along the first column to get

$$\Delta(A, n, k) = \det\left[\frac{1}{(A+i+1)_{j+1}} - \frac{1}{(A+1)_{j+1}}\right]_{\substack{0 \le i \le n-1 \\ 0 \le j \le n-1, j \ne k-1}}.$$

But since

$$\frac{1}{(A+i+1)_{j+1}} - \frac{1}{(A+i)_{j+1}} = \frac{1}{(A+i)_{j+2}}[(A+i) - (A+i+j+1)]$$

$$= \frac{-(j+1)}{(A+i)_{j+2}} = \frac{-(j+1)}{(A+i)(A+i+1)(A+i+2)_j},$$

then

$$\Delta(A, n, k) = \det\left[\frac{-(j+1)}{(A+i)(A+i+1)(A+i+2)_j}\right]_{\substack{0 \le i \le n-2 \\ 0 \le j \le n-1, j \ne k-1}}.$$

Now $-(j+1)$ is a common factor in the jth column and $1/[(A+i)(A+i+1)]$ is common in the ith row and so these factors can be taken outside the determinant. Thus

$$\Delta(A, n, k) = \left[(-1)^{n-1} \prod_{\substack{j=0 \\ j \ne k-1}}^{n-1} (j+1) \middle/ \prod_{i=0}^{n-2}(A+i)(A+i+1)\right]$$

$$\cdot \Delta(A+2, n-1, k-2)$$

$$= \frac{(-1)^{n-1}n!}{k(A)_{n-1}(A+1)_{n-1}}\Delta(A+2, n-1, k-1).$$

Repeat this process k times to get

$$\Delta(A, n, k)$$

$$= \left\{\frac{1}{k!}\prod_{s=0}^{k-1}\frac{(-1)^{n-s-1}(n-s)!}{(A+2s)_{n-s-1}(A+2s+1)_{n-s-1}}\right\}\Delta(A+2k, n-k, 0). \quad (6.3.7)$$

Recall that we had assumed $k \ne 0$. The above equations holds trivially for $k = 0$, provided we assume that the product in the braces is one. In the determinant $\Delta(A+2k, n-k, 0)$, the column index j is not 0 so that j goes from 1 to $n-k$. Therefore,

$$\Delta(A+2k, n-k, 0) = \det\left(\frac{1}{(A+2k+i)_{j+1}}\right)_{\substack{0 \le i \le n-k-1 \\ 0 \le j \le n-k-1}}$$

$$= \det\left(\frac{1}{(A+2k+i)(A+2k+i+1)_j}\right)_{\substack{0 \le i \le n-k-1 \\ 0 \le j \le n-k-1}}$$

$$= \frac{\Delta(A+2k+1, n-k, n-k)}{(A+2k)_{n-k}}.$$

Combine this with (6.3.7). The result after some rearrangement is

$$\Delta(A, n, k) = \frac{(A + n - 1)_k}{k!(n - k)!} \prod_{s=0}^{n-1} \frac{(-1)^{n-s-1}(n - s)!}{(A + 2s)_{n-s-1}(A + 2s + 1)_{n-s-1}}.$$

The product depends only on n so it can be absorbed in C_n. Substitute this value of the determinant in (6.3.6) to arrive at

$$C_n \sum_{k=0}^{n} \frac{(-1)^k}{(\alpha + 1)_k} \frac{(\alpha + \beta + n + 1)_k}{k!(n - k)!} \left(\frac{1 - x}{2} \right)^k$$

$$= \frac{C_n}{n!} \, {}_2F_1 \left(\begin{matrix} -n, n + \alpha + \beta + 1 \\ \alpha + 1 \end{matrix} ; \frac{1 - x}{2} \right).$$

Choose $C_n = (\alpha + 1)_n$. This gives the hypergeometric representation of $P_n^{(\alpha,\beta)}(x)$ and the theorem is proved. ■

An immediate consequence of the hypergeometric representation (6.3.5) of Jacobi polynomials is the formula for the derivative:

$$\frac{d}{dx} \{ P_n^{(\alpha,\beta)}(x) \} = \frac{1}{2}(n + \alpha + \beta + 1) P_{n-1}^{(\alpha+1,\beta+1)}(x). \tag{6.3.8}$$

We have seen in Chapter 3 that $_2F_1(a, b; c; x)$ is a solution of

$$x(1 - x) \frac{d^2y}{dx^2} + \{c - (a + b + 1)x\} \frac{dy}{dx} - aby = 0.$$

Therefore, $u = P_n^{(\alpha,\beta)}(x)$ is a solution of the differential equation

$$(1 - x^2)u'' + \{\beta - \alpha - (\alpha + \beta + 2)x\}u' + n(n + \alpha + \beta + 1)u = 0. \tag{6.3.9}$$

Another straightforward result from (6.3.5) is that the coefficient of x^n in $P_n^{(\alpha,\beta)}(x)$ is

$$\frac{(\alpha + \beta + n + 1)_n}{2^n n!}. \tag{6.3.10}$$

6.4 Generating Functions for Jacobi Polynomials

Generating functions are of great importance in the theory of orthogonal polynomials. We have already used them to study Hermite and Laguerre polynomials. In this chapter, we derive Jacobi's form of the generating function by two methods. One is due to Jacobi and uses Lagrange inversion; the other is due to Hermite. A different generating function, particularly useful in studying ultraspherical polynomials, is also derived.

One way of finding the generating function for Jacobi polynomials is to use Lagrange inversion. We use the following lemma. Its derivation and some applications are given in Appendix E.

Lemma 6.4.1   *Suppose that $\phi(y)$ is analytic in a neighborhood of $y = x$,*

$$r = \frac{y - x}{\phi(y)} = \sum_{n=1}^{\infty} a_n (y - x)^n \quad \text{with } a_1 \neq 0, \tag{6.4.1}$$

and f is analytic in a neighborhood of $y = x$. Then $f(y)$ can be expanded in powers of r:

$$f(y) = f(x) + \sum_{n=1}^{\infty} \frac{r^n}{n!} \frac{d^{n-1}}{dx^{n-1}} (f'(x)(\phi(x))^n). \tag{6.4.2}$$

Theorem 6.4.2   *The generating function for the Jacobi polynomials $P_n^{(\alpha,\beta)}(x)$ is given by*

$$F(x, r) = 2^{\alpha+\beta} R^{-1} (1 - r + R)^{-\alpha} (1 + r + R)^{-\beta}, \tag{6.4.3}$$

when

$$R = (1 - 2xr + r^2)^{1/2}.$$

Proof.   Take $\phi(y) = (y^2 - 1)/2$ in (6.4.1). Then

$$y = \frac{1}{r} - \frac{(1 - 2xr + r^2)^{1/2}}{r} = \frac{1}{r} - \frac{R}{r}.$$

The derivative of (6.4.2) with respect to x is

$$f'(y) \frac{dy}{dx} = f'(x) + \sum_{n=1}^{\infty} \frac{r^n}{n!} \frac{d^n}{dx^n} (f'(x)(\phi(x))^n).$$

For $f'(x) = (1 - x)^\alpha (1 + x)^\beta$ this becomes

$$\frac{(1 - y)^\alpha (1 + y)^\beta}{R} = (1 - x)^\alpha (1 + x)^\beta$$

$$+ \sum_{n=1}^{\infty} \frac{r^n}{n!} \frac{d^n}{dx^n} ((1 - x)^\alpha (1 + x)^\beta (x^2 - 1)^n / 2^n).$$

Use Rodrigues's formula (2.5.13'),

$$(1 - x)^\alpha (1 + x)^\beta P_n^{(\alpha,\beta)}(x) = \frac{(-1)^n}{2^n n!} \frac{d^n}{dx^n} \{(1 - x)^{n+\alpha} (1 + x)^{n+\beta}\},$$

to arrive at

$$\frac{1}{R} \left(\frac{1 - y}{1 - x} \right)^\alpha \left(\frac{1 + y}{1 + x} \right)^\beta = 1 + \sum_{n=1}^{\infty} \frac{r^n}{n!} P_n^{(\alpha,\beta)}(x).$$

A short calculation shows that

$$\frac{1 - y}{1 - x} = \frac{2}{1 - r + R} \quad \text{and} \quad \frac{1 + y}{1 + x} = \frac{2}{1 + r + R}.$$

This completes the proof of the theorem. (See Jacobi [1859].)   ∎

If one were to take the definition of $P_n^{(\alpha,\beta)}(x)$ as the polynomials that satisfy

$$F(x,r) = 2^{\alpha+\beta} R^{-1}(1-r+R)^{-\alpha}(1+r+R)^{-\beta} = \sum_0^\infty P_n^{(\alpha,\beta)}(x)r^n, \quad (6.4.4)$$

then one would face the problem of proving orthogonality. When $\alpha = \beta = 0$, we have the Legendre polynomials $P_n(x)$. For these polynomials Legendre observed that

$$\frac{1}{\sqrt{rs}} \log \frac{1+\sqrt{rs}}{1-\sqrt{rs}} = \int_{-1}^1 \frac{dx}{\sqrt{1-2xr+r^2}\sqrt{1-2xs+s^2}}$$

$$= \int_{-1}^1 \sum_{m,n} P_n(x) P_m(x) r^n s^m dx.$$

This implies orthogonality. Chebyshev [1870] applied the same method to the general Jacobi polynomials. His proof is a *tour de force* and shows his skill in handling formulas. A simpler proof was given by Hermite [1890]. He treated the Legendre case, but his argument extends in an obvious way to Jacobi polynomials.

Theorem 6.4.3 *If $\{P_n^{(\alpha,\beta)}(x)\}$ is given by (6.4.4), then the sequence is orthogonal with respect to the weight function $(1-x)^\alpha(1+x)^\beta$ and conversely.*

Proof. Consider the integral

$$I_m = \int_{-1}^1 x^m F(x,r)(1-x)^\alpha(1+x)^\beta dx$$

$$= \sum_{n=0}^\infty r^n \int_{-1}^1 x^m P_n^{(\alpha,\beta)}(x)(1-x)^\alpha(1+x)^\beta dx.$$

Set

$$(1-2xr+r^2)^{1/2} = 1 - ry$$

or

$$x = y + \frac{r}{2}(1-y^2).$$

Then

$$I_m = \int_{-1}^1 \left[y + \frac{r(1-y^2)}{2} \right]^m (1-y)^\alpha(1+y)^\beta dy.$$

Clearly, I_m is a polynomial in r of degree m. So we must have

$$\int_{-1}^{1} x^m P_n^{(\alpha,\beta)}(x)(1-x)^\alpha(1+x)^\beta dx = 0, \quad n = m+1, m+2, \ldots.$$

This is equivalent to

$$\int_{-1}^{1} P_m^{(\alpha,\beta)}(x) P_n^{(\alpha,\beta)}(x)(1-x)^\alpha(1+x)^\beta dx = 0, \quad m \neq n.$$

For the converse, consider I_m again. The same change of variables gives

$$I_m = \int_{-1}^{1} \left[y + \frac{r(1-y^2)}{2} \right]^m F(x,r) \left[\frac{1-r+1-ry}{2} \right]^\alpha$$

$$\cdot \left[\frac{1+r+1-ry}{2} \right]^\beta (1-ry)(1-y)^\alpha(1+y)^\beta dy.$$

This is clearly a polynomial of degree m in r if

$$F(x,r) = C(1-r+1-ry)^{-\alpha}(1+r+1-ry)^{-\beta}(1-ry)^{-1},$$

where C is a constant. To find C, take $x = 1$ to get

$$F(1,r) = \sum_{n=0}^{\infty} P_n^{(\alpha,\beta)}(1)r^n = \sum_{n=0}^{\infty} \frac{(\alpha+1)_n}{n!} r^n$$

$$= (1-r)^{-\alpha-1} = C2^{-\alpha-\beta}(1-r)^{-\alpha-1},$$

or $C = 2^{\alpha+\beta}$. ∎

Remark 6.4.1   The last theorem gives another way of finding the polynomials orthogonal with respect to the beta distribution $(1-x)^\alpha(1+x)^\beta dx$. These polynomials must have $F(x,r)$ in (6.4.4) as their generating function. Then by Lagrange inversion the polynomials must be

$$\frac{(-1)^n}{2^n n!}(1-x)^{-\alpha}(1+x)^{-\beta} \frac{d^n}{dx^n}\{(1-x)^{n+\alpha}(1+x)^{n+\beta}\}.$$

The ultraspherical polynomials are the important subclass of the Jacobi polynomials $P_n^{(\alpha,\beta)}(x)$, when $\alpha = \beta$. These are defined by a different normalization than $P_n^{(\alpha,\alpha)}(x)$, giving a simpler generating function than (6.4.4) when $\alpha = \beta$. The new generating function is useful in obtaining properties of ultraspherical polynomials. To motivate this generating function consider the Poisson kernel

$$\sum_{k=0}^{\infty} P_k^{(\alpha,\beta)}(x) P_k^{(\alpha,\beta)}(y) r^k / h_k \quad (\alpha, \beta > -1), \tag{6.4.5}$$

where

$$h_k = \int_{-1}^{1} [P_k^{(\alpha,\beta)}(x)]^2 (1-x)^\alpha (1+x)^\beta dx$$

$$= \frac{2^{\alpha+\beta+1}\Gamma(k+\alpha+1)\Gamma(k+\beta+1)}{(2k+\alpha+\beta+1)\Gamma(k+\alpha+\beta+1)\Gamma(k+1)}.$$

Take $y = 1$ in (6.4.5) to get the sum

$$\frac{\Gamma(\alpha+\beta+1)}{2^{\alpha+\beta+1}\Gamma(\alpha+1)\Gamma(\beta+1)} \sum_{k=0}^{\infty} \frac{(2k+\alpha+\beta+1)(\alpha+\beta+1)_k}{(\beta+1)_k} P_k^{(\alpha,\beta)}(x)r^k.$$

Observe that the last series, except for a factor, is obtained from

$$r^{-(\alpha+\beta+1)/2} \sum_{n=0}^{\infty} \frac{(\alpha+\beta+1)_n P_n^{(\alpha,\beta)}(x)r^n}{(\beta+1)_n} \tag{6.4.6}$$

by differentiation. Rewrite this series with the hypergeometric expression for $P_n^{(\alpha,\beta)}(x)$ in powers of $(1+x)/2$. We have

$$\sum_{n=0}^{\infty} \sum_{k=0}^{n} \frac{(1+\alpha+\beta)_{n+k}((x+1)/2)^k(-1)^n r^n}{(n-k)!k!(1+\beta)_k}$$

$$= \sum_{k,n} \frac{(1+\alpha+\beta)_{n+2k}(x+1)^k(-1)^n r^{n+k}}{n!k!(1+\beta)_k 2^k}$$

$$= \sum_{k=0}^{\infty} \sum_{n=0}^{\infty} \frac{(1+\alpha+\beta+2k)_n(-1)^n r^n}{n!} \cdot \frac{(1+\alpha+\beta)_{2k}(x+1)^k r^k}{k!(1+\beta)_k 2^k}$$

$$= \sum_{k=0}^{\infty} \frac{(1+\alpha+\beta)_{2k}(x+1)^k r^k}{k!2^k(1+r)^{\alpha+\beta+1+2k}}$$

$$= \frac{1}{(1+r)^{\alpha+\beta+1}} \sum_{k=0}^{\infty} \frac{((1+\alpha+\beta)/2)_k((2+\alpha+\beta)/2)_k 2^k(x+1)^k r^k}{k!(1+\beta)_k(1+r)^{2k}}.$$

We used the binomial theorem in this computation, along with the fact that

$$(a)_{2k} = 2^{2k}(a/2)_k((a+1)/2)_k.$$

The final result is

$$\sum_{n=0}^{\infty} \frac{(\alpha+\beta+1)_n P_n^{(\alpha,\beta)}(x)r^n}{(\beta+1)_n}$$

$$= \frac{1}{(1+r)^{\alpha+\beta+1}} \, {}_2F_1\left(\begin{matrix} (\alpha+\beta+1)/2, \, (\alpha+\beta+2)/2 \\ 1+\beta \end{matrix} ; 2r(1+x)/(1+r)^2 \right).$$

$$\tag{6.4.7}$$

This is the other generating function we were looking for. When $\alpha = \beta$, this gives

$$\sum_{n=0}^{\infty} \frac{(2\alpha+1)_n P_n^{(\alpha,\alpha)}(x) r^n}{(\alpha+1)_n} = \frac{1}{(1+r)^{2\alpha+1}} \left(1 - \frac{2(1+x)r}{(1+r)^2}\right)^{-\alpha-\frac{1}{2}}$$

$$= (1 - 2xr + r^2)^{-\alpha-\frac{1}{2}}. \tag{6.4.8}$$

It is reasonable to define polynomials

$$C_n^{\lambda}(x) := \frac{(2\lambda)_n}{(\lambda + (1/2))_n} P_n^{(\lambda-(1/2),\lambda-(1/2))}(x) \tag{6.4.9}$$

with the generating function

$$(1 - 2xr + r^2)^{-\lambda} = \sum_{n=0}^{\infty} C_n^{\lambda}(x) r^n. \tag{6.4.10}$$

When $\lambda = 0$, $C_n^{\lambda}(x) \equiv 0$. So the case $\lambda = 0$ has to be considered as $\lambda \to 0$ in the formulas involving $C_n^{\lambda}(x)$. The polynomials $C_n^{\lambda}(x)$ are called ultraspherical polynomials or Gegenbauer polynomials. They are orthogonal with respect to the weight function $(1 - x^2)^{\lambda-(1/2)}$ when $\lambda > -(1/2)$.

An important expression for $C_n^{\lambda}(x)$ follows immediately from the generating function (6.4.10). Let $x = \cos\theta$, factor the left-hand side using

$$1 - 2r\cos\theta + r^2 = (1 - re^{i\theta})(1 - re^{-i\theta}),$$

expand by the binomial theorem, and equate the coefficients of r^n. The result is

$$C_n^{\lambda}(\cos\theta) = \sum_{k=0}^{n} \frac{(\lambda)_k (\lambda)_{n-k}}{k!(n-k)!} e^{i(n-2k)\theta}$$

$$= \sum_{k=0}^{n} \frac{(\lambda)_k (\lambda)_{n-k}}{k!(n-k)!} \cos(n-2k)\theta. \tag{6.4.11}$$

The second equation is true because $C_n^{\lambda}(\cos\theta)$ is real and the real part of $e^{i(n-2k)\theta}$ is $\cos(n-2k)\theta$. When $\lambda > 0$, (6.4.11) implies

$$\left|C_n^{\lambda}(\cos\theta)\right| \leq C_n^{\lambda}(1) = \frac{(2\lambda)_n}{n!}.$$

Another hypergeometric representation for $C_n^{\lambda}(x)$ is obtained by taking a different factorization:

$$(1 - 2xr + r^2)^{-\lambda} = (1+r^2)^{-\lambda} \left(1 - \frac{2xr}{1+r^2}\right)^{-\lambda}$$

$$= \sum_{k=0}^{\infty} \frac{(\lambda)_k}{k!} \frac{(2xr)^k}{(1+r^2)^{k+\lambda}}$$

$$= \sum_{n,k} \frac{(\lambda)_k}{k!} \frac{(\lambda+k)_n}{n!} (-1)^n (2x)^k r^{k+2n}$$

$$= \sum_{n,k} \frac{(\lambda)_{k+n}}{k!n!} (2xr)^k (-1)^n r^{2n}$$

$$= \sum_{n=0}^{\infty} \frac{(\lambda)_n}{n!} r^n (2x)^n {}_2F_1 \left(\begin{array}{c} -n/2, (1-n)/2 \\ 1-n-\lambda \end{array} ; 1/x^2 \right).$$

Thus,

$$C_n^\lambda(x) = \frac{(\lambda)_n}{n!} (2x)^n {}_2F_1 \left(\begin{array}{c} -n/2, (1-n)/2 \\ 1-n-\lambda \end{array} ; \frac{1}{x^2} \right). \tag{6.4.12}$$

Even though ultraspherical polynomials are special cases of Jacobi polynomials, their great importance compels us to note and prove some of their properties. When $\lambda \to 0$, we get the Chebyshev polynomials of the first kind:

$$\lim_{\lambda \to 0} \frac{n+\lambda}{\lambda} C_n^\lambda(x) = \begin{cases} 1, & n = 0, \\ 2T_n(x), & n = 1, 2, \ldots, \end{cases} \tag{6.4.13}$$

$$\lim_{\lambda \to 0} \frac{C_n^\lambda(x)}{C_n^\lambda(1)} = T_n(x). \tag{6.4.13'}$$

When $\lambda \to \infty$, we have

$$\lim_{\lambda \to \infty} \frac{C_n^\lambda(x)}{C_n^\lambda(1)} = x^n.$$

The Rodrigues formula (2.5.13') takes the form

$$(1-x^2)^{\lambda-1/2} C_n^\lambda(x) = \frac{(-2)^n (\lambda)_n}{n!(n+2\lambda)_n} \frac{d^n}{dx^n} (1-x^2)^{\lambda+n-1/2}, \tag{6.4.14}$$

and the formulas for the derivative and the three-term recurrence relation are

$$\frac{d}{dx} C_n^\lambda(x) = 2\lambda C_{n-1}^{\lambda+1}(x) \tag{6.4.15}$$

and

$$nC_n^\lambda(x) = 2(n+\lambda-1)x C_{n-1}^\lambda(x) - (n+2\lambda-2)C_{n-2}^\lambda(x) \tag{6.4.16}$$

for $n \geq 2$ and $C_0^\lambda(x) = 1$, $C_1^\lambda(x) = 2\lambda x$.

It should also be noted that $u(\theta) = (\sin\theta)^\lambda C_n^\lambda(\cos\theta)$ satisfies the differential equation

$$\frac{d^2u}{d\theta^2} + \left\{ (n+\lambda)^2 + \frac{\lambda(1-\lambda)}{\sin^2\theta} \right\} u = 0. \tag{6.4.17}$$

A proof of (6.4.15) from the generating function for $C_n^\lambda(x)$ is as follows: The derivative of (6.4.10) with respect to x gives

$$\sum_{n=0}^{\infty} \frac{d}{dx} C_n^\lambda(x) r^n = 2\lambda r (1 - 2xr + r^2)^{-(\lambda+1)}$$

$$= 2\lambda r \sum_{n=0}^{\infty} C_n^{\lambda+1}(x) r^n$$

and hence (6.4.15). To get (6.4.16) take the derivative of (6.4.10) with respect to r. The result is

$$2\lambda(x - r)(1 - 2xr + r^2)^{-\lambda} = (1 - 2xr + r^2) \sum_{n=0}^{\infty} n C_n^\lambda(x) r^{n-1}.$$

The left side is $2\lambda(x - r) \sum_{n=0}^{\infty} C_n^\lambda(x) r^n$. Relation (6.4.16) follows on equating the coefficients of r^n on each side.

We now state a property of the relative extrema of ultraspherical polynomials. Let $y_{k,n}^{(\alpha)}$, $k = 1, \ldots, n - 1$ denote the zeros of the derivative of $P_n^{(\alpha,\alpha)}(x)$. Order the zeros so that $y_{k,n}(\alpha) < y_{k-1,n}(\alpha)$ and set $y_{0,n}(\alpha) = 1$, $y_{n,n}(\alpha) = -1$. Define

$$\mu_{k,n}(\alpha) = \left| P_n^{(\alpha,\alpha)}(y_{k,n}(\alpha)) \right| / P_n^{(\alpha,\alpha)}(1), \quad k = 0, 1, \ldots, n. \tag{6.4.18}$$

These numbers satisfy the inequality

$$\mu_{k,n}(\alpha) < \mu_{k,n-1}(\alpha), \quad \alpha > -1/2, \quad k = 1, 2, \ldots, n - 1, \quad n = 1, 2, \ldots, \tag{6.4.19}$$

but the inequality is reversed for $-1 < \alpha < -1/2$. For $\alpha = -1/2$, $\mu_{k,n}(\alpha) = 1$. For $\alpha = 0$, (6.4.19) was observed by Todd [1950] after studying graphs of Legendre polynomials. Todd's conjecture was proved by Szegö [1950]. We prove (6.4.19) with $\alpha = 0$ by an argument that generalizes to give (6.4.19). It is left to the reader to prove the general case.

We begin by stating some necessary results about Jacobi polynomials. The following two identities follow directly from Remark 5.6.3:

$$(n + \alpha + 1) P_n^{(\alpha,\beta)}(x) - (n + 1) P_{n+1}^{(\alpha,\beta)}(x)$$

$$= \frac{(2n + \alpha + \beta + 2)(1 - x)}{2} P_n^{(\alpha+1,\beta)}(x), \tag{6.4.20}$$

$$(2n + \alpha + \beta + 1) P_n^{(\alpha,\beta)}(x)$$

$$= (n + \alpha + \beta + 1) P_n^{(\alpha+1,\beta)}(x) - (n + \beta) P_{n-1}^{(\alpha+1,\beta)}(x). \tag{6.4.21}$$

For the reader's convenience, we again note that

$$\frac{d}{dx} P_n^{(\alpha,\beta)}(x) = \frac{n + \alpha + \beta + 1}{2} P_{n-1}^{(\alpha+1,\beta+1)}(x) \tag{6.4.22}$$

and

$$P_n^{(\alpha,\beta)}(-x) = (-1)^n P_n^{(\beta,\alpha)}(x). \tag{6.4.23}$$

Following convention, we denote the Legendre polynomial $P_n^{(0,0)}(x)$ by $P_n(x)$. By (6.4.20), (6.4.23), and then (6.4.21), we have

$$
\begin{aligned}
[P_n(x)]^2 - [P_{n+1}(x)]^2 &= [P_n(x) - P_{n+1}(x)][P_n(x) + P_{n+1}(x)] \\
&= (1-x)P_n^{(1,0)}(x)(1+x)P_n^{(0,1)}(x) \\
&= (1-x^2)\frac{(n+2)P_n^{(1,1)}(x) - (n+1)P_{n-1}^{(1,1)}(x)}{2(n+1)} \\
&\quad \cdot \frac{(n+2)P_n^{(1,1)}(x) + (n+1)P_{n-1}^{(1,1)}(x)}{2(n+1)} \\
&= \frac{(1-x^2)}{(n+1)^2}\left\{\left[\frac{d}{dx}P_{n+1}(x)\right]^2 - \left[\frac{d}{dx}P_n(x)\right]^2\right\},
\end{aligned}
$$

where the last step follows from (6.4.22).

Thus

$$
\begin{aligned}
f(x) &:= [P_n(x)]^2 + \frac{(1-x^2)}{(n+1)^2}\left[\frac{d}{dx}P_n(x)\right]^2 \\
&= [P_{n+1}(x)]^2 + \frac{(1-x^2)}{(n+1)^2}\left[\frac{d}{dx}P_{n+1}(x)\right]^2.
\end{aligned}
$$

The zeros of $f'(x)$ contain the zeros of $P_n'(x) = P_{n-1}^{(1,1)}(x)/2$ as well as the zeros of $P_{n+1}'(x) = P_n^{(1,1)}(x)/2$. Since $f(x)$ is of degree $2n$, we must have

$$f'(x) = \lambda_n P_n^{(1,1)}(x) P_{n-1}^{(1,1)}(x).$$

By comparing coefficients of highest powers, we see that

$$\lambda_n = 1/(4n+4).$$

Because of the separation of the zeros of $P_n^{(1,1)}(x)$ and $P_{n-1}^{(1,1)}(x)$, they take opposite signs for $y_{k,n} < x < y_{k,n+1}$. Thus $f(x)$ decreases in this interval and (6.4.19) is proved for $\alpha = 0$.

It is also easy to prove that

$$\mu_{k,n} < \mu_{k-1,n}, \quad k = 1, 2, \ldots, \lfloor n/2 \rfloor, \tag{6.4.24}$$

where $\mu_{k,n} = \mu_{k,n}(0)$. Consider the function $f(x)$ defined in Exercise 40 with $\alpha = \beta = 0$. Then

$$f(x) = [P_n(x)]^2 + \frac{(1-x^2)[P_n'(x)]^2}{n(n+1)}. \tag{6.4.25}$$

Compute $f'(x)$ and check that $f(x)$ is increasing for $0 < x \le 1$. This implies (6.4.24).

Szász [1950] proved (6.4.19). His argument also works for the case $-1 < \alpha < -1/2$, though he did not make this observation. Later he proved a similar result for Hermite functions. See Szász [1951]. Exercise 10 contains a statement of his result.

We close this section with the observation that the Hermite and Laguerre polynomials are limits of Jacobi polynomials. There are a number of ways to obtain these limits. One uses a generating function. Observe that

$$\lim_{\lambda \to \infty} \left(1 - 2\frac{xr}{\lambda} + \frac{r^2}{\lambda}\right)^{-\lambda} = e^{2xr - r^2},$$

and we conclude that

$$\lim_{\lambda \to \infty} \lambda^{-n/2} C_n^\lambda(x/\lambda) = H_n(x)/n!. \tag{6.4.26}$$

Another method is to use the hypergeometric representations

$$\lim_{\beta \to \infty} P_n^{(\alpha,\beta)}(1 - 2x/\beta) = \lim_{\beta \to \infty} \frac{(\alpha+1)_n}{n!} \sum_{k=0}^{n} \frac{(-n)_k(n+\alpha+\beta+1)_k}{(\alpha+1)_k k! \beta^k} x^k = L_n^\alpha(x).$$

$$\tag{6.4.27}$$

This means that it is possible to derive the properties of Laguerre and Hermite polynomials from those of Jacobi polynomials. However, it is usually easier to deal with these polynomials directly as we did in Sections 6.1 and 6.2.

6.5 Completeness of Orthogonal Polynomials

The problem of expanding an arbitrary function in terms of orthogonal polynomials was briefly considered in the previous chapter. Here we consider expansion by means of Jacobi, Laguerre, and Hermite polynomials. The latter two are the most interesting as they involve integration over infinite intervals. Just as in the case of Fourier series, the result can be nicely stated for functions in $L^2(a, b)$, that is, the Hilbert space of square integrable functions. We reproduce Hewitt's [1954] proof of the completeness of the Hermite and Laguerre polynomials. This proof depends on the uniqueness of Fourier transforms of integrable functions. For completeness, we give a complex variables proof of the latter result due to Bak and Newman [1982, p. 228].

Theorem 6.5.1   *If f is integrable on $(-\infty, \infty)$ and if*

$$\hat{f}(x) = \int_{-\infty}^{\infty} f(t)e^{ixt}dt \equiv 0,$$

then $f = 0$ almost everywhere.

Proof. Clearly,

$$\int_{-\infty}^{\infty} f(t)e^{ix(t-a)}dt \equiv 0.$$

So, if a is real, then

$$\int_{-\infty}^{a} f(t)e^{ix(t-a)}dt = -\int_{a}^{\infty} f(t)e^{ix(t-a)}dt. \qquad (6.5.1)$$

Define two functions of $z = x + iy$:

$$L(z) = \int_{-\infty}^{a} f(t)e^{iz(t-a)}dt, \quad R(z) = -\int_{a}^{\infty} f(t)e^{iz(t-a)}dt.$$

It is clear that $L(z)$ exists for Im $z \leq 0$ and is analytic in Im $z < 0$. Similarly, $R(z)$ exists for Im $z \geq 0$ and is analytic in Im $z > 0$. Moreover, by the dominated convergence theorem and (6.5.1), we have

$$\lim_{y\to 0} L(x + iy) = \int_{-\infty}^{a} f(t)e^{ix(t-a)}dt = \lim_{y\to 0} R(x + iy).$$

This implies that

$$F(z) = \begin{cases} L(z), & \text{Im } z \leq 0, \\ R(z), & \text{Im } z > 0, \end{cases}$$

is a bounded entire function. By Liouville's theorem, $F(z)$ is a constant. Again, by the dominated convergence theorem,

$$\lim_{y\to\infty} F(iy) = \lim_{y\to\infty} -\int_{a}^{\infty} f(t)e^{-y(t-a)}dt = 0.$$

Thus $F(z) \equiv 0$ and, in particular, $F(0) = 0$. This means that for all real a (since a is arbitrary)

$$\int_{-\infty}^{a} f(t)dt = 0.$$

This implies that $f = 0$ almost everywhere and the theorem is proved.   ∎

Let $p(t)$ denote a square integrable function that dies out exponentially at infinity, that is,

$$p(t) = 0(e^{-\alpha|t|}) \quad \text{for some } \alpha > 0 \text{ as } |t| \to \infty. \qquad (6.5.2)$$

Theorem 6.5.2 *Let* $-\infty \leq a < b \leq \infty$. *Let* $p(t) \in L^2(a, b)$, *with* $p(t)$ *different from zero almost everywhere, and let* $p(t)$ *satisfy* (6.5.2), *if* $a = -\infty$ *or* $b = \infty$. *If* $f \in L^2(a, b)$ *and*

$$\int_{a}^{b} t^n f(t)p(t)dt = 0 \quad \text{for } n = 0, 1, 2, \ldots$$

then $f = 0$ *almost everywhere.*

Proof.   Let $z = x + iy$ and define

$$F(z) = \int_a^b e^{izt} p(t) f(t) dt.$$

If $-\infty < a < b < \infty$, then F is an entire function; otherwise, F is analytic in $-\alpha < y < \alpha$. Therefore,

$$F^{(n)}(z) = i^n \int_a^b e^{izt} t^n p(t) f(t) dt.$$

By hypothesis $F^{(n)}(0) = 0$ for $n = 0, 1, 2, \ldots$. This implies $F(z) \equiv 0$ in $-\alpha < y < \alpha$. In particular,

$$F(x) = \int_a^b e^{ixt} p(t) f(t) dt = 0.$$

Since $p(t) f(t)$ is integrable on (a, b), the uniqueness of the Fourier transform gives $p(t) f(t) = 0$. Since $p(t)$ is different from zero almost everywhere, $f(t) = 0$ almost everywhere. This proves the theorem.   ∎

For the next theorem, let $d\alpha(x)$ denote either $x^\alpha e^{-x} dx$ or $e^{-x^2} dx$. In the former case $(a, b) = (0, \infty)$ and in the latter $(a, b) = (-\infty, \infty)$. Let ϕ_n denote either the nth Laguerre or Hermite polynomial, normalized so that

$$\int_a^b \phi_n^2 d\alpha(x) = 1.$$

For any function f such that

$$\int_a^b f(x)^2 d\alpha(x) < \infty,$$

set

$$c_n = \int_a^b f(x) \phi_n(x) d\alpha(x).$$

Theorem 6.5.3   *Suppose $s_n = \sum_{k=0}^n c_k \phi_k$. Then*

$$\int_a^b f(x)^2 d\alpha(x) = \sum_{n=0}^\infty |c_n|^2$$

and

$$\lim_{n \to \infty} \int_a^b [f(x) - s_n(x)]^2 d\alpha(x) = 0.$$

Proof. For $n > m$, it is clear that

$$\int_a^b \left(\sum_{m+1}^n c_k \phi_k \right)^2 d\alpha(x) = \sum_{m+1}^n |c_k|^2. \tag{6.5.3}$$

Also,

$$\int_a^b [f(x) - s_n(x)]^2 d\alpha(x) \geq 0$$

implies that

$$\sum_{k=0}^n |c_k|^2 \leq \int_a^b f(x)^2 d\alpha(x). \tag{6.5.4}$$

It follows from (6.5.3) and (6.5.4) that $\{s_n(x)\}$ is a Cauchy sequence in $L_\alpha^2(a, b)$. (Here L_α^2 is the set of all square integrable functions with respect to the measure $d\alpha$.) There is therefore a function $g \in L_\alpha^2$ such that

$$\lim_{n \to \infty} \int_a^b [g(x) - s_n(x)]^2 d\alpha(x) = 0. \tag{6.5.5}$$

Now for $n > k$,

$$\int_a^b g(x)\phi_k(x)d\alpha(x) - c_k$$

$$= \int_a^b [g(x) - s_n(x)]\phi_k(x)d\alpha(x) \leq \int_a^b [g(x) - s_n(x)]^2 d\alpha(x),$$

by the Cauchy–Schwartz inequality and the fact that the norm of ϕ_k is 1. Let $n \to \infty$ to get

$$c_k = \int_a^b g(x)\phi_k(x)d\alpha(x).$$

By Theorem 6.5.2, it follows that $f = g$ almost everywhere. By (6.5.5), we arrive at

$$\lim_{n \to \infty} \int_a^b s_n^2(x)d\alpha(x) = \int_a^b f(x)^2 d\alpha(x),$$

so that

$$\sum_{n=0}^\infty |c_n|^2 = \int_a^b f(x)^2 d\alpha(x).$$

This proves the theorem. ∎

Remark   In the last theorem, the series $\sum_0^\infty c_n\phi_n$ converged to f in the L^2 sense. Pointwise convergence can be obtained, for example, by assuming that f is smooth or piecewise smooth. In the latter case the series converges to $\frac{1}{2}[f(x+0) + f(x-0)]$, when x is a point of discontinuity.

We can use Theorem 6.5.2 to prove the following result: Suppose $\{p_n(x)\}$ is a sequence of polynomials orthogonal with respect to the weight function $w(x) = O(e^{-c\sqrt{w}})$ on $(0, \infty)$. Then the sequence $\{p_n(x)\}$ is complete.

We can prove this by taking another sequence $\{q_n(x)\}$ orthogonal with respect to $xw(x)$ on $(0, \infty)$. Define a sequence $\{r_n(x)\}$ by

$$r_{2n}(x) = p_n(x^2),$$

$$r_{2n+1}(x) = xq_n(x^2).$$

This sequence is orthogonal with respect to $|x|w(x^2)$ on $(-\infty, \infty)$. This weight function satisfies the conditions for Theorem 6.5.2. Hence the sequence $\{r_n(x)\}$ is complete, which implies the completeness of $\{r_{2n}(x)\}$ for even polynomials. Thus the result is proved.

Compare this result with the comments after Corollary 5.7.6.

6.6 Asymptotic Behavior of $P_n^{(\alpha,\beta)}(x)$ for Large n

Suppose $f(r) = \sum_{n=0}^\infty a_n r^n$ is an analytic function in a neighborhood of zero with only a finite number of singularities on the circle of convergence. Suppose, for convenience, that the radius of convergence is one. We would like to have an estimate of a_n for large n. Let us see how a_n can be found approximately by knowing the singularities of f. Assume that the singularities are poles that we take to be of order one for simplicity. Let the singular part of f on the unit circle be

$$S = \frac{\alpha_1}{1 - \beta_1 r} + \frac{\alpha_2}{1 - \beta_2 r} + \cdots + \frac{\alpha_k}{1 - \beta_k r}.$$

These functions can be expanded by the binomial theorem, so we know the coefficient of r^n, which we call b_n. Now $f - S = \Sigma(a_n - b_n)r^n$ has a larger radius of convergence than f. This means that

$$(a_n - b_n)r^n = o(1)$$

for some $r > 1$, or

$$a_n - b_n = o(r^{-n}).$$

Since b_n is known, we have an estimate of a_n. This is the idea behind what is known as Darboux's method for finding the asymptotic behavior of a_n. The application to orthogonal polynomials is possible because the generating function has the nth polynomial as a coefficient of r^n. In this case, there is also a need to consider algebraic singularities. Take as an example $(1 - 2xr + r^2)^{-1/2}$, the generating function for the Legendre polynomials. With $x = \cos\theta$, we have

$$f(r) = (1 - 2xr + r^2)^{-1/2} = (1 - re^{i\theta})^{-1/2}(1 - re^{-i\theta})^{-1/2}.$$

The singularities are at $r = e^{i\theta}$ and $r = e^{-i\theta}$. In the neighborhood of $r = e^{i\theta}$, the behavior of $f(r)$ is like

$$g = (1 - e^{2i\theta})^{-1/2}(1 - re^{-i\theta})^{-1/2}.$$

In this case $f(r) - g(r) = h(r)(1 - re^{-i\theta})^{1/2}$, where $h(r)$ is continuous at $e^{i\theta}$. So $f - g$ still has an algebraic singularity, but it is now continuous. Thus it is possible to say something about $a_n - b_n$. The precise result is contained in the theorem below. Before stating it, we consider the case

$$f(r) = (1 - re^{i\theta})^{-\lambda}(1 - re^{-i\theta})^{-\lambda}.$$

If $\lambda > 1$, then subtracting a term like g does not make $f - g$ continuous. More terms have to be subtracted. These can be determined by expanding $(1 - re^{i\theta})^{-\lambda}$ in powers of $(1 - re^{-i\theta})$ and vice versa. We have, about $r = e^{i\theta}$,

$$f(r) = (1 - re^{-i\theta})^{-\lambda}\left[1 - \frac{e^{2i\theta}}{e^{2i\theta} - 1}(1 - re^{-i\theta})\right]^{-\lambda}(1 - e^{2i\theta})^{-\lambda}$$

$$= (1 - re^{-i\theta})^{-\lambda}(1 - e^{2i\theta})^{-\lambda}\left[\sum_{k=0}^{\infty} \frac{(\lambda)_k}{k!}\left(\frac{e^{2i\theta}}{e^{2i\theta} - 1}\right)^k(1 - re^{-i\theta})^k\right].$$

If for an integer n, $n - \lambda > 0$, we can take

$$g = (1 - re^{-i\theta})^{-\lambda}(1 - e^{2i\theta})^{-\lambda}\sum_{k=0}^{n} \frac{(\lambda)_k}{k!}\left(\frac{e^{2i\theta}}{e^{2i\theta} - 1}\right)^k(1 - re^{-i\theta})^k.$$

Theorem 6.6.1 *Let $f(z) = \sum_0^{\infty} a_n z^n$ be analytic in $|z| < r$, $r < \infty$, and have a finite number of singularities on $|z| = r$. Assume that $g(z) = \sum_0^{\infty} b_n z^n$ is also analytic in $|z| < r$ and that $f - g$ is continuous on $|z| = r$. Then $a_n - b_n = o(r^{-n})$ as $n \to \infty$.*

Proof. By Cauchy's theorem and the hypothesis on $f - g$,

$$a_n - b_n = \frac{1}{2\pi i}\int_{|z|=r} \frac{f(z) - g(z)}{z^{n+1}}dz = \frac{1}{2\pi r^n}\int_0^{2\pi}[f(re^{i\theta}) - g(re^{i\theta})]e^{-in\theta}d\theta.$$

The Riemann–Lebesgue lemma for Fourier series implies that the last integral tends to zero as $n \to \infty$. This proves the theorem. ∎

In fact, in the above theorem it is not necessary to assume continuity of $f - g$ on $|z| = r$. The same conclusion can be obtained by assuming that $f - g$ has a finite number of singularities on $|z| = r$ and at each singularity z_j, say,

$$f(z) - g(z) = O((z - z_j)^{\sigma_j - 1}), \qquad z \to z_j,$$

where σ_j is a positive constant. For these refinements and further examples, see Olver [1974, §8.9] or Szegö [1975, §8.4].

The generating function for $P_n^{(\alpha, \beta)}(x)$ is

$$2^{\alpha+\beta}(1 - r + \sqrt{1 - 2rx + r^2})^{-\alpha}(1 + r + \sqrt{1 - 2rx + r^2})^{-\beta}(1 - 2rx + r^2)^{-1/2}.$$

Take $x = \cos\theta$. The above function has singularities at $r = e^{\pm i\theta}$. In the neighborhood of $r = e^{i\theta}$, the generating function behaves as

$$2^{\alpha+\beta}(1 - e^{i\theta})^{-\alpha}(1 + e^{i\theta})^{-\beta}(1 - e^{2i\theta})^{-1/2}(1 - re^{-i\theta})^{-1/2}$$

$$= 2^{\alpha+\beta}(1 - e^{i\theta})^{-\alpha-\frac{1}{2}}(1 + e^{i\theta})^{-\beta-\frac{1}{2}}\sum_0^\infty \frac{(1/2)_n}{n!} e^{-in\theta} r^n.$$

This implies that

$$P_n^{(\alpha,\beta)}(\cos\theta) \sim \left[2^{\alpha+\beta}(1 - e^{i\theta})^{-\alpha-\frac{1}{2}}(1 + e^{i\theta})^{-\beta-\frac{1}{2}}e^{-in\theta}\right.$$

$$\left. + \text{ the conjugate}\right]\frac{(1/2)_n}{n!}.$$

Write

$$A = (1 - e^{i\theta})^{-\alpha-\frac{1}{2}}(1 + e^{i\theta})^{-\beta-\frac{1}{2}} = |A|e^{-i\phi},$$

where

$$|A| = \frac{1}{2^{(\alpha+\beta+1)/2}\sqrt{(1 - x)^{\alpha+\frac{1}{2}}(1 + x)^{\beta+\frac{1}{2}}}}, \qquad x = \cos\theta.$$

Therefore,

$$P_n^{(\alpha,\beta)}(\cos\theta) \sim \frac{(1/2)_n}{n!} 2^{\alpha+\beta+1}|A|\cos(n\theta + \phi).$$

Observe that the denominator of $|A|^2$ is $(1 - x)^\alpha(1 + x)^\beta$, except for the factor $2^{\alpha+\beta+1}(1 - x^2)^{1/2}$. This is exactly the weight function for Jacobi polynomials. This is not a coincidence. To state the theorem in a form due to Nevai [1979, pp. 141–143], suppose

$$P_{n+1}(x) = (A_n x + B_n)P_n(x) - C_n P_{n-1}(x), \qquad n = 0, 1, \ldots,$$

$$P_{-1} = 0, \quad P_0 = 1, \quad \text{and} \quad A_{n-1}A_n C_n > 0 \quad \text{for } n = 1, 2, \ldots.$$

Theorem 6.6.2 *If the series*

$$\sum_{n=0}^{\infty} \left\{ \left| \frac{B_n}{A_n} \right| + \left| \left(\frac{C_{n+1}}{A_n A_{n+1}} \right)^{1/2} - \frac{\delta}{2} \right| \right\} \quad converges,$$

then $d\psi$ can be expressed in the form

$$d\psi(x) = \psi'(x)dx + d\psi_j(x).$$

Here $\psi'(x)$ is continuous and positive in $(-\delta, \delta)$, supp $(\psi') = (-\delta, \delta)$, and $\psi_j(x)$ is a step function constant in $(-\delta, \delta)$. Furthermore, the limiting relation

$$\lim_{n\to\infty} \sup\left\{ \psi'(x)\sqrt{\delta^2 - x^2} P_n^2(x)/h_n \right\} = \frac{2}{\pi}$$

holds almost everywhere in supp($d\psi$).

The singularity in the generating function for Laguerre polynomials is more complex. This makes it difficult to apply Darboux's method. Fejér, however, has shown how this can be done. See Szegö [1975, §§8.2–8.3].

6.7 Integral Representations of Jacobi Polynomials

Integral representations for hypergeometric functions imply the existence of such representations for Jacobi polynomials. A few important and useful integral representations are given in this section.

Recall Bateman's [1909] fractional integral formula:

$$_2F_1\left(\begin{matrix} a, b \\ c + \mu \end{matrix}; x \right) = \frac{\Gamma(c + \mu)x^{1-(c+\mu)}}{\Gamma(c)\Gamma(\mu)} \int_0^x t^{c-1}(x - t)^{\mu-1} {}_2F_1\left(\begin{matrix} a, b \\ c \end{matrix}; t \right) dt,$$

where Re $c > 0$, Re $\mu > 0$, and $|x| < 1$, if the series is infinite. This formula is a particular case of (2.2.4) and we use it to prove the next theorem, which is called the Dirichlet–Mehler formula.

Theorem 6.7.1 *For $0 < \theta < \pi$, the nth Legendre polynomial is given by*

$$P_n(\cos\theta) = \frac{2}{\pi} \int_0^\theta \frac{\cos\left(n + \frac{1}{2}\right)\phi}{(2\cos\phi - 2\cos\theta)^{1/2}} d\phi$$

$$= \frac{2}{\pi} \int_\theta^\pi \frac{\sin\left(n + \frac{1}{2}\right)\phi}{(2\cos\theta - 2\cos\phi)^{1/2}} d\phi.$$

Proof. Take $a = -n, b = n+1, c = \mu = 1/2, x = \sin^2(\theta/2)$, and $t = \sin^2(\phi/2)$ in Bateman's formula. This gives

$$P_n(\cos\theta) = \frac{1}{\pi} \int_0^\theta \frac{{}_2F_1(-n, n + 1; 1/2; \sin^2(\phi/2))\cos(\phi/2)d\phi}{(\sin^2(\theta/2) - \sin^2(\phi/2))^{1/2}}.$$

The $_2F_1$ in the integral is the hypergeometric form of the fourth Chebyshev polynomial given by

$$\frac{\cos(n + \frac{1}{2})\phi}{\cos(\phi/2)}.$$

To get the other form of the integral change θ to $\pi - \theta$ and ϕ to $\pi - \phi$. Then use the fact that $P_n(-x) = (-1)^n P_n(x)$. ∎

One way to use this theorem is to show, as Fejér did, that the sum of Legendre polynomials, $\sum_{k=0}^n P_k(x)$, is greater than or equal to 0 for $0 < x < 1$, since

$$\sum_{k=0}^n \frac{P_k^{(1/2,-1/2)}(\cos\theta)}{P_k^{(-1/2,1/2)}(1)} = \sum_{k=0}^n \frac{\sin(k + \frac{1}{2})\theta}{\sin(\theta/2)} = \left(\frac{\sin(n + 1/2)\theta}{\sin(\theta/2)}\right)^2 \geq 0.$$

For the reference to Fejér and other related results, see Askey [1975, Lecture 3].

Theorem 6.7.2 *For $\mu > 0$, $-1 < x < 1$,*

(a) $$(1 - x)^{\alpha+\mu} \frac{P_n^{(\alpha+\mu,\beta-\mu)}(x)}{P_n^{(\alpha+\mu,\beta-\mu)}(1)}$$

$$= \frac{\Gamma(\alpha + \mu + 1)}{\Gamma(\alpha + 1)\Gamma(\mu)} \int_x^1 (1 - y)^\alpha \frac{P_n^{(\alpha,\beta)}(y)}{P_n^{(\alpha,\beta)}(1)}(y - x)^{\mu-1}dy, \quad \alpha > -1;$$

(b) $$(1 + x)^{\beta+\mu} \frac{P_n^{(\alpha-\mu,\beta+\mu)}(x)}{P_n^{(\beta+\mu,\alpha-\mu)}(1)}$$

$$= \frac{\Gamma(\beta + \mu + 1)}{\Gamma(\beta + 1)\Gamma(\mu)} \int_{-1}^x (1 + y)^\beta \frac{P_n^{(\alpha,\beta)}(y)}{P_n^{(\alpha,\beta)}(1)}(x - y)^{\mu-1}dy, \quad \beta > -1;$$

(c) $$\frac{(1 - x)^{\alpha+\mu}}{(1 + x)^{n+\alpha+1}} \frac{P_n^{(\alpha+\mu,\beta)}(x)}{P_n^{(\alpha+\mu,\beta)}(1)} = \frac{2^\mu \Gamma(\alpha + \mu + 1)}{\Gamma(\alpha + 1)\Gamma(\mu)} \int_x^1 \frac{(1 - y)^\alpha}{(1 + y)^{n+\alpha+\mu+1}} \frac{P_n^{(\alpha,\beta)}(y)}{P_n^{(\alpha,\beta)}(1)}$$

$$\times (y - x)^{\mu-1}dy, \quad \alpha > -1;$$

(d) $$\frac{(1 + x)^{\beta+\mu}}{(1 - x)^{n+\beta+1}} \frac{P_n^{(\alpha,\beta+\mu)}(x)}{P_n^{(\beta+\mu,\alpha)}(1)} = \frac{2^\mu \Gamma(\beta + \mu + 1)}{\Gamma(\beta + 1)\Gamma(\mu)} \int_{-1}^x \frac{(1 + y)^\beta}{(1 - y)^{n+\beta+\mu+1}} \frac{P_n^{(\alpha,\beta)}(y)}{P_n^{(\beta,\alpha)}(1)}$$

$$\times (x - y)^{\mu-1}dy, \quad \beta > -1.$$

Proof. To obtain (a) use the hypergeometric representation of the Jacobi polynomials and apply Bateman's formula with an appropriate change of variables.
(b) now follows from (a). Apply $P_n^{(\alpha,\beta)}(-x) = (-1)^n P_n^{(\beta,\alpha)}(x)$.

Finally, (c) and (d) are derived from (a) and (b) respectively by an application of the Pfaff transformation:

$$_2F_1\left(\begin{matrix} a, b \\ c \end{matrix}; x\right) = (1 - x)^{-a} {}_2F_1\left(\begin{matrix} a, c - b \\ c \end{matrix}; x/(x - 1)\right).$$

Note that when this is applied to Bateman's formula we get

$$x^{\mu+c-1}(1 - x)^{a-c} {}_2F_1\left(\begin{matrix} a, b + \mu \\ c + \mu \end{matrix}; x\right)$$

$$= \frac{\Gamma(c + \mu)}{\Gamma(c)\Gamma(\mu)} \int_0^x (x - t)^{\mu-1}(1 - t)^{a-c-\mu} t^{c-1} {}_2F_1\left(\begin{matrix} a, b \\ c \end{matrix}; t\right) dt.$$

The theorem is proved. We also note that Bateman's formula and the results in the theorem are all particular cases of Theorem 2.9.1. ∎

An important integral of Feldheim [1963] and Vilenkin [1958] can be obtained from (c) by using the quadratic transformation

$$\frac{P_{2n}^{(\alpha,\alpha)}(x)}{P_{2n}^{(\alpha,\alpha)}(1)} = \frac{P_n^{(\alpha,-1/2)}(2x^2 - 1)}{P_n^{(\alpha,-1/2)}(1)}$$

and

$$\frac{P_{2n+1}^{(\alpha,\alpha)}(x)}{P_{2n+1}^{(\alpha,\alpha)}(1)} = \frac{x P_n^{(\alpha,1/2)}(2x^2 - 1)}{P_n^{(\alpha,1/2)}(1)},$$

and taking $\beta = \pm 1/2$. The result is

$$\frac{C_n^\nu(\cos\theta)}{C_n^\nu(1)} \frac{\sin^{2\nu-1}\theta}{\cos^{n+2\lambda+1}\theta}$$

$$= \frac{2\Gamma\left(\nu + \frac{1}{2}\right)}{\Gamma\left(\lambda + \frac{1}{2}\right)\Gamma(\nu - \lambda)} \int_0^\theta \sin^{2\lambda}\phi \frac{[\cos^2\phi - \cos^2\theta]^{\nu-\lambda-1}}{\cos^{n+2\nu}\phi} \frac{C_n^\lambda(\cos\phi)}{C_n^\lambda(1)} d\phi,$$

$$0 < \theta < \frac{\pi}{2}, \quad \nu > \lambda > -\frac{1}{2}.$$

A change of variables gives

Corollary 6.7.3 *For* $\nu > \lambda > -1/2$, $0 \le \theta \le \pi$

$$\frac{C_n^\nu(\cos\theta)}{C_n^\nu(1)} = \frac{2\Gamma\left(\nu + \frac{1}{2}\right)}{\Gamma\left(\lambda + \frac{1}{2}\right)\Gamma(\nu - \lambda)} \int_0^{\pi/2} \sin^{2\lambda}\phi \cos^{2\nu-2\lambda-1}\phi[1 - \sin^2\theta \cos^2\phi]^{n/2}$$

$$\cdot \frac{C_n^\lambda(\cos\theta(1 - \sin^2\theta \cos^2\phi)^{-1/2})}{C_n^\lambda(1)} d\phi.$$

Here

$$\frac{C_n^0(\cos\theta)}{C_n^0(1)} = \lim_{\lambda\to 0}\frac{C_n^\lambda(\cos\theta)}{C_n^\lambda(1)} = \cos n\theta.$$

The final theorem of this section is the Laplace integral representation for ultra-spherical polynomials. It is due to Gegenbauer [1875].

Theorem 6.7.4 *For $\lambda > 0$,*

$$C_n^\lambda(\cos\theta) = \frac{\Gamma(n+2\lambda)}{2^{2\lambda-1}n!(\Gamma(\lambda))^2}\int_0^\pi[\cos\theta + i\sin\theta\cos\phi]^n\sin^{2\lambda-1}\phi\,d\phi.$$

Proof. Recall that

$$C_n^\lambda(\cos\theta) = \sum_{k=0}^n\frac{(\lambda)_k(\lambda)_{n-k}}{k!(n-k)!}e^{i(n-2k)\theta}.$$

Rewrite this using the beta integral. Then

$$C_n^\lambda(\cos\theta)$$

$$= \frac{\Gamma(n+2\lambda)}{n!(\Gamma(\lambda))^2}\sum_{k=0}^n\int_0^1 y^{\lambda+k-1}(1-y)^{\lambda+n-k-1}\binom{n}{k}e^{i(n-2k)\theta}dy$$

$$= \frac{\Gamma(n+2\lambda)}{n!(\Gamma(\lambda))^2}\int_0^1 y^{\lambda-1}(1-y)^{\lambda-1}\sum_{k=0}^n\binom{n}{k}y^k e^{i(n-2k)\theta}(1-y)^{n-k}dy$$

$$= \frac{\Gamma(n+2\lambda)}{n!(\Gamma(\lambda))^2}\int_0^1 y^{\lambda-1}(1-y)^{\lambda-1}[ye^{-i\theta}+(1-y)e^{i\theta}]^n dy.$$

Set $y = \sin^2\psi$ to get

$$C_n^\lambda(\cos\theta) = \frac{2\Gamma(n+2\lambda)}{n!(\Gamma(\lambda))^2}\int_0^{\pi/2}[\cos\theta + i\sin\theta(\cos^2\psi - \sin^2\psi)]^n$$

$$\times \sin^{2\lambda-1}\psi\cos^{2\lambda-1}\psi\,d\psi.$$

Now let $\phi = 2\psi$ to get the result in the theorem. ∎

6.8 Linearization of Products of Orthogonal Polynomials

The addition theorem for cosines implies the formula

$$\cos m\theta\cos n\theta = \tfrac{1}{2}\cos(n+m)\theta + \tfrac{1}{2}\cos(n-m)\theta.$$

In the previous chapter, we noted that this result pertains to Chebyshev polynomials of the first kind, $P^{(-1/2,-1/2)}(x)$, where $x = \cos\theta$. This is called a linearization formula because it gives a product of two polynomials as a linear combination of other polynomials of the same kind. More generally, given a sequence of polynomials $\{p_n(x)\}$ one would like to know something about the coefficients $a(k, m, n)$ in

$$p_m(x)p_n(x) = \sum_{k=0}^{m+n} a(k, m, n)p_k(x). \tag{6.8.1}$$

If the $p_n(x)$ are orthogonal with respect to a distribution $d\alpha(x)$, then

$$a(k, m, n) = \frac{1}{h_k} \int_I p_m(x)p_n(x)p_k(x)d\alpha(x). \tag{6.8.2}$$

Thus the problem of the evaluation of the integral of the product of three orthogonal polynomials of the same kind is equivalent to the linearization problem.

As another example of a linerization formula, recall the identity

$$\frac{\sin(n+1)\theta}{\sin\theta} \frac{\sin(m+1)\theta}{\sin\theta} = \sum_{k=0}^{\min(m,n)} \frac{\sin(n+m+1-2k)\theta}{\sin\theta}.$$

This comes from the addition formula for sines. The addition formula is contained in an important special case of (6.8.1), that is,

$$x^m x^n = x^{m+n}$$

when $x = e^{i\theta}$.

One way of obtaining linearization formulas would be to look for those polynomials for which the integral (6.8.2) can be computed. A simpler integral would involve only the product of two polynomials; but this would yield the orthogonality relation. As we have seen, using the generating function is one way of obtaining orthogonality in some cases. The simplest generating function is for a Hermite polynomial, since it involves only the exponential function, which can be multiplied by itself without resulting in something very complicated. For example, to get orthogonality, note that

$$\int_{-\infty}^{\infty} \sum_{m,n} \frac{H_m(x)H_n(x)}{m!n!} r^m s^n e^{-x^2} dx$$

$$= \int_{-\infty}^{\infty} e^{2xr-r^2+2xs-s^2-x^2} dx$$

$$= \int_{-\infty}^{\infty} e^{-(x-r-s)^2} e^{2rs} dx$$

$$= \sqrt{\pi} \, e^{2rs}.$$

Therefore,

$$\int_{-\infty}^{\infty} H_m(x)H_n(x)e^{-x^2} dx = 2^m \sqrt{\pi} \, m! \delta_{mn}.$$

Similarly, to find the integral of the product of three Hermite polynomials, consider

$$\int_{-\infty}^{\infty} \sum_{\ell,m,n} \frac{H_\ell(x) H_m(x) H_n(x)}{\ell! m! n!} r^\ell s^m t^n e^{-x^2} dx$$

$$= \int_{-\infty}^{\infty} e^{2xr - r^2 + 2xs - s^2 + 2xt - t^2 - x^2} dx$$

$$= \int_{-\infty}^{\infty} e^{-(x-r-s-t)^2} dx \, e^{2(rs+rt+st)}$$

$$= \sqrt{\pi} \sum_{a,b,c} \frac{2^{a+b+c} r^{a+b} t^{b+c} s^{a+c}}{a! b! c!}.$$

This shows that

$$\int_{-\infty}^{\infty} H_\ell(x) H_m(x) H_n(x) e^{-x^2} dx = \frac{2^{(\ell+m+n)/2} \ell! m! n! \sqrt{\pi}}{\left(\frac{\ell+m-n}{2}\right)! \left(\frac{m+n-\ell}{2}\right)! \left(\frac{n+\ell-m}{2}\right)!}, \qquad (6.8.3)$$

when $\ell + m + n$ is even and the sum of any two of ℓ, m, n is not less than the third. In all other cases the integral is zero.

Theorem 6.8.1

$$H_m(x) H_n(x) = \sum_{k=0}^{\min(m,n)} \binom{m}{k} \binom{n}{k} 2^k k! H_{m+n-2k}(x).$$

Proof. This follows from the integral formula (6.8.3) and the orthogonality relation for Hermite polynomials. ∎

Remark 6.8.1 An important feature of the coefficients in the linearization formula is their positivity. This property is shared by the integral (6.8.3).

At this point we have linearization formulas for $P_n^{(1/2,1/2)}(x)$, $P_n^{(-1/2,-1/2)}(x)$, and the Hermite polynomials. Symmetric Jacobi polynomials are ultraspherical or Gegenbauer polynomials and the Hermite polynomials are limiting cases, since

$$\frac{H_n(x)}{n!} = \lim_{\lambda \to \infty} \lambda^{-n/2} C_n^\lambda(x/\sqrt{\lambda}).$$

This suggests the possibility of the existence of a linearization formula for the Gegenbauer polynomials. Of course, there are linearization formulas for $P_n^{(1/2,-1/2)}$ ×(x) and $P_n^{(-1/2,1/2)}(x)$, which are essentially the Chebyshev polynomials of the third and fourth kind, but these are related to $P_n^{(1/2,1/2)}(x)$ and $P_n^{(-1/2,-1/2)}(x)$ by quadratic transformations.

The idea of using the generating function $(1 - 2xr + r^2)^{-\lambda}$ to obtain a linearization formula for Gegenbauer polynomials does not appear very hopeful, because the product of three of these does not seem tractable. One useful idea, due to Bailey [1933], is to express $C_n^\lambda(x)$ as a hypergeometric series and then use the Whipple transformation for a very well poised $_7F_6$ to handle the product $C_m^\lambda(x)C_n^\lambda(x)$. The resulting linearization formula is due to Dougall [1919]. It is interesting to note that a more general result was known to Rogers [1895]. See formula (10.11.10).

Theorem 6.8.2

$$C_m^\lambda(x)C_n^\lambda(x) = \sum_{k=0}^{\min(m,n)} a(k,m,n)C_{m+n-2k}^\lambda(x), \tag{6.8.4}$$

when

$$a(k,m,n) = \frac{(m+n+\lambda-2k)(\lambda)_k(\lambda)_{m-k}(\lambda)_{n-k}(2\lambda)_{m+n-k}}{(m+n+\lambda-k)k!(m-k)!(n-k)!(\lambda)_{m+n-k}}$$

$$\cdot \frac{(m+n-2k)!}{(2\lambda)_{m+n-2k}}.$$

Proof.   Recall that

$$C_n^\lambda(\cos\theta) = e^{in\theta}\sum_{k=0}^{n}\frac{(\lambda)_{n-k}(\lambda)_k}{(n-k)!k!}e^{-2ki\theta}$$

$$= \frac{(\lambda)_n}{n!}e^{in\theta}{}_2F_1\left(\begin{array}{c}-n,\lambda\\1-n-\lambda\end{array};e^{-2i\theta}\right). \tag{6.8.5}$$

Apply Euler's transformation (Theorem 2.2.5) to get

$$C_m^\lambda(\cos\theta) = \frac{(\lambda)_m}{m!}e^{im\theta}(1-e^{-2i\theta})^{1-2\lambda}{}_2F_1\left(\begin{array}{c}1-\lambda,1-2\lambda-m\\1-\lambda-m\end{array};e^{-2i\theta}\right). \tag{6.8.6}$$

Multiply the two equations above to obtain

$$C_m^\lambda(\cos\theta)C_n^\lambda(\cos\theta)$$

$$= \frac{(\lambda)_m(\lambda)_n}{m!n!}e^{i(m+n)\theta}(1-e^{-2i\theta})^{1-2\lambda}$$

$$\cdot\sum_{k\geq 0}\frac{(-n)_k(\lambda)_k}{(1-\lambda-n)_kk!}e^{-2k\theta i}\sum_{j\geq 0}\frac{(1-\lambda)_j(1-2\lambda-m)_j}{j!(1-\lambda-m)_j}e^{-2ji\theta}. \tag{6.8.7}$$

When $s = j+k$, the two sums can be rewritten as

$$\sum_{s\geq 0}\frac{(1-\lambda)_s(1-2\lambda-m)_s}{(1-\lambda-m)_s}e^{-2si\theta}\sum_{k\geq 0}\frac{(-n)_k(\lambda)_k(m+\lambda-s)_k}{k!(1-\lambda-n)_k(\lambda-s)_k(2\lambda+m-s)_k}$$

$$= \sum_{s\geq 0}\frac{(1-\lambda)_s(1-2\lambda-m)_s}{(1-\lambda-m)_s}{}_4F_3\left(\begin{array}{c}-n,\lambda,-s,m+\lambda-s\\1-\lambda-n,\lambda-s,2\lambda+m-s\end{array};1\right)e^{-2si\theta}. \tag{6.8.8}$$

This $_4F_3$ is balanced. Recall Whipple's formula (Theorem 2.4.5), which transforms a balanced $_4F_3$ to a very well poised $_7F_6$:

$$_4F_3\left(\begin{matrix} a+1-b-c, d, e, -s \\ a+1-b, a+1-c, d+e-a-s \end{matrix}; 1\right) = \frac{(a+1-d)_s(a+1-e)_s}{(a+1)_s(a+1-d-e)_s}$$

$$\cdot\, _7F_6\left(\begin{matrix} a, 1+a/2, b, c, d, e, -s \\ a/2, a+1-b, a+1-c, a+1-d, a+1-e, a+1+s \end{matrix}; 1\right),$$

when $s = 0, 1, 2, \ldots$. Take $a = -\lambda - m - n$, $b = -m$, $c = 1 - 2\lambda - m - n + s$, $d = \lambda$, and $e = -n$. Then (6.8.8) is transformed to

$$\sum_s \frac{(1-\lambda)_s(1-2\lambda-m-n)_s e^{-2si\theta}}{s!(1-\lambda-m-n)_s}$$

$$\cdot \sum_k \frac{(-\lambda-m-n)_k(1-(\lambda+m+n)/2)_k(-m)_k}{k!(-(\lambda+m+n)/2)_k(1-\lambda-n)_k}$$

$$\cdot \frac{(1-2\lambda-m-n+s)_k(\lambda)_k(-n)_k(-s)_k}{(\lambda-s)_k(1-2\lambda-m-n)_k(1-\lambda-m)_k(1-\lambda-m-n+s)_k}.$$

Reverse the order of summation, set $s = k + \ell$, and simplify. Put this in (6.8.7) and use (6.8.6) for the inner sum to get an identity that reduces to Dougall's identity (6.8.4). ∎

The limit $\lambda \to 0$ gives the identity for $\cos m\theta \cos n\theta$ and $\lambda = 1$ is the identity for $\sin(m+1)\theta \sin(n+1)\theta$. When $\lambda \to \infty$, Dougall's identity reduces to

$$x^{m+n} = x^m \cdot x^n.$$

The next corollary was given by Ferrers [1877] and Adams [1877].

Corollary 6.8.3   *For Legendre polynomials $P_n(x)$,*

$$P_m(x)P_n(x) = \sum_{k=0}^{\min(m,n)} \frac{2m+2n+1-4k}{2m+2n+1-2k}$$

$$\cdot \frac{(1/2)_k(1/2)_{m-k}(1/2)_{n-k}(m+n-k)!}{k!(m-k)!(n-k)!(1/2)_{m+n-k}} P_{m+n-2k}(x). \quad (6.8.9)$$

Proof.   Take $\lambda = 1/2$ in Dougall's identity. ∎

Remark 6.8.2   The coefficients $a(k, m, n)$ in (6.8.4) are positive for $\lambda > 0$. Moreover, Theorem 6.8.2 implies a terminating form of Clausen's formula. See Exercise 3.17(d) for the statement of Clausen's formula.

Corollary 6.8.4   *For $\lambda > -1/2$ and $\lambda \neq 0$*

$$\int_{-1}^{1} C_{\ell}^{\lambda}(x) C_m^{\lambda}(x) C_n^{\lambda}(x)(1-x^2)^{\lambda-1/2} dx$$

$$= \frac{(\lambda)_{s-\ell}(\lambda)_{s-m}(\lambda)_{s-n} s!}{(s-\ell)!(s-m)!(s-n)!(\lambda)_s} \cdot \frac{2^{1-2\lambda}\pi \Gamma(s+2\lambda)}{[\Gamma(\lambda)]^2 s!(s+\lambda)}, \qquad (6.8.10)$$

when $\ell + m + n = 2s$ is even and the sum of any two of ℓ, m, n is not less than the third. The integral is zero in all other cases.

This is straightforward from Dougall's identity. It contains (6.8.3) as a limiting case.

Integrals involving products of some orthogonal polynomials also have combinatorial interpretations. In Section 6.9, we show how (6.8.3) can be computed combinatorially.

The coefficients $a(k, m, n)$ in (6.8.1) can also be computed in terms of gamma functions when $p_n(x) = P_n^{(\alpha,\beta)}(x)$ and α, β differ by one. This once again covers the cases of the third and fourth Chebyshev polynomials.

Hsü [1938] showed how to use the result in Theorem 5.11.6 to go from (6.8.10) to the corresponding integral of Bessel functions:

$$\int_0^{\infty} J_{\alpha}(at) J_{\alpha}(bt) J_{\alpha}(ct) t^{1-\alpha} dt$$

$$= \begin{cases} 0 & \text{if } a, b, c \text{ are not sides of a triangle,} \\ \dfrac{2^{\alpha-1}\Delta^{2\alpha-1}}{\sqrt{\pi}\Gamma(\alpha+1/2)(abc)^{\alpha}} & \text{if } a, b, c \text{ are sides of a triangle of area } \Delta. \end{cases}$$

This integral was evaluated by a different method in Section 4.11.

It is possible to define a second solution of the differential equation for ultraspherical polynomials that converges to the second solution $Y_{\alpha}(x)$ of the Bessel equation. In fact, there is an analog of the second solution for general orthogonal polynomials $\{p_n(x)\}$. For simplicity, suppose that the orthogonality measure (or distribution) $d\alpha(t)$ has support in a finite interval $[a, b]$ and that, on $[a, b]$,

$$d\alpha(t) = \omega(t)dt,$$

where $\omega(t)$ is continuously differentiable and square integrable. Define the function of the second kind q_n outside $[a, b]$ by

$$q_n(z) = \int_a^b \frac{p_n(t)}{z-t}\omega(t)dt, \qquad z \in \mathbb{C}, z \notin [a, b], \qquad (6.8.11)$$

and on the cut (a, b) by

$$q_n(x) = \lim_{y \to 0^+} \frac{1}{2}(q_n(x+iy) + q_n(x-iy)) = \int_a^b \frac{p_n(t)}{x-t}\omega(t)dt, \qquad (6.8.12)$$

$a < x < b$. Note that, on the cut, q_n is the finite Hilbert transform on (a, b) of the function ωp_n. The ultraspherical function of the second kind, $D_n^\lambda(x)$, is defined by

$$(1 - x^2)^{\lambda - 1/2} D_n^\lambda(x) = \frac{1}{\pi} q_n(x) = \frac{1}{\pi} \int_{-1}^1 \frac{C_n^\lambda(t)}{x - t} (1 - t^2)^{\lambda - 1/2} dt. \qquad (6.8.13)$$

It can be shown that

$$\lim_{n \to \infty} n^{1 - 2\lambda} D_n^\lambda \left(1 - \frac{y}{2n^2}\right) = -\frac{\sqrt{\pi}}{\Gamma(\lambda)} 2^{1/2 - \lambda} y^{1/4 - \lambda/2} Y_{\lambda - 1/2}(\sqrt{y}). \qquad (6.8.14)$$

Askey, Koornwinder, and Rahman [1986] have considered the integral

$$\int_{-1}^1 D_n^\lambda(x) C_m^\lambda(x) C_\ell^\lambda(x) (1 - x^2)^{2\lambda - 1} dx. \qquad (6.8.15)$$

It vanishes if the parity of $\ell + m + n$ is even. It also vanishes when $\ell + m + n$ is odd and there is a triangle with sides ℓ, m, n. In the other cases its value is

$$\frac{[\Gamma(\lambda + 1/2)]^2 (2\lambda)_n (2\lambda)_\ell (2\lambda)_m}{[\Gamma(\lambda + 1)]^2 n! \ell! m!}$$
$$\cdot \frac{(\lambda)_{(n + m + \ell + 1)/2} ((n + m - \ell - 1)/2)! (-\lambda)_{(n - m - \ell + 1)/2} ((n - m - \ell + 1)/2)_\ell}{(2\lambda)_{(n + m + \ell + 1)/2} (\lambda + 1)_{(n + m - \ell - 1)/2} (\lambda)_{(n - m + \ell + 1)/2}}.$$

They also evaluate a more general integral in which $D_n^\lambda(x)$ is replaced by a function of a different order, $D_n^\mu(x)$. The proof of these results used Whipple's $_7F_6$ transformation. Corresponding integrals for Bessel functions are also derived. For these results and for references, the reader should see their paper. This work arose from a special case studied by Din [1981].

For general Jacobi polynomials, $P_n^{(\alpha, \beta)}(x)$, the linearization coefficients can not be found as products. Hylleraas [1962] found a three-term recurrence relation for these coefficients from a differential equation satisfied by the product $P_n^{(\alpha, \beta)}(x) P_m^{(\alpha, \beta)}(x)$. In addition to the case $\alpha = \beta$, Hylleraas showed that when $\alpha = \beta + 1$, the linearization coefficients are products. For many problems, only the nonnegativity of these coefficients is necessary. Gasper used Hylleraas's recurrence relation to determine the values of (α, β) with all of the linearization coefficients nonnegative. For many years, the best representation of these coefficients was as a double sum. Finally, Rahman [1981] showed that these coefficients can be written as a very well poised 2-balanced $_9F_8$. These series satisfy three-term contiguous relations and comprise the most general class of hypergeometric series that satisfy a three-term recurrence relation. Although Rahman's result was unexpected, in retrospect it would have been natural to expect such a result.

If Jacobi polynomials are normalized to be positive at $x = 1$, as they are when $\alpha > -1$, then the linearization coefficients are nonnegative when $\alpha \geq \beta \geq -\frac{1}{2}$, when $\alpha + \beta \geq 0$, $-1 < \beta < 1/2$, and for some $\alpha < -\beta$ when $-1 < \beta < -1/2$. For the first two of these regions, there is a general maximum principle for hyperbolic difference equations that implies nonnegativity. See Szwarc [1992].

6.9 Matching Polynomials

Orthogonal polynomials have connections with some combinatorial objects. These connections have been studied intensively in recent years. Here we shall define matching polynomials of graphs and show their relation to Hermite polynomials. This relationship will then be used to evaluate integrals of products of Hermite polynomials; in particular, it will give their orthogonality. We begin with some definitions from graph theory.

Let G be a graph. We can think of G as an ordered pair (V, E), where V is a set of vertices (points) and E is a set of edges that join pairs of vertices. We denote the number of vertices by $|v|$. A graph G is complete if every pair of vertices is joined by an edge. We denote a complete graph on $m = |v|$ vertices by K_m. The complement of a graph G is another graph $\overline{G}$ that has the same set of vertices as G but those and only those edges of $K_{|v|}$ that are not in G. See Figure 6.1. A k-match on G is a set of k disjoint edges of G. By disjoint we mean that no two edges meet at the same vertex.

Example
The set of edges $\{\{1, 2\}, \{3, 4\}\}$ in Figure 6.2 is a 2-match in G. There is no 3-match in this example.

Let $p(G, k)$ denote the number of k-matches in G. We take $p(G, 0) = 1$ and $p(G, -1) = 0$. It is clear that $p(G, 1)$ is the number of edges in G and that for

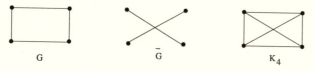

Figure 6.1

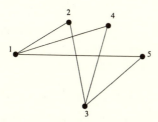

Figure 6.2

$k > [m/2]$ (the greatest integer in $m/2 = |v|/2$), $p(G, k) = 0$. A match that uses every vertex of G is called a complete match and is denoted by $p_m(G)$.

For a graph G, we define the matching polynomial of G, $\alpha(G)$, by

$$\alpha(G) = \alpha(G, x) = \sum_{k=0}^{[m/2]} (-1)^k p(G, k) x^{m-2k}, \qquad (6.9.1)$$

where $m = |v|$.

Theorem 6.9.1 $\alpha(K_m, x) = 2^{-m/2} H_m(x/\sqrt{2})$.

First Proof.   We proved earlier that

$$2^{-m/2} H_m(x/\sqrt{2}) = \sum_{k=0}^{[m/2]} (-1)^k \frac{m!}{k!(m-2k)!} \frac{x^{m-2k}}{2^k}.$$

It is therefore enough to show that

$$p(K_m, k) = \frac{1}{2^k} \frac{m!}{k!(m-2k)!}.$$

We have to find the number of k matchings in a complete graph with m vertices. The number of ways of choosing $2k$ vertices for the k matching from m vertices is $\binom{m}{2k}$. From a given set of $2k$ vertices any particular vertex can be joined to $2k-1$ vertices to give a match. Any vertex of the remaining $2k-2$ unmatched vertices can be matched up with $2k-3$ vertices and so on. This implies that

$$p(K_m, k) = \binom{m}{2k}(2k-1)(2k-3)\cdots 3 \cdot 1$$

$$= \frac{1}{2^k} \frac{m!}{k!(m-2k)!}.$$

This proves the theorem.   ∎

Second Proof.   A more interesting approach is to show that $\alpha(K_m, k)$ satisfies the same recurrence relation and initial conditions as $He_m(x) := 2^{-m/2} H_m(x/\sqrt{2})$. From the recurrence relation (6.1.10) for $H_m(x)$, we find that

$$He_{m+1}(x) = x He_m(x) - m He_{m-1}(x), \qquad He_0(x) = 1, \qquad He_1(x) = x.$$

We first prove that

$$p(K_{m+1}, k) = p(K_m, k) + m p(K_{m-1}, k-1).$$

Take a vertex $v \in K_{m+1}$. This vertex can be a part of a k-match in m ways, and the number of ways to complete each k-match is $p(K_{m-1}, k-1)$. If v is not in a k-match of K_{m+1}, then the number of such k-matches is $p(K_m, k)$. This proves the recurrence relation for p. Now

$$\alpha(K_{m+1}, x) = \sum_{k=0}^{[(m+1)/2]} (-1)^k p(K_{m+1}, k) x^{m+1-2k}$$

$$= \sum_{k=0}^{[(m+1)/2]} (-1)^k p(K_m, k) x^{m+1-2k}$$

$$+ \sum_{k=0}^{[(m+1)/2]} (-1)^k m p(K_{m-1}, k-1) x^{m+1-2k}.$$

If $[(m+1)/2] > [m/2]$ in the first sum, then $p(K_m, [(m+1)/2]) = 0$. So the first sum is

$$x \sum_{k=0}^{[m/2]} (-1)^k p(K_m, k) x^{m-2k} = x\alpha(K_m, x).$$

In the second sum, change k to $k+1$. It becomes

$$\sum_{k=0}^{[(m-1)/2]} (-1)^{k+1} m p(K_{m-1}, k) x^{m-1-2k} = -m\alpha(K_{m-1}, k).$$

This proves the theorem. ■

Our next objective is to give a combinatorial evaluation of the integral

$$I(n_1, n_2, \ldots, n_k) = \int_{-\infty}^{\infty} H_{n_1}(x) H_{n_2}(x) \cdots H_{n_k}(x) e^{-x^2} dx.$$

For the case $k=2$, its evaluation will give the orthogonality of the Hermite polynomials. A related integral is

$$J(n_1, n_2, \ldots, n_k) = \int_{-\infty}^{\infty} He_{n_1}(x) He_{n_2}(x) \cdots He_{n_k}(x) e^{-x^2/2} dx,$$

where $He_m(x) = 2^{-m/2} H_m(x/\sqrt{2})$. A change of variables shows that

$$I(n_1, n_2, \ldots, n_k) = 2^{(n_1 + \cdots + n_k - 1)/2} J(n_1, n_2, \ldots, n_k).$$

It is convenient to shorten the notation by writing $\vec{n} = (n_1, n_2, \ldots, n_k)$. Also, let

$$J_{\vec{n}}^{(i)} = J(n_1, \ldots, n_{i-1}, n_i - 1, n_{i+1}, \ldots, n_k).$$

Here the ith parameter n_i is reduced by 1. Similarly, $J_{\vec{n}}^{(i,j)}$ will mean that the ith and jth parameters are reduced by 1. The next lemma gives a recurrence relation for $J_{\vec{n}}$.

Lemma 6.9.2 $J_{\bar{n}} = \sum_{i=2}^{k} n_i J_{\bar{n}}^{(1,i)}$ and $J_{\bar{0}} = \sqrt{2\pi}$.

Proof. First observe that the Rodrigues-type formula (6.1.3) for Hermite polynomials gives

$$He_m(x) = (-1)^m e^{x^2/2} \frac{d^m}{dx^m} e^{-x^2/2}.$$

Also, (6.1.11) implies

$$H'e_m = \frac{d}{dx} He_m(x) = m He_{m-1}(x).$$

Applying integration by parts, we get

$$J_{\bar{n}} = \int_{-\infty}^{\infty} (-1)^{n_1} \frac{d^{n_1}}{dx^{n_1}} e^{-x^2/2} He_{n_2}(x) \cdots He_{m_k}(x) dx$$

$$= (-1)^{n_1-1} \int_{-\infty}^{\infty} \frac{d^{n_1-1}}{dx^{n_1-1}} e^{-x^2/2} \left\{ \sum_{i=2}^{k} H'e_{n_i}(x) \prod_{\substack{j=2 \\ j\neq i}}^{k} He_{n_j}(x) \right\} dx$$

$$= \sum_{i=2}^{k} n_i \int_{-\infty}^{\infty} He_{n_i-1}(x) He_{n_1-1}(x) \prod_{\substack{j=2 \\ j\neq i}}^{k} He_{n_j}(x) e^{-x^2/2} dx$$

$$= \sum_{i=2}^{k} n_i J_{\bar{n}}^{(1,i)}.$$

Since $J_{\bar{0}}$ is the normal integral,

$$\int_{-\infty}^{\infty} e^{-x^2/2} dx = \sqrt{2\pi},$$

the lemma is proved. ∎

    A combinatorial object that satisfies the same functional relation as in Lemma 6.9.2 is obtained as follows: Let $V_1, V_2, \ldots, V_k$ be a disjoint set of vertices. Let $V = V_1 \cup V_2 \cup \cdots \cup V_k$. Let $|V_i| = n_i$ so that $|V| = \sum_{i=1}^{k} n_i$. Construct a graph G from V by putting an edge between every pair of vertices that does not belong to the same V_i. G is called the complete k-partite graph on $V_1 \cup V_2 \cup \cdots \cup V_k$. Let $P(n_1, n_2, \ldots, n_k) = P_{\bar{n}}$ denote the number of complete matches on G, that is, the number of matches that use all the vertices of G. We set $P_{\bar{0}} = 1$, in accordance with the earlier convention. It is clear that if $\sum_{i=1}^{k} n_i$ is an odd number, then $P_{\bar{n}} = 0$. We also define $P_{\bar{n}}^{(i,j)}$ similarly to $J_{\bar{n}}^{(i,j)}$.

Lemma 6.9.3 $P_{\bar{n}} = \sum_{i=2}^{k} n_i P_{\bar{n}}^{(1,i)}$ and $P_{\bar{0}} = 1$.

Proof. Choose a specific vertex in V_1. This vertex can be matched with any of the n_i vertices in V_i, $i \neq 1$. Once one such match is made, the rest can be completed in $P_{\bar{n}}^{(1,i)}$ ways. This implies that

$$P_{\bar{n}} = \sum_{i=2}^{k} n_i P_{\bar{n}}^{(1,i)},$$

and the proof of the lemma is done. ∎

Since $J_{\bar{n}}$ and $P_{\bar{n}}$ satisfy the same recurrence relation we have the following:

Theorem 6.9.4 $J_{\bar{n}} = \sqrt{2\pi}\, P_{\bar{n}}$, $I_{\bar{n}} = (2^{n_1+n_2+\cdots+n_k}\pi)^{1/2} P_{\bar{n}}$.

Theorem 6.9.5 $P(m, n) = m! \delta_{mn}$.

Proof. In this case $V = V_1 \cup V_2$, $|V_1| = m$, and $|V_2| = n$. If $m \neq n$, then the vertices of V_1 cannot be matched with the vertices of V_2 to give a complete matching. So $P(m, n) = 0$ for $m \neq n$. If $m = n$, then the number of complete matchings is $m!$ and the theorem is proved. ∎

This theorem implies the orthogonality of the Hermite polynomials, for we have the well-known result

$$\int_{-\infty}^{\infty} H_m(x) H_n(x) e^{-x^2} dx = 2^m m! \sqrt{\pi} \delta_{mn}.$$

Now suppose $V = V_1 \cup V_2 \cup V_3$, where V_1, V_2, V_3 have ℓ, m, n elements respectively. If $\ell + m + n$ is odd, or if $\ell > m + n$, then it is easy to see that $P(\ell, m, n) = 0$. The next theorem considers the other situations.

Theorem 6.9.6 *Suppose $\ell + m + n$ is even and $s = (\ell + m + n)/2$. Suppose also that the sum of any two of ℓ, m, n is greater than or equal to the third. Then*

$$P(\ell, m, n) = \frac{\ell! m! n!}{(s - \ell)!(s - m)!(s - n)!}.$$

Proof. Without loss of generality, we assume $m \geq n$. After all vertices in V_1 are matched with vertices in V_2 and V_3, the same number of vertices must be left over in V_2 and V_3 for a complete matching to be possible. This means that there are $m - n$ more matchings of V_1 into V_2 than V_1 into V_3 in a given complete match. So if x denotes the number of V_1, V_2 pairs and y the V_1, V_3 pairs, then $x + y = \ell$ and $x - y = m - n$. Therefore, $x = s - n$ and $y = s - m$. This implies that there are $(s - \ell)$ V_2, V_3 pairs. There are $\binom{\ell}{s-n}$ ways of choosing elements in V_1 to pair with elements in V_2. The remaining elements in V_1 then pair with elements in V_3. Moreover, for any given $2(s - n)$ elements, taking $s - n$ from V_1 and $s - n$ from

V_2, there are $(s - n)!$ ways of doing the pairing. All this means that

$$P(\ell, m, n) = \binom{\ell}{s-n}\binom{m}{s-\ell}\binom{n}{s-m}(s-n)!(s-\ell)!(s-m)!$$

$$= \frac{\ell!m!n!}{(s-\ell)!(s-m)!(s-n)!}.$$

The theorem is proved. ∎

The theorem implies that

$$\int_{-\infty}^{\infty} H_\ell(x)H_m(x)H_n(x)e^{-x^2}dx = 2^{(\ell+m+n)/2}\frac{\ell!m!n!\sqrt{\pi}}{\left(\frac{\ell+m-n}{2}\right)!\left(\frac{m+n-\ell}{2}\right)!\left(\frac{n+\ell-m}{2}\right)!},$$

when $\ell + m + n$ is even and the sum of any two of ℓ, m, n is not smaller than the third. Otherwise the above integral is zero. It is possible to compute $P(k, \ell, m, n)$ as well, but the result is a single series rather than a product. The reader should read the paper of Azor, Gillis, and Victor [1982] for this and other results. For further results on matching polynomials see Godsil [1981].

We now give a different approach to the theorem that $J_{\tilde{n}} = \sqrt{2\pi} P_{\tilde{n}}$. First observe that the matching polynomial of G, $\alpha(G, x)$, can be written as

$$\alpha(G) = \alpha(G, x) = \sum_\alpha (-1)^{|\alpha|}x^{m-2|\alpha|},$$

where α runs through all the matchings of G and $|\alpha| = $ the number of edges in the matching α. Let the disjoint union of two graphs G_1 and G_2 be denoted by $G_1 \cup G_2$. See Figure 6.3.

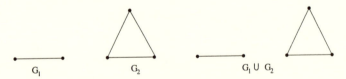

$$\underset{G_1}{\qquad} \qquad \underset{G_2}{\qquad} \qquad \underset{G_1 \cup G_2}{\qquad}$$

Figure 6.3

Lemma 6.9.7 $\alpha(G_1 \cup G_2) = \alpha(G_1)\alpha(G_2).$

Proof.   Suppose G_1 has m vertices and G_2 has n vertices. Then

$$\alpha(G_1)\alpha(G_2) = \left(\sum_\alpha (-1)^{|\alpha|}x^{m-2|\alpha|}\right)\left(\sum_\beta (-1)^{|\beta|}x^{m-2|\beta|}\right)$$

$$= \sum_\gamma (-1)^{|\gamma|}x^{m+n-2|\gamma|}.$$

The last relation follows because every matching γ breaks up uniquely into a matching α of G_1 and β of G_2. The lemma is proved. ∎

Let ϕ be a linear operator on polynomials defined by

$$\phi(x^n) = \frac{1}{\sqrt{2\pi}} \int_{-\infty}^{\infty} x^n e^{-x^2/2} dx.$$

If n is odd, then $\phi(x^n) = 0$. When n is even, say $n = 2m$, then

$$\phi(x^n) = (2m - 1)(2m - 3) \cdots 5 \cdot 3 \cdot 1.$$

We have seen that this is also the number of perfect matchings on K_m. We denote this quantity by $pm(K_n)$.

Let V_i $(i = 1, 2, \ldots, k)$ have n_i vertices and $V = V_1 \cup V_2 \cup \cdots \cup V_k$ be their disjoint union. Let K_V be the complete graph on V. An edge of K_V is called homogeneous if it joins two vertices in the same V_i; otherwise it is inhomogeneous. With this terminology, $P_{\bar{n}}$ is the number of perfect matchings of K_n with no homogeneous edges.

Lemma 6.9.8

$$\frac{1}{\sqrt{2\pi}} J_{\bar{n}} =: L_{\bar{n}} = \sum_{\alpha} (-1)^{|\alpha|} pm(K_{n-2|\alpha|}),$$

where α runs over all matchings of $G = K_{V_1} \cup K_{V_2} \cup \cdots \cup K_{V_k}$. Here $n = \sum_{i=1}^{k} n_i$.

Proof. By the previous lemma and the above remarks

$$L_{\bar{n}} = \phi(He_{n_1}(x) He_{n_2}(x) \cdots He_{n_k}(x))$$

$$= \phi(\alpha(K_{n_1}) \alpha(K_{n_2}) \cdots \alpha(K_{n_k}))$$

$$= \phi(\alpha(G, x)) = \phi\left(\sum_{\alpha} (-1)^{|\alpha|} x^{n-2|\alpha|} \right)$$

$$= \sum_{\alpha} (-1)^{|\alpha|} pm(K_{n-2|\alpha|}).$$

This proves the lemma. ∎

The expression $\sum_{\alpha} (-1)^{|\alpha|} pm(K_{n-2|\alpha|})$ can also be written as

$$\sum_{\alpha_1, \ldots, \alpha_k} (-1)^{|\alpha_1| + \cdots + |\alpha_k|} pm(K_{n-2|\alpha|}),$$

where α_i is a matching in K_{V_i}. Finally, we can rewrite this as

$$\sum_{\alpha_1, \ldots, \alpha_k, \gamma} (-1)^{|\alpha_1| + \cdots + |\alpha_k|},$$

where γ runs through all the complete matchings of $K_{n-2|\alpha|}$ with $|\alpha| = |\alpha_1| + \cdots + |\alpha_k|$. The matchings $\alpha_1, \alpha_2, \ldots, \alpha_k, \gamma$, taken together, give a complete matching of K_V.

To complete the final step, we need one more lemma that uses the concept of a colored complete matching of K_V. For each matching $\alpha_1, \alpha_2, \ldots, \alpha_k, \gamma$, color the edges in each α_i red and the edges in γ blue. Thus all the red edges are homogeneous and the blue edges are either homogeneous or inhomogeneous. The set of all matchings $\alpha_1, \ldots, \alpha_k, \gamma$ in the summation is the set X of all matchings of K_V in which only the homogeneous edges are red. Let $Y \subseteq X$, where Y is the set of all matchings in which there are no red edges and all the blue edges are inhomogeneous. These are the complete matchings in the k-partite subgraph in K_V. If $r(\alpha)$ denotes the number of red edges in α, then by Lemma 6.9.8 we have shown that

$$L_{\bar{n}} = \sum_{\alpha \in X} (-1)^{r(\alpha)}.$$

Also, by definition

$$P_{\bar{n}} = \sum_{\alpha \in Y} 1.$$

The next lemma will complete the proof of $P_{\bar{n}} = L_{\bar{n}}$.

Lemma 6.9.9

$$\sum_{\alpha \in X - Y} (-1)^{r(\alpha)} = 0.$$

Proof. First define an involution θ on $X - Y$. Number the edges of K_V arbitrarily. For any $\alpha \in X - Y$, consider the set of all homogeneous edges of α. This set is nonempty. Consider the smallest edge in this set and change its color from red to blue or from blue to red. This gives a new matching $\alpha' = \theta(\alpha)$ in $X - Y$. Clearly $\theta(\theta(\alpha)) = \alpha$. It is also clear that $(-1)^{r(\alpha)} + (-1)^{r(\theta(\alpha))} = 0$. This proves the lemma and the theorem. $\blacksquare$

The above proof follows DeSainte-Catherine and Viennot [1983]. Also see Viennot [1983].

6.10 The Hypergeometric Orthogonal Polynomials

The hypergeometric representations of the Jacobi, Laguerre, and Hermite polynomials, which we have extensively studied in this chapter, are respectively given by

$$\frac{(\alpha+1)_n}{n!}{}_2F_1\left(\begin{matrix} -n, n+\alpha+\beta+1 \\ \alpha+1 \end{matrix} ; \frac{1-x}{2}\right), \quad \frac{(\alpha+1)_n}{n!}{}_1F_1\left(\begin{matrix} -n \\ \alpha+1 \end{matrix} ; x\right),$$

and

$$(2x)^n {}_2F_0\left(\begin{matrix} -n/2, -(n-1)/2 \\ - \end{matrix} ; -\frac{1}{x^2}\right).$$

In Chapter 3, the Wilson polynomials were introduced. These polynomials can be represented as ${}_4F_3$ hypergeometric functions:

$$\frac{W_n(x^2; a, b, c, d)}{(a+b)_n(a+c)_n(a+d)_n}$$

$$= {}_4F_3\left(\begin{matrix} -n, n+a+b+c+d-1, a+ix, a-ix \\ a+b, a+c, a+d \end{matrix} ; 1\right). \qquad (6.10.1)$$

We saw that Jacobi polynomials are limiting cases of Wilson polynomials and in turn the Laguerre and Hermite polynomials are limits of Jacobi polynomials. A question arises as to whether there are hypergeometric orthogonal polynomials at the ${}_3F_2$ level. In fact there are such polynomials. A few are treated in this section and others are given in the exercises. For a more complete treatment the reader should see Koekoek and Swarttouw [1998].

It is easily seen that

$$\lim_{d\to\infty} \frac{W_n(x^2; a, b, c, d)}{(a+d)_n} = (a+b)_n(a+c)_n {}_3F_2\left(\begin{matrix} -n, a+ix, a-ix \\ a+b, a+c \end{matrix} ; 1\right)$$

$$=: S_n(x^2; a, b, c) \qquad (6.10.2)$$

and

$$\lim_{t\to\infty} \frac{W_n((x+t)^2; a-it, b-it, c+it, d+it)}{(-2)^n n!}$$

$$= i^n \frac{(a+c)_n(a+d)_n}{n!} {}_3F_2\left(\begin{matrix} -n, n+a+b+c+d-1, a+ix \\ a+c, a+d \end{matrix} ; 1\right)$$

$$=: p_n(x; a, b, c, d). \qquad (6.10.3)$$

The polynomials $S_n(x^2; a, b, c)$ and $p_n(x; a, b, c, d)$ are called the continuous dual Hahn and continuous Hahn polynomials respectively. Their orthogonality and recurrence relations can be obtained from those of Wilson polynomials, which were derived in Chapter 3. We restate them here for convenience. When $\mathrm{Re}(a, b, c, d) > 0$ and the nonreal parameters occur in conjugate pairs, the orthogonality is given by

$$\frac{1}{2\pi} \int_0^\infty \left| \frac{\Gamma(a+ix)\Gamma(b+ix)\Gamma(c+ix)\Gamma(d+ix)}{\Gamma(2ix)} \right|^2$$

$$\cdot W_m(x^2; a, b, c, d)W_n(x^2; a, b, c, d)dx$$

$$= (n+a+b+c+d-1)_n n!$$

$$\cdot \frac{\Gamma(n+a+b)\Gamma(n+a+c)\Gamma(n+a+d)\Gamma(n+b+c)\Gamma(n+b+d)\Gamma(n+c+d)}{\Gamma(2n+a+b+c+d)} \delta_{mn}.$$

$$(6.10.4)$$

The recurrence relation is

$$-(a^2 + x^2)\widetilde{W}_n(x^2) = A_n \widetilde{W}_{n+1}(x^2) - (A_n + C_n)\widetilde{W}_n(x^2) + C_n \widetilde{W}_{n-1}(x^2),$$

$$(6.10.5)$$

where

$$\widetilde{W}_n(x^2) = \widetilde{W}_n(x^2, a, b, c, d) = \frac{W_n(x^2; a, b, c, d)}{(a+b)_n(a+c)_n(a+d)_n},$$

$$A_n = \frac{(n+a+b+c+d-1)(n+a+b)(n+a+c)(n+a+d)}{(2n+a+b+c+d-1)(2n+a+b+c+d)},$$

and

$$C_n = \frac{n(n+b+c-1)(n+b+d-1)(n+c+d-1)}{(2n+a+b+c+d-2)(2n+a+b+c+d-1)}.$$

These polynomials also satisfy a difference equation that is a dual of the recurrence relation. This is given by

$$n(n+a+b+c+d-1)y(x) = B(x)y(x+i) - [B(x)+D(x)]y(x)$$
$$+ D(x)y(x-i),$$

$$(6.10.6)$$

where

$$y(x) = W_n(x^2; a, b, c, d),$$

$$B(x) = \frac{(a-ix)(b-ix)(c-ix)(d-ix)}{2ix(2ix-1)},$$

and

$$D(x) = \frac{(a+ix)(b+ix)(c+ix)(d+ix)}{2ix(2ix+1)}.$$

The corresponding results for the continuous dual Hahn polynomials are

$$\frac{1}{2\pi} \int_0^\infty \left| \frac{\Gamma(a+ix)\Gamma(b+ix)\Gamma(c+ix)}{\Gamma(2ix)} \right|^2 S_m(x^2)S_n(x^2)dx$$

$$= \Gamma(n+a+b)\Gamma(n+a+c)\Gamma(n+b+c)n!\delta_{mn}.$$

$$(6.10.7)$$

Here $S_n(x^2) = S_n(x^2; a, b, c)$ and a, b, c are either all positive or one is positive and the other two are complex conjugates with positive real parts,

$$-(a^2 + x^2)\widetilde{S}_n(x^2) = A_n\widetilde{S}_{n+1}(x^2) - (A_n + C_n)\widetilde{S}_n(x^2) + C_n\widetilde{S}_{n-1}(x^2), \quad (6.10.8)$$

where

$$\widetilde{S}_n(x^2) = S_n(x^2)/[(a + b)_n(a + c)_n],$$

$$A_n = (n + a + b)(n + a + c),$$

and

$$C_n = n(n + b + c - 1);$$

$$ny(x) = B(x)y(x + i) - [B(x) + D(x)]y(x) + D(x)y(x - i), \quad (6.10.9)$$

where

$$y(x) = S_n(x^2),$$

$$B(x) = \frac{(a - ix)(b - ix)(c - ix)}{2ix(2ix - 1)},$$

and

$$D(x) = \frac{(a + ix)(b + ix)(c + ix)}{2ix(2ix + 1)}.$$

In the case of the continuous Hahn polynomials, the results are

$$\frac{1}{2\pi} \int_{-\infty}^{\infty} \Gamma(a + ix)\Gamma(b + ix)\Gamma(c - ix)\Gamma(d - ix) p_m(x) p_n(x) dx$$

$$= \frac{\Gamma(n + a + c)\Gamma(n + a + d)\Gamma(n + b + c)\Gamma(n + b + d)}{(2n + a + b + c + d - 1)\Gamma(n + a + b + c + d - 1)} \delta_{mn}, \quad (6.10.10)$$

when $\operatorname{Re}(a, b, c, d) > 0$, $c = \bar{a}$, and $d = \bar{b}$;

$$(a + ix)\tilde{p}_n(x) = A_n\tilde{p}_{n+1}(x) - (A_n + C_n)\tilde{p}_n(x) + C_n\tilde{p}_{n-1}(x), \quad (6.10.11)$$

where

$$\tilde{p}_n(x) = \frac{n!}{i^n(a + c)_n(a + d)_n} p_n(x; a, b, c, d),$$

$$A_n = -\frac{(n + a + b + c + d - 1)(n + a + c)(n + a + d)}{(2n + a + b + c + d - 1)(2n + a + b + c + d)},$$

$$C_n = \frac{n(n + b + c - 1)(n + b + d - 1)}{(2n + a + b + c + d - 2)(2n + a + b + c + d - 1)};$$

and

$$n(n + a + b + c + d - 1)y(x)$$
$$= B(x)y(x + i) - [B(x) + D(x)]y(x) + D(x)y(x - i), \quad (6.10.12)$$

where

$$y(x) = p_n(x; a, b, c, d),$$

$$B(x) = (c - ix)(d - ix),$$

and

$$D(x) = (a + ix)(b + ix).$$

We observed earlier that Jacobi polynomials are limits of Wilson polynomials. In this case, however, the difference equation for the Wilson polynomials becomes the differential equation for Jacobi polynomials. See the exercises for other examples of hypergeometric orthogonal polynomials. For recent developments on some polynomials considered here and their extensions and applications, see Nevai [1990].

6.11 An Extension of the Ultraspherical Polynomials

The generating function of the ultraspherical polynomials is given by the product of $(1 - re^{i\theta})^{-\lambda}$ and its conjugate. More generally, Fejér [1925] studied a sequence of polynomials defined as follows.

Let $f(z) = \sum_{n=0}^{\infty} a_n z^n$ be a function that is analytic in a neighborhood of $z = 0$, with real coefficients. The generalized Legendre polynomials or the Legendre–Fejér polynomials are defined by

$$|f(re^{i\theta})|^2 = \sum_{n=0}^{\infty} r^n \sum_{k=0}^{n} a_k a_{n-k} e^{i(n-2k)\theta}$$

$$= \sum_{n=0}^{\infty} r^n \sum_{k=0}^{n} a_k a_{n-k} \cos(n - 2k)\theta$$

$$:= \sum_{n=0}^{\infty} p_n(\cos\theta) r^n. \quad (6.11.1)$$

Feldheim [1941a] and Lanzewizky [1941] independently asked whether the $p_n(\cos\theta)$ give rise to orthogonal polynomials other than the Gegenbauer polynomials. We know that if the $p_n(x)$ are orthogonal with respect to some positive measure, then they must satisfy

$$x p_n(x) = A_n p_{n+1}(x) + B_n p_n(x) + C_n p_{n-1}(x), \quad n = 0, 1, 2, \ldots \quad (6.11.2)$$

with $A_n C_{n+1} > 0$ and A, B_n, C_{n+1} real. We can normalize to take $p_{-1}(x) = 0$, $p_0(x) = 1$. The converse is also true, although we did not prove it. So, to find polynomials $p_n(x)$ that are orthogonal, it is enough to derive those that satisfy the three-term recurrence relation.

Note that if

$$p_n(\cos\theta) = \sum_{k=0}^{n} a_k a_{n-k} \cos(n - 2k)\theta,$$

then by $\theta \to \theta + \pi$ we obtain

$$p_n(-\cos\theta) = (-1)^n p_n(\cos\theta).$$

Therefore, if $p_n(x)$ satisfies (6.11.2), it must, in fact, satisfy

$$2x p_n(x) = A_n p_{n+1}(x) + C_n p_{n-1}(x),$$

with A_n, C_n real, $A_n C_{n+1} > 0$, $n = 0, 1, 2, \ldots$. This implies that

$$2\cos\theta \sum_{k=0}^{n} a_k a_{n-k} \cos(n - 2k)\theta = A_n \sum_{k=0}^{n+1} a_k a_{n+1-k} \cos(n + 1 - 2k)\theta$$

$$+ C_n \sum_{k=0}^{n-1} a_k a_{n-1-k} \cos(n - 1 - 2k)\theta.$$

$$(6.11.3)$$

Now use the trigonometric identity

$$2\cos\theta \cos(n - 2k)\theta = \cos(n + 1 - 2k)\theta + \cos(n - 1 - 2k)\theta$$

to write the left side of (6.11.3) as

$$\sum_{k=0}^{n} a_k a_{n-k} \cos(n + 1 - 2k)\theta + \sum_{k=0}^{n} a_k a_{n-k} \cos(n - 1 - 2k)\theta.$$

Substitute this in (6.11.3) and equate the coefficient of $\cos(n + 1)\theta$ to get

$$A_n a_0 a_{n+1} = a_0 a_n$$

or

$$A_n = \frac{a_n}{a_{n+1}}.$$

The coefficient of $\cos(n - 1 - 2k)\theta$ gives

$$C_n = \frac{a_{k+1}}{a_k} + \frac{a_{n-k}}{a_{n-k-1}} - \frac{a_n}{a_{n+1}} \frac{a_{k+1}}{a_k} \frac{a_{n-k}}{a_{n-k-1}}.$$

Take $k = 0$ and 1 to obtain an equation for the variables a. To simplify this equation, set $s_n = a_n/a_{n-1}$. We obtain the nonlinear difference equation

$$s_1 + s_n - \frac{s_1 s_n}{s_{n+1}} = s_2 + s_{n-1} - \frac{s_2 s_{n-1}}{s_{n+1}},$$

or

$$s_{n+1}(s_n - s_{n-1} + s_1 - s_2) = s_1 s_n - s_2 s_{n-1}.$$

For further simplification, set $s_n = t_n + s_1$. The equation becomes

$$t_{n+1}(t_n - t_{n-1} - t_2) = -t_2 t_{n-1}, \quad t_1 = 0.$$

Write $t_n = t_2 u_n$; then

$$u_{n+1}(u_n - u_{n-1} - 1) = -u_{n-1}, \quad u_1 = 0.$$

For linear difference equations we get polynomial solutions $\sum A_n q^n$. Such a solution is not possible here and as there is no general method for solving nonlinear equations, we try the simplest rational expression as a possible solution, keeping in mind that $u_1 = 0$. Set

$$u_n = \frac{A(1 - q^{n-1})}{1 - Bq^n},$$

with $|q| \le 1$, for $u_n(q, A, B) = u_n(q^{-1}, A/(Bq), B^{-1})$. Then

$$\frac{A(1 - q^n)}{1 - Bq^{n+1}} \left[\frac{A(1 - q^{n-1}) - (1 - Bq^n)}{1 - Bq^n} \right]$$

$$= \left[\frac{A(1 - q^n) - (1 - Bq^{n+1})}{1 - Bq^{n+1}} \right] \frac{A(1 - q^{n-2})}{1 - Bq^{n-1}}.$$

For this to be true, we must have $B = 1$ and

$$(1 - q^{n-1})(A - 1 - (A - q)q^{n-1}) = (1 - q^{n-2})(A - 1 - (A - q)q^n).$$

This is identically true for $A - 1 = q$. Thus

$$u_n = \frac{(1 + q)(1 - q^{n-1})}{1 - q^n}. \qquad (6.11.4)$$

So

$$s_n = \frac{(1 + q)(1 - q^{n-1})(s_2 - s_1)}{1 - q^n} + s_1,$$

which shows that for some α and β

$$s_n = \frac{\alpha(1 - \beta q^{n-1})}{1 - q^n}.$$

This gives

$$A_n = \frac{1 - q^{n+1}}{\alpha(1 - \beta q^n)} \tag{6.11.5}$$

and after some simplification

$$C_n = \frac{\alpha(1 - \beta^2 q^{n-1})}{1 - \beta q^n}. \tag{6.11.6}$$

The recurrence relation for $p_n(x)$ is given by

$$2x\alpha(1 - \beta q^n)p_n = (1 - q^{n+1})p_{n+1} + \alpha^2(1 - \beta^2 q^{n-1})p_{n-1},$$

with

$$\frac{(1 - q^{n+1})(1 - \beta^2 q^n)}{(1 - \beta q^n)(1 - \beta q^{n+1})} > 0, \quad n = 0, 1, 2, \dots.$$

We have taken $|q| \le 1$. If $q = 1$, then the value of u_n in (6.11.4) is defined by the limit as $q \to 1$. This case also gives rise to orthogonal polynomials. For example, if $\beta = q^\lambda$ in (6.11.5) and (6.11.6), then $A_n = (n + 1)/(n + \lambda)$, $C_n = (n + 2\lambda - 1)/(n + \lambda)$ (with $\alpha = 1$), and the recurrence relation for ultraspherical polynomials is obtained. Clearly, there are other cases where division by zero may be involved in (6.11.5) and (6.11.6). These do not lead to orthogonal polynomials of all degrees unless q is a root of unity. Consider the situation where this problem does not arise and let us see what polynomials we get.

We need an expression for a_n. We have

$$\frac{a_n}{a_0} = \frac{1}{A_{n-1}A_{n-2}\cdots A_0} = \alpha^n \frac{(1 - \beta)(1 - \beta q)\cdots(1 - \beta q^{n-1})}{(1 - q)(1 - q^2)\cdots(1 - q^n)}.$$

Therefore,

$$f(re^{i\theta}) = a_0 \sum_0^\infty \frac{(1 - \beta)(1 - \beta q)\cdots(1 - \beta q^{n-1})}{(1 - q)(1 - q^2)\cdots(1 - q^n)}\alpha^n r^n e^{in\theta},$$

which suggests that we take $a_0 = \alpha = 1$. In this case, the polynomial is

$$p_n(\cos\theta) = \sum_{k=0}^n \frac{(1 - \beta)\cdots(1 - \beta q^{k-1})(1 - \beta)\cdots(1 - \beta q^{n-k-1})}{(1 - q)\cdots(1 - q^k)(1 - q)\cdots(1 - q^{n-k})} \cos(n - 2k)\theta.$$

This expression may appear a little strange at this point. As pointed out before, taking $\beta = q^\lambda$ and letting $q \to 1$ gives the ultraspherical polynomials. One should keep this procedure in mind. In Chapter 10, we give an introduction to objects of this kind. By the methods developed there, it can be shown that the generating function is given by

$$|f(re^{i\theta})|^2 = \prod_{n=0}^{\infty} \frac{(1 - \beta re^{i\theta}q^n)(1 - \beta re^{-i\theta}q^n)}{(1 - re^{i\theta}q^n)(1 - re^{-i\theta}q^n)}.$$

This may appear like a much more complicated expression than the generating function for the ultraspherical polynomials that it is supposed to extend. But there is one sense in which it is simpler. Recall that the singularities of the generating function can be used to get information about the asymptotic behavior of the polynomials and the weight function (see Theorem 6.6.2). The generating function for the ultraspherical polynomials has algebraic singularities, whereas the singularities here are simple poles. These are easier to deal with. The poles closest to the origin are at $r = e^{i\theta}$ and $r = e^{-i\theta}$. Near $r = e^{i\theta}$, the generating function behaves like

$$\prod_{n=0}^{\infty} \frac{(1 - \beta e^{2i\theta}q^n)(1 - \beta q^n)}{(1 - e^{2i\theta}q^n)(1 - q^{n+1})} \cdot \frac{1}{(1 - re^{-i\theta})}.$$

So,

$$p_n(\cos\theta) \approx \prod_{n=0}^{\infty} \frac{(1 - \beta e^{2i\theta}q^n)(1 - \beta q^n)}{(1 - e^{2i\theta}q^n)(1 - q^{n+1})} e^{-in\theta} + \text{conjugate}, \quad n \to \infty.$$

Write the infinite product as $Re^{i\phi}$. Then

$$p_n(\cos\theta) \approx \prod_{n=0}^{\infty} \frac{(1 - \beta q^n)}{(1 - q^{n+1})} \left[\prod_{n=0}^{\infty} \frac{(1 - \beta e^{2i\theta}q^n)(1 - \beta e^{-2i\theta}q^n)}{(1 - e^{2i\theta}q^n)(1 - e^{-2i\theta}q^n)} \right]^{1/2} 2\cos(n\theta - \phi),$$

as $n \to \infty$.

By Theorem 6.6.2, we expect the weight function to be

$$\omega_p(\cos\theta) = \prod_{n=0}^{\infty} \frac{(1 - e^{2i\theta}q^n)(1 - e^{-2i\theta}q^n)}{(1 - \beta e^{2i\theta}q^n)(1 - \beta e^{-2i\theta}q^n)}$$

$$= \prod_{n=0}^{\infty} \frac{(1 - 2\cos 2\theta q^n + q^{2n})}{(1 - 2\beta \cos 2\theta q^n + \beta^2 q^{2n})}.$$

These infinite products are well known in the theory of elliptic functions. In Chapter 10, we hope to convince the reader that they are quite natural and tractable, unwieldy though they may appear now.

Exercises

1. Evaluate the integral

$$I = \int_{-\infty}^{\infty} e^{-x^2} e^{2ixt} dx = 2 \int_0^{\infty} e^{-x^2} \cos 2xt dx$$

(a) by contour integration, (b) by expanding $\cos 2xt$ in powers of x and integrating term by term, and (c) by showing that the integral satisfies the differential equation

$$\frac{dI}{dt} = -2t I.$$

2. Prove that $F(x, r) := e^{2xr - r^2} = \sum_{n=0}^{\infty} \frac{H_n(x)}{n!} r^n$ by showing that

$$\left[\frac{\partial^n F}{\partial r^n} \right]_{r=0} = H_n(x).$$

3. Prove that

$$\lim_{\alpha \to \infty} \left(\frac{2}{\alpha} \right)^{n/2} L_n^\alpha(\alpha + \sqrt{2\alpha}x) = (-1)^n \frac{H_n(x)}{n!}.$$

Hint: You can use generating functions, recurrence relations, or Rodrigues's formula.

4. Prove that $u_n = e^{-x^2/2} H_n(x)$ satisfies the equation

$$u_n'' + (2n + 1 - x^2) u_n = 0.$$

Deduce that

$$\frac{d}{dx} \left(u_n' u_m - u_m' u_n \right) + 2(n - m) u_m u_n = 0.$$

Hence prove the orthogonality of Hermite polynomials, that is,

$$\int_{-\infty}^{\infty} u_m u_n dx = 0 \quad \text{for } m \neq n.$$

5. Use the generating function for Hermite polynomials in Exercise 2 to prove that

$$H_n(x \cos u + y \sin u) = n! \sum_{k=0}^{n} \frac{H_k(x) H_{n-k}(y)}{k!(n-k)!} \cos^k u \sin^{n-k} u.$$

6. Let n be a nonnegative integer. Show that

$$x^{2n} = \frac{(2n)!}{2^{2n}} \sum_{k=0}^{n} \frac{H_{2k}(x)}{(2k)!(n-k)!}$$

and

$$x^{2n+1} = \frac{(2n+1)!}{2^{2n+1}} \sum_{k=0}^{n} \frac{H_{2k+1}(x)}{(2k+1)!(n-k)!}.$$

7. Define

$$\mathrm{sgn}\, x = \begin{cases} 1, & x > 0, \\ -1, & x < 0. \end{cases}$$

Show that

$$\mathrm{sgn}\, x = \frac{1}{\sqrt{\pi}} \sum_{n=0}^{\infty} \frac{(-1)^n}{2^{2n}(2n+1)n!} H_{2n+1}(x).$$

8. Use the generating function (6.2.4) for Laguerre polynomials to prove that

$$\sum_{n=0}^{\infty} \frac{H_n(x)}{\lfloor n/2 \rfloor!} r^n = (1 + 4r^2)^{-3/2}(1 + 2xr + 4r^2)e^{4x^2r^2/(1+4r^2)}.$$

9. Obtain the generating function (6.2.4) from the integral representation (6.2.15) of Laguerre polynomials.

10. Let $\phi_n(x) = e^{-x^2/2} H_n(x)/\sqrt{2^n n!}$, $n = 0, 1, 2, \ldots$. Denote the relative maxima of $|\phi_n(x)|$, as x decreases from $+\infty$ to 0, by $\mu_{0,n}, \mu_{1,n}, \mu_{2,n}, \ldots$. Prove that

$$\mu_{r,n} > \mu_{r,n+1}, \quad n \geq r \geq 0.$$

Deduce that $|\phi_n(x)| \leq \max |\phi_0(x)| = 1$. (See Szász [1951].)

11. Show that the Fourier transform of $u_n(x) = e^{-x^2/2} H_n(x)$ is $i^n u_n(x)$ by filling in and completing the following steps:

$$\int_{-\infty}^{\infty} u_n(x)e^{ixy}dx = \int_{-\infty}^{\infty} e^{-x^2} \frac{d^n}{dx^n} e^{ixy+x^2/2}dx$$

$$= (-i)^n e^{y^2/2} \int_{-\infty}^{\infty} e^{-x^2} \frac{d^n}{dy^n} e^{(x+iy)^2/2}dx$$

$$= (i)^n e^{y^2/2} \frac{d^n}{dy^n} \int_{-\infty}^{\infty} e^{-(x^2/2)+ixy-y^2/2}dx$$

$$= i^n \sqrt{2\pi}\, u_n(y).$$

12. Let $\psi_n(x) = \frac{1}{(2^n \cdot 1/2 n!)^{1/2}} H_n(\sqrt{2\pi}x)e^{-\pi x^2}$. Suppose f is square integrable on $(-\infty, \infty)$ and g is its Fourier transform. Let

$$f(x) \sim \sum_{n=0}^{\infty} a_n \psi_n(x), \quad g(x) \sim \sum_{n=0}^{\infty} b_n \psi_n(x),$$

$$xf(x) \sim \sum_{n=0}^{\infty} c_n \psi_n(x), \quad xg(x) \sim \sum_{n=0}^{\infty} d_n \psi_n(x).$$

(a) Show that $a_n = i^n b_n$.

(b) Use the recurrence relation for Hermite polynomials to obtain

$$\sqrt{4\pi} x\, \psi_n(x) = \sqrt{n+1}\, \psi_{n+1}(x) + \sqrt{n}\, \psi_{n-1}(x).$$

(c) Use (a) and (b) to show that

$$\sqrt{4\pi}\, c_n = \sqrt{n+1}\, a_n + \sqrt{n}\, a_{n-1},$$

$$\sqrt{4\pi}\, d_n = i^{-n-1}[\sqrt{n+1}\, a_n - \sqrt{n}\, a_{n-1}].$$

(d) Deduce that

$$\int_{-\infty}^{\infty} x^2 |f(x)|^2 dx + \int_{-\infty}^{\infty} x^2 |g(x)|^2 dx \geq \frac{1}{2\pi} \int_{-\infty}^{\infty} |f(x)|^2 dx$$

with equality only if $f(x)$ is almost everywhere equal to a constant multiple of $\exp(-\pi x^2)$.

(e) Rescale to show that (d) implies that (for $p > 0$)

$$p^2 \int_{-\infty}^{\infty} x^2 |f(x)|^2 dx + p^{-2} \int_{-\infty}^{\infty} x^2 |g(x)|^2 dx \geq \frac{1}{2\pi} \int_{-\infty}^{\infty} |f(x)|^2 dx.$$

(f) Show that (e) implies Heisenberg's inequality:

$$\left[\int_{-\infty}^{\infty} x^2 |f(x)|^2 dx \right]^{1/2} \left[\int_{-\infty}^{\infty} x^2 |g(x)|^2 dx \right]^{1/2} \geq \frac{1}{4\pi} \int_{-\infty}^{\infty} |f(x)|^2 dx.$$

(See de Bruin [1967].)

13. Show that

$$\int_0^{\infty} e^{-st} t^{\alpha} L_n^{\alpha}(t) dt = \frac{\Gamma(n+\alpha+1)}{n!} \frac{(s-1)^n}{s^{\alpha+n+1}}.$$

14. Show that

$$L_n^{\alpha}(x) = \frac{(-1)^n \Gamma(n+\alpha+1)}{\sqrt{\pi}\, \Gamma(\alpha+1/2)(2n)!} \int_{-1}^{1} (1-t^2)^{\alpha-1/2} H_{2n}(\sqrt{xt}) dt, \quad \alpha > -1/2.$$

15. Prove that

$$L_n(x^2 + y^2) = \frac{(-1)^n}{2^{2n}} \sum_{k=0}^{n} \frac{H_{2k}(x) H_{2n-2k}(y)}{k!(n-k)!}.$$

16. Prove that

$$L_n^{\alpha+\beta+1}(x+y) = \sum_{k=0}^{n} L_k^{\alpha}(x) L_{n-k}^{\beta}(y).$$

17. For $\mathrm{Re}(\alpha+1, \beta) > 0$, prove that

$$\int_0^1 t^{\alpha} (1-t)^{\beta-1} L_n^{\alpha}(xt) dt = \frac{\Gamma(\beta)\Gamma(\alpha+n+1)}{\Gamma(\beta+\alpha+n+1)} L_n^{\alpha+\beta}(x)$$

and

$$\frac{\Gamma(\alpha+1)\Gamma(\alpha+\beta+2)}{\Gamma(\beta+1)} \int_0^1 t^{\alpha} \frac{L_n^{\alpha}(xt)}{L_n^{\alpha}(0)} (1-t)^{\beta} \frac{L_m^{\beta}(x(1-t))}{L_m^{\beta}(0)} dt = \frac{L_{m+n}^{\alpha+\beta+1}(x)}{L_{m+n}^{\alpha+\beta+1}(0)}.$$

18. Prove

(a)
$$\sum_{k=0}^{n} \binom{n}{k} \frac{L_k^\alpha(x)}{L_k^\alpha(0)} y^{n-k} = (y+1)^n \frac{L_n^\alpha(x/(y+1))}{L_n^\alpha(0)}.$$

(b)
$$\sum_{k=0}^{n} \binom{n}{k} H_k(x)(xy)^{n-k} = H_n(x+y).$$

19. Prove the identity

$$\sum_{k=0}^{n} \binom{n}{k} \frac{C_k^\lambda(x)}{C_k^\lambda(1)} y^k = (1 + 2xy + y^2)^{n/2} C_n^\lambda \left(\frac{1+xy}{(1+2xy+y^2)^{1/2}} \right) \Big/ C_n^\lambda(1).$$

20. Show that for $\alpha > -1, r > 0$, and $x > 0$,

$$J_\alpha(2\sqrt{rx}) = (rx)^{\alpha/2} e^{-r} \sum_{n=0}^{\infty} \frac{r^n}{\Gamma(n+\alpha+1)} L_n^\alpha(x).$$

21. Prove that for Legendre polynomials $P_n(x)$,

$$\sum_{n=0}^{\infty} \frac{P_n(x)}{n!} r^n = e^{xr} J_0(\sqrt{1-x^2}r).$$

More generally,

$$\sum_{n=0}^{\infty} \frac{C_n^\lambda(x)}{C_n^\lambda(1)} \frac{r^n}{n!} = 2^{\lambda-1/2} \Gamma(\lambda+1/2) e^{xr} (\sqrt{1-x^2}r)^{-\lambda+1/2} \cdot J_\lambda(\sqrt{1-x^2}r),$$

$$\lambda > -1/2.$$

22. Suppose $P_n(x)$ is the Legendre polynomial of degree n. Then Turán's inequality states that

$$[P_n(x)]^2 - P_{n-1}(x)P_{n+1}(x) \geq 0, \quad n \geq 1, \quad -1 \leq x \leq 1.$$

This exercise sketches a proof of Turán's inequality. See Szegö [1948] on which this is based.

(a) Show that if the polynomial

$$S_n(y) = u_0 + \binom{n}{1} u_1 y + \binom{n}{2} u_2 y^2 + \cdots + \binom{n}{n} u_n y^n$$

has all real roots, then

$$u_{n-1}^2 - u_n u_{n-2} \geq 0.$$

(b) The following is a result from entire function theory: Suppose

$$f(y) = \lim_{n \to \infty} S_n(y/n) = \sum_{n=0}^{\infty} \frac{u_n}{n!} y^n$$

is an entire function with the factorization

$$f(y) = e^{-\alpha y^2 + \beta y} \prod_{n=0}^{\infty} (1 + \beta_n y) e^{-\beta_n y},$$

where $\alpha \geq 0$, β and β_n are real, and $\sum_{n=0}^{\infty} \beta_n^2$ is convergent. Then $S_n(y)$ has all real roots.

To obtain Turán's inequality, use Exercise 21 and (4.14.3).

23. (a) Use Exercise 19 to show that $S_n(y)$ in Exercise 22(a) with

$$u_k = P_k(x)$$

has all real roots and thus obtain another proof of Turán's inequality.

(b) Extend Turán's inequality to the polynomials $H_n(x)$, $L_n^\alpha(x)$, and $C_n^\lambda(x)$. Prove these inequalities by two different methods.

24. For the Legendre polynomial $P_n(x)$, prove the following results:

(a)
$$\int_{-1}^{1} P_n(x) e^{-itx} dx = i^{-n} \sqrt{\frac{2\pi}{t}} J_{n+1/2}(t).$$

(b)
$$\int_{-\infty}^{\infty} t^{-1/2} J_{n+1/2}(t) e^{itx} dt = \begin{cases} \sqrt{2\pi} i^n P_n(x) & \text{if } -1 < x < 1, \\ 0 & \text{if } x > 1 \text{ or } x < -1. \end{cases}$$

(c) For a nonnegative integer k,

$$\int_{-1}^{1} P_n(x) x^{n+2k} dx = \frac{(2k+1)_n}{2^n (k+1/2)_{n+1}}.$$

25. Suppose $\alpha, \beta > -1$. Show that

$$(1 - x^2) \frac{d^2 y}{dx^2} + [\beta - \alpha - (\alpha + \beta + 2)x] \frac{dy}{dx} + \lambda y = 0$$

has a nontrivial polynomial solution if and only if λ has the form $n(n + \alpha + \beta + 1)$, where n is a nonnegative integer. This solution is $C P_n^{(\alpha, \beta)}(x)$, where C is a constant.

26. Prove the following results for ultraspherical polynomials:

(a)
$$\lim_{x \to \infty} x^{-n} C_n^\lambda(x) = 2^n \frac{(\lambda)_n}{n!}.$$

(b)
$$\sum_{n=0}^{\infty} \frac{(\lambda + 1/2)_n}{(2\lambda)_n} C_n^\lambda(x) r^n = 2^{\lambda - 1/2} R^{-1} \{1 - xr + R\}^{-\lambda + 1/2},$$

where

$$R = (1 - 2xr + r^2)^{1/2}.$$

(c)
$$\sum_{k=0}^{n} (k + \lambda) C_k^\lambda(x) = \frac{(n + 2\lambda) C_n^\lambda(x) - (n + 1) C_{n+1}^\lambda(x)}{2(1 - x)}.$$

27. Use Rodrigues's formula to prove

(a) $P_n^{(\alpha,\beta)}(x) = \dfrac{1}{2\pi i} \displaystyle\int_{(x+)} \left(\dfrac{t^2-1}{2(t-x)}\right)^n \left(\dfrac{1-t}{1-x}\right)^\alpha \left(\dfrac{1+t}{1+x}\right)^\beta dt$

when $x \neq \pm 1$, and when the contour of integration is a simple closed curve, around $t = x$ in the positive direction, that does not contain $t = \pm 1$.

(b)
$$2n \int_x^1 (1-y)^\alpha (1+y)^\beta P_n^{(\alpha,\beta)}(y)dy$$
$$= (1-x)^{\alpha+1}(1+x)^{\beta+1} P_{n-1}^{(\alpha+1,\beta+1)}(x).$$

28. Prove that

(a)
$$C_{2m}^\lambda(x) = \dfrac{(\lambda)_m}{(1/2)_m} P_m^{(\lambda-1/2,-1/2)}(2x^2-1)$$
$$= \dfrac{(2\lambda)_{2m}}{(2m)!} {}_2F_1\left(\begin{matrix} -m, m+\lambda \\ \lambda+1/2 \end{matrix}; 1-x^2\right).$$

(b)
$$C_{2m+1}^\lambda(x) = \dfrac{(\lambda)_{m+1}}{(1/2)_{m+1}} x P_m^{(\lambda-1/2,1/2)}(2x^2-1)$$
$$= \dfrac{(2\lambda)_{2m+1}}{(2m+1)!} x {}_2F_1\left(\begin{matrix} -m, m+\lambda+1 \\ \lambda+1/2 \end{matrix}; 1-x^2\right).$$

The following problems define some important hypergeometric orthogonal polynomials. For a given nonnegative integer N appearing in the definition of a discrete orthogonal polynomial, we use the notation

$${}_p\tilde{F}_q(a_1,\ldots,a_p; b_1,\ldots,b_q; x) := \sum_{k=0}^N \dfrac{(a_1)_k \cdots (a_p)_k}{k!(b_1)_k \cdots (b_q)_k} x^k.$$

29. The Racah polynomials are defined by

$$R_n(\lambda(x)) := R_n(\lambda(x); \alpha, \beta, \gamma, \delta)$$
$$:= {}_4\tilde{F}_3\left(\begin{matrix} -n, n+\alpha+\beta+1, -x, x+\gamma+\delta+1 \\ \alpha+1, \beta+\delta+1, \gamma+1 \end{matrix}; 1\right)$$

for $n = 0, 1, 2, \ldots, N$, and where
$$\lambda(x) = x(x+\gamma+\delta+1)$$
and one of the bottom parameters is $-N$. Show that the orthogonality relation is given by

$$\sum_{x=0}^N \dfrac{(\gamma+\delta+1)_x((\gamma+\delta+3)/2)_x(\alpha+1)_x(\beta+\delta+1)_x(\gamma+1)_x}{x!((\gamma+\delta+1)/2)_x(\gamma+\delta-\alpha+1)_x(\gamma-\beta+1)_x(\delta+1)_x}$$

$$\cdot R_m(\lambda(x)) R_n(\lambda(x))$$

$$= M \dfrac{(n+\alpha+\beta+1)_n(\beta+1)_n(\alpha-\delta+1)_n(\alpha+\beta-\gamma+1)_n n!}{(\alpha+\beta+2)_{2n}(\alpha+1)_n(\beta+\delta+1)_n(\gamma+1)_n} \delta_{mn},$$

where

$$M = \begin{cases} \dfrac{(\gamma + \delta + 2)_N (-\beta)_N}{(\gamma - \beta + 1)_N (\delta + 1)_N} & \text{if } \alpha + 1 = -N, \\[3mm] \dfrac{(\gamma + \delta + 2)_N (\delta - \alpha)_N}{(\gamma + \delta - \alpha + 1)_N (\delta + 1)_N} & \text{if } \beta + \delta + 1 = -N, \\[3mm] \dfrac{(-\delta)_N (\alpha + \beta + 2)_N}{(\alpha - \delta + 1)_N (\beta + 1)_N} & \text{if } \gamma + 1 = -N. \end{cases}$$

Show that the recurrence relation is given by

$$\lambda(x) R_n(\lambda(x)) = A_n R_{n+1}(\lambda(x)) - (A_n + C_n) R_n(\lambda(x)) + C_n R_{n-1}(\lambda(x)),$$

where

$$A_n = \frac{(n + \alpha + \beta + 1)(n + \alpha + 1)(n + \beta + \delta + 1)(n + \gamma + 1)}{(2n + \alpha + \beta + 1)(2n + \alpha + \beta + 2)}$$

and

$$C_n = \frac{n(n + \beta)(n + \alpha + \beta - \gamma)(n + \alpha - \delta)}{(2n + \alpha + \beta)(2n + \alpha + \beta + 1)}.$$

Show that the difference equation satisfied by $y(x) = R_n(\lambda(x))$ is

$$n(n + \alpha + 1)y(x) = B(x)y(x + 1) - [B(x) + D(x)]y(x) + D(x)y(x - 1),$$

with

$$B(x) = \frac{(x + \alpha + 1)(x + \beta + \delta + 1)(x + \gamma + 1)(x + \gamma + \delta + 1)}{(2x + \gamma + \delta + 1)(2x + \gamma + \delta + 2)}$$

and

$$D(x) = \frac{x(x + \delta)(x - \beta + \gamma)(x - \alpha + \gamma + \delta)}{(2x + \gamma + \delta)(2x + \gamma + \delta + 1)}.$$

30. The Hahn polynomials are limits of Racah polynomials defined by

$$\lim_{\delta \to \infty} R_n(\lambda(x); \alpha, \beta, -N - 1, \delta) = Q_n(x; \alpha, \beta, N)$$

or

$$\lim_{\gamma \to \infty} R_n(\lambda(x); \alpha, \beta, \gamma, -\beta - N - 1) = Q_n(x; \alpha, \beta, N).$$

Show that

(a) $Q_n(x; \alpha, \beta, N) = {}_3\tilde{F}_2 \left(\begin{matrix} -n, n + \alpha + \beta + 1, -x \\ \alpha + 1, -N \end{matrix} ; 1 \right), \quad n = 0, 1, \ldots, N.$

(b) $\displaystyle\sum_{x=0}^{N} \frac{(\alpha + 1)_x (\beta + 1)_{N-x}}{x!(N - x)!} Q_m(x; \alpha, \beta, N) Q_n(x; \alpha, \beta, N)$

$$= \frac{(-1)^n n!(\beta + 1)_n (n + \alpha + \beta + 1)_{N+1}}{N!(2n + \alpha + \beta + 1)(-N)_n (\alpha + 1)_n} \delta_{mn}.$$

(c) $-xQ_n(x) = A_n Q_{n+1}(x) - (A_n + C_n)Q_n(x) + C_n Q_{n-1}(x)$, where

$$A_n = \frac{(n + \alpha + \beta + 1)(n + \alpha + 1)(N - n)}{(2n + \alpha + \beta + 1)(2n + \alpha + \beta + 2)},$$

and

$$C_n = \frac{n(n + \beta)(n + \alpha + \beta + N + 1)}{(2n + \alpha + \beta)(2n + \alpha + \beta + 1)}.$$

(d) $Q_n(x)$ satisfies the difference equation

$$n(n + \alpha + \beta + 1)y(x) = B(x)y(x + 1)$$
$$- [B(x) + D(x)]y(x) + D(x)y(x - 1),$$

where

$$B(x) = (x - N)(x + \alpha + 1),$$
$$D(x) = x(x - \beta - N - 1).$$

(e) Use (b) to show that $Q_n(x; \alpha, \beta, N) = C_n Q_n(N - x; \beta, \alpha, N)$, where $C_n = (-1)^n (\alpha + 1)_n / (\beta + 1)_n$. Deduce Corollary 3.3.2.

31. Define the dual Hahn polynomials by

$$R_n(\lambda(x); \gamma, \delta, N) := \lim_{\beta \to \infty} R_n(\lambda(x); -N - 1, \beta, \gamma, \delta)$$

and deduce properties corresponding to (a) through (d) in the previous problem. Observe that dual Hahn is obtained from the Hahn by interchanging n and x.

32. Show that, for the Hahn polynomials $Q_n(x)$ defined in Exercise 31, the following limit formula holds:

$$\lim_{N \to \infty} Q_n(Nx; \alpha, \beta, N) = \frac{P_n^{(\alpha,\beta)}(1 - 2x)}{P_n^{(\alpha,\beta)}(1)}.$$

33. The Meixner polynomials can be defined by

$$\lim_{N \to \infty} Q_n\left(x; b - 1, \frac{N(1 - c)}{c}, N\right) =: M_n(x; b, c).$$

Show that

(a) $M_n(x; b, c) = {}_2F_1(-n, -x; b; 1 - 1/c).$

(b) $\displaystyle\sum_{x=0}^{\infty} \frac{(b)_x}{x!} c^x M_m(x; b, c) M_n(x; b, c) = \frac{c^{-n} n!}{(b)_n (1 - c)^b} \delta_{mn},$

$$b > 0 \quad \text{and} \quad 0 < c < 1.$$

(c) $(c - 1)x M_n(x) = c(n + b)M_{n+1}(x) - [n + (n + b)c]M_n(x) + n M_{n-1}(x).$

(Note that an application of the Pfaff transformation (Theorem 2.2.5) shows that Meixner polynomials satisfy a three-term recurrence, which, by Favard's theorem, implies that they are also orthogonal with respect to a positive measure when $c > 1$. The reader is encouraged to find this orthogonality relation, which is obtainable from (b).)

(d) $n(c - 1)M_n(x) = c(x + b)M_n(x + 1)$

$$- [x + (x + b)c]M_n(x) + x M_n(x - 1).$$

Observe the duality in n and x exhibited by relations (c) and (d).

34. A way of defining Krawtchouk polynomials is given by

$$K_n(x; p, N) := \lim_{t \to \infty} Q_n(x; pt, (1 - p)t, N).$$

Prove the following relations:

(a) $K_n(x; p, N) = {}_2\tilde{F}_1(-n, -x; -N; 1/p), \quad n = 0, 1, \ldots, N.$

(b)
$$\sum_{x=0}^{N} \binom{N}{x} p^x (1 - p)^{N-x} K_m(x; p, N) K_n(x; p, N)$$
$$= \frac{(-1)^n n!}{(-N)_n} \left(\frac{1 - p}{p}\right)^n \delta_{mn}.$$

(c) $-x K_n(x) = p(N - n)K_{n+1}(x) - [p(N - n) + n(1 - p)]K_n(x)$
$$+ n(1 - p)K_{n-1}(x).$$

(d) $-n K_n(x) = p(N - x)K_n(x + 1) - [p(N - x) + x(1 - p)]K_n(x)$
$$+ x(1 - p)K_n(x - 1).$$

Note the relationship of (c) and (d) as in the case of the Meixner polynomials given in the previous exercise. In fact

$$K_n(x; p, N) = M_n(x; -N, p/(p - 1)).$$

35. Define the Charlier polynomials by

$$C_n(x; a) := \lim_{b \to \infty} M_n(x; b, a/(a + b))$$

or

$$C_n(x; a) = \lim_{N \to \infty} K_n(x; a/N, N).$$

Deduce that

(a) $C_n(x; a) = {}_2F_0(-n, -x; -; -1/a).$

(b) $\sum_{x=0}^{\infty} \frac{a^x}{x!} C_m(x; a) C_n(x; a) = n! a^{-n} e^a \delta_{mn}, a > 0.$

(c) $-x C_n(x) = a C_{n+1}(x) - (n + a)C_n(x) + n C_{n-1}(x).$

(d) $-nC_n(x) = aC_n(x+1) - (x+a)C_n(x) + xC_n(x-1).$

Compare (c) and (d) as in the previous two exercises.

36. Prove that
$$\lim_{a \to \infty} (2a)^{n/2} C_n(\sqrt{2ax} + a; a) = (-1)^n H_n(x).$$

37. The hypergeometric representation of the Meixner–Pollaczek polynomials is
$$P_n^\lambda(x; \phi) = \frac{(2\lambda)_n}{n!} e^{in\phi} {}_2F_1\left(\begin{matrix} -n, \lambda + ix \\ 2\lambda \end{matrix}; 1 - e^{-2i\phi}\right).$$

These polynomials can be obtained as limits of continuous dual Hahn (or continuous Hahn) polynomials. Show that

(a) $P_n^\lambda(x; \phi) = \lim\limits_{t \to \infty} \dfrac{S_n((x-t)^2; \lambda + it, \lambda - it, t\cot\phi)}{n!(t/\sin\phi)_n}.$

(b) $\displaystyle\int_{-\infty}^{\infty} e^{(2\phi - \pi)x} |\Gamma(\lambda + ix)|^2 P_m^\lambda(x; \phi) P_n^\lambda(x; \phi) dx$

$$= \frac{2\pi \Gamma(n + 2\lambda)}{(2\sin\phi)^{2\lambda} n!} \delta_{mn}, \quad \lambda > 0, \quad 0 < \phi < \pi.$$

(c) $(n+1)P_{n+1}^\lambda(x) - 2[x\sin\phi + (n+\lambda)\cos\phi]P_n^\lambda(x)$

$$+ (n + 2\lambda - 1)P_{n-1}^\lambda(x) = 0.$$

(d) $e^{i\phi}(\lambda - ix)P_n^\lambda(x + i) + 2i[x\cos\phi - (n+\lambda)\sin\phi]P_n^\lambda(x)$

$$- e^{-i\phi}(\lambda + ix)P_n^\lambda(x - i) = 0.$$

38. This problem gives generating functions for some orthogonal polynomials.
    (a) Wilson polynomials:

$${}_2F_1\left(\begin{matrix} a + ix, b + ix \\ a + b \end{matrix}; t\right) {}_2F_1\left(\begin{matrix} c - ix, d - ix \\ c + d \end{matrix}; t\right) = \sum_{n=0}^{\infty} \frac{W_n(x^2; a, b, c, d)t^n}{(a+b)_n(c+d)_n n!}$$

and

$$(1 - t)^{1 - a - b - c - d}$$

$$\cdot {}_4F_3\left(\begin{matrix} (a+b+c+d-1)/2, (a+b+c+d)/2, a+ix, a-ix \\ a+b, a+c, a+d \end{matrix}; \frac{-4t}{(1-t)^2}\right)$$

$$= \sum_{n=0}^{\infty} \frac{(a+b+c+d-1)_n}{(a+b)_n(a+c)_n(a+d)_n n!} W_n(x^2; a, b, c, d).$$

A simple corollary is that

$$\frac{d^k}{dx^k} {}_2F_1\left(\begin{matrix} a+x, b+x \\ a+b \end{matrix}; t\right) {}_2F_1\left(\begin{matrix} c-x, d-x \\ c+d \end{matrix}; t\right) \geq 0$$

for $-\infty < x < \infty$ when $a, b, c, d > 0$ and $0 < t < 1$.

(b) Continuous dual Hahn:

$$(1-t)^{-c+ix} {}_2F_1\left(\begin{matrix} a+ix, b+ix \\ a+b \end{matrix}; t\right) = \sum_{n=0}^{\infty} \frac{S_n(x^2; a, b, c)t^n}{(a+b)_n n!}$$

and

$$e^t {}_2F_2\left(\begin{matrix} a+ix, a-ix \\ a+b, a+c \end{matrix}; -t\right) = \sum_{n=0}^{\infty} \frac{S_n(x^2; a, b, c)t^n}{(a+b)_n(a+c)_n n!}.$$

(c) Continuous Hahn:

$$ {}_1F_1\left(\begin{matrix} a+ix \\ a+c \end{matrix}; -it\right) {}_1F_1\left(\begin{matrix} d-ix \\ b+d \end{matrix}; it\right) = \sum_{n=0}^{\infty} \frac{p_n(x; a, b, c, d)t^n}{(a+c)_n(b+d)_n}$$

and

$$(1-t)^{1-a-b-c-d}$$

$$\cdot {}_3F_2\left(\begin{matrix} (a+b+c+d-1)/2, (a+b+c+d)/2, a+ix \\ a+c, a+d \end{matrix}; \frac{-4t}{(1-t)^2}\right)$$

$$= \sum_{n=0}^{\infty} \frac{(a+b+c+d-1)_n}{(a+c)_n(a+d)_n i^n} p_n(x; a, b, c, d)t^n.$$

(d) Meixner–Pollaczek:

$$(1 - e^{i\phi}t)^{-\lambda+ix}(1 - e^{-i\phi}t)^{-\lambda-ix} = \sum_{n=0}^{\infty} P_n^\lambda(x; \phi)t^n$$

and

$$e^t {}_1F_1\left(\begin{matrix} \lambda+ix \\ 2\lambda \end{matrix}; (e^{-2i\phi}-1)t\right) = \sum_{n=0}^{\infty} \frac{P_n^\lambda(x; \phi)}{(2\lambda)_n e^{in\phi}} t^n.$$

(e) Meixner:

$$\left(1 - \frac{t}{c}\right)^x (1-t)^{-x-b} = \sum_{n=0}^{\infty} \frac{(b)_n}{n!} M_n(x; b, c)t^n$$

and

$$e^t {}_1F_1\left(\begin{matrix} -x \\ b \end{matrix}; \frac{(1-c)t}{c}\right) = \sum_{n=0}^{\infty} \frac{M_n(x; b, c)}{n!} t^n.$$

(f) Charlier:

$$e^t (1 - t/a)^x = \sum_{n=0}^{\infty} \frac{C_n(x; a)}{n!} t^n.$$

For more examples of orthogonal polynomials and their properties, see
Chihara [1978, Chapter 6].

39. Prove the inequalities in (6.4.19) for $\alpha > -1/2$. Prove the corresponding
result for $-1/2 > \alpha > -1$.

40. Suppose $\alpha, \beta > -1$. Show that

$$\max_{-1 \le x \le 1} \left| P_n^{(\alpha,\beta)}(x) \right| = \begin{cases} \frac{(q+1)_n}{n!} \sim n^q & \text{when } q = \max(\alpha, \beta) \ge -1/2, \\ \left| P_n^{(\alpha,\beta)}(x_1) \right| \sim 1/\sqrt{n} & \text{when } \max(\alpha, \beta) < -1/2. \end{cases}$$

Here x_1 is one of the two maximum points nearest $(\beta - \alpha)/(\alpha + \beta + 1) = x_0$.
(Compare with Exercise 4.18. Take

$$n(n + \alpha + \beta + 1) f(x) = n(n + \alpha + \beta + 1) \left\{ P_n^{(\alpha,\beta)}(x) \right\}^2$$

$$+ (1 - x^2) \left\{ \frac{d}{dx} P_n^{(\alpha,\beta)}(x) \right\}^2,$$

and show that $f'(x)$ can change sign only at x_0.)

41. Show that

$$\max_{-1 \le x \le 1} \left| C_n^\lambda(x) \right| = \begin{cases} \frac{(2\lambda)_n}{n!} & \text{if } \lambda > 0, \\ \left| C_n^\lambda(x_1) \right| & \text{if } \lambda < 0, \lambda \text{ nonintegral.} \end{cases}$$

Here x_1 is one of the two maximum points nearest 0 if n is odd; $x_1 = 0$ if n is
even.

42. Show that for a fixed c and $n \to \infty$,

$$P_n^{(\alpha,\beta)}(\cos \theta) = \begin{cases} \theta^{-\alpha-1/2} O(n^{-1/2}) & \text{if } c/n \le \theta \le \pi/2, \\ O(n^\alpha) & \text{if } 0 \le \theta \le c/n. \end{cases}$$

[Use the result on the asymptotic behavior of $P_n^{(\alpha,\beta)}(x)$ in Section 6.6. Also
apply Exercise 4.18, with

$$y(x) = (\sin(x/2))^{\alpha+1/2} (\cos(x/2))^{\beta+1/2} P_n^{(\alpha,\beta)}(\cos x)$$

and

$$\phi(x) = \frac{1/4 - \alpha^2}{4 \sin^2(x/2)} + \frac{1/4 - \beta^2}{4 \cos^2(x/2)} + (n + (\alpha + \beta + 1)/2)^2.]$$

43. Show that the sequence formed by the relative maxima of $|L_n^\alpha(x)|$ and $|L_n^\alpha(0)|$
is decreasing for $x < \alpha + 1/2$ and increasing for $x > \alpha + 1/2$. (Consider the
function $n\{L_n^\alpha(x)\}^2 + x\{\frac{d}{dx} L_n^\alpha(x)\}^2$.)

44. Prove that the successive relative maxima of $|H_n(x)|$ is a decreasing or increasing sequence according as $x \le 0$ or $x \ge 0$.

45. Use Theorem 6.7.2(c) and Gegenbauer's integral in Theorem 6.7.4 to obtain Koornwinder's Laplace-type integral for Jacobi polynomials:

$$\frac{P_n^{(\alpha,\beta)}(x)}{P_n^{(\alpha,\beta)}(1)} = \frac{2\Gamma(\alpha+1)}{\Gamma(\beta+1/2)\Gamma(\alpha-\beta)\sqrt{\pi}}$$
$$\cdot \int_0^1 \int_0^\pi \left[\frac{1+x-(1-x)u^2}{2} + i\sqrt{1-x^2}u\cos\theta \right]^n$$
$$\cdot u^{2\beta+1}(1-u^2)^{\alpha-\beta-1}(\sin 2\theta)^{2\beta}\, d\theta\, du,$$

$$\alpha > \beta > -1/2.$$

46. Note that $y(x) = e^{-x^2/2}H_n(x)$ is a solution of

$$y'' + (2n+1)y = x^2 y.$$

(a) Hence verify that for $s = \sqrt{2n+1}$,

$$y(x) = A_n[\cos(sx - n\pi/2) + R_n(x)],$$

where

$$R_n(x) = \frac{1}{sA_n} \int_0^x t^2 y(t)\sin(s(x-t))dt,$$

and

$$A_n = \begin{cases} n!/k!, & n = 2k, \\ (n!/k!)(2/s), & n = 2k+1. \end{cases}$$

(b) Use Schwartz's inequality to prove that

$$|R_n(x)| \le C|x|^{5/2}n^{-1/4},$$

where C is a constant.

(c) Deduce that

$$H_n(x) \sim 2^{(n+1)/2}n^{n/2}e^{-n/2}e^{x^2/2}\cos\left(sx - \frac{n\pi}{2}\right) \quad \text{as } n \to \infty.$$

47. Show that $y(x) = e^{-x/2}L_n^\alpha(x)$ satisfies the equation

$$xy'' + (\alpha+1)y' + \left(n + \frac{\alpha+1}{2}\right)y = \frac{xy}{4}.$$

(a) Deduce that, for $N = (\alpha+2n+1)/2$,

$$y(x) = \frac{\Gamma(n+\alpha+1)}{n!}y_1(x)$$
$$+ \frac{\pi}{4N}\int_0^x (Nt)^{\alpha+1}y(t)[y_1(t)y_2(x) - y_1(x)y_2(t)]dt,$$

where

$$y_1(x) = J_\alpha(2\sqrt{Nx})/(\sqrt{Nx})^\alpha, \qquad y_2(x) = Y_\alpha(2\sqrt{Nx})/(\sqrt{Nx})^\alpha.$$

(b) As in Exercise 46, but after much more work, it can be shown that the integral divided by $\Gamma(n+\alpha+1)/n!$ tends to zero as $n \to \infty$. Thus prove that

$$L_n^\alpha(x) \sim \frac{\Gamma(n+\alpha+1)}{n!} e^{x/2}(Nx)^{-\alpha/2} J_\alpha(2\sqrt{Nx}) \quad \text{as } n \to \infty.$$

(c) Use (4.8.5) to conclude that

$$L_n^\alpha(x) \sim \frac{n^{(\alpha-1/2)/2}e^{x/2}}{\sqrt{\pi}x^{(\alpha+1/2)/2}} \cos\left(2\sqrt{nx} - \frac{\alpha\pi}{2} - \frac{\pi}{4}\right) \quad \text{as } n \to \infty.$$

48. Let r_k denote the number of ways in which k rooks can be placed in different rows and columns on an $m \times n$ chessboard. Let $R_{m,n} = \sum_{k=0}^{m\wedge n} r_k x^k$. Prove that, when α is an integer, $R_{n,n+\alpha} = n!x^n L_n^\alpha(-1/x)$.

49. Use the methods of Section 6.9 to evaluate the integral

$$\frac{1}{\sqrt{2\pi}} \int_{-\infty}^{\infty} He_a(x)He_b(x)He_c(x)He_d(x)e^{-x^2/2}dx.$$

Then evaluate it by a different method.

50. Suppose $f(x)$ is expandable in terms of Jacobi polynomials. Let

$$f(x) = \sum_{n=0}^{\infty} a(n)h_n^{-1} P_n^{(\alpha,\beta)}(x),$$

where h_n is defined in (6.2.5). Let $g(x)$ be an average of $f(x)$ defined by

$$g(x) = (1-x)^{-\alpha-1}(1-x)^{-\beta} \int_x^1 f(t)(1-t)^\alpha (1+t)^\beta dt.$$

Suppose for simplicity that $a(0)=0$. If $b(n)$ and $a(n)$ are the Jacobi coefficients of g and f respectively, show that

$$b(n) = \frac{a(n)}{n+\alpha+\beta+1} + \frac{\Gamma(n+\alpha+1)}{\Gamma(n+1)}$$

$$\cdot \sum_{k=n+1}^{\infty} \frac{a(k)[2k+\alpha+\beta+1]}{k[k+\alpha+\beta+1]} \frac{\Gamma(k+1)}{\Gamma(k+\alpha+1)},$$

for $n = 0, 1, \ldots$.

A sufficient condition for the result to hold is the convergence of the series $\Sigma |a(n)| n^{1+\alpha}$. Similar results hold for Laguerre and Hermite polynomials.

51. This exercise gives a proof of the following theorem of Hardy [1933]: If f and its Fourier transform g are both $O(|x|^m e^{-x^2/2})$ for large x and some m, then each is a finite linear combination of functions of the form $e^{-x^2/2} H_n(x)$, where $H_n(x)$ is the Hermite polynomial of degree n.

Note that it is enough to prove the theorem for self-reciprocal ($f = g$) and skew-reciprocal functions ($f = -g$). Take f to be self-reciprocal and show that:

(a) For Re $s = \sigma > -1$,

$$\lambda(s) = \int_0^\infty e^{-sx^2/2} f(x)\,dx$$

satisfies

$$\lambda(s) = s^{-1/2}\lambda(1/s).$$

(b) When $\mu(s) = \sqrt{(s+1)}\lambda(s)$,

$$\mu(s) = \mu(1/s).$$

The function $\mu(s)$ may have a singularity at $s = -1$ but is analytic at all other points including infinity. Hence,

$$\lambda(s) = \sum_{n=0}^\infty a_n/(s+1)^{n+1/2}.$$

(c) $\lambda(s) = O(|s+1|^{-p})$ for some p, near $s = -1$. This can be proven by the argument below. Let $\sigma + 1 = \tau$. On the unit circle $|s| = 1$, $\tau = |s+1|^2/2$. For $|s| \leq 1$

$$\lambda(s) = O\left(\int_0^\infty e^{-\tau x^2/2} x^m dx\right) = O(|s+1|^{-m-1}).$$

(d)
$$\int_0^\infty e^{-sx^2/2} e^{-x^2/2} H_{2n}(x)\,dx = (-1)^n \frac{(2n)!}{n!}\sqrt{\frac{\pi}{2}}\frac{(s-1)^n}{(s+1)^{n+1/2}}.$$

(e)
$$\int_0^\infty e^{-sx^2/2}(f - \Phi)\,dx = 0,$$

where Φ is an appropriate linear combination of functions of the form $e^{-x^2/2} H_n(x)$.

52. Hardy's theorem given in the previous exercise extends to the following result of Roosenraad [1969]:

Set

$$c_n^\alpha = [n!/\Gamma(n+\alpha+1)]^{1/2}$$

and define generalized Laguerre functions by

$$\mathcal{L}_{2n}^\alpha(x) = c_n^\alpha |x|^{\alpha+1/2} e^{-x^2/2} L_n^\alpha(x^2),$$

$$\mathcal{L}_{2n+1}^\alpha(x) = c_n^{\alpha+1} |x|^{\alpha+1/2} e^{-x^2/2} x L_n^{\alpha+1}(x^2), \quad n = 0, 1, 2, \ldots.$$

For a function f defined for all real numbers, set

$$(\mathcal{R}_\alpha f)(t) = \frac{1}{2} \int_{-\infty}^{\infty} f(x) \left\{ |xt|^{1/2} J_\alpha(|xt|) + i \frac{(xt)}{|xt|^{1/2}} J_{\alpha+1}(|xt|) \right\} dx.$$

Note that $\mathcal{R}_\alpha$ is the sum of an even and an odd Hankel transform. Check that $\mathcal{L}_m^\alpha$ are eigenfunctions of $\mathcal{R}_\alpha$, that is,

$$\mathcal{R}_\alpha \mathcal{L}_m^\alpha = i^m \mathcal{L}_m^\alpha, \quad m = 0, 1, 2, \ldots.$$

Theorem *If f and $g_\alpha = \mathcal{R}_\alpha f$ are both $O(x^{m+\alpha+1/2} e^{-x^2/2})$ for large x and some $m \geq 0$, then each is a finite linear combination of the functions $\mathcal{L}_n^\alpha(x)$.*

As before, it is sufficient to consider the cases where $\mathcal{R}_\alpha f = \pm f$. Take $\mathcal{R}_\alpha f = f$. Show that, for Re $s > -1$,

$$\lambda(s) = \int_0^\infty x^{\alpha+1/2} e^{-sx^2/2} f(x) dx$$

satisfies

$$\lambda(s) = s^{-\alpha-1} \lambda(1/s).$$

So, if $\mu(s) = (1+s)^{\alpha+1} \lambda(s)$, then $\mu(s) = \mu(1/s)$. Now complete the proof as in the previous exercise.

7

Topics in Orthogonal Polynomials

As we have seen before, we can gain insight into Jacobi polynomials by using the fact that they are hypergeometric functions. In this chapter, we reverse our procedure and see that Jacobi polynomials can shed light on some aspects of hypergeometric function theory. Thus, we discuss the connection coefficient problem for Jacobi polynomials. We also discuss the positivity of sums of Jacobi polynomials. We mention several methods but here, too, there are situations in which the hypergeometric function plays an important role. Finally, for its intrinsic interest, we present Beukers's use of Legendre polynomials to prove the irrationality of $\zeta(3)$, a result first proved by Apéry.

It is evident that the Jacobi polynomial $P_n^{(\gamma,\delta)}(x)$ can be expressed as a sum: $\sum_{k=0}^{n} c_{nk} P_k^{(\alpha,\beta)}(x)$. The significant point is that the connection coefficient c_{nk} is expressible as a $_3F_2$ hypergeometric function. This $_3F_2$ can be evaluated in terms of shifted factorials under conditions on the parameters α, β, γ, and δ. Surprisingly, this leads to an illuminating proof of Whipple's $_7F_6$ transformation. We have seen that, with the exception of Gauss's $_2F_1$, most summable hypergeometric series are either balanced or well poised. A puzzling fact is that at the $_5F_4$ and higher levels, the series are very well poised. The above-mentioned proof of Whipple's transformation sheds light on this fact by showing that very well poisedness arises from the orthogonality relation for Jacobi polynomials.

Fejér used the positivity of the series $\sum_{k=0}^{n} \sin\left(k + \frac{1}{2}\right)\theta$ to prove his famous theorem on the Cesàro summability of Fourier series. The positivity of some other trigonometric series have also been important in mathematics. It turns out that these inequalities are generalizable to inequalities for sums of Jacobi polynomials. The soundness of this generalization is illustrated by the usefulness of the inequalities. A dramatic example is an inequality proved by Gasper that played an unexpected but significant role in de Branges's proof of the Bieberbach conjecture. For this interesting story, see Baernstein et al. [1986]. We also state and prove a trigonometric inequality due to Vietoris.

7.1 Connection Coefficients

Suppose V is the vector space of all polynomials over the real or complex numbers and V_m is the subspace of polynomials of degree $\leq m$. Suppose $p_0(x)$, $p_1(x)$, $p_2(x)$, ... is a sequence of polynomials such that $p_n(x)$ is of exact degree n; let $q_0(x), q_1(x), q_2(x), \ldots$ be another such sequence. Clearly, these sequences form a basis for V. It is also evident that $p_0(x), \ldots, p_m(x)$ and $q_0(x), \ldots, q_m(x)$ give two bases of V_m. It is often necessary in working with finite-dimensional vector spaces to find the matrix that transforms a basis of a given space to another basis. This means that one is interested in the coefficients c_{nk} that satisfy

$$q_n(x) = \sum_{k=0}^{n} c_{nk} p_k(x). \tag{7.1.1}$$

The choice of p_n or q_n depends on the situation. For example, suppose

$$p_n(x) = x^n, \quad q_n(x) = x(x-1)\cdots(x-n+1).$$

Then the coefficients c_{nk} are Stirling numbers of the first kind. If the roles of these p_n and q_n are interchanged, then we get Stirling numbers of the second kind. These numbers are useful in some combinatorial problems and were defined by Stirling [1730].

Usually, little can be said about these connection coefficients. However, there are some cases where simple formulas can be obtained. For example,

$$C_n^\lambda(\cos\theta) = \sum_{k=0}^{n} \frac{(\lambda)_{n-k}(\lambda)_k}{(n-k)!k!} \cos(n-2k)\theta \tag{7.1.2}$$

gives an expansion of $P_n^{(\alpha,\alpha)}(x)$ in terms of $P_k^{(-1/2,-1/2)}(x)$. This formula was derived in the previous chapter from the generating function for $C_n^\lambda(x)$. See (6.4.11). Another example is

$$L_n^\beta(x) = \sum_{k=0}^{n} \frac{(\beta-\alpha)_{n-k}}{(n-k)!} L_k^\alpha(x). \tag{7.1.3}$$

This can be obtained from the generating function for $L_n^\beta(x)$. We have

$$\sum_{n=0}^{\infty} L_n^\beta(x) r^n = (1-r)^{-\beta-1} \exp(-xr/(1-r))$$

$$= (1-r)^{-\alpha-1} \exp(-xr/(1-r))(1-r)^{-(\beta-\alpha)}$$

$$= \left(\sum_{k=0}^{\infty} L_k^\alpha(x) r^k \right) \left(\sum_{s=0}^{\infty} \frac{(\beta-\alpha)_s}{s!} r^s \right)$$

$$= \sum_{n=0}^{\infty} \left(\sum_{k=0}^{n} \frac{(\beta-\alpha)_{n-k}}{(n-k)!} L_k^\alpha(x) \right) r^n.$$

Notice that in both these cases the polynomials are similar, in that they are orthogonal on the same interval and their weight functions are closely related.

The next lemma is a basic result of this section and gives the connection coefficient c_{nk} when $q_n(x) = P_n^{(\gamma,\delta)}(x)$ and $p_k(x) = P_k^{(\alpha,\beta)}(x)$.

Lemma 7.1.1 *Suppose* $P_n^{(\gamma,\delta)}(x) = \sum_{k=0}^n c_{nk} P_k^{(\alpha,\beta)}(x)$. *Then*

$$c_{nk} = \frac{(n+\gamma+\delta+1)_k (k+\gamma+1)_{n-k}(2k+\alpha+\beta+1)\Gamma(k+\alpha+\beta+1)}{(n-k)!\,\Gamma(2k+\alpha+\beta+2)}$$

$$\cdot\, {}_3F_2\left(\begin{array}{c} -n+k, n+k+\gamma+\delta+1, k+\alpha+1 \\ k+\gamma+1, 2k+\alpha+\beta+2 \end{array}; 1\right).$$

Proof. From the orthogonality of Jacobi polynomials,

$$c_{nk} = a_{nk}/h_k,$$

where

$$h_k = \int_{-1}^1 \left[P_k^{(\alpha,\beta)}(x)\right]^2 (1-x)^\alpha (1+x)^\beta dx$$

$$= \frac{2^{\alpha+\beta+1}\Gamma(k+\alpha+1)\Gamma(k+\beta+1)}{(2k+\alpha+\beta+1)\Gamma(k+\alpha+\beta+1)\Gamma(k+1)} \qquad (7.1.4)$$

and

$$a_{nk} = \int_{-1}^1 P_n^{(\gamma,\delta)}(x) P_k^{(\alpha,\beta)}(x)(1-x)^\alpha (1+x)^\beta dx$$

$$= \frac{(-1)^k}{2^k k!}\int_{-1}^1 P_n^{(\gamma,\delta)}(x)\frac{d^k}{dx^k}[(1-x)^{\alpha+k}(1+x)^{\beta+k}]dx$$

$$= \frac{1}{2^k k!}\int_{-1}^1 \frac{d^k}{dx^k}\left[P_n^{(\gamma,\delta)}(x)\right](1-x)^{\alpha+k}(1+x)^{\beta+k}dx.$$

We have seen earlier that

$$\frac{d}{dx}P_n^{(\gamma,\delta)}(x) = \frac{n+\gamma+\delta+1}{2}P_{n-1}^{(\gamma+1,\delta+1)}(x). \qquad (6.3.8)$$

Therefore,

$$\frac{d^k}{dx^k}P_n^{(\gamma,\delta)}(x) = \frac{(n+\gamma+\delta+1)_k}{2^k}P_{n-k}^{(\gamma+k,\delta+k)}(x).$$

Use this in the integral for a_{nk} to get

$$
\begin{aligned}
a_{nk} &= \frac{(n+\gamma+\delta+1)_k}{2^{2k}k!} \int_{-1}^{1} P_{n-k}^{(\gamma+k,\delta+k)}(x)(1-x)^{\alpha+k}(1+x)^{\beta+k}dx \\
&= \frac{(n+\gamma+\delta+1)_k(\gamma+k+1)_{n-k}}{2^{2k}k!(n-k)!} \sum_{j=0}^{n-k} \frac{(-n+k)_j(n+k+\gamma+\delta+1)_j}{(k+\gamma+1)_j j! 2^j} \\
&\quad \cdot \int_{-1}^{1}(1-x)^{\alpha+k+j}(1+x)^{\beta+k}dx \\
&= \frac{(n+\gamma+\delta+1)_k(k+\gamma+1)_{n-k}\Gamma(k+\alpha+1)\Gamma(k+\beta+1)2^{\alpha+\beta+1}}{k!(n-k)!\Gamma(\alpha+\beta+2k+2)} \\
&\quad \cdot {}_3F_2\left(\begin{array}{c} -n+k, n+k+\gamma+\delta+1, k+\alpha+1 \\ k+\gamma+1, 2k+\alpha+\beta+2 \end{array}; 1\right).
\end{aligned}
$$

This is equivalent to the claim in Lemma 7.1.1.   ■

In general, the ${}_3F_2$ in the lemma cannot be summed. If we take $\gamma = \alpha$, then the ${}_3F_2$ reduces to a terminating ${}_2F_1$, which can be evaluated by the Chu–Vandermonde formula (Corollary 2.2.3). The ${}_3F_2$ can again be summed if $\delta = \beta$. For in this case we get a balanced ${}_3F_2$ whose value is given by the Pfaff–Saalschütz identity (Theorem 2.2.6). Finally, the ${}_3F_2$ can be summed by Watson's identity if $\alpha = \beta$ and $\gamma = \delta$ (Theorem 3.5.5). It is, however, sufficient to do the $\alpha = \gamma$ case in Lemma 7.1.1 as the other two cases are consequences of this one.

Theorem 7.1.2

$$
P_n^{(\alpha,\delta)}(x) = \frac{(\alpha+1)_n}{(\alpha+\beta+2)_n} \cdot \sum_{k=0}^{n}
$$
$$
\cdot \frac{(-1)^{n-k}(\delta-\beta)_{n-k}(\alpha+\beta+1)_k(\alpha+\beta+2k+1)(\alpha+\delta+n+1)_k}{(n-k)!(\alpha+1)_k(\alpha+\beta+1)(\alpha+\beta+n+2)_k} P_k^{(\alpha,\beta)}(x).
$$

Proof. Take $\gamma = \alpha$ in Lemma 7.1.1. The ${}_3F_2$ reduces to

$$
{}_2F_1\left(\begin{array}{c} -n+k, n+k+\alpha+\delta+1 \\ 2k+\alpha+\beta+2 \end{array}; 1\right) = \frac{(\beta-\delta-n+k+1)_{n-k}}{(\alpha+\beta+2k+2)_{n-k}}.
$$

Use this in Lemma 7.1.1 and simplify to get the result.   ■

The case $\delta = \beta$ is a corollary.

Theorem 7.1.3

$$
P_n^{(\gamma,\beta)}(x) = \frac{(\beta+1)_n}{(\alpha+\beta+2)_n}
$$
$$
\cdot \sum_{k=0}^{n} \frac{(\gamma-\alpha)_{n-k}(\alpha+\beta+1)_k(\alpha+\beta+2k+1)(\beta+\gamma+n+1)_k}{(n-k)!(\beta+1)_k(\alpha+\beta+1)(\alpha+\beta+n+2)_k} P_k^{(\alpha,\beta)}(x).
$$

Proof. Use Theorem 7.1.2 and the fact that $P_n^{(\alpha,\beta)}(x) = (-1)^n P_n^{(\beta,\alpha)}(-x)$. ∎

Theorem 7.1.4

$$P_m^{(\gamma,\gamma)}(x) = \frac{(\gamma+1)_m}{(2\gamma+1)_m}$$

$$\cdot \sum_{k=0}^{[m/2]} \frac{(2\alpha+1)_{m-2k}(\gamma+1/2)_{m-k}(\alpha+3/2)_{m-2k}(\gamma-\alpha)_k}{(\alpha+1)_{m-2k}(\alpha+3/2)_{m-k}(\alpha+1/2)_{m-2k}k!} P_{m-2k}^{(\alpha,\alpha)}(x).$$

Proof. Replace x with $2x^2-1$ and $\beta = \pm 1/2$ in Theorem 7.1.3. Then

$$P_n^{(\gamma,-1/2)}(2x^2-1) = \frac{(1/2)_n}{(\alpha+3/2)_n}$$

$$\cdot \sum_{k=0}^{n} \frac{(\gamma-\alpha)_{n-k}(\alpha+1/2)_k(2k+\alpha+1/2)(n+\gamma+1/2)_k}{(n-k)!(1/2)_k(\alpha+1/2)(n+\alpha+3/2)_k} P_k^{(\alpha,-1/2)}(2x^2-1)$$

$$(7.1.5)$$

and

$$x P_n^{(\gamma,1/2)}(2x^2-1) = \frac{(3/2)_n}{(\alpha+5/2)_n}$$

$$\cdot \sum_{k=0}^{n} \frac{(\gamma-\alpha)_{n-k}(\alpha+3/2)_k(2k+\alpha+3/2)(n+\gamma+3/2)_k}{(n-k)!(3/2)_k(\alpha+3/2)(n+\alpha+5/2)_k} x P_k^{(\alpha,1/2)}(2x^2-1).$$

$$(7.1.6)$$

Now use the quadratic transformation formula from Chapter 3 (see (3.1.1)) to obtain

$$P_{2n}^{(\alpha,\alpha)}(x) = \frac{(\alpha+1)_{2n}n!}{(\alpha+1)_n(2n)!} P_n^{(\alpha,-1/2)}(2x^2-1) \qquad (7.1.7)$$

and

$$P_{2n+1}^{(\alpha,\alpha)}(x) = \frac{(\alpha+1)_{2n+1}n!}{(\alpha+1)_n(2n+1)!} x P_n^{(\alpha,1/2)}(2x^2-1). \qquad (7.1.8)$$

Take $m = 2n$ and combine (7.1.5) and (7.1.7) and simplify to get

$$P_m^{(\gamma,\gamma)}(x) = \frac{(\gamma+1)_m}{(2\gamma+1)_m}$$

$$\cdot \sum_{k=0}^{m/2} \frac{(\gamma-\alpha)_{\frac{m}{2}-k}(2\alpha+1)_{2k}(\alpha+3/2)_{2k}(\gamma+1/2)_{\frac{m}{2}+k}}{(\frac{m}{2}-k)!(\alpha+1)_{2k}(\alpha+3/2)_{\frac{m}{2}+k}(\alpha+1/2)_{2k}} P_{2k}^{(\alpha,\alpha)}(x).$$

The formula in the theorem for even m follows from this by reversing the order of summation, that is, by changing k to $\frac{m}{2} - k$. The odd case is obtained similarly from (7.1.6) and (7.1.8). ∎

We can also write the previous theorem in terms of ultraspherical polynomials:

Theorem 7.1.4′

$$C_n^\lambda(x) = \sum_{k=0}^{[n/2]} \frac{(\lambda)_{n-k}(\lambda - \mu)_k(n + \mu - 2k)}{(\mu + 1)_{n-k}k!\mu} C_{n-2k}^\mu(x). \qquad (7.1.9)$$

Remark 7.1.1 The relation in (7.1.2) is obtained from the above formula by letting $\mu \to 0$. Surprisingly, it is easy to obtain Theorem 7.1.4′ directly from (7.1.2).

Proof of Theorem 7.1.4′   Note that

$$\frac{d}{dx}C_n^\lambda(x) = 2\lambda C_{n-1}^{\lambda+1}(x).$$

Differentiate (7.1.2) with respect to θ and divide by $-\sin\theta$ to get

$$2\lambda C_{n-1}^{\lambda+1}(\cos\theta) = \sum_{k=0}^{n} \frac{(\lambda)_{n-k}(\lambda)_k(n - 2k)}{(n - k)!k!} \frac{\sin(n - 2k)\theta}{\sin\theta}. \qquad (7.1.10)$$

If k is replaced by $n - k$, the expression in the summation does not change. So for n odd the terms of the sum can be paired and for n even the same can be done, because the term when $n = k$ is zero. Thus, after changing $n - 1$ to n and λ to $\lambda - 1$ in (7.1.10), we get

$$C_n^\lambda(x) = \sum_{k=0}^{[n/2]} \frac{(\lambda)_{n-k}(\lambda - 1)_k}{(2)_{n-k}k!} \frac{n + 1 - 2k}{1} C_{n-2k}^1(x).$$

Here we used the fact that

$$C_n^1(\cos\theta) = \frac{\sin(n + 1)\theta}{\sin\theta}.$$

Repeat this process, that is, differentiate with respect to x and change $n - 1$ to n and λ to $\lambda - 1$ to get

$$C_n^\lambda(x) = \sum_{k=0}^{[n/2]} \frac{(\lambda)_{n-k}(\lambda - 2)_k}{(3)_{n-k}k!} \frac{n + 2 - 2k}{2} C_{n-2k}^2(x).$$

By induction it follows that

$$C_n^\lambda(x) = \sum_{k=0}^{[n/2]} \frac{(\lambda)_{n-k}(\lambda - \mu)_k}{(\mu + 1)_{n-k}k!} \frac{n + \mu - 2k}{\mu} C_{n-2k}^\mu(x) \qquad (7.1.11)$$

for $\mu = 1, 2, 3, \ldots$. It is evident from (7.1.2) that $C_{n-2k}^\mu(x)$ is a polynomial in μ. Hence the right side of (7.1.11) is a rational function of μ. Since (7.1.11) is true for infinitely many values of μ, it is identically true. This completes another proof of Theorem 7.1.4'. ∎

It is possible to obtain Dougall's sum of a very well poised $_7F_6$ from Theorem 7.1.3. This evaluation of Dougall's identity gives an insight different from those suggested by earlier evaluations. To start, take $x = 1$ in Theorem 7.1.3 to get

$$\frac{(\gamma + 1)_n (\alpha + \beta + 2)_n}{(\beta + 1)_n n!} = \frac{(\gamma - \alpha)_n}{n!}$$

$$\cdot \sum_{k=0}^{n} \frac{(-n)_k (\alpha + \beta + 1)_k (2k + \alpha + \beta + 1)(n + \gamma + \beta + 1)_k (\alpha + 1)_k}{(-n - \gamma + \alpha + 1)_k (\beta + 1)_k (\alpha + \beta + 1)(n + \alpha + \beta + 2)_k k!}.$$

Note that

$$\frac{2k + \alpha + \beta + 1}{\alpha + \beta + 1} = \frac{(1 + (\alpha + \beta + 1)/2)_k}{((\alpha + \beta + 1)/2)_k}. \tag{7.1.12}$$

So

$$_5F_4 \left(\begin{matrix} \alpha + \beta + 1, (\alpha + \beta + 3)/2, n + \gamma + \beta + 1, \alpha + 1, -n \\ (\alpha + \beta + 1)/2, \alpha - n - \gamma + 1, \beta + 1, n + \alpha + \beta + 2 \end{matrix}; 1 \right)$$

$$= \frac{(\gamma + 1)_n (\alpha + \beta + 2)_n}{(\beta + 1)_n (\gamma - \alpha)_n}.$$

Set $\alpha + \beta + 1 = a$, $\alpha + 1 = b$, and $n + \gamma + \beta + 1 = c$ to rewrite this formula as

$$_5F_4 \left(\begin{matrix} a, \ a/2 + 1, \ b, \ c, \ -n; \\ a/2, a + 1 - b, a + 1 - c, a + 1 + n \end{matrix}; 1 \right) = \frac{(a + 1)_n (a + 1 - b - c)_n}{(a + 1 - b)_n (a + 1 - c)_n}.$$

This identity gives the sum of a terminating very well poised $_5F_4$. It is interesting that after Kummer's sum of the well-poised $_2F_1$ at $x = -1$ and Dixon's sum of the well-poised $_3F_2$ at $x = 1$, most well-poised series that can be summed have the additional feature of a numerator and denominator parameter differing by one. This makes the series very well poised. In the above series, this followed from (7.1.12), which in turn came from the orthogonality relation for Jacobi polynomials (7.1.4). This partially explains why the summable well-poised series after $_3F_2$ are very well poised.

The above $_5F_4$ is only a particular case of Dougall's $_7F_6$. To find the general case, note that evaluation of a function at a point is an example of a linear operator. This was the operator applied to the identity in Theorem 7.1.3 to get the $_5F_4$. A more general operator is an integral with respect to a measure. To obtain attractive formulas, one chooses the measure suitably. The Jacobi polynomials can be written as $_2F_1$ hypergeometric series and we know that the integration of a hypergeometric function with respect to a beta distribution introduces two new independent parameters into the series. This shows how we may obtain the generalization of the $_5F_4$ formula. Write Theorem 7.1.3 as

$$\frac{(\gamma+1)_n}{n!}\,{}_2F_1\left(\begin{array}{c}-n,\gamma+\beta+n+1\\\gamma+1\end{array};ut\right)$$

$$=\frac{(\beta+1)_n}{(\alpha+\beta+2)_n}\sum_{k=0}^{n}\frac{(\gamma-\alpha)_{n-k}(\alpha+\beta+1)_k}{(n-k)!}$$

$$\cdot\frac{(\alpha+\beta+2k+1)(\gamma+\beta+n+1)_k(\alpha+1)_k}{(\beta+1)_k(\alpha+\beta+1)(\alpha+\beta+n+2)_k k!}$$

$$\cdot\,{}_2F_1\left(\begin{array}{c}-k,\alpha+\beta+k+1\\\alpha+1\end{array};ut\right). \tag{7.1.13}$$

Integrate this with respect to two independent beta distributions. The result is

$$\frac{(\alpha+\beta+2)_n(\gamma+1)_n}{(\beta+1)_n n!}\,{}_4F_3\left(\begin{array}{c}-n,\gamma+\beta+n+1,a,b\\\gamma+1,c,d\end{array};1\right)=\frac{(\beta+1)_n}{(\alpha+\beta+2)_n}$$

$$\cdot\sum_{k=0}^{n}\frac{(\gamma-\alpha)_{n-k}(\alpha+\beta+1)_k(\alpha+\beta+2k+1)(\gamma+\beta+n+1)_k(\alpha+1)_k}{(n-k)!(\beta+1)_k(\alpha+\beta+1)(\alpha+\beta+n+2)_k k!}$$

$$\cdot\,{}_4F_3\left(\begin{array}{c}-k,\alpha+\beta+k+1,a,b\\\alpha+1,c,d\end{array};1\right). \tag{7.1.14}$$

The $_4F_3$s here cannot be summed without restriction on the parameters. Observe that the $_4F_3$ on the left-hand side is balanced if $a+b+\beta+1=c+d$. This is also the condition for the $_4F_3$s on the right to be balanced. So assume that the parameters are chosen to balance the $_4F_3$s. If we further let $b=\alpha+1$, then the $_4F_3$ on the right side is reduced to a balanced $_3F_2$, which can be summed by the Pfaff–Saalschütz formula. The result is the formula

$$_7F_6\left(\begin{array}{c}-n,\alpha+\beta+1,(\alpha+\beta+3)/2,\gamma+\beta+n+1,\alpha+1,\alpha+\beta-c+2,c-a\\\alpha-\gamma-n+1,(\alpha+\beta+1)/2,\alpha+\beta+n+2,\beta+1,c,a+\alpha+\beta-c+2\end{array};1\right)$$

$$=\frac{(\gamma+1)_n(\alpha+\beta+2)_n}{(\beta+1)_n(\gamma-\alpha)_n}\,{}_4F_3\left(\begin{array}{c}-n,\gamma+\beta+n+1,a,\alpha+1\\\gamma+1,c,d\end{array};1\right), \tag{7.1.15}$$

when $\alpha+\beta+a+2=c+d$. This is just Whipple's transformation formula from which Dougall's identity can be obtained, as we have seen earlier.

Remark 7.1.2　The coefficients in Theorem 7.1.4 are nonnegative when $\gamma>\alpha>-1$. This fact is useful in the proof of the positivity of a certain $_3F_2$ function. This played a significant role in the first proof of the Bieberbach conjecture. We prove the inequality in a later section. The nonnegativity of the coefficients in Theorem 7.1.3 holds under the same condition, that is, $\gamma>\alpha>-1$. In Theorem 7.1.2, nonnegativity occurs when $\delta-\beta=-1,-2,\ldots,\delta>-1$. For general $\alpha,\beta,\gamma,\delta$

the problem of the nonnegativity of the coefficients reduces by Lemma 7.1.1 to that of a certain $_3F_2$. One way to deal with this is to use three-term contiguous relations for $_3F_2$s. For details the reader may look to Askey and Gasper [1971].

7.2 Rational Functions with Positive Power Series Coefficients

We start by showing that

$$\sum_{k=0}^{\min(m,n)} \binom{m-k+\alpha}{m-k}\binom{n-k+\alpha}{n-k}\binom{k-\alpha-2}{k} \geq 0 \qquad (7.2.1)$$

for $\alpha \geq 0$. Lorentz and Zeller [1964] used this to obtain a new proof of a theorem of Hardy and Bohr. The above inequality is not directly related to orthogonal polynomials but its proof gives a nice introduction to the method of generating functions. This method will be used again in this section to prove an inequality involving Laguerre polynomials. Moreover, (7.2.1) is really an inequality for a $_3F_2$, a topic we discussed in the previous section. Note that (7.2.1) can be written as

$$\frac{(\alpha+1)_m(\alpha+1)_n}{m!n!} {}_3F_2\left(\begin{matrix} -m, -n, -\alpha-1 \\ -m-\alpha, -n-\alpha \end{matrix}; 1\right) \geq 0 \qquad (7.2.2)$$

for $\alpha \geq 0$.

We prove a more general result, which is the content of the next theorem, due to Askey, Gasper, and Ismail [1975].

Theorem 7.2.1 If $0 \leq \alpha \leq \min(\beta, \gamma)$, then

$$\sum_{k=0}^{\min(m,n)} \binom{m-k+\beta}{m-k}\binom{n-k+\gamma}{n-k}\binom{k-\alpha-2}{k} \geq 0,$$

$$m, n = 0, 1, 2, \ldots . \qquad (7.2.3)$$

Proof. Observe that

$$(1-x)^{-(\alpha+1)} = \sum_{k=0}^{\infty} \binom{k+\alpha}{k} x^k.$$

Thus (7.2.3) must be the coefficient of some term in an expansion obtained as the product of three binomial expansions. In fact, (7.2.3) is the coefficient of $r^m s^n$ in the product

$$\sum_{k=0}^{\infty} \binom{k-\alpha-2}{k} (rs)^k \sum_{m=k}^{\infty} \binom{m-k+\beta}{m-k} r^{m-k} \sum_{n=k}^{\infty} \binom{n-k+\gamma}{n-k} s^{n-k}.$$

Verify that we can rewrite this product as

$$\sum_{m,n=0}^{\infty} \sum_{k=0}^{\min(m,n)} \binom{m-k+\beta}{m-k} \binom{n-k+\gamma}{n-k} \binom{k-\alpha-2}{k} r^{m-k}s^{n-k}(rs)^k$$

$$= \frac{(1-rs)^{\alpha+1}}{(1-r)^{\beta+1}(1-r)^{\gamma+1}}$$

$$= (1-r)^{\alpha-\beta}(1-s)^{\alpha-\gamma} \frac{(1-rs)^{\alpha+1}}{(1-r)^{\alpha+1}(1-s)^{\alpha+1}}.$$

The two factors $(1-r)^{\alpha-\beta}$ and $(1-s)^{\alpha-\gamma}$ have nonnegative power series coefficients when $\beta \geq \alpha$ and $\gamma \geq \alpha$. This shows that it is sufficient to prove the case where $\alpha = \beta = \gamma$. Note that

$$\frac{1-rs}{(1-r)(1-s)} = \frac{1}{1-r} + \frac{1}{1-s} - 1$$

$$= 1 + \sum_{n=1}^{\infty}(r^n + s^n).$$

Thus the expansion of $(1-rs)/[(1-r)(1-s)]$ has positive coefficients and it follows that any positive integer power of this rational function also has positive power series coefficients. Now write

$$\left[\frac{1-rs}{(1-r)(1-s)}\right]^{\alpha+1} = \left[\frac{1-rs}{(1-r)(1-s)}\right]^{[\alpha]+1} \left[\frac{1-rs}{(1-r)(1-s)}\right]^{\alpha-[\alpha]}.$$

Since $0 \leq \alpha - [\alpha] < 1$, we need only consider the case $\alpha = \beta = \gamma$ and $0 \leq \alpha < 1$. Observe that the $_3F_2$ in (7.2.2) when written out is

$$1 - \frac{mn(\alpha+1)}{(m+\alpha)(n+\alpha)1!} + \frac{m(m-1)n(n-1)(\alpha+1)\alpha}{(m+\alpha)(m+\alpha-1)(n+\alpha)(n+\alpha-1)2!}$$

$$- \frac{m(m-1)(m-2)n(n-1)(n-2)(\alpha+1)\alpha(\alpha-1)}{(m+\alpha)(m+\alpha-1)(m+\alpha-2)(n+\alpha)(n+\alpha-1)(n+\alpha-2)3!} + \cdots.$$

There are both positive and negative terms in this series, so the sign of the sum is not immediately evident. To show that the series is positive, we transform it into another series, all of whose terms are positive. For this purpose apply to (7.2.2) Thomae's formula (Corollary 3.3.4),

$$_3F_2\left(\begin{matrix} a, b, c \\ d, e \end{matrix}; 1\right) = \frac{\Gamma(d)\Gamma(e)\Gamma(s)}{\Gamma(a)\Gamma(s+b)\Gamma(s+c)} {}_3F_2\left(\begin{matrix} d-a, e-a, s \\ s+b, s+c \end{matrix}; 1\right),$$

where $s = d + e - a - b - c$. The result, after a little simplification, is

$$\frac{\alpha(\alpha+1)}{(m+\alpha)(n+\alpha)} {}_3F_2\left(\begin{matrix} 1-m, 1-n, 1-\alpha \\ 1-m-\alpha, 1-n-\alpha \end{matrix}; 1\right).$$

It is clear that every term of this $_3F_2$ is positive when $0 < \alpha < 1$. This proves the theorem. ∎

Remark Theorem 7.2.1 is equivalent to the statement that

$$_3F_2\left(\begin{matrix} -m, -n, -\alpha - 1 \\ -m - \beta, -n - \gamma \end{matrix}; 1\right) \geq 0, \quad m, n = 0, 1, 2, \ldots \tag{7.2.4}$$

when $0 \leq \alpha \leq \min(\beta, \gamma)$. The condition $0 \leq \alpha \leq \min(\beta, \gamma)$ is necessary. Take $m = 1$ in (7.2.4) to get

$$_3F_2\left(\begin{matrix} -1, -n, -\alpha - 1 \\ -1 - \beta, -n - \gamma \end{matrix}; 1\right) = 1 - \frac{(\alpha + 1)n}{(1 + \beta)(n + \gamma)} \geq 0.$$

Let $n \to \infty$ to see that $\alpha \leq \beta$. By symmetry $\alpha \leq \gamma$.

The next problem is to show the positivity of the coefficients $A(k, m, n)$ in the power series expansion of the rational function

$$\frac{1}{(1-r)(1-s) + (1-r)(1-t) + (1-s)(1-t)} = \sum_{k,m,n=0}^{\infty} A(k, m, n) r^k s^m t^n.$$

$$\tag{7.2.5}$$

The $A(k, m, n)$ satisfy a finite-difference equation that approximates a two-dimensional wave equation. Friedrichs and Lewy wished to use the positivity of $A(k, m, n)$ to prove the convergence of solutions of finite-difference approximations to solutions of the wave equation. Szegö [1933] gave a proof using Bessel functions. He also translated this problem into an equivalent problem about the positivity of integrals of products of Laguerre polynomials. We follow this direction here.

Rewrite the left-hand side of (7.2.5) as

$$\frac{1}{(1-r)(1-s)(1-t)} \cdot \frac{1}{\frac{1}{1-r} + \frac{1}{1-s} + \frac{1}{1-t}}$$

$$= \int_0^\infty \frac{e^{-x/(1-r)}}{1-r} \cdot \frac{e^{-x/(1-s)}}{1-s} \cdot \frac{e^{-x/(1-t)}}{1-t} dx. \tag{7.2.6}$$

Recall the generating function for the Laguerre polynomials $L_n^\alpha(x)$,

$$\frac{e^{-xr/(1-r)}}{(1-r)^{\alpha+1}} = \sum_{n=0}^{\infty} L_n^\alpha(x) r^n. \tag{6.2.4}$$

Then

$$\frac{e^{-x/(1-r)}}{1-r} = e^{-x} \cdot \frac{e^{-xr/(1-r)}}{1-r} = e^{-x} \sum_{n=0}^{\infty} L_n(x) r^n.$$

Thus (7.2.6) is equal to

$$\sum_{k,m,n} \int_0^\infty L_k(x)L_m(x)L_n(x)e^{-3x}dx\, r^k s^m t^n$$

and

$$A(k,m,n) = \int_0^\infty L_k(x)L_m(x)L_n(x)e^{-3x}dx.$$

A more general situation can be treated in a similar way. Note that if $f(x) = (x-r)(x-s)(x-t)$ then the left-hand side of (7.2.5) is $1/f'(1)$. Write

$$\frac{1}{[f'(1)]^{\alpha+1}} = \sum_{k,m,n=0}^\infty A^\alpha(k,m,n)r^k s^m t^n,$$

so that

$$A^\alpha(k,m,n) = \frac{1}{\Gamma(\alpha+1)}\int_0^\infty L_k^\alpha(x)L_m^\alpha(x)L_n^\alpha(x)x^\alpha e^{-3x}dx. \qquad (7.2.7)$$

Theorem 7.2.2   For $\alpha \geq -1/2$, $A^\alpha(k,m,n) \geq 0$. For $\alpha \geq 0$, the inequality is strict, that is, $A^\alpha(k,m,n) > 0$.

Proof.   In Chapter 6, we computed integrals of the products of three Hermite or three ultraspherical polynomials. This also gave their nonnegativity. These integrals were obtained from corresponding linearization formulas. That method does not work here. But recall that

$$\lim_{\beta \to \infty} P_n^{(\alpha,\beta)}(1 - 2x/\beta) = L_n^\alpha(x).$$

Thus it is reasonable to consider the positivity of the integral

$$\int_{-1}^1 P_k^{(\alpha,\alpha+j)}(x)P_m^{(\alpha,\alpha+j)}(x)P_n^{(\alpha,\alpha+j)}(x)(1-x)^\alpha(1+x)^{\alpha+3j}dx. \qquad (7.2.8)$$

We already know that

$$\int_{-1}^1 P_k^{(\alpha,\alpha)}(x)P_m^{(\alpha,\alpha)}(x)P_n^{(\alpha,\alpha)}(x)(1-x^2)^\alpha dx \geq 0 \quad \text{for } \alpha \geq -1/2. \qquad (7.2.9)$$

The question is: Can one increase the second parameter β in $P_\ell^{(\alpha,\beta)}(x)$ and still retain positivity in (7.2.8)? In fact, we have the formula

$$(1+x)P_n^{(\alpha,\beta+1)}(x) = \frac{2(n+1)}{2n+\alpha+\beta+2}P_{n+1}^{(\alpha,\beta)}(x) + \frac{2(n+\beta+1)}{2n+\alpha+\beta+2}P_n^{(\alpha,\beta)}(x). \qquad (7.2.10)$$

Verify this by noting that the right side vanishes when $x = -1$ and that Jacobi polynomials are orthogonal. Since the coefficients in (7.2.10) are positive, we get the nonnegativity of (7.2.8) from (7.2.9) and (7.2.10). This in turn implies the nonnegativity of $A^\alpha(k, m, n)$ for $\alpha \geq -1/2$.

The strict positivity for $\alpha \geq 0$ comes from

$$\frac{1}{[f'(1)]^{\alpha+1}} = \left[\frac{1}{[f'(1)]^{(\alpha-1)/2+1}}\right]^2.$$

This implies

$$[\Gamma((\alpha+1)/2)]^2 \int_0^\infty L_k^\alpha(x)L_m^\alpha(x)L_n^\alpha(x)x^\alpha e^{-3x}dx$$

$$= \Gamma(\alpha+1)\sum_{a=0}^k\sum_{b=0}^m\sum_{c=0}^n I(k-a, m-b, n-c)I(a, b, c), \quad (7.2.11)$$

where

$$I(i, j, k) = \int_0^\infty L_i^{(\alpha-1)/2}(x)L_j^{(\alpha-1)/2}(x)L_k^{(\alpha-1)/2}(x)x^{(\alpha-1)/2}e^{-3x}dx.$$

When $\alpha \geq 0$, all terms in (7.2.11) are nonnegative. So it is enough to find one strictly positive term. The positivity of the term $a = k, b = m, c = 0$ follows from the next lemma, which proves the theorem. ∎

Lemma 7.2.3 *For $\alpha > -1, \epsilon > 0$, we have*

$$\int_0^\infty e^{-\epsilon x}L_n^\alpha(x)L_m^\alpha(x)x^\alpha e^{-x}dx > 0. \quad (7.2.12)$$

Proof. Consider the generating function

$$\sum_{m,n}^\infty r^n s^m \int_0^\infty L_n^\alpha(x)L_m^\alpha(x)x^\alpha e^{-(1+\epsilon)x}dx$$

$$= \int_0^\infty \frac{x^\alpha e^{-xr/(1-r)-xs/(1-s)-(1+\epsilon)x}}{(1-r)^{\alpha+1}(1-s)^{\alpha+1}}dx$$

$$= \frac{\Gamma(\alpha+1)}{(1+\epsilon)^{\alpha+1}}\left[1 - \left(\frac{\epsilon}{1+\epsilon}(r+s) + \frac{1-\epsilon}{1+\epsilon}rs\right)\right]^{-(\alpha+1)}.$$

From the last expression it is clear that if $0 < \epsilon < 1$, then the coefficient of $r^n s^m$ is positive. The result may be extended to larger values of ϵ by iteration as follows. Since $e^{-\epsilon x}L_n^\alpha(x)$ is smooth and integrable, we can expand it in terms of Laguerre polynomials. Let

$$e^{-\epsilon x}L_n^\alpha(x) = \sum_{k=0}^\infty C_k(\epsilon)L_k^\alpha(x).$$

For $0 < \epsilon < 1$, $C_k(\epsilon) > 0$ by our previous remarks. Now

$$e^{-2\epsilon x} L_n^\alpha(x) = \sum_{k=0}^{\infty} C_k(\epsilon) e^{-\epsilon x} L_k^\alpha(x).$$

So $e^{-2\epsilon x} L_n^\alpha(x)$ can be written as a sum with positive coefficients. Iteration of this process completes the proof of the lemma. Another proof is in Exercises 6 and 7. ∎

A consequence of Lemma 7.2.3 is the next result. The proof is left to the reader.

Corollary 7.2.4 *Let* $\alpha > -1$, *and suppose that* $f(x)$ *is represented by its Laguerre–Fourier expansion. Suppose also that the coefficients of the expansion are positive, that is,*

$$a_n = \int_0^\infty f(x) L_n^\alpha(x) x^\alpha e^{-x} dx \geq 0, \quad n = 0, 1, \dots.$$

Then

$$a_n(\epsilon) = \int_0^\infty f(x) e^{-\epsilon x} L_n^\alpha(x) x^\alpha e^{-x} dx > 0, \quad n = 0, 1, 2, \dots, \epsilon > 0,$$

unless $f(x) = 0$, $x \geq 0$.

Theorem 7.2.2 is surprising when considered from a different point of view, as we shall see below. Consider the following result of Sarmanov [1968].

Theorem 7.2.5 *If*

$$f(x, y) = \sum_{n=0}^{\infty} a_n L_n^\alpha(x) L_n^\alpha(y) / L_n^\alpha(0) \geq 0, \quad 0 \leq x, y < \infty, \tag{7.2.13}$$

then

$$a_n = \int_0^1 r^n d\mu(r),$$

where $d\mu(r)$ *is a positive measure.*

In fact, the positivity of (7.2.13) for $a_n = r^n$, $0 \leq r < 1$, is a consequence of (6.2.28). Now Laguerre polynomials satisfy a differential equation in x and a difference equation in n. It frequently happens that a dual result can be obtained by interchanging n and x. For example, Lemma 7.2.3 is the dual of the positivity of (7.2.13) when $a_n = r^n$. A dual of Szegö's positive integral (7.2.7) would be

$$\sum_{n=0}^{\infty} a_n L_n^\alpha(x) L_n^\alpha(y) L_n^\alpha(z) \geq 0, \quad 0 \leq x, y, z < \infty,$$

for some sequence a_n. However, Theorem 7.2.5 shows that

$$a_n L_n^\alpha(z) L_n^\alpha(0) = \int_0^1 r^n d\mu(r, z),$$

with $d\mu(r, z) \geq 0$ for all z, $0 \leq z < \infty$. This is possible only when $d\mu(r, z)$ is a point mass at $r = 0$, and when $a_0 = c \geq 0$, $a_n = 0$, $n = 1, 2, \ldots$, since $L_n^\alpha(z)$ is negative for some $z > 0$.

Sarmanov's paper contains a proof of Theorem 7.2.5. Another proof that makes more explicit use of special functions is in Askey [1970].

There are some extensions of Theorem 7.2.2 for some α.

Theorem 7.2.6   *If* $\alpha \geq \alpha_0 = (-5 + \sqrt{17})/2$, *then*

$$\int_0^\infty x^a L_k^\alpha(x) L_m^\alpha(x) L_n^\alpha(x) e^{-2x} dx \geq 0, \qquad \alpha \geq \alpha_0,$$

$k, m, n = 0, 1, \ldots$ *. The only case of equality occurs when* $k = m = n = 1$ *and* $\alpha = \alpha_0$.

For a proof, see Askey and Gasper [1977]. The case $\alpha = 0, 1, \ldots$ is outlined in Exercise 10. The fact that Theorem 7.2.6 implies Theorem 7.2.2 when $\alpha \geq \alpha_0$ follows from Lemma 7.2.3.

Theorem 7.2.7   *If* $0 < a < 1$, $a + b = 1$, *and* $\alpha \geq 0$, *then*

$$\int_0^\infty x^a L_k^\alpha(ax) L_m^\alpha(bx) L_n^\alpha(x) e^{-x} dx \geq 0,$$

$k, m, n = 0, 1, \ldots$.

For a proof, see Koornwinder [1978]. We sketch a proof of Koornwinder's inequality in Theorem 7.2.7 for nonnegative integer values of α after a discussion of MacMahon's Master Theorem.

The theorem of MacMahon [1917–1918, pp. 93–98], known as the Master Theorem, makes it possible to give combinatorial interpretations of coefficients of series expansions of rational functions in several variables. MacMahon's Master Theorem can be stated as follows: Suppose

$$V_n = (-1)^n x_1 \cdots x_n \begin{vmatrix} a_{11} - 1/x_1 & a_{12} & \cdots & a_{1n} \\ a_{21} & a_{22} - 1/x_2 & \cdots & a_{2n} \\ \vdots & \vdots & & \vdots \\ a_{n1} & a_{n2} & \cdots & a_{nn} - 1/x_n \end{vmatrix}.$$

Then the coefficient of $x_1^{k_1} x_2^{k_2} \cdots x_n^{k_n}$ in the expansion of $1/V_n$ is the same as the coefficient of the same term in

$$(a_{11}x_1 + \cdots + a_{1n}x_n)^{k_1} \cdots (a_{n1}x_1 + \cdots + a_{nn}x_n)^{k_n}.$$

As an application of this theorem, consider the following example. An easy calculation shows that

$$1 - \frac{1}{2}(r+s+t) + \frac{1}{2}rst = -rst \begin{vmatrix} 1/2 - 1/r & -1/2 & -1/2 \\ -1/2 & 1/2 - 1/s & -1/2 \\ -1/2 & -1/2 & 1/2 - 1/t \end{vmatrix}.$$

The Master Theorem implies that the coefficient of $r^k s^m t^n$ in the series expansion $[1 - (r+s+t)/2 + rst/2]^{-1}$ is the same as the coefficient of $r^k s^m t^n$ in $(r-s-t)^k(-r+s-t)^m(-r-s+t)^n/2^{k+m+n}$. The combinatorial interpretation of this result is as follows. Take three boxes with $k, m,$ and n distinguishable objects in them. Rearrange these objects among the boxes so that the number of objects in each box remains the same. Then the coefficient of $r^k s^m t^n$ in $(r-s-t)^k(-r+s-t)^m(-r-s+t)^n$ represents the number of rearrangements where an even number of objects has been moved from one box to a different box minus the number of rearrangements where an odd number of objects has been moved to a different box. By Exercise 10 (where this coefficient has been obtained as an integral of a product of three Laguerre polynomials), this coefficient must be positive. Thus, we see that in the above combinatorial situation, the number of "even" rearrangements exceed the number of "odd" rearrangements.

As another application of the Master Theorem, we outline Ismail and Tamhankar's [1979] proof of Koornwinder's result in Theorem 7.2.7 for $\alpha = 0, 1, 2, \ldots$. Let

$$B^\alpha(k, m, n) = \int_0^\infty L_k^\alpha(x)L_m^\alpha((1-\lambda)x)L_n^\alpha(\lambda x)x^\alpha e^{-x}dx.$$

A simple calculation, using the generating function for Laguerre polynomials, shows that

$$\sum_{k,m,n} B^\alpha(k, m, n)r^k s^m t^n = \frac{\Gamma(\alpha+1)}{[1 - (1-\lambda)r - \lambda s - \lambda rt - (1-\lambda)st + rst]^{\alpha+1}}.$$

Since α is a nonnegative integer, $B^0(k, m, n) \geq 0$ implies that $B^\alpha(k, m, n) \geq 0$. So we take $\alpha = 0$. To apply the Master Theorem observe that

$$1 - (1-\lambda)r - \lambda s - \lambda rt - (1-\lambda)st + rst$$

$$= -rst \begin{vmatrix} (1-\lambda) - 1/r & -\sqrt{\lambda(1-\lambda)} & -\sqrt{\lambda} \\ -\sqrt{\lambda(1-\lambda)} & \lambda - 1/s & -\sqrt{1-\lambda} \\ -\sqrt{\lambda} & -\sqrt{1-\lambda} & -1/t \end{vmatrix}.$$

By the Master Theorem, $B^0(k, m, n)$ is the coefficient of $r^k s^m t^n$ in

$$[(1 - \lambda)r - \sqrt{\lambda(1 - \lambda)}s - \sqrt{\lambda}t]^k[-\sqrt{\lambda(1 - \lambda)}r$$
$$+ \lambda s - \sqrt{1 - \lambda}t]^m[-\sqrt{\lambda}r - \sqrt{1 - \lambda}s]^n.$$

By applications of the binomial theorem, Ismail and Tamhankar show that

$$B^0(k, m, n) = \lambda^{2k+m-n}(1 - \lambda)^{n-k}\frac{(k + m - n)!n!}{k!m!}$$

$$\cdot \left[\sum_i (-1)^i\{(1 - \lambda)/\lambda\}^i \binom{k}{i}\binom{m}{n - k + i}\right]^2$$

$$\geq 0. \tag{7.2.14}$$

This proves Koornwinder's theorem for nonnegative integer values of α.

Note that Koornwinder's inequality implies that for $0 < \lambda < 1, \alpha \geq 0$,

$$L_m^\alpha(\lambda x)L_n^\alpha((1 - \lambda)x) = \sum_{k=0}^{m+n} a_{k,m,n}L_k^\alpha(x) \tag{7.2.15}$$

with $a_{k,m,n} \geq 0$. This relation can be iterated to give

$$L_{n_1}^\alpha(\lambda_1 x) \cdots L_{n_j}^\alpha(\lambda_j x) = \sum_{k=0}^{n_1+\cdots+n_j} a_k L_k^\alpha(x), \tag{7.2.16}$$

with $a_k \geq 0$ when $\alpha \geq 0$, $\sum_{i=1}^j \lambda_i = 1$, and $\lambda_i \geq 0, i = 1, 2, \ldots, j$.

Several proofs of the Master Theorem now exist. Perhaps the proof that best explains its combinatorial significance is due to Foata [1965]. This proof was later simplified by Cartier and Foata [1969]. A readily accessible treatment of this argument is given by Brualdi and Ryser [1991]. A short proof using a multiple complex variables integral was given by Good [1962].

7.3 Positive Polynomial Sums from Quadrature and Vietoris's Inequality

Fejér used the inequality

$$\sum_{k=0}^n \sin(k + 1/2)\theta = \frac{1 - \cos(n + 1)\theta}{2 \sin \theta/2} = \frac{\sin^2((n + 1)\theta/2)}{\sin \theta/2} \geq 0, \quad 0 \leq \theta \leq 2\pi,$$

to prove that the Fourier series of a continuous function is $(C, 1)$ summable to the function. This inequality can be expressed as

$$\sum_{k=0}^n \frac{P_k^{(1/2,-1/2)}(\cos \theta)}{P_k^{(-1/2,1/2)}(1)} = \left(\frac{\sin((n + 1)\theta/2)}{\sin \theta/2}\right)^2 \geq 0, \quad 0 \leq \theta \leq 2\pi. \tag{7.3.1}$$

In Section 6.7, we saw that a similar inequality holds for Legendre polynomials, that is,

$$\sum_{k=0}^{n} P_k(x) = \sum_{k=0}^{n} \frac{P_k(x)}{P_k(1)} > 0, \quad -1 < x \le 1. \tag{7.3.2}$$

Fejér used this to study the summability of a series of spherical functions. He also conjectured that

$$\sum_{k=0}^{n} \frac{\sin(k+1)\theta}{k+1} > 0, \quad 0 < \theta < \pi. \tag{7.3.3}$$

These sums are partial sums of the Fourier series

$$\frac{\pi - \theta}{2} = \sum_{k=0}^{\infty} \frac{\sin(k+1)\theta}{k+1}, \quad 0 < \theta \le \pi,$$

which was studied because it illustrates the Gibbs phenomenon. It is possible that the graphs of the partial sums suggested the conjecture to Fejér. We can write (7.3.3) as

$$\sum_{k=0}^{n} \frac{P_k^{(1/2,1/2)}(\cos\theta)}{P_k^{(1/2,1/2)}(1)} > 0, \quad 0 < \theta < \pi. \tag{7.3.4}$$

The earliest proofs of (7.3.3) are due to Jackson [1911] and Gronwall [1912]. Recall that (7.3.2) was obtained from (7.3.1) by using Mehler's integral in Section 6.7. There are other integrals in that section that give extensions to sums involving $P_n^{(\alpha,\beta)}(x)$. It is possible to obtain positive sums with terms of either the form $P_n^{(\alpha,\beta)}(x)/P_n^{(\alpha,\beta)}(1)$ or $P_n^{(\alpha,\beta)}(x)/P_n^{(\beta,\alpha)}(1)$. Without some applications in mind, it is difficult to determine which extension is going to be useful. The inequality in (7.3.1) suggests that sums of $P_n^{(\alpha,\beta)}(x)/P_n^{(\beta,\alpha)}(1)$ may be important. Here we consider a problem in quadrature that provides some confirmation of this.

Let $\{P_n(x)\}$ be a sequence of polynomials orthonormal with respect to the distribution $d\alpha(x)$ on (a, b). As in Gauss quadrature discussed in Chapter 5, interpolation is done at the zeros of the polynomials $P_n(x)$, but now the integration may be with respect to a different distribution. Let x_ν, $\nu = 1, \ldots, n$, denote the zeros of $P_n(x)$. Let $f(x)$ be a continuous function and let the interpolation polynomial be given by

$$\sum_{\nu=1}^{n} \ell_\nu(x) f(x_\nu) := \sum_{\nu=1}^{n} \frac{P_n(x) f(x_\nu)}{P_n'(x_\nu)(x - x_\nu)}. \tag{7.3.5}$$

Then we have the approximate formula

$$\int_a^b f(x)d\beta(x) \approx \sum_{v=1}^n \lambda_v f(x_v), \tag{7.3.6}$$

where

$$\lambda_v = \int_a^b \frac{P_n(x)d\beta(x)}{P_n'(x_v)(x - x_v)} \tag{7.3.7}$$

$$= \int_a^b \frac{P_n(x)P_{n+1}(x_v) - P_n(x_v)P_{n+1}(x)}{P_{n+1}(x_v)P_n'(x_v)(x - x_v)} d\beta(x)$$

$$= -\frac{k_{n+1}}{k_n} \cdot \frac{1}{P_n'(x_v)P_{n+1}(x_v)} \int_a^b \sum_{k=0}^n P_k(x_v)P_k(x)d\beta(x).$$

Here k_n is the coefficient of x^n in $P_n(x)$ and the last equation follows from the Christoffel–Darboux formula (Theorem 5.2.4). Now write

$$K(x) = \sum_{k=0}^n \left[P_k(x) \int_a^b P_k(t)d\beta(t) \right]. \tag{7.3.8}$$

Then

$$\lambda_v = -\frac{k_{n+1}}{k_n} \cdot \frac{K(x_v)}{P_n'(x_v)P_{n+1}(x_v)}. \tag{7.3.9}$$

If λ_v is positive, then it can be shown that the sum on the right-hand side of (7.3.6) converges to the integral as $n \to \infty$. The proof given for Gaussian quadrature in Chapter 5 works here as well.

Take the case in which $P_n(x) = P_n^{(\alpha,\beta)}(x)$, $d\beta(x) = dx$, and $(a, b) = (-1, 1)$. Then

$$K(x) = \sum_{k=0}^n \left[P_k^{(\alpha,\beta)}(x) \int_{-1}^1 P_k^{(\alpha,\beta)}(t)dt \right] \bigg/ h_k^{\alpha,\beta},$$

where $h_k^{\alpha,\beta}$ is given by (7.1.4). Write $P_k^{(\alpha,\beta)}(t)$ in hypergeometric form and integrate term by term to get

$$\frac{-(\alpha)_{k+1}}{(k + \alpha + \beta)(1)_{k+1}} \sum_{j=0}^k \frac{(-k - 1)_{j+1}(k + \alpha + \beta)_{j+1}}{(1)_{j+1}(\alpha)_{j+1}}$$

$$= \frac{-(\alpha)_{k+1}}{(k + \alpha + \beta)(1)_{k+1}} \left[{}_2F_1\left(\begin{matrix} -k - 1, k + \alpha + \beta \\ \alpha \end{matrix}; 1 \right) - 1 \right].$$

This $_2F_1$ can be evaluated by the Pfaff–Saalschütz identity (Theorem 2.2.6). The result is

$$K(x) = \sum_{k=0}^{n} \frac{(2k + \alpha + \beta + 1)\Gamma(k + \alpha + \beta)k!}{\Gamma(k + \alpha + 1)\Gamma(k + \beta + 1)}$$

$$\cdot \left[\frac{\Gamma(k + \alpha + 1)}{\Gamma(\alpha)\Gamma(k + 2)} + \frac{(-1)^k \Gamma(k + \beta + 1)}{\Gamma(\beta)\Gamma(k + 2)} \right] P_k^{(\alpha,\beta)}(x).$$

This sum is intractable when written as

$$\sum_{k=0}^{n} a_k \frac{P_k^{(\alpha,\beta)}(x)}{P_k^{(\alpha,\beta)}(1)}.$$

It can, however, be written in the form

$$K_n(x) = \frac{1}{\Gamma(\alpha + 1)\Gamma(\beta + 1)} \sum_{k=0}^{n} \frac{(2k + \alpha + \beta + 1)\Gamma(k + \alpha + \beta)}{(k + 1)!}$$

$$\cdot \left[\alpha \frac{P_k^{(\alpha,\beta)}(x)}{P_k^{(\beta,\alpha)}(1)} + \beta \frac{P_k^{(\beta,\alpha)}(-x)}{P_k^{(\alpha,\beta)}(1)} \right]. \tag{7.3.10}$$

Assume $k_n > 0$ in (7.3.9). By Corollary 5.2.6, it follows that

$$P_n'(x_\nu) P_{n+1}(x_\nu) < 0.$$

Thus to show that $\lambda_\nu > 0$ in (7.3.9), it is sufficient to prove that

$$K_n(x) \geq 0, \quad -1 \leq x \leq 1.$$

It is easy to check that

$$c_k = \frac{(2k + \alpha + \beta + 1)\Gamma(k + \alpha + \beta)}{(k + 1)!}$$

satisfies $0 \leq c_{k+1} \leq c_k$ when $0 < \alpha + \beta \leq 1$. So the nonnegativity of $K_n(x)$ in (7.3.10) will follow for $\alpha, \beta \geq 0, \alpha + \beta \leq 1$ upon summation by parts, provided that

$$D_n^{(\alpha,\beta)}(x) = \sum_{k=0}^{n} \frac{P_k^{(\alpha,\beta)}(x)}{P_k^{(\beta,\alpha)}(1)} \geq 0, \quad -1 \leq x \leq 1. \tag{7.3.11}$$

Observe that $D_n^{(\alpha,\beta)}(-1) = 0$ when n is odd. Thus (7.3.11) is sharp in the sense that equality holds at some point in $[-1, 1]$ for infinitely many values of n. Proof of the inequality (7.3.11) for some values of (α, β) is given in the next section.

Now suppose that $d\beta(x) = (1 - x)^{\alpha-\gamma}(1 + x)^{\beta-\delta}dx$. We shall look at λ_ν for some specific $\alpha, \beta, \gamma, \delta$. When $\alpha = 1/2, \beta = -1/2, \gamma = 1, \delta = 0$, the positivity of λ_v, using the expression (7.3.9), reduces to the positivity of the sum

$$\sum_{k=0}^{n} \sin(k + 1/2)\theta.$$

The two cases

$$\alpha = \beta = -1/2, \quad \gamma = 1/4, \quad \delta = -1/4$$

and

$$\alpha = \beta = 1/2, \quad \gamma = 3/4, \quad \delta = 1/4$$

lead to the respective sums

$$\sum_{k=0}^{n} c_k \cos kx, \quad \sum_{k=1}^{n} c_k \sin kx, \tag{7.3.12}$$

where

$$c_{2k} = c_{2k+1} = \frac{(1/2)_k}{k!}, \quad k = 0, 1, 2, \dots. \tag{7.3.13}$$

Vietoris [1958] proved the strict positivity of these sums for $0 < x < \pi$. A proof of these inequalities is given below. The inequality

$$\sum_{k=1}^{n} c_k \sin kx > 0, \quad 0 < x < \pi, \tag{7.3.14}$$

extends the inequality (7.3.4) of Jackson and Gronwall. To see this let $\theta \to \pi$ in (7.3.4). The result is

$$1 - 1 + 1 - 1 + \cdots + (-1)^{n+1},$$

which vanishes when n is even. So it might appear that the inequality cannot be improved. Now suppose that all we assume about the series in (7.3.14) is that

$$1 = c_1 \geq c_2 \geq c_3 \geq \cdots.$$

Divide the series (7.3.14) by $\sin x$ and let $x \to \pi$. We obtain

$$1 - 2c_2 + 3c_3 - 4c_4 + \cdots + (-1)^{n+1} n c_n. \tag{7.3.15}$$

For nonnegativity of this series for all n, we require that $c_2 \leq 1/2$. Take the largest value of $c_2 = 1/2$. Then $c_3 \leq c_2 = 1/2$ and the largest value of $c_3 = 1/2$. With these values of c_1, c_2, and c_3, we have $4c_4 \leq 3/2$ or $c_4 \leq 3/8$. So take $c_4 = 3/8$. If we continue in this manner, we get the sequence c_k as defined in (7.3.13).

As a first step in the proof of Vietoris's inequality, we show that the two sums in (7.3.12) are the partial sums of a Fourier series just as (7.3.3) is.

Proposition 7.3.1   *If c_k is the sequence defined by (7.3.13), then*

$$\sum_{k=1}^{\infty} c_k \sin kx = \sum_{k=0}^{\infty} c_k \cos kx = \left(\frac{1}{2}\cot(x/2)\right)^{1/2} \quad \text{for } 0 < x < \pi. \quad (7.3.16)$$

Proof.   For $|z| \le 1, z \ne 0$, we have

$$(1-z)^{-1/2} = \sum_{k=0}^{\infty} c_{2k} z^k.$$

It follows that

$$(1+z)(1-z^2)^{-1/2} = \sum_{k=0}^{\infty} c_k z^k, \quad |z| \le 1, z \ne \pm 1.$$

Set $z = e^{ix}, 0 < x < \pi$, and take the real and imaginary parts to get the result of the proposition.   ∎

Proposition 7.3.2   *For $m \ge 1$,*

$$\binom{2m}{m} < \frac{2^{2m}}{\sqrt{\pi m}}. \quad (7.3.17)$$

Proof.   Set

$$a_m = \frac{\sqrt{m}}{2^{2m}} \binom{2m}{m}.$$

It is easily seen that $a_m < a_{m+1}$ for $m \ge 1$. Now observe that

$$\lim_{m \to \infty} a_m = \lim_{m \to \infty} \frac{(1/2)_m \sqrt{m}}{m!}$$

$$= \frac{1}{\Gamma(1/2)} = \frac{1}{\sqrt{\pi}}.   ∎$$

The proof of the next proposition is from Brown and Hewitt [1984].

Proposition 7.3.3   *For c_k defined by (7.3.13), we have*

$$2\sin(\theta/2) \sum_{k=0}^{n} c_k \cos k\theta \ge \sqrt{\sin \theta} - 2c_{n+1}, \quad (7.3.18)$$

$$2\sin(\theta/2) \sum_{k=1}^{n} c_k \sin k\theta \ge \sqrt{\sin \theta} - 2c_{n+1}. \quad (7.3.19)$$

Proof.   Observe that for $m > n$

$$2\sin(\theta/2) \sum_{k=n+1}^{m} c_k \cos k\theta = \sum_{k=n+1}^{m} c_k [\sin(k+1/2)\theta - \sin(k-1/2)\theta]$$

$$= -c_{n+1} \sin(n+1/2)\theta + \sum_{k=n+1}^{m-1} (c_k - c_{k+1}) \sin(k+1/2)\theta$$

$$+ c_m \sin(m+1/2)\theta$$

$$\leq c_{n+1}(1 - \sin(n+1/2)\theta) - c_m(1 - \sin(m+1/2)\theta)$$

$$\leq 2c_{n+1}.$$

By Proposition 7.3.1,

$$\sqrt{\sin\theta} = 2\sin(\theta/2) \sum_{k=0}^{\infty} c_k \cos k\theta$$

$$= 2\sin(\theta/2) \left(\sum_{k=0}^{n} c_k \cos k\theta + \sum_{k=n+1}^{\infty} c_k \cos k\theta \right)$$

$$\leq 2\sin(\theta/2) \sum_{k=0}^{n} c_k \cos k\theta + 2c_{n+1}.$$

This proves (7.3.18), and the proof of (7.3.19) is similar. The proposition is proved. ∎

We are now in a position to prove Vietoris's inequalities, which are explicitly stated in the next theorem.

Theorem 7.3.4   *If*

$$c_{2k} = c_{2k+1} = \frac{1}{2^{2k}} \binom{2k}{k}, \quad k \geq 0,$$

then

$$\sigma_n(x) = \sum_{k=1}^{n} c_k \sin kx > 0, \quad 0 < x < \pi, \tag{7.3.20}$$

and

$$r_n(x) = \sum_{k=0}^{n} c_k \cos kx > 0, \quad 0 < x < \pi. \tag{7.3.21}$$

Proof.   Consider (7.3.20) first. The result is clearly true for $n = 1$. So let $n \geq 2$. We need a separate argument for each of the three intervals: $0 < x \leq \pi/n, \pi/n < x < \pi - \pi/n$, and $\pi - \pi/n \leq x < \pi$.

The positivity of $\sigma_n(x)$ for $0 < x \le \pi/n$ is obvious since each term in the sum is nonnegative and the first term is strictly positive. When $\pi - \pi/n \le x < \pi$, set $x = \pi - y$ so that $0 < y \le \pi/n$. Suppose n is even, say $n = 2m$. Then

$$\sigma_n(x) = \sum_{k=1}^{2m}(-1)^{k-1}c_k \sin ky = \sum_{k=1}^{m}[c_{2k-1}\sin(2k-1)y - c_{2k}\sin 2ky]$$

$$= \sum_{k=1}^{m}(2k-1)c_{2k-1}\left[\frac{\sin(2k-1)y}{2k-1} - \frac{\sin 2ky}{2k}\right].$$

The last term in square brackets is positive because $\sin t/t$ is decreasing in $(0, \pi]$ and $2ky \le 2my = ny \le \pi$. So $\sigma_n(x) > 0$. When n is odd there is an extra term in the sum, $c_n = \sin ny$, which is positive for $0 < y < \pi/n$. Thus $\sigma_n(x) > 0$ whether n is even or odd.

Now note that

$$\sin u \ge u - u^3/6.$$

For the interval $\pi/n < x < \pi - \pi/n$, which is nontrivial for $n \ge 3$, we then have

$$\sin x > \sin(\pi/n) \ge (\pi/n)(1 - \pi^2/6n^2).$$

By (7.3.19)

$$2\sin(x/2)\sigma_n(x) \ge [(\pi/n)(1 - \pi^2/6n^2)]^{1/2} - 2c_{n+1}. \tag{7.3.22}$$

An easy calculation shows that the term in square brackets decreases for $n \ge \pi/\sqrt{2}$. So for $n \ge 3$, and by the definition of c_n, the right-hand side of (7.3.22) is positive for $n = 2m - 1$, if it is positive for $n = 2m$. For the latter value of n, (7.3.17) implies that the right-hand side of (7.3.22) is at least equal to

$$\frac{1}{\sqrt{2\pi m}}[\pi(1 - \pi^2/24m^2)^{1/2} - 2\sqrt{2}].$$

A simple computation shows that this expression is positive for $m \ge 2$. This proves the inequality in (7.3.20) .

The inequality (7.3.21) is clearly true for $n = 0$ and 1. Moreover,

$$r_2(x) = \frac{1}{2}\cos 2x + \cos x + 1$$

$$= \cos^2 x + \cos x + \frac{1}{2} = \left(\cos x + \frac{1}{2}\right)^2 + \frac{1}{4} > 0.$$

Assume that $n \ge 3$. For $0 < x \le \pi/n$,

$$\frac{dr_n}{dx} = -\sum_{k=1}^{n}kc_k \sin kx < 0, \quad 0 < x < \pi/n.$$

So $r_n(x)$ is decreasing in $0 < x < \pi/n$, and its value at π/n is positive. Note that

$$r_n(\pi/n) = \sum_{k=0}^{[n/2]} (c_k - c_{n-k}) \cos(k\pi/n) > 01.$$

Thus $r_n(x) > 0$ for $0 < x \leq \pi/n$. Now let $\pi - \pi/(n+1) < x < \pi$ and set $y = \pi - x$ so that

$$r_n(x) = \sum_{k=0}^{[(n-1)/2]} c_{2k}[\cos 2ky - \cos(2k+1)y] + \epsilon_n,$$

where $\epsilon_n = 0$ if $n = 2m - 1$ and $\epsilon_n = c_{2m} \cos 2my$ if $n = 2m$. The expression in the sum is positive because $\cos x$ is decreasing in $0 \leq x \leq \pi$. This implies that for $n = 2m - 1$, $r_n(x) > 0$ for $0 < y < \pi/n$. When $n = 2m$, we have

$$r_n(x) \geq c_{2m}(1 - \cos y + \cos 2y - \cos 3y + \cdots + \cos 2my)$$
$$= c_{2m}(1 + \cos x + \cos 2x + \cdots + \cos 2mx)$$
$$= c_{2m} \operatorname{Re}\left[\frac{e^{i(2m+1)x} - 1}{e^{ix} - 1}\right]$$
$$= c_{2m} \operatorname{Re}\left[e^{imx} \frac{e^{i(m+1/2)x} - e^{-i(m+1/2)x}}{e^{ix/2} - e^{-ix/2}}\right]$$
$$= c_{2m} \frac{\sin(m + 1/2)x \cos mx}{\sin(x/2)}$$
$$= c_{2m} \frac{\cos(m + 1/2)y \cos my}{\cos(y/2)}.$$

It follows that $r_n(x) > 0$ for $0 < (m+1/2)y < \pi/2$, that is, for $0 < y < \pi/(n+1)$. The rest of the argument can be completed as before. Suppose that $n \geq 3$ and $\pi/(n+1) \leq x \leq \pi - \pi/(n+1)$. As in the case of $\sigma_n(x)$ on $\pi/n < x < \pi - \pi/n$, it is sufficient to show that

$$\left[\frac{\pi}{n+1}\left(1 - \frac{\pi^2}{6(n+1)^2}\right)\right]^{1/2} - 2c_{n+1} > 0.$$

Again it suffices to consider even values of n, say $n = 2m$. The inequality can be directly checked for $m = 2$ and 3. For $m \geq 4$, apply (7.3.17) to see that the following inequality is stronger:

$$\left[\frac{\pi}{2m+1}\left(1 - \frac{\pi^2}{6(2m+1)^2}\right)\right]^{1/2} - \frac{2}{\sqrt{\pi m}} > 0.$$

This is true for $m = 4$, and when the left side is multiplied by $\sqrt{m}$, it is an increasing function of m. Thus the inequality holds for $m \geq 4$, and the theorem is proved. ∎

The next theorem, which is apparently a generalization of Theorem 7.3.4, is in fact equivalent to it. It is also due to Vietoris.

Theorem 7.3.5   *If $a_0 \geq a_1 \geq \cdots \geq a_n > 0$ and $2ka_{2k} \leq (2k-1)a_{2k-1}, k \geq 1$, then*

$$s_n(x) = \sum_{k=1}^{n} a_k \sin kx > 0, \quad 0 < x < \pi, \qquad (7.3.23)$$

and

$$t_n(x) = \sum_{k=0}^{n} a_k \cos kx > 0, \quad 0 < x < \pi. \qquad (7.3.24)$$

Proof.   For c_k as defined in Theorem 7.3.4, let $a_k = c_k d_k$. Then $d_0 \geq d_1 \geq d_2 \geq \cdots \geq d_n > 0$ and summation by parts gives

$$s_n(x) = \sum_{k=1}^{n} c_k d_k \sin kx$$
$$= \sum_{k=1}^{n-1} (d_k - d_{k+1})\sigma_k(x) + d_n \sigma_n(x) > 0, \quad 0 < x < \pi,$$

by (7.3.20). This proves the theorem since (7.3.24) can be done in a similar way.   ■

The Jackson–Gronwall inequality is a consequence of (7.3.23). Just take $a_k = 1/k$.

There is a nice application of these inequalities of Vietoris to the problem of finding sufficient conditions on the coefficients of trigonometric polynomials to force all the zeros to be real, and then also yields information about the distribution of these zeros. Szegö [1936] proved the following theorem.

Theorem 7.3.6   *If $\lambda_0 > \lambda_1 \geq \lambda_2 \geq \cdots \geq \lambda_n > 0$, and s_k and t_k denote the zeros of*

$$p(\theta) = \sum_{k=0}^{n} \lambda_k \cos(n-k)\theta$$

and

$$q(\theta) = \sum_{k=1}^{n-1} \lambda_k \sin(n-k)\theta$$

respectively, with their order such that they are increasing in size on $(0, \pi)$, then

$$\left(k - \frac{1}{2}\right)\pi \Big/ \left(n + \frac{1}{2}\right) < s_k < \left(k + \frac{1}{2}\right)\pi \Big/ \left(n + \frac{1}{2}\right), \quad k = 1, \ldots, n \quad (7.3.25)$$

and

$$k\pi \Big/ \left(n + \frac{1}{2}\right) < t_k < (k + 1)\pi \left(n + \frac{1}{2}\right), \quad k = 1, \ldots, n - 1. \quad (7.3.26)$$

If the λ_k are not only increasing but satisfy the following convexity-type condition:

$$2\lambda_0 - \lambda_1 > \lambda_1 - \lambda_2 \geq \lambda_2 - \lambda_3 \geq \cdots \geq \lambda_{n-1} - \lambda_n \geq \lambda_n \geq 0, \quad (7.3.27)$$

then the right-hand sides of (7.3.25) and (7.3.26) can be replaced by $k\pi/n$ and $(k + 1/2)\pi/n$ respectively.

The Vietoris inequalities can be used to obtain two other trigonometric inequalities. See Exercise 17. These two inequalities along with the conditions

$$(2k - 1)\lambda_{k-1} \geq 2k\lambda_k > 0, \quad k = 1, 2, \ldots, \quad (7.3.28)$$

lead to the following different improvements in (7.3.25) and (7.3.26) :

$$\left(k - \frac{1}{2}\right)\pi \Big/ \left(n + \frac{1}{4}\right) < s_k < k\pi \Big/ \left(n + \frac{1}{2}\right), \quad k = 1, \ldots, n, \quad (7.3.29)$$

$$k\pi \Big/ \left(n + \frac{1}{4}\right) < t_k < \left(k + \frac{1}{2}\right)\pi \Big/ \left(n + \frac{1}{4}\right), \quad k = 1, \ldots n - 1. \quad (7.3.30)$$

For a proof of these inequalities see Askey and Steinig [1974].

7.4 Positive Polynomial Sums and the Bieberbach Conjecture

In the previous section, we saw the significance of showing the positivity of the sums

$$\sum_{k=0}^{n} \frac{P_k^{(\alpha,\beta)}(x)}{P_k^{(\beta,\alpha)}(1)}. \quad (7.4.1)$$

The positivity of some of these sums has turned out to be important. We illustrate this for some specific α and β, though much more is known. For more information, see Askey [1975].

The strict positivity of (7.4.1) for $\alpha = \beta = 0, -1 < x \leq 1$ was proved in Chapter 6, Section 6.7. This implies, after summation by parts, that for Legendre polynomials $P_k(x)$,

$$\sum_{k=0}^{n} a_k P_k(x) > 0, \quad -1 < x \leq 1, \tag{7.4.2}$$

when $a_k \geq a_{k+1} \geq 0, a_0 > 0, k = 0, 1, \ldots, n - 1$. The next result is due to Feldheim [1963] and gives the positivity of (7.4.1) when $\alpha = \beta \geq 0$.

Theorem 7.4.1   *For $0 \leq \theta < \pi$ and $v \geq 1/2$, we have*

$$\sum_{k=0}^{n} \frac{C_k^v(\cos\theta)}{C_k^v(1)} > 0. \tag{7.4.3}$$

Proof.   By the Feldheim–Vilenkin integral (Corollary 6.7.3), we have for $v > 1/2$,

$$\sum_{k=0}^{n} \frac{C_k^v(\cos\theta)}{C_k^v(1)} = \frac{2\Gamma(v + 1/2)}{\Gamma(v - 1/2)} \int_0^{\pi/2} \sin\phi \cos^{2v-2}\phi$$
$$\cdot \sum_{k=0}^{n} [1 - \sin^2\theta \cos^2\phi]^{k/2} P_k(\cos\theta(1 - \sin^2\theta \cos^2\phi)^{-1/2})d\phi.$$

Take $a_k = [1 - \sin^2\theta \cos^2\phi]^{k/2}$. Then $a_k \geq a_{k+1} \geq 0$ and $a_0 = 1$. So by (7.4.2) the integral is positive and the theorem is proved.   ■

The Jackson–Gronwall inequality

$$\sum_{k=1}^{n} \frac{\sin k\theta}{k} > 0, \quad 0 < \theta < \pi, \tag{7.3.3}$$

is a corollary of Theorem 7.4.1 when $v = 1$.

The positivity of (7.4.1) for $\beta = 0$ and $\alpha = 0, 1, 2, \ldots$ was needed in the first proof of the Bieberbach conjecture on univalent functions. See de Branges [1985]. More generally, take $\alpha > -1$.

Since $P_n^{(0,\alpha)}(1) = 1$, the sum we are interested in is

$$\sum_{k=0}^{n} P_k^{(\alpha,0)}(x), \quad \alpha > -1. \tag{7.4.4}$$

The first step in the proof, due to Gasper, of the positivity of (7.4.4) is to express it as a hypergeometric series. Thus

$$\sum_{k=0}^{n} P_k^{(\alpha,0)}(x) = \sum_{k=0}^{n} \frac{(\alpha+1)_k}{k!} \sum_{j=0}^{k} \frac{(-k)_j (k+\alpha+1)_j}{(\alpha+1)_j j!} \left(\frac{1-x}{2}\right)^j$$

$$= \sum_{j=0}^{n} \frac{(-1)^j ((1-x)/2)^j}{j!(\alpha+1)_j} \sum_{k=j}^{n} \frac{(\alpha+1)_{k+j}}{(k-j)!}$$

$$= \sum_{j=0}^{n} \frac{(\alpha+1)_{2j}((x-1)/2)^j}{j!(\alpha+1)_j} \sum_{k=0}^{n-j} \frac{(\alpha+2j+1)_k}{k!}.$$

Since the inner sum is

$$\frac{(\alpha+2j+2)_{n-j}}{(n-j)!},$$

we have

$$\sum_{k=0}^{n} P_k^{(\alpha,0)}(x) = \sum_{j=0}^{n} \frac{(\alpha+1)}{(\alpha+2j+1)} \cdot \frac{(\alpha+2)_{n+j}(-n)_j}{(\alpha+1)_j j! n!} \left(\frac{1-x}{2}\right)^j$$

$$= \frac{(\alpha+2)_n}{n!} \, {}_3F_2\left(\begin{matrix} -n, n+\alpha+2, (\alpha+1)/2 \\ (\alpha+3)/2, \alpha+1 \end{matrix}; \frac{1-x}{2}\right). \quad (7.4.5)$$

There is a formula of Clausen that gives the square of a ${}_2F_1$ as a ${}_3F_2$ (see Exercise 3.17). The formula is

$$ {}_3F_2\left(\begin{matrix} 2a, 2b, a+b \\ a+b+1/2, 2a+2b \end{matrix}; x\right) = \left[{}_2F_1\left(\begin{matrix} a, b \\ a+b+1/2 \end{matrix}; x\right)\right]^2. \quad (7.4.6)$$

This ${}_3F_2$ is nonnegative because it is a square. The ${}_3F_2$ in (7.4.5) is fairly close to this but different in one numerator and one denominator parameter. We have seen before that by fractional intregration of a ${}_pF_q$, it is possible to get a ${}_{p+1}F_{q+1}$ with the necessary extra parameters. So use formula (2.2.2) to write the ${}_3F_2$ in (7.4.5) as

$$ {}_3F_2\left(\begin{matrix} -n, n+\alpha+2, (\alpha+1)/2 \\ (\alpha+3)/2, \alpha+1 \end{matrix}; t\right)$$

$$= \frac{\Gamma(\alpha+1)}{[\Gamma((\alpha+1)/2)]^2} \int_0^1 {}_2F_1\left(\begin{matrix} -n, n+\alpha+2 \\ (\alpha+3)/2 \end{matrix}; st\right) s^{(\alpha-1)/2}(1-s)^{(\alpha-1)/2} ds,$$

$$(7.4.7)$$

for $\alpha > -1$. The ${}_2F_1$ in the integral is really the ultraspherical polynomial

$$C_n^{\alpha/2}(1-2st)/C_n^{\alpha/2}(1).$$

This has zeros in $(0, 1)$ and so we do not have a positive integrand in (7.4.7). To get an idea about what should be done, write the ${}_3F_2$ in Clausen's formula with

$2a = -k, 2b = k + \alpha + 1$. Then

$$_3F_2\left(\begin{array}{c} -k, k + \alpha + 1, (\alpha + 1)/2 \\ (\alpha + 2)/2, \alpha + 1 \end{array}; t\right)$$

$$= \frac{\Gamma(\alpha + 1)}{[\Gamma((\alpha + 1)/2)]^2} \int_0^1 \frac{C_k^{(\alpha-1)/2}(1 - 2st)}{C_k^{(\alpha-1)/2}(1)} s^{(\alpha-1)/2}(1 - s)^{(\alpha-1)/2} ds.$$

$$(7.4.8)$$

The proof would be complete if it is possible to write $C_n^{\alpha/2}(x)$ in terms of $C_k^{(\alpha-1)/2}(x)$ using a positive coefficient. However, this is obtainable from the connection coefficient formula (7.1.9). Thus we have proved the next theorem.

Theorem 7.4.2   *For $\alpha > -1$,*

$$\sum_{k=0}^n P_k^{(\alpha,0)}(x) > 0, \quad -1 < x \le 1. \tag{7.4.9}$$

The integral in Theorem 6.7.2(b) can be applied to (7.4.9) to give the positivity of (7.4.1) for $-1 < x \le 1$ when $\beta \ge 0$, $\alpha + \beta > -1$. For $\alpha \ge -\beta, -1/2 \le \beta < 0$, Gasper [1977] has shown the positivity for $-1 < x \le 1$, except when $\alpha = -\beta = -1/2$, in which case this sum reduces to (7.3.1), when there are cases of equality as well as nonnegativity.

7.5 A Theorem of Turán

In the last section we proved the Jackson–Gronwall inequality. There is a theorem of Turán [1952] that shows another way of doing this.

Theorem 7.5.1   *If $\sum_{j=0}^\infty |a_j|$ converges and*

$$\sum_{j=0}^\infty a_j \sin(j + 1/2)\phi \ge 0, \quad 0 \le \phi \le \pi, \tag{7.5.1}$$

then

$$\sum_{j=0}^\infty a_j \frac{\sin(j + 1)\theta}{j + 1} > 0, \quad 0 < \theta < \pi, \tag{7.5.2}$$

unless $a_j \equiv 0, j = 0, 1, 2, \ldots$.

Proof.   In the integral formula given by Theorem 6.7.2(d), take $\alpha = 1/2$, $\beta = -1/2$, and $\mu = 1$. This gives

$$\frac{\sin(n + 1)\theta}{2(n + 1)(\sin(\theta/2))^{2n+2}} = \int_{\theta/2}^{\pi/2} \frac{\sin(2n + 1)\phi}{(\sin \phi)^{2n+3}} d\phi. \tag{7.5.3}$$

Multiply by $2a_n(\sin(\theta/2))^{2n+2}$ and sum to obtain

$$\sum_{n=0}^{\infty} a_n \frac{\sin(n+1)\theta}{n+1} = 2 \int_{\theta/2}^{\pi/2} \sum_{n=0}^{\infty} \left(\frac{\sin(\theta/2)}{\sin \phi}\right)^{2n+2} \frac{a_n \sin(n+1/2)(2\phi) d\phi}{\sin \phi}.$$

Write the integrand as

$$\frac{\sin^2(\theta/2)}{\sin^3 \phi} \sum_{n=0}^{\infty} a_n r^n \sin(n+1/2)(2\phi), \quad 0 \le \frac{1}{2}\theta \le \phi < \frac{1}{2}\pi,$$

$$r = \frac{\sin^2(\theta/2)}{\sin^2 \phi} \le 1.$$

The strict positivity of this sum for $\frac{1}{2}\theta < \phi < \frac{1}{2}\pi$ follows from the fact that

$$\sum_{n=0}^{\infty} a_n r^n \sin\left(n+\frac{1}{2}\right)\varphi$$

$$= \frac{2}{\pi} \int_0^{\pi} \sum_n r^n \sin\left(n+\frac{1}{2}\right)\phi \sin\left(n+\frac{1}{2}\right)\psi \sum_m a_m \sin\left(m+\frac{1}{2}\right)\psi d\psi.$$

(7.5.4)

This formula is obtained by using the orthogonality of the sine function. Now note that the closed form of the Poisson kernel,

$$\sum_{n=0}^{\infty} r^n \sin\left(n+\frac{1}{2}\right)\phi \sin\left(n+\frac{1}{2}\right)\psi$$

$$= \frac{(1-r)\sin\frac{1}{2}\phi \sin\frac{1}{2}\psi \left[(1-r)^2 + 4r\left(1 - \cos\frac{1}{2}(\phi+\psi)\cos\frac{1}{2}(\phi-\psi)\right)\right]}{\left[1 - 2r\cos\frac{1}{2}(\phi+\psi) + r^2\right]\left[1 - 2r\cos\frac{1}{2}(\phi-\psi) + r^2\right]},$$

(7.5.5)

gives its strict positivity for $0 \le r < 1$. This and (7.5.1) make the integrand in (7.5.4) nonnegative, and the theorem is proved. ∎

There is a generalization of (7.5.5) to Jacobi polynomials due to Watson. The result states that

$$\sum_{n=0}^{\infty} \frac{r^k P_k^{(\alpha,\beta)}(\cos 2\phi) P_k^{(\alpha,\beta)}(\cos 2\theta)}{h_k}$$

$$= \frac{\Gamma(\alpha+\beta+2)(1-r)}{2^{\alpha+\beta+1}\Gamma(\alpha+1)\Gamma(\beta+1)(1+r)^{\alpha+\beta+2}}$$

$$\cdot \sum_{m,n}^{\infty} \frac{((\alpha+\beta+2)/2)_{m+n}((\alpha+\beta+3)/2)_{m+n}}{(\alpha+1)_m(\beta+1)_n m! n!}$$

$$\cdot \frac{(4r \sin^2 \phi \sin^2 \theta)^m (4r \cos^2 \phi \cos^2 \theta)^n}{(1+r)^{2m+2n}},$$

(7.5.6)

where h_k is given by (7.1.4). We base our proof on a result of Bailey. For the result, see Bailey [1935, p. 81] and for the reference to Watson see Bailey, p. 102, Example 19. Bailey's formula is the following:

$$\begin{aligned}
&{}_2F_1(\alpha, \beta; \gamma; x)\,{}_2F_1(\alpha, \beta; \alpha + \beta + 1 - \gamma; y) \\
&= \sum_{m,n} \frac{(\alpha)_{m+n}(\beta)_{m+n}[x(1-y)]^m[y(1-x)]^n}{(\gamma)_m(\alpha + \beta + 1 - \gamma)_n m!n!}.
\end{aligned} \tag{7.5.7}$$

We sketch a proof of (7.5.7) leaving some details to the reader. Start with the double series

$$(1-s)^{-\alpha}(1-t)^{-\beta} \sum_{j,k} \frac{(\alpha)_{j+k}(\beta)_{j+k}}{(\gamma)_j(\gamma')_k j!k!} \left[\frac{-s}{(1-s)(1-t)}\right]^j \left[\frac{-t}{(1-s)(1-t)}\right]^k. \tag{7.5.8}$$

Expand in powers of s and t and show that the coefficient of $s^m t^n$ is

$$\frac{(\alpha)_m(\beta)_n(1 + \alpha - \gamma')_m(1 + \beta - \gamma)_n(\gamma - \beta)_{m-n}}{m!n!(\gamma)_m(\gamma')_n(1 + \alpha - \gamma')_{m-n}}.$$

When $\gamma + \gamma' = \alpha + \beta + 1$, the last factor in the numerator cancels the last factor in the denominator and so (7.5.8) is equal to

$$\begin{aligned}
\sum_{m=0}^{\infty}\sum_{n=0}^{\infty} &\frac{(\alpha)_m(\beta)_n(\gamma - \beta)_m(\gamma' - \alpha)_n}{m!n!(\gamma)_m(\gamma')_n} s^m t^n \\
&= {}_2F_1(\alpha, \gamma - \beta; \gamma; s)\,{}_2F_1(\beta, \gamma' - \alpha; \gamma'; t) \\
&= (1-s)^{-\alpha}(1-t)^{-\beta}\,{}_2F_1(\alpha, \beta; \gamma; -s/(1-s))\,{}_2F_1(\alpha, \beta; \gamma'; -t/(1-t)).
\end{aligned}$$

The last step follows from Pfaff's transformation given in Theorem 2.2.5. This proves (7.5.7).

The left side of (7.5.6) can be written as

$$\sum_{k=0}^{\infty} \frac{k!(\alpha + \beta + 1)_k}{(\alpha + 1)_k(\beta + 1)_k}(2k + \alpha + \beta + 1)P_k^{(\alpha,\beta)}(\cos 2\phi)P_k^{(\alpha,\beta)}(\cos 2\theta)\,r^k. \tag{7.5.9}$$

First consider the simpler series

$$\sum_{k=0}^{\infty} \frac{k!(\alpha + \beta + 1)_k}{(\alpha + 1)_k(\beta + 1)_k}P_k^{(\alpha,\beta)}(\cos 2\phi)P_k^{(\alpha,\beta)}(\cos 2\theta)\,r^k. \tag{7.5.10}$$

Replace the Jacobi polynomials by their hypergeometric representations and apply

(7.5.7) to get

$$\sum_{k=0}^{\infty} \frac{(\alpha + \beta + 1)_k}{k!} (-r)^k {}_2F_1 \left(\begin{matrix} -k, k + \alpha + \beta + 1 \\ \alpha + 1 \end{matrix} ; \sin^2 \theta \right)$$

$$\cdot {}_2F_1 \left(\begin{matrix} -k, k + \alpha + \beta + 1 \\ \beta + 1 \end{matrix} ; \cos^2 \phi \right)$$

$$= \sum_{k=0}^{\infty} \frac{(\alpha + \beta + 1)_k}{k!} (-r)^k \sum_{m,n} \frac{(-k)_{m+n} (k + \alpha + \beta + 1)_{m+n}}{(\alpha + 1)_m (\beta + 1)_n}$$

$$\cdot (\sin \theta \sin \phi)^{2m} (\cos \theta \cos \phi)^{2n}$$

$$= \sum_{m,n} \frac{(\sin^2 \theta \sin^2 \phi)^m (\cos^2 \theta \cos^2 \phi)^n}{(\alpha + 1)_m (\beta + 1)_n m! n!} r^{m+n} \sum_{k=0}^{\infty} \frac{(\alpha + \beta + 1)_{k+2m+2n}}{k!} (-r)^k$$

$$= \sum_{m,n} \frac{(r \sin^2 \theta \sin^2 \phi)^m (r \cos^2 \theta \cos^2 \phi)^n}{(\alpha + 1)_m (\beta + 1)_n m! n!}$$

$$\cdot (\alpha + \beta + 1)_{2m+2n} (1 + r)^{-2m-2n-\alpha-\beta-1}.$$

Now multiply (7.5.10) and the last expression by $r^{(\alpha+\beta+1)/2}$ and take the derivative. This introduces the factor $(2k + \alpha + \beta + 1)$ needed on the left side of (7.5.9). The right side of Watson's formula follows after an easy calculation. This proves (7.5.6).

Theorem 7.5.2 *If $\beta > \alpha > -1$ and*

$$f(x) = \sum_{k=0}^{n} a_k \frac{P_k^{(\alpha,\alpha)}(x)}{P_k^{(\alpha,\alpha)}(1)} \geq 0, \quad -1 \leq x \leq 1,$$

then

$$g(y) = \sum_{k=0}^{n} a_k \frac{P_k^{(\beta,\beta)}(y)}{P_k^{(\beta,\beta)}(1)} > 0, \quad -1 < y < 1,$$

unless $a_k \equiv 0$, $k = 0, 1, \ldots, n$.

Proof. As in the proof of Theorem 7.4.1, use the Feldheim–Vilenkin formula to get

$$g(\cos \theta) = \frac{2 \Gamma(\beta + 1)}{\Gamma(\beta - \alpha) \Gamma(\alpha + 1)} \int_0^{\pi/2} \sin^{2\alpha+1} \phi \cos^{2\beta-2\alpha-1} \phi$$

$$\cdot \sum_{k=0}^{n} a_k [1 - \sin^2 \theta \cos^2 \phi]^{k/2} \frac{P_k^{(\alpha,\alpha)}(\cos \theta (1 - \sin^2 \theta \cos^2 \phi)^{-1/2})}{P_k^{(\alpha,\alpha)}(1)} d\phi.$$

Let $r = (1 - \sin^2 \theta \cos^2 \phi)^{1/2}$ and $u = \cos \theta (1 - \sin^2 \theta \cos^2 \phi)^{-1/2}$. Then for $0 < \theta < \pi$ and $0 \le \phi < \pi/2$, we have $0 \le r < 1$ and $|u| < 1$. We can conclude from (7.5.6) that the sum inside the integral is strictly positive unless $f(x) \equiv 0$. We know this because

$$\sum_{k=0}^{n} a_k r^k \frac{P_k^{(\alpha,\alpha)}(u)}{P_k^{(\alpha,\alpha)}(1)} = \int_{-1}^{1} \sum_{0}^{\infty} r^k \frac{P_k^{(\alpha,\alpha)}(u) P_k^{(\alpha,\alpha)}(y)}{h_k^{\alpha,\alpha}} f(y) dy.$$

Now observe that orthogonality of the Jacobi polynomials implies that $f(x) \equiv 0$ for $-1 \le x \le 1$ if and only if $a_k \equiv 0$, $k = 0, 1, 2, \ldots, n$. This proves the theorem. ∎

7.6 Positive Summability of Ultraspherical Polynomials

The $(C, 1)$ means of the formal series $1 + 2 \sum_{k=1}^{\infty} \cos k\theta$ are

$$\sigma_n^1 = 1 + 2 \sum_{k=1}^{n} \left(1 - \frac{k}{n+1}\right) \cos k\theta = \frac{1}{n+1} \left\{ \frac{\sin \frac{1}{2}(n+1)\theta}{\sin \frac{1}{2}\theta} \right\} \ge 0. \quad (7.6.1)$$

Fejér used this positivity to prove the $(C, 1)$ summability of the Fourier series of a continuous function. The generating function for the sequence σ_n^1 in (7.6.1) is

$$(1 - r)^{-2} \left(1 + 2 \sum_{n=1}^{\infty} \cos n\theta \; r^n\right) = \frac{1+r}{1-r} \cdot \frac{1}{1 - 2r \cos \theta + r^2}. \quad (7.6.2)$$

The last equality follows from (5.1.16). Now observe that

$$\lim_{\nu \to 0} \frac{n+\nu}{\nu} C_n^\nu(\cos \theta) = \begin{cases} 2 \cos n\theta, & n > 0, \\ 1, & n = 0, \end{cases}$$

and the generating function for $\{\frac{n+\nu}{\nu} C_n^\nu(\cos \theta)\}$ is

$$\sum_{n=0}^{\infty} \frac{n+\nu}{\nu} C_n^\nu(\cos \theta) r^n = \frac{1 - r^2}{(1 - 2r \cos \theta + r^2)^{\nu+1}}. \quad (7.6.3)$$

This follows from the generating function for ultraspherical polynomials given in Chapter 6, namely,

$$\sum_{n=0}^{\infty} C_n^\nu(\cos \theta) r^n = \frac{1}{(1 - 2r \cos \theta + r^2)^\nu}. \quad (7.6.4)$$

Kogbetliantz [1924] proved the following generalization of (7.6.1). For a discussion of Cesàro summability of infinite series, see Appendix B.

Theorem 7.6.1   *The $(C, 2v + 1)$ means of the formal series*

$$\sum_{n=0}^{\infty} \frac{n+v}{v} C_n^v(x), \quad -1 \le x \le 1, v > 0,$$

are positive. That is,

$$\sum_{k=0}^{n} \frac{(2v+2)_{n-k}}{(n-k)!} \frac{(k+v)}{v} C_k^v(x) \ge 0, \quad -1 \le x \le 1, v > 0. \tag{7.6.5}$$

The proof depends on the following lemma.

Definition 7.6.2   *A function is called absolutely monotonic if its power series has nonnegative coefficients.*

Lemma 7.6.3   *The function $1/[(1-r)^{2v}(1-2xr+r^2)^v]$ is absolutely monotonic for $-1 \le x \le 1$.*

Proof.   Denote the function by $[g(r)]^v$ and let $x = \cos\theta$. Then

$$h(r) := \log g(r) = -2\log(1-r) - \log(1-re^{i\theta}) - \log(1-re^{-i\theta})$$

$$= \sum_{n=1}^{\infty} \frac{(2 + 2\cos n\theta)}{n} r^n.$$

Thus $h(r)$ and

$$[g(r)]^v = \sum_{n=0}^{\infty} \frac{v^n [h(r)]^n}{n!}$$

are absolutely monotonic.   ∎

Proof of Theorem 7.6.1   The generating function for (7.6.5) is

$$\frac{1-r^2}{(1-r)^{2v+2}(1-2xr+r^2)^{v+1}}$$

$$= \frac{1-r^2}{(1-r)^2(1-2xr+r^2)} \cdot \frac{1}{(1-r)^{2v}(1-2xr+r^2)^v}.$$

The first factor is absolutely monotonic by (7.6.1) and (7.6.2). The second factor is absolutely monotonic by Lemma 7.6.3. This proves the inequality (7.6.5) and the theorem.   ∎

The lemma has the following corollary.

Corollary 7.6.4   *For $v > 0, -1 \le x \le 1$,*

$$\sum_{k=0}^{n} \frac{(2v)_{n-k}}{(n-k)!} C_k^v(x) \ge 0. \tag{7.6.6}$$

Proof.   This follows from (7.6.4) and Lemma 7.6.3.   ∎

Similar results have been obtained for Jacobi series. The results are very important, but the proofs involve quite complicated formulas. See Gasper [1977] for many of these inequalities. For example, Gasper proved the following extension of the inequality in Exercise 14:

$$\sum_{k=0}^{n} \frac{(\lambda+1)_{n-k}(\lambda+1)_k}{(n-k!)k!} \frac{P_k^{(\alpha,\beta)}(x)}{P_k^{(\beta,\alpha)}(1)} \geq 0, \quad -1 \leq x \leq 1, \tag{7.6.7}$$

$0 \leq \lambda \leq \alpha + \beta$ and $\alpha \geq \beta \geq 0$ or $\alpha + \beta \geq 0$ and $\beta \geq -1/2$. Proofs of this inequality use many of the formulas given here for hypergeometric series, as well as some others. Note the particular case $\lambda = 0$:

$$\sum_{k=0}^{n} \frac{P_k^{(\alpha,\beta)}(x)}{P_k^{(\beta,\alpha)}(1)} \geq 0, \quad -1 \leq x \leq 1 \tag{7.6.8}$$

for $\alpha + \beta \geq 0$, $\beta \geq -1/2$. Inequality is strict in $-1 < x \leq 1$, when $\alpha = 1/2$, $\beta = -1/2$ are excluded.

These inequalities have implications for Bessel functions. Using Theorem 5.11.6, it is easy to see that for $\alpha > \beta - 1$

$$\lim_{n\to\infty} \left(\frac{\theta}{n}\right)^{\alpha+1-\beta} \sum_{k=0}^{n} \frac{P_k^{(\alpha,\beta)}\left(\cos\frac{\theta}{n}\right)}{P_k^{(\beta,\alpha)}(1)} = 2^\alpha \Gamma(\beta+1) \int_0^\theta t^{-\beta} J_\alpha(t)dt. \tag{7.6.9}$$

The condition $\alpha > \beta - 1$ is needed for convergence of the integral at $t = 0$, but it is not needed for the sum. In an appendix to the paper of Feldheim [1963], Szegö considered the limit when $\alpha = \beta$, and also the resulting integral. He showed that

$$\int_0^x t^{-\alpha} J_\alpha(t)dt > 0, \quad x > 0 \tag{7.6.10}$$

when $\alpha > \bar{\alpha}$ and when $\bar{\alpha}$ is the solution of

$$\int_0^{j_{\alpha,2}} t^{-\alpha} J_\alpha(t)dt = 0 \tag{7.6.11}$$

with $j_{\alpha,2}$ the second zero of $J_\alpha(t)$. A similar result holds for

$$\int_0^x t^{-\beta} J_\alpha(t)dt, \quad -1 < \alpha < 1/2.$$

From (7.6.8), we have

$$\int_0^x t^{-\beta} J_\alpha(t)dt \geq 0, \quad x > 0 \tag{7.6.12}$$

when $\alpha + \beta \geq 0$, $-1/2 \leq \beta \leq 0$. In fact, (7.6.8) also holds when $\alpha + \beta \geq -1$, $\beta \geq 0$. Thus (7.6.12) holds when $\alpha > \beta - 1$ and $\beta \geq 0$. Gasper [1975] has shown that the inequality (7.6.12) is strict for these α, β except when $\alpha = 1/2$, $\beta = -1/2$.

Gasper [1975] derived the following interesting identity when $\beta = -1/2$ and $\alpha > -3/2$ in (7.6.12):

$$\int_0^x t^{1/2} J_\alpha(t)dt = \frac{\left[\Gamma\left(\frac{\alpha}{2} + \frac{5}{4}\right)\right]^2}{\left(\alpha + \frac{3}{2}\right)\Gamma(\alpha + 1)} 2^{\alpha+1} x$$

$$\cdot \sum_{n=0}^\infty \frac{\left(\frac{1}{2}\right)_n \left(\frac{\alpha}{2} - \frac{1}{4}\right)_n \left(\alpha + \frac{3}{2}\right)_n \left(2n + \alpha + \frac{1}{2}\right)}{(\alpha + 1)_n \left(\frac{\alpha}{2} + \frac{7}{4}\right)_n n! \left(n + \alpha + \frac{1}{2}\right)} \left[J_{n + \frac{\alpha}{2} + \frac{1}{4}}\left(\frac{x}{2}\right)\right]^2 .$$

$$(7.6.13)$$

This series is clearly positive when $\alpha > 1/2$, nonnegative when $\alpha = 1/2$, and negative when x is a zero of $J_{(2\alpha+1)/4}(x/2)$, where $-3/2 < \alpha < 1/2$. Gasper also has a similar result for

$$\int_0^x (x - t)^{\alpha - 1/2} t^\alpha J_\alpha(t)dt, \quad \alpha > -1/2.$$

See Exercise 25. This and (7.6.13) are derived by using the identity

$$x^{2\nu} = \frac{\Gamma^2(\nu + 1)2^{2\nu+1}}{\Gamma(2\nu + 1)} \sum_{n=0}^\infty \frac{(n + \nu)\Gamma(n + 2\nu)}{n!} J_{n+\nu}^2(x). \qquad (7.6.14)$$

See Watson [1944, §5.1] for (7.6.14). An approach to (7.6.12) via differential equations can be found in Makai [1974]. For more references, see Askey [1975] or Gasper [1975].

7.7 The Irrationality of $\zeta(3)$

The topic of this section is unrelated to the earlier parts of this chapter. However, it involves an interesting application of Legendre polynomials and so has been included here.

In Chapter 1, we gave Euler's formula for $\zeta(2n)$ when n is an integer ≥ 1. This showed that $\zeta(2n)$ is irrational. In spite of repeated attempts, Euler was unable to evaluate $\zeta(2n + 1)$. Later mathematicians have not had better luck. It was only as recently as 1978 that R. Apéry proved that $\zeta(3)$ is irrational. A simpler proof using Legendre polynomials was given by Beukers [1979]. We follow Beukers's exposition.

The basic lemma is the following:

Lemma 7.7.1 *There exist two sequences of integers $\{A_n\}$ and $\{B_n\}$ such that*

$$0 < |A_n + B_n \zeta(3)| < 3(9/10)^n. \qquad (7.7.1)$$

This immediately implies the theorem.

Theorem 7.7.2 $\zeta(3)$ *is irrational.*

Proof. If $\zeta(3) = p/q$, where p and q are integers, then the sequence of nonzero rational numbers $|A_n + B_n \zeta(3)| \geq 1/q$. The second inequality in (7.7.1), however, implies that this sequence becomes arbitrarily small as n increases. This contradiction implies that $\zeta(3)$ is irrational. ∎

The proof of Lemma 7.7.1 depends on the following lemmas.

Lemma 7.7.3 *For nonnegative integers r and s,*

$$\int_0^1 \int_0^1 -\frac{\log xy}{1-xy} x^r y^s dx dy = 2\left\{\zeta(3) - \sum_{k=1}^r \frac{1}{k^3}\right\} \quad \text{when} \quad r = s$$

$$= \text{rational number whose denominator divides } d_r^3,$$

$$\text{where} \quad d_r = \ell cm(1, 2, \ldots, r), \quad \text{when} \quad r > s.$$

Proof. Note that for $\sigma > 0$ and $r > s$

$$\int_0^1 \int_0^1 \frac{x^{r+\sigma} y^{s+\sigma}}{1-xy} dx dy = \sum_{k=0}^\infty \frac{1}{(k+r+\sigma+1)(k+s+\sigma+1)}$$

$$= \sum_{k=0}^\infty \frac{1}{r-s}\left\{\frac{1}{k+s+\sigma+1} - \frac{1}{k+r+\sigma+1}\right\}$$

$$= \frac{1}{r-s}\left\{\frac{1}{s+1+\sigma} + \cdots + \frac{1}{r+\sigma}\right\}. \qquad (7.7.2)$$

Differentiate with respect to σ and then set $\sigma = 0$ to get

$$\int_0^1 \int_0^1 \frac{\log xy}{1-xy} x^r y^s dx dy = \frac{-1}{r-s}\left\{\frac{1}{(s+1)^2} + \cdots + \frac{1}{r^2}\right\}.$$

The second equality of the lemma follows from this. Now take $r = s$ in the first equation of (7.7.2) and again differentiate with respect to σ and set $\sigma = 0$. The result is

$$\int_0^1 \int_0^1 \frac{\log xy}{1-xy} x^r y^r dx dy = -\sum_{k=0}^\infty \frac{2}{(k+r+1)^3}.$$

This is the first equation of the lemma, which is now completely proved. ∎

For the next lemma, let

$$p_n(x) = \frac{1}{n!}\frac{d^n}{dx^n}\{x^n(1-x)^n\},$$

which is essentially the Legendre polynomial on $(0, 1)$.

Lemma 7.7.4 *There exist integers A_n and B_n such that*

$$0 \neq \int_0^1 \int_0^1 -\frac{\log xy}{1-xy} p_n(x) p_n(y) dx dy = (A_n + B_n \zeta(3)) d_n^{-3}.$$

Here $d_n = \ell cm(1, 2, \ldots, n)$.

Proof. Since $p_n(x)$ is a polynomial with integer coefficients, the equality follows from Lemma 7.7.3. Now observe that

$$-\frac{\log xy}{1-xy} = \int_0^1 \frac{1}{1-(1-xy)z} dz.$$

Then the integral in the lemma can be written as

$$\int_0^1 \int_0^1 \int_0^1 \frac{p_n(x) p_n(y)}{1-(1-xy)z} dx dy dz.$$

Integrate by parts n times with respect to x to see that this integral is equal to

$$\int_0^1 \int_0^1 \int_0^1 \frac{(xyz)^n (1-x)^n p_n(y)}{(1-(1-xy)z)^{n+1}} dx dy dz.$$

Set

$$\omega = \frac{1-z}{1-(1-xy)z}$$

and rewrite the last integral as

$$\int_0^1 \int_0^1 \int_0^1 (1-x)^n (1-\omega)^n \frac{p_n(y)}{1-(1-xy)\omega} dx dy d\omega$$

$$= \int_0^1 \int_0^1 \int_0^1 \frac{\{x(1-x)y(1-y)\omega(1-\omega)\}^n}{\{1-(1-xy)\omega\}^{n+1}} dx dy d\omega. \qquad (7.7.3)$$

The last equality uses integration by parts n times. It is clear that the final integral is nonzero and so the lemma is proved. ∎

Lemma 7.7.5 *For A_n and B_n as in Lemma 7.7.4,*

$$0 < |A_n + B_n \zeta(3)| d_n^{-3} < 2(\sqrt{2}-1)^{4n} \zeta(3).$$

Proof. The integral in Lemma 7.7.4 is equal to integral (7.7.3). We find the maximum value of the integrand in (7.7.3). Let

$$f(x, y, \omega) = \frac{x(1-x)y(1-y)\omega(1-\omega)}{1-(1-xy)\omega}.$$

By solving the equations

$$\frac{\partial f}{\partial x} = \frac{\partial f}{\partial y} = \frac{\partial f}{\partial \omega} = 0,$$

it is easy to see that at the maximum, $x = y$ and $\omega = 1/(1+x)$. Thus f is bounded by $x^2(1-x)^2/(1+x)^2$, which is maximized at $x = \sqrt{2}-1$. Thus integral (7.7.3) is bounded above by

$$
(\sqrt{2}-1)^{4n} \int_0^1 \int_0^1 \int_0^1 \frac{1}{1-(1-xy)\omega} dx\,dy\,d\omega
$$
$$
= (\sqrt{2}-1)^{4n} \int_0^1 \int_0^1 -\frac{\log xy}{1-xy} dx\,dy
$$
$$
= 2(\sqrt{2}-1)^{4n}\zeta(3).
$$

This proves the lemma.   ■

We need the following result from elementary number theory due to Chebyshev. Let $\pi(n)$ denote the number of primes less than n. Then

$$
\pi(n) < 1.06n/\log n. \tag{7.7.4}
$$

See Ingham [1932, p. 15].

Proof of Lemma 7.7.1.   Observe that

$$
d_n = \prod_{\substack{p \leq n \\ p=\text{prime}}} p^{[\log n/\log p]} < \prod_{p \leq n} p^{\log n/\log p} = n^{\pi(n)}.
$$

By (7.7.4), it follows that

$$
d_n < n^{1.06n/\log n} = e^{1.06n} < 3^n.
$$

An upper bound for $\zeta(3)$ is given by

$$
\zeta(3) < 1 + \int_1^\infty \frac{dx}{x^3} = \frac{3}{2}.
$$

Therefore, Lemma 7.7.5 gives

$$
0 < |A_n + B_n\zeta(3)| < 3\left(\frac{27}{(\sqrt{2}+1)^4}\right)^n < 3\left(\frac{9}{10}\right)^n. \quad ■
$$

Apéry's proof used some sequences satisfying three-term recurrence relations. For a lively account of Apéry's proof, see van der Poorten [1979]. These sequences were analyzed in terms of contiguous relations by Askey and Wilson [1984]. See Exercises 28 and 29.

1. Let

$$x(x-1)\cdots(x-n+1) = \sum_{r=0}^{n} s(n,r)x^r$$

and

$$x^n = \sum_{r=0}^{n} S(n,r)x(x-1)\cdots(x-r+1).$$

The integers $s(n,r)$ and $S(n,r)$ are called Stirling numbers of the first and second kind respectively. Show that

$$\sum_r S(n,r)s(r,m) = \delta_{nm}.$$

Use this to prove that

$$a_n = \sum_r s(n,r)b_r, \quad n \geq 1,$$

if and only if

$$b_n = \sum_r S(n,r)a_r, \quad n \geq 1.$$

2. Derive formulas (7.1.14) and (7.1.15) as indicated in the text.

3. Verify formula (7.2.7).

4. Prove that

$$P_n^{(\alpha+1,\beta)}(x) = \frac{2}{2n+\alpha+\beta+2} \frac{(n+\alpha+1)P_n^{(\alpha,\beta)}(x) - (n+1)P_{n+1}^{(\alpha,\beta)}(x)}{1-x}.$$

Deduce (7.2.10).

5. Suppose $m(x)$ is a positive integrable function on $(-1,1)$; $\{p_n(x)\}$ a sequence of polynomials orthogonal with respect to $m(x)$; and $\{q_n(x)\}$ orthogonal with respect to $(1+x)m(x)$. Let $p_n(1) > 0$ and $q_n(1) > 0$. Prove that

$$(1+x)q_n(x) = A_n p_{n+1}(x) + B_n p_n(x),$$

where A_n and B_n are positive. Determine A_n and B_n in the case illustrated in (7.2.10).

6. Let $\{p_m(x)\}$ be a sequence of polynomials orthogonal with respect to a distribution $d\alpha(x)$ in $(0, \infty)$. Let $\xi_1, \xi_2, \ldots, \xi_k$ be zeros of $p_m(x)$ and let

$$f_k(t) = \int_0^\infty e^{-xt}\{(\xi_1 - x)\cdots(\xi_k - x)\}^{-1}\{p_m(x)\}^2 d\alpha(x), \quad t > 0.$$

(a) Show that

$$f_k(t) = e^{-\xi_k t} \int_0^t e^{\xi_k s} f_{k-1}(s)ds.$$

(b) Deduce that $f_k(t) > 0$ for $t > 0$.

7. With $\{p_m(x)\}$ as in the previous problem, assume that $p_m(0) > 0$. Prove that

$$\int_0^\infty e^{-xt} p_m(x) p_n(x) d\alpha(x) > 0, \quad t > 0.$$

For Exercises 6 and 7, see Karlin and McGregor [1957, pp. 507–509].

8. Deduce Lemma 7.2.3 from the previous problem.

9. Prove Corollary 7.2.4.

10. (a) Prove that

$$\sum_{k,m,n=0}^\infty r^k s^m t^n \frac{\int_0^\infty L_k^\alpha(x) L_m^\alpha(x) L_n^\alpha(x) x^\alpha e^{-2x} dx}{\Gamma(\alpha+1)}$$

$$= [2 - (r+s+t) + rst]^{-\alpha-1}.$$

(b) Use the multinomial theorem to prove that, except for a constant factor, the right side of (a) also equals

$$\sum_{k,m,n\geq0} r^k s^m t^n \frac{(\alpha+1)_{k+m+n} 2^{-k-m-n}}{k!m!n!}$$

$$\cdot {}_3F_2\left(\begin{matrix} -k, -m, -n \\ (-\alpha-k-m-n)/2, (1-\alpha-k-m-n)/2 \end{matrix}; 1\right).$$

(c) Let $k \leq \min(m, n)$. Reverse the order of summation in the above ${}_3F_2$ to get

$$\frac{(-1)^k m! n! 2^{2k} \Gamma(m+n+\alpha+1-k)}{(m-k)!(n-k)! \Gamma(m+n+k+\alpha+1)}$$

$$\cdot {}_3F_2\left(\begin{matrix} -k, (\alpha+1+m+n-k)/2, (\alpha+2+m+n-k)/2 \\ m+1-k, n+1-k \end{matrix}; 1\right).$$

(d) Apply Kummer's transformation (Corollary 3.3.3) to the ${}_3F_2$ in (c) to show that the expression in (c) is positive for $\alpha = 0, 1, 2, \ldots$. This proves (7.2.13) for these values of α.

11. Prove that

(a) $$\sum_{n=0}^\infty r^n \sum_{k=0}^n \sin(k+1/2)\theta = \frac{(1+r)\sin(\theta/2)}{(1-r)(1-2r\cos\theta+r^2)},$$

(b) $$\sum_{n=0}^\infty r^n \sum_{k=0}^n (n+1-k)\sin(k+1)\theta = \frac{\sin\theta}{(1-r)^2(1-2r\cos\theta+r^2)}.$$

12. (a) Deduce from (7.3.1) that

$$2\sum_{k=1}^n \sin k\theta + \sin(n+1)\theta \geq 0 \quad \text{for } 0 \leq \theta \leq \pi.$$

(b) Use the results in the previous problem to prove that

$$\sum_{k=0}^n (n+1-k)\sin(k+1)\theta \geq 0 \quad \text{for } 0 \leq \theta \leq \pi.$$

13. Let $S_n(x) = \sum_{k=1}^{n} \frac{\sin kx}{k}$. Use induction and the fact that the extrema of $S_n(x)$ lie at $2k\pi/n$, where k is an integer, to prove that $S_n(x) > 0$ for $0 < x < \pi$.

14. Use Theorem 7.5.2 and Corollary 7.6.4 to prove that

$$\sum_{k=0}^{n} \frac{(2\nu)_{n-k}(2\nu)_k}{(n-k)!k!} \frac{C_k^\lambda(x)}{C_k^\lambda(1)} \geq 0, \quad -1 \leq x \leq 1, \ \lambda \geq \nu > 0.$$

Deduce that

$$\sum_{k=0}^{n} \frac{(a)_{n-k}(a)_k}{(n-k)!k!} \frac{\sin(k+1)\theta}{(k+1)\sin\theta} \geq 0, \quad 0 < a \leq 2, \ 0 \leq \theta \leq \pi.$$

Consider the cases $a = 1, 2$.

15. Define the difference operator Δ_h by $(\Delta_h f)(t) = (f(t+h) - f(t))/h$. Show that $(-h)^k(\Delta_h^k f)(t) = \sum_{\ell=0}^{k}(-1)^\ell \binom{k}{\ell} f(t+\ell h)$, where Δ_h^k is the kth iterate of Δ_h.

Suppose f is continuous on $[0, 1]$ and $f(1) = 1$. Prove that the following statements are equivalent:

(a) f is absolutely monotonic in $[0,1]$, that is, $f(t) = \sum_0^\infty a_n t^n$, $a_n \geq 0$, $t \in [0, 1]$;

(b) $f \in C^\infty(0, 1)$ and $f^{(k)}(t) \geq 0$ for $k = 0, 1, 2, \ldots$ and $t \in (0, 1)$;

(c) $(\Delta_{1/n}^m f)(0) \geq 0$ for $n = 1, 2, \ldots$ and $0 \leq m \leq n$.

One way to show that $(c) \Rightarrow (a)$ is to use the Bernstein polynomials $B_n(t; f) = \sum_{k=0}^{n} \binom{n}{k} t^k (1-t)^{n-k} f(\frac{k}{n}) = \sum_{k=0}^{n} \binom{n}{k}(\frac{t}{n})^k (\Delta_{1/n}^k f)(0)$, which uniformly approximate f in $[0, 1]$. To prove the last equality, use the result at the beginning of this problem.

16. Let $a_0 \geq a_1 \geq \cdots \geq a_m > 0$ and $1 \leq n \leq m$. Show that

$$\sum_{k=0}^{m} a_k \cos k\theta \sin(\theta/2) \geq \sum_{k=0}^{n-1} a_k \cos k\theta \sin(\theta/2) - (a_n/2)(1 + \sin(n - 1/2)\theta)$$

and

$$\sum_{k=0}^{m} a_k \sin k\theta \sin(\theta/2) \geq \sum_{k=0}^{n-1} a_k \sin k\theta \sin(\theta/2) - (a_n/2)(1 - \cos(n - 1/2)\theta).$$

(Vietoris)

17. Let $c_k = 2^{-2k} \binom{2k}{k}$. Show that for $0 < x < 2\pi$

$$\sum_{k=0}^{n} c_k \sin(k + 1/4)x > 0 \quad \text{and} \quad \sum_{k=0}^{n} c_k \cos(k + 1/4)x > 0.$$

Deduce that the inequalities hold if c_k is replaced by α_k, satisfying $(2k - 1)\alpha_{k-1} \geq 2k\alpha_k > 0$ for $k \geq 1$.

18. Let α_k satisfy the conditions given in the previous problem. Prove that if $0 \leq \nu \leq 1/4$ and $0 < x < 2\pi$ or if $-1/4 \leq \nu \leq 1/4$ and $0 < x < \pi$, then

$$\sum_{k=0}^{n} \alpha_k \cos(k + \nu)x > 0.$$

19. With α_k as in Exercise 17, show that if $1/4 \leq \nu \leq 1/2$ and $0 < x < 2\pi$ or if $1/4 \leq \nu \leq 3/4$ and $0 < x < \pi$, then

$$\sum_{k=0}^{n} \alpha_k \sin(k+\nu)x \geq 0.$$

20. Show that

$$1 + \sum_{k=1}^{n} \frac{\cos kx}{k} > 0, \quad 0 \leq x < \pi.$$

21. With c_k as in Vietoris's inequalities, show that

$$\sum_{k=0}^{n} c_k \frac{C_k^{\nu}(x)}{C_k^{\nu}(1)} > 0,$$

for $\nu > 0$ and $0 < x < 1$.

22. Show that if

$$\sum_{k=0}^{n} a_k \frac{P_k^{(\alpha,\beta)}(x)}{P_k^{(\beta,\alpha)}(1)} \geq 0, \quad -1 \leq x \leq 1,$$

then

$$\sum_{k=0}^{n} a_k \frac{P_k^{(\alpha,\gamma)}(y)}{P_k^{(\gamma,\alpha)}(1)} \geq 0, \quad -1 \leq y \leq 1, \ \gamma > \beta.$$

Observe that this implies the terminating form of Theorem 7.5.1.

23. Prove that if

$$\sum_{k=0}^{n} a_k \frac{P_k^{(\alpha,\beta)}(x)}{P_k^{(\beta,\alpha)}(1)} \geq 0, \quad -1 \leq x \leq 1,$$

then

$$\sum_{k=0}^{n} a_k \frac{P_k^{(\alpha-\mu,\beta+\mu)}(y)}{P_k^{(\beta+\mu,\alpha-\mu)}(1)} \geq 0, \quad -1 \leq y \leq 1, \ \mu > 0.$$

24. (a) Show that $[_0F_1(c; x)]^2 = {}_1F_2(c - 1/2; 2c - 1, c; 4x)$.

   (b) Prove the formula of Bailey that

$$\int_0^{2x} J_{2\alpha}(t)dt = 2x \int_0^{\pi/2} [J_\alpha(x \sin \phi)]^2 \sin \phi d\phi, \quad \alpha > -1/2$$

   (c) Show that $\int_0^x J_\alpha(t)dt > 0, \alpha > -1$.

25. (a) Use (6.14) to show that

$$\int_0^x (x-t)^\lambda t^\mu J_\alpha(t)dt$$

$$= \frac{\Gamma(\lambda+1)\Gamma(\alpha+\mu+1)\Gamma^2(\nu+1)}{\Gamma(\alpha+1)\Gamma(\alpha+\lambda+\mu+2)} 2^{4\nu-\alpha} x^{\alpha+\lambda+\mu+1-2\nu}$$

$$\cdot \sum_{n=0}^{\infty} {}_5F_4 \left[\begin{matrix} -n, n+2\nu, \nu+1, (\alpha+\mu+1)/2, (\alpha+\mu+2)/2 \\ \nu+1/2, \alpha+1, (\alpha+\lambda+\mu+2)/2, (\alpha+\lambda+\mu+3)/2 \end{matrix} ; 1 \right]$$

$$\cdot \frac{(2\nu+1)_n}{n!} \frac{2n+2\nu}{n+2\nu} J_{n+\nu}^2 \left(\frac{x}{2} \right),$$

when $\alpha + \mu > -1$, $\lambda > -1$, and $2\nu \neq -1, -2, \ldots$ and the factor $(2n + 2\nu)/(n + 2\nu)$ is replaced by 1 at $n = 0$.

(b) Take $\mu = \lambda + 1/2$ and set $\nu = (\alpha + \lambda + 1/2)/2$ so that the $_5F_4$ reduces to a balanced $_4F_3$.

(c) Take $\lambda = 0$ to get (6.13).

(d) Take $\lambda = \alpha - 1/2$ to get

$$\int_0^x (x - t)^{\alpha - 1/2} t^\alpha J_\alpha(t)dt$$

$$= \frac{\Gamma(\alpha + 1/2)\Gamma(2\alpha + 1)\Gamma(\alpha + 1)}{\Gamma(3\alpha + 3/2)} 2^{3\alpha} x^{\alpha + 1/2}$$

$$\cdot \sum_{n=0}^\infty \frac{((2\alpha + 1)/4)_n ((2\alpha - 1)/4)_n}{((6\alpha + 3)/4)_n ((6\alpha + 5)/4)_n} \frac{(2\alpha + 1)_n}{n!} \frac{2n + 2\alpha}{n + 2\alpha} J_{n+2\alpha}^2 \left(\frac{x}{2}\right),$$

for $\alpha > -1/2$.   (Gasper)

26. Suppose that $\{p_n(x)\}$ and $\{q_n(x)\}$ are orthonormal polynomials associated with the weights $\omega(x)$ and $\omega_1(x)$ respectively. Prove that if

$$q_n(x) = \sum_{k=0}^n c_{k,n} p_k(x),$$

then

$$\omega(x)p_k(x) = \sum_{n=k}^\infty c_{k,n} q_n(x)\omega_1(x).$$

27. Use Exercise 26 and Theorem 7.1.4′ to show that for $-1 < x < 1$ and $\mu > (\lambda - 1)/2$,

$$(1 - x^2)^{\mu - 1/2} C_n^\mu(x) = \sum_{k=0}^\infty c_{k,n}^{\mu,\lambda} C_{n+2k}^\lambda(x)(1 - x^2)^{\lambda - 1/2},$$

where

$$c_{k,n}^{\mu,\lambda}$$

$$= \frac{\Gamma(\lambda)2^{2\lambda - 2\mu}(n + 2k + \lambda)(n + 2k)!\Gamma(n + 2\mu)\Gamma(n + k + \lambda)\Gamma(k + \lambda - \mu)}{\Gamma(\lambda - \mu)\Gamma(\nu)n!k!\Gamma(n + k + \mu + 1)\Gamma(n + 2k + 2\lambda)}.$$

Note that $c_{k,n}^{\mu,\lambda} > 0$ for $(\lambda - 1)/2 < \mu < \lambda$. Deduce the special case

$$(\sin\theta)^{2\lambda - 1} C_n^\mu(\cos\theta) = \sum_{k=0}^\infty c_{k,n}^\mu \sin(n + 2k + 1)\theta$$

when $\mu > 0$, $\mu \neq 1, 2, \ldots$, and

$$c_{k,n}^\mu = \frac{2^{2-2\mu}(n + k)!\Gamma(n + 2\mu)\Gamma(k + 1 - \mu)}{\Gamma(\mu)\Gamma(1 - \mu)k!n!\Gamma(n + k + \mu + 1)}.$$

28. Show that

$$b_n = \sum_{k=0}^n \binom{n}{k}^2 \binom{n+k}{k}^2$$

satisfies the three-term recurrence relation

$$n^3 b_n + (n-1)^3 b_{n-2} = (34n^3 - 51n^2 + 27n - 5)b_{n-1}.$$

(Apéry)

29. If $a + d = b + c$,

$$g_n = \sum_{k=0}^{n} \binom{n}{k}\binom{n+a+d}{k+d}\binom{n+k+b+\ell}{k+\ell}\binom{n+k+c+f}{k+f}$$

satisfies a three-term recurrence relation. Find it.

30. Complete the proof of Bailey's formula (7.5.7).

31. Use (7.5.7) to prove Brafman's [1951] generating-function formula for Jacobi polynomials,

$$\sum_{n=0}^{\infty} \frac{(\gamma)_n (\alpha + \beta - \gamma + 1)_n P_n^{(\alpha,\beta)}(x) r^n}{(\alpha + 1)_n (\beta + 1)_n}$$

$$= {}_2F_1(\gamma, \alpha + \beta - \gamma + 1; \alpha + 1; (1 - r - R)/2)$$

$$\cdot {}_2F_1(\gamma, \alpha + \beta - \gamma + 1; \beta + 1; (1 + r - R)/2),$$

where

$$R = (1 - 2xr + r^2)^{1/2}.$$

The case $\gamma = \alpha$ is interesting.

32. Complete the proof of (7.2.14).

33. Show that, if $\alpha > -1$, $k, m, n = 0, 1, \ldots$, then

$$(-1)^{k+m+n} \int_0^{\infty} L_k^\alpha(x) L_m^\alpha(x) L_n^\alpha(x) x^\alpha e^{-x} dx \geq 0.$$

34. If $\alpha \geq 0$, $j, k, m, n = 0, 1, 2, \ldots$, then prove that

$$\int_0^{\infty} L_j^\alpha(x) L_k^\alpha(x) L_m^\alpha(x) L_n^\alpha(x) x^\alpha e^{-2x} dx \geq 0.$$

8

The Selberg Integral and Its Applications

Dirichlet's straightforward though useful multidimensional generalization of the beta integral was presented in Chapter 1. In the 1940s, more than 100 years after Dirichlet's work, Selberg found a more interesting generalized beta integral in which the integrand contains a power of the discriminant of the n variables of integration. Recently, Aomoto evaluated a yet slightly more general integral. An important feature of this evaluation is that it provides a simpler proof of Selberg's formula, reminiscent of Euler's evaluation of the beta integral by means of a functional equation. The depth of Selberg's integral formula may be seen in the fact that in two dimensions it implies Dixon's identity for a well-poised $_3F_2$. Bressoud observed that Aomoto's extension implies identities for nearly poised $_3F_2$.

After presenting Aomoto's proof, we give another proof of Selberg's formula due to Anderson. This proof is similar to Jacobi's or Poisson's evaluation of Euler's beta integral in that it depends on the computation of a multidimensional integral in two different ways. The basis for Anderson's proof is Dirichlet's multidimensional integral mentioned above. A very significant aspect of Anderson's method is that it applies to the finite-field analog of Selberg's integral as well. We give a brief treatment of this analog at the end of the chapter.

Stieltjes posed and solved an electrostatic problem that is equivalent to obtaining the maximum of an n variable function very closely related to the integrand in Selberg's formula. Stieltjes's remarkable solution showed that the maximum is attained when the n variables are zeros of a certain Jacobi polynomial of degree n. We devote a section of this chapter to Stieltjes's work and show how his result can be combined with Selberg's formula to derive the discriminants of Jacobi, Laguerre, and Hermite polynomials.

Siegel used the discriminant of the Laguerre polynomial to extend the arithmetic and geometric mean inequality. Siegel's result, which we include, contains an interesting inequality of Schur relating the arithmetic mean and the discriminant.

8.1 Selberg's and Aomoto's Integrals

The theorem given below contains Selberg's [1944] extension of the beta integral.

Theorem 8.1.1   *If n is a positive integer and α, β, γ are complex numbers such that* $\operatorname{Re}\alpha > 0$, $\operatorname{Re}\beta > 0$, *and* $\operatorname{Re}\gamma > -\min\{1/n, (\operatorname{Re}\alpha)/(n-1), (\operatorname{Re}\beta)/(n-1)\}$, *then*

$$S_n(\alpha, \beta, \gamma) = \int_0^1 \cdots \int_0^1 \prod_{i=1}^n \left\{ x_i^{\alpha-1}(1-x_i)^{\beta-1} \right\} |\Delta(x)|^{2\gamma} dx_1 \cdots dx_n$$

$$= \prod_{j=1}^n \frac{\Gamma(\alpha + (j-1)\gamma)\Gamma(\beta + (j-1)\gamma)\Gamma(1+j\gamma)}{\Gamma(\alpha+\beta+(n+j-2)\gamma)\Gamma(1+\gamma)},$$

where

$$\Delta(x) = \prod_{1 \le i < j \le n} (x_i - x_j).$$

The conditions on α, β, γ are needed for the convergence of the integral. For a discussion of this, see Selberg [1944]. Note, however, that the condition on γ is related to the first occurrence of a pole of the function on the right-hand side of the integral formula. Selberg's proof of this formula appeared in 1944, but for more than three decades it was not well known. More recently, Aomoto [1987] found a simpler proof which depends on a recurrence relation satisfied by a slightly more general integral.

Theorem 8.1.2   *With the same conditions on the parameter α, β, γ and with $k \le n$,*

$$\int_0^1 \cdots \int_0^1 \prod_{i=1}^k x_i \prod_{i=1}^n x_i^{\alpha-1}(1-x_i)^{\beta-1} \prod_{1 \le i < j \le n} |x_i - x_j|^{2\gamma} dx_1 \cdots dx_n$$

$$= \prod_{j=1}^k \frac{(\alpha + (n-j)\gamma)}{(\alpha+\beta+(2n-j-1)\gamma)} S_n(\alpha, \beta, \gamma).$$

Anderson [1990] proved a finite-field analog of Selberg's formula and then noted (Anderson [1991]) that the idea could be carried over to the continuous case.

8.2 Aomoto's Proof of Selberg's Formula

To motivate this proof, recall the two basic steps of the proof of the formula $B(\alpha, \beta) = \Gamma(\alpha)\Gamma(\beta)/\Gamma(\alpha+\beta)$.

Step 1. Obtain the functional equation $B(\alpha, \beta) = \frac{\alpha+\beta}{\beta} B(\alpha, \beta+1)$. Though we did this differently in Chapter 1, it can be done as follows: For Re $\alpha > 0$, Re $\beta > 0$,

$$0 = \int_0^1 \frac{\partial}{\partial x}[x^\alpha(1-x)^\beta]dx$$

$$= \alpha \int_0^1 x^{\alpha-1}(1-x)^\beta dx - \beta \int_0^1 x^\alpha(1-x)^{\beta-1}dx$$

$$= (\alpha+\beta) \int_0^1 x^{\alpha-1}(1-x)^\beta dx - \beta \int_0^1 x^{\alpha-1}(1-x)^{\beta-1}dx$$

$$= (\alpha+\beta)B(\alpha, \beta+1) - \beta B(\alpha, \beta)$$

Step 2. Iterate the first step n times to get

$$B(\alpha, \beta) = \frac{(\alpha+\beta)_n}{(\beta)_n} B(\alpha, \beta+n).$$

Then apply a change of variables in the integral for $B(\alpha, \beta+n)$ and let $n \to \infty$ to obtain the necessary result.

We apply the same basic procedure to prove a generalization of Selberg's formula. Let

$$w(x) = w(x; \alpha, \beta, \gamma) = \prod_{i=1}^n x_i^{\alpha-1}(1-x_i)^{\beta-1} \prod_{1 \le i < j \le n} |x_i - x_j|^{2\gamma} \quad (8.2.1)$$

and

$$I_k = \int_{C_n} \prod_{i=1}^k x_i w(x; \alpha, \beta, \gamma)dx, \quad (8.2.2)$$

where C_n is the n-dimensional cube and $dx = dx_1 dx_2 \cdots dx_n$. By symmetry the product $\prod_{i=1}^k x_i$ may be replaced by the product of any k distinct variables without changing the value of the integral. Let I_0 denote the integral without the factor $\prod_{i=1}^k x_i$.

To obtain the functional equation start with

$$0 = \int_{C_n} \frac{\partial}{\partial x_1}\left[(1-x_1) \prod_{i=1}^k x_i w(x)\right] dx$$

$$= \alpha \int_{C_n} (1-x_1) \prod_{i=2}^k x_i w(x)dx - \beta \int_{C_n} \prod_{i-1}^k x_i w(x)dx$$

$$+ 2\gamma \sum_{j=2}^n \int_{C_n} (1-x_1)\frac{\prod_{i=1}^k x_i w(x)dx}{x_1 - x_j}. \quad (8.2.3)$$

The third term in (8.2.3) is derived from the fact that

$$\frac{d}{dx}|x|^c = c|x|^{c-1}\operatorname{sgn} x = \frac{c|x|^c}{x} \qquad \text{if } x \neq 0.$$

The next lemma shows that the third integral can be written in terms of I_k and I_{k-1}.

Lemma 8.2.1

(a)
$$\int_{C_n} \frac{\prod_{i=1}^k x_i w(x)dx}{x_1 - x_j} = \begin{cases} 0 & \text{if } 2 \leq j \leq k, \\ \frac{1}{2}I_{k-1} & \text{if } k < j \leq n. \end{cases}$$

(b)
$$\int_{C_n} \frac{x_1 \prod_{i=1}^k x_i w(x)dx}{x_1 - x_j} = \begin{cases} \frac{1}{2}I_k & \text{if } 2 \leq j \leq k, \\ I_k & \text{if } k < j \leq n. \end{cases}$$

Proof.

(a) In the first case, where $2 \leq j \leq k$, the transposition $x_1 \leftrightarrow x_j$ changes the sign of the integrand, so the integral vanishes. In the second case the same transposition leads to

$$\frac{x_1}{x_1 - x_j} \longrightarrow \frac{x_j}{x_j - x_1} = 1 - \frac{x_1}{x_1 - x_j},$$

so

$$2\int_{C_n} \frac{\prod_1^k x_i w(x)dx}{x_1 - x_j} = \int_{C_n} \prod_{i=2}^k x_i w(x)dx = I_{k-1}.$$

(b) For $2 \leq j \leq k$, the transposition $x_1 \leftrightarrow x_j$ gives

$$\frac{x_1^2 x_j}{x_1 - x_j} \rightarrow \frac{x_1 x_j^2}{x_j - x_1} = x_1 x_j - \frac{x_1^2 x_j}{x_1 - x_j},$$

which proves the first part of (b). For the second part observe that

$$\frac{x_1^2}{x_1 - x_j} = x_1 + \frac{x_1 x_j}{x_1 - x_j}$$

and the last term changes sign in the transposition $x_1 \leftrightarrow x_j$. So its presence in the integral makes that part zero. The other part coming from x_1 gives the necessary result. Thus the lemma is proved. ∎

Use this lemma to rewrite (8.2.3) as

$$0 = \alpha I_{k-1} - (\alpha + \beta)I_k + \gamma(n-k)I_{k-1} - \gamma(2n-k-1)I_k.$$

This gives

$$I_k = \frac{\alpha + (n-k)\gamma}{\alpha + \beta + (2n-k-1)\gamma} I_{k-1}. \tag{8.2.4}$$

Iterate this functional relation to arrive at

$$\int_{C_n} \prod_{i=1}^{k} x_i w(x; \alpha, \beta, \gamma) dx$$

$$= \prod_{i=1}^{k} \frac{\alpha + (n - i)\gamma}{\alpha + \beta + (2n - i - 1)\gamma} \int_{C_n} w(x; \alpha, \beta, \gamma) dx.$$

The last integral is Selberg's integral, which we write as $S_n(\alpha, \beta, \gamma)$. The problem now is to evaluate this integral. Fortunately, it is possible to apply the functional equation (8.2.4) to Selberg's integral itself. Note that

$$S_n(\alpha + 1, \beta, \gamma) = \prod_{j=1}^{n} \frac{\alpha + (n - j)\gamma}{\alpha + \beta + (2n - j - 1)\gamma} S_n(\alpha, \beta, \gamma). \qquad (8.2.5)$$

Symmetry in α and β and iteration give

$$S_n(\alpha, \beta, \gamma) = \prod_{j=1}^{n} \frac{(\alpha + \beta + (2n - j - 1)\gamma)_k}{(\beta + (n - j)\gamma)_k} S_n(\alpha, \beta + k, \gamma)$$

$$= \prod_{j=1}^{n} \frac{(\alpha + \beta + (2n - j - 1)\gamma)_k}{(\beta + (n - j)\gamma)_k}$$

$$\cdot \int_0^k \cdots \int_0^k \prod_{i=1}^{n} \left(\frac{x_i}{k}\right)^{\alpha-1} \left(1 - \frac{x_i}{k}\right)^{\beta+k-1} \prod_{1 \le i < j \le n} \left|\frac{x_i - x_j}{k}\right|^{2\gamma} \frac{dx}{k^n}.$$

Let $k \to \infty$ and use the limit definition of the gamma function to get

$$S_n(\alpha, \beta, \gamma) = \prod_{j=1}^{n} \frac{\Gamma(\beta + (n - j)\gamma)}{\Gamma(\alpha + \beta + (2n - j - 1)\gamma)}$$

$$\cdot \int_0^\infty \cdots \int_0^\infty \prod_{i=1}^{n} x_i^{\alpha-1} e^{-x_i} \prod_{1 \le i < j \le n} |x_i - x_j|^{2\gamma} dx. \qquad (8.2.6)$$

Denote the integral in (8.2.6) by $G_n(\alpha, \gamma)$. Then by symmetry in α and β, relation (8.2.6) implies that

$$\frac{G_n(\alpha, \gamma)}{\prod_{j=1}^{n} \Gamma(\alpha + (n - j)\gamma)} = \frac{G_n(\beta, \gamma)}{\prod_{j=1}^{n} \Gamma(\beta + (n - j)\gamma)} =: D_n(\gamma).$$

Thus we can write

$$S_n(\alpha, \beta, \gamma) = \prod_{j=1}^{n} \frac{\Gamma(\alpha + (n - j)\gamma)\Gamma(\beta + (n - j)\gamma)}{\Gamma(\alpha + \beta + (2n - j - 1)\gamma)} D_n(\gamma). \qquad (8.2.7)$$

To compute $D_n(\gamma)$, first note that by the symmetry in the variables $x_1, x_2, \ldots, x_n$

$$\int_{C_n} w(x; \alpha, \beta, \gamma)dx = n! \int_0^1 \int_{x_n}^1 \cdots \int_{x_2}^1 w(x; \alpha, \beta, \gamma)dx_1 \cdots dx_n. \qquad (8.2.8)$$

Since

$$\lim_{\alpha \to 0^+} \alpha \int_0^1 t^{\alpha-1} f(t)dt = f(0)$$

for a continuous function f (for a proof see Exercise 1 at the end of this chapter), multiply (8.2.8) by α, let $\alpha \to 0^+$, and apply (8.2.7) to get

$$n \int_{C_{n-1}} \prod_{i=1}^{n-1} [x_i^{2\gamma-1}(1-x_i)^{\beta-1}] \prod_{1 \le i < j \le n-1} |x_i - x_j|^{2\gamma} dx$$

$$= D_n(\gamma) \prod_{j=1}^n \frac{\Gamma(\beta + (n-j)\gamma)}{\Gamma(\beta + (2n-j-1)\gamma)} \prod_{j=2}^n \Gamma((j-1)\gamma)$$

$$= D_n(\gamma) \prod_{j=1}^n \frac{\Gamma(\beta + (j-1)\gamma)}{\Gamma(\beta + (n+j-2)\gamma)} \prod_{j=2}^n \Gamma((j-1)\gamma).$$

Again by (8.2.7), the last integral also equals

$$D_{n-1}(\gamma) \prod_{j=1}^{n-1} \frac{\Gamma(2\gamma + (j-1)\gamma)\Gamma(\beta + (j-1)\gamma)}{\Gamma(2\gamma + \beta + (n+j-3)\gamma)}.$$

This gives the functional relation

$$D_n(\gamma) = \frac{n\Gamma(n\gamma)}{\Gamma(\gamma)} D_{n-1}(\gamma) = \frac{\Gamma(n\gamma+1)}{\Gamma(\gamma+1)} D_{n-1}(\gamma).$$

This implies

$$D_n(\gamma) = \prod_{j=1}^n \frac{\Gamma(1+j\gamma)}{\Gamma(1+\gamma)}.$$

Thus, Selberg's formula and Aomoto's extension are both proved.

Applying the change of variables and limiting procedure for obtaining (8.2.6) to Aomoto's integral formula, we get:

Corollary 8.2.2   *With the conditions on the parameters α and γ as in Theorem 8.1.1,*

$$\int_0^\infty \cdots \int_0^\infty \prod_{i=1}^k x_i \prod_{i=1}^n x_i^{\alpha-1} e^{-x_i} \prod_{1 \le i < j \le n} |x_i - x_j|^{2\gamma} dx$$

$$= \prod_{j=1}^k (\alpha + (n-j)\gamma) \prod_{j=1}^n \left(\frac{\Gamma(\alpha + (j-1)\gamma)\Gamma(1+j\gamma)}{\Gamma(1+\gamma)} \right).$$

To derive another corollary from Selberg's formula, take $\alpha = \beta$ and $x_i = (1 + t_i/\sqrt{2\alpha})/2$, let $\alpha \to \infty$, and apply Stirling's formula.

Corollary 8.2.3 *For* $\mathrm{Re}\, \gamma > -1/n$,

$$\int_{-\infty}^{\infty} \cdots \int_{-\infty}^{\infty} \exp\left(-\frac{1}{2}\sum_{i=1}^{n} x_i^2\right) \prod_{1 \leq i < j \leq n} |x_i - x_j|^{2\gamma} dx = (2\pi)^{n/2} \prod_{j=1}^{n} \frac{\Gamma(\gamma j + 1)}{\Gamma(\gamma + 1)}.$$

Remark 8.2.1 One can use Carlson's theorem to prove (8.2.7) without having to let β go to infinity. It follows from (8.2.5) that (8.2.7) is true when α and β are positive integers. Moreover, both sides of (8.2.7) are analytic functions of α and β for $\mathrm{Re}\,\alpha > 0$ and $\mathrm{Re}\,\beta > 0$ and they are bounded for $\mathrm{Re}\,\alpha \geq 1$, $\mathrm{Re}\,\beta \geq 1$.

8.3 Extensions of Aomoto's Integral Formula

Aomoto's formula involves the introduction of the extra factors $\prod_{i=1}^{k} x_i$, $k \leq n$. One may ask whether extra factors of the type $\Pi(1 - x_j)$ can be inserted in the integrand. The simplest integral of this type occurs when there is no common variable among the two different kinds of factors. Let

$$B(j, k) := \int_{C_n} \prod_{i=1}^{j} x_i \prod_{i=j+1}^{j+k} (1 - x_i)w(x)dx,$$

where $j + k \leq n$ and

$$w(x) = w(x; \alpha, \beta, \gamma) = \prod_{i=1}^{n} x_i^{\alpha-1}(1 - x_i)^{\beta-1} \prod_{1 \leq i < j \leq n} |x_i - x_j|^{2\gamma}.$$

The formula is

$$B(j, k) = \prod_{i=1}^{n} \frac{\Gamma(\alpha + (n - i)\gamma)\Gamma(\beta + (n - i)\gamma)\Gamma(i\gamma + 1)}{\Gamma(\alpha + \beta + (2n - i - 1)\gamma)\Gamma(\gamma + 1)}$$
$$\cdot \frac{\prod_{i=1}^{j}[\alpha + (n - i)\gamma]\prod_{i=1}^{k}[\beta + (n - i)\gamma]}{\prod_{i=1}^{j+k}[\alpha + \beta + (2n - i - 1)\gamma]}. \tag{8.3.1}$$

This is easy to verify. Denote the right side of (8.3.1) by $C(j, k)$. Observe that Aomoto's integral implies

$$B(j, 0) = C(j, 0) \quad \text{and} \quad B(0, k) = C(0, k); \tag{8.3.2}$$

moreover, both B and C satisfy the same recurrence relation

$$B(j - 1, k - 1) - B(j, k - 1) = B(j - 1, k). \tag{8.3.3}$$

This proves (8.3.1). Now let

$$B(j, k, \ell) = \int_{C_n} \prod_{i=1}^{j} x_i \prod_{i=j+1-\ell}^{j+k-\ell} (1 - x_i) w(x) dx,$$

so that j represents the number of extra x_i factors, k the number of extra $(1 - x_i)$ factors, and ℓ the number of variables that overlap among the extra factors. Here $\ell \leq j, k \leq n$ and $j + k - \ell \leq n$.

Theorem 8.3.1

$$B(j, k, \ell) = \prod_{i=1}^{\ell} \frac{[\alpha + \beta + (n - i - 1)\gamma]}{[\alpha + \beta + 1 + (2n - i - 1)\gamma]}$$

$$\cdot \frac{\prod_{i=1}^{j}[\alpha + (n - i)\gamma] \prod_{i=1}^{k}[\beta + (n - i)\gamma]}{\prod_{i=1}^{j+k}[\alpha + \beta + (2n - i - 1)\gamma]} S_n(\alpha, \beta, \gamma),$$

where $S_n(\alpha, \beta, \gamma)$ is the value of the Selberg integral.

Proof.   First consider the case where $k = \ell$. This integral, after renumbering variables, can be written as

$$B(j, k, k) = \int_{C_n} \prod_{i=1}^{j} x_i \prod_{i=1}^{k} (1 - x_i) w(x) dx.$$

It satisfies the functional relation that generalizes (8.2.4):

$$(\alpha + \beta + (2n - j - k - 1)\gamma) B(j + 1, k - 1, k - 1)$$
$$= (\alpha + (n - j - 1)\gamma) B(j, k - 1, k - 1).   \tag{8.3.4}$$

To prove (8.3.4), start with

$$0 = \int_{C_n} \frac{\partial}{\partial x_1} \prod_{i=1}^{j} x_i \prod_{i=1}^{k} (1 - x_i) w(x) dx$$

$$= \alpha \int_{C_n} \prod_{i=2}^{j} x_i \prod_{i=1}^{k} (1 - x_i) w(x) dx - \beta \int_{C_n} \prod_{i=1}^{j} x_i \prod_{i=2}^{k} (1 - x_i) w(x) dx$$

$$+ 2\gamma \sum_{m=2}^{n} \int_{C_n} \frac{\prod_{i=1}^{j} x_i \prod_{i=1}^{k} (1 - x_i)}{x_1 - x_m} w(x) dx.$$

The functional equation (8.3.4) now follows from the following lemma.

Lemma 8.3.2

(a)
$$\sum_{m=2}^{k} \int_{C_n} \frac{\prod_{i=1}^{j} x_i \prod_{i=1}^{k}(1 - x_i)}{x_1 - x_m} w(x)dx = 0.$$

(b)
$$\sum_{m=k+1}^{j} \int_{C_n} \frac{\prod_{i=1}^{j} x_i \prod_{i=1}^{k}(1 - x_i)}{x_1 - x_m} w(x)dx = \frac{(k - j)}{2} B(j, k - 1, k - 1).$$

(c)
$$\sum_{m=j+1}^{n} \int_{C_n} \frac{\prod_{i=1}^{j} x_i \prod_{i=1}^{k}(1 - x_i)}{x_1 - x_m} w(x)dx$$

$$= (n - j)\left[\frac{B(j - 1, k - 1, k - 1)}{2} - B(j, k - 1, k - 1)\right].$$

The proof of this lemma is left to the reader.
A direct consequence of (8.3.4) is that

$$B(j, k, k) = \prod_{i=j+1}^{n} \frac{[\alpha + \beta + (2n - i - k - 1)\gamma]}{[\alpha + (n - i)\gamma]} B(n, k, k)$$

$$= \prod_{i=j+1}^{n} \frac{(\alpha + \beta + (2n - i - k - 1)\gamma)}{(\alpha + (n - i)\gamma)}$$

$$\cdot \prod_{i=1}^{k} \frac{(\beta + (n - i)\gamma)}{(\alpha + \beta + 1 + (2n - i - 1)\gamma)}$$

$$\cdot \prod_{i=1}^{n} \frac{(\alpha + (n - i)\gamma)}{(\alpha + \beta + (2n - i - 1)\gamma)} S_n(\alpha, \beta, \gamma). \qquad (8.3.5)$$

The second equation follows from (8.3.1) and the fact that $B(n, k, k)$ is the same as $B(0, k, k)$ except that α is replaced by $\alpha + 1$. Equation (8.3.5) is, in fact, equivalent to

$$B(j, k, k) = \prod_{i=1}^{k} \frac{(\alpha + \beta + (n - i - 1)\gamma)}{(\alpha + \beta + 1 + (2n - i - 1)\gamma)}$$

$$\cdot \frac{\prod_{i=1}^{j}(\alpha + (n - i)\gamma) \prod_{i=1}^{k}(\beta + (n - i)\gamma)}{\prod_{i=1}^{j+k}(\alpha + \beta + (2n - i - 1)\gamma)} S_n(\alpha, \beta, \gamma). \qquad (8.3.6)$$

This proves Theorem 8.3.1 for the case $k = \ell$. By writing x_1 as $1 - (1 - x_1)$ in the integral $B(j, k, \ell)$ one verifies that

$$B(j + 1, k, \ell) = B(j, k, \ell) - B(j, k + 1, \ell) \quad \text{for } j \geq \ell. \qquad (8.3.7)$$

The right side of Theorem 8.3.1 also satisfies this recurrence relation. Thus Theorem 8.3.1 is proved. The above proof is due to Shaun Cooper. ∎

We end this section with the statement of another beta integral of Selberg.

Theorem 8.3.3 *Let $D_n = \{(x_1, \ldots, x_n) \mid x_i \geq 0, \sum_{i=1}^{n} x_i \leq 1\}$. Then*

$$\int_{D_n} \prod_{i=1}^{k} x_i \prod_{i=1}^{n} x_i^{\alpha-1} \left(1 - \sum_{i=1}^{n} x_i\right)^{\beta-1} \prod_{1 \leq i < j \leq n} |x_i - x_j|^{2\gamma} dx$$

$$= \frac{\Gamma(\beta)}{\Gamma(\beta + k + n\alpha + (n-1)n\gamma)} \prod_{j=1}^{n} \frac{\Gamma(\alpha + (n-j)\gamma)\Gamma(j\gamma + 1)}{\Gamma(\gamma + 1)}.$$

A proof is outlined in Exercise 5.

Historical Remark   Hardy and Pólya independently proved the following theorem on entire functions:

Let $f(z)$ be an entire function of exponential type less than $\log 2$. If $f(n)$ is an integer for $n = 0, 1, 2, \ldots$, then $f(z)$ is a polynomial.

This theorem can be restated as follows: 2^z is the smallest transcendental entire function taking integral values at the positive integers.

Gelfond generalized this theorem as follows:

If $f(z)$ is an entire function such that

$$f(n), f'(n), \ldots, f^{(p-1)}(n)$$

are all integers and f is of exponential type less than $p \log(1 + e^{(1-p)/p})$, then f is a polynomial. The case $p = 1$ is the result of Hardy and Pólya.

Selberg discovered his integral formula when he generalized Gelfond's theorem. He proved that $p \log(1 + e^{(1-p)/p})$ can be replaced by $\log \{\min \prod_{i=1}^{p}(1 + y_i)\}$, where $y_i > 0$, $y_1 y_2 \cdots y_p = e^{1-p}$, and $|\prod_{1 \leq i < j \leq p}(\frac{1}{y_i} - \frac{1}{y_j})| = 1$. To see that this is a generalization, note that since the y_j are distinct,

$$\prod_{i=1}^{p}(1 + y_i) > (1 + \sqrt[p]{y_1 y_2 \cdots y_p})^p = \left(1 + e^{\frac{1-p}{p}}\right)^p.$$

For references, see Boas [1954].

Corollary 8.2.3 was conjectured by Mehta and Dyson in the mid-1960s. They considered a gas of N point charges at $x_1, x_2, \ldots, x_N$, which are free to move on the infinite straight line $-\infty < x < \infty$. The potential energy of this gas is given by

$$W = \frac{1}{2} \sum_{i=1}^{n} x_i^2 - \sum_{1 \leq i < j \leq n} \log |x_i - x_j|,$$

where the first term represents a harmonic potential that attracts each charge independently toward the point $x = 0$ and the second term represents an electrostatic repulsion between each pair of charges. An important role in the thermodynamical study of this system is played by the partition function

$$\psi_n(\beta) = \int_{-\infty}^{\infty} \cdots \int_{-\infty}^{\infty} e^{-\beta W} dx_1 \cdots dx_n.$$

It was the value of this integral that was conjectured by Dyson and Mehta. See Mehta [1991]. Mehta's book also contains Selberg's original proof of Selberg's integral formula.

8.4 Anderson's Proof of Selberg's Formula

In Chapter 1, we gave two essentially different proofs of the formula $B(\alpha, \beta) = \Gamma(\alpha)\Gamma(\beta)/\Gamma(\alpha + \beta)$. One was done by constructing a functional equation and the other by evaluating a suitable double integral in two different ways. The second method applied to the finite field analogs of the beta and gamma integrals, that is, to the Jacobi and Gauss sums. Anderson [1991] found a proof of Selberg's formula which involves the computation of a $(2n - 1)$-dimensional integral in two ways. This proof carries over to a formula for the finite field analog of the Selberg integral. In fact, Anderson [1990] obtained a proof of the latter result first.

Anderson's proof depends on Dirichlet's generalization of the beta integral given in Chapter 1: For $\operatorname{Re}\alpha_i > 0$,

$$\iint \cdots \int_V \rho_0^{\alpha_0 - 1} \rho_1^{\alpha_1 - 1} \cdots \rho_n^{\alpha_n - 1} d\rho_0 \cdots d\rho_{n-1} = \frac{\prod_{i=0}^n \Gamma(\alpha_i)}{\Gamma(\Sigma\alpha_i)}, \qquad (8.4.1)$$

where V is the set $\rho_i \geq 0$, $\sum_{i=0}^n \rho_i = 1$. The formula is used after a change of variables. To see this, first consider Selberg's integral, which may be written as

$$n! A_n(\alpha, \beta, \gamma) := n! \int_0^1 \int_0^{x_n} \cdots \int_0^{x_2} |F(0)|^{\alpha - 1} |F(1)|^{\beta - 1} |\Delta_F|^\gamma dx_1 \cdots dx_n,$$

where $0 < x_1 < x_2 < \cdots < x_n < 1$,

$$F(t) = (t - x_1)(t - x_2) \cdots (t - x_n) = t^n - F_{n-1}t^{n-1} + \cdots + (-1)^n F_0,$$

and Δ_F is the discriminant of F, so that

$$|\Delta_F| = \left| \prod_{i=1}^n F'(x_i) \right| = \left| \prod_{i<j} (x_i - x_j) \right|^2 .$$

We now change the variables from $x_1, x_2, \ldots, x_n$ to $F_0, F_1, \ldots, F_{n-1}$, which are the elementary symmetric functions of the x_i s.

Lemma 8.4.1

$$A_n(\alpha, \beta, \gamma) = \int |F(0)|^{\alpha - 1} |F(1)|^{\beta - 1} |\Delta_F|^{\gamma - \frac{1}{2}} dF_0 dF_1 \cdots dF_{n-1},$$

where the integration is over all points $(F_0, F_1, \ldots, F_{n-1})$ in which the F_i are elementary symmetric functions of $x_1, \ldots, x_n$ with $0 < x_1 < \cdots < x_n$.

Proof.   It is sufficient to prove that the Jacobian

$$\left|\left(\frac{\partial F_i}{\partial x_j}\right)\right| = |\Delta_F|^{1/2}.$$ (8.4.2)

Observe that two columns of the Jacobian are equal when $x_i = x_j$. Thus $\prod_{i<j}(x_i - x_j)$ is a factor of the determinant. Moreover, the Jacobian and $\prod_{i<j}(x_i - x_j)$ are homogeneous and of the same degree. This proves (8.4.2) and the lemma.   ∎

We make a similar change of variables in (8.4.1). To accomplish this, set

$$Z(t) = (t - \zeta_0)(t - \zeta_1)\cdots(t - \zeta_n) \quad (0 \le \zeta_0 < \zeta_1 < \cdots < \zeta_n < 1)$$

and let

$$\mathcal{D} = \{(t - x_1)(t - x_2)\cdots(t - x_n) \mid \zeta_{i-1} < x_i < \zeta_i; i = 1, \ldots, n\}.$$ (8.4.3)

Lemma 8.4.2   *For all* $F(t) = t^n - F_{n-1}t^{n-1} + \cdots + (-1)^n F_0 \in \mathcal{D}$, *the map*

$$(F_0, F_1, \ldots, F_{n-1}) \mapsto \left(\frac{F(\zeta_0)}{Z'(\zeta_0)}, \ldots, \frac{F(\zeta_n)}{Z'(\zeta_n)}\right) = (\rho_0, \rho_1, \ldots, \rho_n) \in R^{n+1},$$

where $Z'(t)$ *denotes the derivative of* $Z(t)$, *is a bijection and* $\rho_j > 0$ *with* $\sum_{i=0}^{n} \rho_j = 1$.

Proof.   Observe that

$$\rho_j = \frac{F(\zeta_j)}{Z'(\zeta_j)} = \frac{(\zeta_j - x_1)(\zeta_j - x_2)\cdots(\zeta_j - x_n)}{(\zeta_j - \zeta_0)\cdots(\zeta_j - \zeta_{j-1})(\zeta_j - \zeta_{j+1})\cdots(\zeta_j - \zeta_n)} > 0$$

since the numerator and denominator have exactly $n - j$ negative factors.
Now let

$$Z_j(t) = \frac{Z(t)}{t - \zeta_j}.$$

By Lagrange's interpolation formula

$$F(t) = \sum_{j=0}^{n} \rho_j Z_j(t) \equiv \sum_{j=0}^{n} \frac{Z_j(t)}{Z'(\zeta_j)} F(\zeta_j).$$ (8.4.4)

One can directly verify this by checking that both sides of the equation are polynomials of degree n and are equal at $n + 1$ points $t = \zeta_0, \zeta_1, \ldots, \zeta_n$. Equate the coefficients of t^n on both sides to get

$$1 = \sum_{j=0}^{n} \rho_j.$$

Now for a given point $(\rho_0, \rho_1, \ldots, \rho_n)$ with $\Sigma \rho_j = 1$ and $\rho_j > 0$, $j = 0, \ldots, n$, define $F(t)$ by (8.4.4). The expressions

$$F(\zeta_i) = \rho_i Z_i(\zeta_i) = \rho_i(\zeta_i - \zeta_0) \cdots (\zeta_i - \zeta_{i-1})(\zeta_i - \zeta_{i+1}) \cdots (\zeta_i - \zeta_n)$$

and

$$F(\zeta_{i+1}) = \rho_{i+1} Z_{i+1}(\zeta_{i+1})$$
$$= \rho_{i+1}(\zeta_{i+1} - \zeta_0) \cdots (\zeta_{i+1} - \zeta_i)(\zeta_{i+1} - \zeta_{i+2}) \cdots (\zeta_{i+1} - \zeta_n)$$

show that $F(\zeta_i)$ and $F(\zeta_{i+1})$ have different signs and F vanishes at some point x_{i+1} between ζ_i and ζ_{i+1}. Thus $F \in \mathcal{D}$ by (8.4.3). This proves the bijection. ∎

We can now restate Dirichlet's formula (8.4.1).

Lemma 8.4.3 *With the notation of Lemma 8.4.2 and* $\operatorname{Re} \alpha_i > 0$,

$$\int_{F(t) \in \mathcal{D}} \prod_{i=0}^{n} |F(\zeta_i)|^{\alpha_i - 1} dF_0 \cdots dF_{n-1} = \frac{\prod_{i=0}^{n} |Z'(\zeta_i)|^{\alpha_i - \frac{1}{2}} \Gamma(\alpha_i)}{\Gamma\left(\sum_{i=0}^{n} \alpha_i\right)}.$$

Proof. We need to verify that the Jacobian

$$\left| \frac{\partial(\rho_0, \ldots, \rho_{n-1})}{\partial(F_0, \ldots, F_{n-1})} \right| = \prod_{i=0}^{n} |Z'(\zeta_i)|^{-1/2}.$$

Since

$$\rho_i = \frac{F(\zeta_i)}{Z'(\zeta_i)} = \frac{1}{Z'(\zeta_i)} \left(\zeta_i^n - F_{n-1} \zeta_i^{n-1} + \cdots + (-1)^n F_0 \right),$$

the Jacobian is

$$\left| \left(\frac{\partial \rho_i}{\partial F_j} \right) \right| = \frac{|(\zeta_i^j)|}{\Pi |Z'(\zeta_i)|}.$$

The numerator is a Vandermonde determinant and the result follows. ∎

The final step is to obtain the $(2n - 1)$-dimensional integral. Let $F(t)$ and $G(t)$ be two polynomials such that

$$\begin{aligned} F(t) &= (t - x_1)(t - x_2) \cdots (t - x_{n-1}), \\ G(t) &= (t - y_1)(t - y_2) \cdots (t - y_n), \end{aligned} \tag{8.4.5}$$

and

$$0 < y_1 < x_1 < y_2 < \cdots < x_{n-1} < y_n < 1.$$

The resultant of F and G, denoted $R(F, G)$, is given by

$$|R(F, G)| = \left| \prod_{\substack{i=1,\ldots,n-1 \\ j=1,\ldots,n}} (x_i - y_j) \right| = \left| \prod_{j=1}^{n} F(y_j) \right| = \left| \prod_{i=1}^{n-1} G(x_i) \right|. \qquad (8.4.6)$$

The absolute value of the discriminant of F can be written as $|R(F, F')|$. The $(2n - 1)$-dimensional integral is

$$\int_{(F,G)} |G(0)|^{\alpha-1} |G(1)|^{\beta-1} |R(F, G)|^{\gamma-1} dF_0 \cdots dF_{n-2} dG_0 \cdots dG_{n-1}$$

$$= \int_{(F,G)} |G(0)|^{\alpha-1} |G(1)|^{\beta-1} \left| \prod_{j=1}^{n} F(y_i) \right|^{\gamma-1} dF_0 \cdots dF_{n-2} dG_0 \cdots dG_{n-1}.$$

$$(8.4.7)$$

Here the integration is over all F and G defined by (8.4.6).

Lemma 8.4.4   *Selberg's integral* $A_n(\alpha, \beta, \gamma)$ *of Lemma 8.4.1 satisfies the recurrence relation*

$$A_n(\alpha, \beta, \gamma) = \frac{\Gamma(\alpha)\Gamma(\beta)\Gamma(n\gamma)}{\Gamma(\alpha + \beta + (n - 1)\gamma)} A_{n-1}(\alpha + \gamma, \beta + \gamma, \gamma).$$

Proof.   Integrate the $(2n - 1)$-dimensional integral (8.4.7) with respect to $dF_0 \cdots dF_{n-2}$ and use Lemma 8.4.3 with $G(t)$ instead of $Z(t)$ to get

$$\int_G |G(0)|^{\alpha-1} |G(1)|^{\beta-1} \left| \prod_{j=1}^{n} G'(y_j) \right|^{\gamma-\frac{1}{2}} dG_0 \cdots dG_{n-1} \frac{\Gamma(\gamma)^n}{\Gamma(n\gamma)}$$

$$= A_n(\alpha, \beta, \gamma) \frac{\Gamma(\gamma)^n}{\Gamma(n\gamma)}.$$

To compute (8.4.7) in another way, set $\tilde{F}(t) = t(t - x_1)(t - x_2) \cdots (t - x_n)$, $\alpha_0 = \alpha, \alpha_n = \beta, \alpha_j = \gamma$ for $j = 1, \ldots, n - 1, x_0 = 0$, and $x_n = 1$ so that (8.4.7) is equal to

$$\int_{(F,G)} |G(0)|^{\alpha-1} |G(1)|^{\beta-1} \left| \prod_{j=1}^{n-1} G(x_j) \right|^{\gamma-1} dG_0 \cdots dG_{n-1} dF_0 \cdots dF_{n-2}$$

$$= \int_{(F,G)} \prod_{j=0}^{n} |G(x_i)|^{\alpha_i - 1} dG_0 \cdots dG_{n-1} dF_0 \cdots dF_{n-2}. \qquad (8.4.8)$$

Now integrate (8.4.8) with respect to $dG_0 \cdots dG_{n-1}$ and use $\tilde{F}(t)$ instead of $Z(t)$ in Lemma 8.4.3 to obtain

$$\int_F \left| \prod_{j=1}^{n-1} \tilde{F}'(x_j) \right|^{\gamma-1/2} |\tilde{F}'(0)|^{\alpha-1/2} |\tilde{F}'(1)|^{\beta-1/2} dF_0 \cdots dF_{n-2}$$

$$\cdot \frac{\Gamma(\gamma)^{n-1}\Gamma(\alpha)\Gamma(\beta)}{\Gamma(\alpha+\beta+(n-1)\gamma)}.$$

Since

$$|\tilde{F}'(0)| = |x_1 x_2 \cdots x_{n-1}|,$$
$$|\tilde{F}'(1)| = |(1-x_1)\cdots(1-x_{n-1})|,$$

and

$$\prod_{j=1}^{n} |F'(x_j)| = \prod_{j=1}^{n-1} |x_j| \prod_{j=1}^{n-1} |(1-x_j)| |\Delta_F|,$$

the last integral can be written as

$$\frac{\Gamma(\gamma)^{n-1}\Gamma(\alpha)\Gamma(\beta)}{\Gamma(\alpha+\beta+(n-1)\gamma)} \int_F \prod_{j=1}^{n-1} x_j^{\alpha+\gamma-1} \prod_{j=1}^{n-1} (1-x_j)^{\beta+\gamma-1} |\Delta_F|^{\gamma-\frac{1}{2}} dF_0 \cdots dF_{n-2}$$

$$= \frac{\Gamma(\gamma)^{n-1}\Gamma(\alpha)\Gamma(\beta)}{\Gamma(\alpha+\beta+(n-1)\gamma)} A_{n-1}(\alpha+\gamma, \beta+\gamma, \gamma).$$

Equate the two different evaluations of the $(2n-1)$-dimensional integral to obtain the result in Lemma 8.4.4. ∎

Selberg's formula is obtained by iterating Lemma 8.4.4 $n-1$ times.

R. J. Evans has shown that Aomoto's extension (8.3.1) can also be proved by this method. The idea is sketched in Exercise 21 at the end of this chapter.

8.5 A Problem of Stieltjes and the Discriminant of a Jacobi Polynomial

There is a problem of Stieltjes that connects up Jacobi polynomials, the hypergeometric differential equation and Selberg's integral in a very interesting way. It may be stated as a two-dimensional electrostatics problem. Let $p > 0$ and $q > 0$ be fixed. Suppose there are charges of size p at 0 and q at 1 and unit charges at $x_1, x_2, \ldots, x_n$, where $0 < x_i < 1$, $i = 1, \ldots, n$. Assume the potential is logarithmic to get the energy of the system as

$$T(x_1, x_2, \ldots, x_n) = -p \sum_{i=1}^{n} \log x_i - q \sum_{i=1}^{n} \log(1-x_i) - \sum_{1 \le i < j \le n} \log |x_i - x_j|.$$

$$(8.5.1)$$

The problem is to find the location of the charges so that they are in electrostatic equilibrium. The latter occurs when the energy is a minimum, so we either minimize (8.5.1) or maximize

$$H(x_1, x_2, \ldots, x_n) := \prod_{i=1}^{n} x_i^p (1 - x_i)^q \prod_{1 \le i < j \le n} |x_i - x_j|. \qquad (8.5.2)$$

Theorem 8.5.1   *The maximum of (8.5.2) occurs when $x_1, x_2, \ldots, x_n$ are the zeros of the Jacobi polynomial $P_n^{(2p-1, 2q-1)}(1 - 2x)$.*

Proof.   Since H is a continuous function of $x_1, x_2, \ldots, x_n$ for $0 \le x_i \le 1$, $i = 1, \ldots, n$, it has a maximum value at some point. If any of the x_i is 0 or 1, the value of H is 0. So the minimum of T (or the maximum of H) occurs where

$$\frac{\partial T}{\partial x_i} = 0, \quad i = 1, 2, \ldots, n.$$

Therefore

$$\frac{p}{x_i} - \frac{q}{1 - x_i} + \sum_{j \ne i} \frac{1}{x_i - x_j} = 0, \quad i = 1, 2, \ldots, n. \qquad (8.5.3)$$

This is a set of n nonlinear equations in n unknowns. Stieltjes introduces a polynomial f whose zeros x_i, $i = 1, \ldots, n$ satisfy (8.5.3), and he shows that f satisfies a specific hypergeometric differential equation. Set

$$f(x) = \prod_{i=1}^{n} (x - x_i)$$

so that the discriminant Δ of f is given by

$$\Delta = \prod_{1 \le i < j \le n} (x_i - x_j)^2 = \prod_{i=1}^{n} f'(x_i).$$

Take the logarithmic derivative to obtain

$$\frac{\partial}{\partial x_i} \log \Delta = 2 \sum_{j \ne i} \frac{1}{x_i - x_j} = \frac{f''(x_i)}{f'(x_i)}, \quad i = 1, \ldots, n.$$

Write (8.5.3) as

$$\frac{1}{2} \frac{f''(x_i)}{f'(x_i)} + \frac{p}{x_i} - \frac{q}{1 - x_i} = 0, \quad i = 1, \ldots, n,$$

or

$$x_i(1 - x_i) f''(x_i) + 2[p - (p + q)x_i] f'(x_i) = 0, \quad i = 1, \ldots, n.$$

Now consider the expression

$$x(1 - x) f''(x) + 2[p - (p + q)x] f'(x).$$

This is a polynomial of degree $\leq n$, since f is a polynomial of degree n, and it is zero at $x_i, i = 1, 2, \ldots, n$. Thus the expression is a constant multiple of $f(x)$ and we deduce that $y = f(x)$ satisfies the differential equation

$$x(1 - x)y'' + [2p - 2(p + q)x]y' + \lambda y = 0 \qquad (8.5.4)$$

for some constant λ. Compare (8.5.4) with the hypergeometric equation

$$x(1 - x)y'' + [c - (a + b + 1)x]y' - aby = 0,$$

which has the two independent solutions

$$_2F_1\left(\begin{matrix} a, b \\ c \end{matrix}; x\right) \quad \text{and} \quad x^{1-c}\,_2F_1\left(\begin{matrix} a + 1 - c, b + 1 - c \\ 2 - c \end{matrix}; x\right).$$

We see that these are also the independent solutions of (8.5.4) with $c = 2p$, $a + b = 2p + 2q - 1$, and $ab = -\lambda$. We get a polynomial solution of degree n only when a or b is $-n$. So if $b = -n$, then $a = n + 2p + 2q - 1$ and $\lambda = -n(n + 2p + 2q - 1)$ and

$$f(x) = k_2F_1\left(\begin{matrix} -n, n + 2p + 2q - 1 \\ 2p \end{matrix}; x\right). \qquad (8.5.5)$$

To find k, note that the coefficient of x^n in $f(x)$ is 1. So

$$k = \frac{(-1)^n (2p)_n}{(n + 2p + 2q - 1)_n}.$$

Except for a constant factor, the hypergeometric polynomial in (8.5.5) is the Jacobi polynomial $P_n^{(2p-1, 2q-1)}(1 - 2x)$. This proves the theorem. ∎

It is actually possible to find the maximum value of H from Selberg's integral.

Theorem 8.5.2

$$\max_{0 \leq x_i \leq 1} H^2(x_1, x_2, \ldots, x_n) = \prod_{j=1}^{n} \frac{(2p + j - 1)^{2p+j-1}(2q + j - 1)^{2q+j-1} j^j}{(2p + 2q + n + j - 2)^{2p+2q+n+j-2}},$$

where H is defined by (8.5.2).

Proof. Recall that if μ is a positive measure on a measure space X and

$$\|f\|_k = \left(\int_X |f|^k d\mu\right)^{1/k} < \infty, \quad \text{for some } k \text{ in } 0 < k < \infty,$$

then

$$\|f\|_k \to \|f\|_\infty \quad \text{as } k \to \infty. \qquad (8.5.6)$$

In Selberg's integral, take $\alpha - 1 = 2pk$, $\beta - 1 = 2qk$, and $\gamma = k$ and then apply (8.5.6) together with Stirling's formula to obtain the theorem. ∎

Since the expression for $H(x_1, \ldots, x_n)$ involves the discriminant we have the next theorem.

Theorem 8.5.3   *The discriminant of the Jacobi polynomial $P_n^{(\alpha,\beta)}(x)$ is*

$$2^{n(n-1)} \prod_{j=1}^{n} \frac{j^j (\alpha+j)^{j-1}(\beta+j)^{j-1}}{(\alpha+\beta+n+j)^{n+j-2}}.$$

Proof.   We have

$$\prod_{i=0}^{n} x_i^{2p}(1-x_i)^{2q} = \prod_{i=1}^{n}((-1)^n f(0))^{2p}(f(1))^{2q},$$

$$(-1)^n f(0) = (-1)^n k = \frac{(2p)_n}{(n+2p+2q-1)_n},$$

and

$$f(1) = k {}_2F_1\left(\begin{matrix} -n, n+2p+2q-1 \\ 2p \end{matrix}; 1\right)$$

$$= \frac{(2q)_n}{(n+2p+2q-1)_n}.$$

The Chu–Vandermonde identity (Corollary 2.2.3) was used to sum the ${}_2F_1$. Combine these results with Theorem 8.5.2 to see that the discriminant of $P_n^{(2p-1,2q-1)}$ $(1-2x)$ is

$$\prod_{j=1}^{n} \frac{(2p+j-1)^{j-1}(2q+j-1)^{j-1} j^j}{(2p+2q+n+j-2)^{n+j-2}}.$$

Theorem 8.5.3 follows after an appropriate change of variables.   ∎

Similar theorems can be stated for Laguerre and Hermite polynomials.

Theorem 8.5.4   *The maximum of*

$$U(x_1, x_2, \ldots, x_n) = \prod_{i=1}^{n} x_i^p e^{-x_i} \prod_{1 \le i < j \le n} |x_i - x_j|$$

is obtained when $x_1, x_2, \ldots, x_n$ are the zeros of the Laguerre polynomial $L_n^{(2p-1)}(2x)$. The discriminant of $L_n^\alpha(x)$ is

$$\prod_{j=1}^{n} j^j (\alpha+j)^{j-1}.$$

Theorem 8.5.5   *The maximum of*

$$V(x_1, x_2, \ldots, x_n) = \prod_{i=1}^{n} e^{-x_i^2/2} \prod_{1 \le i < j \le n} |x_i - x_j|$$

is attained when $x_1, \ldots, x_n$ are the zeros of the Hermite polynomial $H_n(x)$. The discriminant of $H_n(x)$ is

$$\frac{1}{2^{2n(n-1)}} \prod_{j=1}^{n} j^j.$$

The proofs of these theorems require Corollaries 8.2.2 and 8.2.3 and are left to the reader as exercises. Other formulations of Theorems 8.5.4 and 8.5.5 are also possible. These are given in the exercises at the end of the chapter. For references to the work of Stieltjes and to other methods for the calculation of the discriminants of the classical orthogonal polynomials, see Szegö [1975, §§ 6.7–6.71].

Remark 8.5.1 Stieltjes's interpretation that the zeros of the Jacobi polynomials are the equilibrium positions of charges placed in the interval $[0,1]$ is useful for guessing theorems about these zeros. For example, if the zeros of $P_n^{(\alpha,\beta)}(1-2x)$ are written in increasing order $0 < x_1 < x_2 < \cdots < x_n < 1$, then $\frac{\partial x_\nu}{\partial \alpha} > 0$ and $\frac{\partial x_\nu}{\partial \beta} < 0$. Observe that if α is increased the unit charges in $(0,1)$ are pushed up toward 1, and if β is increased they are pushed down toward 0.

8.6 Siegel's Inequality

Siegel [1945] derived an inequality that refines the arithmetic and geometric mean inequality. He used it to obtain results about traces of algebraic integers, all of whose conjugates are real and positive.

The idea of the proof is as follows: Let s and p be two positive numbers such that $s^n > p$. Find the maximum of $\Delta = \prod_{1 \le i < j \le n}(x_i - x_j)^2$ considered as a function of $x_1, \ldots, x_n$ and subject to the conditions that $x_1 + \cdots + x_n = ns$ and $x_1 x_2 \cdots x_n = p^n$, $x_i > 0$. It turns out that the maximum lies at the zeros of a certain Laguerre polynomial. In fact, the method by which this is determined is the same as that used in the solution of the Stieltjes problem of the previous section. Siegel's inequality drops out easily after this.

Before stating Siegel's inequality we prove the arithmetic and geometric mean inequality.

Lemma 8.6.1 Let $x_1, x_2, \ldots, x_n$ be n positive numbers. Then

$$\sqrt[n]{x_1 x_2 \cdots x_n} \le \frac{x_1 + x_2 + \cdots + x_n}{n},$$

where equality holds if and only if $x_1 = x_2 = \cdots = x_n$.

Proof. Start with the inequality in Exercise 1.6. For α, β, u, and v nonnegative and $\alpha + \beta = 1$ write the inequality as

$$u^\alpha v^\beta \le \alpha u + \beta v. \tag{8.6.1}$$

Here equality holds if and only if $u = v$. We prove the lemma by induction. It holds for $n = 1$. Assume the result true up to n. Then, by (8.6.1) and the inductive hypothesis,

$$(x_1 x_2 \cdots x_{n+1})^{1/(n+1)} = \left(x_1^{1/n} \cdots x_n^{1/n}\right)^{n/(n+1)} x_{n+1}^{1/(n+1)}$$

$$\leq \frac{n}{n+1} \sqrt[n]{x_1 \cdots x_n} + \frac{1}{n+1} x_{n+1}$$

$$\leq \frac{n}{n+1} \left(\frac{x_1 + \cdots + x_n}{n}\right) + \frac{x_{n+1}}{n+1}$$

$$= \frac{x_1 + \cdots + x_{n+1}}{n+1}.$$

It is clear from the proof that equality holds if and only if all the x_i are equal. This proves the lemma. ■

To state Siegel's extension, set

$$P(t) = P_n(t) = \frac{1}{n!} \prod_{k=0}^{n-2} \left(\frac{t+k}{n-k}\right)^{n-k-1}, \tag{8.6.2}$$

$$Q(t) = Q_n(t) = \prod_{k=1}^{n-1} \left(1 + \frac{n-k}{t+k-1}\right), \tag{8.6.3}$$

and

$$\Delta = \prod_{1 \leq i < j \leq n} (x_i - x_j)^2,$$

where the x_i are positive. For $n \geq 2$ and $\Delta \neq 0$, let μ denote the unique positive root of the algebraic equation

$$P_n(\mu) = (x_1 \cdots x_n)^{n-1} \Delta^{-1}. \tag{8.6.4}$$

(The polynomial $P(t)$ has only positive coefficients and $P(0) = 0$, so μ is uniquely determined.) Siegel's inequality is contained in the next theorem.

Theorem 8.6.2   *Let $n \geq 2$, $\Delta \neq 0$, and μ be the positive root of (8.6.4); then*

$$\left(\frac{x_1 + \cdots + x_n}{n}\right)^n \geq x_1 \cdots x_n Q_n(\mu),$$

where Q is given by (8.6.3).

Remark 8.6.1   Since $Q_n(\mu) > 1$ for positive μ, Theorem 8.6.2 is a refinement of the arithmetic and geometric mean inequality.

We first prove the following lemma.

Lemma 8.6.3 *Let s and p be two positive numbers such that $s^n > p$. The maximum of $\Delta = \prod(x_i - x_j)^2$, subject to the conditions*

$$x_1 + \cdots + x_n = ns \quad \text{and} \quad x_1 \cdots x_n = p^n,$$

occurs when $x_1, \ldots, x_n$ are the zeros of a certain Laguerre polynomial.

Proof. To show the existence of the maximum, observe that for any positive set of values $x_3, x_4, \ldots, x_n$ the equations

$$x_1 + x_2 = ns - (x_3 + \cdots + x_n), \qquad x_1 x_2 = p(x_3 \cdots x_n)^{-1}$$

have a unique solution that is positive if and only if

$$x_3 + \cdots + x_n < ns \quad \text{and} \quad \{ns - (x_3 + \cdots + x_n)\}^2 \geq 4p(x_3 \cdots x_n)^{-1}. \quad (8.6.5)$$

Moreover, $x_1 = x_2$ if and only if equality holds in (8.6.5). The conditions (8.6.5) define a compact domain D in a $(n-2)$-dimensional space whose points have positive coordinates $x_3, \ldots, x_n$. The boundary of D consists of the surface $x_1 = x_2$ so that the maximum of Δ is inside D. By Lagrange's method of undetermined multipliers, at the maximum point

$$\frac{\partial \phi}{\partial x_k} = 0, \quad k = 1, \ldots, n,$$

where

$$\phi(x_1, \ldots, x_n) = \frac{1}{2} \log \Delta - \lambda(x_1 + \cdots + x_n) + \mu \log(x_1 \cdots x_n)$$

and λ and μ are some constants. In fact, μ will be seen to be the positive root of (8.6.4) when the x_i maximize Δ.

As in the solution of Stieltjes's problem of the previous section, show that $f(x) = \prod_{i=1}^{n}(x - x_i)$ satisfies

$$\frac{f''(x_k)}{f'(x_k)} - \lambda + \frac{\mu}{x_k} = 0, \quad k = 1, 2, \ldots, n.$$

So $x f''(x) - (\lambda x - \mu) f'(x)$ is a polynomial of degree n that vanishes at $x_1, \ldots, x_n$ and hence is a constant multiple of $f(x)$. This constant has to be $-\lambda n$. Therefore, f satisfies the differential equation

$$xy'' - (\lambda x - \mu)y' + \lambda n y = 0.$$

Set $t = \lambda x$ to transform this equation to

$$ty'' - (t - \mu)y' + ny = 0.$$

This equation is satisfied by the Laguerre polynomial $L_n^{(\mu-1)}(t)$. So we get

$$f(x) = kL_n^{(\mu-1)}(\lambda x) \tag{8.6.6}$$

for some constant k. This proves the lemma. ∎

We make a few observations necessary to complete the proof of Siegel's inequality. First note that the discriminate of the polynomial in (8.6.6) is, according to Theorem 8.5.4,

$$\lambda^{-n(n-1)} \prod_{j=1}^{n-1} \{(\mu + j)^j (j + 1)^{j+1}\}. \tag{8.6.7}$$

Also observe that since

$$L_n^{(\mu-1)}(\lambda x) = \frac{(\mu)_n}{n!} {}_1F_1\left(\begin{matrix} -n \\ \mu \end{matrix}; \lambda x\right),$$

the constant k in (8.6.6) is $(-1)^n n! / \lambda^n$. So

$$s = \frac{x_1 + \cdots + x_n}{n} = \frac{\mu + n - 1}{\lambda}$$

and

$$p = x_1 x_2 \cdots x_n = \frac{(\mu)_n}{\lambda^n}.$$

This implies

$$s^n p^{-1} = \frac{(\mu + n - 1)^n}{(\mu)_n} = \prod_{k=1}^{n-1} \left(1 + \frac{n - k}{\mu + k - 1}\right) = Q_n(\mu) \tag{8.6.8}$$

and (using (8.6.7))

$$p^{n-1} \Delta^{-1} = \frac{1}{n!} \prod_{k=0}^{n-2} \left(\frac{\mu + k}{n - k}\right)^{n-k-1} = P_n(\mu). \tag{8.6.9}$$

Proof of Theorem 8.6.2   Suppose that $y_1, y_2, \ldots, y_n$ are n positive numbers such that

$$\frac{y_1 + \cdots + y_n}{n} = s \quad \text{and} \quad y_1 \cdots y_n = p.$$

Let μ_0 be the solution of equation (8.6.4) with these y in place of the x and let Δ_0 denote the discriminant using the y. Since $P_n(t)$ is an increasing function and

$$P_n(\mu) = p^{n-1} \Delta^{-1} \le p^{n-1} \Delta_0^{-1} = P_n(\mu_0)$$

we have $\mu \le \mu_0$. However, $Q_n(t)$ is a decreasing function and so

$$s^n p^{-1} = Q_n(\mu) \ge Q_n(\mu_0).$$

This proves Siegel's inequality. ∎

The following corollary is due to Schur [1918].

Corollary 8.6.4 *For positive numbers* $x_1, \ldots, x_n$

$$\left(\frac{x_1 + x_2 + \cdots + x_n}{n}\right)^{n(n-1)} \geq \frac{(n-1)^{n(n-1)}}{\prod_{k=2}^{n} k^k (k-1)^{k-1}} \prod_{1 \leq i < j \leq n} (x_i - x_j)^2.$$

Proof. Set

$$R(t) = P(t) Q^{n-1}(t) = \prod_{k=1}^{n} \frac{(t+n-1)^{n-1}}{k^k (t+k-1)^{k-1}}.$$

Use (8.6.4) and the inequality in Theorem 8.6.2 to eliminate $p = x_1 \cdots x_n$ and obtain

$$s^{n(n-1)} \geq \Delta R(\mu). \tag{8.6.10}$$

Show that $R(t)$ is an increasing function of $t > 0$ by proving $\frac{d}{dt} \log R(t) > 0$. Thus $R(\mu) \geq R(0)$ and

$$s^{n(n-1)} \geq \Delta R(0).$$

This is Schur's inequality. ∎

Siegel applied his inequality to the derivation of some results on traces of positive algebraic integers with positive conjugates. We include statements of two of these results as they are quite interesting. A proof of one of them is sketched. The details are left to the reader, since they involve an easy application of the Euler–MacLaurin summation formula.

Before we state the next theorem, observe that the equation

$$g(v) := (1+v)^2 \log(1 + (1/v)) + \log v - v - 1 = 0 \tag{8.6.11}$$

has exactly one positive root, say, θ. This is true since $g(0) = -1$, $g(\infty) = \infty$, and

$$g'(v) = 2(1+v) \log(1 + (1/v)) - 2 > 0 \quad \text{for } v > 0.$$

Theorem 8.6.5 *Suppose the algebraic equation with integer coefficients*

$$x^n + a_1 x^{n-1} + \cdots + a_n = 0$$

has all positive roots $x_1, x_2, \ldots, x_n$. *Let* θ *be the unique positive root of (8.6.11) and let* $\lambda_0 = e(1 + (1/\theta))^{-\theta}$; *then for any* $\lambda < \lambda_0$, *there exists a number* $N = N(\lambda)$ *such that*

$$x_1 + \cdots + x_n > \lambda n$$

for all $n > N$.

The proof depends on the following lemma.

Lemma 8.6.6   *If t is any positive number satisfying $\Delta P_n(t) \geq 1$, then*

$$s^n \geq Q_n(t).$$

Proof.   Since $p = x_1 x_2 \cdots x_n = (-1)^n a_n$ is a positive integer, Theorem 8.6.2 and Corollary 8.6.4 imply that

$$s^{n(n-1)} \geq \max(Q^{n-1}(\mu), \Delta R(\mu)).$$

We have seen that $Q(t)$ is decreasing and $R(t)$ is increasing for $t > 0$. It follows for $t > 0$ that

$$\max(Q^{n-1}(\mu), \Delta R(\mu)) \geq \min(Q^{n-1}(t), \Delta R(t)) = Q^{n-1}(t) \min(1, \Delta P(t)).$$

This proves the lemma.   ■

Proof of Theorem 8.6.5   Start with the Euler–Maclaurin summation formula (see Appendix D)

$$\sum_{k=1}^n f(k) = \frac{f(n) - f(0)}{2} + \int_0^n f(x)dx + \int_0^n f'(x)B_1(x - [x])dx. \quad (8.6.12)$$

(Recall that $B_1(x) = x - \frac{1}{2}$.) Take $f(x) = (n - x)\log(x + vn - 1)$ in (8.6.12) and show that

$$\log P(vn) = \sum_{k=1}^{n-1}(n - k)\log(k + vn - 1) - \sum_{k=2}^n k\log k$$

$$= \frac{1}{2}g(v)n^2 + O(n\log n).$$

Since

$$\lim_{v \to \theta}\{1 - v\log(1 + (1/v))\} = \log \lambda_0,$$

it follows that $s = (x_1 + \cdots + x_n)/n > \lambda$, for any $\lambda < \lambda_0$ and all n greater than some suitably large $N(\lambda)$. This proves the theorem.   ■

Siegel computed the value of $\lambda_0 = 1.7336105\ldots$. The best possible value of the constant that could replace λ_0 in the theorem is certainly ≤ 2. This follows from the fact that $4\cos^2\frac{\pi}{p}$, for odd prime p, has trace equal to $2n - 1 < 2n$, where $n = (p - 1)/2$. The following theorem was also proved by Siegel, but because its proof is longer we omit it.

Theorem 8.6.7 *Suppose ξ is an algebraic integer $\neq 1$ or $\frac{1}{2}(3 \pm \sqrt{5})$. Suppose all the conjugates $x_1, x_2, \ldots, x_n$ of ξ are positive. Then*

$$x_1 + x_2 + \cdots + x_n > \frac{3}{2}n.$$

8.7 The Stieltjes Problem on the Unit Circle

In Section 8.5 we considered Stieltjes's problem for the unit interval. Dyson and others have looked at the situation where the freely moving charges lie on a thin circular conductor of unit radius. See Mehta [1991] for references. We work out the case in which n unit charges are placed on the unit circle, although the more general case with a charge of size q at $\theta = 0$ and one of size p at $\theta = \pi$ can also be treated similarly.

The potential energy of the system with n unit charges on the unit circle is

$$W = - \sum_{1 \leq j < k \leq n} \log|e^{i\theta_k} - e^{i\theta_j}|. \tag{8.7.1}$$

In the equilibrium position, W is a minimum.

Theorem 8.7.1 *The minimum value of W is $-\frac{n}{2}\log n$ and is attained when $e^{i\theta_k}$ for $k = 1, \ldots, n$ are roots of the equation $x^n \pm 1 = 0$.*

Proof. Write $|e^{i\theta_k} - e^{i\theta_j}|$ as follows:

$$|e^{i\theta_k} - e^{i\theta_j}| = 2\sin\left(\frac{\theta_k - \theta_j}{2}\right) = \frac{1}{i}(e^{i\theta_k} - e^{i\theta_j})e^{-i(\theta_k+\theta_j)/2} \quad \text{for } \theta_k \geq \theta_j.$$

This shows that at a minimum

$$0 = \frac{\partial W}{\partial \theta_k} = -i\left(\frac{n-1}{2} - \sum_{j \neq k} \frac{e^{i\theta_k}}{e^{i\theta_k} - e^{i\theta_j}}\right), \quad k = 1, \ldots, n.$$

As in the Stieltjes problem, define

$$f(x) = \prod_{j=1}^{n}(x - e^{i\theta_j})$$

so that

$$\frac{f''(e^{i\theta_k})}{2f'(e^{i\theta_k})} = \sum_{j \neq k} \frac{1}{e^{i\theta_k} - e^{i\theta_j}}.$$

Thus f satisfies the equation

$$xy'' - (n-1)y' = 0$$

or

$$f(x) = Cx^n + D.$$

Since the coefficient of x^n in f is 1 and the roots lie on the unit circle, $C = 1$ and $D = \pm 1$. Thus

$$f(x) = x^n \pm 1.$$

The least value of W is $-\frac{1}{2} \log \Delta$, where Δ is the discriminant of $f(x)$. Since

$$\Delta = \left| \prod_{k=1}^{n} f'(e^{i\theta_k}) \right| = \left| \prod_{k=1}^{n} n e^{i(n-1)\theta_k} \right| = n^n,$$

this minimum is $-\frac{n}{2} \log n$.   ■

Remark 8.7.1   The partition function for this charge distribution is given by

$$\psi_n(\beta) = \frac{1}{(2\pi)^n} \int_{-\pi}^{\pi} \cdots \int_{-\pi}^{\pi} e^{-\beta W} d\theta_1 \cdots d\theta_n$$

$$= \frac{1}{(2\pi)^n} \int_{-\pi}^{\pi} \cdots \int_{-\pi}^{\pi} \prod_{1 \le j < k \le n} |e^{i\theta_j} - e^{i\theta_k}| d\theta_1 \cdots d\theta_n.$$

This integral is sometimes called Dyson's integral, but it is a special case of Selberg's integral, as can be seen after a suitable transformation. Its value is $\Gamma(1 + (\beta N/2))/(\Gamma(1 + \beta/2))^N$. Incidentally, one may use the last integral to evaluate the discriminant of $x^n \pm 1$. Forrester and Rogers [1986] have considered a different distribution of charges on the unit circle, which leads to Jacobi polynomials. See Exercise 13.

8.8 Constant-Term Identities

Take $\beta = 2k$, where k is a positive integer, in Dyson's integral (8.7.2). Its value is equal to the constant term in the Laurent expansion of the product

$$\prod_{\substack{\ell, j \\ \ell \ne j}} \left(1 - \frac{z_\ell}{z_j} \right)^k. \tag{8.8.1}$$

To prove this, set $z_j = e^{i\theta_j}$, so that

$$|z_j - z_\ell|^{2k} = (z_j - z_\ell)^k \left(z_j^{-1} - z_\ell^{-1} \right)^k = \left(1 - \frac{z_\ell}{z_j} \right)^k \left(1 - \frac{z_j}{z_\ell} \right)^k.$$

Now observe that any power other than 0 of z_j vanishes on integration. From the value of Dyson's integral given in our last remark, we have

$$\text{C.T.} \prod_{\substack{\ell, j \\ \ell \ne j}} \left(1 - \frac{z_\ell}{z_j} \right)^k = \frac{\Gamma(1 + kn)}{\Gamma(1 + k)^n} = \frac{(nk)!}{(k!)^n}, \tag{8.8.2}$$

where C.T. stands for "constant term of." More generally, Dyson conjectured the result contained in the next theorem, which was first independently proved by Gunson and Wilson. The elegant proof given below is due to Good [1970].

Theorem 8.8.1   *If $a_1, a_2, \ldots, a_n$ are nonnegative integers, then*

$$\text{C.T.} \prod_{\substack{j,\ell \\ j \neq \ell}} \left(1 - \frac{z_j}{z_\ell}\right)^{a_j} = \frac{(a_1 + a_2 + \cdots + a_n)!}{\prod_{j=1}^{n}(a_j)!}.$$

Proof.   Let $p(x) = \prod_{i=1}^{n}(x - z_i)$. Then

$$\sum_{j=1}^{n} \frac{p(x)}{(x - z_j)p'(z_j)} \equiv 1,$$

since the left side is a polynomial of degree $\leq n - 1$ that is equal to 1 at the n points z_i, $i = 1, 2, \ldots, n$. Rewrite the identity as

$$\sum_{j=1}^{n} \prod_{\substack{1 \leq k \leq n \\ j \neq k}} \frac{x - z_k}{z_j - z_k} = 1.$$

Let $x = 0$ to get

$$\sum_{j=1}^{n} \prod_{\substack{1 \leq k \leq n \\ j \neq k}} \frac{1}{1 - z_j/z_k} = 1. \tag{8.8.3}$$

Now put

$$F_n(a_1, a_2, \ldots, a_n) = \prod_{j \neq \ell} \left(1 - \frac{z_j}{z_\ell}\right)^{a_j}$$

and multiply (8.8.3) by $F_n(a_1, \ldots, a_n)$ to arrive at the recurrence relation

$$F_n(a_1, \ldots, a_n) = \sum_{j=1}^{n} F_n(a_1, \ldots, a_j - 1, \ldots, a_n).$$

Obviously, C.T. $F_n(a_1, \ldots, a_n)$ must satisfy the same relation:

$$\text{C.T. } F_n(a_1, \ldots, a_n) = \sum_{j=1}^{n} \text{C.T. } F(a_1, \ldots, a_{j-1}, a_j - 1, a_{j+1}, \ldots, a_k).$$

Also C.T. $F_n(0, 0, \ldots, 0) = 1$ and if $a_k = 0$, then

$$\text{C.T. } F_n(a_1, \ldots, a_n) = \text{C.T. } F_{n-1}(a_1, \ldots, a_{k-1}, a_{k+1}, \ldots, a_n).$$

The last relation holds because $a_k = 0$ implies that only nonpositive powers of z_k appear in $F_n(a_1, \ldots, a_n)$. It is easy to check that $(a_1 + a_2 + \cdots + a_n)!/$

$(a_1! \, a_2! \cdots a_n!)$ also satisfies the same recurrence relations and initial conditions. This proves Theorem 8.8.1 by induction. ∎

Morris [1984] derived the following constant-term identity from Selberg's integral.

Theorem 8.8.2 *Suppose p, q, and r are nonnegative integers. Then*

$$\text{C.T.} \prod_{i=1}^{n} (1 - z_i)^p \left(1 - \frac{1}{z_i} \right)^q \prod_{1 \le j \ne k \le n} \left(1 - \frac{z_j}{z_k} \right)^r$$

$$= \prod_{j=1}^{n} \frac{(p + q + (j - 1)r)! \, (jr)!}{(p + (j - 1)r)! \, (q + (j - 1)r)! \, r!}.$$

The proof of this identity is left as an exercise.

8.9 Nearly Poised $_3F_2$ Identities

In Chapters 2 and 3 we gave several derivations of Dixon's sum for a well-poised $_3F_2$. A terminating form of this identity also follows from Selberg's formula. The more general result of Aomoto allows us to sum a few nearly poised $_3F_2$. Recall that a series is nearly poised if all but one of the pairs of upper and lower parameters have the same sum.

To see that Selberg's formula gives the terminating Dixon sum, take $n = 2$ and $\gamma = y$ to be positive integers in Selberg's integral. We get

$$\int_0^1 \int_0^1 (x_1 x_2)^{\alpha - 1} [(1 - x_1)(1 - x_2)]^{\beta - 1} (x_1 - x_2)^{2y} dx_1 dx_2$$

$$= \int_0^1 \int_0^1 x_1^{\alpha - 1} x_2^{\alpha - 1} (1 - x_1)^{\beta - 1} (1 - x_2)^{\beta - 1} x_1^{2y} \left(1 - \frac{x_2}{x_1} \right)^{2y} dx_1 dx_2.$$

Expand $(1 - x_2/x_1)^{2y}$ by the binomial theorem and integrate term by term with respect to x_1 to get

$$\int_0^1 \sum_{r=0}^{2y} \frac{\Gamma(2y + 1)}{\Gamma(r + 1)\Gamma(2y - r + 1)} \frac{\Gamma(2y + \alpha - r)\Gamma(\beta)}{\Gamma(2y + \alpha + \beta - r)} x_2^{\alpha + r - 1} (1 - x_2)^{\beta - 1} dx_2$$

$$= \sum_{r=0}^{2y} \frac{\Gamma(2y + 1)\Gamma(2y + \alpha - r)\Gamma(\beta)\Gamma(\alpha + r)\Gamma(\beta)}{\Gamma(2r + 1)\Gamma(2y - r + 1)\Gamma(2y + \alpha + \beta - r)\Gamma(\alpha + \beta + r)}$$

$$= \frac{\Gamma(\alpha)\Gamma(\beta)^2\Gamma(\alpha + 2y)}{\Gamma(\alpha + \beta)\Gamma(\alpha + \beta + 2y)} \, _3F_2 \left(\begin{matrix} -2y, \alpha, -\alpha - \beta - 2y + 1 \\ -\alpha - 2y + 1, \alpha + \beta \end{matrix} ; 1 \right).$$

This, together with the value of Selberg's integral for $n = 2$, gives

$$_3F_2\left(\begin{array}{c} -2y, \alpha, -\alpha - \beta - 2y + 1 \\ -\alpha - 2y + 1, \alpha + \beta \end{array}; 1\right) = \frac{\Gamma(1 + 2y)\Gamma(\alpha + \beta)\Gamma(\beta + y)\Gamma(\alpha + \beta)}{\Gamma(\beta)\Gamma(\alpha + 2y)\Gamma(1 + y)\Gamma(\alpha + \beta + y)}.$$

This is the terminating form of Dixon's identity we mentioned earlier.

The next theorem gives the values of some terminating nearly poised series that can be derived from Aomoto's formula and its extension. This theorem was also obtained by Bressoud [1987] in the more general setting of basic hypergeometric series.

Theorem 8.9.1

(a) $\qquad _3F_2\left(\begin{array}{c} -2k, \alpha, -\alpha - \beta - 2k \\ -\alpha - 2k, \alpha + \beta \end{array}; 1\right) = \dfrac{(\alpha + 1)_k(\beta)_k(1)_{2k}}{(\alpha + \beta)_k(1)_k(\alpha)_{2k}}$,

(b) $_3F_2\left(\begin{array}{c} -2k, \alpha, 1 - \alpha - \beta - 2k \\ 1 - \alpha - 2k, \alpha + \beta + 1 \end{array}; 1\right) = \dfrac{(\alpha)_k(\beta + 1)_k(1)_{2k}(\alpha + \beta)}{(\alpha + \beta)_k(1)_k(\alpha)_{2k}(\alpha + \beta + 2k)}$,

(c) $_3F_2\left(\begin{array}{c} -2k, \alpha + 1, 1 - \alpha - \beta - 2k \\ 1 - \alpha - 2k, \alpha + \beta + 1 \end{array}; 1\right) = \dfrac{(\alpha + 1)_k(\beta)_k(1)_{2k}(\alpha + \beta)}{(\alpha + \beta)_k(1)_k(\alpha)_{2k}(\alpha + \beta + 2k)}$,

(d) $\qquad _3F_2\left(\begin{array}{c} -2k, \alpha, -\alpha - \beta - 2k \\ 1 - \alpha - 2k, \alpha + \beta \end{array}; 1\right) = \dfrac{(\alpha)_k(\beta + 1)_k(1)_{2k}}{(\alpha + \beta)_k(1)_k(\alpha)_{2k}}$,

(e) $\qquad _3F_2\left(\begin{array}{c} -2k, \alpha + 1, 1 - \alpha - \beta - 2k \\ 1 - \alpha - 2k, \alpha + \beta + 2 \end{array}; 1\right)$

$$= \frac{(\alpha + 1)_k(\beta + 1)_k(1)_{2k}(\alpha + \beta)_{2k}}{(\alpha + \beta + 1)_k(1)_k(\alpha)_{2k}(\alpha + \beta + 2)_{2k}}.$$

Proof. Consider the integral

$$I = \int_0^1 \int_0^1 x_1^{\alpha_1 - 1} x_2^{\alpha_2 - 1}(1 - x_1)^{\beta_1 - 1}(1 - x_2)^{\beta_2 - 1}(x_1 - x_2)^{2k} dx_1 dx_2,$$

when k is a nonnegative integer. Expand $(x_1 - x_2)^{2k}$ by the binomial theorem and integrate. The result is

$$I = \frac{\Gamma(\alpha_1 + 2k)\Gamma(\beta_1)\Gamma(\alpha_2)\Gamma(\beta_2)}{\Gamma(\alpha_1 + \beta_1 + 2k)\Gamma(\alpha_2 + \beta_2)} {}_3F_2\left(\begin{array}{c} -2k, 1 - \alpha_1 - \beta_1 - 2k, \alpha_2 \\ 1 - \alpha_1 - 2k, \alpha_2 + \beta_2 \end{array}; 1\right).$$

The value of I from Aomoto's formula can be found in the following cases:

(a) $\alpha_1 = \alpha_2 + 1 (= \alpha + 1)$, $\beta_2 = \beta_1 (= \beta)$,
(b) $\alpha_1 = \alpha_2 (= \alpha)$, $\beta_2 = \beta_1 + 1 (= \beta + 1)$,
(c) $\alpha_2 = \alpha_1 + 1 (= \alpha + 1)$, $\beta_1 = \beta_2 (= \beta)$,
(d) $\alpha_1 = \alpha_2 (= \alpha)$, $\beta_1 = \beta_2 + 1 (= \beta + 1)$.

These give the first four cases of the theorem. Theorem 8.3.1, which extends Aomoto's formula, can be applied when

(e) $\alpha_2 = \alpha_1 + 1 (= \alpha + 1), \beta_2 = \beta_1 + 1 (= \beta + 1)$.

Thus the theorem is proved. ∎

The reader may also use Theorem 8.3.1 to work out the case

$$\alpha_1 = \alpha_2 + 1 (= \alpha + 1), \qquad \beta_1 = \beta_2 + 1 (= \beta + 1). \qquad (8.9.1)$$

There are nonterminating extensions of the sums in Theorem 8.3.1. See Exercise 15 for one example. It is possible to sum similar, almost very well poised $_5F_4$ series. See Bressoud [1987] for some terminating cases, again done there for basic hypergeometric series.

8.10 The Hasse–Davenport Relation

The finite-field analogs of the gamma and beta integrals are the Gauss and Jacobi sums respectively. It is worthwhile to look into the analog of Selberg's integral. Anderson discovered his idea for the evaluation of the integrals by studying the finite-field analog. Evans [1981] conjectured formulas that are analogs of Selberg's integral formula, and Anderson [1990] proved a special case of it. Later Evans [1991] used Anderson's ideas to obtain the complete result. In this section, we prove the Hasse–Davenport relation, which can be viewed as another finite-field analog of the multivariable beta integral of Dirichlet, which was used in Anderson's proof of the Selberg integral. In the next section, we give the statement and proof of the analog of Selberg's formula due to Anderson. Though this is a particular case of the more general known result, it contains the essential ideas that help explain the origin of the concepts used in Section 8.3. In this and the next section some knowledge of finite fields is assumed.

Every finite field is a finite-dimensional vector space over the field $\mathbb{Z}/p\mathbb{Z} \equiv \mathbb{Z}(p)$, for some prime p. A finite field F thus has $q = p^m$ elements for some integer $m \geq 0$ and a prime p. Denote by F_s the finite field with q^s elements. For $\alpha \in F_s$ define the trace and norm of α from F_s to F as

$$\mathrm{Tr}_{F_s/F}(\alpha) = \alpha + \alpha^q + \alpha^{q^2} + \cdots + \alpha^{q^{s-1}}$$

and

$$N_{F_s/F}(\alpha) = \alpha \cdot \alpha^q \cdots \cdot \alpha^{q^{s-1}}.$$

Check that the trace and norm of α belong to F and that

$$N_{F_s/F}(\alpha + \beta) = N_{F_s/F}(\alpha) + N_{F_s/F}(\beta)$$

and

$$N_{F_s/F}(\alpha\beta) = N_{F_s/F}(\alpha)N_{F_s/F}(\beta).$$

We omit F_s/F in the notation for trace and norm if the context is clear. For $\alpha \in F$, $\alpha^q = \alpha$. This implies that $\mathrm{Tr}(\alpha) = s\alpha$ and $N(\alpha) = \alpha^s$ when α is viewed as an element of F_s. More generally, $\mathrm{Tr}_{F_s/F}(\alpha) = \frac{s}{d}\mathrm{Tr}_{F(\alpha)/F}(\alpha)$ and $N_{F_s/F}(\alpha) = [N_{F(\alpha)/F}(\alpha)]^{s/d}$, where $d = [F(\alpha):F] = $ the dimension of $F(\alpha)$ over F.

Suppose that α is a root of the monic irreducible polynomial

$$f(x) = x^d - c_1x^{d-1} + c_2x^{d-2}\cdots + (-1)^dc_d \in F[x]. \tag{8.10.1}$$

Lemma 8.10.1

(a) *The trace and norm of α from $F(\alpha)$ to F are given by*

$$\mathrm{Tr}(\alpha) = c_1 \quad and \quad N(\alpha) = c_d.$$

(b) *If α is viewed as an element of $F_s \supseteq F(\alpha)$, then*

$$\mathrm{Tr}(\alpha) = \frac{s}{d}c_1 \quad and \quad N(\alpha) = c_d^{s/d}.$$

Proof.

(a) In this case, $\mathrm{Tr}(\alpha) = \alpha + \alpha^q + \cdots + \alpha^{q^{d-1}}$ and $N(\alpha) = \alpha\alpha^q\cdots\alpha^{q^{d-1}}$. Since the automorphism $\beta \to \beta^q$ of any finite field, which contains the field F with q elements, fixes F, it follows that $0 = (f(\alpha))^q = f(\alpha^q)$. Thus if α is a root of f, so is α^q. This implies (a).

(b) This follows from (a) and the discussion preceding Lemma 8.10.1. ∎

We now extend the idea of a Gauss sum, introduced in Chapter 1 for $\mathbb{Z}(p)$, to any finite field. For convenience in writing, denote $\mathbb{Z}(p)$ by F_p (not to be confused with F_s, the finite field of dimension s over F.)

Let ψ denote the additive character on F defined by

$$\psi(\alpha) = \zeta_p^{\mathrm{Tr}_{F/F_p}(\alpha)}, \quad where \quad \zeta_p = e^{2\pi i/p}. \tag{8.10.2}$$

A multiplicative character on F is a homomorphism χ from $F - \{0\}$ to the complex numbers, that is,

$$\chi(\alpha\beta) = \chi(\alpha)\chi(\beta). \tag{8.10.3}$$

By convention $\chi(0) = 0$. In what follows, let χ denote a nontrivial multiplicative character. Define the Gauss sum $g(\chi)$ by

$$g(\chi) = \sum_{\alpha \in F}\chi(\alpha)\psi(\alpha). \tag{8.10.4}$$

We are interested in the connection between $g(\chi)$ and

$$g(\chi') = \sum_{\beta \in F_s} \chi'(\beta)\psi'(\beta), \tag{8.10.5}$$

where

$$\chi' = \chi \circ N_{F_s/F} \quad \text{and} \quad \psi' = \psi \circ \text{Tr}_{F_s/F}. \tag{8.10.6}$$

Check that χ' and ψ' are characters of the appropriate kind on F_s. A relationship between $g(\chi)$ and $g(\chi')$ was given by Davenport and Hasse [1934]. We state and prove it after a couple of lemmas.

Lemma 8.10.2   *Suppose $\alpha \in F_s$ is a root of the irreducible polynomial (8.10.1). Then*

$$\chi'(\alpha)\psi'(\alpha) = [\chi(c_d)\psi(c_1)]^{s/d},$$

where χ' and ψ' are defined by (8.10.6).

Proof.

$$\begin{aligned}
\chi'(\alpha)\psi'(\alpha) &= \chi(N(\alpha))\psi(\text{Tr}(\alpha)) \\
&= \chi\left(c_d^{s/d}\right)\psi\left(\frac{s}{d}c_1\right) \quad \text{(by Lemma 8.10.1)} \\
&= [\chi(c_d)\psi(c_1)]^{s/d}.
\end{aligned}$$

The next lemma is a well-known result on finite fields.   ∎

Lemma 8.10.3   *The polynomial $x^{q^s} - x$ is the product of all monic irreducible polynomials in $F[x]$ of degrees that divide s. (F has q elements.)*

The proof is left to the reader.

The essential idea in the proof of the Hasse–Davenport relation given here is due to Weil [1949]. Weil's proof is much simpler than the original. We follow the account given by Ireland and Rosen [1991, §11.4], which contains a further simplification of Weil's argument by P. Monsky.

The Hasse–Davenport relation is given in the next theorem.

Theorem 8.10.4 $-g(\chi') = (-g(\chi))^s.$

In order to make the proof more manageable, we break it down into two lemmas followed by the completion of the proof. Suppose f is the monic polynomial $x^n - c_1 x^{n-1} + c_2 x^{n-2} + \cdots + (-1)^n c_n$, where $c_i \in F$. Define a complex-valued function λ on the set of monic polynomials with coefficients in F by the equation $\lambda(f) = \chi(c_n)\psi(c_1)$.

Lemma 8.10.5 *The function λ is multiplicative, that is, if f and g are monic polynomials in $F[x]$, then $\lambda(fg) = \lambda(f)\lambda(g)$.*

Proof. Suppose $f(x) = x^n - c_1 x^{n-1} + c_2 x^{n-2} + \cdots + (-1)^n c_n$ and $g(x) = x^m - d_1 x^{m-1} + \cdots + (-1)^m d_m$. Then

$$f(x)g(x) = x^{m+n} - (c_1 + d_1)x^{m+n-1} + \cdots + (-1)^{m+n} c_n d_m.$$

By definition

$$\lambda(fg) = \chi(c_n d_m)\psi(c_1 + d_1)$$
$$= \chi(c_n)\psi(c_1)\chi(d_m)\psi(d_1) = \lambda(f)\lambda(g). \qquad \blacksquare$$

Lemma 8.10.6 $g(\chi') = \sum_f (\deg f)\lambda(f)^{s/\deg f}$, *where the sum is over all monic irreducible polynomials in $F[x]$ whose degrees divide s.*

Proof. Suppose $\alpha \in F_s$ satisfies an irreducible polynomial f of degree d. Then $\chi'(\alpha)\psi'(\alpha) = [\chi(c_n)\psi(c_1)]^{s/d} = \lambda(f)^{s/d}$ by Lemma 8.10.2. This implies that

$$\sum_{\text{conjugates}} \chi'(\alpha)\psi'(\alpha) = (\deg f)\lambda(f)^{s/\deg f},$$

where the summation is over all the conjugates of α. Since every element of F_s is the root of a monic irreducible polynomial whose degree divides s and conversely (in fact, these polynomials are the only irreducible factors of $x^{q^s} - x$), the result is proved. $\blacksquare$

Proof of Theorem 8.10.4. Consider the L-function given by the formal power series $L(\lambda, t) = \sum_{f \text{ monic}} \lambda(f) t^{\deg f}$, where the summation is over all monic polynomials in $F[x]$. The following identity is easy to verify:

$$L(\lambda, t) = \sum_{f \text{ monic}} \lambda(f) t^{\deg f} = \prod_{f \text{ irred.}} (1 - \lambda(f) t^{\deg f})^{-1},$$

where the product is over all monic irreducible polynomials. Here it is understood that $\lambda(1) = 1$. Write the L-function as

$$L(\lambda, t) = \sum_{n=0}^{\infty} \left(\sum_{\deg f = n} \lambda(f) \right) t^n.$$

The coefficient of t is

$$\sum_{a \in F} \lambda(x - a) = \sum_{a \in F} \chi(a)\psi(a) = g(\chi),$$

and the coefficient of t^n for $n > 1$ is

$$\sum_{c_1, c_n \in F} \chi(c_n)\psi(c_1) = q^{n-2} \sum_{c_n \in F} \chi(c_n) \sum_{c_1 \in F} \psi(c_1) = 0.$$

The factor q^{n-2} arises because, for a given pair $c_1, c_n \in F$, there are q^{n-2} ways of writing the other coefficients of a polynomial. We have shown that

$$L(\lambda, t) = 1 + g(\chi)t = \prod_{f \text{ irred.}} (1 - \lambda(f)t^{\deg f})^{-1}.$$

Take the logarithmic derivative and multiply the equation by t. Then

$$\frac{tg(\chi)}{1 + g(\chi)t} = \sum_{f \text{ irred.}} \frac{\lambda(f)(\deg f)t^{\deg f}}{1 - \lambda(f)t^{\deg f}},$$

or

$$\sum_{s=1}^{\infty} (-1)^{s-1} g(\chi)^s t^s = \sum_{s=1}^{\infty} \left(\sum_{f \text{ irred.}} (\deg f)\lambda(f)^{s/\deg f} \right) t^s.$$

Equate the coefficients of t^s on each side and use Lemma 8.10.6 to complete the proof of the theorem. ∎

Remark 8.10.1   In Exercise 1.50, the Dirichlet L-function $\prod(1 - \chi(p)p^{-s})^{-1}$ was introduced. The prime ideals of the ring of integers p are the ideals generated by prime numbers p. The number of elements in the field $\mathbb{Z}(p)$ is p. For the ring of polynomials $F[x]$, the prime ideals are generated by irreducible polynomials. If F has q elements, then $F[x]/(f(x))$, where f is irreducible, has $q^{\deg f}$ elements. Thus, when $t = q^{-s}$,

$$\prod_{f \text{ irred.}} (1 - \lambda(f)q^{-s \deg f})^{-1} = \prod_{f \text{ irred.}} (1 - \lambda(f)t^{\deg f})^{-1}$$

is called an L-function. Note that λ is multiplicative on $F[x]$ just as χ is multiplicative on $\mathbb{Z}$.

8.11   A Finite-Field Analog of Selberg's Integral

We begin by recalling the definition and some elementary properties of the resultant of two polynomials. Suppose K is a field and let

$$f(x) = a_0 x^n + a_1 x^{n-1} + \cdots + a_n,$$
$$g(x) = b_0 x^m + b_1 x^{m-1} + \cdots + b_m$$

be two polynomials in $K[x]$.

The resultant of f and g, $R(f, g)$, is defined by the $(m+n) \times (m+n)$ determinant

$$R(f, g) = \begin{vmatrix} a_0 \ a_1 \dots a_n & & & \\ & a_0 \ a_1 \dots a_n & & \\ & \dots\dots\dots & & \\ & & a_0 \ a_1 \dots a_n & \\ b_0 \ b_1 \dots b_m & & & \\ & b_0 \ b_1 \dots b_m & & \\ & \dots\dots\dots & & \\ & & b_0 \ b_1 \dots b_m & \end{vmatrix}. \tag{8.11.1}$$

It can be shown (an exercise for the reader) that the resultant vanishes if and only if either f and g have a common nonconstant factor or the leading coefficients of f and g are zero.

Suppose that f and g can be factored as

$$f = a_0(x - x_1)(x - x_2) \cdots (x - x_n),$$
$$g = b_0(x - y_1)(x - y_2) \cdots (x - y_m).$$

Then

$$R(f, g) = a_0^m b_0^n \prod_i \prod_j (x_i - y_j) = a_0^m \prod_i g(x_i) = (-1)^{mn} b_0^n \prod_j f(y_j).$$
$$\tag{8.11.2}$$

It follows that

$$R(f, gh) = R(f, g)R(f, h). \tag{8.11.3}$$

It is also clear from the definition that if either g or f is a constant, then

$$R(f, b_0) = b_0^n, \quad R(a_0, g) = a_0^m. \tag{8.11.4}$$

Now let F be a field with $q = p^n$ elements where p is an odd prime. Let χ and ψ be the multiplicative and additive characters defined in the previous section. For all positive integers α, define the Gauss sum

$$g(\chi^\alpha) = g(\alpha) = \sum_{x \in F} \chi(x)^\alpha \psi(x). \tag{8.11.5}$$

Extend the definition to all integers α by the requirement

$$g(\alpha + q - 1) = g(\alpha), \quad \text{and let} \quad g^*(\alpha) = q/g(-\alpha). \tag{8.11.6}$$

For all positive integers α, β, γ, n, the Selberg sum is defined by

$$S_n(\alpha, \beta, \gamma) = \sum_f \chi \left((-1)^{n\alpha} f(0)^\alpha f(1)^\beta \Delta_f^\gamma \right) \delta(\Delta_f), \tag{8.11.7}$$

where the sum is over all monic polynomials f of degree n in $F[x]$, Δ_f is the discriminant of f, and $\delta = \chi^{(q-1)/2}$.

To evaluate this sum we need a property of the following L-function: For a monic polynomial $V \in F[x]$, let

$$L(t, V) := \sum_W \chi(R(V, W))t^{\deg W},$$

where W ranges over the monic polynomials in $F[x]$. Let G be the product of all distinct monic irreducible polynomials that divide V. Denote the multiplicity of the monic irreducible polynomial f as a factor of V by $\mathrm{ord}_f V$. We say that V is primitive if for every factor f of V, $q - 1$ does not divide $\mathrm{ord}_f V$.

The next lemma plays a role in the evaluation of Selberg's sum similar to the role of Lemma 8.4.3, which is a consequence of Dirichlet's beta integral formula, in the evaluation of Selberg's integral. The proof of this lemma employs the Hasse–Davenport relation (proved in the last section).

Lemma 8.11.1   *Suppose V is primitive and of positive degree. Then*

$$\deg L(t, V) \le \deg G - 1$$

and the coefficient $\epsilon(V)$ of $t^{\deg G-1}$ is

$$\epsilon(V) = \delta(\Delta_G)\chi(R(V, G'))g^*(\deg(V))^{-1} \prod_{f|G} g^*(\mathrm{ord}_f V)^{\deg f}.$$

Proof.   It follows from (8.11.2) that $R(V, W)$ depends only on the value of $W \bmod G$. Suppose m is an integer $\ge \deg G$. For a given polynomial S of degree $\le \deg G - 1$, there are $q^{m-\deg G}$ polynomials Q of $\deg m - \deg G$ such that $GQ + S$ is of $\deg m$. Write

$$L(t, V) = \sum_{m=0}^{\infty} \left(\sum_{\deg W = m} \chi(R(V, W)) \right) t^m.$$

Since V is primitive, $\chi(R(V, S)) \ne 1$ for some polynomial S, and hence

$$\sum_{\deg W=m} \chi(R(V, W)) = q^{m-\deg G} \sum_S \chi(R(V, S)) = 0.$$

Thus $L(t, V)$ is a polynomial of degree $\le \deg G - 1$.

To find $\epsilon(V)$, consider the double sum

$$\mu = \sum_U \sum_W \psi\left(-\mathrm{Res}_\infty \frac{UW}{G} dx \right) \bar\chi(R(V, U)),$$

where W ranges over monic polynomials of $\deg G - 1$ and U ranges over polynomials of degree $< \deg G$ and Res_∞ denotes the residue at infinity. When U is a constant, say $a \in F$, then

$$-\text{Res}_\infty \frac{UW}{G} dx = \text{coefficient of } 1/x \text{ in } \frac{U(1/x)W(1/x)}{G(1/x)} \cdot \frac{1}{x^2} = a.$$

(The factor $1/x^2$ comes in because the transformation $x \to 1/x$ changes dx to $-dx/x^2$.) By (8.11.4)

$$\bar{\chi}(R(V,a)) = \bar{\chi}(a)^{\deg V}.$$

When U is not a constant, then, by an argument similar to the one given before, the sum over all W vanishes. Thus

$$\mu = q^{\deg G - 1} \sum_{a \in F} \psi(a)\bar{\chi}(a)^{\deg V} = q^{\deg G - 1} g(-\deg(V)). \tag{8.11.8}$$

To evaluate μ in another way, start with the fact that the sum of the residues of UW/G at finite points plus the residue at inifinity is 0. The finite poles are at the zeros of G. This implies

$$-\text{Res}_\infty \frac{UW}{G} = \sum_{\text{Res } \eta_f} \frac{UW}{G}.$$

Here η_f denotes a root of G that comes from the irreducible factor f. The residue at η_f is given by $U(\eta_f)W(\eta_f)/G'(\eta_f)$, where G' denotes the formal derivative of G. The sum of residues at all the roots of f is then

$$\text{Tr}\left(\frac{U(\eta_f)W(\eta_f)}{G'(\eta_f)}\right).$$

Use this to write μ as

$$\mu = \sum_U \sum_W \prod_{f|G} \psi\left(\text{Tr}\left(\frac{U(\eta_f)W(\eta_f)}{G'(\eta_f)}\right)\right) \bar{\chi}(N(U(\eta_f)))^{\text{ord}_f V}.$$

Here N denotes the norm, and we used (8.11.2) to obtain the expression involving N. To simplify μ, observe that when $R(V,W) \neq 0$,

$$\psi\left(\text{Tr}\left(\frac{U(\eta_f)W(\eta_f)}{G'(\eta_f)}\right)\right) \bar{\chi}(N(U(\eta_f)))^{\text{ord}_f V}$$

$$= \psi\left(\text{Tr}\left(\frac{U(\eta_f)W(\eta_f)}{G'(\eta_f)}\right)\right) \bar{\chi}\left(N\left(\frac{U(\eta_f)W(\eta_f)}{G'(\eta_f)}\right)\right)^{\text{ord}_f V}$$

$$\cdot \chi\left(N\left(\frac{W(\eta_f)}{G'(\eta_f)}\right)\right)^{\text{ord}_f V}. \tag{8.11.9}$$

It is clear that since V is primitive, the sum over U in μ vanishes when $R(V,W) = 0$. Thus the terms in the sum indexed by W that have $R(V,W) = 0$ can be dropped.

This observation, together with (8.11.9), (8.11.2), and (8.11.3), implies that

$$\mu = \sum_W \chi \left(\frac{R(V, W)}{R(V, G')} \right) \prod_{f | G} g_f(-\mathrm{ord}_f V). \qquad (8.11.10)$$

In the above expression the Gauss sum g_f is over the field $F(\eta_f)$, where $[F(\eta_f) : F] =$ degree of f. By the Hasse–Davenport relation,

$$g_f(-\mathrm{ord}_f V) = (-1)^{\deg f - 1}(g(-\mathrm{ord}_f V))^{\deg f}.$$

Now observe that $\prod_{f | G}(-1)^{\deg f - 1}$ is equal to the sign of the permutation of the roots of G effected by the qth power automorphism of the algebraic closure of the field F. Since q is odd, this value is equal to $\delta(\Delta_G)$. Thus

$$\mu = \epsilon(V)\delta(\Delta_G)\bar{\chi}(R(V, G')) \prod_{f | G} g(-\mathrm{ord}_f V)^{\deg f}. \qquad (8.11.11)$$

Compare the two expressions for μ, (8.11.11) and (8.11.8), to prove the lemma. ∎

Theorem 8.11.2 *Suppose that the numbers* $\alpha, \alpha + \gamma, \ldots, \alpha + (n-1)\gamma$; $\beta, \beta + \gamma, \ldots, \beta + (n-1)\gamma$; $\gamma, 2\gamma, \ldots, n\gamma$ *are not divisible by* $q - 1$. *Then the Selberg sum can be written as*

$$S_n(\alpha, \beta, \gamma) = \prod_{j=0}^{n-1} \frac{g^*(\alpha + j\gamma)g^*(\beta + j\gamma)g^*((j+1)\gamma)}{g^*(\alpha + \beta + (n-1+j)\gamma)g^*(\gamma)}.$$

Proof.   The proof is by induction. It is sufficient to prove that

$$S_n(\alpha, \beta, \gamma) = \frac{g^*(\alpha)g^*(\beta)g^*(n\gamma)}{g^*(\alpha + \beta + (n-1)\gamma)g^*(\gamma)} \cdot S_{n-1}(\alpha + \gamma, \beta + \gamma, \gamma). \quad (8.11.12)$$

For this purpose, consider the double sum

$$S = \sum_P \sum_Q \chi(Q(0)^\alpha Q(1)^\beta R(P, Q)^\gamma),$$

where the sum on P ranges over monic polynomials of degree $n - 1$ and Q is over monic polynomials of degree n. First note that

$$\chi(Q(0)^\alpha Q(1)^\beta R(P, Q)^\gamma) = \chi(Q(0)^\alpha Q(1)^\beta R(Q^\gamma, P))$$
$$= \chi(R(x^\alpha(1 - x)^\beta P^\gamma, Q)).$$

So we can take $V = Q^\gamma$ or $V = x^\alpha(x - 1)^\beta P^\gamma$ in Lemma 8.11.1, since the hypothesis of the theorem implies that V is primitive. The lemma also implies that the sum over P (respectively Q) is zero if Q (respectively P) is not square-free. Therefore, summing over P with Q square-free, Lemma 8.11.1 implies that

$$S = \sum_Q \chi(Q(0)^\alpha Q(1)^\beta)\delta(\Delta_Q)\chi(R(Q, Q')^\gamma)\frac{g^*(\gamma)^n}{g^*(n\gamma)}.$$

This follows since, if $Q = Q_1 \cdots Q_s$ is a factorization into irreducible polynomials, then $\deg V = \deg Q^\gamma = n\gamma$ and $\mathrm{ord}_{Q_i} V = \gamma$. Thus by (8.11.2),

$$S = \chi(-1)^{n\alpha + n(n-1)\gamma/2} S_n(\alpha, \beta, \gamma) g^*(\gamma)^n / g^*(n\gamma). \qquad (8.11.13)$$

Similarly, summation over Q with P square-free and $V = x^\alpha (x-1)^\beta P^\gamma$ gives

$$S = \sum_P \delta(\Delta_{x(x-1)P}) \chi(R(x^\alpha (x-1)^\beta P^\gamma, \frac{d}{dx} x(x-1)P)) \cdot \frac{g^*(\alpha)g^*(\beta)g^*(\gamma)^{n-1}}{g^*(\alpha + \beta + (n-1)\gamma)}.$$

Set

$$T = \frac{d}{dx} x(x-1)P = (x-1)P + xP + x(x-1)P'$$

and observe that

$$\chi(R(x^\alpha (x-1)^\beta P^\gamma, T)) = \chi(T(0)^\alpha T(1)^\beta R(P, T)^\gamma)$$
$$= \chi((-1)^\alpha P(0)^\alpha P(1)^\beta R(P, x(x-1)P')^\gamma)$$

and

$$R(P, x(x-1)P') = R(P, x)R(P, x-1)R(P, P')$$
$$= P(0)P(1)R(P, P').$$

These relations and the fact that $\delta(\Delta_{x(x-1)P}) = \delta(\Delta_P)$ give

$$S = \chi(-1)^{n\alpha + n(n-1)\gamma/2} S_{n-1}(\alpha + \gamma, \beta + \gamma, \gamma) \frac{g^*(\alpha)g^*(\beta)g^*(\gamma)^{n-1}}{g^*(\alpha + \beta + (n-1)\gamma)}. \qquad (8.11.14)$$

A comparison of (8.11.13) and (8.11.14) shows that (8.11.12) is true, which proves the theorem. ∎

Exercises

1. Prove that if f is continuous on $[0, 1]$, then

$$\lim_{\alpha \to 0+} \alpha \int_0^1 t^{\alpha - 1} f(t) \, dt = f(0).$$

The case in which f is differentiable is easier and can be done by integration by parts.

2. Suppose that f is a complex measurable function on a measure space X with a positive measure μ and that $\|f\|_\infty > 0$. Prove that if $\|f\|_p < \infty$ for some $0 < p < \infty$, then

$$\|f\|_p \to \|f\|_\infty \quad \text{as} \quad p \to \infty.$$

3. Work out the details of the proofs of Corollaries 8.2.2 and 8.2.3, that is, prove the formulas

$$\int_0^\infty \cdots \int_0^\infty \prod_{i=1}^k x_i \prod_{i=1}^n x_i^{\alpha-1} e^{-x_i} \prod_{1\le i<j\le n} |x_i - x_j|^{2\gamma} dx$$

$$= \prod_{j=1}^k (\alpha + (n-j)\gamma) \prod_{j=1}^n \frac{\Gamma(\alpha + (j-1)\gamma)\Gamma(1+j\gamma)}{\Gamma(1+\gamma)}$$

and

$$\int_{-\infty}^\infty \cdots \int_{-\infty}^\infty \exp\left(-\frac{1}{2}\sum_{i=1}^n x_i^2\right) \prod_{1\le i<j\le n} |x_i - x_j|^{2\gamma} dx = (2\pi)^{n/2} \prod_{j=1}^n \frac{\Gamma(\gamma j + 1)}{\Gamma(\gamma + 1)}.$$

4. Denote the first integral in Exercise 3 as $G_n(k; \alpha, \gamma)$ and employ the method used in proving Aomoto's formula to show that

$$G_n(k; \alpha, \gamma) = [\alpha + (n-k)\gamma]G_n(k-1; \alpha, \gamma).$$

Obtain another evaluation of $G_n(k; \alpha, \gamma)$ from this recurrence relation.

5. Here is one proof of Selberg's alternative beta integral formula (Theorem 8.3.3). With $R_+ = [0, \infty)$, set

$$G(\lambda) = \int_{R_+^n} \left(\prod_{i=1}^k x_i\right) \prod_{1\le i<j\le n} |x_i - x_j|^{2\gamma} \prod_{i=1}^n x_i^{\alpha-1} e^{-\lambda x_i} dx.$$

(a) Show that $\lambda^{\beta+k+n\alpha+(n-1)n\gamma-1} e^{-\lambda} G(\lambda) = G(1)\lambda^{\beta-1} e^{-\lambda}$.

(b) Integrate (a) with respect to λ over $[0, \infty)$ and show that

$$\int_{R_+^n} \frac{\left(\prod_{i=1}^k x_i\right) \prod_{1\le i<j\le n} |x_i - x_j|^{2\gamma}}{\left[1 + \sum_1^n x_i\right]^{\beta+k+n\alpha+n(n-1)\gamma}} \prod_{i=1}^n x_i^{\alpha-1} dx$$

$$= \frac{\Gamma(\beta) \prod_{j=1}^k (\alpha + (n-j)\gamma)}{\Gamma(\beta + k + n\alpha + (n-1)n\gamma)} \prod_{j=1}^n \frac{\Gamma(\alpha + (n-j)\gamma)\Gamma(j\gamma + 1)}{\Gamma(\gamma + 1)}.$$

(c) Change variables so that

$$x_i = y_i \left(1 - \sum_1^n y_i\right)^{-1};$$

show that

$$\left|\frac{\partial(x_1, \ldots, x_n)}{\partial(y_1, \ldots, y_n)}\right| = \left(1 - \sum_1^n y_i\right)^{-n-1}$$

and obtain Selberg's formula.

6. Prove Lemma 8.3.2.
7. Verify Equations (8.3.6) and (8.3.7) in the proof of Theorem 8.3.1.

8. Prove Theorems 8.5.4 and 8.5.5.

   The next five problems are similar to Steiltjes's maximum problem and have similar solutions.

9. Let there be a positive mass p at the fixed point $x = 0$ and unit masses at the variable points $x_1, x_2, \ldots, x_n$ in $[0, \infty)$ such that

$$x_1 + \cdots + x_n \le nK,$$

   where K is a given positive number. Show that the maximum of

$$U(x_1, \ldots, x_n) = \prod_{j=1}^{n} x_j^p \prod_{1 \le i < j \le n} |x_i - x_j|$$

   is attained if and only if the $\{x_j\}$ are zeros of the Laguerre polynomial $L_n^{(\alpha)}(cx)$, where $\alpha = 2p - 1$ and $c = (n + \alpha)/K$.

10. Suppose there are unit charges at the variable points $x_1, \ldots, x_n$ in the interval $(-\infty, \infty)$ such that

$$x_1^2 + \cdots + x_n^2 \le nL,$$

    for a given positive number L. Show that the maximum of

$$V(x_1, \ldots, x_n) = \prod_{1 \le i < j \le n} |x_i - x_j|$$

    is attained if and only if the $\{x_i\}$ are the zeros of the Hermite polynomials $H_n(Cx)$, $C = \sqrt{(n-1)/(2L)}$.

11. Suppose there are n unit masses, $n \ge 2$, at the points $x_1, x_2, \ldots, x_n$ in $[-1, 1]$. Find the positions of these points for which $\prod_{1 \le i < j \le n} |x_i - x_j|$ is a maximum.

12. Suppose that the n unit masses exist at $x_1, \ldots, x_n$ in $[0, \infty)$ and satisfy the condition

$$x_1 + x_2 + \cdots + x_n \le nK,$$

    where K is a given positive number. Find the positions of these points for which $\prod_{1 \le i < j \le n} |x_i - x_j|$ is a maximum.

13. Suppose charges are distributed on a unit circle. At $\theta = 0$, fix a charge $+q$ and at, $\theta = \pi$, fix a charge $+p$. Distribute $2N$ freely moving unit charges at $\theta_1, \ldots, \theta_{2N}$ so that

$$0 < \theta_j < \pi, j = 1, \ldots, N \quad \text{and} \quad \pi < \theta_j < 2\pi, j = N+1, \ldots, 2N.$$

    Show that the potential

$$T = -q \sum_{k=1}^{2N} \log|1 - e^{i\theta_k}| - p \sum_{k=1}^{2N} \log|1 + e^{i\theta_k}| - \sum_{1 \le k < j \le 2N} \log|e^{i\theta_k} - e^{i\theta_j}|$$

    is a miminum when the θ_is are zeros of the Jacobi polynomial $P_N^{(q-1/2, p-1/2)}(\cos\theta)$, $0 < \theta < 2\pi$. See Forrester and Rogers [1986].

14. Prove that, with appropriate conditions on the parameters for convergence,

$$\int_0^\infty \cdots \int_0^\infty \prod_{j=1}^n x_j^{\alpha-1}(1+x_j)^{-\alpha-\beta-2\gamma(n-1)} \prod_{1\le i<j\le n} |(x_i-x_j)|^{2\gamma} dx_1 \ldots dx_n$$

$$= \prod_{j=1}^n \frac{\Gamma(\alpha+(j-1)\gamma)\Gamma(\beta+(j-1)\gamma)\Gamma(1+j\gamma)}{\Gamma(\alpha+\beta+(n+j-2)\gamma)\Gamma(1+\gamma)}$$

and

$$\frac{1}{(2\pi)^n} \int_{-\infty}^\infty \cdots \int_{-\infty}^\infty \prod_{j=1}^n (a+ix_j)^{-\alpha}(b-ix_j)^{-\beta}$$

$$\cdot \prod_{1\le i<j\le n} |(x_i-x_j)|^{2\gamma} dx_1 \ldots dx_n$$

$$= \frac{1}{(a+b)^{(\alpha+\beta)n-\gamma n(n-1)-n}}$$

$$\cdot \prod_{j=1}^n \frac{\Gamma(\alpha+\beta-(n+j-2)\gamma-1)\Gamma(1+j\gamma)}{\Gamma(\alpha-(j-1)\gamma)\Gamma(\beta-(j-1)\gamma)\Gamma(1+\gamma)}.$$

For the second integral, begin with Cauchy's beta integral.

15. Show that

$$_3F_2\left(\begin{matrix} a-1, b, c \\ a+1-b, a+1-c \end{matrix}; 1\right)$$

$$= \frac{(a-1)\Gamma(a+1-b)\Gamma(a+2-c)\Gamma(a/2+1)\Gamma(a/2+2-b-c)}{(b-1)(c-1)\Gamma(a/2+1-b)\Gamma(a/2+1-c)\Gamma(a+1)\Gamma(a+1-b-c)}$$

$$- \frac{\Gamma(a+1-b)\Gamma(a+1-c)\Gamma(a/2+1/2)\Gamma(a/2+3/2-b-c)}{(b-1)(c-1)\Gamma(a/2+1/2-b)\Gamma(a/2+1/2-c)\Gamma(a)\Gamma(a+1-b-c)}.$$

Note that, if $a = 2n$, the second term vanishes and the first can be evaluated by setting $a = -2n - \epsilon$ and letting $\epsilon \to 0$. When $a = -2n - 1$, the first term vanishes and the second can be evaluated by a similar limit.

16. Evaluate the integral in Remark 8.7.1, that is,

$$\frac{1}{(2\pi)^n} \int_{-\pi}^\pi \cdots \int_{-\pi}^\pi \prod_{1\le i<j\le n} |e^{i\theta_j} - e^{i\theta_k}| d\theta_1 \cdots d\theta_n.$$

One approach is to set $x_i = \tan(\theta_i/2)$ in the second integral in Exercise 14.

17. Prove Morris's constant-term identity contained in Theorem 8.8.2.

18. Complete the missing steps in the proof of Theorem 8.6.5.

19. Prove the arithmetic and geometric mean inequality as follows: Let $x_1 \le x_2 \le x_3 \cdots \le x_n$. Show that with $s = (x_1 + \cdots + x_n)/n$, $s(x_1 + x_n - s) \ge x_1 x_n$. By the inductive hypothesis the result holds for the $n-1$ numbers $x_2, \ldots, x_{n-1}, x_1 + x_n - s$. Now obtain the necessary result.

20. Fill in the details in the proof of Theorem 8.9.1. In particular, work out case (8.9.1).

21. Consider $S_n(\alpha, \beta, \gamma; u) = \int |F(0)|^{\alpha-1}|F(1)|^{\beta-1}|\Delta_F|^{\gamma-2}F(u)dF_0dF_1\cdots dF_{n-1}$, where u is a parameter.

(a) Show that

$$\int_{F(t)\in\mathcal{D}} F(u)\prod_{i=0}^{n}|F(\zeta)|^{\alpha_i-1}dF_0\cdots dF_{n-1}$$

$$= \prod_{i=0}^{n}\frac{|Z'(\zeta_i)|^{\alpha_i-1/2}\Gamma(\alpha_i)}{\Gamma\left(\sum_0^n \alpha_i\right)}\cdot\frac{\sum_{k=0}^n \alpha_k \prod_{i\neq k}(u-\zeta_i)}{\sum_0^n \alpha_i}.$$

(This follows from Lemma 8.4.3.)

(b) Let

$$I_n(\alpha, \beta, \gamma; u) = \int_{(F,G)}|G(0)|^{\alpha-1}|G(1)|^{\beta-1}|R(F,G)|^{\gamma-1}$$

$$\cdot F(u)dF_0\cdots dF_{n-2}dG_0\cdots dG_{n-1}.$$

Show that

$$I_n(\alpha, \beta, \gamma; u) = \frac{1}{n}\frac{d}{du}S_n(\alpha, \beta, \gamma; u)\cdot\frac{\Gamma(\gamma)^n}{\Gamma(n\gamma)}$$

$$= S_{n-1}(\alpha+\gamma, \beta+\gamma, \gamma; u)\frac{\Gamma(\alpha)\Gamma(\beta)\Gamma(\gamma)^{n-1}}{\Gamma(\alpha+\beta+(n-1)\gamma)}.$$

(c) Now prove by induction that

$$S_n(\alpha, \beta, \gamma; u)$$

$$= S_n(\alpha, \beta, \gamma)\left\{\sum_{m=0}^{n}(-1)^m\binom{n}{m}u^{n-m}\prod_{i=0}^{m-1}\frac{(\alpha+(n-1-i)\gamma)}{(\alpha+\beta+(2n-2-i)\gamma)}\right\}$$

To prove this, let $T_n(\alpha, \beta, \gamma; u)$ denote the sum inside the braces show that

$$T_n(\alpha, \beta, \gamma; u) = nT_{n-1}(\alpha+\gamma, \beta+\gamma, \gamma; u).$$

(d) Use (c) to show that the introduction of the factor

$$x_1 x_2\cdots x_m(1-x_{m+1})\cdots(1-x_{m+\ell}); \quad m+\ell\leq n,$$

inside the Selberg integral, multiplies its value by

$$\sum_{j=0}^{\ell}(-1)^j\binom{\ell}{j}\prod_{i=0}^{m+j-1}\frac{\alpha+(n-1-i)\gamma}{\alpha+\beta+(2n-2-i)\gamma}.$$

(e) Use the Chu–Vandermonde identity to sum the expression in (d).

The following problems on finite fields use the notation and definitions given in Section 8.10.

22. For $\alpha \in F_s$ show that

(a) $Tr_{F_s/F}(\alpha) \in F$, $N_{F_s/F}(\alpha) \in F$,

(b) $Tr_{F_s/F}$ maps F_s onto F,

(c) There exists a $\beta \in F$ such that $\psi(\beta) \neq 1$,

(d) $\sum_{\beta\in F}\psi(\beta) = 0$.

23. Define $g_\alpha(\chi) = \sum_{t \in F} \chi(t) \psi(\alpha t)$. Show that

(a) $|g_\alpha(\chi)| = q^{1/2}$,

(b) If $\chi \neq id$, then $g_\alpha(\chi) g_\alpha(\chi^{-1}) = \chi(-1)q$.

24. (a) Prove Lemma 8.10.3,

(b) Take $F = \mathbb{Z}/p\mathbb{Z} = \mathbb{Z}(p)$ in (a). Let N_d denote the number of monic irreducible polynomials of degree d in $F[x]$. Use (a) to show that

$$p^s = \sum_{d|s} d N_d.$$

(c) Prove that

$$N_s = s^{-1} \sum_{d|s} \mu\left(\frac{s}{d}\right) p^d.$$

25. Prove that

$$\frac{1}{1 - qt} = \sum_{n=0}^{\infty} q^n t^n$$

$$= \sum_{n=0}^{\infty} (\text{\# of monic polynomials of degree } n) t^n$$

$$= \prod_{f \text{ irreducible}} (1 - t^{\deg f})^{-1}$$

$$= \prod_{d=1}^{\infty} (1 - t^d)^{-N_d},$$

where N_d has the same meaning as in the previous problem. Now deduce the result in Exercise 24(b).

26. Verify formulas (8.11.4).

27. Prove that $R(f, g) = 0$ if and only if either f and g have a common nonconstant factor or the leading coefficients of f and g are 0. (Note that f and g have a common factor if and only if there exist polynomials f_1 and g_1 such that $\deg f_1 \leq \deg f - 1$, $\deg g_1 \leq \deg g - 1$, and $f g_1 = f_1 g$.)

28. Prove that if $f = a_0(x - x_1) \cdots (x - x_n)$ and $g = b_0(y - y_1) \cdots (y - y_m)$, then

$$R(f, g) = a_0^m b_0^n \prod_i \prod_j (x_i - y_j) = a_0^m \prod_i g(x_i) = (-1)^{mn} b_0^n \prod_j f(y_j).$$

9

Spherical Harmonics

The aim of this chapter is to introduce the basic functions necessary for Fourier analysis in higher dimensions. One way to view $\cos n\theta$ is as a restriction to the unit circle of the homogeneous polynomial $[(x + iy)^n + (x - iy)^n]/2$, which is a solution of the two-dimensional Laplace equation. Spherical harmonics are restrictions to the sphere $x_1^2 + x_2^2 + \cdots + x_n^2 = 1$ of homogeneous polynomials that are solutions of the n-dimensional Laplace equation. These functions are related to the ultraspherical polynomials studied in Chapter 6.

An important result in this chapter is the addition theorem for ultraspherical polynomials, which generalizes the addition formula for the cosine. A useful tool in the proof of this result is a theorem of Funk and Hecke on an integral of a product of a continuous function and a spherical harmonic. Our presentation owes much to Müller [1966]. We also employ the Funk–Hecke formula to obtain the Fourier transform of a function on R^n, which is the product of a radial function and a spherical harmonic.

The final six sections of the chapter show how spaces of spherical harmonics of a given degree give irreducible representations of $SU(2)$, the group of all 2×2 matrices

$$\begin{pmatrix} a & -\bar{b} \\ b & \bar{a} \end{pmatrix}$$

of determinant one. Representation theory provides a very important approach to the study of special functions. Unfortunately, we do not have space for it here. The reader may consult Vilenkin [1968] or Miller [1972]. Here we content ourselves with showing the manner in which Jacobi polynomials appear in representations of $SU(2)$ and deriving an addition theorem.

9.1 Harmonic Polynomials

Solutions of the Laplace equation

$$\sum_{i=1}^{n} \frac{\partial^2 u}{\partial x_i^2} = 0 \tag{9.1.1}$$

are called harmonic functions. We are particularly interested in polynomial solutions. Since any polynomial in the variables $x_1, x_2, \ldots, x_n$ is a sum of a finite
number of homogeneous polynomials of different degrees, we concentrate on
these.

Definition 9.1.1   *A polynomial $H_m(x)$, which is homogeneous of degree m and
satisfies (9.1.1), is called a harmonic polynomial.*

   Interest in these polynomials arises from the fact that they are useful in finding
harmonic functions with given boundary values. In particular, one would like to
determine such functions in the unit ball $\sum_{i=1}^{n} x_i^2 \leq 1$. To motivate the study
of these polynomials and functions, consider the case $n = 2$. In this situation we
usually write $u = u(x, y)$, where x and y are real. It is clear that $z = x + iy$ can
be used to determine harmonic functions. Since

$$\frac{\partial^2(x+iy)^n}{\partial x^2} = n(n-1)(x+iy)^{n-2}$$

and

$$\frac{\partial^2(x+iy)^n}{\partial y^2} = -n(n-1)(x+iy)^{n-2},$$

it follows that $z^n = (x+iy)^n$ is a harmonic polynomial. Similarly, $\bar{z}^n = (x-iy)^n$
is also a harmonic polynomial. There is a technical problem in using $(x+iy)^n$ and
$(x-iy)^n$ to find harmonic functions with a given boundary value on $x^2 + y^2 = 1$,
because it is hard to see some of the important properties of these polynomials
when they are given in terms of $(x+iy)^n$ and $(x-iy)^n$. This problem can be
resolved by using polar coordinates.

   Set $x = r\cos\theta$ and $y = r\sin\theta$ so that $z = re^{i\theta}$ and $z^n = r^n e^{in\theta}$. Then

$$u_n(x, y) = r^n \cos n\theta = [(x+iy)^n + (x-iy)^n]/2$$

and

$$v_n(x, y) = r^n \sin n\theta = [(x+iy)^n - (x-iy)^n]/(2i)$$

are harmonic polynomials of degree n. The Poisson integral

$$v(r, \theta) = \frac{1}{2\pi} \int_0^{2\pi} f(\phi) \frac{1-r^2}{1-2r\cos(\theta-\phi)+r^2} d\phi$$

then solves the problem of obtaining a harmonic function $u(x, y) = v(r, \theta)$ with
$\lim_{r \to 1^-} v(r, \theta) = f(\theta), 0 \leq \theta \leq 2\pi$.

   Note that the Poisson kernel

$$\frac{1-r^2}{1-2r\cos\theta+r^2} = 1+2\sum_{n=1}^{\infty} \cos n\theta r^n = 1+2\sum_{n=1}^{\infty} u_n(x, y).$$

Thus the kernel in the above integral is an infinite sum of harmonic polynomials.

The polynomials $u_n(x, y)$ and $v_n(x, y)$ can also be expressed as

$$u_n(x, y) = \frac{z^n + \bar{z}^n}{2} = \frac{1}{2} \sum_{k \text{ even}} \binom{n}{k} x^{n-k} (iy)^k$$

$$= \frac{1}{2} \sum_{j=0}^{[n/2]} (-1)^j \binom{n}{2j} x^{n-2j} y^{2j}$$

and

$$v_n(x, y) = \frac{1}{2} \sum_{j=0}^{[\frac{n-1}{2}]} (-1)^j \binom{n}{2j+1} x^{n-2j-1} y^{2j+1}.$$

On the circle $x^2 + y^2 = 1$, it is simpler to view these polynomials as $\cos n\theta$ and $\sin n\theta$ respectively. Note that, for $n = 1, 2, 3, \ldots$, there are two independent harmonic polynomials of degree n. For $n = 0$, there is just one. This is reflected in the coefficients in the expansion of the Poisson kernel.

9.2 The Laplace Equation in Three Dimensions

We have seen that it is convenient to use polar coordinates to study the harmonic polynomials on the unit ball in two dimensions. We could have employed polar coordinates right in the beginning by writing the Laplace equation as

$$\frac{\partial^2 v}{\partial r^2} + \frac{1}{r^2} \frac{\partial^2 v}{\partial \theta^2} + \frac{1}{r} \frac{\partial v}{\partial r} = 0. \tag{9.2.1}$$

A set of solutions of the form $U(r, \theta) = R(r)T(\theta)$ can be obtained by separation of variables. We have

$$r^2 R'' T + r R' T + R T'' = 0$$

or

$$\frac{r^2 R'' + r R'}{R} = -\frac{T''}{T} = c,$$

where c is a constant. So

$$r^2 R'' + r R' - cR = 0,$$

which is Euler's equation with a solution of the form $R = r^\lambda$. The exponent λ satisfies

$$\lambda(\lambda - 1) + \lambda - c = 0$$

or

$$\lambda = \pm\sqrt{c}.$$

For polynomial solutions we must have $\lambda = 0, 1, 2, \ldots$, and $c = n^2$ for an integer n. Thus T satisfies the equation

$$T'' + n^2 T = 0.$$

Two independent solutions are $\cos n\theta$ and $\sin n\theta$.

The above calculations can be carried over to three dimensions. We write Laplace's equation in spherical coordinates as

$$\frac{\partial}{\partial r}\left(r^2 \frac{\partial v}{\partial r}\right) + \frac{1}{\sin\theta}\frac{\partial}{\partial\theta}\left(\sin\theta\frac{\partial v}{\partial\theta}\right) + \frac{1}{\sin^2\theta}\frac{\partial^2 v}{\partial\phi^2} = 0.$$

The ranges for θ and ϕ are $0 \le \theta \le \pi$ and $0 \le \phi \le 2\pi$. Separating variables $U(r, \phi, \theta) = R(r)F(\phi)T(\theta)$ gives

$$\frac{(r^2 R')'}{R} + \frac{1}{\sin\theta}\frac{(\sin\theta T')'}{T} + \frac{1}{\sin^2\theta}\frac{F''}{F} = 0.$$

Therefore,

$$\frac{(r^2 R')'}{R} = c$$

and

$$\frac{1}{\sin\theta}\frac{(\sin\theta T')'}{T} + \frac{1}{\sin^2\theta}\frac{F''}{F} = -c.$$

The first equation can be rewritten as

$$r^2 R'' + 2r R' - cR = 0.$$

Again, $R = r^\lambda$ gives $\lambda(\lambda - 1) + 2\lambda - c = 0$ or $\lambda(\lambda + 1) - c = 0$. Write $c = n(n + 1)$, so that $\lambda = n$ and $\lambda = -n - 1$ are the solutions. Again, for polynomial solutions take $\lambda = n$, a nonnegative integer. Then

$$\sin\theta\frac{(\sin\theta T')'}{T} + n(n + 1)\sin^2\theta + \frac{F''}{F} = 0.$$

So, $F'' - dF = 0$ for a constant, say, $d = -m^2$. Then T satisfies the equation

$$\frac{1}{\sin\theta}\frac{d}{d\theta}\left(\sin\theta\frac{dT}{d\theta}\right) + \left\{n(n + 1) - \frac{m^2}{\sin^2\theta}\right\}T = 0.$$

Set $x = \cos\theta$ and $T(\theta) = y(x)$ so that this equation becomes

$$(1 - x^2)\frac{d^2 y}{dx^2} - 2x\frac{dy}{dx} + \left\{n(n + 1) - \frac{m^2}{1 - x^2}\right\}y = 0.$$

This differential equation was mentioned in Chapter 3 as a second-order equation with regular singularities at -1, 1, and ∞ and no other singularities. In Riemann's notation, its set of solutions was given by

$$P \left\{ \begin{array}{ccc} -1 & \infty & 1 \\ m/2 & n+1 & m/2 \\ -m/2 & -n & -m/2 \end{array} \; x \right\}. \tag{3.9.5}$$

Notice that the indices associated with 1 and -1 are the same. From the discussion in Chapter 2, and in particular from Equation (2.3.6), it follows that the set of solutions can also be written as

$$(1-x^2)^{m/2} P \left\{ \begin{array}{ccc} -1 & \infty & 1 \\ 0 & n+m+1 & 0 \\ -m & m-n & -m \end{array} \; x \right\} \equiv (1-x^2)^{m/2} v.$$

By Theorem 2.3.1, v satisfies the equation

$$(1-x^2)y'' - 2(1+m)xy' + (n-m)(n+m+1)y = 0.$$

This equation has polynomial solutions when $n - m \geq 0$ is an integer. In fact, the polynomial solution is the ultraspherical polynomial $C_{n-m}^{m+\frac{1}{2}}(x)$. The reader may verify this by comparing the equation with the differential equation for Jacobi polynomials in Chapter 6. See Exercise 6.25.

We have shown that the Laplace equation is satisfied by $r^n(1-x^2)^{m/2}C_{n-m}^{m+\frac{1}{2}}(x)$ $\cos m\phi$ and $r^n(1-x^2)^{m/2}C_{n-m}^{m+\frac{1}{2}}(x)\sin m\phi$, where $x = \cos\theta$ and $0 \leq m \leq n$. These are $2n + 1$ independent solutions. We will show that these span harmonic polynomials of degree n in three dimensions. Observe that only one solution, namely $r^n C_n^{1/2}(x)$, does not depend on ϕ.

9.3 Dimension of the Space of Harmonic Polynomials of Degree *k*

Let V_k denote the vector space of homogeneous polynomials of degree k in n variables. Each polynomial $p \in V_k$ has the form

$$p(x) = \sum_{|\alpha|=k} c_\alpha x^\alpha,$$

where $\alpha = (\alpha_1, \alpha_2, \ldots, \alpha_n)$, $x = (x_1, x_2, \ldots, x_n)$, $c_\alpha = c_{\alpha_1, \alpha_2, \ldots, \alpha_n}$, $x^\alpha = x_1^{\alpha_1} \cdots x_n^{\alpha_n}$, and $|\alpha| = \sum_{i=1}^n \alpha_i$, with α_i nonnegative integers. The dimension of V_k is the number of n-tuples $(\alpha_1, \alpha_2, \ldots, \alpha_n)$ with $\sum_{i=1}^n \alpha_i = k$. Represent a given n-tuple $(\alpha_1, \alpha_2, \ldots, \alpha_n)$ by a sequence of k dots and $n - 1$ vertical lines as in the example given below:

$$\cdots \mid \cdots \mid \cdots \mid \cdots .$$

There are α_1 dots before the first line, then α_2 dots between the first and the second line, and so on. Clearly, there is a one-to-one correspondence between such sequences and the n-tuples. The total number of dots and lines is $n + k - 1$ and so the number of different arrangements of dots and lines is

$$d_{k,n} = \binom{n+k-1}{n-1} = \binom{n+k-1}{k}. \tag{9.3.1}$$

A second way of seeing this is that $d_{k,n}$ is the coefficient of x^k in the expansion of $(1-x)^{-n} = (1-x)^{-1} \cdots (1-x)^{-1}$, for this coefficient is the number of solutions of $\Sigma \alpha_i = k$ with α_i nonnegative. Yet another argument in a special case is given by Anno and Mori [1986]. It is evident that not all the homogeneous polynomials are harmonic. For instance, in the case of two variables, the dimension of the space of harmonic polynomials of degree $k > 0$ is 2 but $d_{k,2} = k + 1$.

To find the number of independent harmonic polynomials of degree k in n variables, write the homogeneous polynomial

$$p(x) = \sum_{j=0}^{k} x_n^j A_{k-j}(x_1, \ldots, x_{n-1}),$$

where A_{k-j} is homogeneous of degree $k - j$ in $x_1, \ldots, x_{n-1}$. Apply the operator

$$\Delta = \Delta_n = \sum_{i=1}^{n-1} \frac{\partial^2}{\partial x_i^2} + \frac{\partial^2}{\partial x_n^2}$$

to $p(x)$ to get

$$\Delta p(x) = \sum_{j=2}^{k} j(j-1)x_n^{j-2} A_{k-j}(x_1, \ldots, x_{n-1})$$

$$+ \sum_{j=0}^{k-2} x_n^j \Delta_{n-1} A_{k-j}(x_1, \ldots, x_{n-1}).$$

If $p(x)$ is a harmonic polynomial, then we must have

$$\Delta_{n-1} A_{k-j} = -(j+2)(j+1) A_{k-j-2}, \quad j = 0, 1, \ldots, k-2.$$

So once A_k and A_{k-1} are given, the remaining A_i are determined. Therefore, the number of linearly independent harmonic polynomials of degree k in n variables is

$$c_{k,n} = d_{k,n-1} + d_{k-1,n-1}$$

$$= \binom{k+n-2}{k} + \binom{k+n-3}{k-1} = (2k+n-2)\frac{(k+n-3)!}{k!(n-2)!}.$$

Observe that $c_{k,2} = 2$ for $k > 0$ and $c_{k,3} = 2k + 1$, which reconfirm the statements at the end of the first two sections.

9.4 Orthogonality of Harmonic Polynomials

The harmonic polynomials of different degrees in two variables are orthogonal when integration is over the unit circle; the same is true in three variables over the unit sphere. This is clear by looking at the polynomials given in the earlier sections that formed a basis for the space of harmonic polynomials in two and three dimensions. This result continues to hold in higher dimensions. Before proving this generalization, note that polar coordinates in n dimensions, $(r, \theta_1, \ldots, \theta_{n-2}, \phi)$ can be defined by the equations

$$
\begin{aligned}
x_1 &= r \cos \theta_1, \\
x_2 &= r \sin \theta_1 \, \cos \theta_2, \\
x_3 &= r \sin \theta_1 \, \sin \theta_2 \cos \theta_3, \\
&\;\;\vdots \qquad \vdots \\
x_{n-1} &= r \sin \theta_1 \cdots \sin \theta_{n-2} \cos \phi, \\
x_n &= r \sin \theta_1 \cdots \sin \theta_{n-2} \sin \phi,
\end{aligned} \tag{9.4.1}
$$

where $0 \le \theta_i \le \pi$ and $0 \le \phi \le 2\pi$.

Let $H_k(x)$ and $H_j(x)$ be homogeneous harmonic polynomials in n variables of degrees k and j respectively with $j \ne k$. By Green's theorem,

$$
\begin{aligned}
0 &= \int_{|x| \le 1} [H_j(x) \Delta H_k(x) - H_k(x) \Delta H_j(x)] dV \\
&= \int_{|\xi|=1} \left[H_j(\xi) \frac{\partial}{\partial r} H_k(r\xi) \Big|_{r=1} - H_k(\xi) \frac{\partial}{\partial r} H_j(r\xi) \Big|_{r=1} \right] d\omega(\xi), \quad (9.4.2)
\end{aligned}
$$

where $\xi = x/|x|$, $|x| = r$, and $d\omega(\xi)$ is the invariant measure on the surface of the sphere. We used the fact that the normal derivative on the sphere is in the radial direction. The homogeneity of $H_k(x)$ gives

$$
\frac{\partial}{\partial r} H_k(r\xi) \Big|_{r=1} = \frac{\partial}{\partial r} r^k H_k(\xi) \Big|_{r=1} = k H_k(\xi).
$$

Substitute this in (9.4.2) to arrive at

$$
(k - j) \int_{|\xi|=1} H_j(\xi) H_k(\xi) d\omega(\xi) = 0. \tag{9.4.3}
$$

The functions $H_k(\xi)$, which are restrictions of homogeneous harmonic polynomials to the surface of the sphere in R^n, are called spherical harmonics. Sometimes they are also called surface spherical harmonics and the $H_k(x)$ are called solid spherical harmonics. A notation sometimes used when expressing spherical harmonics in polar coordinates is

$$H_k(\xi) = Y_k(\theta, \phi),$$

where $\theta = (\theta_1, \theta_2, \ldots, \theta_{n-2})$.

Let F be the space of real-valued continuous functions on the sphere $|\xi|^2 = \xi_1^2 + \xi_2^2 + \cdots + \xi_n^2 = 1$. An inner product on this space can be defined by

$$\langle f, g \rangle = \int_{|\xi|=1} f(\xi)g(\xi)d\omega(\xi), \quad \text{for } f, g \in F. \tag{9.4.4}$$

The conjugate of g is taken in the integral, if complex-valued functions are used. The result contained in (9.4.3) is stated in the next theorem.

Theorem 9.4.1 *Spherical harmonics of different degrees are orthogonal with respect to the inner product (9.4.4).*

Spherical harmonics of the same degree may or may not be orthogonal. For example, $\cos k\theta$ and $\sin k\theta$ are two independent spherical harmonics in two dimensions and are orthogonal, but $\cos k\theta + \sin k\theta$ and $\cos k\theta$ are not orthogonal. Using Gram–Schmidt orthogonalization, it is possible to choose an orthonormal basis of $c_{k,n} = (2k + n - 2)(k + n - 3)!/(k!(n - 2)!)$ spherical harmonics of degree k in n variables. Denote the members of this basis by

$$S_{k,j}(\xi), \quad j = 1, 2, \ldots, c_{k,n}.$$

Theorem 9.4.1 and the definition of $S_{k,j}(\xi)$ imply

$$\langle S_{k,j}, S_{k',j'} \rangle = \delta_{kk'}\delta_{jj'}.$$

9.5 Action of an Orthogonal Matrix

Recall the definition of an orthogonal matrix.

Definition 9.5.1 *An $n \times n$ matrix O is called orthogonal if $OO^t = I$, where O^t is the transpose of O.*

It is clear that if $OO^t = I$ then $O^tO = I$, that is, O^t is also orthogonal. Write a vector $x \in R^N$ as

$$x = \begin{pmatrix} x_1 \\ x_2 \\ \vdots \\ x_n \end{pmatrix},$$

and write the inner product of two vectors x and y in R^n as

$$(x, y) = x'y = x_1 y_1 + x_2 y_2 + \cdots + x_n y_n.$$

If O is an orthogonal matrix, then

$$(Ox, Oy) = x'O'Oy = x'y = (x, y). \tag{9.5.1}$$

So, if we write $(x, x) = \|x\|^2$, then $\|Ox\| = \|x\|$. In particular, if $\|\xi\| = 1$, then $\|O\xi\| = 1$.

We now show that the Laplace equation remains invariant under the action of an orthogonal matrix. Write

$$\left(\frac{\partial}{\partial x}\right) = \left(\frac{\partial}{\partial x_1}, \ldots, \frac{\partial}{\partial x_n}\right)^t.$$

Then the Laplace operator is given by $(\frac{\partial}{\partial x})^t(\frac{\partial}{\partial x})$. For a change of variables $x' = Ox$, we have $(\frac{\partial}{\partial x'}) = O^t(\frac{\partial}{\partial x})$. So

$$\left(\frac{\partial}{\partial x'}\right)^t\left(\frac{\partial}{\partial x'}\right) = \left(\frac{\partial}{\partial x}\right)^t OO^t\left(\frac{\partial}{\partial x}\right) = \left(\frac{\partial}{\partial x}\right)^t\left(\frac{\partial}{\partial x}\right). \tag{9.5.2}$$

This proves our claim.

Consider the action of the orthogonal matrix O on a spherical harmonic $H_k(\xi)$ defined by the mapping $H_k(\xi) \to H_k(O\xi)$. This action transforms an orthonormal basis $S_{k,j}(\xi)$, $j = 1, 2, \ldots, c_{k,n}$ to another orthonormal basis $S_{k,j}(O\xi)$, $j = 1, 2, \ldots, c_{k,n}$ for the space of spherical harmonics of degree k in n variables, since

$$\int_{|\xi|=1} S_{k,j}(O\xi)S_{k,j'}(O\xi)d\omega(\xi)$$

$$= \int_{|\xi|=1} S_{k,j}(\xi)S_{k,j'}(\xi)d\omega(\xi) = \delta_{jj'}.$$

Moreover,

$$S_{k,j}(O\xi) = \sum_{\ell=1}^{c_{k,n}} A_{j\ell}^k S_{k,\ell}(\xi) \tag{9.5.3}$$

and the coefficients $(A_{j\ell}^k)$ form an orthogonal matrix. This follows from the relation

$$\sum_{\ell=1}^{c_{k,n}} A_{j\ell}^k A_{j'\ell}^k = \int_{|\xi|=1} S_{k,j}(O\xi)S_{k,j'}(O\xi)d\omega(\xi) = \delta_{jj'}.$$

9.6 The Addition Theorem

The spherical harmonic of degree k in two variables, $\cos k\theta$, has the addition formula

$$\cos k\theta \cos k\theta_0 + \sin k\theta \sin k\theta_0 = \cos k(\theta - \theta_0). \qquad (9.6.1)$$

We note two interesting properties of this spherical harmonic. The only orthogonal transformation that fixes the point $(\cos \theta_0, \sin \theta_0)$ on the circle is the reflection about the line $\theta = \theta_0$; the unique independent spherical harmonic of degree k, which is invariant under this transformation is $\cos k(\theta - \theta_0)$. Moreover, since $\sin k\theta = \cos(k\theta - \frac{\pi}{2})$, we can view $\cos k\theta$ as the basic spherical harmonic in two variables. This uniqueness continues to hold in higher dimensions but with a different basic function. The goal of this section is to obtain these generalizations.

There is no nonconstant harmonic polynomial that is invariant under all orthogonal transformations. But there is exactly one, modulo a constant factor, invariant under transformations that leave one point fixed. To prove this, start with the following preliminary lemma.

Lemma 9.6.1 *Up to a constant factor, there exists at most one harmonic polynomial of degree k that is invariant under all those orthogonal transformations that leave one point η of the unit sphere fixed.*

Proof. Since the inner product $(Ox, \eta) = (x, \eta)$ for all orthogonal transformations O that leave η fixed, it is sufficient to prove that there is, at most, one (up to a constant factor) harmonic polynomial $H_k(x)$ that depends only on r and (x, η).

From the homogeneity of $H_k(x)$, we have

$$H_k(x) = c_0(x, \eta)^k + c_1 r(x, \eta)^{k-1} + c_2 r^2(x, \eta)^{k-2} + \cdots.$$

It is an easy calculation that

$$\Delta[r^\ell(x, \eta)^m] = m(m - 1)(\eta, \eta)r^\ell(x, \eta)^{m-2} + \ell(\ell + 2m + n - 2)r^{\ell-2}(x, \eta)^m.$$

Then $\Delta H_k = 0$ implies that the coefficients c_k satisfy the relations

$$(k - m)(k - m - 1)c_m + (m + 2)(2k - m - 2 + n - 2)c_{m+2} = 0$$

for $m = 0, 1, 2, \ldots$ and $c_1 = 0$. This shows that c_0 determines $H_k(x)$, and the lemma is proved. ∎

We now show the existence of the harmonic polynomial of Lemma 9.6.1. Let $S_{k,j}(\xi)$, $j = 1, 2, \ldots, c_{k,n}$, be the orthonormal basis of the space of harmonic polynomials of degree k in n variables. The form of (9.6.1) and the remarks after that suggest the consideration of the function $F(\xi, \eta)$ defined below. First define the vector function

$$S(\xi) = \begin{pmatrix} S_{k,1}(\xi) \\ \vdots \\ S_{k,c_{k,n}}(\xi) \end{pmatrix};$$

then set $F(\xi, \eta) = S(\xi)' S(\eta)$, where ξ and η are points on the sphere. From (9.5.3) and the orthogonality of the matrix $C = (A^k_{j\ell})$, we have

$$F(O\xi, O\eta) = S(O\xi)' S(O\eta)$$
$$= S(\xi)' C' C S(\eta) = F(\xi, \eta),$$

when O is an orthogonal transformation. It is clear that F, as a function of ξ, is the restriction to the unit sphere of a harmonic polynomial of degree k. Moreover, if O is any orthogonal transformation that fixes η, then $F(O\xi, \eta) = F(\xi, \eta)$. This proves the existence of the function we were looking for. Its uniqueness was proved in Lemma 9.6.1.

It is clear that the function $F(\xi, \eta)$ depends only on the inner product (ξ, η). Denote it $b_k P_k((\xi, \eta))$ and normalize by taking $P_k((\eta, \eta)) = P_k(1) = 1$.

Definition 9.6.2 *The function $F(\xi, \eta) = b_k P_k((\xi, \eta))$ is called the zonal harmonic of degree k with pole η.*

To determine b_k, take $\xi = \eta$ to get

$$b_k = \sum_{j=1}^{c_{k,n}} (S_{k,j}(\eta))^2;$$

then integration over the sphere with respect to $d\omega(\eta)$ gives $b_k \omega_n = c_{k,n}$ or $b_k = c_{k,n}/\omega_n$, where $\omega_n = 2(\pi)^{n/2} \Gamma(n/2)$ is the surface area of the unit sphere $x_1^2 + x_2^2 + \cdots + x_n^2 = 1$. Now multiply the equation

$$b_k P_k((\xi, \eta)) = \sum_{j=1}^{c_{k,n}} S_{k,j}(\xi) S_{k,j}(\eta)$$

by itself and integrate with respect to η. The result is

$$\frac{c_{k,n}^2}{\omega_n^2} \int_{|\eta|=1} [P_k((\xi, \eta))]^2 d\omega(\eta)$$
$$= \sum_j \sum_\ell S_{k,j}(\xi) S_{k,\ell}(\xi) \int_{|\eta|=1} S_{k,j}(\eta) S_{k,\ell}(\eta) d\omega(\eta)$$
$$= \sum S_{k,j}^2(\xi) = \frac{c_{k,n}}{\omega_n}.$$

Therefore,

$$\int_{|\eta|=1} P_k((\xi, \eta)) P_j((\xi, \eta)) d\omega(\eta) = \frac{\omega_n}{c_{k,n}} \delta_{kj}. \qquad (9.6.2)$$

This orthogonality relation for P_k will help us identify the function. Rotate ξ to $\epsilon_1 = (1, 0, \ldots, 0)$ and take $\eta = t\epsilon_1 + \sqrt{1 - t^2}\eta'$, where $|\eta'| = 1$ and the first component of η' is zero. A change to polar coordinates gives

$$t = (\eta, \epsilon_1) = \cos\theta_1. \qquad (9.6.3)$$

The Jacobian is given by

$$dx_1 dx_2 \cdots dx_n = r^{n-1} \sin^{n-2}\theta_1 \cdots \sin^2\theta_{n-3} \sin\theta_{n-2} dr d\theta_1 \cdots d\theta_{n-2} d\phi,$$

which implies that, on the sphere,

$$d\omega_n = \sin^{n-2}\theta_1 d\theta_1 d\omega_{n-1} = (1 - t^2)^{\frac{n-3}{2}} dt d\omega_{n-1}. \qquad (9.6.4)$$

The orthogonality relation (9.6.2) can be written as

$$\int_{-1}^{1} P_k(t) P_j(t) (1 - t^2)^{\frac{n-3}{2}} dt = \frac{\omega_n}{\omega_{n-1} c_{k,n}} \delta_{kj}. \qquad (9.6.5)$$

Thus $P_k(t) = A C_k^{(n-2)/2}(t)$ is an ultraspherical polynomial. Since $P_k(1) = 1$, the constant is given by $A = 1/C_k^{(n-2)/2}(1)$.

We have, therefore, proved the following addition theorem.

Theorem 9.6.3   *Let* $S_{k,j}(\xi)$, $j = 1, 2, \ldots, c_{k,n}$, *be an orthonormal set of spherical harmonics of degree* k. *Then*

$$\sum_{j=1}^{c_{k,n}} S_{k,j}(\xi) S_{k,j}(\eta) = \frac{c_{k,n}}{\omega_n} \frac{C_k^{(n-2)/2}((\xi, \eta))}{C_k^{(n-2)/2}(1)}. \qquad (9.6.6)$$

Remark 9.6.1   This result contains (9.6.1) as a limiting case. Recall that

$$\lim_{n \to 2} \frac{C_k^{(n-2)/2}(t)}{C_k^{(n-2)/2}(1)} = \cos k\theta, \quad \text{where } t = \cos\theta.$$

Remark 9.6.2   Since the integral (9.6.5) can be directly evaluated from the properties of ultraspherical polynomials, the value of $c_{k,n}$, the dimension of the space of spherical harmonics of degree k, can be computed from (9.6.5).

To see what the addition formula (9.6.6) looks like for $n = 3$, consider the set of independent spherical harmonics in three variables listed at the end of Section 9.2. Rewrite them in terms of the associated Legendre function defined by

$$P_k^m(x) = (-1)^m \frac{(2m)!}{2^m m!} (1 - x^2)^{m/2} C_{k-m}^{m+\frac{1}{2}}(x). \qquad (9.6.7)$$

It is easily verified that

$$\int_{-1}^{1} \left[P_k^m(x) \right]^2 dx = \frac{2}{2k+1} \cdot \frac{(k+m)!}{(k-m)!},$$

so that an orthonormal set of spherical harmonics of degree k for $n = 3$ is given by

$$\sqrt{\frac{2k+1}{4\pi}} P_k(x), \qquad A_m \cos m\phi \, P_n^m(x), \qquad A_m \sin m\phi \, P_n^m(x), \quad m = 1, \ldots, 2k,$$

where

$$A_m = \sqrt{\frac{(k-m)!(2k+1)}{(k+m)!2\pi}}.$$

Now take

$$\xi = (\cos\alpha, \, \sin\alpha\cos\phi_1, \, \sin\alpha\sin\phi_1)$$

and

$$\eta = (\cos\beta, \, \sin\beta\cos\phi_2, \, \sin\beta\sin\phi_2)$$

so that $(\xi, \eta) = \cos\alpha\cos\beta + \sin\alpha\sin\beta\cos\phi$ when $\phi = \phi_1 - \phi_2$. Since

$$\frac{c_{k,3}}{\omega_3} = \frac{2k+1}{4\pi} \quad \text{and} \quad P_k(1) = 1 = C_k^{1/2}(1),$$

(9.6.6) gives

$$P_k(\cos\alpha\cos\beta + \sin\alpha\sin\beta\cos\phi)$$

$$= P_k(\cos\alpha) P_k(\cos\beta) + 2 \sum_{m=1}^{k} \frac{(k-m)!}{(k+m)!} P_k^m(\cos\alpha) P_k^m(\cos\beta) \cos m\phi,$$

$$(9.6.8)$$

where $\phi = \phi_1 - \phi_2$.

The addition formula (9.6.6) shows that the ultraspherical function $C_k^{(n-2)/2}$ $((\xi, \eta))$ is the basic spherical harmonic in n dimensions, analogous to $\cos k\theta$ in two dimensions. Observe that it is possible to find $c_{k,n}$ points $\eta_1, \eta_2, \ldots, \eta_{c_{k,n}}$ on the sphere $x_1^2 + x_2^2 + \cdots + x_n^2 = 1$ such that the matrix

$$\begin{pmatrix} S_{k,1}(\eta_1) & S_{k,2}(\eta_1) & \cdots & S_{k,c_{k,n}}(\eta_1) \\ \vdots & & & \\ S_{k,1}(\eta_{c_{k,n}}) & S_{k,2}(\eta_{c_{k,n}}) & \cdots & S_{k,c_{k,n}}(\eta_{c_{k,n}}) \end{pmatrix} \qquad (9.6.9)$$

is invertible. Now consider the system of $c_{k,n}$ linear equations by choosing $\eta = \eta_1, \eta_2, \ldots, \eta_{c_{k,n}}$ in (9.6.6). This system of equations can be solved uniquely for $S_{k,j}(\xi)$ in terms of $C_k^{(n-2)/2}((\xi, \eta_\ell))$. We have proved the following:

Theorem 9.6.4 *It is possible to choose points $\eta_1, \eta_2, \ldots, \eta_{c_{k,n}}$ such that every spherical harmonic can be expressed in the form*

$$S_k(\xi) = \Sigma a_\ell C_k^{(n-2)/2}((\xi, \eta_\ell)).$$

9.7 The Funk–Hecke Formula

In this section we prove the Funk–Hecke formula, which will be useful in finding a basis for the space of spherical harmonics of degree k. This leads to the addition formula for ultraspherical polynomials.

In the following material we write the inner product of two vectors α and β in R^n as $\alpha \cdot \beta$ instead of (α, β). Observe that any continuous function f on the interval $[-1, 1]$ extends to a continuous function of two variables $g(\alpha, \beta)$ on the sphere defined by $g(\alpha, \beta) = f(\alpha \cdot \beta)$. If either α or β is kept fixed, then we have a function on the sphere. Now consider the integral

$$F(\alpha, \beta) = \int_{|\eta|=1} f(\alpha \cdot \eta) C_k^{(n-2)/2}(\beta \cdot \eta) d\omega(\eta). \tag{9.7.1}$$

For an orthogonal transformation O,

$$F(O\alpha, O\beta) = \int_{|\eta|=1} f(O\alpha \cdot \eta) C_k^{(n-2)/2}(O\beta \cdot \eta) d\omega(\eta)$$

$$= \int_{|\eta|=1} f(\alpha \cdot O^t \eta) C_k^{(n-2)/2}(\beta \cdot O^t \eta) d\omega(\eta).$$

By the invariance of measure under orthogonal transformations, it follows that

$$F(O\alpha, O\beta) = F(\alpha, \beta). \tag{9.7.2}$$

As a function of β, F is a spherical harmonic and depends only on $\alpha \cdot \beta$. The argument of the previous section implies that $F(\alpha, \beta)$ is a constant multiple of $C_k^{(n-2)/2}(\alpha \cdot \beta)$. Therefore,

$$\int_{|\eta|=1} f(\alpha \cdot \eta) C_k^{(n-2)/2}(\beta \cdot \eta) d\omega(\eta) = \lambda_k C_k^{(n-2)/2}(\alpha \cdot \beta). \tag{9.7.3}$$

To find λ_k, take $\alpha = \beta = \epsilon_1 = (1, 0, \ldots, 0)$ and set $\eta = t\epsilon_1 + \sqrt{1 - t^2}\eta'$ where the first component of η' is zero. A calculation similar to the one used to derive (9.6.5) gives

$$\lambda_k C_k^{(n-2)/2}(1) = \int_{|\tilde\eta'|=1} \int_{-1}^{1} f(t) C_k^{(n-2)/2}(t)(1 - t^2)^{(n-3)/2} dt \, d\omega(\tilde\eta')$$

$$= \omega_{n-1} \int_{-1}^{1} f(t) C_k^{(n-2)/2}(t)(1 - t^2)^{(n-3)/2} dt, \tag{9.7.4}$$

where $\tilde\eta'$ is obtained from η' by removing the first component.

The Funk–Hecke formula is contained in the next theorem. It was first published by Funk [1916] and a little later by Hecke [1918].

Theorem 9.7.1 *Let $f(t)$ be continuous on $[-1, 1]$ and $S_k(\xi)$ be any surface harmonic of degree k. Then for a unit vector α,*

$$\int_{|\eta|=1} f(\alpha \cdot \eta) S_k(\eta) d\omega(\eta) = \lambda_k S_k(\alpha), \qquad (9.7.5)$$

where λ_k is given by (9.7.4).

Proof 1. By (9.7.3), the result is true when $S_k(\eta)$ is replaced by $C_k^{(n-2)/2}(\beta \cdot \eta)$. Theorem 9.6.4 says that any $S_k(\eta)$ is a linear combination of $C_k^{(n-2)/2}(\beta_\ell \cdot \eta)$. Thus (9.7.5) follows. ∎

Proof 2. An integrated form of the addition formula (9.6.6) is

$$S_{k,j}(\xi) = \frac{c_{k,n}}{\omega_n C_k^{(n-2)/2}(1)} \int_{|\eta|=1} C_k^{(n-2)/2}(\xi \cdot \eta) S_{k,j}(\eta) d\omega(\eta). \qquad (9.7.6)$$

Now $S_{k,j}(\xi)$ can be replaced by any spherical harmonic $S_k(\xi)$, because the $S_{k,j}(\xi)$ form a basis for the space of such functions. Multiply (9.7.3) across by $S_k(\beta) d\omega(\beta)$ and integrate with respect to β. The theorem follows after an application of (9.7.6). ∎

Remark 9.7.1 Formula (9.7.6) suggests another way of arriving at the zonal harmonic function. The map $\phi : S_{k,j} \to S_{k,j}(\xi)$ is a linear functional on the finite-dimensional space of spherical harmonics of degree k. So ϕ is given by an inner product, that is, there exists a function g_ξ such that

$$\phi(S_{k,j}) = (S_{k,j}, g_\xi).$$

Written out in full, we have

$$S_{k,j}(\xi) = \int_{|\eta|=1} S_{k,j}(\eta) g_\xi(\eta) d\eta. \qquad (9.7.7)$$

One can then prove that $g_\xi(\eta) = \Sigma S_{k,j}(\xi) S_{k,j}(\eta)$.

9.8 The Addition Theorem for Ultraspherical Polynomials

To motivate the technique used to find a basis for the space of spherical harmonics, we start with an integral formula for ultraspherical polynomials obtained in Theorem 6.7.3. The formula can be written in the form

$$\frac{C_k^{(n-2)/2}(t)}{C_k^{(n-2)/2}(1)} = \frac{\omega_{n-2}}{\omega_{n-1}} \int_{-1}^{1} [t + i\sqrt{1 - t^2} s]^k (1 - s^2)^{(n-4)/2} ds, \quad n \geq 3. \qquad (9.8.1)$$

To obtain a proof different from the one given in Chapter 6, consider the integral

$$g(x) = \int_{|\tilde{\eta}_{n-1}|=1} [x \cdot \epsilon_1 + ix \cdot \eta_{n-1}]^k \, d\omega(\tilde{\eta}_{n-1}),$$

where $x = (x_1, \ldots, x_n)$, $\epsilon_1 = (1, 0, \ldots, 0)$, $\eta = t\epsilon_1 + \sqrt{1-t^2}\eta_{n-1}$, $t = \eta \cdot \epsilon_1$, and $\tilde{\eta}_{n-1}$ is obtained from η_{n-1} by removing the first entry, which is zero. In what follows we write η_{n-1} instead of $\tilde{\eta}_{n-1}$. We first show that g is a harmonic function. Observe that

$$\frac{\partial^2 g}{\partial x_1^2} = k(k-1) \int_{|\eta_{n-1}|=1} [x \cdot \epsilon_1 + ix \cdot \eta_{n-1}]^{k-2} d\omega(\eta_{n-1}),$$

and for $j = 2, \ldots, n$,

$$\frac{\partial^2 g}{\partial x_j^2} = -k(k-1) \int_{|\eta_{n-1}|=1} [x \cdot \epsilon_1 + ix \cdot \eta_{n-1}]^{k-2} (\epsilon_j \cdot \eta_{n-1})^2 d\omega(\eta_{n-1}),$$

where $\epsilon_j = (0, \ldots, 1, \ldots, 0)$ with 1 in the jth position. So

$$\Delta g = k(k-1) \int_{|\eta_{n-1}|=1} [x \cdot \epsilon_1 + ix \cdot \eta_{n-1}]^{k-2} \left[1 - \sum_{j=2}^n (\epsilon_j \cdot \eta_{n-1})^2 \right] d\omega(\eta_{n-1}) = 0$$

since

$$\sum_{j=2}^n (\epsilon_j \cdot \eta_{n-1})^2 = |\eta_{n-1}|^2 = 1.$$

Thus $g(x)$ is a solid spherical harmonic of degree k. It is clear that g is also invariant under all orthogonal transformations that fix ϵ_1. If we take $x = \epsilon_1$, then the value of $g(x)$ is ω_{n-1}, the volume of the unit sphere in $n-1$ dimensions.

Set $\xi = x/|x|$ and then $\xi = t\epsilon_1 + \sqrt{1-t^2}\xi_{n-1}$. From the previous remarks, $g(\xi)$ is a multiple of $C_k^{(n-2)/2}(t)$. The normalization at $t = 1$ gives

$$\frac{C_k^{(n-2)/2}(t)}{C_k^{(n-2)/2}(1)} = \frac{1}{\omega_{n-1}} \int_{|\eta_{n-1}|=1} [t + i\sqrt{1-t^2}\xi \cdot \eta_{n-1}]^k \, d\omega(\eta_{n-1}). \qquad (9.8.2)$$

Now apply the procedure used in the derivation of (9.6.5) to the above integral to get (9.8.1).

The next step in the derivation of the addition theorem for ultraspherical polynomials is to take any spherical harmonic of degree k in n variables and express it in terms of a spherical harmonic of a different degree in $n-1$ variables. For this purpose consider the integral

$$g(\xi) = \int_{|\tilde{\eta}_{n-1}|=1} [\xi \cdot \epsilon_1 + i\xi \cdot \eta_{n-1}]^k S_j(\tilde{\eta}_{n-1}) d\omega(\tilde{\eta}_{n-1}). \qquad (9.8.3)$$

It is immediately obvious from the argument used above that g is a spherical harmonic of degree k in n variables. Moreover, by the Funk–Hecke formula, the integral can be written as

$$g(\xi) = \int_{|\tilde{\eta}|=1} [t + i\sqrt{1-t^2}\xi_{n-1} \cdot \tilde{\eta}_{n-1}]^k S_j(\tilde{\eta}_{n-1}) d\omega(\tilde{\eta}_{n-1})$$

$$= S_j(\xi_{n-1})\omega_{n-2} \int_{-1}^{1} (t + i\sqrt{1-t^2}s)^k C_j^{(n-3)/2}(s)(1-s^2)^{(n-4)/2} ds.$$

Rodrigues's formula for ultraspherical polynomials,

$$(1-x^2)^{\lambda-\frac{1}{2}} C_k^{\lambda}(x) = \frac{(-1)^k \Gamma(k+\lambda)\Gamma(k+2\lambda)}{k!\Gamma(\lambda)\Gamma(2k+2\lambda)} \frac{d^k}{dx^k}(1-x^2)^{k+\lambda-1/2},$$

implies that the term $(1-s^2)^{(n-4)/2} C_j^{(n-3)/2}(s)$ in the integrand is a constant times

$$\frac{d^j}{ds^j}(1-s^2)^{j+(n-4)/2}.$$

Integration by parts j times then gives

$$g(\xi) = K S_j(\xi_{n-1})(1-t^2)^{j/2} \int_{-1}^{1} (t + i\sqrt{1-t^2}s)^{k-j}(1-s^2)^{j+(n-4)/2} ds,$$

where K is a constant. By (9.8.1) the last integral is proportional to $C_{k-j}^{j+(n-2)/2}(t)$. We have, therefore, proved that $S_j(\xi_{n-1})(1-t^2)^{j/2} C_{k-j}^{j+(n-2)/2}(t)$ is a spherical harmonic of degree k in n variables. Recall that t and ξ_{n-1} are related by $t\epsilon_1 + \sqrt{1-t^2}\xi_{n-1} = \xi$, a point on the n-dimensional unit sphere. With the notation of (9.4.5), $S_{j,\ell}(\xi_{n-1})$, $\ell = 1, 2, \ldots, c_{j,n-1}$, form a basis for the surface harmonics of degree j in $n-1$ variables. Thus the set of functions

$$S_{j,\ell}(\xi_{n-1})(1-t^2)^{j/2} C_{k-j}^{j+(n-2)/2}(t), \quad j = 0, \ldots, k, \ \ell = 1, \ldots, c_{j,n-1}, \quad (9.8.4)$$

forms an orthogonal basis for the vector space of spherical harmonics of degree k in n variables. The orthogonality is easily verified, and since $c_{k,n} = \sum_{j=0}^{k} c_{j,n-1}$, there are the correct number of vectors to form a basis.

Observe that, since

$$A_j = \int_{-1}^{1} \left[C_{k-j}^{j+(n-2)/2}(t)\right]^2 (1-t^2)^{j+(n-3)/2} dt$$

$$= \frac{\pi \Gamma(k+j+n-2)}{2^{2j+n-3}[\Gamma(j+(n-2)/2)]^2(k-j)!(k+(n-2)/2)},$$

the functions in (9.8.4) form an orthonormal basis when multiplied by $1/\sqrt{A_j} = B_j$. If this basis is used in (9.6.6) with

$$\xi = t\epsilon_1 + \sqrt{1-t^2}\xi_{n-1}$$

and

$$\eta = s\epsilon_1 + \sqrt{1-s^2}\eta_{n-1},$$

then we have

$$\frac{2k+n-2}{(n-2)\omega_n}C_k^{(n-2)}(ts + \sqrt{1-t^2}\sqrt{1-s^2}\xi_{n-1}\cdot\eta_{n-1})$$

$$= \sum_{j=0}^{k} B_j^2 (1-t^2)^{j/2} C_{k-j}^{j+(n-2)/2}(t)(1-s^2)^{j/2}C_{k-j}^{j+(n-2)/2}(s)$$

$$\cdot \sum_{\ell} S_{j,\ell}(\xi_{n-1})S_{j,\ell}(\eta_{n-1}).$$

By (9.6.6) the inner sum is

$$\frac{2j+n-3}{\omega_{n-1}(n-3)}C_j^{(n-3)/2}(\xi_{n-1}\cdot\eta_{n-1}).$$

This gives the next theorem due to Gegenbauer [1875].

Theorem 9.8.1   *For an integer $n > 3$,*

$$C_k^{(n-2)/2}(st + \sqrt{1-s^2}\sqrt{1-t^2}\xi_{n-1}\cdot\eta_{n-1})$$

$$= \sum_{j=0}^{k} a_j(1-s^2)^{j/2}C_{k-j}^{j+(n-2)/2}(s)(1-t^2)^{j/2}$$

$$\cdot C_{k-j}^{j+(n-2)/2}(t)C_j^{(n-3)/2}(\xi_{n-1}\cdot\eta_{n-1}), \qquad (9.8.5)$$

where

$$a_j = \frac{\Gamma(n-3)2^{2j}(k-j)![\Gamma(j+(n-2)/2)]^2(2j+n-3)}{[\Gamma((n-2)/2)]^2\Gamma(j+k+n-2)}.$$

Remark 9.8.1   The result is true for $n = 2$ and $n = 3$ but a limit has to be taken. The case $n = 2$ is the addition formula for the cosine function and $n = 3$ gives the addition theorem for Legendre polynomials, (9.6.8).

Formula (9.8.5) is often written as

$$C_k^{(n-2)/2}(\cos\alpha\cos\beta + \sin\alpha\sin\beta\cos\phi)$$

$$= \sum_{j=0}^{k} a_j(\sin\alpha)^j C_{k-j}^{j+(n-2)/2}(\cos\alpha)(\sin\beta)^j C_{k-j}^{j+(n-2)/2}(\cos\beta)C_j^{(n-3)/2}(\cos\phi).$$

$$(9.8.5')$$

The addition theorem can be extended to any spherical polynomial $C_k^\lambda(x)$, $\lambda > 0$ by analytic continuation, since both sides of (9.8.5') are rational functions of

$\lambda = (n - 2)/2$. This identity holds for complex λ as long as no poles occur. It is also worth noting that (9.8.5′) can be obtained from (9.6.8) by differentiating with respect to ϕ.

Remark 9.8.2 The addition formulas for trigonometric functions have generalizations to elliptic functions that are different from those given by (9.8.5). Elliptic functions satisfy an addition formula of the type

$$f(u + v) = A(f(u), f(v)), \tag{9.8.6}$$

where $A(x, y)$ is an algebraic function. Weierstrass proved that the only solutions of (9.8.6) are algebraic functions, algebraic functions of e^{ciu} for some constant c, or algebraic functions of elliptic functions. Apparently, Weierstrass never published this result though he mentioned it in his lectures. See Copson [1935, p. 363].

9.9 The Poisson Kernel and Dirichlet Problem

The solution of the Dirichlet problem for the unit disk is given by the integral

$$u(r, \theta) = \frac{1}{2\pi} \int_0^{2\pi} f(\phi) \frac{1 - r^2}{1 - 2r \cos(\theta - \phi) + r^2} d\phi. \tag{9.9.1}$$

The Poisson kernel,

$$\frac{1 - r^2}{1 - 2r \cos(\theta - \phi) + r^2} = 1 + 2 \sum_{n=1}^{\infty} r^n \cos n(\theta - \phi),$$

is the sum of all the two-dimensional zonal harmonics with poles at ϕ. In n dimensions, the zonal harmonic of degree k with pole at η is, according to (9.6.6), given by

$$\frac{2k + n - 2}{(n - 2)\omega_n} C_k^{(n-2)/2}(\xi \cdot \eta).$$

In Chapter 6, we saw that

$$\sum_{k=0}^{\infty} \frac{2k + n - 2}{(n - 2)\omega_n} C_k^{(n-2)/2}(\xi \cdot \eta) r^k = \frac{1}{\omega_n} \frac{1 - r^2}{(1 - 2r \cos(\xi \cdot \eta) + r^2)^{n/2}}. \tag{9.9.2}$$

The generalization of (9.9.1) is contained in the next theorem.

Theorem 9.9.1 *Suppose f is a continuous function on the n-dimensional sphere. Let α be a point on the sphere and $0 < r < 1$. Then*

$$u(r\alpha) = \frac{1}{\omega_n} \int_{|\eta|=1} f(\eta) \frac{1 - r^2}{(1 - 2r \cos(\alpha \cdot \eta) + r^2)^{n/2}} d\omega(\eta)$$

is harmonic inside the sphere and $u(\alpha) = f(\alpha)$ on the sphere.

This theorem is proved using the same method as we used in the two-dimensional case.

9.10 Fourier Transforms

In Chapter 4, we considered the Fourier transform of functions of two variables. If the point (x, y) is identified with the complex number $x + iy = z = re^{i\theta}$, then we can write the Fourier series of an integrable function $f(re^{i\theta})$ as

$$f(re^{i\theta}) \sim \sum_{-\infty}^{\infty} f_k(r)e^{ik\theta}. \qquad (9.10.1)$$

The Fourier transform of (9.10.1) was seen to be expressible in terms of Bessel functions. This continues to be true in higher dimensions as well. First, we need a definition.

Definition 9.10.1 *A function $f : R^n \to R$ is called radial if there is a function $f_0(u)$ on $0 < u < \infty$ such that $f(x) = f_0(|x|)$.*

We generalize the functions $f_k(r)e^{ik\theta}$ in (9.10.1) to functions expressible as the product of a radial function and a harmonic polynomial in higher dimensions. The main result of this section concerns the Fourier transform of such functions. The presentation is based on Bochner [1955], though the proofs of the basic results are different.

Lemma 9.10.2 *For any spherical harmonic $S_k(\xi)$ of degree k in n variables,*

$$\int_{|\xi|=1} e^{-2\pi i t(\eta \cdot \xi)} S_k(\xi)d\omega(\xi) = 2\pi i^k S_k(\eta)\frac{J_{k+(n-2)/2}(2\pi t)}{t^{(n-2)/2}}. \qquad (9.10.2)$$

Proof. By the Funk–Hecke formula, integral (9.10.2) equals

$$\omega_{n-1} S_k(\eta) \int_{-1}^{1} e^{-2\pi i t s} C_k^{(n-2)/2}(s)(1 - t^2)^{\frac{n-3}{2}} dt.$$

Now apply Gegenbauer's formula:

$$J_{v+k}(x) = \frac{(-i)^k \Gamma(2v)k!(x/2)^v}{\Gamma(v + 1/2)\Gamma(1/2)\Gamma(2v + k)} \int_{-1}^{1} e^{ixs}(1 - s^2)^{v-\frac{1}{2}} C_k^v(s)ds. \qquad (4.7.7)$$

The result follows. ∎

Let $f \in L_1(R^n)$ and let Tf be its Fourier transform:

$$Tf(y) = \int_{R^n} e^{-2\pi i(y \cdot x)} f(x)dx. \qquad (9.10.3)$$

Theorem 9.10.3 *Suppose $f \in L_1(R^n)$ is of the form $f(x) = f_0(|x|) S_k(\xi)$. Then*

$$Tf(y) = F_0(|y|) S_k(\eta),$$

where

$$F_0(t) = 2\pi i^k t^{1-n/2} \int_0^\infty f_0(s) J_{k-1+n/2}(2\pi st) s^{n/2} ds \qquad (9.10.4)$$

and $y = |y| \eta$.

Proof. It is easy to see that with $x = (x_1, \ldots, x_n)$ and $x_j = s\xi_j$, we have

$$Tf(y) = \int_0^\infty F_0(s) s^{n-1} \left(\int_{|\xi|=1} e^{-2\pi i s |y|(\eta \cdot \xi)} S_k(\xi) d\omega(\xi) \right) ds.$$

The result now follows from Lemma 9.10.2. ∎

An interesting consequence is the next result, which is obtained by combining Theorem 9.10.3 with the fact that for $\operatorname{Re}(\mu + \nu) > 0$,

$$\lim_{\epsilon \to 0^+} \int_0^\infty e^{-\epsilon t} J_\nu(t) t^{\mu-1} dt = \frac{2^{\mu-1} \Gamma((\mu + \nu)/2)}{\Gamma(1 + (\nu - \mu)/2)}. \qquad (9.10.5)$$

To obtain (9.10.5), apply Pfaff's transformation (Theorem 2.2.5) to the $_2F_1$ in (4.11.4) and then take the limit.

Corollary 9.10.4 *For a spherical harmonic $S_k(\xi)$,*

$$\lim_{\epsilon \to 0} \int_{R^n} e^{-\epsilon|x|} (\sqrt{2\pi}|x|)^\alpha S_k(\xi) e^{-2\pi i (y \cdot x)} dx$$

$$= \frac{i^k S_k(\eta)}{(\sqrt{2\pi}|y|)^{n+\alpha}} \frac{2^{\alpha+n/2} \Gamma((n + k + \alpha)/2)}{\Gamma((k - \alpha)/2)} \qquad (9.10.6)$$

when $|y| \neq 0$ and $y = |y| \eta$.

Proof. Take $F_0(t) = (\sqrt{2\pi} t)^\alpha e^{-\epsilon t}$ in Theorem 9.10.3 and then use (9.10.5). ∎

A particular case of (9.10.6) is worthy of note. Take $\alpha = -n/2$. We have

$$\lim_{\epsilon \to 0} \int_{R^n} \frac{S_k(\xi)}{|x|^{n/2}} e^{-\epsilon|x|} e^{-2\pi i (x \cdot y)} dx = i^k \frac{S_k(\eta)}{|y|^{n/2}}. \qquad (9.10.7)$$

This implies that $|x|^{-n/2} S_k(\xi)$ is an eigenfunction of the Fourier transform with eigenvalue i^n. Observe that this transform is actually the Abel mean of the Fourier transform.

One may restate Theorem 9.10.3 for the case where $f \in L_1(R^n)$ is of the form $f(x) = f_0(|x|) S_k(x)$ and $S_k(x)$ is a homogeneous harmonic polynomial of degree k. It is sufficient to remark that $S_k(x) = |x|^n S_k(\xi)$.

Theorem 9.10.5   *Supposing f is integrable and of the form $f_0(|x|)S_k(x)$; then*

$$Tf(y) = F_0(|y|)S_k(y),$$

where

$$F_0(t) = 2\pi i^k t^{-k-\frac{1}{2}n+1} \int_0^\infty f_0(s)J_{k+\frac{1}{2}n-1}(2\pi st)s^{\frac{1}{2}n+k}ds.$$

It is well known that the Fourier transform of $e^{-\pi|x|^2}$ is $e^{-\pi|y|^2}$. A similar result holds when these exponential functions are multiplied by spherical harmonics.

Theorem 9.10.6   *For any homogeneous harmonic polynomial $S_k(x)$ of degree k,*

$$\int_{R^n} e^{-2\pi i(y\cdot x)}e^{-\pi|x|^2}S_k(x)dx = i^k e^{-\pi|y|^2}S_k(y).$$

Thus $e^{-\pi|x|^2}S_k(x)$ is an eigenfunction of the Fourier transform with eigenvalue i^k.

Proof.   This can be derived from Theorem 9.10.5 and the following formula of Sonine:

$$\int_0^\infty J_\nu(st)e^{-t^2}t^{\nu+1}dt = \frac{s^\nu}{2^{\nu+1}}e^{-s^2/4}. \tag{9.10.8}$$

∎

9.11 Finite-Dimensional Representations of Compact Groups

Representation theory provides an important and powerful approach to special functions. Unfortunately, we can do no more than devote a few sections to this topic. After giving basic definitions, we show how Jacobi polynomials appear in representations of $SU(2)$. We also explain how spaces of spherical harmonics give irreducible representations of $SU(2)$.

Suppose G is a group and V a finite-dimensional vector space. Let $GL(V)$ denote the group of linear transformations from V onto V.

Definition 9.11.1   *A finite-dimensional representation of G in V is a homomorphism from G to $GL(V)$. If G is a topological group, we assume that the homomorphism is continuous.*

Thus if

$$U : G \mapsto GL(V)$$

is a representation, then $U(g_1g_2) = U(g_1)U(g_2)$. The linear mappings $U(g_1)$, $U(g_2)$, and $U(g_1g_2)$ can be represented by matrices if we choose a basis for V.

Suppose dim $V = n$ and $\{x_1, x_2, \ldots, x_n\}$ is a basis for V. It is clear that the matrix entries $U_{ij}(g_1 g_2)$, $1 \leq i, j \leq n$, satisfy the relation

$$U_{ij}(g_1 g_2) = \sum_{k=1}^{n} U_{ik}(g_1) U_{kj}(g_2), \quad 1 \leq i, \; j \leq n. \tag{9.11.1}$$

If a different basis $\{x_1', x_2', \ldots, x_n'\}$ is chosen, then there exists a matrix P such that

$$U(g) = P U'(g) P^{-1}$$

for all $g \in G$. Here $U'(g)$ is the matrix representation of U corresponding to the new basis.

Suppose U_1 and U_2 are two representations of G in the vector spaces V_1 and V_2. The two representations are called isomorphic, or more briefly, V_1 and V_2 are isomorphic, if there is a linear isomorphism

$$T : V_1 \to V_2$$

such that

$$T \circ U_1(g) = U_2(g) \circ T \quad \text{for all } g \in G.$$

Now suppose that the vector space V has an inner product $\langle x, y \rangle$ defined on it. Let $\{x_1, x_2, \ldots, x_n\}$ be an orthonormal basis of V with respect to this inner product. Then

$$U_{ij}(g) = \langle U(g) x_j, x_i \rangle. \tag{9.11.2}$$

Definition 9.11.2 *A representation $U : G \mapsto GL(V)$ is called unitary if*

$$\langle U(g)x, U(g)y \rangle = \langle x, y \rangle$$

for all $x, y \in V$ and $g \in G$.

In this section, we study the representations of the compact group $SU(2)$. For compact groups, it is always possible to define an inner product on V such that a representation U of G in V is unitary. This is easy to prove, if the existence of an invariant measure on G is assumed. In fact, there exists a unique measure dg on G such that, for a continuous function f,

(a)
$$\int_G f(g) dg = \int_G f(gh) dg,$$

where h and $g \in G$, and

(b)
$$\int_G dg = 1.$$

The first condition gives invariance under right translation. It can be shown that the measure dg is also invariant under left translation. This invariant measure is often called the Haar measure. See Halmos [1950]. The only group we study in detail here is $SU(2)$. For this group the invariant measure is easy to construct. $SU(2)$ is the group of all matrices with complex entries of the form

$$g = \begin{pmatrix} a & b \\ -\bar{b} & \bar{a} \end{pmatrix}$$

with determinant one, that is, $|a|^2 + |b|^2 = 1$. Suppose

$$h = \begin{pmatrix} a_0 & b_0 \\ -\bar{b}_0 & \bar{a}_0 \end{pmatrix}$$

is also in $SU(2)$. It is easily checked that gh is constructed by the parameters a_1 and b_1 where

$$a_1 = a_0 a - \bar{b}_0 b,$$
$$b_1 = b_0 a + \bar{a}_0 b.$$

Thus

$$dg = da \wedge d\bar{a} \wedge db \wedge d\bar{b} \tag{9.11.3}$$

is an invariant measure on $SU(2)$. A simple calculation shows that

$$dgh = da_1 \wedge d\bar{a}_1 \wedge db_1 \wedge d\bar{b}_1$$
$$= (|a_0|^2 + |b_0|^2)^2 da \wedge d\bar{a} \wedge db \wedge d\bar{b}$$
$$= dg. \tag{9.11.4}$$

Suppose that $\langle \ \rangle_1$ is an inner product on the vector space V. For a given representation U of G in V, define a new inner product by

$$\langle x, y \rangle = \int_G \langle U(g)x, U(g)y \rangle_1 dg. \tag{9.11.5}$$

That U is unitary with respect to this inner product follows from property (a) of the Haar measure. Note that

$$\langle U(h)x, U(h)y \rangle = \int_G \langle U(g)U(h)x, U(g)U(h)y \rangle_1 dg$$
$$= \int_G \langle U(gh)x, U(gh)y \rangle_1 dg$$
$$= \int_G \langle U(g)x, U(g)y \rangle_1 dg$$
$$= \langle x, y \rangle.$$

Let $U : G \to GL(V)$ be a representation of G, and let W be a subspace of V. We say that W is invariant under the action of G if $U(g)$ maps W onto itself. This gives another representation of G, namely $U_W : G \to GL(W)$. Note that $U_W(g) = U(g)|_W$. U_W is called a subrepresentation of G. If U has no nontrivial subrepresentations, then U is called irreducible.

Suppose G is a compact group; then there is an inner product on V with respect to which U is unitary. If W is an invariant subspace of V, then the orthogonal complement of W is also an invariant subspace. This is easily verified. In this situation, the matrix for $U(g)$ has the form

$$U(g) = \begin{pmatrix} U_W(g) & 0 \\ 0 & U_{W^\perp}(g) \end{pmatrix}.$$

By continuing this process, we see that

$$V = W_1 \oplus W_2 \oplus \cdots \oplus W_k, \tag{9.11.6}$$

where $W_1, \ldots, W_k$ are irreducible. It can be shown that this decomposition is unique up to isomorphism. Thus, if

$$V = W_1' \oplus W_2' \oplus \cdots \oplus W_\ell',$$

then $k = \ell$ and, after renumbering if necessary, $W_i \simeq W_i'$.

9.12 The Group $SU(2)$

Recall that $SU(2)$ is defined as the group of matrices of the form

$$\begin{pmatrix} a & b \\ -\bar{b} & \bar{a} \end{pmatrix}, \qquad |a|^2 + |b|^2 = 1. \tag{9.12.1}$$

Thus, the group is defined by three parameters, which one may choose to be $|a|$, arg a, and arg b. When $ab \neq 0$, one can uniquely choose another set of three parameters ϕ, ψ, and θ called Euler's angles. These are obtained from the relations

$$a = e^{i(\phi+\psi)/2} \cos \frac{1}{2}\theta, \qquad b = i e^{i(\phi-\psi)/2} \sin \frac{1}{2}\theta, \tag{9.12.2}$$

where $0 \leq \phi < 2\pi$, $0 < \theta < \pi$, and $-2\pi \leq \psi < 2\pi$. When $ab = 0$, the correspondence between a, b and θ, ϕ, ψ is not one to one. Another way of writing the relation (9.12.2) in terms of matrices is

$$\begin{pmatrix} a & b \\ -\bar{b} & \bar{a} \end{pmatrix} = \begin{pmatrix} e^{i\phi/2} & 0 \\ 0 & e^{-i\phi/2} \end{pmatrix} \begin{pmatrix} \cos \frac{1}{2}\theta & i \sin \frac{1}{2}\theta \\ i \sin \frac{1}{2}\theta & \cos \frac{1}{2}\theta \end{pmatrix} \begin{pmatrix} e^{i\psi/2} & 0 \\ 0 & e^{-i\psi/2} \end{pmatrix}.$$

$$\tag{9.12.3}$$

Set

$$g(\phi, \theta, \psi) = \begin{pmatrix} a & b \\ -\bar{b} & \bar{a} \end{pmatrix},$$

with a and b as in (9.12.2). Then (9.12.3) is equivalent to

$$g(\phi, \theta, \psi) = g(\phi, 0, 0)g(0, \theta, 0)g(0, 0, \psi). \tag{9.12.4}$$

It also follows from (9.12.2) that

$$\cos\theta = 2|a|^2 - 1, \quad e^{i\phi} = -\frac{abi}{|a||b|}, \quad \text{and} \quad e^{i\psi} = \frac{ia}{b}\frac{|b|}{|a|}. \tag{9.12.5}$$

A question that arises here is the following: If $g(\phi, \theta, \psi) = g_1(\phi_1, \theta_1, \psi_1)$ $g_2(\phi_2, \theta_2, \psi_2)$, then what is the relation of ϕ, θ, ψ to ϕ_1, θ_1, ψ_1 and ϕ_2, θ_2, ψ_2? This general case follows from the consideration of the particular case when $\phi_1 = \psi_1 = \psi_2 = 0$. Then

$$g(\phi, \theta, \psi) = \begin{pmatrix} a & b \\ -\bar{b} & \bar{a} \end{pmatrix}$$

$$= \begin{pmatrix} \cos\frac{1}{2}\theta_1 & i\sin\frac{1}{2}\theta_1 \\ i\sin\frac{1}{2}\theta_1 & \cos\frac{1}{2}\theta_1 \end{pmatrix} \begin{pmatrix} \cos\frac{1}{2}\theta_2 e^{i\phi_2/2} & i\sin\frac{1}{2}\theta_2 e^{i\phi_2/2} \\ i\sin\frac{1}{2}\theta_2 e^{-i\phi_2/2} & \cos\frac{1}{2}\theta_2 e^{-i\phi_2/2} \end{pmatrix}.$$

So

$$a = \cos\frac{1}{2}\theta_1 \cos\frac{1}{2}\theta_2 e^{i\phi_2/2} - \sin\frac{1}{2}\theta_1 \sin\frac{1}{2}\theta_2 e^{-i\phi_2/2},$$

$$b = i\left(\cos\frac{1}{2}\theta_1 \sin\frac{1}{2}\theta_2 e^{i\phi_2/2} + \cos\frac{1}{2}\theta_2 \sin\frac{1}{2}\theta_1 e^{-i\phi_2/2} \right). \tag{9.12.6}$$

From (9.12.5) and (9.12.6), we get

$$\cos\theta = \cos\theta_1 \cos\theta_2 - \sin\theta_1 \sin\theta_2 \cos\phi_2,$$

$$e^{i\phi} = \frac{\sin\theta_1 \cos\theta_2 + \cos\theta_1 \sin\theta_2 \cos\phi_2 + i\sin\theta_2 \sin\phi_2}{\sin\theta}, \tag{9.12.7}$$

$$e^{i(\phi+\psi)/2} = \frac{\cos\frac{1}{2}\theta_1 \cos\frac{1}{2}\theta_2 e^{i\phi_2/2} - \sin\frac{1}{2}\theta_1 \sin\frac{1}{2}\theta_2 e^{-i\phi_2/2}}{\cos\frac{1}{2}\theta}.$$

This gives the formulas for the product $g(0, \theta_1, 0)g(\phi_2, \theta_2, 0)$. To obtain the general case the following remarks are sufficient. Observe that $g(\phi_1, 0, 0)g(\phi, \theta, \psi) = g(\phi_1+\phi, \theta, \psi)$, $g(\phi, \theta, \psi)g(0, 0, \psi_2) = g(\phi, \theta, \psi+\psi_2)$, $g(0, 0, \psi)g(\psi, 0, 0) = g(0, \theta, \psi)$, and $g(0, 0, \psi_1)g(\phi_2, 0, 0) = g(\psi_1 + \phi_2, 0, 0)$. Apply these relations to

$$g(\phi_1, \theta_1, \psi_1)g(\phi_2, \theta_2, \psi_2)$$
$$= g(\phi_1, 0, 0)g(0, \theta_1, 0)g(0, 0, \psi_1)g(\phi_2, 0, 0)g(0, \theta_2, 0)g(0, 0, \psi_2).$$

The general case follows immediately.

Remark The invariant measure dg for $G = SU(2)$ defined by (9.11.3) when written in terms of Euler angles is

$$dg = \frac{1}{2} \sin\theta \, d\theta \, d\varphi \, d\psi.$$

Usually the measure is normalized to

$$d\tilde{g} = \frac{1}{16\pi^2} \sin\theta \, d\theta \, d\varphi \, d\psi$$

so that

$$\int_G d\tilde{g} = 1.$$

Observe that $d\tilde{g}$ is half the product of the normalized measure on the sphere, $\sin\theta \, d\theta \, d\varphi / 4\pi$, and the normalized measure on the circle, $d\psi / 2\pi$.

9.13 Representations of $SU(2)$

Let V_{N+1} denote the $(N+1)$-dimensional vector space consisting of homogeneous polynomials of degree N in two complex variables with complex coefficients. If $P \in V_{N+1}$, then

$$P(x_1, x_2) = \sum_{k=0}^{N} r_k x_1^k x_2^{N-k}, \tag{9.13.1}$$

where x_1 and x_2 are complex variables and r_k, $k = 0, \ldots, N$, are complex constants. We also write $P(x) \equiv P(x_1, x_2)$ with

$$x = \begin{pmatrix} x_1 \\ x_2 \end{pmatrix}.$$

A representation of $SL_2(\mathbb{C})$ in V_{N+1} can be defined by

$$U(g)P(x) = P(g^t x), \tag{9.13.2}$$

where

$$g = \begin{pmatrix} a & b \\ c & d \end{pmatrix}, \quad ad - bc = 1, a, b, c, d \in \mathbb{C}.$$

Note that $P(g^t x) = P(ax_1 + cx_2, bx_1 + dx_2)$. It is easy to check that

$$U(g_1 g_2) = U(g_1)U(g_2). \tag{9.13.3}$$

It can be shown that U gives an irreducible representation of $SL_2(\mathbb{C})$. In fact, all the finite-dimensional irreducible representations of $SL_2(\mathbb{C})$ are of this form. The restrictions to the compact subgroup $SU(2)$ of these representations give all the finite-dimensional irreducible representations of $SU(2)$.

It is standard practice to write the polynomial in (9.13.1) in a slightly different form. Let $N = 2\ell$, so that ℓ is an integer multiple of $1/2$. Write

$$P(x_1, x_2) = \sum_{n=-\ell}^{\ell} r_n x_1^{\ell-n} x_2^{\ell+n}. \qquad (9.13.4)$$

Here $\ell + n$ takes integer values from 0 to $N = 2\ell$.

Associate with P a nonhomogeneous polynomial Q given by $P(x, 1) = Q(x)$. Thus,

$$Q(x) = \sum_{n=-\ell}^{\ell} r_n x^{\ell-n}, \qquad (9.13.5)$$

and the homogeneous polynomial corresponding to Q is obtained from

$$P(x_1, x_2) = x_2^{2\ell} Q(x_1/x_2).$$

Denote the space of all nonhomogeneous polynomials of the form (9.13.5) by H_ℓ and the representation of $SU(2)$ corresponding to U in the space H_ℓ by T_ℓ. This implies that

$$T_\ell(g)Q(x) = (bx + d)^{2\ell} Q((ax + c)/(bx + d)). \qquad (9.13.6)$$

Since $SU(2)$ is compact, the inner product (9.11.5) shows that it is possible to choose an inner product in H_ℓ such that T_ℓ is unitary. With respect to this inner product, the basis $\{1, x, \ldots, x^{2\ell}\}$ of H_ℓ is orthogonal. In fact, we have the lemma given below. Since the above inner product is defined up to a constant factor, assume that $\langle 1, 1 \rangle = (2\ell)!$.

Lemma 9.13.1 $\langle x^{\ell-m}, x^{\ell-n} \rangle = (\ell - n)!(\ell + n)!\delta_{mn}, \ -\ell \le m, n \le \ell.$

Proof.   Let

$$g = \begin{pmatrix} e^{it/2} & 0 \\ 0 & e^{-it/2} \end{pmatrix}.$$

From (9.13.6), it follows that $T_\ell(g)x^{\ell-k} = e^{-ikt} x^{\ell-k}$. Since $T_\ell(g)$ is unitary,

$$\langle x^{\ell-m}, x^{\ell-n} \rangle = \langle T_\ell(g)x^{\ell-m}, T_\ell(g)x^{\ell-n} \rangle$$
$$= e^{-i(m-n)t} \langle x^{\ell-m}, x^{\ell-n} \rangle.$$

This implies that $\langle x^{\ell-m}, x^{\ell-n} \rangle = 0$ for $m \ne n$.

The case $m = n$ requires a little more work. Take

$$g = \begin{pmatrix} \cos \frac{1}{2}t & -\sin \frac{1}{2}t \\ \sin \frac{1}{2}t & \cos \frac{1}{2}t \end{pmatrix},$$

and observe that, by the first part of the theorem,

$$0 = \langle x^{\ell-n}, x^{\ell-n+1} \rangle - \langle T_\ell(g)x^{\ell-n}, T_\ell(g)x^{\ell-n+1} \rangle$$
$$= \langle u^{\ell+n} v^{\ell-n}, u^{\ell+n-1} v^{\ell-n+1} \rangle, \qquad (9.13.7)$$

where

$$u = \left(\sin \frac{1}{2}t \right) x + \cos \frac{1}{2}t \quad \text{and} \quad v = \left(\cos \frac{1}{2}t \right) x - \sin \frac{1}{2}t.$$

Take the derivative of (9.13.7) with respect to t and set $t = 0$. The result is

$$(\ell + n)\langle x^{\ell-n+1}, x^{\ell-n+1} \rangle - (\ell - n + 1)\langle x^{\ell-n}, x^{\ell-n} \rangle = 0.$$

The theorem follows when the condition $\langle 1, 1 \rangle = (2\ell)!$ is used. ∎

9.14 Jacobi Polynomials as Matrix Entries

Choose the functions

$$\psi_n(x) = \frac{x^{\ell-n}}{\sqrt{(\ell-n)!(\ell+n)!}}, \quad -\ell \le n \le \ell, \qquad (9.14.1)$$

as an orthonormal basis for the space H_ℓ. Then, for $g \in SL_2(C)$,

$$T_\ell(g)\psi_n(x) = \frac{(ax+c)^{\ell-n}(bx+d)^{\ell+n}}{\sqrt{(\ell-n)!(\ell+n)!}} = \sum_{-\ell}^{\ell} t_{mn}^\ell(g)\psi_m(x). \quad (9.14.2)$$

Now use Taylor's formula to see that the coefficient $t_{mn}^\ell(g)$ is given by

$$t_{mn}^\ell(g) = \sqrt{\frac{(\ell+m)!}{(\ell-n)!(\ell+n)!(\ell-m)!}} \frac{d^{\ell-m}}{dx^{\ell-m}}[(ax+c)^{\ell-n}(bx+d)^{\ell+n}]_{x=0}.$$

Set $y + 1 = a(bx + d)$. Since $ad - bc = 1$, we have $ax + c = y/b$ and

$$t_{mn}^\ell(g) = \sqrt{\frac{(\ell+m)!}{(\ell-n)!(\ell+n)!(\ell-m)!}} \frac{b^{n-m}}{a^{n+m}} \frac{d^{\ell-m}}{dy^{\ell-m}}[y^{\ell-m}(y+1)^{\ell+n}]_{y=bc}.$$
$$(9.14.3)$$

If $g \in SU(2)$, then the use of Euler's angles provides a simpler formula for $t_{mn}^\ell(g)$. Consider the decomposition

$$T_\ell(g(\phi, \theta, \psi)) = T_\ell(g(\phi, 0, 0))T_\ell(g(0, \theta, 0))T_\ell(g(0, 0, \psi)). \quad (9.14.4)$$

It follows from (9.12.3) and (9.14.2) that

$$T_\ell(g(\phi, 0, 0))\psi_n = e^{-in\phi}\psi_n, \quad -\ell \le n \le \ell. \qquad (9.14.5)$$

This means that $T_\ell(g(\phi, 0, 0))$ is a diagonal matrix given by

$$T_\ell(g(\phi, 0, 0)) = \begin{pmatrix} e^{i\ell\phi} & & & 0 \\ & e^{i(\ell-1)\phi} & & \\ & & \ddots & \\ 0 & & & e^{-i\ell\phi} \end{pmatrix}. \qquad (9.14.6)$$

The matrix for $T_\ell(g(0, 0, \psi))$ is similar. Write $t^\ell_{mn}(g(0, \theta, 0)) = t^\ell_{mn}(\theta)$; then Equation (9.14.4) gives

$$t^\ell_{mn}(g) = e^{-i(m\phi + n\psi)} t^\ell_{mn}(\theta).$$

Denote $t^\ell_{mn}(\theta)$, $0 < \theta < \pi$, by $P^\ell_{mn}(\cos\theta)$, so that

$$t^\ell_{mn}(g) = e^{-i(m\phi + n\psi)} P^\ell_{mn}(\cos\theta). \qquad (9.14.7)$$

Since

$$g(0, \theta, 0) = \begin{pmatrix} \cos\frac{\theta}{2} & i\sin\frac{\theta}{2} \\ i\sin\frac{\theta}{2} & \cos\frac{\theta}{2} \end{pmatrix} \equiv \begin{pmatrix} a & b \\ c & d \end{pmatrix},$$

the quantity bc in (9.14.3) equals $-\sin^2\frac{\theta}{2} = (\cos\theta - 1)/2$. Thus, replacing y with $(z-1)/2$ in (9.14.3), we obtain

$$P^\ell_{mn}(z) = \frac{(-1)^{\ell-n} i^{n-m}}{2^\ell} \sqrt{\frac{(\ell+m)!}{(\ell-n)!(\ell+n)!(\ell-m)!}} (1+z)^{-(m+n)/2}$$
$$\cdot (1-z)^{(n-m)/2} \frac{d^{\ell-m}}{dz^{\ell-m}}[(1-z)^{\ell-n}(1+z)^{\ell+n}]. \qquad (9.14.8)$$

This shows that $P^\ell_{mn}(z)$ can be written in terms of a Jacobi polynomial. It is a constant multiple of

$$(1-z)^{(m-n)/2}(1+z)^{(m+n)/2} P^{(m-n, m+n)}_{\ell-m}(z).$$

9.15 An Addition Theorem

It follows from (9.12.7) that when

$$g(\phi, \theta, \psi) = g(0, \theta_1, 0)g(\phi_2, \theta_2, 0)$$

we have

$$\cos\theta = \cos\theta_1 \cos\theta_2 - \sin\theta_1 \sin\theta_2 \cos\phi_2. \qquad (9.15.1)$$

Use this in (9.11.1), (9.14.4), and (9.14.7) to get an addition formula:

$$e^{-i(m\phi + n\psi)} P^\ell_{mn}(\cos\theta) = \sum_{k=-\ell}^{\ell} e^{-ik\phi_2} P^\ell_{mk}(\cos\theta_1) P^\ell_{kn}(\cos\theta_2), \qquad (9.15.2)$$

where θ, ϕ, and ψ are given by (9.12.7). When $\phi_2 = 0$ and $\theta = \theta_1 + \theta_2 < \pi$, we have $\phi = \psi = 0$. In this case

$$P_{mn}^{\ell}(\cos(\theta_1 + \theta_2)) = \sum_{k=-\ell}^{\ell} P_{mk}^{\ell}(\cos\theta_1) P_{kn}^{\ell}(\cos\theta_2). \tag{9.15.3}$$

The addition formula (9.15.2) is the analog of Graf's addition formula (4.10.6) for Bessel functions. Formula (9.8.5) is an analog of Gegenbauer's addition formula for Bessel functions. An analog of this for Jacobi polynomials was found when $\beta = 0$ by Săpiro [1968] and in the general case by Koornwinder [1972, 1975].

Koornwinder used the addition formula to derive the Laplace-type integral for Jacobi polynomials given in Exercise 6.45 and an integral formula for a product of Jacobi polynomials. Koornwinder [1974] observed that it is possible to obtain the product formula from the Laplace-type integral by using a result of Bateman [1932, pp. 392–393]. Bateman's result is

$$\frac{P_n^{(\alpha,\beta)}(s)}{P_n^{(\alpha,\beta)}(1)} \frac{P_n^{(\alpha,\beta)}(t)}{P_n^{(\alpha,\beta)}(1)} = \sum_{k=0}^{n} b_{k,n}(s+t)^k \frac{P_k^{(\alpha,\beta)}((1+st)/(s+t))}{P_k^{(\alpha,\beta)}(1)}, \tag{9.15.4}$$

where $b_{k,n}$ is defined by (9.15.4) when $t = 1$,

$$\frac{P_n^{(\alpha,\beta)}(s)}{P_n^{(\alpha,\beta)}(1)} = \sum_{k=0}^{n} b_{k,n}(s+1)^k. \tag{9.15.5}$$

Bateman proved (9.15.4) by showing that both sides of the equation are solutions of the partial differential equation

$$\left[\frac{\partial^2}{\partial \xi^2} + ((2\alpha+1)\cot\xi - (2\beta+1)\tan\xi)\frac{\partial}{\partial\xi} \right.$$
$$\left. + \frac{\partial^2}{\partial\eta^2} + ((2\alpha+1)\coth\eta + (2\beta+1)\tanh\eta)\frac{\partial}{\partial\eta} \right] F = 0, \tag{9.15.6}$$

when

$$s = \cos 2\xi, \quad t = \cosh 2\eta.$$

Now recall the formula in Exercise 6.45:

$$\frac{P_n^{(\alpha,\beta)}(x)}{P_n^{(\alpha,\beta)}(1)} = \int_0^1 \int_0^{\pi} \left[\frac{1+x-(1-x)u^2}{2} + i\sqrt{1-x^2}\, u\cos\theta \right]^n dm_{\alpha,\beta}(u,\theta), \tag{9.15.7}$$

where $\alpha > \beta > -1/2$ and

$$dm_{\alpha,\beta}(u,\theta) = \frac{2\Gamma(\alpha+1)}{\sqrt{\pi}\Gamma(\alpha-\beta)\Gamma(\beta+1/2)}(1-u^2)^{\alpha-\beta-1}u^{2\beta+1}(\sin\theta)^{2\beta}\,du\,d\theta.$$

This implies

$$(x+y)^n \frac{P_n^{(\alpha,\beta)}((1+xy)/(x+y))}{P_n^{(\alpha,\beta)}(1)}$$

$$= \int_0^1 \int_0^\pi \left[\frac{(1+x)(1+y)+(1-x)(1-y)u^2}{2} \right.$$

$$\left. + \sqrt{(1-x^2)(1-y^2)} u \cos\theta \right]^n dm_{\alpha,\beta}(u,\theta).$$

When this formula is combined with (9.15.4) and (9.15.5) the result is the product formula for Jacobi polynomials given by

$$\frac{P_n^{(\alpha,\beta)}(x)P_n^{(\alpha,\beta)}(y)}{P_n^{(\alpha,\beta)}(1)P_n^{(\alpha,\beta)}(1)}$$

$$= \frac{1}{P_n^{(\alpha,\beta)}(1)} \int_0^1 \int_0^\pi P_n^{(\alpha,\beta)}[\{(1+x)(1+y)+(1-x)(1-y)\}/2$$

$$+ \sqrt{(1-x^2)(1-y^2)} u \cos\theta - 1] dm_{\alpha,\beta}(u,\theta). \qquad (9.15.8)$$

9.16 Relation of $SU(2)$ to the Rotation Group $SO(3)$

The description of elements of $SU(2)$ in terms of Euler angles suggests a connection of $SU(2)$ with the group of rotations in three dimensions. The explicit relationship is given below. An interesting consequence of this connection is that the spaces of spherical harmonics in three variables are seen as the irreducible representation spaces of $SU(2)$.

The rotation group $SO(3)$ consists of 3×3 matrices g with real entries and determinant 1 such that the transpose of g is also its inverse. These are the orientation-preserving linear mappings g from R^3 to R^3 such that for $x \in R^3$, $|x|^2 = |gx|^2$. To define a homomorphism ϕ from $SU(2)$ to $SO(3)$, first identify the points $x = (x_1, x_2, x_3) \in R^3$ with 2×2 Hermitian matrices of trace 0:

$$u_x = \begin{pmatrix} -x_3 & x_1 - ix_2 \\ x_1 + ix_2 & x_3 \end{pmatrix}.$$

Note that $\det u_x = -|x|^2$. For $g \in SU(2)$, define

$$\phi(g)x = gu_xg^{-1} \quad \text{for } x \in R^3.$$

It is clear that $\phi(g)x$ is a Hermitian matrix with trace zero and $\det[\phi(g)x] = -|x|^2$. Moreover, $\phi(g)$ is an orientation-preserving linear mapping. Thus $\phi(g)$ can be identified with an element of $SO(3)$, and we have a homomorphism ϕ from $SU(2)$ to $SO(3)$ whose kernel is easily seen to be $\{\pm I\}$. It can also be shown, though we do not do so here, that ϕ is an onto mapping. Thus ϕ gives an isomorphism of $SU(2)/\{\pm I\}$ onto $SO(3)$. Recall that the Euler angle ψ in (9.12.2) ranges over $[-2\pi, 2\pi)$ whereas the range for ψ in a rotation is $[0, 2\pi)$. This is related to the fact that $\pm g \in SU(2)$ are associated with the same rotation in $SO(3)$.

It is easy to check that $\omega_1(t), \omega_2(t), \omega_3(t) \in SU(2)$, defined by

$$\omega_1(t) = \begin{pmatrix} \cos t & i\sin t \\ i\sin t & \cos t \end{pmatrix}, \quad \omega_2(t) = \begin{pmatrix} \cos t & -\sin t \\ \sin t & \cos t \end{pmatrix},$$

$$\omega_3(t) = \begin{pmatrix} e^{it} & 0 \\ 0 & e^{-it} \end{pmatrix},$$

correspond to rotations (by an angle of $2t$) about the $x_1, x_2,$ and x_3 axes respectively. Thus by (9.12.3) and (9.12.4) we can view any rotation with Euler angles ϕ, θ, ψ as a product of a rotation by the angle ψ about x_3, a rotation by the angle θ about x_1, and a rotation by the angle ϕ about x_3.

The object of the remainder of this section is to show that $H_k(x)$, the space of harmonic polynomials of degree k in three variables provides an irreducible representation of $SO(3)$ and hence of $SU(2)$ as well.

Let ϕ be the homomorphism from $SU(2)$ onto $SO(3)$. We have seen that $\phi(g_1) = \phi(g_2)$ if and only if $g_1 = \pm g_2$. It is clear that if U is a (irreducible) representation of $SO(3)$ in a vector space V, then $\phi \cdot U$ is a (irreducible) representation of $SU(2)$ in V. Conversely, if T is a (irreducible) representation of $SU(2)$ in V such that T (-identity) = identity, then T gives rise to a (irreducible) representation of $SO(3)$.

Recall that in the notation of Section 9.13, the proof of Lemma 9.13.1 shows that the functions $x^{\ell-k} \in H_\ell$ (not to be confused with $H_k(x)$) for $k = -\ell, \ldots, \ell$ are eigenvectors of $T_\ell(g)$ corresponding to the eigenvalues e^{2ikt} ($k = -\ell, \ldots, \ell$) when $g = \omega_3(-t)$. This fact may be used to prove that if U is a representation of $SU(2)$ in V with $\dim V = 2\ell + 1$ and $e^{2i\ell t}$ occurs as an eigenvalue of $U(\omega_3(t))$, then (U, V) is isomorphic to (T_ℓ, H_ℓ). We have

$$V = H_{k_1} \oplus H_{k_2} \oplus \cdots \oplus H_{k_p}$$

for some integers $k_1, \ldots, k_p$. If $p = 1$ and $k_1 = \ell$, then we are done. If $p \neq 1$, then $k_i < \ell$ and the eigenvalues of $U(\omega_3(t))$ are of the form e^{2imt} with $|m| < \ell$. Thus $V \simeq H_\ell$ and the result is proved.

Now define a representation U of $SO(3)$ in $H_k(x)$ by

$$U(g)p(x) = p(g^{-1}x)$$

for $x \in R^3$, $p \in H_k(x)$, and $g \in SO(3)$. This in turn gives a representation $U \cdot \phi$ of $SU(2)$. Now observe that $(x_1 + ix_2)^k \in H_k(x)$ and that $\omega_3(t)$ maps to a rotation by an angle $2t$ about x_3. Thus $U(\phi(\omega_3(t)))$ has $(x_1 + ix_2)^k$ as an eigenvector with eigenvalue e^{2ikt}. Thus by the result of the previous paragraph, $H_k(x)$, the space of harmonic polynomials of degree k, is an irreducible representation of $SU(2)$.

Exercises

1. Verify that the only polynomial solution of the differential equation
$$(1 - x^2)y'' - 2(1 + m)xy' + (n - m)(n + m + 1)y = 0,$$
where $m, n \in \mathbb{Z}$, $n - m \geq 0$, is the polynomial $K C_{n-m}^{m+\frac{1}{2}}(x)$.

2. Let V_k denote the vector space of all homogeneous polynomials of degree k in n variables. For $\alpha = (\alpha_1, \ldots, \alpha_n)$, let $x^\alpha = x_1^{\alpha_1} x_2^{\alpha_2} \cdots x_n^{\alpha_n}$ and
$$D^\alpha = \partial^{\alpha_1 + \alpha_2 + \cdots + \alpha_n} / \partial x_1^{\alpha_1} \partial x_2^{\alpha_2} \cdots \partial x_n^{\alpha_n}.$$
For a polynomial P in n variables, let $P(D)$ denote the differential operator obtained by replacing x^α by D^α in $P(x)$. For $P, Q \in V_k$, define $\langle P, Q \rangle = P(D)Q$ (or $P(D)\bar{Q}$ if complex coefficients are used). Prove that $\langle \, , \, \rangle$ is an inner product on V_k.

3. Let Δ be the Laplace operator and $k \geq 2$. Show that $\Delta : V_k \to V_{k-2}$ is mapping by proving that there is no nonzero vector in V_{k-2} that is orthogonal (with respect to the inner product in Exercise 2) to the range of Δ.

4. Let $H_k \subset V_k$ denote the subspace of harmonic polynomials. Suppose $L_k = \{P \in V_k \mid P(x) = |x|^2 Q(x), Q \in V_{k-2}\}$. Prove that
$$V_k = H_k \oplus L_k.$$

5. Use the result of Exercise 4 to prove that if $P \in V_k$, then
$$P(x) = P_0(x) + |x|^2 P_1(x) + \cdots + |x|^{2\ell} P_\ell(x),$$
where P_j is a homogeneous harmonic polynomial of degree $k - 2j$, $j = 0, 1, \ldots, \ell$. Deduce that any polynomial in n variables and restricted to the unit sphere is a sum of spherical harmonics.

6. Use the results of the previous problems to show that $c_{k,n} = $ dimension of the space of spherical harmonics of degree k in n variables $= \dim V_k - \dim V_{k-2}$
$$= \binom{n + k - 1}{k} - \binom{n + k - 3}{k - 2}.$$

7. Prove that it is possible to choose points $\eta_1, \eta_2, \ldots, \eta_{c_{kn}}$ such that the matrix (9.6.9) is invertible.

8. Show that the function $g_\xi(\eta)$ in (9.7.7) is given by $\Sigma S_{k,j}(\xi) S_{k,j}(\eta)$.

9. Derive the addition formula (9.8.5) from (9.6.8) by differentiation.

10. Prove Theorem 9.9.1.

11. Use (9.8.5′) to derive:
   (a) Gegenbauer's product formula
$$\frac{C_n^\lambda(\cos \theta) C_n^\lambda(\cos \phi)}{C_n^\lambda(1)}$$
$$= c_\lambda \int_0^\pi C_n^\lambda(\cos \theta \cos \phi + \sin \theta \sin \phi \cos \psi)(\sin \psi)^{2\lambda - 1} d\psi,$$

where $\lambda > 0$ and

$$c_\lambda^{-1} = \int_0^\pi (\sin \psi)^{2\lambda-1} d\psi.$$

(b) The integral formula (9.8.1) for ultraspherical polynomials.

12. Prove that when one lets $\lambda \to 0$ in formulas (a) and (b) of Exercise 11, one obtains the well-known formulas

$$\cos n\theta \cos n\phi = \frac{1}{2}[\cos n(\theta + \phi) + \cos n(\theta - \phi)],$$

$$\cos n\theta = \frac{1}{2}[e^{in\theta} + e^{-in\theta}].$$

13. Let $x, y \in R^n$, $x = R\xi$, $y = r\eta$, $|\xi| = |\eta| = 1$, and $R > r$, and note that

$$|x - y|^{2-n} = R^{2-n}\left(1 - 2\left(\frac{r}{R}\right)\xi \cdot \eta + \left(\frac{r}{R}\right)^2\right)^{\frac{2-n}{2}}$$

$$= R^{2-n} \sum_{k=0}^\infty C_k^{(n-2)/2}(\xi \cdot \eta)\left(\frac{r}{R}\right)^k.$$

(a) Let $\sum_{j=1}^n y_j \frac{\partial}{\partial x_j} = r(\eta \cdot \nabla)$. Use Taylor's theorem to prove that

$$|x - y|^{2-n} = \sum_{k=0}^\infty \frac{(-1)^k}{k!} r^k (\eta \cdot \nabla)^k |x|^{2-n}.$$

(b) Deduce Maxwell's formula that

$$(\eta \cdot \nabla)^k |x|^{2-n} = \frac{(-1)^k}{k!} \frac{C_k^{(n-2)/2}(\xi \cdot \eta)}{|x|^{n+k-2}}.$$

14. Deduce the following formulas from (9.15.2):

(a) $P_{mk}^\ell(\cos \theta_1) P_{kn}^\ell(\cos \theta_2) = \frac{1}{2\pi} \int_{-\pi}^\pi e^{i(k\phi_2 - m\phi - n\psi)} P_{mn}^\ell(\cos \theta) d\phi_2.$

(b) $P_\ell(\cos \theta_1) P_\ell(\cos \theta_2) = \frac{1}{2\pi} \int_{-\pi}^\pi P_\ell(\cos \theta_1 \cos \theta_2 - \sin \theta_1 \sin \theta_2 \cos \phi) d\phi,$

where ϕ, ψ, and θ are as defined in the text. In (b) ℓ is an integer.

15. This problem gives a generating function for spherical harmonics in three dimensions. Let $x = (x_1, x_2, x_3)$ and $u_n = (-2t, 1 - t^2, i + it^2)$. Then define $H_n^k(x)$ by

$$(u \cdot x)^n = [x_2 + ix_3 - 2x_1 t - (x_2 - ix_3)t^2]^n = t^n \sum_{k=-n}^n H_n^k(x) t^k.$$

(a) Show that $\bar{H}_n^k = (-1)^k H_n^{-k}$.

(b) Prove that $H_n^k(x)$ is a homogeneous harmonic polynomial by showing that $\nabla^2(u \cdot x)^n = 0$.

(c) Let $v = (-2s, 1 - s^2, i + is^2)$ and define the polynomial $\phi(u, v)$ by

$$\phi(u, v) = \int_{|\xi|=1} (u \cdot \xi)^n (\bar{v} \cdot \xi)^n d\omega(\xi).$$

Use the technique of Lemma 9.6.2 to show that $\phi(u, v)$ is a constant times $(u \cdot \bar{v})^n$. [Note that $u \cdot u = 0$ and $v \cdot v = 0$.]

(d) Define $S_n^k(\xi) = |x|^{-n} H_n^k(x)$, where $\xi = x/|x|$. Show that

$$\int_{|\xi|=1} S_n^k(\xi) \bar{S}_n^\ell(\xi) d\omega(\xi) = \begin{cases} 0, & k \neq \ell, \\ 2\pi \cdot \frac{\Gamma(1/2)\Gamma(n+1)}{\Gamma(n+(3/2))} \binom{2n}{k+n}, & k = \ell. \end{cases}$$

(e) Also prove that

$$S_n^k(\xi) = (-1)^{n+k} \frac{2^n n!}{(n+k)!} (\xi_2 - i\xi_2)^k (1 - \xi_1^2)^{-k/2} P_n^k(\xi_1).$$

16. Show that the associated Legendre functions $P_k^m(x)$ defined by (9.6.7) satisfy the recurrence

$$P_k^{m+2}(x) + \frac{2(m+1)x}{(x^2-1)^{1/2}} P_k^{m+1}(x) - (k-m)(k+m+1)P_k^m(x) = 0$$

for $m = 0, 1, 2, \ldots$.

17. Derive the formulas

(a) $$P_k^m(x) = \frac{\Gamma(k+m+1)}{\pi \Gamma(k+1)} \int_0^\pi (x + \sqrt{x^2 - 1} \cos\psi)^k \cos m\psi \, d\psi,$$

(b) $$P_k^m(\cos\theta) = \frac{(-1)^m 2\Gamma(k+m+1)}{\sqrt{\pi}\Gamma(m+(1/2))\Gamma(k-m+1)} \frac{1}{(2\sin\theta)^m}$$
$$\cdot \int_0^\theta \frac{\cos(k+(1/2))\phi \, d\phi}{[2\cos\phi - 2\cos\theta]^{-m+(1/2)}},$$

where $\text{Re} x > 0, 0 < \theta < \pi$, and m is a nonnegative integer.

18. Show that

$$P_k^m(\cos\theta) = \frac{(-1)^m}{(k-m)!} \int_0^\infty u^k e^{-u\cos\theta} J_m(u \sin\theta) du.$$

19. Show that

$$\left[P_k^m(\cos\theta) \right]^2$$
$$= \frac{(k+m)!}{(k-m)!} \sum_{j=m}^k (-1)^{j+m} \frac{(2j)!}{2^{2j}(j-m)!(j+m)!(j!)^2} \frac{(k+j)!}{(k-j)!} \sin^{2j}\theta.$$

20. Prove the relation (9.10.5), that is,

$$\lim_{\epsilon \to 0^+} \int_0^\infty e^{-\epsilon t} J_\nu(t) t^{\mu-1} dt = \frac{2^{\mu-1}\Gamma((\mu+\nu)/2)}{\Gamma(1+(\nu-\mu)/2)}.$$

10

Introduction to q-Series

When one is counting, one may use a generating function to keep track of the number of objects being counted. This is yet another way in which hypergeometric series arise. For example, the finite binomial theorem is usually written as

$$(x + y)^n = \sum_{k=0}^{n} \binom{n}{k} x^{n-k} y^k.$$

The coefficient $\binom{n}{k}$ counts the number of ways $n - k$ x and k y can be arranged. The usual argument is to observe that the first y can be put in any of the n places, the second in $n - 1$ places, and so on, until the

$$n(n - 1) \cdots (n - k + 1)$$

ways are obtained. However, note that the first y could have been in any of the k spots, the second in any of the remaining $k - 1$ spots, through the kth, so that $k!$ of those arrangements are the same. Thus we can represent the number of combinations as

$$\binom{n}{k} = \frac{n(n - 1) \cdots (n - k + 1)}{k!} = \frac{n!}{k!(n - k)!}.$$

The last expression makes the symmetry in k and $n - k$ as clear as it is by counting the x first rather than the y.

Observe that since $x = y = 1$ gives

$$2^n = \sum_{k=0}^{n} \binom{n}{k},$$

the binomial coefficients provide a refinement to the cruder result that 2^n is the total number of arrangements of x and y in n places.

A further refinement is possible. One nice way to illustrate this is to consider lattice paths on the first quadrant, starting at $(0, 0)$ and moving to $(n - k, k)$ by n steps, each either one unit to the right or one up. Consider the case of two moves

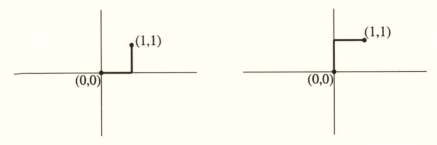

Figure 10.1

shown in Figure 10.1. In the first case there is no area under the path; in the second the area is one. We will split the $\binom{n}{k}$ paths according to the area under the curve. Since

$$(x + y)^2 = xx + xy + yx + yy$$

we can keep track of the area by rewriting each term with the xs first and adding a unit area whenever yx is changed to xy. We will use a parameter q to do the counting, requiring that

$$yx = qxy.$$

Since we wish to collect the qs together, also assume that

$$yq = qy,$$
$$xq = qx.$$

As an exercise, work out $(x + y)^4$: The coefficient of x^2y^2 comes from six pictures, and the generating function for these six pictures collected by areas under the graph is

$$1 + q + 2q^2 + q^3 + q^4.$$

A little reflection shows that this is

$$(1 + q^2)(1 + q + q^2),$$

demonstrating that there is some structure to this coefficient. It is possible to rewrite this in a form that suggests a general form for the coefficients, but there is a more elegant way to derive this formula. Recall Pascal's triangle property for binomial coefficients:

$$\binom{n + 1}{k} = \binom{n}{k} + \binom{n}{k - 1}. \qquad (10.0.1)$$

This can be explained by a combinatorial argument, since the $(n + 1)$-th spot could contain an x in $\binom{n}{k}$ ways or a y in $\binom{n}{k-1}$ ways.

However (10.0.1) also comes from

$$(x + y)^{n+1} = (x + y)^n (x + y).$$

We use this method for finding the q-binomial coefficients $\begin{bmatrix} n \\ k \end{bmatrix}_q$ that are defined by

$$(x + y)^n = \sum_{k=0}^{n} \begin{bmatrix} n \\ k \end{bmatrix}_q x^{n-k} y^k \qquad (10.0.2)$$

when

$$yx = qxy,$$
$$xq = qx, \quad yq = qy.$$

First

$$(x + y)^{n+1} = (x + y)^n (x + y)$$

gives

$$\sum_{k=0}^{n} \begin{bmatrix} n + 1 \\ k \end{bmatrix}_q x^{n+1-k} y^k = \sum_{k=0}^{n} \begin{bmatrix} n \\ k \end{bmatrix}_q x^{n-k} y^k (x + y).$$

Since

$$y^k x = q^k x y^k$$

we have

$$\begin{bmatrix} n + 1 \\ k \end{bmatrix}_q = \begin{bmatrix} n \\ k \end{bmatrix}_q q^k + \begin{bmatrix} n \\ k - 1 \end{bmatrix}_q. \qquad (10.0.3)$$

This is a q-extension of (10.0.1). In the case when $q = 1$ there is only one Pascal triangle relation. In the q-case there is a second. Using the same argument,

$$(x + y)^{n+1} = (x + y)(x + y)^n$$

gives

$$\begin{bmatrix} n + 1 \\ k \end{bmatrix}_q = \begin{bmatrix} n \\ k \end{bmatrix}_q + q^{n+1-k} \begin{bmatrix} n \\ k - 1 \end{bmatrix}_q. \qquad (10.0.4)$$

Relations (10.0.3) and (10.0.4) can be combined to give

$$\begin{bmatrix} n \\ k \end{bmatrix}_q = \frac{(1 - q^{n+1-k})}{(1 - q^k)} \begin{bmatrix} n \\ k - 1 \end{bmatrix}_q.$$

Iteration leads to

$$\begin{bmatrix} n \\ k \end{bmatrix}_q = \frac{(1 - q^{n+1-k}) \cdots (1 - q^n)}{(1 - q^k) \cdots (1 - q)} \begin{bmatrix} n \\ 0 \end{bmatrix}_q.$$

But

$$\begin{bmatrix} n \\ 0 \end{bmatrix}_q = 1$$

and so

$$\begin{bmatrix} n \\ k \end{bmatrix}_q = \frac{(1-q)\cdots(1-q^n)}{(1-q)\cdots(1-q^k)(1-q)\cdots(1-q^{n-k})} = \frac{(q;q)_n}{(q;q)_k(q;q)_{n-k}},$$

$$(10.0.5)$$

where

$$(q;q)_n = \prod_{j=1}^n (1-q^j).$$

$$(10.0.6)$$

Another way to write (10.0.5) is

$$\begin{bmatrix} n \\ k \end{bmatrix}_q = \frac{n!_q}{k!_q(n-k)!_q},$$

$$(10.0.7)$$

where

$$n!_q = (1+q)\cdots(1+q+\cdots+q^{n-1}) = (q;q)_n(1-q)^{-n}.$$

$$(10.0.8)$$

The question naturally arises whether there is a commutative extension of the binomial theorem that uses q-binomial coefficients. To derive this result, replace y with xy in (10.0.2). This is possible since $y(xy) = (yx)y = q(xy)y$. Then

$$(x+xy)^n = \sum_{k=0}^n \begin{bmatrix} n \\ k \end{bmatrix}_q x^{n-k}(xy)^k.$$

Observe that $(xy)^k = xyxy\cdots xy = x^k y^k q^{k(k-1)/2}$. Also,

$$(x+xy)(x+xy)\cdots(x+xy)$$
$$= x(1+y)\cdots x(1+y)x(1+y)$$
$$= x(1+y)\cdots x(1+y)x^2(1+qy)(1+y)$$
$$= x^n(1+y)(1+qy)\cdots(1+q^{n-1}y).$$

Therefore,

$$(1+y)(1+qy)\cdots(1+q^{n-1}y) = \sum_{k=0}^n q^{k(k-1)/2}\begin{bmatrix} n \\ k \end{bmatrix}_q y^k.$$

$$(10.0.9)$$

Replace y with y/x to obtain

$$(x+y)(x+qy)\cdots(x+q^{n-1}y) = \sum_{k=0}^n \begin{bmatrix} n \\ k \end{bmatrix}_q q^{k(k-1)/2} x^{n-k} y^k.$$

$$(10.0.10)$$

This is a q-extension of the binomial theorem. The noncommutative binomial theorem is due to Schützenberger [1953]. The q-binomial theorem was independently known to several mathematicians of the nineteenth century. The interpretation of the q-binomial coefficient in terms of areas under lattice paths is due to Pólya [1984, Vol. 4, p. 444].

The infinite q-binomial theorem can be seen as an analog of the formula for the beta integral on $(0, 1)$ in terms of gamma functions. To show this, we introduce a q-integral. This was explicitly done by Thomae and Jackson, but the essential idea was discovered by Fermat. We also develop the q-extensions of the gamma and beta functions.

A generalization of the q-binomial theorem is the $_1\psi_1$ formula of Ramanujan. This can be considered a q-extension of the beta integral on $(0, \infty)$. Ramanujan's formula and one of its consequences, the Jacobi triple product identity, are very important in number theory. We show how they imply results on representations of numbers as sums of squares.

The remainder of the chapter is devoted to the study of a few other important q-beta integrals and to developing the elementary theory of basic hypergeometric (or q-hypergeometric) series. We also give a very short exposition of the theory of q-ultraspherical polynomials. We note that some of these infinite series and products are also modular functions.

10.1 The q-Integral

Even before the systematic development of calculus by Leibniz and Newton in the latter half of the seventeenth century, mathematicians from many parts of the world attempted to evaluate the integral

$$\int_0^a x^\alpha dx.$$

For example, Archimedes computed the case $\alpha = 2$. He did this in two different ways, one using the value of $1^2 + 2^2 + \cdots + n^2$, which was familiar to the Babylonians in 1700 B.C., and the other using the sum of a finite geometric series. In the early seventeenth century this integral was computed for other small values of α (up to nine according to some accounts). The difficulty experienced by those mathematicians was the problem of treating the sums $1^k + 2^k + \cdots + n^k$ in a general way. In the 1650s, Fermat, Pascal, and others found a method for this. Fermat also gave an easier way of computing the integral, using a geometric series. In his studies of the Greeks, Fermat must have noted that Archimedes also used a geometric series in the quadrature of a parabola. For more history and references, see A. Edwards [1987] or C. Edwards [1979].

Decompose the interval $[0, a]$ into subintervals using a geometric dissection, that is, subintervals with endpoints $\{x_n\}_0^\infty$ where $x_n = aq^n$, $0 < q < 1$. In this case the sum approximating the integral is

$$\sum_{n=0}^\infty x_n^\alpha (x_n - x_{n+1}) = \sum_{n=0}^\infty (aq^n)^\alpha (aq^n - aq^{n+1})$$

$$= a^{\alpha+1}(1-q) \sum_{n=0}^\infty q^{(\alpha+1)n}$$

$$= \frac{a^{\alpha+1}(1-q)}{1-q^{\alpha+1}}. \tag{10.1.1}$$

Fermat considered the case $\alpha = \ell/m$, where ℓ and m are positive integers. Set $t = q^{1/m}$ and write (10.1.1) as

$$\frac{a^{(\ell+m)/m}(1-t^m)}{1-t^{m+n}} = a^{(\ell+m)/m} \frac{1 + t + \cdots + t^{m-1}}{1 + t + \cdots + t^{m+n-1}}$$

$$\rightarrow \frac{m}{m+n} a^{(\ell+m)/m} \quad \text{as } t \rightarrow 1. \tag{10.1.2}$$

Thus, Fermat evaluated the integral when α is rational.

Thomae [1869] and later Jackson [1910] introduced the q-integral defined by

$$\int_0^a f(x) d_q x = \sum_{n=0}^\infty f(aq^n)(aq^n - aq^{n+1}). \tag{10.1.3}$$

We call $d_q x$ the Fermat measure. Jackson also defined an integral on $(0, \infty)$ by

$$\int_0^\infty f(x) d_q x = (1-q) \sum_{n=-\infty}^\infty f(q^n) q^n. \tag{10.1.4}$$

Notice that

$$\lim_{N \to \infty} \int_0^{q^{-N}} f(x) d_q x = \int_0^\infty f(x) d_q x.$$

The idea here is that on $(1, \infty)$ the division points are at $q^{-1}, q^{-2}, q^{-3}, \ldots$ when $0 < q < 1$. It is easy to see that, when $f(x)$ is continuous on $(0, a)$,

$$\lim_{q \to 1^-} \int_0^a f(x) d_q x = \int_0^a f(x) dx. \tag{10.1.5}$$

It is clear that we can write the q-integral for any continuous function, but it is important to keep in mind that the resulting sum should be interesting and manageable. After all, Fermat used the q-integral because a geometric series can be summed.

Suppose that $f(x) = x^{\alpha-1}(1-x)^{\beta-1}$; we are interested in obtaining an analog of the beta integral. From (10.1.3), the sum corresponding to this $f(x)$ is

$$\sum_{n=0}^{\infty} q^{n(\alpha-1)}(1-q^n)^{\beta-1}(q^n - q^{n+1}).$$

We are unable to sum this series because of the term $(1-q^n)^{\beta-1}$. So we look for a function $f_q(x)$ such that

$$f_q(x) \to x^{\alpha-1}(1-x)^{\beta-1} \quad \text{as } q \to 1^-,$$

and for which the q-integral $\int_0^1 f_q(x) d_q x$ can be evaluated in an appropriate form. It seems very likely that $x^{\alpha-1}$ should be retained as it is. Then we might deal with $(1-x)^{\beta-1}$ by expressing it as a power series in x and then deciding what to do with the coefficients. By the binomial theorem,

$$(1-x)^{-\alpha} = \sum_{k=0}^{\infty} \frac{(\alpha)_k}{k!} x^k \quad \text{for } |x| < 1.$$

We now need the q-analogs of $k!$ and more generally of $(\alpha)_k$ and finally an analog of the binomial theorem itself.

10.2 The q-Binomial Theorem

To define $k!_q$, the q-analog of $k!$, note that (10.1.2) indicates that we should replace an integer m with $1 + q + \cdots + q^{m-1} = (1-q^m)/(1-q)$. Thus,

$$k!_q = \frac{(1-q)(1-q^2)\cdots(1-q^k)}{(1-q)(1-q)\cdots(1-q)},$$

and we can replace the shifted factorial $(\alpha)_k$ with

$$\frac{(1-q^\alpha)(1-q^{\alpha+1})\cdots(1-q^{\alpha+k-1})}{(1-q)^k}.$$

Now write

$$(a; q)_k = (1-a)(1-aq)\cdots(1-aq^{k-1}). \tag{10.2.1}$$

We see that the series corresponding to $\sum_0^\infty (\alpha)_k x^k / k!$ is

$$\sum_{k=0}^{\infty} \frac{(q^\alpha; q)_k}{(q; q)_k} x^k. \tag{10.2.2}$$

This series can be summed and its evaluation in terms of infinite products is given in the next theorem, called the q-binomial theorem. To see how to sum it, consider the following proof of the binomial theorem. Let

$$g_\alpha(x) = \sum_{k=0}^\infty \frac{(\alpha)_k}{k!} x^k, \quad |x| < 1.$$

First differentiate

$$g_\alpha'(x) = \sum_{k=1}^\infty \frac{(\alpha)_k}{(k-1)!} x^{k-1} = \alpha g_{\alpha+1}(x).$$

To remove $g_{\alpha+1}(x)$, consider

$$g_\alpha(x) - g_{\alpha+1}(x) = \sum_{k=1}^\infty \frac{(\alpha)_k - (\alpha+1)_k}{k!} x^k$$

$$= \sum_{k=1}^\infty \frac{(\alpha+1)_{k-1}[\alpha - (\alpha+k)]}{k!} x^k = -x g_{\alpha+1}(x).$$

Eliminate $g_{\alpha+1}(x)$ from the two equations to get

$$\frac{g_\alpha'(x)}{g_\alpha(x)} = \frac{\alpha}{1-x},$$

which implies

$$g_\alpha(x) = (1-x)^{-\alpha}.$$

Application of this idea to the evaluation of (10.2.2) requires the q-difference operator. This operator is defined by

$$\Delta_q f(x) = \frac{f(x) - f(qx)}{x - qx} = \frac{f(x) - f(qx)}{(1-q)x}. \tag{10.2.3}$$

We now state and prove the q-binomial theorem.

Theorem 10.2.1   *For $|x| < 1, |q| < 1$,*

$$\sum_{k=0}^\infty \frac{(a;q)_k}{(q;q)_k} x^k = \frac{(ax;q)_\infty}{(x;q)_\infty},$$

where $(a;q)_\infty = \prod_{k=0}^\infty (1 - aq^k)$.

First Proof.   Let

$$f_a(x) = \sum_{k=0}^\infty \frac{(a;q)_k}{(q;q)_k} x^k.$$

Apply the q-difference operator Δ_q to both sides. Then

$$\frac{f_a(x) - f_a(qx)}{x} = \sum_{k=0}^{\infty} \frac{(a; q)_k}{(q; q)_k}(1 - q^k)x^{k-1}$$

$$= (1 - a)\sum_{k=1}^{\infty} \frac{(aq; q)_{k-1}}{(q; q)_{k-1}}x^{k-1}$$

$$= (1 - a)\sum_{k=0}^{\infty} \frac{(aq; q)_k}{(q; q)_k}x^k = (1 - a)f_{aq}(x),$$

or

$$f_a(x) - f_a(qx) = (1 - a)xf_{aq}(x).$$

Now consider

$$f_a(x) - f_{aq}(x) = \sum_{k=0}^{\infty} \frac{(aq; q)_{k-1}}{(q; q)_k}(1 - a - 1 + aq^k)x^k$$

$$= -axf_{aq}(x),$$

or

$$f_a(x) = (1 - ax)f_{aq}(x).$$

Eliminate $f_{aq}(x)$ from the two equations to get

$$f_a(x) = \frac{1 - ax}{1 - x}f_a(qx).$$

Iterate this relation n times and let $n \to \infty$ to arrive at

$$f_a(x) = \frac{(ax; q)_n}{(x; q)_n}f_a(q^nx) = \frac{(ax; q)_\infty}{(x; q)_\infty}f_a(0) = \frac{(ax; q)_\infty}{(x; q)_\infty}.$$

This proves the theorem. ∎

Second Proof. The infinite product $(ax; q)_\infty/(x; q)_\infty$ is uniformly and absolutely convergent for fixed a and q in $|x| \leq 1 - \epsilon$ and so represents an analytic function in $|x| < 1$. Consider its Taylor expansion in $|x| < 1$,

$$F(x) = \frac{(ax; q)_\infty}{(x; q)_\infty} = \sum_{n=0}^{\infty} A_n x^n.$$

Clearly,

$$F(x) = \frac{(1 - ax)}{(1 - x)}F(qx).$$

This implies

$$(1 - x) \sum_{n=0}^{\infty} A_n x^n = (1 - ax) \sum_{n=0}^{\infty} A_n q^n x^n.$$

Equate the coefficients of x^n on both sides. Then

$$A_n = \frac{(1 - aq^{n-1})}{1 - q^n} A_{n-1}$$

$$= \frac{(a; q)_n}{(q; q)_n}.$$

This completes the second proof. ∎

Remark 10.2.1   The infinite product in the q-binomial theorem also arises naturally when we look for the analog of $(1 - x)^{-\alpha}$. To see this, suppose α is a positive integer n. A possible q-analog of $(1 - x)^{-n}$ is

$$\frac{1}{(1 - x)(1 - qx) \cdots (1 - q^{n-1}x)} = \frac{(1 - q^n x)(1 - q^{n+1}x) \cdots}{(1 - x)(1 - qx) \cdots} = \frac{(q^n x; q)_\infty}{(x; q)_\infty}.$$

The last expression is meaningful even if n is not an integer, so more generally we consider $(ax; q)_\infty / (x; q)_\infty$.

There are many interesting special cases of Theorem 10.2.1.

Corollary 10.2.2

(a)
$$\sum_{n=0}^{\infty} \frac{x^n}{(q; q)_n} = \frac{1}{(x; q)_\infty}, \qquad |x| < 1, |q| < 1. \tag{Euler}$$

(b)
$$\sum_{n=0}^{\infty} \frac{(-1)^n q^{\binom{n}{2}} x^n}{(q; q)_n} = (x; q)_\infty, \qquad |q| < 1. \tag{Euler}$$

(c)
$$\sum_{k=0}^{N} \begin{bmatrix} N \\ k \end{bmatrix}_q (-1)^k q^{\binom{k}{2}} x^k = (x; q)_N = (1 - x) \cdots (1 - xq^{N-1}). \tag{Rothe}$$

(d)
$$\sum_{k=0}^{\infty} \begin{bmatrix} N + k - 1 \\ k \end{bmatrix}_q x^k = \frac{1}{(x; q)_N} = \frac{1}{(1 - x) \cdots (1 - xq^{N-1})}, \qquad |x| < 1,$$

where the q-binomial coefficient is

$$\begin{bmatrix} n \\ k \end{bmatrix}_q = (q; q)_n / (q; q)_k (q; q)_{n-k}.$$

Proof.

(a) Set $a = 0$ in Theorem 10.2.1.
(b) Replace a with $1/a$, and x with ax and then set $a = 0$.

(c) Set $a = q^{-N}$.

(d) Set $a = q^N$.   ∎

Remark 10.2.2   The q-binomial theorem was apparently discovered independently by several mathematicians including Gauss [1866], Cauchy [1843], and Heine [1847]. It seems that the first statement of the q-binomial theorem in approximately the form given in Corollary 10.2.2(c) was published by Rothe [1811]. (He stated this as (10.0.10), but with a misprint.) For the references to Euler in Corollary 10.2.2, see Andrews [1976, p. 30].

In most cases we do not give detailed proofs to justify limiting processes. It is, however, interesting and important to know what is involved here. Hence we end this section with Koornwinder's [1990] proof of the fact that the ordinary binomial theorem is obtained from the q-binomial theorem as $q \to 1^-$.

The proof of the theorem depends on the following lemma; we omit the proof.

Lemma 10.2.3   *Suppose μ, λ, k are real; $0 \le \mu - \lambda \le k$; $\mu + \lambda \ge 1$; and*

$$f(t) = \frac{e^{-\mu t} - e^{-(\lambda+k)t}}{1 - e^{(k+1)t}}, \quad t > 0.$$

Then $f'(t) \le 0$ if $t > 0$.

Theorem 10.2.4   *Suppose λ and μ are real. Then*

$$\lim_{q \to 1^-} \frac{(q^\lambda x; q)_\infty}{(q^\mu x; q)_\infty} = (1 - x)^{\mu - \lambda},$$

uniformly on $\{x \in \mathbb{C} : |x| \le 1\}$, if $\mu \ge \lambda$, $\mu + \lambda \ge 1$, and uniformly on compact subsets of $\{x \in \mathbb{C} : |x| \le 1, x \ne 1\}$ for other choices of λ and μ.

Proof.   First observe that since

$$\frac{(q^\lambda x; q)_\infty}{(q^\mu x; q)_\infty} = \frac{(q^\lambda x; q)_\ell}{(q^\mu x; q)_m} \frac{(q^{\lambda+\ell} x; q)_\infty}{(q^{\mu+m} x; q)_\infty} \tag{10.2.4}$$

we can choose ℓ and m appropriately so that $\mu + m \ge \lambda + \ell$ and $\mu + \lambda + \ell + m \ge 1$. Moreover, the first quotient on the right-hand side of (10.2.4) tends to $(1 - x)^{\ell - m}$ uniformly on compact subsets of $\{x \in \mathbb{C} : |x| \le 1, x \ne 1\}$ as $q \to 1^-$. Consequently, we need only consider the case where $\mu > \lambda$ and $\mu + \lambda \ge 1$. By the q-binomial theorem, the left side of (10.2.4) is

$$\frac{(q^{\lambda-\mu} q^\mu x; q)_\infty}{(q^\mu x; q)_\infty} = 1 + \sum_{n=1}^\infty \frac{q^\mu - q^\lambda}{1 - q} \cdot \frac{q^\mu - q^{\lambda+1}}{1 - q^2} \cdots \frac{q^\mu - q^{\lambda+n-1}}{1 - q^n} \cdot x^n.$$

$$\tag{10.2.5}$$

It is easy to check that $(q^\mu - q^{\lambda+k})/(1 - q^{k+1})$ increases with q for $\lambda + k \geq \mu$ and so

$$\frac{q^\mu - q^{\lambda+k}}{1 - q^{k+1}} \leq \lim_{q \to 1^-} \frac{q^\mu - q^{\lambda+k}}{1 - q^{k+1}} = \frac{\lambda - \mu + k}{k + 1}.$$

Let m be the largest integer such that $\lambda + m - 1 < \mu$. Then for $|x| \leq 1$

$$\frac{q^\mu - q^\lambda}{1 - q} \cdots \frac{q^\mu - q^{\lambda+m-1}}{1 - q^m} \cdot \frac{q^\mu - q^{\lambda+m}}{1 - q^{m+1}} \cdots \frac{q^\mu - q^{\lambda+n-1}}{1 - q^n} \cdot x^n$$
$$\leq \frac{q^\mu - q^\lambda}{1 - q} \cdots \frac{q^\mu - q^{\lambda+m-1}}{1 - q^m} \cdot \frac{\lambda - \mu + m}{m + 1} \cdots \frac{\lambda - \mu + n - 1}{n}.$$

Thus the series (10.2.5) from the mth term onward is majorized by the convergent series

$$\sup_{0<q<1} \left| \frac{q^\mu - q^\lambda}{1 - q} \cdots \frac{q^\mu - q^{\lambda+m-1}}{1 - q^m} \right| \sum_{n=m}^{\infty} \frac{(\lambda - \mu + m)_{n-m}}{(m + 1)_{n-m}} \cdot x^n.$$

This implies that we can take the termwise limit in (10.2.5) to get

$$\lim_{q \to 1^-} \frac{(q^\lambda x; q)_\infty}{(q^\mu x; q)_\infty} = 1 + \sum_{n=1}^{\infty} \frac{(\mu - \lambda)_n}{n!} x^n = (1 - x)^{\mu - \lambda}. \tag{10.2.6}$$

This proves the theorem. ∎

In a similar way, if we replace x with $(1 - q)x$ in Corollary 10.2.2(a) and let $q \to 1^-$, the series converges to e^x:

$$\lim_{q \to 1^-} \frac{1}{((1 - q)x; q)_\infty} = e^x, \tag{10.2.7}$$

and we have a q-analog of the exponential function. The sum in Corollary 10.2.2(b) also gives a series that converges to the exponential series, but the infinite product form is equivalent to (10.2.7). Similarly,

$$\lim_{q \to 1^-} (-(1 - q)x; q)_\infty = e^x. \tag{10.2.8}$$

The functions used in (10.2.7) and (10.2.8) are occasionally given other names:

$$e_q(x) := \frac{1}{((1 - q)x; q)_\infty}, \tag{10.2.9}$$

$$E_q(x) := (-(1 - q)x; q)_\infty. \tag{10.2.10}$$

10.3 The q-Gamma Function

We return to the problem of finding an analog of the beta integral over $(0, 1)$. By
Theorem 10.2.4 it is reasonable to replace

$$x^{\alpha-1}(1-x)^{\beta-1}$$

with

$$x^{\alpha-1}(qx; q)_\infty/(q^\beta x; q)_\infty.$$

We write the q-binomial theorem as

$$\sum_{n=0}^{\infty} \frac{(q^{n+1}; q)_\infty x^n}{(q^n a; q)_\infty} = \frac{(ax; q)_\infty (q; q)_\infty}{(x; q)_\infty (a; q)_\infty}. \tag{10.3.1}$$

Replace x with q^α and a with q^β in (10.3.1) to get

$$\int_0^1 x^{\alpha-1} \frac{(qx; q)_\infty}{(q^\beta x; q)_\infty} d_q x = \frac{(1-q)(q^{\alpha+\beta}; q)_\infty (q; q)_\infty}{(q^\alpha; q)_\infty (q^\beta; q)_\infty}. \tag{10.3.2}$$

This is the q-extension of

$$\int_0^1 x^{\alpha-1}(1-x)^{\beta-1} dx = \frac{\Gamma(\alpha)\Gamma(\beta)}{\Gamma(\alpha+\beta)}.$$

To write (10.3.2) in this form we need the q-version of the gamma function.
The existence of a useful $n!_q$ indicates the possibility of a convenient analog of
the gamma function. We follow Euler's procedure and look for an interpolation
formula by using infinite products. Now

$$n!_q = \frac{(q; q)_n}{(1-q)^n} = \frac{(q; q)_\infty}{(1-q)^n (q^{n+1}; q)_\infty}, \quad 0 < q < 1.$$

The last expression does not require n to be a positive integer, so we set

$$\Gamma_q(x) := \frac{(q; q)_\infty}{(q^x; q)_\infty} (1-q)^{1-x} \quad \text{when } |q| < 1. \tag{10.3.3}$$

Here we take the principal values of q^x and $(1-q)^{1-x}$. Then $\Gamma_q(x)$ is a meromorphic function with obvious poles at $x = -n \pm 2\pi ik/\log q$, where k and n are
nonnegative integers. It is not difficult to see that the residue at $x = -n$ is

$$\frac{(1-q)^{n+1}}{(q^{-n}; q)_n \log q^{-1}}.$$

Because $\Gamma_q(x)$ has no zeros, its reciprocal is an entire function. It is left to the
reader to check these facts. These properties of $\Gamma_q(x)$ are similar to those of $\Gamma(x)$.
We can now write (10.3.2), which is just another form of the q-binomial theorem,
as follows:

Theorem 10.3.1

$$B_q(\alpha, \beta) := \int_0^1 x^{\alpha-1} \frac{(qx; q)_\infty}{(q^\beta x; q)_\infty} d_q x = \frac{\Gamma_q(\alpha)\Gamma_q(\beta)}{\Gamma_q(\alpha + \beta)}.$$

Remark 10.3.1   At first sight it might appear that we could have replaced $(1 - x)^{\beta-1}$ with $(q^{1-\beta}x; q)_\infty/(x; q)_\infty$, but this is not as useful. The function $(1-x)^{\beta-1}$ is positive in the interval of integration $(0, 1)$ and $1 - x$ vanishes at 1. In the q-integral the set of points over which the summation is carried out is q^n, $n = 0, 1, 2, \ldots$, that is, a discrete set of points in $[0, 1]$. The first point to the right of the interval where we want our function to vanish is q^{-1}. To get such a function replace $xq^{-\beta}$ by x in both infinite products. The result is $(qx; q)_\infty/(q^\beta x; q)_\infty$. The function $(q^{1-\beta}x; q)_\infty/(x, q)_\infty$ could have been used, but the q-integral would be on $[0, q^\beta]$.

Theorem 10.3.1 gives us one reason to accept $\Gamma_q(x)$ as the natural q-analog of $\Gamma(x)$. Another reason is the following: The Bohr–Mollerup theorem states that $\Gamma(x)$ is the unique function satisfying the functional equation

$$f(x + 1) = xf(x), \quad f(1) = 1,$$

and is also logarithmically convex. It can be shown that $\Gamma_q(x)$ is the only function that satisfies the functional equation

$$f_q(x + 1) = \frac{1 - q^x}{1 - q} f_q(x), \quad f_q(1) = 1,$$

and is also logarithmically convex. The proof of the latter result is identical with that of the Bohr–Mollerup theorem and was included in Chapter 1. See Exercise 10.

From the definition of $\Gamma_q(x)$ and the Bohr–Mollerup theorem, one may suspect that $\lim_{q \to 1^-} \Gamma_q(x) = \Gamma(x)$, and this is indeed true. We derive it as a consequence of the next theorem, the proof of which requires the following lemma.

Lemma 10.3.2   If $f(1) = f(2) = g(1) = g(2) = 0$ and $0 \le f''(x) \le g''(x)$ for $x > 0$, then $f(x) \le g(x)$ in $[0, 1] \cup [2, \infty)$ and $f(x) \ge g(x)$ in $[1, 2]$.

Proof. For $x \in [1, 2]$, it is easy to see that

$$f(x) = \int_1^2 h(x, t) f''(t) dt$$

with

$$h(x, t) = \begin{cases} (x - 2)(t - 1), & 1 \le t < x \le 2, \\ (t - 2)(x - 1), & 1 \le x < t \le 2. \end{cases}$$

Since $h(x, t)$ is negative for x and t in $[1, 2]$, it follows that $f(x) \ge g(x)$ in the interval $[1, 2]$. We may assume without loss of generality that $f(x) \equiv 0$. We have just shown that, in that case, $g(x) \leq 0$ in $[1, 2]$ and $g = 0$ at the endpoints. By the mean-value theorem, g' is zero somewhere in $(1, 2)$, and since g' is increasing we must have $g'(x) \leq 0$ for $0 < x < 1$ and $g'(x) \geq 0$ for $x \geq 2$. So g is decreasing in $(0, 1)$ and increasing in $(2, \infty)$. This implies the result. ∎

Theorem 10.3.3 *For $0 < r < q < 1$, we have*

$$\Gamma_r(x) \leq \Gamma_q(x) \leq \Gamma(x), \qquad 0 < x \leq 1 \quad \text{or} \quad x \geq 2,$$

and

$$\Gamma(x) \leq \Gamma_q(x) \leq \Gamma_r(x), \qquad 1 \leq x \leq 2.$$

Proof. Observe that

$$\frac{d^2}{dx^2} \log \Gamma_q(x) = (\log q)^2 \sum_{n=0}^{\infty} \frac{q^{x+n}}{(1 - q^{x+n})^2} > 0.$$

We show that each term of the series

$$h(q) = \frac{(\log q)^2 q^{x+n}}{(1 - q^{x+n})^2}$$

is increasing in $(0, 1)$. Set $a = n + x$; then

$$h'(q) = \frac{aq^{a-1}(\log q)(1 + q^a)}{(1 - q^a)^3} \left[\frac{2(1 - q^a)}{a(1 + q^a)} + \log q \right].$$

To prove that $h' > 0$, it is sufficient to demonstrate that the expression within the square brackets, which we denote by $g(q)$, is negative. A simple calculation gives

$$g'(q) = \frac{(1 - q^a)^2}{q(1 + q^a)^2} \geq 0, \qquad q > 0.$$

Since $g(1) = 0$, we have $g(q) \leq 0$ in $(0, 1]$. Thus $h(q)$ is increasing and so $\frac{d^2}{dx^2} \log \Gamma_q(x)$ is increasing for $0 < q < 1$. This means that

$$\frac{d^2}{dx^2} \log \Gamma_r(x) \leq \frac{d^2}{dx^2} \log \Gamma_q(x), \qquad 0 < r < q < 1, \quad x > 0.$$

Moreover,

$$\log \Gamma_r(1) = \log \Gamma_q(1) = \log \Gamma_r(2) = \log \Gamma_q(2) = 0.$$

The theorem is now a consequence of Lemma 10.3.2. ∎

Corollary 10.3.4 $\lim_{q \to 1^-} \Gamma_q(x) = \Gamma(x).$

Proof. Theorem 10.3.3 implies that $\lim_{q \to 1^-} \Gamma_q(x) = \lambda(x)$ exists. Moreover, $\lambda(x)$ satisfies the conditions of the Bohr–Mollerup theorem. Thus $\lambda(x) = \Gamma(x)$ and the corollary is proved for $0 < x < \infty$. For real nonintegral values, the functional equations then give the same result. For complex x, the Stieltjes–Vitali theorem completes the proof. See Hille [1962, p. 251] for this theorem. (For Gosper's proof of Corollary 10.3.4, see Andrews [1986, p. 109].) ∎

There are several results about $\Gamma_q(x)$ that are analogs of corresponding statements about $\Gamma(x)$. The analogs of the Legendre duplication formula and the Gauss multiplication theorem are given in the theorem below.

Theorem 10.3.5

(a) $\Gamma_q(2x)\Gamma_{q^2}(1/2) = (1+q)^{2x-1}\Gamma_{q^2}(x)\Gamma_{q^2}(x+1/2)$.
(b) Let $r = q^n$. Then

$$\Gamma_q(nx)\Gamma_r(1/n)\Gamma_r(2/n)\cdots\Gamma_r((n-1)/n)$$
$$= (1+q+\cdots+q^{n-1})^{nx-1}\Gamma_r(x)\Gamma_r(x+1/n)\cdots\Gamma_r(x+(n-1)/n).$$

The proofs are straightforward and left to the reader as exercises. The next two formulas can be regarded as asymptotic formulas for $\Gamma_q(x)$ for large x, but they are of a different nature from Stirling's formula. They follow from Corollary 10.2.2:

$$\Gamma_q(x) = (q;q)_\infty(1-q)^{1-x}\sum_{n=0}^{\infty}\frac{q^{nx}}{(q;q)_n}, \qquad \text{Re } x > 0, \tag{10.3.4}$$

$$\frac{1}{\Gamma_q(x)} = \frac{(1-q)^{x-1}}{(q;q)_\infty}\sum_{n=0}^{\infty}\frac{(-1)^n q^{\binom{n}{2}}q^{nx}}{(q;q)_n}, \qquad \text{Re } x > 0. \tag{10.3.5}$$

Now recall that $1/\Gamma(x)$ has zeros at $x = 0, -1, -2, \ldots$ whereas $1/\Gamma(1-x)$ has zeros at $x = 1, 2, 3, \ldots$. Thus their product has a zero at each integer. This fact is among the properties we see reflected in Euler's formula

$$\Gamma(x)\Gamma(1-x) = \pi/\sin\pi x.$$

Similarly, $1/\Gamma_q(x)$ has zeros at $x = -n$ due to the factor $(q^x; q)_\infty$. (In fact, $1/\Gamma_q(x)$ has zeros at $x = -n \pm 2\pi i k/\log q$ where k and n are nonnegative integers.) Hence the function with the full range of integer points as zeros may be taken as $(q^x; q)_\infty(q^{1-x}; q)_\infty = (q^x, q)_\infty(q/q^x; q)_\infty$. Replace q^x with y to write the function as $(y; q)_\infty(q/y; q)_\infty$. We expect this function to have interesting properties. In fact, it is one of the theta functions discovered by Gauss and Jacobi and is the topic of the next section.

10.4 The Triple Product Identity

The triple product identity expresses $(x; q)_\infty(q/x; q)_\infty(q; q)_\infty$ as a Laurent series in $0 < |x| < \infty$. One proof of this identity follows from the terminating

q-binomial theorem due to Rothe given in Corollary 10.2.2. This proof was known to Gauss [1866a] and Cauchy [1843a]. At the end of the previous section we remarked that the infinite product $(x; q)_\infty (q/x; q)_\infty$ arises naturally when we look for a q-analog of the Euler reflection formula. We shall see that the Laurent-series side of the identity also appears naturally from a particular Riemann sum approximation of the normal integral. This raises the question of whether the triple product identity could be approached from the series side. The answer is yes and this leads to another proof of the identity. We end the section with a number of identities that are important but simple consequences of the triple product identity. In later sections we give applications to number theory and combinatorics.

Theorem 10.4.1   *For $|q| < 1$ and $x \in \mathbb{C} - \{0\}$,*

$$(x; q)_\infty (q/x; q)_\infty (q; q)_\infty = \sum_{k=-\infty}^{\infty} (-1)^k q^{\binom{k}{2}} x^k.$$

Proof.   Take $N = 2n$ in Corollary 10.2.2(c) to obtain

$$(x; q)_{2n} = \sum_{k=-n}^{n} \begin{bmatrix} 2n \\ n+k \end{bmatrix}_q (-1)^{k+n} q^{(k+n)(k+n-1)/2} x^{k+n}.$$

Then replace x by xq^{-n} and rewrite $(xq^{-n}; q)_{2n}$ as

$$(xq^{-n}; q)_n (x; q)_n = (-1)^n x^n q^{-n^2+n(n-1)/2} (q/x; q)_n (x; q)_n.$$

The above identity then becomes

$$(q/x; q)_n (x; q)_n = \sum_{k=-n}^{n} \frac{(q; q)_{2n} (-1)^k q^{k(k-1)/2} x^k}{(q; q)_{n+k} (q; q)_{n-k}}.$$

When $n \to \infty$, this gives

$$(x; q)_\infty (q/x; q)_\infty = \sum_{k=-\infty}^{\infty} \frac{(-1)^k q^{\binom{k}{2}} x^k}{(q; q)_\infty}.$$

This limiting process can be justified by Tannery's theorem. The result in Theorem 10.4.1 is called the triple product identity.   ■

Remark 10.4.1   Replace q with e^{-2t} and x with $-e^{-t} e^{i\theta}$ in the identity. The result is

$$\sum_{k=-\infty}^{\infty} e^{-n^2 t} e^{in\theta} = \prod_{n=0}^{\infty} \left(1 + 2e^{-(2n+1)t} \cos\theta + e^{-(4n+2)t}\right) \left(1 - e^{-2(n+1)t}\right).$$

The left side is a solution of the heat equation

$$\frac{\partial^2 u}{\partial \theta^2} = \frac{\partial u}{\partial t}.$$

The right side is positive for $t > 0$ since

$$1 + 2r\cos\theta + r^2 \geq 1 - 2r + r^2 = (1 - r)^2 > 0,$$

when $0 \leq r < 1$. The positivity is not evident from the left side. Thus the two sides give different properties of the function. Clearly, the right side also gives the zeros of the function.

In view of the importance of the triple product identity, we look at it from another point of view. First replace q with q^2 and x with $-qx$ to get

$$(-qx; q^2)_\infty(-q/x; q^2)_\infty(q^2; q^2)_\infty = \sum_{n=-\infty}^{\infty} q^{n^2} x^n. \qquad (10.4.1)$$

The sum on the right of (10.4.1) comes from an important integral. Consider the normal integral

$$\sqrt{\pi} = \int_{-\infty}^{\infty} e^{-x^2} dx.$$

Shift x by $a/2$ to get

$$e^{a^2/4}\sqrt{\pi} = \int_{-\infty}^{\infty} e^{-x^2 - ax} dx.$$

Replace this integral with the approximation formed by summing over a discrete one-dimensional lattice with space size δ:

$$\delta \sum_{n=-\infty}^{\infty} e^{-\delta^2 n^2 - a\delta n}.$$

It is natural to ask here how close the sum is to the integral when δ is small. To answer this, consider the formula in Exercise 2.26:

$$\sum_{n=-\infty}^{\infty} e^{-\pi t(n+\alpha)^2} = \frac{1}{\sqrt{t}} \sum_{n=-\infty}^{\infty} e^{-\pi n^2/t} e^{2\pi i n\alpha}. \qquad (10.4.2)$$

When t is small the terms on the left are close to one for small values of n but all the terms on the right with one exception are very small. So the expression on the right works very well for numerical calculation of the series for small δ. Later we shall see that there is a deeper reason for the importance of the transformation formula (10.4.2). It shows that $\sum_{-\infty}^{\infty} q^{n^2}$ is a modular form.

Let us now return to the sum in Theorem 10.4.1 when $x = 1$. The sum is $A = \sum_{-\infty}^{\infty}(-1)^n q^{\binom{n}{2}}$. Apply the two changes in variables $n \to -n$ and $n \to n+1$ to get

$$A = \sum_{n=-\infty}^{\infty} (-1)^n q^{n(n+1)/2} = \sum_{n=-\infty}^{\infty} (-1)^{n+1} q^{n(n+1)/2} = -A.$$

Thus A is zero. Write

$$H(x) = \sum_{n=-\infty}^{\infty} (-1)^n q^{\frac{n(n-1)}{2}} x^n.$$

Then

$$H(qx) = -\frac{1}{x} \sum_{n=-\infty}^{\infty} (-1)^{n+1} q^{\frac{n(n+1)}{2}} x^{n+1} = -\frac{1}{x} H(x),$$

or

$$H(x) = -xH(qx). \tag{10.4.3}$$

This equation implies that if x is a root of $H(x) = 0$, then so are qx and x/q. Since we know that $x = 1$ is a root, it follows that q^n for every integer n is a root. Thus $H(x)$ has $(x; q)_\infty (q/x; q)_\infty = T(x)$ as a factor. Without knowledge of Theorem 10.4.1 we cannot be sure that $H(x)$ has no other zeros. However, it can be shown by a simple calculation that

$$T(x) = -xT(qx).$$

That is, $T(x)$ satisfies the same functional equation as H. It follows that the Laurent-series expansion of (10.4.1) is uniquely determined in a deleted neighborhood of $x = 0$ up to a constant factor. Consequently,

$$T(x) = C_0(q)H(x).$$

Replace q with q^2 and x with $-qx$ to get

$$(-qx; q^2)_\infty (-q/x; q^2)_\infty = C_0(q^2) \sum_{n=-\infty}^{\infty} q^{n^2} x^n. \tag{10.4.4}$$

There are a number of ways of finding $C_0(q^2)$. We apply a device due to Gauss and Jacobi. Another method, also due to Gauss [1866b], using the arithmetic–geometric mean is given in Exercise 13. There also exist some combinatorial methods but we shall not describe them here. One such method is given in the next chapter.

Let $x = i$ in (10.4.4). The result is

$$(-iq; q^2)_\infty (iq; q^2)_\infty = C_0(q^2) \left[\sum_{n=-\infty}^{\infty} (-1)^n q^{4n^2} + i \sum_{n=-\infty}^{\infty} (-1)^n q^{(2n+1)^2} \right].$$

Since the left side is real for q real, we get

$$(-iq; q^2)_\infty (iq; q^2)_\infty = C_0(q^2) \sum_{n=-\infty}^{\infty} (-1)^n q^{4n^2}.$$

The left side is identical to $(-q^2; q^4)_\infty$. Now set $x = -1$ and replace q with q^4 in (10.4.4) to get

$$(q^4; q^8)_\infty^2 = C_0(q^8) \sum_{n=-\infty}^{\infty} (-1)^n q^{4n^2}.$$

The last two identities imply

$$\frac{C_0(q^2)}{C_0(q^8)} = \frac{(-q^2; q^4)_\infty}{(q^4; q^8)_\infty^2} \cdot \frac{(q^2; q^4)_\infty}{(q^2; q^4)_\infty} = \frac{(q^8; q^8)_\infty}{(q^2; q^2)_\infty}.$$

This gives

$$C_0(q^2)(q^2; q^2)_\infty = C_0(q^{2 \cdot 2^2})(q^{2 \cdot 2^2}; q^{2 \cdot 2^2})_\infty$$
$$= C_0\left(q^{2 \cdot 2^n}\right)\left(q^{2 \cdot 2^n}; q^{2 \cdot 2^n}\right)_\infty.$$

Note that $(q; q)_\infty$ and $\sum_{k=-\infty}^{\infty} q^{k^2} x^k$ are continuous functions of q in $|q| < 1$. Thus $C_0(q^2)$ is also continuous, which implies that $C_0(0) = 1$. Let $n \to \infty$ to get

$$C_0(q^2)(q^2; q^2)_\infty = 1.$$

This gives $C_0(q^2) = 1/(q^2; q^2)_\infty$ and we have another proof of the triple product identity.

Corollary 10.4.2

$$\sum_{n=-\infty}^{\infty} q^{n^2} = \prod_{n=1}^{\infty} (1 - q^{2n})(1 + q^{2n-1})^2, \qquad \text{(Gauss)} \quad (10.4.5)$$

$$\sum_{n=-\infty}^{\infty} (-1)^n q^{n^2} = \prod_{n=1}^{\infty} (1 - q^{2n})(1 - q^{2n-1})^2, \qquad \text{(Gauss)} \quad (10.4.6)$$

$$\sum_{n=-\infty}^{\infty} (-1)^n q^{n(3n+1)/2} = \prod_{n=1}^{\infty} (1 - q^n), \qquad \text{(Euler)} \quad (10.4.7)$$

$$\sum_{n=0}^{\infty} q^{n(n+1)/2} = \prod_{n=1}^{\infty} [(1 - q^{2n})/(1 - q^{2n+1})], \qquad \text{(Gauss)} \quad (10.4.8)$$

$$\sum_{n=0}^{\infty} (-1)^n (2n+1) q^{n(n+1)/2} = \prod_{n=1}^{\infty} (1 - q^n)^3. \qquad \text{(Jacobi)} \quad (10.4.9)$$

Proof. The identities in (10.4.5) and (10.4.6) are obvious from (10.4.4). Note that (10.4.6) also follows from (10.4.5) by changing q to $-q$. To get (10.4.7) replace q with $q^{3/2}$ and set $x = -\sqrt{q}$. For (10.4.8) and (10.4.9), write the identity in

Theorem 10.4.1, with x replaced by $-qx$, as

$$\sum_{n=0}^{\infty} q^{n(n+1)/2}(x^n + x^{-n-1}) = (1 + 1/x)\prod_{n=1}^{\infty}(1 - q^n)(1 + q^n/x)(1 + q^n x).$$

$$(10.4.10)$$

Set $x = 1$. Then

$$\sum_{n=0}^{\infty} q^{n(n+1)/2} = \prod_{n=1}^{\infty}[(1 - q^{2n})^2/(1 - q^n)]$$

$$= \prod_{n=1}^{\infty}[(1 - q^{2n})/(1 - q^{2n+1})].$$

This is (10.4.8). Now divide (10.4.10) by $x + 1$ and let $x \to -1$ to get (10.4.9). The corollary is proved. ∎

The sequences of numbers $\{n^2\}$, $\{n(n + 1)/2\}$, $\{n(3n \pm 1)/2\}$ are the square, triangular, and pentagonal numbers, respectively. These are of number theoretic interest and their appearance as powers of q in the series make the above identities useful in combinatorial number theory.

Remark 10.4.2 Gauss and Jacobi independently discovered the triple product identity. Jacobi's results including (10.4.9) are contained in his famous book, Fundamenta Nova (Jacobi [1829]). See Gauss [1866] for (10.4.1), (10.4.5), and (10.4.6). Identity (10.4.8) was published in Gauss [1808], a paper noted for containing the first evaluation of the quadratic Gauss sum. See Exercises 5 and 6. Euler's result (10.4.7) is the famous pentagonal number theorem. A discussion of Euler's proof and references are given in Weil [1983, p. 281]. See also Euler [1748].

10.5 Ramanujan's Summation Formula

The q-binomial theorem evaluated a q-analog of the beta integral over $(0, 1)$. We have seen that the q-integral over $(0, \infty)$ is a bilateral series. Therefore, a q-analog of the beta integral over $(0, \infty)$, that is,

$$\int_0^{\infty} \frac{x^{\alpha-1}}{(1 + x)^{\alpha+\beta}} dx,$$

should be a bilateral series, but similar to the series in the q-binomial theorem. The correct generalization was found by Ramanujan. He considered the bilateral sum

$$\sum_{n=-\infty}^{\infty} \frac{(a; q)_n}{(b; q)_n} x^n,$$

$$(10.5.1)$$

which, as we shall see later, is a q-integral analog of

$$\frac{B(\alpha, \beta)}{c^\alpha} = \int_0^\infty \frac{x^{\alpha-1}}{(1+cx)^{\alpha+\beta}} dx.$$

We should first clarify the meaning of $(a; q)_n$ in (10.5.1) for negative n. Since

$$(a; q)_n = (a; q)_\infty / (aq^n; q)_\infty \tag{10.5.2}$$

for $n > 0$, and the right side is meaningful for negative n as well, we take (10.5.2) as the definition of $(a; q)_n$ for all n. If $n = -m$, then

$$(a; q)_{-m} = \frac{1}{(aq^{-m}; q)_m} = \frac{(-1)^m q^{\binom{m}{2}}}{a^m (a^{-1}q; q)_m}.$$

Ramanujan's evaluation of (10.5.1) in terms of infinite products is given next. It contains the q-binomial theorem and the triple product identity as special cases.

Theorem 10.5.1   *For $|q| < 1$ and $|ba^{-1}| < |x| < 1$,*

$$\sum_{n=-\infty}^\infty \frac{(a; q)_n}{(b; q)_n} x^n = \frac{(ax; q)_\infty (q/ax; q)_\infty (q; q)_\infty (b/a; q)_\infty}{(x; q)_\infty (b/ax; q)_\infty (b; q)_\infty (q/a; q)_\infty}. \tag{10.5.3}$$

First Proof.   Use (10.5.2) to write the series as

$$\sum_{n=0}^\infty \frac{(a; q)_n}{(b; q)_n} x^n + \sum_{n=1}^\infty \frac{(b^{-1}q; q)_n}{(a^{-1}q; q)_n} \left(\frac{b}{ax}\right)^n.$$

The first series converges for $|x| < 1$ and the second for $|b/ax| < 1$. So the bilateral series converges when $|ba^{-1}| < |x| < 1$.

Observe that

$$\sum_{n=-\infty}^\infty \frac{(a; q)_n}{(b; q)_n} x^n = \frac{(a; q)_\infty}{(b; q)_\infty} \sum_{n=-\infty}^\infty \frac{(bq^n; q)_\infty}{(aq^n; q)_\infty} x^n$$

$$= \frac{(a; q)_\infty}{(b; q)_\infty} f(b).$$

This proof of Ramanujan's identity depends on a functional relation satisfied by $f(b)$. To find it, note that

$$f(b) = \sum_{n=-\infty}^\infty \frac{(bq^{n+1}; q)_\infty}{(aq^n; q)_\infty} x^n [1 - b(q^n - a^{-1}) - ba^{-1}]$$

$$= \left(1 - \frac{b}{a}\right) f(bq) + \frac{b}{ax} \sum_{n=-\infty}^\infty \frac{(bq^{n+1}; q)_\infty}{(aq^{n+1}; q)_\infty} x^{n+1}$$

$$= (1 - b/a) f(bq) + (b/ax) f(b).$$

This gives the desired functional equation, namely

$$f(b) = \frac{(1-b/a)}{(1-b/ax)} f(bq).$$

Now $f(b)$ is an analytic function of b for $|b|$ sufficiently small. Iteration gives

$$f(b) = \frac{(b/a; q)_\infty}{(b/ax; q)_\infty} f(0). \qquad (10.5.4)$$

It is not easy to sum $f(0)$, but $f(q)$ can be obtained from the q-binomial theorem. Note that

$$f(q) = \sum_{n=-\infty}^{\infty} \frac{(q^{n+1}; q)_\infty}{(aq^n; q)_\infty} x^n = \sum_{n=0}^{\infty} \frac{(q^{n+1}; q)_\infty}{(aq^n; q)_\infty} x^n$$

$$= \frac{(q; q)_\infty}{(a; q)_\infty} \sum_{n=0}^{\infty} \frac{(a; q)_n}{(q; q)_n} x^n = \frac{(q; q)_\infty}{(a; q)_\infty} \cdot \frac{(ax; q)_\infty}{(x; q)_\infty}.$$

From (10.5.4)

$$f(0) = \frac{(q/ax; q)_\infty}{(q/a; q)_\infty} f(q)$$

$$= \frac{(q/ax; q)_\infty (q; q)_\infty (ax; q)_\infty}{(q/a; q)_\infty (a; q)_\infty (x; q)_\infty}.$$

Use this in (10.5.4) to get $f(b)$ and then

$$\sum_{n=-\infty}^{\infty} \frac{(a; q)_n}{(b; q)_n} x^n = \frac{(ax; q)_\infty (q/ax; q)_\infty (q; q)_\infty (b/a; q)_\infty}{(x; q)_\infty (b/ax; q)_\infty (b; q)_\infty (q/a; q)_\infty}.$$

This argument assumed that we could take $b = q$. To prove the general case apply analytic continuation on b and x. ■

Second Proof (Venkatachaliengar). As in the case of the q-binomial theorem, we can start with the infinite product and get the Laurent series. Suppose that

$$F(x) = \frac{(ax; q)_\infty (q/ax; q)_\infty}{(x; q)_\infty (b/ax; q)_\infty} = \sum_{n=-\infty}^{\infty} A_n x^n. \qquad (10.5.5)$$

The Laurent series is defined in $|x| < 1$ and $|b/ax| < 1$, that is, $|b/a| < |x| < 1$. We consider the Laurent expansion of $F(qx)$, which we want to exist for $|b/aq| < |x| < 1$; so assume for the present that $|b/aq| < |x| < 1$. Both $F(x)$ and $F(qx)$ are defined in $|b/aq| < |x| < 1$. Thus it is possible to look for a functional relation between $F(x)$ and $F(qx)$ in this region. Now

$$F(qx) = \frac{(ax;q)_\infty (q/ax;q)_\infty (1-1/ax)(1-x)}{(x;q)_\infty (b/ax;q)_\infty (1-ax)(1-b/aqx)} = \sum_{n=-\infty}^{\infty} A_n q^n x^n. \quad (10.5.6)$$

Therefore,

$$q(1-x) \sum_{n=-\infty}^{\infty} A_n x^n = (b-aqx) \sum_{n=-\infty}^{\infty} A_n q^n x^n.$$

Equate the coefficients of x^n to get

$$q(A_n - A_{n-1}) = bA_n q^n - aA_{n-1} q^n,$$

or

$$A_n = \frac{1-aq^{n-1}}{1-bq^{n-1}} A_{n-1}$$

$$= \frac{(a;q)_n}{(b;q)_n} A_0.$$

This implies that

$$\frac{(ax;q)_\infty (q/ax;q)_\infty}{(x;q)_\infty (b/ax;q)_\infty} = A_0 \sum_{n=-\infty}^{\infty} \frac{(a;q)_n}{(b;q)_n} x^n. \quad (10.5.7)$$

Multiply both sides by $(1-x)$ and let $x \to 1^-$ to get

$$\frac{(a;q)_\infty (q/a;q)_\infty}{(q;q)_\infty (b/a;q)_\infty} = \frac{(a;q)_\infty}{(b;q)_\infty} A_0. \quad (10.5.8)$$

Substitute this value of A_0 in (10.5.7). Ramanujan's identity has the restriction on b that $|b/aq| < |x| < 1$. This can be removed by analytic continuation. In the derivation of (10.5.8), Abel's continuity theorem was used in the last step. This theorem states that if $\lim_{n\to\infty} a_n = a$ then

$$\lim_{x\to 1^-} (1-x) \sum_{n=0}^{\infty} a_n x^n = a. \quad \blacksquare$$

See Appendix B for a discussion of this and related matters.

Third Proof (Ismail [1977]).　　Rewrite the q-binomial theorem as follows

$$\frac{(ax;q)_\infty}{(x;q)_\infty} = \sum_{n=0}^{\infty} \frac{(a;q)_n}{(q;q)_n} x^n = \sum_{n=-N}^{\infty} \frac{(a;q)_{n+N}}{(q;q)_{n+N}} x^{n+N}$$

$$= \frac{(a;q)_N}{(q;q)_N} x^N \sum_{n=-N}^{\infty} \frac{(aq^N;q)_n}{(q^{N+1};q)_n} x^n.$$

Replace a with aq^{-N} to obtain

$$\sum_{n=-\infty}^{\infty} \frac{(a;q)_n}{(q^{N+1};q)_n} x^n = \frac{(aq^{-N}x;q)_\infty x^{-N}(q;q)_N}{(aq^{-N};q)_N(x;q)_\infty}$$

$$= \frac{(aq^{-N}x;q)_N}{(aq^{-N};q)_N} x^{-N} \frac{(q;q)_\infty(ax;q)_\infty}{(q^{N+1};q)_\infty(x;q)_\infty} = \frac{(q/ax;q)_N(q;q)_\infty(ax;q)_\infty}{(q/a;q)_N(q^{N+1};q)_\infty(x;q)_\infty}$$

$$= \frac{(ax;q)_\infty(q/ax;q)_\infty(q;q)_n(q^{N+1}/a;q)_\infty}{(x;q)_\infty(q^{N+1}/ax;q)_\infty(q^{N+1};q)_\infty(q/a;q)_\infty}.$$

If q^{N+1} is replaced by b, both sides of this equality are analytic in b for b close to 0, and they agree when $b = q^{N+1}$. Zero is the limit point of this sequence, and two functions analytic in an open set around $b = 0$ and that are equal at infinitely many points in this set must be identically equal there. Analytic continuation gives Theorem 10.5.1 for $|b/a| < |x| < 1$. ∎

Theorem 10.5.1 is called Ramanujan's $_1\psi_1$ formula because bilateral q-series are denoted by ψ and there is one upper and one lower parameter.

Remark 10.5.1 If the argument in the second proof is generalized to

$$F(x) = \frac{(ax;q)_\infty(b/x;q)_\infty}{(cx;q)_\infty(d/x;q)_\infty},$$

a second-order difference equation arises. This reduces to a first-order equation for $d = q/c$; the coefficients can then be determined. Unfortunately, the resulting series is incorrect. This problem occurs because the change from $F(x)$ to $F(qx)$ moves one region of the analyticity of F to an adjacent one, since the poles of $F(x)$ are at $x = c^{-1}q^n$, $n = 0, \pm 1, \ldots$. Details are given in Askey [1987].

Remark 10.5.2 As remarked earlier, Ramanujan's formula is a q-analog of a beta integral. Rewrite the formula as

$$\sum_{n=-\infty}^{\infty} \frac{(bq^n;q)_\infty}{(aq^n;q)_\infty} x^n = \frac{(ax;q)_\infty(q/ax;q)_\infty(q;q)_\infty(b/a;q)_\infty}{(a;q)_\infty(q/a;q)_\infty(x;q)_\infty(b/ax;q)_\infty}.$$

Set $x = q^\alpha$, $a = -c$, and $b = -cq^{\alpha+\beta}$ to get

$$\int_0^\infty \frac{(-cq^{\alpha+\beta}x;q)_\infty}{(-cx;q)_\infty} x^{\alpha-1} d_q x = \frac{(-cq^\alpha;q)_\infty(-c^{-1}q^{1-\alpha};q)_\infty \Gamma_q(\alpha)\Gamma_q(\beta)}{(-c;q)_\infty(-c^{-1}q;q)_\infty \Gamma_q(\alpha+\beta)}.$$

Finally, by Theorem 10.2.4,

$$\lim_{q\to 1^-} \frac{(-cq^\alpha;q)_\infty(-c^{-1}q^{1-\alpha};q)_\infty}{(-c;q)_\infty(-c^{-1}q;q)_\infty} = (1+c)^{-\alpha}(1+1/c)^\alpha = c^{-\alpha}.$$

10.6 Representations of Numbers as Sums of Squares

Ramanujan's formula for $\sum_{n=-\infty}^{\infty}(a;q)_n x^n/(b;q)_n$ is useful in obtaining simple and direct derivations of some results on the number of representations of a number as a sum of squares. Three cases are worked out here: two squares (x^2+y^2), three squares where two are equal (x^2+2y^2), and four squares $(x^2+y^2+u^2+v^2)$. For other results of this type consult Fine [1988]. At the heart of the proofs is the following special case of Ramanujan's identity. Take $b=aq$ in Theorem 10.5.1 and divide by $1-a$ to get

$$\sum_{n=-\infty}^{\infty}\frac{x^n}{1-aq^n}=\frac{(ax;q)_\infty(q/ax;q)_\infty(q;q)_\infty^2}{(x;q)_\infty(q/x;q)_\infty(a;q)_\infty(q/a;q)_\infty},\qquad |q|<|x|<1.$$

$$(10.6.1)$$

Now observe that

$$\left(\sum_{n=-\infty}^{\infty}q^{n^2}\right)^s=\sum_{n=0}^{\infty}r_s(n)q^n,\tag{10.6.2}$$

where $r_s(n)$ is the number of ways n can be written as the sum of s squares. Observe also that $r_s(n)$ gives the number of points with integer coordinates on the sphere in s-dimensional space. Here the counting distinguishes between the order, so that $25=3^2+4^2=4^2=4^2+3^2$ are counted, and negative integers are also used; hence, for example, $(-3)^2+4^2$ is counted. We consider $s=2$ and $s=4$. Now note that

$$\sum_{n=-\infty}^{\infty}q^{n^2}\sum_{m=-\infty}^{\infty}q^{2m^2}=\sum_{n=0}^{\infty}s(n)q^n,\tag{10.6.3}$$

where $s(n)$ is the number of ways of writing n as x^2+2y^2.

The strategy for finding simple expressions for $r_s(n)$ and $s(n)$ is to use the triple product identity to express the left sides of (10.6.2) and (10.6.3) as products and then use (10.6.1) to express these products as a different sum.

Let $d_{i,j}(n)$ denote the number of divisors of n congruent to i mod j.

Theorem 10.6.1

(a) $r_2(n)=4[d_{1,4}(n)-d_{3,4}(n)]$,

(b) $s(n)=2[d_{1,8}(n)+d_{3,8}(n)-d_{5,8}(n)-d_{7,8}(n)]$,

(c) $r_4(n)=8\sum_{d\mid n,4\nmid d}d$.

Proof.

(a) By (10.4.5) and (10.6.2),

$$\sum_{n=0}^{\infty}r_2(n)q^n=(q^2;q^2)_\infty^2(-q;q^2)_\infty^4.$$

To apply (10.6.1) rewrite the product as

$$\frac{(q^2;q^2)_\infty^2(-q;q^2)_\infty^2(q^2;q^4)_\infty^2}{(q;q^2)_\infty^2} = \frac{(q^2;q^2)_\infty^2(-q;q^2)_\infty^2(q^2;q^2)_\infty}{(q;q^2)_\infty^2(q^4;q^4)_\infty}$$

$$= \frac{(q^2;q^2)_\infty^2(-q;q^2)_\infty^2}{(q;q^2)_\infty^2(-q^2;q^2)_\infty^2}. \qquad (10.6.4)$$

Replace q with q^2 in (10.6.1) and set $x = q$ and $a = -1$. Then

$$2\sum_{n=-\infty}^\infty \frac{q^n}{1+q^{2n}} = \frac{(-q;q^2)_\infty^2(q^2;q^2)_\infty^2}{(q;q^2)_\infty^2(-q^2;q^2)_\infty^2}.$$

This implies

$$\sum_{n=0}^\infty r_2(n)q^n = 2\sum_{n=-\infty}^\infty \frac{q^n}{1+q^{2n}} = 1 + 4\sum_{n=1}^\infty q^n \sum_{m=0}^\infty (-1)^m q^{2mn}$$

$$= 1 + 4\sum_{n=1}^\infty \sum_{m=0}^\infty \left(q^{(4m+1)n} - q^{(4m+3)n}\right)$$

$$= 1 + 4\sum_{n=1}^\infty (d_{1,4}(n) - d_{3,4}(n))q^n. \qquad (10.6.5)$$

This proves (a).

(b) As in (a), one can show that

$$\sum_{n=-\infty}^\infty q^{n^2} \sum_{n=-\infty}^\infty q^{2m^2} = (q^2;q^2)_\infty(-q;q^2)_\infty^2(q^4;q^4)_\infty(-q^2;q^4)_\infty^2$$

$$= \frac{(q^4;q^4)_\infty(-q;q^4)_\infty(-q^3;q^4)_\infty(q^4;q^4)_\infty}{(q;q^4)_\infty(q^3;q^4)_\infty(-q^4;q^4)_\infty(-q^4;q^4)_\infty}.$$

This time replace q with q^4 in (10.6.1) and then set $a = -1$ and $x = q$ to get

$$\sum_{n=0}^\infty s(n)q^n = 2\sum_{n=-\infty}^\infty \frac{q^n}{1+q^{4n}} = 1 + 2\sum_{n=1}^\infty \left(\frac{q^n}{1+q^{4n}} + \frac{q^{3n}}{1+q^{4n}}\right)$$

$$= 1 + 2\sum_{n=1}^\infty \sum_{m=0}^\infty (-1)^m \left[q^{(4m+1)n} + q^{(4m+3)n}\right]$$

$$= 1 + 2\sum_{n=1}^\infty \sum_{m=0}^\infty \left[q^{(8m+1)n} + q^{(8m+3)n} - q^{(8m+5)n} - q^{(8m+7)n}\right].$$

This proves (b).

(c) Rewrite (10.6.1) as follows:

$$\sum_{n=-\infty}^{\infty} \frac{x^n}{1-aq^n} = \frac{1}{1-a} + \sum_{n=1}^{\infty} \frac{x^n(1-aq^n)+aq^n x^n}{1-aq^n} - \sum_{n=1}^{\infty} \frac{a^{-1}q^n x^{-n}}{1-a^{-1}q^n}$$

$$= \frac{1-ax}{(1-x)(1-a)} + \frac{1}{a}\sum_{n=1}^{\infty} x^{-n}q^n[a^2 x^{2n}(1-aq^n)^{-1}$$

$$- (1-q^n/a)^{-1}].$$

Let $a = -1$ and combine with (10.6.1) to get

$$\frac{(-qx;q)_\infty(-q/x;q)_\infty(q;q)_\infty^2}{(xq;q)_\infty(q/x;q)_\infty(-q;q)_\infty^2} = 1 + \frac{2(1-x)}{1+x}\sum_{n=1}^{\infty}\frac{(q/x)^n[1-x^{2n}]}{1+q^n}.$$

Then $x \to -1$ gives

$$\left[\frac{(q;q)_\infty}{(-q;q)_\infty}\right]^4 = 1 + 8\sum_{n=1}^{\infty}\frac{n(-q)^n}{1+q^n}. \qquad (10.6.6)$$

It follows from (10.4.6) that

$$\sum_{n=-\infty}^{\infty}(-1)^n q^{n^2} = \frac{(q;q)_\infty}{(-q;q)_\infty}.$$

Change q to $-q$ in (10.6.6) to arrive at

$$\sum_{n=0}^{\infty} r_4(n)q^n = 1 + 8\sum_{n=1}^{\infty}\frac{nq^n}{1+(-q)^n}$$

$$= 1 + 8\left[\sum_{n=1}^{\infty}\frac{nq^n}{1-q^n} - \sum_{n=1}^{\infty}2nq^{2n}\left(\frac{1}{1-q^{2n}} - \frac{1}{1+q^{2n}}\right)\right]$$

$$= 1 + 8\left[\sum_{n=1}^{\infty}\frac{nq^n}{1-q^n} - \sum_{n=1}^{\infty}\frac{4nq^{4n}}{1-q^{4n}}\right]. \qquad (10.6.7)$$

The result follows as before. ∎

The identities (10.6.5) and (10.6.7) and their interpretations were first discovered by Jacobi [1829]. The number theoretic result in (b) was stated and proved by Gauss for n prime.

10.7 Elliptic and Theta Functions

The theory of elliptic functions with its ramifications has been studied for two centuries. Its founders are Euler, Gauss, Abel, and Jacobi. In this century the theory of elliptic functions has been largely incorporated within the theory of elliptic curves, which was recently used by Wiles to prove Fermat's last theorem. In this section we will simply give some definitions and show how Ramanujan obtained the Fourier-series expansions of Jacobi elliptic functions from (10.6.1).

There are four theta functions of Jacobi. They are all really the same function, just as $\sin z$ and $\cos z$ are the same, but it is useful to consider all four:

$$\theta_1(z, q) = 2 \sum_{n=0}^{\infty} (-1)^n q^{(n+1/2)^2} \sin(2n+1)z$$

$$= -i \sum_{n=-\infty}^{\infty} (-1)^n q^{(n+1/2)^2} e^{(2n+1)iz}, \tag{10.7.1}$$

$$\theta_2(z, q) = 2 \sum_{n=0}^{\infty} q^{(n+1/2)^2} \cos(2n+1)z$$

$$= \sum_{n=-\infty}^{\infty} q^{(n+1/2)^2} e^{(2n+1)iz} = \theta_1(z + \pi/2; q), \tag{10.7.2}$$

$$\theta_3(z, q) = 1 + 2 \sum_{n=1}^{\infty} q^{n^2} \cos 2nz = \sum_{n=-\infty}^{\infty} q^{n^2} e^{2niz}, \tag{10.7.3}$$

$$\theta_4(z, q) = 1 + 2 \sum_{n=1}^{\infty} (-1)^n q^{n^2} \cos 2nz$$

$$= \sum_{n=-\infty}^{\infty} (-1)^n q^{n^2} e^{2niz} = \theta_3\left(z + \frac{\pi}{2}, q\right). \tag{10.7.4}$$

Here $q = e^{\pi i \tau}$ with $\operatorname{Im} \tau > 0$ so that $|q| < 1$ for a given τ. The value of q^λ for some λ is determined from $e^{\pi i \lambda \tau}$. The following relations, satisfied by the theta functions, follow immediately from the definitions:

$$\theta_1(z + \pi, q) = -\theta_1(z, q),$$
$$\theta_2(z + \pi, q) = -\theta_2(z, q),$$
$$\theta_3(z + \pi, q) = \theta_3(z, q),$$
$$\theta_3(z + \pi, q) = \theta_4(z, q); \tag{10.7.5}$$

and

$$\theta_1(z + \pi\tau, q) = -q^{-1}e^{-2iz}\theta_1(z, q),$$
$$\theta_2(z + \pi\tau, q) = q^{-1}e^{-2iz}\theta_2(z, q),$$
$$\theta_3(z + \pi\tau, q) = q^{-1}e^{-2iz}\theta_3(z, q),$$
$$\theta_4(z + \pi\tau, q) = q^{-1}e^{-2iz}\theta_4(z, q). \tag{10.7.6}$$

The following formulas are obtained from the triple product identity:

$$
\begin{aligned}
\theta_1(z, q) &= 2q^{1/4} \sin z (q^2, q^2)_\infty (q^2 e^{2iz}; q^2)_\infty (q^2 e^{-2iz}; q^2)_\infty \\
&= -iq^{1/4} e^{iz} (q^2; q^2)_\infty (q^2 e^{2iz}; q^2)_\infty (e^{-2iz}; q^2)_\infty, \\
\theta_2(z, q) &= 2q^{1/4} \cos z (q^2, q^2)_\infty (-q^2 e^{2iz}; q^2)_\infty (-q^2 e^{-2iz}; q^2)_\infty \\
&= q^{1/4} e^{iz} (q^2, q^2)_\infty (-q^2 e^{2iz}; q^2)_\infty (-e^{-2iz}; q^2)_\infty, \\
\theta_3(z, q) &= (q^2; q^2)_\infty (-qe^{2iz}; q^2)_\infty (-qe^{-2iz}; q^2)_\infty, \\
\theta_4(z, q) &= (q^2; q^2)_\infty (qe^{2iz}; q^2)_\infty (qe^{-2iz}; q^2)_\infty.
\end{aligned}
\tag{10.7.7}
$$

For $z = 0$, these reduce to

$$
\begin{aligned}
\theta_2 &= \theta_2(0) = \theta_2(0, q) = 2q^{1/4}(q^2; q^2)_\infty (-q^2; q^2)_\infty^2, \\
\theta_3 &= \theta_3(0) = \theta_3(0, q) = (q^2; q^2)_\infty (-q; q^2)_\infty^2, \\
\theta_4 &= \theta_4(0)\theta_4(0, q) = (q^2, q^2)_\infty (q; q^2)_\infty^2,
\end{aligned}
\tag{10.7.8}
$$

and

$$
\theta_1' = \theta_1'(0, q) = \lim_{z \to 0} \frac{\theta_1(z, q)}{z} = 2q^{1/4}(q^2; q^2)_\infty^3.
$$

Proposition 10.7.1 $\theta_1' = \theta_2 \theta_3 \theta_4$.

Proof. The right side equals

$$
2q^{1/4}(q^2; q^2)_\infty^3 (-q; q^2)_\infty^2 (-q^2; q^2)_\infty^2 (q, q^2)_\infty^2.
$$

The factor $(-q; q^2)_\infty (-q^2; q^2)_\infty (q; q^2)$ equals one because it can be written as

$$
\begin{aligned}
(-q; q)_\infty (q; q^2) &= \frac{(-q; q)_\infty (q, q^2)(q^2; q^2)_\infty}{(q^2; q^2)_\infty} \\
&= \frac{(-q; q)_\infty (q; q)_\infty}{(q^2; q^2)_\infty} = \frac{(q^2; q^2)_\infty}{(q^2; q^2)_\infty} = 1.
\end{aligned}
$$

This proves the proposition, which, it should be noted, is equivalent to (10.4.9). ∎

A meromorphic function is called an elliptic function if it has two periods whose ratio is not a real number. To define the Jacobi elliptic functions, set

$$
k^{1/2} = \theta_2/\theta_3,
\tag{10.7.9}
$$

and let

$$
\operatorname{sn}(u, k) = \frac{\theta_3 \, \theta_1 \left(u / \theta_3^2\right)}{\theta_2 \, \theta_4 \left(u / \theta_3^2\right)}.
\tag{10.7.10}
$$

It follows from (10.7.5) and (10.7.6) that $\text{sn}(u, k)$ is periodic in u with periods $2\pi\theta_3^2$ and $\pi\tau\theta_3^2$, and so it is an elliptic function. Other Jacobi functions are

$$\text{cn}(u, k) = \frac{\theta_2}{\theta_3} \frac{\theta_2(u/\theta_3^2)}{\theta_4(u/\theta_3^2)}$$

and

$$\text{dn}(u, k) = \frac{\theta_4}{\theta_3} \frac{\theta_3(u/\theta_3^2)}{\theta_4(u/\theta_3^2)}.$$

There are also functions that correspond to $\csc z$, $\tan z$, and so on:

$$\text{ns}(u, k) = \frac{1}{\text{ns}(u, k)}, \quad \text{nc}(u, k) = \frac{1}{\text{cn}(u, k)}, \quad \text{nd}(u, k) = \frac{1}{\text{dn}(u, k)},$$

$$\text{sc}(u, k) = \frac{\text{sn}(u, k)}{\text{cn}(u, k)}, \quad \text{cs}(u, k) = \frac{\text{cn}(u, k)}{\text{sn}(u, k)}, \quad \text{cd}(u, k) = \frac{\text{cn}(u, k)}{\text{dn}(u, k)}, \quad (10.7.11)$$

$$\text{sd}(u, k) = \frac{\text{sn}(u, k)}{\text{dn}(u, k)}, \quad \text{ds}(u, k) = \frac{\text{dn}(u, k)}{\text{sn}(u, k)}, \quad \text{dc}(u, k) = \frac{\text{dn}(u, k)}{\text{cn}(u, k)}.$$

Exercise 2.10 required us to show that

$$\frac{\pi}{2}\theta_3^2 = \int_0^1 \frac{dt}{\sqrt{(1-t^2)(1-k^2t^2)}} \equiv K.$$

Set $u = 2Kx/\pi$. Then

$$\text{sn}(u, k) = \frac{\theta_3}{\theta_2} \cdot \frac{\theta_1(x)}{\theta_4(x)}, \quad (10.7.12)$$

and similar relations hold for the other functions. We omit the modulus k in $\text{sn}(u, k)$ etc., and simply write $\text{sn}\, u$ etc. It is easy to see that

$$\text{sn}\, u = 2q^{1/4}k^{-1/2}\sin x \prod_{n=1}^{\infty}\left\{\frac{1 - 2q^{2n}\cos 2x + q^{4n}}{1 - 2q^{2n-1}\cos 2x + q^{4n-2}}\right\},$$

$$\text{cn}\, u = 2q^{1/4}k'^{1/2}k^{-1/2}\cos x \prod_{n=1}^{\infty}\left\{\frac{1 + 2q^{2n}\cos 2x + q^{4n}}{1 - 2q^{2n-1}\cos 2x + q^{4n-2}}\right\}, \quad (10.7.13)$$

and

$$\text{dn}\, u = k'^{1/2}\prod_{n=1}^{\infty}\left\{\frac{1 + 2q^{2n-1}\cos 2x + q^{4n-2}}{1 - 2q^{2n-1}\cos 2x + q^{4n-2}}\right\}.$$

Here $k' > 0$ is defined by $k'^2 = 1 - k^2$. Again, by the exercise mentioned above, $k'^{1/2} = \theta_4/\theta_3$. There we took $0 < k < 1$, but the results hold more generally. One can also deduce from Exercise 2.11 that $\text{sn}\, u$ has periods $4K$ and $2iK'$.

The Fourier expansion of the Jacobian functions is given in the next theorem.

Theorem 10.7.2   *With $u = 2Kx/\pi$, we have*

$$\operatorname{sn} u = \frac{2\pi}{Kk} \sum_{n=0}^{\infty} \frac{q^{n+1/2} \sin(2n+1)x}{1 - q^{2n+1}}, \tag{10.7.14}$$

$$\operatorname{dn} u = \frac{\pi}{2K} + \frac{2\pi}{K} \sum_{n=1}^{\infty} \frac{q^n \cos 2nx}{1 + q^{2n}}, \tag{10.7.15}$$

$$\operatorname{cn} u = \frac{2\pi}{Kk} \sum_{n=0}^{\infty} \frac{q^{n+1/2} \cos(2n+1)x}{1 + q^{2n+1}}. \tag{10.7.16}$$

Proof. We prove (10.7.14) and leave the other two parts as exercises. In Ramanujan's formula (10.6.1), first replace q with q^2 and then set $x = qe^{-2ix}$ and $a = 1/q$. Then

$$\sum_{0}^{\infty} \frac{q^n e^{2nix}}{1 - q^{2n+1}} - \sum_{1}^{\infty} \frac{q^{n-1} e^{-2nix}}{1 - q^{2n-1}} = \frac{(q^2 e^{2ix}; q^2)_\infty (e^{-2ix}; q^2)_\infty (q^2; q^2)_\infty^2}{(qe^{2ix}; q^2)_\infty (qe^{-2ix}; q^2)_\infty (q; q^2)_\infty^2}.$$

Change n to $n + 1$ in the second sum; then combine with the first to get

$$e^{-ix} 2i \sum_{0}^{\infty} \frac{q^n \sin(2n+1)x}{1 - q^{2n+1}}.$$

Now multiply both sides by $-ie^{ix}q^{1/2}\pi/Kk$ and show that the product reduces to

$$\frac{\theta_3}{\theta_2} \theta_1(x)\theta_4(x) = \operatorname{sn} u.$$

A calculation almost identical with the one in Proposition 10.7.1 is needed. This proves the theorem. ■

There are similar series for cd, sd, and nd. The series for the other Jacobi functions such as ns u are somewhat different. For example,

$$\operatorname{ns} u = \frac{\pi}{2K \sin x} + \frac{2\pi}{K} \sum_{n=0}^{\infty} \frac{q^{2n+1} \sin(2n+1)x}{1 - q^{2n+1}}. \tag{10.7.17}$$

This can be proved directly or by first showing that

$$\operatorname{sn}(u + iK') = k^{-1}\operatorname{dc}(u - K) = k^{-1}\operatorname{ns} u.$$

See Whittaker and Watson [1940, p. 511].

10.8 q-Beta Integrals

We have seen that beta integrals and their extensions and analogs are very useful and important. There are several more extensions and it is worthwhile to look at a few here. One is quite important as it is associated with a set of orthogonal polynomials that is beginning to find applications in several areas of mathematics.

We noted earlier that Ramanujan's $_1\psi_1$ could be viewed as the integral of $x^{\alpha-1}(-ax; q)_\infty/(-x; q)_\infty$ with respect to the Fermat measure. Ramanujan also integrated this function using the usual measure. He used a general interpolation-type theorem to evaluate it. An account of this is given later. For now, we give an evaluation using a functional relation since that fits in with the methods used earlier.

Let $0 < q < 1$, Re $\alpha > 0$, and $|aq^{-\alpha}| < 1$. Assume for the moment that $0 < \alpha < 1$. Define

$$f(a) = \int_0^\infty \frac{x^{\alpha-1}(-ax; q)_\infty}{(-x; q)_\infty} dx.$$

Then

$$f(a) = \int_0^\infty x^{\alpha-1} \frac{(-axq; q)_\infty}{(-x; q)_\infty} (1 - a + a(1 + x)) dx$$

$$= (1 - a) f(aq) + aq^{-\alpha} f(a).$$

This iterates to

$$f(a) = \frac{1 - a}{1 - aq^{-\alpha}} f(aq) = \frac{(a; q)_\infty}{(aq^{-\alpha}; q)_\infty} f(0).$$

As in the evaluation of Ramanujan's $_1\psi_1$, it is hard to find $f(0)$. But

$$f(q) = \int_0^\infty \frac{x^{\alpha-1}}{1 + x} dx = \frac{\pi}{\sin \pi \alpha}.$$

Thus

$$f(a) = \frac{(a, q)_\infty (q^{1-\alpha}; q)_\infty}{(aq^{-\alpha}; q)_\infty (q; q)_\infty} \frac{\pi}{\sin \pi \alpha}, \tag{10.8.1}$$

when $0 < \alpha < 1$. The general case follows by analytic continuation in α, and by continuity when α is a positive integer. The limiting case when $\alpha = k$ is a positive integer is

$$\int_0^\infty x^{k-1} \frac{(-ax; q)_\infty}{(-x; q)_\infty} dx = \frac{(-1)^{k+1}(q/a)^k (q; q)_{k-1}(\log q)}{(a^{-1}q; q)_k}.$$

To see that (10.8.1) extends

$$\int_0^\infty \frac{x^{\alpha-1}}{(1+x)^{\alpha+\beta}} dx = \frac{\Gamma(\alpha)\Gamma(\beta)}{\Gamma(\alpha+\beta)},$$

take $a = q^{\alpha+\beta}$ in (10.8.1) to get

$$\int_0^\infty x^{\alpha-1} \frac{(-xq^{\alpha+\beta}; q)_\infty}{(-x; q)_\infty} dx = \frac{\Gamma_q(\beta)\Gamma(\alpha)\Gamma(1-\alpha)}{\Gamma_q(\alpha+\beta)\Gamma_q(1-\alpha)}.$$

This formula lacks symmetry in α and β. Symmetry is restored by the following integral:

$$f(a, b) = \int_0^\infty x^{c-1} \frac{(-ax; q)_\infty(-qb/x; q)_\infty}{(-x; q)_\infty(-q/x; q)_\infty} dx.$$

The details of the evaluation are left to the reader. First show that

$$f(a, b) = f(aq, b) + aq^{-c} f(a, bq)$$

and then by a similar procedure that

$$f(a, b) = f(a, bq) + bq^c f(aq, b).$$

The two equations together give

$$f(a, b) = \frac{1-ab}{(1-bq^c)} f(a, bq) = \frac{(ab; q)_\infty}{(bq^c; q)_\infty} f(a, 0).$$

The value of $f(a, 1)$ is given by (10.8.1), and the final result is

$$f(a, b) = \frac{(ab; q)_\infty(q^c; q)_\infty(q^{1-c}, q)_\infty \pi}{(bq^c; q)_\infty(aq^{-c}; q)_\infty(q; q)_\infty \sin \pi c}.$$

When $a = q^{\alpha+c}, b = q^{\beta-c}$, then

$$\int_0^\infty x^{c-1} \frac{(-xq^{\alpha+c}; q)_\infty(-q^{\beta+1-c}/x; q)_\infty}{(-x; q)_\infty(-q/x; q)_\infty} dx$$

$$= \frac{\Gamma(c)\Gamma(1-c)}{\Gamma_q(c)\Gamma_q(1-c)} \cdot \frac{\Gamma_q(\alpha)\Gamma_q(\beta)}{\Gamma_q(\alpha+\beta)}.$$

To obtain other q-beta integrals, let us again consider Ramanujan's $_1\psi_1$ formula (see (10.5.3)). When $|b/a| < x < 1$ and $|q| < 1$,

$$\sum_{n=-\infty}^\infty \frac{(a; q)_n x^n}{(b; q)_n} = \frac{(ax; q)_\infty(q/ax; q)_\infty(b/q; q)_\infty(q; q)_\infty}{(x; q)_\infty(b/ax; q)_\infty(q/a; q)_\infty(b; q)_\infty}.$$

Set $ax = q^{1/2}e^{i\theta}$ and $x = \alpha e^{i\theta}$, $|\alpha| < 1$. Then

$$\int_{-\pi}^{\pi} \frac{(q^{1/2}e^{i\theta}; q)_\infty (q^{1/2}e^{-i\theta}; q)_\infty}{(\alpha e^{i\theta}; q)_\infty (bq^{-1/2}e^{-i\theta}; q)_\infty} d\theta$$

$$= \frac{(b; q)_\infty (\alpha q^{1/2}; q)_\infty}{(q; q)_\infty (\alpha bq^{-1/2}; q)_\infty} \int_{-\pi}^{\pi} \sum_{n=-\infty}^{\infty} \frac{(q^{1/2}/\alpha; q)_n}{(b; q)_n} e^{in\theta} d\theta$$

$$= \frac{(b; q)_\infty (\alpha q^{1/2}; q)_\infty}{(q; q)_\infty (\alpha bq^{-1/2}; q)_\infty} 2\pi. \qquad (10.8.2)$$

To obtain a positive kernel in (10.8.2), take $b = \alpha q^{1/2}$. Then the integrand is an even function of θ, and becomes

$$\int_0^{\pi} \prod_{n=0}^{\infty} \frac{1 - 2q^{(2n+1)/2} \cos\theta + q^{2n+1}}{1 - 2\alpha q^n \cos\theta + \alpha^2 q^{2n}} d\theta = \frac{(\alpha\sqrt{q}; q)_\infty (\alpha\sqrt{q}; q)_\infty}{(\alpha^2; q)_\infty (q; q)_\infty} \pi. \quad (10.8.3)$$

To get a perspective on these integrals, replace α with $q^{\alpha-1/2}$ and b with q^β in (10.8.1). The result is

$$\int_{-\pi}^{\pi} \frac{(q^{1/2}e^{i\theta}; q)_\infty (q^{1/2}e^{-i\theta}; q)_\infty}{(q^{\alpha-1/2}e^{i\theta}; q)_\infty (q^{\beta-1/2}e^{-i\theta}; q)_\infty} d\theta = 2\pi \frac{\Gamma_q(\alpha + \beta - 1)}{\Gamma_q(\alpha)\Gamma_q(\beta)}, \qquad (10.8.4)$$

and the limit when $q \to 1$ is

$$\int_{-\pi}^{\pi} (1 - e^{i\theta})^{\alpha-1} (1 - e^{-i\theta})^{\beta-1} d\theta = 2\pi \frac{\Gamma(\alpha + \beta - 1)}{\Gamma(\alpha)\Gamma(\beta)}. \qquad (10.8.5)$$

In a similar fashion, (10.8.3) gives the special case $\alpha = \beta$ of (10.8.5), that is,

$$\int_0^{\pi} (2 - 2\cos\theta)^{\alpha-1} d\theta = 2^{\alpha-1} \int_{-1}^{1} (1 - x)^{\alpha-1} (1 - x^2)^{-1/2} dx = \frac{\pi\Gamma(2\alpha - 1)}{\Gamma(\alpha)\Gamma(\alpha)}.$$

One simple extension of (10.8.2) can be seen when we multiply by $e^{-ik\theta}$ before integrating. A more important extension is obtained when we reconsider the argument given in (10.8.2) and realize that the numerator could be replaced by other products. The product

$$(-q^{1/2}e^{i\theta}; q)_\infty (-q^{1/2}e^{-i\theta}; q)_\infty$$

is one obvious choice. Two others are

$$(e^{i\theta}; q)_\infty (e^{-i\theta}; q)_\infty \quad \text{and} \quad (-e^{i\theta}; q)_\infty (-e^{-i\theta}; q)_\infty.$$

Each of these has an extra factor from the theta product in the numerator of the $_1\psi_1$ sum, and there is no problem in evaluating a similar integral. One could get greedy and ask whether it is possible to evaluate the integral when all of these factors are present simultaneously. The integral is

$I(a, b, c, d)$

$$= \int_0^\pi \frac{(e^{i\theta}; q)_\infty (e^{-i\theta}; q)_\infty (q^{1/2}e^{i\theta}; q)_\infty (q^{1/2}e^{-i\theta}; q)_\infty (-e^{i\theta}; q)_\infty (-e^{-i\theta}; q)_\infty}{(ae^{i\theta}; q)_\infty (ae^{-i\theta}; q)_\infty (be^{i\theta}; q)_\infty (be^{-i\theta}; q)_\infty (ce^{i\theta}; q)_\infty (ce^{-i\theta}; q)_\infty}$$

$$\cdot \frac{(-q^{1/2}e^{i\theta}; q)_\infty (-q^{1/2}e^{-i\theta}; q)_\infty}{(de^{i\theta}; q)_\infty (de^{i\theta}; q)_\infty} d\theta$$

$$= \int_{-1}^1 \frac{h(x, 1)h(x, q^{1/2})h(x, -1)h(x, -q^{1/2})dx}{h(x, a)h(x, b)h(x, c)h(x, d)\sqrt{1 - x^2}},$$

where

$$h(x, a) = \prod_{n=0}^\infty (1 - 2axq^n + a^2q^{2n}).$$

The condition on the parameters is that $\max(|a|, |b|, |c|, |d|, |q|) < 1$. Instead of directly evaluating the integral, let us first try to guess what its value should be. We would like to discover some functional relations. Observe that

$$\frac{b}{h(x, aq)h(x, b)} - \frac{a}{h(x, a)h(x, bq)} = \frac{b(1 - 2ax + a^2) - a(1 - 2bx + b^2)}{h(x, a)h(x, b)}$$

$$= \frac{(b - a)(1 - ab)}{h(x, a)h(x, b)}.$$

The reason we put the factors b and a on the left is to remove the x in the numerator. This leads to

$$bI(aq, b, c, d) - aI(a, bq, c, d) = (1 - ab)(b - a)I(a, b, c, d). \qquad (10.8.6)$$

This suggests that $I(a, b, c, d)$ is a function of ab and the symmetric products. So we try

$$I(a, b, c, d) = f(ab)f(ac)f(ad)f(bc)f(bd)f(cd). \qquad (10.8.7)$$

Apply this to (10.8.6) with $c = d = 0$ to get

$$(b - a)(1 - ab)f(ab) = (b - a)f(abq).$$

This gives

$$f(ab) = \frac{f(abq)}{1 - ab} = \cdots = \frac{f(abq^n)}{(ab; q)_n} = \frac{f(0)}{(ab; q)_\infty}.$$

To see if this is correct, use it in (10.8.7) and then put this conjectured value of the integral in (10.8.6). The result is equivalent to

$$\frac{\dfrac{b}{(abq)_\infty(acq)_\infty(adq)_\infty(bc)_\infty(bd)_\infty(cd)_\infty}}{} - \frac{a}{(abq)_\infty(ac)_\infty(ad)_\infty(bcq)_\infty(bdq)_\infty(cd)_\infty}$$
$$= \frac{b-a}{(abq)_\infty(ac)_\infty(ad)_\infty(bc)_\infty(bd)_\infty(cd)_\infty},$$

where we have written $(x)_\infty$ instead of $(x;q)_\infty$ for convenience. This equation is equivalent to

$$b(1-ac)(1-ad) - a(1-bc)(1-bd) = b - a.$$

This is not true because the left side is $(b-a)(1-abcd)$. However, it suggests that the error is a function of $abcd$. So we write

$$I(a,b,c,d) = \frac{h(abcd)}{(ab)_\infty(ac)_\infty(ad)_\infty(bc)_\infty(bd)_\infty(cd)_\infty}.$$

Now use this in (10.8.6) to get

$$\frac{bh(abcdq)}{(abq)_\infty(acq)_\infty(adq)_\infty(bc)_\infty(bd)_\infty(cd)_\infty}$$
$$- \frac{ah(abcdq)}{(abq)_\infty(ac)_\infty(ad)_\infty(bcq)_\infty(bdq)_\infty(cd)_\infty}$$
$$= \frac{(b-a)(1-ab)h(abcd)}{(ab)_\infty(ac)_\infty(ad)_\infty(bc)_\infty(bd)_\infty(cd)_\infty}.$$

Simplify to get

$$h(abcd) = (1-abcd)h(abcdq).$$

Iterate this to arrive at

$$h(abcd) = (abcd;q)_\infty h(0) = (abcd;q)_\infty M(q). \tag{10.8.8}$$

To identify $M(q)$ we need to find specific values of $a, b, c,$ and d where we can evaluate both the integral and the infinite product in (10.8.8). The obvious values $(a,b,c,d) = (1, \sqrt{q}, -1, -\sqrt{q})$ work for the integral as well as the product. The integral is

$$\int_0^\pi d\theta = \pi,$$

and

$$\frac{(q;q)_\infty M(q)}{(q^{1/2};q)_\infty(-1;q)_\infty(-q^{1/2};q)_\infty(-q^{1/2};q)_\infty(-q;q)_\infty(q^{1/2}q)_\infty}$$

$$= \frac{1}{2}(q;q)_\infty M(q).$$

This implies that $M(q) = 2\pi/(q;q)_\infty$. The conjectured value of $I(a,b,c,d)$ is in fact correct and we have the theorem:

Theorem 10.8.1   *For* $\max(|a|,|b|,|c|,|d|,|q|) < 1$,

$$\int_{-1}^1 \frac{h(x,1)h(x,\sqrt{q})h(x,-1)h(x,-\sqrt{q})}{h(x,a)h(x,b)h(x,c)h(x,d)} \frac{dx}{\sqrt{1-x^2}}$$

$$= \frac{2\pi(abcd;q)_\infty}{(q;q)_\infty(ab;q)_\infty(ac;q)_\infty(ad;q)_\infty(bc;q)_\infty(bd;q)_\infty(cd;q)_\infty},$$

when

$$h(x,q) = \prod_{n=0}^\infty (1 - 2axq^n + a^2 q^{2n}).$$

Proof. We have already seen that both sides of this relation satisfy the functional equation (10.8.6). It is also clear that both sides are analytic functions of a,b,c, and d when $\max(|a|,|b|,|c|,|d|) < 1$. By the uniqueness of analytic functions, it will be sufficient to show that they are equal for $(a,b,c,d) = (q^j, q^{k+1/2}, -q^\ell, -q^{m+1/2})$, $j,k,\ell,m = 0,1,2,\ldots$. We first show that they are equal for $a = 1, c = -1, d = \sqrt{q}$, and all b, $|b| < 1$. In this case the integral is simply

$$\int_{-1}^{+1} \frac{h(x;\sqrt{q})}{h(x,b)} \frac{dx}{\sqrt{1-x^2}},$$

and this was evaluated in (10.8.3). Thus we have to show that

$$\frac{\pi(bq^{1/2};q)_\infty(bq^{1/2};q)_\infty}{(b^2;q)_\infty(q;q)_\infty}$$

$$= \frac{2\pi(bq^{1/2};q)_\infty}{(q;q)_\infty(b;q)_\infty(-b;q)_\infty(-bq^{1/2};q)_\infty(-1;q)_\infty(-q^{1/2};q)_\infty(q^{1/2};q)_\infty}.$$

But the right side is

$$\frac{2\pi(bq^{1/2};q)_\infty(bq^{1/2};q)_\infty}{2(q;q)_\infty(b^2;q^2)_\infty(b^2q;q^2)_\infty(-q;q)_\infty(q;q^2)_\infty} = \frac{\pi(bq^{1/2};q)_\infty^2}{(b^2;q)_\infty(q;q)_\infty}.$$

This gives us a starting value, and the functional equation (10.8.6) then shows that if the integral can be computed for some (a,b,c,d), then it can be computed when one of the parameters is multiplied by q. This proves the formula for

$$(a, b, c, d) = (q^j, b, -q^\ell, -q^{m+1/2}), \quad j, \ell, m = 0, 1, 2, \ldots,$$

and all b, $|b| < 1$, which completes the proof. ∎

If the restriction to $\max(|a|, |b|, |c|, |d|) < 1$ is removed, then some discrete mass points appear. This case was first treated directly via Cauchy's theorem by Askey and Wilson [1985]. A direct extension from the case in Theorem 10.8.1 was provided by Gasper and Rahman [1990, §6.6].

An extension of Theorem 10.8.1 is given below.

Theorem 10.8.2 *For* $\max_{1 \leq i \leq 5}(|a_i|, |q|) < 1$,

$$\int_{-1}^{1} \frac{h(x, 1)h(x, \sqrt{q})h(x, -1)h(x, -\sqrt{q})h\left(x, \prod_1^5 a_i\right)}{\prod_1^5 h(x, a_i)} \frac{dx}{\sqrt{1 - x^2}}$$

$$= \frac{2\pi (a_1a_2a_3a_4)_\infty (a_1a_2a_3a_5)_\infty}{(a_1a_2)_\infty (a_1a_3)_\infty (a_1a_4)_\infty (a_1a_5)_\infty (a_2a_3)_\infty}$$

$$\cdot \frac{(a_2a_3a_4a_5)_\infty (a_1a_2a_4a_5)_\infty (a_1a_3a_4a_5)_\infty}{(a_2a_4)_\infty (a_2a_5)_\infty (a_3a_4)_\infty (a_3a_5)_\infty (a_4a_5)_\infty (q)_\infty},$$

where $(x)_\infty = (x; q)_\infty$.

This is proved in the same way we proved Theorem 10.8.1, except that a more elementary initial result can be used. A more general integral was evaluated by Nassrallah and Rahman [1985]. This special case was noted by Rahman [1986], and the proof outlined here was given by Askey [1988].

Theorem 10.8.1 has an important limiting case, as we saw earlier. Write

$$(e^{i\theta}; q)_\infty = (q^{ix}; q)_\infty$$

and

$$(ae^{i\theta}; q)_\infty = (q^{a+ix}; q)_\infty$$

in the integral and let $q \to 1^-$. The result is the de Branges-Wilson integral (3.6.1):

$$\frac{1}{2\pi} \int_0^\infty \left| \frac{\Gamma(a + ix)\Gamma(b + ix)\Gamma(c + ix)\Gamma(d + ix)}{\Gamma(2ix)} \right|^2 dx$$

$$= \frac{\Gamma(a + b)\Gamma(a + c)\Gamma(a + d)\Gamma(b + c)\Gamma(b + d)\Gamma(c + d)}{\Gamma(a + b + c + d)},$$

$$\text{Re}(a, b, c, d) > 0, \quad (10.8.9)$$

when either all the parameters are real or any complex ones appear in conjugate pairs. When a is complex, $|\Gamma(a + ix)|^2$ should be replaced by $\Gamma(a + ix)\Gamma(a - ix)$.

We saw that the integrand of (10.8.9) is the weight function for the orthogonality of the Wilson polynomials

$$_4F_3\left(\begin{matrix} -n, n+a+b+c+d-1, a-ix, a+ix \\ a+b, a+c, a+d \end{matrix} ; 1\right).$$

There is a q-extension of these polynomials for which the integrand in Theorem 10.8.1 plays the same role. See Gasper and Rahman [1990].

10.9 Basic Hypergeometric Series

The series in the q-binomial theorem takes the form $\sum c_n$ with c_{n+1}/c_n a rational function of q^n. In this case

$$\frac{c_{n+1}}{c_n} = \frac{1-aq^n}{1-q^{n+1}}x$$

and

$$c_n = \frac{(a;q)_n}{(q;q)_n}x^n, \quad c_0 = 1.$$

Earlier, we started with the q-binomial theorem and then proceeded to consider the triple product and its generalization, the $_1\psi_1$. This introduced us to theta functions and elliptic functions, among other things. The above remarks on the q-binomial series suggest another direction. As a generalization of this series we consider a series $\sum c_n$ for which

$$\frac{c_{n+1}}{c_n} = \frac{(1-aq^n)(1-bq^n)}{(1-cq^n)(1-q^{n+1})}x, \quad c_0 = 1.$$

This can be solved as

$$c_n = \frac{(a;q)_n(b;q)_n}{(c;q)_n(q;q)_n}x^n.$$

The function $\sum_{n=0}^{\infty} c_n$ is denoted by

$$_2\phi_1\left(\begin{matrix} a, b \\ c \end{matrix} ; q, x\right)$$

and is an example of a basic hypergeometric series. If we set $a = q^\alpha, b = q^\beta$, and $c = q^\gamma$ and let $q \to 1^-$, we have the ordinary hypergeometric function $_2F_1(\alpha, \beta; \gamma; x)$. The $_2\phi_1$ series was studied by Heine [1847], who proved the following theorem, which is an analog of Euler's integral formula for the hypergeometric function:

Theorem 10.9.1 *For $|q| < 1$, $|x| < 1$, and $|b| < 1$,*

$$_2\phi_1\left(\begin{matrix} a, b \\ c \end{matrix}; q, x\right) = \frac{(b; q)_\infty (ax; q)_\infty}{(c; q)_\infty (x; q)_\infty} {}_2\phi_1\left(\begin{matrix} c/b, x \\ ax \end{matrix}; q, b\right).$$

Proof. As one might suspect from the proof of Euler's integral formula, a proof of the above transformation requires the q-binomial theorem and a change of the order of summation. Once again we write $(a)_n$ instead of $(a; q)_n$ for convenience. Then

$$_2\phi_1\left(\begin{matrix} a, b \\ c \end{matrix}; q, x\right) = \frac{(b)_\infty}{(c)_\infty} \sum_{n=0}^\infty \frac{(a)_n (cq^n)_\infty}{(q)_n (bq^n)_\infty} x^n$$

$$= \frac{(b)_\infty}{(c)_\infty} \sum_{n=0}^\infty \frac{(a)_n x^n}{(q)_n} \sum_{m=0}^\infty \frac{(c/b)_m}{(q)_m} (bq^n)^m$$

$$= \frac{(b)_\infty}{(c)_\infty} \sum_{m=0}^\infty \frac{(c/b)_m (axq^m)_\infty}{(q)_m (xq^m)_\infty} b^m$$

$$= \frac{(b)_\infty (ax)_\infty}{(c)_\infty (x)_\infty} {}_2\phi_1\left(\begin{matrix} c/b, x \\ ax \end{matrix}; q, b\right),$$

which completes the proof. ∎

To help understand this transformation, Thomae [1869] observed that it is a q-integral analog of Euler's integral representation of the $_2F_1$ hypergeometric function. Set $a = q^\alpha$, $b = q^\beta$, and $c = q^\gamma$ in Theorem 10.9.1 to get

$$\sum_{n=0}^\infty \frac{(q^\alpha)_n (q^\beta)_n}{(q)_n (q^\gamma)_n} x^n = \frac{(q^\beta; q)_\infty (q^\alpha x; q)_\infty}{(q^\gamma; q)_\infty (x; q)_\infty} \sum_{m=0}^\infty \frac{(q^{\gamma-\beta}; q)_m (x; q)_m q^{\beta m}}{(q; q)_m (q^\alpha x; q)_m}$$

$$= \frac{(q^\beta; q)_\infty (q^{\gamma-\beta}; q)_\infty}{(q^\gamma; q)_\infty (q; q)_\infty} \sum_{m=0}^\infty \frac{(q^{m+1}; q)_\infty (xq^{\alpha+m}; q)_\infty q^{m\beta}}{(q^{m+1+\gamma-\beta-1}; q)_\infty (xq^m; q)_\infty}.$$

$$(10.9.1)$$

Use the notation

$$(a; q)_\alpha = \frac{(a; q)_\infty}{(aq^\alpha; q)_\infty}$$

to write (10.9.1) as the q-integral formula

$$_2\phi_1\left(\begin{matrix} q^\alpha, q^\beta \\ q^\gamma \end{matrix}; q, x\right) = \frac{\Gamma_q(\gamma)}{\Gamma_q(\beta)\Gamma_q(\gamma-\beta)} \int_0^1 \frac{t^{\beta-1}(qt; q)_{\gamma-\beta-1}}{(xt; q)_\alpha} d_q t.$$

Euler's integral representation is one way of evaluating a $_2F_1$ at $x = 1$. Similarly, we have the q-analog of Gauss's sum due to Heine.

Corollary 10.9.2   *For* $|c/ab| < 1$,

$$_2\phi_1\left(\begin{matrix}a,b\\c\end{matrix};q,c/ab\right) = \frac{(c/a;q)_\infty(c/b;q)_\infty}{(c;q)_\infty(c/ab;q)_\infty}.$$

Proof.   Take $x = c/ab$ in Theorem 10.10.1, assuming that $|b| < 1$ and $|c/ab| < 1$. Then

$$_2\phi_1\left(\begin{matrix}a,b\\c\end{matrix};q,c/ab\right) = \frac{(b;q)_\infty(c/b;q)_\infty}{(c;q)_\infty(c/ab;q)_\infty}\sum_{m=0}^\infty \frac{(c/ab;q)_m b^m}{(q)_m}$$

$$= \frac{(b;q)_\infty(c/b;q)_\infty}{(c;q)_\infty(c/ab;q)_\infty}\cdot\frac{(c/a;q)_\infty}{(b;q)_\infty}.$$

The last step follows from the q-binomial theorem. This proves the corollary except for the assumption on b that $|b| < 1$. This can be removed by analytic continuation. ■

Bailey [1941] found an analog of Kummer's theorem that evaluates $_2F_1(a, b; a + 1 - b; -1)$.

Theorem 10.9.3   *If* $|q| < \min(1, |b|)$, *then*

$$_2\phi_1\left(\begin{matrix}a,b\\aq/b\end{matrix};q,-q/b\right) = \frac{(aq;q^2)_\infty(-q;q)_\infty(aq^2/b^2;q^2)_\infty}{(aq/b;q)_\infty(-q/b;q)_\infty}.$$

Proof.   First assume that $|a| < 1$. Interchange a and b in Heine's transformation to get

$$_2\phi_1\left(\begin{matrix}a,b\\aq/b\end{matrix};q,-q/b\right) = \frac{(a;q)_\infty(-q;q)_\infty}{(aq/b;q)_\infty(-q/b;q)_\infty}\sum_{m=0}^\infty \frac{(q/b;q)_m(-q/b;q)_m}{(q;q)_m(-q;q)_m}a^m$$

$$= \frac{(a;q)_\infty(-q;q)_\infty}{(aq/b;q)_\infty(-q/b;q)_\infty}\sum_{m=0}^\infty \frac{(q^2/b^2;q^2)_m}{(q^2;q^2)_m}a^m$$

$$= \frac{(a;q)_\infty(-q;q)_\infty}{(aq/b;q)_\infty(-q/b;q)_\infty}\cdot\frac{(aq^2/b^2;q^2)_\infty}{(a;q^2)_\infty}$$

$$= \frac{(aq;q^2)_\infty(-q;q)_\infty(aq^2/b^2;q^2)_\infty}{(aq/b;q)_\infty(-q/b;q)_\infty}.$$

Now remove the condition on a and the result is proved.   ■

Corollary 10.9.4

$$\sum_{n=0}^\infty \frac{q^{n(n-1)}x^n}{(q;q)_n(x;q)_n} = \frac{1}{(x;q)_\infty},\qquad \text{(Cauchy)},\tag{10.9.2}$$

$$\sum_{n=0}^\infty \frac{(a;q)_n q^{n(n+1)/2}}{(q;q)_n} = (aq;q^2)_\infty(-q;q)_\infty.\tag{10.9.3}$$

Proof. For (10.9.2) set $a = 1/A$, $b = 1/B$, and $c = x$ in Heine's formula (Corollary 10.9.2) and then let $A \to 0$, $B \to 0$. To obtain (10.9.3), set $b = 1/B$ in Bailey's formula (Theorem 10.9.3) and let $B \to 0$. This proves the corollary. ∎

The case $x = q$ in (10.9.2) gives

$$\sum_{n=0}^{\infty} \frac{q^{n^2}}{(1-q)^2(1-q^2)^2 \cdots (1-q^n)^2} = \prod_{n=1}^{\infty}(1-q^n)^{-1}.$$

This particular case of (10.9.2) was known to Euler. Similarly, the special case $a = q$ in (10.9.3) gives Gauss's formula

$$1 + \sum_{n=1}^{\infty} q^{n(n+1)/2} = \prod_{m=1}^{\infty} \frac{1-q^{2m}}{1-q^{2m-1}}.$$

We obtained this earlier from the triple product identity.

It should be clear by now that many results on hypergeometric series have extensions to basic hypergeometric series. We develop this theme a little more in the next section, where we derive some basic hypergeometric identities. For much more on this subject, see Gasper and Rahman [1990]. Identity (10.9.2) is in Cauchy [1843].

We terminate this section with a generalization of the $_2\phi_1$ series that will be needed in the next section and later. An $_r\phi_s$ basic hypergeometric series is defined by

$$_r\phi_s \left(\begin{matrix} a_1, a_2, \ldots, a_r \\ b_1, \ldots, b_s \end{matrix} ; q, x \right)$$

$$= \sum_{n=0}^{\infty} \frac{(a_1; q)_n (a_2; q)_n \cdots (a_r; q)_n}{(q; q)_n (b_1, q)_n \cdots (b_s; q)_n} \left((-1)^n q^{n(n-1)/2} \right)^{s+1-r} x^n, \quad (10.9.4)$$

where $q \neq 0$, when $r > s + 1$. An $_{r+1}\phi_r$ series is called k-balanced if $x = q$ and

$$b_1 b_2 \cdots b_r = q^k a_1 a_2 \cdots a_{r+1}. \quad (10.9.5)$$

When $k = 1$, the series is called balanced. The series (10.9.4) is called well poised if $s = r - 1$ and

$$qa_1 = b_1 a_2 = \cdots = b_{r-1} a_r. \quad (10.9.6)$$

10.10 Basic Hypergeometric Identities

An iteration of Heine's transformation in Theorem 10.9.1 gives

$$_2\phi_1 \left(\begin{matrix} a, b \\ c \end{matrix} ; q, x \right) = \frac{(c/b; q)_\infty (bx; q)_\infty}{(c; q)_\infty (x; q)_\infty} {}_2\phi_1 \left(\begin{matrix} abx/c, b \\ bx \end{matrix} ; q, c/b \right), \quad (10.10.1)$$

and a second iteration gives

Theorem 10.10.1

$$_2\phi_1\left(\begin{matrix} a,b \\ c \end{matrix}; q, x\right) = \frac{(abx/c; q)_\infty}{(x; q)_\infty}\,_2\phi_1\left(\begin{matrix} c/a, c/b \\ c \end{matrix}; q, abx/c\right). \qquad (10.10.2)$$

This is a q-analog of Euler's transformation formula

$$_2F_1\left(\begin{matrix} a, b \\ c \end{matrix}; x\right) = (1 - x)^{c-a-b}\,_2F_1\left(\begin{matrix} c - a, c - b \\ c \end{matrix}; x\right).$$

In (10.10.2), expand the infinite product on the right by the q-binomial theorem and equate the coefficients of x^n. This gives

$$\frac{(a; q)_n(b; q)_n}{(c; q)_n(q; q)_n} = \sum_{k=0}^{n} \frac{(c/a; q)_k(c/b; q)_k}{(c; q)_k(q; q)_k}(ab/c)^k \frac{(ab/c; q)_{n-k}}{(q; q)_{n-k}}$$

$$= \frac{(ab/c; q)_n}{(q; q)_n} \sum_{k=0}^{n} \frac{(q^{-n}; q)_k(c/a; q)_k(c/b; q)_k q^k}{(q^{1-n}c/ab; q)_k(c; q)_k(q; q)_k}.$$

After renaming parameters, we get

$$_3\phi_2\left(\begin{matrix} q^{-n}, a, b, \\ c, q^{1-n}ab/c \end{matrix}; q, q\right) = \frac{(c/a; q)_n(c/b; q)_n}{(c; q)_n(c/ab; q)_n}. \qquad (10.10.3)$$

This extends the Pfaff–Saalschütz identity for the balanced $_3F_2$. The $_3\phi_2$ in (10.10.3) is balanced because the product of the numerator parameters $q^{-n}ab$ times the power-series variable q equals the product of the denominator parameters. Recall that the q used in the definition of balanced appears as the power-series variable. For Heine's $_2\varphi_1$ sum in Corollary 10.9.2, the variable is c/ab, and the same type of condition holds, that is, the product of the numerator parameters times the power-series variable is the denominator parameter $(a \cdot b \cdot (c/ab) = c)$.

The following is a more general result due to Sears [1951]. It gives the transformation between two terminating balanced $_4\phi_3$ series and extends Whipple's $_4F_3$ transformation.

Theorem 10.10.2   *For a positive integer n,*

$$_4\phi_3\left(\begin{matrix} q^{-n}, a, b, c, \\ d, e, f \end{matrix}; q, q\right) = a^n \frac{(e/a; q)_n(f/a; q)_n}{(e; q)_n(f; q)_n}$$

$$\cdot\,_4\phi_3\left(\begin{matrix} q^{-n}, a, d/b, d/c \\ d, aq^{-n+1}/e, aq^{-n+1}/f \end{matrix}; q, q\right) \qquad (10.10.4)$$

when $def = abcq^{1-n}$.

Proof.   Take the transformation (10.10.2) twice each time with different parameters; then take their product such that the function on the left is multiplied by the function on the right in the other identity. This yields

$$
{}_2\phi_1\left(\begin{matrix} a, b \\ c \end{matrix}; q, x\right)\frac{(dex/f; q)_\infty}{(x; q)_\infty}{}_2\phi_1\left(\begin{matrix} f/d, f/e \\ f \end{matrix}; q, \frac{dex}{f}\right)
$$

$$
= {}_2\phi_1\left(\begin{matrix} d, e \\ f \end{matrix}; q, x\right)\frac{(abx/c; q)_\infty}{(x; q)_\infty}{}_2\phi_1\left(\begin{matrix} c/a, c/b \\ c \end{matrix}; q, \frac{abx}{c}\right).
$$

Reduce this to the product of two series by taking

$$
\frac{ab}{c} = \frac{de}{f}.
$$

The equality of the coefficients of x^n gives

$$
\sum_{k=0}^n \frac{(a; q)_k(b; q)_k}{(c; q)_k(q; q)_k}\frac{(f/d; q)_{n-k}(f/e; q)_{n-k}}{(f; q)_{n-k}(q; q)_{n-k}}(de/f)^{n-k}
$$

$$
= \sum_{k=0}^n \frac{(d; q)_k(e; q)_k}{(f; q)_k(q; q)_k}\cdot\frac{(c/a; q)_{n-k}(c/b; q)_{n-k}}{(c; q)_{n-k}(q; q)_{n-k}}(ab/c)^{n-k}.
$$

After some rearrangement and renaming of parameters, it can be seen that this is equivalent to (10.10.4). ∎

It should be observed that both sides of (10.10.4) are balanced series. There are many interesting limiting cases of this transformation given in the next theorem. Before we state it, note that if $c = d$ in (10.10.4), we get (10.10.3).

Theorem 10.10.3

$$
{}_3\phi_2\left(\begin{matrix} q^{-n}, a, b \\ d, e \end{matrix}; q, q\right) = \frac{(e/a; q)_n a^n}{(e; q)_n}{}_3\phi_2\left(\begin{matrix} q^{-n}, a, d/b \\ d, q^{1-n}a/e \end{matrix}; q, bq/e\right), \tag{10.10.5}
$$

$$
{}_3\phi_2\left(\begin{matrix} q^{-n}, a, b \\ e, f \end{matrix}; q, q\right) = \frac{(e/a; q)_n(f/a; q)_n}{(e; q)_n(f; q)_n}a^n{}_3\phi_2\left(\begin{matrix} q^{-n}, a, abq^{1-n}/ef \\ q^{1-n}a/e, q^{-n+1}a/f \end{matrix}; q, q\right), \tag{10.10.6}
$$

$$
{}_3\phi_2\left(\begin{matrix} a, b, c \\ d, e \end{matrix}; q, de/abc\right) = \frac{(e/a; q)_\infty(de/bc; q)_\infty}{(e; q)_\infty(de/abc; q)_\infty}{}_3\phi_2\left(\begin{matrix} a, d/b, d/c \\ d, de/bc \end{matrix}; q, e/a\right), \tag{10.10.7}
$$

$$
{}_3\phi_2\left(\begin{matrix} a, b, c \\ d, e \end{matrix}; q, de/abc\right) = \frac{(a; q)_\infty(de/ab; q)_\infty(de/ac; q)_\infty}{(d; q)_\infty(e; q)_\infty(de/abc; q)_\infty}
$$

$$
\cdot {}_3\phi_2\left(\begin{matrix} d/a, e/a, de/abc \\ de/ab, de/ac \end{matrix}; q, a\right). \tag{10.10.8}
$$

Proof.   To prove (10.10.5), let f and c tend to zero in (10.10.4), keeping f/c fixed. Similarly, for (10.10.6), let d and c tend to zero with $d/c = q^{-n+1}ab/ef$. Finally, to prove (10.10.7), set $f = q^\lambda$ and let $n \to \infty$ with $n + \lambda$ fixed so that $q^{n+\lambda} = abcq/de$.

Now (10.10.7) is a generalization of Kummer's transformation

$$_3F_2\left(\begin{matrix} a, b, c \\ d, e \end{matrix}; 1\right) = \frac{\Gamma(e)\Gamma(d+e-a-b-c)}{\Gamma(e-a)\Gamma(d+e-b-c)} {_3F_2}\left(\begin{matrix} a, d-b, d-c \\ d, d+e-b-c \end{matrix}; 1\right).$$

(10.10.9)

When (10.10.7) is applied to itself, the result is (10.10.8), with a and c interchanged.   ∎

Remark 10.10.1   It is worth mentioning that if we set $x = c/e$ and let $c \to \infty$, Kummer's transformation becomes

$$_2F_1\left(\begin{matrix} a, b \\ d \end{matrix}; x\right) = (1-x)^{-a} {_2F_1}\left(\begin{matrix} a, d-b \\ d \end{matrix}; \frac{x}{x-1}\right).$$

(10.10.10)

In contrast, if $x = a/e$ and $a \to \infty$ it becomes

$$_2F_1\left(\begin{matrix} b, c \\ d \end{matrix}; x\right) = (1-x)^{d-b-c} {_2F_1}\left(\begin{matrix} d-b, d-c \\ d \end{matrix}; x\right).$$

(10.10.11)

We have already seen a q-extension of the last formula. Let us determine a q-extension of (10.10.10). In formula (10.10.7), let $de/abc = x$ and c and e tend to zero. The result is

$$_2\phi_1\left(\begin{matrix} a, b \\ d \end{matrix}; q, x\right) = \frac{(ax; q)_\infty}{(x; q)_\infty} \sum_{n=0}^{\infty} \frac{(a, q)_n (d/b; q)_n q^{\binom{n}{2}}(-1)^n}{(d; q)_n (ax; q)_n (q; q)_n}(bx)^n.$$

(10.10.12)

When a, b, and c are replaced by q^a, q^b, and q^c respectively and we let $q \to 1^-$, this reduces to (10.10.10). This is an instance where the q-extension is nicer than the $q = 1$ case in the following sense: The left side of (10.10.10) is analytic for $|x| < 1$, while the right side has two factors, the first analytic for x in $\mathbb{C} - [1, \infty)$ and the second analytic for $\text{Re}\, x < 1/2$. In (10.10.12), the $_2\phi_1$ on the left is analytic for $|x| < 1$, while on the right, $1/(x; q)_\infty$ has poles when $x = 1, q^{-1}, q^{-2}, \ldots$. The other factors are entire functions since $(ax; q)_\infty/(ax; q)_n = (axq^n; q)_\infty$ is entire while the series converges uniformly for $|x| \le A$ for each A.

10.11 q-Ultraspherical Polynomials

In Section 6.11, we introduced the (continuous) q-ultraspherical polynomials defined by

$$
C_n(x; \beta \mid q) = \sum_{k=0}^{n} \frac{(\beta; q)_n (\beta; q)_{n-k}}{(q; q)_k (q; q)_{n-k}} \cos(n - 2k)\theta
$$

$$
= \frac{(\beta; q)_n}{(q; q)_n} e^{in\theta} {}_2\phi_1 \left(\begin{matrix} q^{-n}, \beta \\ q^{1-n}\beta^{-1} \end{matrix}; q, q\beta^{-1} e^{-2i\theta} \right). \quad (10.11.1)
$$

(Because in this section we discuss only the continuous q-ultraspherical polynomials, the word continuous will be dropped.) When the generating function found in Chapter 6 is combined with the q-binomial theorem, the result is

$$
\sum_{n=0}^{\infty} C_n(\cos\theta; \beta \mid q) r^n = \frac{(\beta r e^{i\theta}; q)_\infty (\beta r e^{-i\theta}; q)_\infty}{(r e^{i\theta}; q)_\infty (r e^{-\theta}; q)_\infty}, \quad 0 < r < 1. \quad (10.11.2)
$$

It is also easy to conclude from (6.11.5) and (6.11.6) that the polynomials $C_n(x; \beta \mid q)$ satisfy the recurrence relation

$$
2(1 - \beta q^n) x C_n(x; \beta \mid q) = (1 - q^{n+1}) C_{n+1}(x; \beta \mid q)
$$
$$
+ (1 - \beta^2 q^{n-1}) C_{n-1}(x; \beta \mid q). \quad (10.11.3)
$$

An implication of Theorem 6.6.2 is that $\{C_n(x; \beta \mid q)\}, x = \cos\theta$ is an orthogonal polynomial sequence with respect to the distribution

$$
\omega_\beta(\cos\theta) d\theta = \left| \frac{(e^{2i\theta}; q)_\infty}{(\beta e^{2i\theta}; q)_\infty} \right|^2 d\theta. \quad (10.11.4)
$$

Since we did not prove Theorem 6.6.2, we verify this fact directly here.

Theorem 10.11.1 When $|q| < 1$ and $|\beta| < 1$,

$$
\int_0^\pi C_n(\cos\theta; \beta \mid q) C_m(\cos\theta; \beta \mid q) \omega_\beta(\cos\theta) d\theta
$$
$$
= \frac{2\pi(1 - \beta)}{(1 - \beta q^n)} \cdot \frac{(\beta^2; q)_n}{(q; q)_n} \cdot \frac{(\beta; q)_\infty (\beta q; q)_\infty}{(\beta^2; q)_\infty (q; q)_\infty} \delta_{mn}, \quad (10.11.5)
$$

where $\omega_\beta(\cos\theta)$ is given by (10.11.4).

Proof. Take β and q real. Since the integrand in (10.11.5) is unchanged by the transformation $\theta \to -\theta$ and $C_n(\cos\theta; \beta \mid q)$ has an expansion in terms of $e^{i(n-2k)\theta}$, start with the integral

$$
\int_{-\pi}^{\pi} e^{ik\theta} \omega_\beta(\cos\theta) d\theta = \int_{-\pi}^{\pi} e^{ik\theta} \frac{(e^{2i\theta}; q)_\infty (e^{-2i\theta}; q)_\infty}{(\beta e^{2i\theta}; q)_\infty (\beta e^{-2i\theta}; q)_\infty} d\theta.
$$

For $|\beta| < 1$, the q-binomial theorem gives

$$\int_{-\pi}^{\pi} e^{ik\theta} \omega_\beta(\cos\theta)d\theta = \sum_{n=0}^{\infty} \frac{(\beta^{-1};q)_n}{(q;q)_n}\beta^n \sum_{m=0}^{\infty} \frac{(\beta^{-1};q)_m}{(q;q)_m}\beta^m \int_{-\pi}^{\pi} e^{i(k+2n-2m)\theta} d\theta.$$

When k is odd, the integral is zero. When $k = 2\ell$,

$$\int_{-\pi}^{\pi} e^{2i\ell\theta} \omega_\beta(\cos\theta)d\theta = 2\pi \sum_{n=0}^{\infty} \frac{(\beta^{-1};q)_\infty(\beta^{-1};q)_{\ell+n}}{(q;q)_n(q;q)_{\ell+n}}\beta^{2n+\ell}$$

$$= 2\pi \frac{(\beta^{-1};q)_\ell}{(q;q)_\ell}\beta^\ell \, {}_2\phi_1\left(\begin{matrix} \beta^{-1}, \beta^{-1}q^\ell \\ q^{\ell+1} \end{matrix}; q, \beta^2\right).$$

Apply (10.10.1) to the above ${}_2\phi_1$ to get

$$\frac{(\beta q^\ell;q)_\infty(\beta q;q)_\infty}{(\beta^2;q)_\infty(q^{\ell+1};q)_\infty}{}_2\phi_1\left(\begin{matrix} q^{-1}, \beta^{-1}q^\ell \\ \beta q^\ell \end{matrix}; q, \beta q\right).$$

This ${}_2\phi_1$ has only two terms, and so we have

$$\int_{-\pi}^{\pi} e^{i2\ell\theta} \omega_\beta(\cos\theta) = \frac{2\pi\beta^\ell(\beta^{-1};q)_\ell(1+q^\ell)}{(\beta q;q)_\ell} \frac{(\beta q;q)_\infty(\beta;q)_\infty}{(\beta^2;q)_\infty(q;q)_\infty}.$$

A second way to evaluate this integral is to use Ramanujan's ${}_1\psi_1$ sum (10.5.3).

Let $m \le n$. Now consider the integral

$$\int_0^\pi \cos(m-2k)\theta \, C_n(\cos\theta; \beta \mid q)\omega_\beta(\cos\theta)d\theta$$

$$= \frac{1}{2}\int_{-\pi}^{\pi} e^{i(m-2k)\theta} \sum_{\ell=0}^{n} \frac{(\beta;q)_\ell(\beta;q)_{n-\ell}}{(q;q)_\ell(q;q)_{n-\ell}} e^{i(n-2\ell)\theta} \omega_\beta(\cos\theta)d\theta.$$

Since $m + n - 2k - 2\ell$ has to be even for the contribution of the integral to be nonzero, let $m - n - 2k = -2s$ or $m - 2k = n - 2s$. Then the integral is

$$\frac{1}{2}\sum_{\ell=0}^{n} \frac{(\beta;q)_\ell(\beta;q)_{n-\ell}}{(q;q)_\ell(q;q)_{n-\ell}} \int_{-\pi}^{\pi} e^{2i(n-\ell-s)\theta} \omega_\beta(\cos\theta)d\theta$$

$$= \frac{1}{2}\sum_{\ell=0}^{n} \frac{(\beta;q)_\ell(\beta;q)_{n-\ell}}{(q;q)_\ell(q;q)_{n-\ell}} \int_{-\pi}^{\pi} e^{2i(s-\ell)\theta} \omega_\beta(\cos\theta)d\theta$$

$$= \pi \frac{(\beta q;q)_\infty(\beta;q)_\infty}{(\beta^2;q)_\infty(q;q)_\infty} \sum_{\ell=0}^{n} \frac{(\beta;q)_\ell(\beta;q)_{n-\ell}}{(q;q)_\ell(q;q)_{n-\ell}}(1+q^{s-\ell})\frac{(\beta^{-1};q)_{s-\ell}}{(\beta q;q)_{s-\ell}}$$

$$= \pi \frac{(\beta q;q)_\infty(\beta;q)_\infty}{(\beta^2;q)_\infty(q;q)_\infty} \cdot \frac{(\beta;q)_n(\beta^{-1};q)_s}{(q;q)_n(\beta q;q)_s} \cdot \beta^s(1+q^s)$$

$$\cdot \sum_{\ell=0}^{n} \frac{(q^{-n};q)_\ell(q^{-s}\beta^{-1};q)_\ell(\beta;q)_\ell(-q^{1-s};q)_\ell q^\ell}{(q;q)_\ell(q^{1-n}\beta^{-1};q)_\ell(q^{1-s}\beta;q)_\ell(-q^{-s};q)_\ell}.$$

The second equation in the above calculation is obtained by changing ℓ to $n - \ell$ and θ to $-\theta$. The last sum, which is a balanced ${}_4\phi_3$, when transformed by Sears's formula (Theorem 10.10.2) becomes

$$\frac{\beta^n(q^{1-s};q)_n(q^{1-n}/\beta^2;q)_n}{(q^{1-s}\beta;q)_n(q^{1-n}/\beta;q)_n} {}_4\phi_3\left(\begin{matrix} q^{-n}, \beta, -\beta, q^{-1} \\ -q^{-s}, q^{s-n}, \beta^2 \end{matrix}; q, q\right).$$

This ${}_4\phi_3$ has only two nonzero terms and the above expression becomes, after a simple calculation,

$$\frac{(\beta^2;q)_n(q^{1-s};q)_{n-1}(1-q^{n-2s})}{(\beta;q)_n(q^{1-s}\beta;q)_n(1+q^{-s})}.$$

The factor $(q^{1-s};q)_{n-1}$ is zero for $s = 1, 2, \ldots, n-1$. This gives

$$\int_0^\pi \cos(n-2s)\theta C_n(\cos\theta; \beta \mid q)\omega_\beta(\cos\theta)d\theta$$

$$= \begin{cases} 0 & \text{for } s = 1, 2, \ldots, n-1, \\ \pi\dfrac{(\beta;q)_\infty(\beta q;q)_\infty(\beta^2;q)_n}{(\beta^2;q)_\infty(q;q)_\infty(\beta q;q)_n} & \text{for } s = 0 \quad \text{or} \quad n. \end{cases}$$

By (10.11.1) and the above relation we obtain the orthogonality

$$\int_0^\pi C_m(\cos\theta; \beta \mid q)C_n(\cos\theta; \beta \mid q)\omega_\beta(\cos\theta)d\theta$$

$$= 2\pi\frac{(\beta;q)_n}{(q;q)_n}\frac{(\beta^2;q)_n(\beta;q)_\infty(\beta q;q)_\infty}{(\beta q;q)_n(\beta^2;q)_\infty(q;q)_\infty}\delta_{mn}$$

$$= 2\pi\frac{(1-\beta)}{(1-\beta q^n)}\cdot\frac{(\beta^2;q)_n}{(q;q)_n}\cdot\frac{(\beta;q)_\infty(\beta q;q)_\infty}{(q;q)_\infty(\beta^2;q)_\infty}\delta_{mn}.$$

The theorem is proved. ∎

The polynomials $C_n(\cos\theta; \beta|q)$ satisfy a difference equation. To state it, we need the *q*-difference operator D_q defined by

$$D_q f(x) = \frac{\delta_q f(x)}{\delta_q x} \quad \text{with} \quad \delta_q g(e^{i\theta}) = g(q^{1/2}e^{i\theta}) - g(q^{-1/2}e^{i\theta}), \quad x = \cos\theta.$$

$$(10.11.6)$$

It can be shown from the generating function for $C_n(\cos\theta; \beta \mid q)$ that

$$D_q C_n(x; \beta \mid q) = \frac{2(1-\beta)}{(1-q)}q^{(1-n)/2}C_{n-1}(x; \beta q \mid q).$$ (10.11.7)

The *q*-difference equation is

$$(1-q^2)D_q[\omega_{\beta q}(x)D_q y(x)] + 4q^{1-n}(1-q^n)(1-\beta^2 q^n)\omega_\beta(x)y(x) = 0,$$

$$(10.11.8)$$

with $y(x) = C_n(x; \beta \mid q)$. As a first step to the proof of (10.11.8), one can show that

$$D_q(\omega_\beta(x)C_n(x; \beta \mid q))$$

$$= -\frac{2q^{-n/2}(1 - q^{n+1})(1 - \beta^2 q^{n-1})}{(1 - q)(1 - \beta q)} w_{\beta/q}(x)C_{n+1}(x; \beta q \mid q).$$

One can use (10.11.7) to give a proof of the connection coefficient formula:

$$C_n(x; \gamma \mid q) = \sum_{k=0}^{[n/2]} \beta^k \frac{(\gamma \beta^{-1}; q)_k (\gamma; q)_{n-k}}{(q; q)_k (\beta q; q)_{n-k}} \frac{1 - \beta q^{n-2k}}{1 - \beta} C_{n-2k}(x; \beta \mid q).$$

$$(10.11.9)$$

The proof follows the same steps as that of Theorem 7.1.4$'$. Formula (10.11.9) was first given by Rogers [1895]. Rogers also found the following linearization formula:

$$C_m(x; \beta \mid q)C_n(x; \beta \mid q)$$

$$= \sum_{k=0}^{\min(m,n)} \frac{(q; q)_{m+n-2k}(\beta; q)_{m-k}(\beta; q)_{n-k}(\beta; q)_k(\beta^2; q)_{m+n-k}}{(\beta^2; q)_{m+n-2k}(q; q)_{m-k}(q; q)_{n-k}(q; q)_k(\beta q; q)_{m+n-k}}$$

$$\cdot \frac{(1 - \beta q^{m+n-2k})}{(1 - \beta)} C_{m+n-2k}(x; \beta \mid q).$$

$$(10.11.10)$$

Proving this by induction is easy. It is likely that Rogers first computed the formula for a few small values of m and then guessed the general result. The simplest direct evaluation of the formula may be the one similar to the proof of Theorem 6.8.2, which uses the q-analog of Whipple's transformation given in Chapter 12. For details see Gasper [1985]. The proofs of (10.11.7) to (10.11.10) are left to the reader.

When we set $\beta = 0$ in $C_n(x; \beta \mid q)$ we get the (continuous) q-Hermite polynomials. They are defined by

$$H_n(x \mid q) = (q; q)_n C_n(x; 0 \mid q). \tag{10.11.11}$$

The following properties of the q-Hermite polynomials are now immediate:

$$H_n(\cos\theta \mid q) = \sum_{k=0}^{n} \frac{(q; q)_n}{(q; q)_k (q; q)_{n-k}} \cos(n - 2k)\theta$$

$$= \sum_{k=0}^{n} \frac{(q; q)_n e^{i(n-2k)\theta}}{(q; q)_k (q; q)_{n-k}}. \tag{10.11.12}$$

They satisfy the orthogonality relation

$$\int_0^{\pi} H_m(\cos\theta \mid q) H_n(\cos\theta \mid q) |(e^{2i\theta}; q)_\infty|^2 d\theta = \frac{2\pi \delta_{mn}}{(q^{n+1}; q)_\infty}. \tag{10.11.13}$$

Since the weight function for the Hermite polynomials is e^{-x^2} and

$$\lim_{q \to 1^-} \frac{H_n(x((1-q)/2)^{1/2} \mid q)}{((1-q)/2)^{n/2}} = H_n(x), \tag{10.11.14}$$

we may regard the integral

$$\int_0^\pi |(e^{2i\theta}; q)_\infty|^2 d\theta = \frac{2\pi}{(q; q)_\infty}$$

as an extension of the normal integral

$$\int_{-\infty}^\infty e^{-x^2} dx = \sqrt{\pi}.$$

The generating function for $H_n(x \mid q)$ is given by

$$\sum_{n=0}^\infty \frac{H_n(x \mid q)}{(q; q)_n} r^n = \frac{1}{(re^{i\theta}; q)_\infty (re^{-i\theta}; q)_\infty}, \qquad x = \cos\theta; \tag{10.11.15}$$

and the three-term recurrence relation is

$$2x H_n(x \mid q) = H_{n+1}(x \mid q) + (1 - q^n) H_{n-1}(x \mid q). \tag{10.11.16}$$

By (10.11.10), the linearization formula is easily seen to be

$$\frac{H_m(x \mid q) H_n(x \mid q)}{(q; q)_m (q; q)_n} = \sum_{k=0}^{\min(m,n)} \frac{H_{m+n-2k}(x \mid q)}{(q; q)_k (q; q)_{n-k} (q; q)_{m-k}}. \tag{10.11.17}$$

A direct proof is indicated in Exercise 41. From (10.11.17) and the q-binomial theorem, it is possible to derive a formula for the Poisson kernel of the q-Hermite polynomials:

$$\sum_{n=0}^\infty \frac{H_n(\cos\theta \mid q) H_n(\cos\phi \mid q)}{(q; q)_n} r^n = \frac{(r^2; q)_\infty}{\left|\left(re^{i(\theta+\phi)}; q\right)_\infty \left(re^{-i(\theta-\phi)}; q\right)_\infty\right|^2}, \tag{10.11.18}$$

where r is real, $-1 < r < 1$. The derivation is left to the reader as an exercise.

Finally we note that the integral $I(a, b, c, d)$ of Section 10.8 can be written in terms of q-Hermite polynomials as

$$I(a, b, c, d) = \sum_{k,\ell,m,j \geq 0} \frac{a^k b^\ell c^m d^j}{(q; q)_j (q; q)_k (q; q)_\ell (q; q)_m}$$

$$\cdot \int_0^\pi H_k(x \mid q) H_\ell(x \mid q) H_m(x \mid q) H_j(x \mid q) |(e^{2i\theta}; q)_\infty|^2 d\theta, \tag{10.11.19}$$

where $x = \cos\theta$. This follows easily from the generating function (10.11.15). By means of the linearization formula (10.11.17), another evaluation of the integral is obtained. This observation is due to Ismail and Stanton [1988]. They also pointed out that (10.11.17) and (10.11.18) are equivalent. Al-Salam and Ismail [1988] used the connection coefficients and linearization for (continuous) q-ultraspherical and Hermite polynomials to prove Theorem 10.8.2.

10.12 Mellin Transforms

The integral of $x^{\alpha-1}(-ax; q)_\infty/(-x, q)_\infty$, evaluated in Section 10.8 by a functional equation, is also a particular case of an interesting Mellin transformation formula. This formula, given by Ramanujan, has many important applications, some of which will be presented in the exercises. The Mellin transform connects up in an important way the transformations of some q-series with functional equations satisfied by certain Dirichlet series. Earlier we mentioned a transformation for

$$1 + 2\sum_{n=1}^{\infty} q^{n^2}, \quad q = e^{-\pi x},$$

that was useful in computing a Riemann sum approximation of the normal integral $\int_{-\infty}^{\infty} e^{-ct^2} dt$. In Exercise 2.28, the reader was asked to use this transformation to obtain the functional equation for the Riemann zeta function, $\zeta(s)$. This relationship between q-series, which arise from elliptic functions, and Dirichlet series is particularly important in number theory. We discuss a few examples, especially those involving q-series, as considered in the previous sections.

Ramanujan stated his formula as

$$\int_0^\infty x^{s-1}\{\phi(0) - x\phi(1) + x^2\phi(2) - \cdots\}dx = \frac{\pi}{\sin s\pi}\phi(-s).$$

In this form, one must put some strong restrictions on $\phi(s)$. Hardy [1940, pp. 189–190] gave fairly general conditions for the validity of this formula. See also Berndt [1985, p. 299]. Hardy's theorem is stated below.

Let $H(\delta)$ denote the half plane $u = \sigma + it$, $\sigma \geq -\delta$, $0 < \delta < 1$. Suppose $A < \pi$; let $K(A, P, \delta)$ denote the set of all functions ϕ, holomorphic on $H(\delta)$, that satisfy

$$|\phi(u)| = O(e^{P\sigma + A|t|}). \tag{10.12.1}$$

Take $0 < c < \delta$, and define

$$\Phi(x) = \frac{1}{2\pi i}\int_{c-i\infty}^{c+i\infty} \frac{\pi}{\sin \pi u}\phi(-u)x^{-u}du. \tag{10.12.2}$$

The integrand is

$$O\left(e^{-(\pi-A)|t|}e^{-Pc}x^{-c}\right);$$

thus that the integral converges uniformly in any interval $0 < x_0 \leq x \leq X$. Therefore, the integral represents an analytic function $\Phi(x)$ for $x > 0$. An application
of Cauchy's theorem gives

$$\Phi(x) - \frac{1}{2\pi i} \int_{-N-\frac{1}{2}-i\infty}^{-N-\frac{1}{2}+i\infty} \frac{\pi}{\sin \pi u} \phi(-u) x^{-u} du$$

$$= \phi(0) - x\phi(1) + \cdots + (-1)^N x^N \phi(N). \tag{10.12.3}$$

If we take $0 < x < e^{-P}$, then the series $\sum_0^\infty (-1)^n \phi(n) x^n$ converges and the
integral in (10.12.3) goes to 0 as $N \to \infty$. So, for $0 < x < e^{-P}$,

$$\Phi(x) = \phi(0) - \phi(1)x + \phi(2)x^2 \ldots . \tag{10.12.4}$$

Theorem 10.12.1   Let $0 < \operatorname{Re} s < \delta$. If $\Phi(x)$ is given by (10.12.2), then

$$\int_0^\infty x^{s-1} \Phi(x) dx = \frac{\pi}{\sin s\pi} \phi(-s).$$

Proof.   Choose c_1 and c_2 so that $0 < c_1 < \operatorname{Re} s = \sigma < c_2 < \delta$. Compute the
absolutely convergent double integral

$$\frac{1}{2\pi i} \int_0^1 \int_{c_1-i\infty}^{c_1+i\infty} \frac{\pi}{\sin \pi u} \phi(-u) x^{s-u-1} du dx$$

in two different ways. One way it equals

$$\frac{1}{2\pi i} \int_0^1 x^{s-1} \left(\int_{c_1-i\infty}^{c_1+i\infty} \frac{\pi}{\sin \pi u} \phi(-u) x^{-u} du \right) dx = \int_0^1 x^{s-1} \Phi(x) dx.$$

The other way it is

$$\frac{1}{2\pi i} \int_{c_1-i\infty}^{c_1+i\infty} \frac{\pi}{\sin \pi u} \phi(-u) \left(\int_0^1 x^{s-u-1} dx \right) du = \frac{1}{2\pi i} \int_{c_1-i\infty}^{c_1+i\infty} \frac{\pi}{\sin \pi u} \frac{\phi(-u)}{s-u} du.$$

Therefore,

$$\int_0^1 x^{s-1} \Phi(x) dx = \frac{1}{2\pi i} \int_{c_1-i\infty}^{c_1+i\infty} \frac{\pi}{\sin \pi u} \frac{\phi(-u)}{s-u} du.$$

Similarly, use the double integral

$$\frac{1}{2\pi i} \int_1^\infty \int_{c_2-i\infty}^{c_2+i\infty} \frac{\pi}{\sin \pi u} \phi(-u) x^{s-u-1} du dx$$

to get

$$\int_1^\infty x^{s-1} \Phi(x) dx = -\frac{1}{2\pi i} \int_{c_2-i\infty}^{c_2+i\infty} \frac{\pi}{\sin \pi u} \frac{\phi(-u)}{s-u} du.$$

An application of Cauchy's residue theorem gives

$$\int_0^\infty x^{s-1}\Phi(x)dx = \frac{1}{2\pi i}\left(\int_{c_1-i\infty}^{c_1+i\infty} - \int_{c_2-i\infty}^{c_2+i\infty}\right)\frac{\pi}{\sin \pi u}\frac{\phi(-u)}{s-u}du$$

$$= \frac{\pi}{\sin \pi s}\phi(-s). \qquad \blacksquare$$

Corollary 10.12.2 *If* $0 < q < 1$, $s > 0$, *and* $0 < a < q^s$, *then*

$$\int_0^\infty x^{s-1}\frac{(-ax;q)_\infty}{(x;q)_\infty}dx = \frac{\pi}{\sin s\pi}\frac{(q^{1-s};q)_\infty(a;q)_\infty}{(q;q)_\infty(aq^{-s};q)_\infty}.$$

Proof. Take

$$\phi(u) = \frac{(a;q)_\infty(q^{u+1};q)_\infty}{(q;q)_\infty(aq^u;q)_\infty},$$

and check that ϕ satisfies the conditions of Theorem 10.12.1. $\blacksquare$

Another corollary is Carlson's theorem, which we proved and used in previous chapters for ϕ bounded.

Corollary 10.12.3 *Suppose* $\phi(u)$ *is analytic in a half plane* $H(\delta) = \{u \mid \operatorname{Re} u \geq -\delta\}$, $0 < \delta < 1$ *satisfying (10.12.1), and* $\phi(n) = 0$ *for* $n = 0, 1, 2, \ldots$. *Then* $\phi = 0$.

This corollary shows that Ramanujan's formula is actually an interpolation formula. In fact, Newton's interpolation formula

$$\lambda(-s) = \lambda(0) + \frac{s}{1!}\Delta\lambda(0) + \frac{s(s+1)}{2!}\Delta^2\lambda(0) + \cdots,$$

where

$$\Delta\lambda(0) = \lambda(1) - \lambda(0), \quad \Delta^2\lambda(0) = \lambda(2) - 2\lambda(1) + \lambda(0), \ldots,$$

can be derived from Theorem 10.12.1, under conditions on λ. For this and other applications the reader should see Exercises 34–36.

When we discussed Mellin transforms earlier, it was mentioned that Mellin transforms are just Fourier transforms with a change of variables. Thus theorems about Fourier transforms have analogs for Mellin transforms. Therefore, there is a uniqueness theorem for the Mellin transform. Suppose $x^{\sigma-1}f(x)$ is integrable on $(0, \infty)$ for $\alpha < \sigma < \beta$. Then

$$F(s) = \int_0^\infty x^{s-1}f(x)dx \qquad (10.12.5)$$

exists and is analytic for $\alpha < \operatorname{Re} s < \beta$.

Lemma 10.12.4 *If $F(\sigma_0 + it) = 0$ for all $t \in (-\infty, \infty)$ and fixed σ_0 in (α, β), then $f(x) \equiv 0$ almost everywhere.*

Proof. Let $x = e^u$ in (10.12.5). Then

$$F(\sigma_0 + it) = \int_{-\infty}^{\infty} e^{itu} e^{\sigma_0 u} f(e^u) du$$

is the Fourier transform of $e^{\sigma_0 u} f(e^u) = 0$ almost everywhere. Since $e^{\sigma_0 u}$ never vanishes, the lemma is proved. ∎

Note that if, in the lemma, f is continuous at a point, then f must be zero at that point. This implies that if the Mellin transforms of two continuous functions are the same, then the two functions are equal. Here is an application of this fact:

Theorem 10.12.5 *For $u > 0$*

$$\int_0^{\infty} e^{-(ut + x^2/4t)} \frac{dt}{\sqrt{t}} = \frac{\sqrt{\pi} e^{-x\sqrt{u}}}{\sqrt{u}}.$$

Proof. It is sufficient to prove this for $u = 1$. Denote the integral by $f(x)$. For $\operatorname{Re} s > 1$, the double integral

$$\int_0^{\infty} \int_0^{\infty} x^{s-1} e^{-t-(x^2/4t)} \frac{dt}{\sqrt{t}} dx$$

is absolutely convergent. Thus it is equal to

$$\int_0^{\infty} x^{s-1} \left[\int_0^{\infty} e^{-t-(x^2/4t)} \frac{dt}{\sqrt{t}} \right] dx = \int_0^{\infty} x^{s-1} f(x) dx.$$

Now change the order of integration to get

$$\int_0^{\infty} \frac{e^{-t}}{\sqrt{t}} \left[\int_0^{\infty} x^{s-1} e^{-x^2/4t} dx \right] dx = 2^{s-1} \Gamma(s/2) \int_0^{\infty} t^{(s-1)/2} e^{-t} dt$$

$$= 2^{s-1} \Gamma(s/2) \Gamma((s+1)/2)$$

$$= \sqrt{\pi}\, \Gamma(s).$$

The last step follows from Legendre's duplication formula. We have

$$\int_0^{\infty} x^{s-1} f(x) dx = \sqrt{\pi}\, \Gamma(s).$$

This means that $f(x)$ and $\sqrt{\pi} e^{-x}$ have the same Mellin transform and hence they are equal. This concludes the proof of the theorem. ∎

This proof follows Bellman [1961, p. 30], which contains an interesting discussion of theta functions.

There is another related integral transform that is important, the Laplace transform, defined by

$$F(s) = \int_0^\infty e^{-st} f(t) dt.$$

If f is integrable and satisfies $|f(t)| = O(e^{bt})$ as $t \to \infty$, then $F(s)$ is analytic for Re $s > b$. There are uniqueness theorems for Laplace transforms as well. We can use this uniqueness to prove the transformation formula for theta functions mentioned earlier.

Theorem 10.12.6

$$\sum_{n=-\infty}^{\infty} e^{-n^2 x} = \sqrt{\frac{\pi}{x}} \sum_{n=-\infty}^{\infty} e^{-n^2\pi^2/x}, \quad \operatorname{Re} x > 0.$$

(A more general result can be proved by the Poisson summation formula, but this result shows that the theta function defined by the series is a modular form. See Remarks 10.12.1 and 10.12.2 at the end of the section.)

Proof.   For Re $s > 0$,

$$\int_0^\infty e^{-sx} \sum_{n=-\infty}^{\infty} e^{-n^2 x} dx = \frac{1}{s} + 2 \sum_{n=1}^{\infty} \frac{1}{n^2 + s}.$$

By Theorem 10.12.5, when Re $s > 0$

$$\sqrt{\pi} \int_0^\infty e^{-sx} \left[\frac{1}{\sqrt{x}} + 2 \sum_{n=1}^{\infty} \frac{e^{-n^2\pi^2/x}}{\sqrt{x}} \right] dx = \frac{\pi}{\sqrt{s}} + \frac{2\pi}{\sqrt{s}} \sum_{n=1}^{\infty} e^{-2n\pi\sqrt{s}}$$

$$= \frac{\pi}{\sqrt{s}} + \frac{2\pi}{\sqrt{s}} \frac{e^{-2\pi\sqrt{s}}}{1 - e^{-2\pi\sqrt{s}}} = \frac{\pi}{\sqrt{s}} \frac{1 + e^{-2\pi\sqrt{s}}}{1 - e^{-2\pi\sqrt{s}}}.$$

According to (1.2.5), we have

$$\pi \cot \pi x = \frac{1}{x} + 2 \sum_{n=1}^{\infty} \frac{x}{x^2 - n^2}.$$

After proper identification, this shows that

$$\frac{\pi}{\sqrt{s}} \frac{1 + e^{-2\pi\sqrt{s}}}{1 - e^{-2\pi\sqrt{s}}} = \frac{1}{s} + 2 \sum_{n=1}^{\infty} \frac{1}{n^2 + s},$$

and the lemma is proved by the uniqueness of the Laplace transform. This proof is due to Hamburger [1922]. ∎

When we take the Mellin transform of $\sum_{-\infty}^{\infty} e^{-n^2 x}$, the Riemann zeta function is obtained. This gives a connection between q-series and Dirichlet series. An exercise in Chapter 2 asks the reader to use the result of Theorem 10.12.6 to obtain the functional equation of $\zeta(s)$. We give the details here because the result and proof are important.

Theorem 10.12.7 *The expression $\pi^{-s/2}\Gamma(s/2)\zeta(s)$ is invariant under $s \to 1-s$.*

The following proof goes back to Riemann [1859].

Proof. For Re $s > 1$, and $\psi(x) = \sum_{n=1}^{\infty} e^{-n^2\pi x}$,

$$\pi^{-s/2}\Gamma(s/2)\zeta(s) = \int_0^\infty x^{(s/2)-1} \sum_{n=1}^{\infty} e^{-n^2\pi x} dx$$

$$= \int_0^1 x^{(s/2)-1}\psi(x)dx + \int_1^\infty x^{(s/2)-1}\psi(x)dx.$$

Change x to $1/x$ in the first integral to get

$$\int_1^\infty x^{-(1/2)-1}\psi(1/x)dx. \tag{10.12.6}$$

By Theorem 10.12.6

$$1 + 2\psi(x) = \frac{1}{\sqrt{x}}(1 + 2\psi(1/x)).$$

Thus (10.12.6) becomes

$$\int_1^\infty x^{-(s/2)-1}\left[\frac{\sqrt{x}}{2} - \sqrt{x}\,\psi(x) - \frac{1}{2}\right]dx$$

$$= \frac{-1}{1-s} - \frac{1}{s} + \int_1^\infty x^{-(s/2)-(1/2)}\psi(x)dx.$$

We can conclude that

$$\pi^{-s/2}\Gamma(s/2)\zeta(s) + \frac{1}{s} + \frac{1}{1-s} = \int_1^\infty \psi(x)\left(x^{(s/2)-1} + x^{-(s/2)-(1/2)}\right)dx.$$

$$\tag{10.12.7}$$

Because the integral is an entire function of s, the expression on the left is defined for all s. In particular, $\zeta(s)$ is defined for all s except $s = 1$ where it has a pole of order 1. The integral is invariant under $s \to 1 - s$. This proves the theorem. ∎

We have shown that the functional equation for the zeta function, the transformation formula for the theta function, and the partial fraction expansion of the cotangent function are equivalent results. Of course, we still need to show that the theta transformation is a consequence of the functional equation. This can be done by Mellin inversion. To illustrate the technique, we apply it to a different though related function, that is,

$$\eta(\tau) = q^{1/24} \prod_{n=1}^{\infty} (1 - q^n), \quad |q| < 1, \qquad (10.12.8)$$

where $q = e^{2\pi i \tau}$. This is called the Dedekind η-function.

Before we state the theorem, note that the integral in (10.12.7) is bounded in vertical strips, $\alpha < \mathrm{Re}\, s < \beta$. So the same is true for

$$\pi^{-s/2} \Gamma(s/2) \zeta(s) + \frac{1}{s} + \frac{1}{1-s}.$$

Theorem 10.12.8   *The Dedekind function $\eta(\tau)$ satisfies the relation*

$$\eta(-1/\tau) = \sqrt{\frac{\tau}{i}}\, \eta(\tau).$$

Proof.   We follow Weil [1968]. Since $\eta(\tau)$ has no zeros for $\mathrm{Im}\, \tau > 0$, we prove that

$$\log \eta(-1/\tau) = \log \eta(\tau) + \frac{1}{2} \log(\tau/i). \qquad (10.12.9)$$

From (10.12.8)

$$\log \eta(\tau) = \pi i \tau / 12 - \sum_{m,n \geq 1} \frac{e^{2\pi i m n \tau}}{m}.$$

Let

$$f(\tau) = \sum_{m,n=1}^{\infty} \frac{e^{2\pi i m n \tau}}{m}.$$

The Dirichlet series corresponding to f can be found by taking its Mellin transform as before:

$$\int_0^{\infty} x^{s-1} \sum_{m,n} \frac{e^{-2\pi m n x}}{m}\, dx = (2\pi)^{-s} \Gamma(s) \sum_{m,n} \frac{1}{m(mn)^s}.$$

Now

$$\sum_{m,n} \frac{1}{m(mn)^s} = \sum_{m=1}^{\infty} \frac{1}{m^{s+1}} \sum_{n=1}^{\infty} \frac{1}{n^s} = \zeta(s+1)\zeta(s).$$

It follows from Theorem 10.12.7 that

$$\Lambda(s) = (2\pi)^{-s} \Gamma(s) \zeta(s) \zeta(s+1)$$

remains invariant under $s \to -s$. It is clear that $\Lambda(s)$ has simple poles at $s = \pm 1$ with residues $\pm \pi/12$ respectively, since $\zeta(-1) = -1/12$. This last fact can be verified from the functional equation for $\zeta(s)$. At $s = 0$, however, $\Lambda(s)$ has a double pole. These are the only poles of $\Lambda(s)$. Since $\zeta'(0) = -1/2$, we conclude that

$$\Lambda(s) - \frac{\pi}{12(s-1)} + \frac{\pi}{12(s+1)} + \frac{1}{2s^2} \qquad (10.12.10)$$

is entire and bounded on every vertical strip. The last observation follows from the remarks made before the theorem. By Cauchy's theorem, the Mellin inverse of $\Gamma(s)$ is e^{-y}, that is,

$$e^{-y} = \frac{1}{2\pi i} \int_{c-i\infty}^{c+i\infty} \Gamma(s)y^{-s}ds, \quad c > 1.$$

From this we get

$$f(iy) = \frac{1}{2\pi i} \int_{c-i\infty}^{c+i\infty} \Lambda(s)y^{-s}ds, \quad c > 1. \qquad (10.12.11)$$

Observe that $\zeta(s)\zeta(s+1)$ is bounded on the line Re $s = c > 1$, because the series for $\zeta(s)$ converges absolutely. By Stirling's formula

$$\Gamma(s) \sim \sqrt{2\pi}|t|^{\sigma-(1/2)}e^{-\pi|t|/2}, \qquad s = \sigma + it, \quad |t| \to \infty,$$

in any vertical strip $\alpha < \sigma < \beta$. These two facts imply that for any $\mu > 0$

$$|\Lambda(s)| = O(|t|^{-\mu}), \qquad |t| \to \infty, \quad \text{Re } s = c > 1. \qquad (10.12.12)$$

Now choose c_1 so that $-c_1 > 1$. Thus, for any $\mu > 0$,

$$|\Lambda(s)| = |\Lambda(-s)| = O(|t|^{-\mu}), \qquad |t| \to \infty, \quad \text{Re } s = c_1 < -1.$$

Since the expression in (10.12.10) is bounded in every vertical strip, it follows from the Phragmén–Lindelöf theorem that (10.12.12) holds in (c_1, c) for large t. All this implies that we can move the line of integration in (10.12.11) to $-c < -1$, while picking up the residues at $s = \pm 1$ and $s = 0$. The expansion for y^{-s} is

$$1 - s \log y + \cdots,$$

so the residue at $s = 0$ is $(\log y)/2$. At $s = \pm 1$, we have residues $\pi/12y$ and $-\pi y/12$. The result is

$$f(iy) = \frac{1}{2\pi i} \int_{-c-i\infty}^{-c+i\infty} \Lambda(s)y^{-s}ds + \pi/12y + (\log y)/2 - \pi y/12$$

$$= \frac{1}{2\pi i} \int_{-c-i\infty}^{-c+i\infty} \Lambda(-s)y^{-s}ds + \pi/12y + (\log y)/2 - \pi y/12$$

$$= \frac{1}{2\pi i} \int_{c-i\infty}^{c+i\infty} \Lambda(s)y^{s}ds + \pi/12y + (\log y)/2 - \pi y/12$$

$$= f(-1/iy) + \pi/12y + (\log y)/2 - \pi y/12.$$

We may therefore conclude that the relation (10.12.9) holds on the imaginary axis. Since the functions involved are analytic, the result is proved. ∎

Remark 10.12.1   Note that $g(\tau) = \eta^8(\tau)$ has the following two properties:

$$g(\tau + 1) = g(\tau),$$
$$g(-1/\tau)1/\tau^4 = g(\tau).$$

The interest in the two transformations

$$\tau \to \tau + 1, \quad \tau \to -1/\tau$$

stems from the fact that they generate the modular group G, which consists of all the fractional linear transformations $(a\tau + b)/(c\tau + d)$, where a, b, c, and d are integers with $ad - bc = 1$. A function f on the upper half plane is called a modular form if for some integer k

$$f(A\tau)A'(\tau)^k = f(\tau)$$

for all $A \in G$. This holds for $g(\tau) = \eta^8(\tau)$ with $k = 2$. Subgroups of G of finite index also play a very important role. Consider the theta function

$$\theta(\tau) = \sum_{-\infty}^{\infty} q^{n^2},$$

where $q = e^{\pi i \tau}$. The function $h(\tau) = \theta^4(\tau)$ satisfies

$$h(\tau + 2) = h(\tau),$$
$$h(-1/\tau)1/\tau^4 = h(\tau).$$

The transformations $\tau \to \tau + 2$ and $\tau \to -1/\tau$ generate a subgroup of the modular group. A problem of interest to number theorists is to find properties of the Fourier coefficients of modular forms. For example, the coefficient of q^n in the expansion of $h(\tau)$ is the number of representations of n as a sum of four squares, which was studied in Section 10.6 by nonmodular methods. For a proof using modular forms and Hecke operators, see Koblitz [1984, p. 174].

Remark 10.12.2   Theorem 10.12.8 has a simple proof that considers the integral of $\cot z \cot(z/\tau)$ over a suitable contour in the complex plane; it was given by Siegel [1954]. Rademacher [1955] obtained a more general result by a similar method. Let h, h', and k be integers and let

$$\tau = (h + iz)/k, \quad \tau' = (h' + iz^{-1})/k$$

with $\gcd(h, k) = 1$, $k > 0$, $\operatorname{Re} z > 0$, and $hh' \equiv -1 \pmod{k}$. Then Rademacher's result states that

$$\log \eta \left(\frac{h' + iz^{-1}}{k} \right) = \log \eta \left(\frac{h + iz}{k} \right) + \frac{1}{2} \log z + \frac{\pi i}{12k}(h' - h) + \pi i s(h, k),$$

where the Dedekind sum $s(h, k)$ is given by

$$s(h, k) = \sum_{\ell=1}^{k-1} \left(\frac{\ell}{k} - \frac{1}{2} \right) \left(\frac{\ell h}{k} - \left[\frac{\ell h}{k} \right] - \frac{1}{2} \right).$$

It is, however, worth noting here that the transformation formula for $\eta(\tau)$ is implied by a similar formula for theta functions, a particular case of which is contained in Theorem 10.12.6. We have

$$\sum_{n=-\infty}^{\infty} e^{-s(n+x)^2} = s^{-1/2} \sum_{n=-\infty}^{\infty} e^{\pi n^2/s} e^{2\pi i n x} \qquad (10.12.13)$$

(Appendix D, Equation (D.4.2)). Set $q = e^{\pi i \tau}$ and $p = e^{-\pi i/\tau}$. By the definition of θ_3 in (10.7.3), the above formula implies

$$\theta_3 \left(\frac{\pi z}{\tau}, p \right) = \sqrt{\frac{\tau}{i}} e^{\pi i (\pi z)^2/\tau} \theta_3(\pi z, q). \qquad (10.12.14)$$

Now the infinite product for θ_3 contained in (10.7.7) gives

$$\lim_{\lambda \to 1} \frac{\theta_3((\tau + \lambda)\pi/2, q)}{1 + e^{-\lambda \pi i}} = \prod_{n=1}^{\infty} (1 - q^n)^3 \qquad (10.12.15)$$

and

$$\lim_{\lambda \to 1} \frac{\theta_3((\tau + \lambda)\pi/2\tau, p)}{1 - p^{1-\lambda}} = \prod_{n=1}^{\infty} (1 - p^n)^3. \qquad (10.12.16)$$

An easy calculation shows that (10.12.14), (10.12.15), and (10.12.16) imply

$$\eta(-1/\tau)^3 = \frac{\tau}{i} \sqrt{\frac{\tau}{i}} \, \eta(\tau)^3,$$

or

$$\eta(-1/\tau) = c \sqrt{\frac{\tau}{i}} \, \eta(\tau),$$

where c is a cube root of unity. Set $\tau = i$ to see that $c = 1$.

Before closing this chapter, we also note that quadratic Gauss sums are related to theta functions. Cauchy evaluated these sums from the formula in Theorem 10.12.6 by taking $x = \epsilon + 2\pi i/N$ and letting $\epsilon \to 0$, where N is an odd integer. A more general reciprocity relation for Gaussian sums can also be obtained by this method.

Bellman [1961, p. 40] attributed this proof of the reciprocity relation to Landsberg, but it was published earlier by Henry Smith in 1859. Smith's [1859–1866] report contained the reference to Cauchy. For the connection of the reciprocity of Gaussian sums to Tauberian theory, see Bochner [1952]. The reciprocity formula is given in Exercise 43.

Exercises

1. Let $p_n(k)$ denote the number of permutations of $1, 2, \ldots, n$ with k inversions. Prove that

$$n!_q = \sum_{k \geq 0} p_n(k) q^k. \qquad \text{(Rodrigues)}$$

   Then find the total number of inversions for all the permutations of $1, 2, \ldots, n$.
   (Stern)

   The problem on the total number of inversions was posed by Stern and a solution was given by Rodrigues. An interesting history of results on inversions is given in W. Johnson's unpublished notes, *some old and new results on inversions*.

2. Give a proof of (10.0.10) similar to the proof of the noncommutative binomial theorem given in the first part of the chapter.

3. Let $p(m, n, k) = $ number of partitions of k into $\leq n$ parts, each part $\leq m$. Show that

$$\begin{bmatrix} m+n \\ n \end{bmatrix}_q = \sum_{k \geq 0} p(m, n, k) q^k.$$

   Deduce that

$$p(m, n, k) = p(n, m, k).$$

4. Prove the following identities:

   (a) $$\begin{bmatrix} n+m+1 \\ m+1 \end{bmatrix}_q = \sum_{j=0}^{n} q^j \begin{bmatrix} m+j \\ m \end{bmatrix}_q,$$

   (b) $$\begin{bmatrix} m+n \\ \ell \end{bmatrix}_q = \sum_{k=0}^{\ell} q^{(n-k)(\ell-k)} \begin{bmatrix} n \\ k \end{bmatrix}_q \begin{bmatrix} m \\ \ell - k \end{bmatrix}_q.$$

   The last identity is a q-analog of the Chu–Vandermonde identity.

5. Let

$$f(q, m) \equiv \sum_{j=0}^{m} (-1)^j \begin{bmatrix} m \\ j \end{bmatrix}_q, \qquad f(q, 0) = 1.$$

   Prove that

$$f(q, m) = (1 - q^{m-1}) f(q, m-2).$$

   Deduce that

$$f(q, m) = \begin{cases} (1-q)(1-q^3) \cdots (1-q^{m-1}), & m \text{ even}, \\ 0, & m \text{ odd}. \end{cases} \qquad \text{(Gauss)}$$

6. Let n be an odd integer and x a primitive nth root of unity.

(a) Use the result in Exercise 5 to show that

$$\sum_{k=1}^{n} x^{k(k-1)} = (1 - x^{-2})(1 - x^{-6}) \cdots (1 - x^{-2(n-2)}).$$

(b) Deduce that

$$G = \sum_{k=0}^{n-1} x^{k^2} = (x - x^{-1})(x^3 - x^{-3}) \cdots (x^{n-2} - x^{-n+2})$$

$$= (-1)^{\frac{n-1}{2}} (x^2 - x^{-2})(x^4 - x^{-4}) \cdots (x^{n-1} - x^{-n+1}).$$

(c) Show that

$$G = \begin{cases} \pm\sqrt{n} & \text{for } n = 4k + 1, \\ \pm i\sqrt{n} & \text{for } n = 4k + 3. \end{cases}$$

(d) Set $x = e^{2\pi i/n}$ in (b) to obtain

$$G = (2i)^{\frac{n-1}{2}} \sin \frac{2\pi}{n} \sin \frac{6\pi}{n} \cdots \sin \frac{2(n-2)\pi}{n}.$$

Note that when $n = 4k+1$, there are k negative factors in the sine product. Conclude that

$$G = \sqrt{n} \quad \text{for } n = 4k + 1.$$

Do a similar analysis for $n = 4k + 3$ to show that, in this case,

$$G = i\sqrt{n}. \tag{Gauss}$$

7. Define $\text{inv}(m_1, m_2, \ldots, m_r; n)$ to be the number of permutations $x_1 x_2, \ldots,$ $x_{m_1+m_2+\cdots+m_n}$ of $\{1^{m_1} 2^{m_2}, \ldots, r^{m_r}\}$ in which there are exactly n inversions. Show that

(a) $\text{inv}(m_1, \ldots, m_r; n) = \sum_{j=0}^{n} \text{inv}(m_1 + \cdots + m_{r-1}, m_r; j) \text{inv}(m_1, \ldots,$ $m_{r-1}; n-j)$

(b) Use induction and (a) to show that, for $r \geq 1$,

$$\sum_{n \geq 0} \text{inv}(m_1, m_2, \ldots, m_r; n) q^n = \begin{bmatrix} m_1 + m_2 + \cdots + m_r \\ m_1, m_2, \ldots, m_r \end{bmatrix}_q$$

$$= \frac{(q; q)_{m_1+m_2+\cdots+m_r}}{(q; q)_{m_1} \cdots (q; q)_{m_r}}.$$

This is a result of MacMahon. See Andrews [1976, p. 41].

8. Prove that the Gaussian polynomial

$$G(m, n; q) = \begin{bmatrix} m + n \\ n \end{bmatrix}_q$$

is reciprocal, that is, $G(m, n; q) = q^{mn} G(m, n; 1/q)$. Deduce that

$$p(m, n, k) = p(m, n, mn - k).$$

9. Prove the following version of the q-binomial theorem:

$$(ab; q)_n = \sum_{k=0}^{n} \begin{bmatrix} n \\ k \end{bmatrix}_q b^k (a; q)_k (b; q)_{n-k}.$$

State and prove a multinomial extension of this formula.

10. (a) Prove the following analog of the Bohr–Mollerup theorem: For $0 < q < 1$, $\Gamma_q(x)$ is the unique logarithmically convex function that satisfies the functional equation

$$f_q(x+1) = \frac{1-q^x}{1-q} f_q(x) \quad \text{with} \quad f_q(1) = 1.$$

(b) Let $g(x)$ be defined for $x > 0$, and let

$$\lim_{x \to \infty} \frac{g(x)}{x} = 0.$$

Prove that any two convex solutions of

$$f(x+1) - f(x) = g(x)$$

differ at most by a constant. Use this to derive the result in (a). (See John [1938].)

11. Prove the q-analogs of Legendre's duplication formula and Gauss's multiplication formula contained in Theorem 10.3.5.

12. For $q > 1$, define

$$\Gamma_q(x) = \frac{(q^{-1}; q^{-1})_\infty}{(q^{-x}; q^{-1})_\infty} (q-1)^{1-x} q^{x(x-1)/2}.$$

Show that

(a) $\Gamma_q(x)$ satisfies the functional equation in Exercise 10.

(b) $\lim_{q \to 1+} \Gamma_q(x) = \Gamma(x)$.

(c) The residue of $\Gamma_q(x)$ at $x = -n$ is

$$\frac{(q-1)^{n+1} q^{n(n+1)/2}}{(q; q)_n \log q}.$$

(d) If $q > 1$ and

$$f(x+1) = \frac{q^x - 1}{q - 1} f(x), \quad f(1) = 1, \quad \text{and} \quad \frac{d^3}{dx^3}(\log f(x)) \le 0$$

for $x > 0$, then

$$f(x) = \Gamma_q(x) \quad \text{for } x > 0.$$

(e) If $q > 1$ and

$$f(x+1) = \frac{q^x - 1}{q - 1} f(x), \quad f(1) = 1, \quad \text{and} \quad \frac{d^2}{dx^2}(\log f(x)) \ge \log q$$

for $x > 0$, then

$$f(x) = \Gamma_q(x) \quad \text{for } x > 0.$$

(See Moak [1980].)

13. Compute $C_0(q^2)$ in (10.4.4) as follows:

(a) Note that

$$C_0(q^2)\theta_3(q) = \prod_{n=1}^{\infty}(1+q^{2n-1})^2$$

and

$$C_0(q^2)\theta_4(q) = \prod_{n=1}^{\infty}(1-q^{2n-1})^2.$$

(b) Show that

$$C_0(q^2)\theta_4(q^2) = \prod_{n=1}^{\infty}(1-q^{4n-2}).$$

(Hint: Use $\sqrt{\theta_3(q)\theta_4(q)} = \theta_4(q^2)$.)

(c) Show that

$$\frac{C_0(q^4)}{C_0(q^2)} = \prod_{n=1}^{\infty}(1-(q^2)^{2n-1}).$$

Deduce that

$$C_0(q^2)^{-1} = \prod_{n=1}^{\infty}(1-q^{2n}). \qquad \text{(Gauss)}$$

14. Prove the quintuple product identity:

$$H(x) \equiv \prod_{n=1}^{\infty}(1-q^n)(1-xq^n)(1-q^{n-1}/x)(1-x^2q^{2n-1})(1-q^{2n-1}/x^2)$$

$$= \sum_{n=-\infty}^{\infty}(x^{3n}-x^{-3n-1})q^{n(3n+1)/2}.$$

Deduce that

$$\prod_{n=1}^{\infty}(1-q^n)^3(1-q^{2n-1})^2 = \sum_{n=-\infty}^{\infty}(6n+1)q^{n(3n+1)/2}.$$

One method is to take $H(x) = \sum_{n=-\infty}^{\infty} c(n)x^n$. To find $c(0)$, compute $\frac{H(qx)}{H(x)}$ and $\frac{H(1/x)}{H(x)}$ to determine $c(n)$ in terms of $c(0)$. Then specialize x.

15. Prove that the quintuple product identity in the previous problem is equivalent to the identity

$$\prod_{n=1}^{\infty}(1-q^{2n})(1-q^{2n-1}x)(1-q^{2n-1}/x)(1-q^{4n-4}x^2)(1-q^{4n-4}/x^2)$$

$$= \sum_{n=-\infty}^{\infty} q^{3n^2-2n}\left[(x^{3n}+x^{-3n})-(x^{3n-2}+x^{-(3n-2)})\right].$$

For Exercises 16–24 and related results and references, see Gasper and Rahman [1990].

16. Prove the following q-analog of Dougall's formula for a 2-balanced, very well poised $_7F_6$:

$$_8\phi_7\left(\begin{array}{c} a, q\sqrt{a}, -q\sqrt{a}, b, c, d, e, q^{-N} \\ \sqrt{a}, -\sqrt{a}, aq/b, aq/c, aq/d, aq/e, aq^{N+1} \end{array}; q, q\right)$$

$$= \frac{(aq; q)_N (aq/cd; q)_N (aq/bd; q)_N (aq/bc; q)_N}{(aq/b; q)_N (aq/c; q)_N (aq/d; q)_N (aq/bcd; q)_N},$$

when $bcde = a^2 q^{N+1}$ and N is a positive integer. One proof goes as follows:

(a) Write f instead of q^{-N} and express the formula as

$$_8\phi_7\left(\begin{array}{c} a, q\sqrt{a}, -q\sqrt{a}, b, c, d, e, f \\ \sqrt{a}, -\sqrt{a}, aq/b, aq/c, aq/d, aq/e, aq/f \end{array}; q, q\right)$$

$$= \frac{(aq)_\infty (aq/cd)_\infty (aq/bd)_\infty (aq/bc)_\infty (aq/bf)_\infty}{(aq/b)_\infty (aq/c)_\infty (aq/d)_\infty (aq/f)_\infty (aq/bcd)_\infty}$$

$$\cdot \frac{(aq/cf)_\infty (aq/df)_\infty (aq/bcdf)_\infty}{(aq/bcf)_\infty (aq/bdf)_\infty (aq/cdf)_\infty},$$

when $a^2 q = bcdef$.

(b) Suppose the formula is true for $f = 1, q^{-1}, q^{-2}, \ldots, q^{-N+1}$ and take $f = q^{-N}$. By symmetry the result is true if c or $d = a^2 q/bcef$ is equal to $1, q^{-1}, \ldots, q^{-N+1}$.

(c) Observe that if the original formula is multiplied by $(aq/c; q)_N$ and $(aq/bcd; q)_N$, then the formula gives the identity of two polynomials in c of degree $2N$. From (b) the two sides are equal for $2N$ values of c. Now set $c = aq^N$ and verify the equality in this case. (Jackson)

17. Use the formula in Exercise 16 and that method of proof to show that

$$_{10}\phi_9\left(\begin{array}{c} a, q\sqrt{a}, -q\sqrt{a}, c, d, e, f, g, h, j \\ \sqrt{a}, -\sqrt{a}, aq/c, aq/d, aq/e, aq/f, aq/g, aq/h, aq/j \end{array}; q, q\right)$$

$$= \frac{(aq)_\infty (aq/fg)_\infty (aq/fh)_\infty (aq/fj)_\infty (aq/gh)_\infty (aq/gj)_\infty (aq/hj)_\infty (aq/fghj)_\infty}{(aq/f)_\infty (aq/g)_\infty (aq/h)_\infty (aq/j)_\infty (aq/ghj)_\infty (aq/hjf)_\infty (aq/jfg)_\infty (aq/fgh)_\infty}$$

$$\cdot {}_{10}\phi_9\left(\begin{array}{c} k, q\sqrt{k}, -q\sqrt{k}, kc/a, kd/a, ke/a, f, g, h, j \\ \sqrt{k}, -\sqrt{k}, aq/c, aq/d, aq/e, kq/f, kq/g, kq/h, kq/j \end{array}; q, q\right),$$

when $k = a^2 q/cde$ and $cdefghj = a^3 q^2$, and $f, g, h,$ or j is of the form q^{-N} where N is a nonnegative integer. (Bailey)

18. Derive the following formula from Bailey's formula given in the previous exercise:

$$_8\phi_7\left(\begin{matrix} a, q\sqrt{a}, -q\sqrt{a}, c, d, e, f, g \\ \sqrt{a}, -\sqrt{a}, aq/c, aq/d, aq/e, aq/f, aq/g \end{matrix}; q, a^2q^2/cdefg\right)$$

$$= \frac{(aq)_\infty(aq/fg)_\infty(aq/ge)_\infty(aq/ef)_\infty}{(aq/e)_\infty(aq/f)_\infty(aq/g)_\infty(aq/efg)_\infty}$$

$$\cdot _4\phi_3\left(\begin{matrix} aq/cd, e, f, g \\ efg/a, aq/c, aq/d \end{matrix}; q, q\right),$$

when e, f, or g is of the form q^{-N}. This is a q-analog of Whipple's transformation.

Deduce Sears's transformation formula in Theorem 10.11.2. (Watson)

19. Let c, d, e, f, and $g = q^{-N}$ tend to ∞ in Watson's formula given in Exercise 18 to get

$$1 + \sum_{n=1}^{\infty}(-1)^n a^{2n} q^{n(5n-1)/2}(1 - aq^{2n})\frac{(aq; q)_{n-1}}{(q; q)_n}$$

$$= (aq; q)_\infty \sum_{n=0}^{\infty} \frac{q^{n^2}a^n}{(q; q)_n}.$$

Simplify when $a = 1$ and $a = q$.

20. Derive the following identities from Jackson's formula in Exercise 16:

(a)
$$_6\phi_5\left(\begin{matrix} a, q\sqrt{a}, -q\sqrt{a}, b, c, d \\ \sqrt{a}, -\sqrt{a}, aq/b, aq/c, aq/d \end{matrix}; q, aq/bcd\right)$$

$$= \frac{(aq; q)_\infty(aq/bc; q)_\infty(aq/bd; q)_\infty(aq/cd; q)_\infty}{(aq/b; q)_\infty(aq/c; q)_\infty(aq/d; q)_\infty(aq/bcd; q)_\infty},$$

provided $|aq/bcd| < 1$.

(b) The q-Dixon formula

$$_4\phi_3\left(\begin{matrix} a, -q\sqrt{a}, b, c \\ -\sqrt{a}, aq/b, aq/c \end{matrix}; q, q\sqrt{a}/bc\right)$$

$$= \frac{(aq; q)_\infty(aq/bc; q)_\infty(q\sqrt{a}/b; q)_\infty(q\sqrt{a}/c; q)_\infty}{(aq/b; q)_\infty(aq/c; q)_\infty(q\sqrt{a}; q)_\infty(q\sqrt{a}/bc; q)_\infty},$$

provided $|q\sqrt{a}/bc| < 1$.

21. Prove the following identities and also deduce (b) from (a):

(a) $_3\phi_2\left(\begin{matrix} q^{-n}, b, c \\ d, e \end{matrix}; q, q\right) = \frac{(de/bc; q)_n}{(e; q)_n}\left(\frac{bc}{d}\right)^n {}_3\phi_2\left(\begin{matrix} q^{-n}, d/b, d/c \\ d, de/bc \end{matrix}; q, q\right).$

(b) $\quad {}_2\phi_1\left(\begin{matrix} q^{-n}, d/b \\ d \end{matrix}; q, bq/e\right) = (-1)^n q^{-\binom{n}{2}}(e; q)_n e^{-n} {}_3\phi_2\left(\begin{matrix} q^{-n}, b, 0 \\ d, e \end{matrix}; q, q\right).$

(c) $\quad {}_3\phi_2\left(\begin{matrix} q^{-n}, a, b \\ d, e \end{matrix}; q, \dfrac{deq^n}{ab}\right) = \dfrac{(e/a; q)_n}{(e; q)_n} {}_3\phi_2\left(\begin{matrix} q^{-n}a, d/b \\ d, aq^{1-n}/e \end{matrix}; q, q\right).$

22. Prove Watson's q-Barnes integral formula

$$
{}_2\phi_1\left(\begin{matrix} a, b \\ c \end{matrix}; q, z\right)
$$
$$
= \frac{(a; q)_\infty(b; q)_\infty}{(q; q)_\infty(c; q)_\infty}\left(-\frac{1}{2\pi i}\right)\int_{-i\infty}^{i\infty}\frac{(q^{1+s}; q)_\infty(cq^s; q)_\infty \pi(-z)^s}{(aq^s; q)_\infty(bq^s; q)_\infty \sin \pi s}ds,
$$

$|z| < 1, |\arg(-z)| < \pi.$

23. Let $\operatorname{Re} c > 0$, $\operatorname{Re} d > 0$, and $\operatorname{Re}(x + y) > 1$. Prove the following analog of Cauchy's beta integral:

$$
\frac{1}{2\pi i}\int_{-i\infty}^{i\infty}\frac{(-csq^x; q)_\infty(dsq^y; q)_\infty}{(-cs; q)_\infty(ds; q)_\infty}ds
$$
$$
= \frac{\Gamma_q(x + y - 1)}{\Gamma_q(x)\Gamma_q(y)}\frac{(-cq^x/d; q)_\infty(-dq^y/c; q)_\infty}{(c + d)(-cq/d; q)_\infty(-dq/c; q)_\infty}
$$

when $0 < q < 1.$ (Wilson)

24. Verify the following analogs of Barnes's first and second lemmas.

(a) $\quad \dfrac{1}{2\pi i}\displaystyle\int_{i\infty}^{i\infty}\dfrac{(q^{1-c+s}; q)_\infty(q^{1-d+s}; q)_\infty}{(q^{a+s}; q)_\infty(q^{b+s}; q)_\infty}\dfrac{\pi q^s ds}{\sin \pi(c - s)\sin \pi(d - s)}$

$$
= \frac{q^c}{\sin \pi(c - d)} \cdot \frac{(q; q)_\infty(q^{1+c-d}; q)_\infty(q^{d-c}; q)_\infty(q^{a+b+c+d}; q)_\infty}{(q^{a+c}; q)_\infty(q^{a+d}; q)_\infty(q^{b+c}; q)_\infty(q^{b+d}; q)_\infty},
$$

(Watson)

(b) $\quad \dfrac{1}{2\pi i}\displaystyle\int_{-i\infty}^{i\infty}\dfrac{(q^{1+s}; q)_\infty(q^{d+s}; q)_\infty(q^{e+s}; q)_\infty}{(q^{a+s}; q)_\infty(q^{b+s}; q)_\infty(q^{c+s}; q)_\infty}\dfrac{\pi q^s ds}{\sin \pi s \sin \pi(d + s)}$

$$
= \csc \pi d\, \frac{(q; q)_\infty(q^d; q)_\infty(q^{1-d}; q)_\infty(q^{e-a}; q)_\infty(q^{e-b}; q)_\infty(q^{e-c}; q)_\infty}{(q^a; q)_\infty(q^b; q)_\infty(q^c; q)_\infty(q^{1+a-d}; q)_\infty(q^{1+b-d}; q)_\infty(q^{1+c-d}; q)_\infty}
$$

when $d + e = 1 + a + b + c, 0 < q < 1.$ (Agarwal)

25. Prove the theta relations (10.7.5), (10.7.6), and (10.7.7).

26. Prove the product formulas for the Jacobi elliptic functions in (10.7.13).

27. Prove that when $u = 2Kx/\pi,$

(a) $\qquad\qquad \operatorname{dn} u = \dfrac{\pi}{2K} + \dfrac{2\pi}{K}\displaystyle\sum_{n=1}^{\infty}\dfrac{q^n \cos 2nx}{1 + q^n},$

(b)
$$\text{cn } u = \frac{2\pi}{K} \sum_{n=0}^{\infty} \frac{q^{n+1/2} \cos(2n+1)x}{1+q^{2n-1}},$$

(c)
$$\text{ns } u = \frac{\pi}{2K \sin x} + \frac{2\pi}{K} \sum_{n=0}^{\infty} \frac{q^{2n+1} \sin(2n+1)x}{1-q^{2n+1}}.$$

28. Show that for $k \neq 0, \pm 1$,
$$\frac{d}{du} \text{sn}(u, k) = \text{cn}(u, k)\text{dn}(u, k).$$

Also prove that
$$\text{sn}^2(u, k) + \text{cn}^2(u, k) = 1$$

and
$$k^2 \text{sn}^2(u, k) + \text{dn}^2(u, k) = 1.$$

Note that
$$\left(\frac{d \text{ sn}(u, k)}{du}\right)^2 = (1 - \text{sn}^2(u, k))(1 - k^2 \text{sn}^2(u, k)).$$

29. Use the ideas of Section 10.8 to prove that
$$\int_0^{\infty} x^{c-1} \frac{(-ax; q)_{\infty}(-qb/x; q)_{\infty}}{(-x; q)_{\infty}(-q/x; q)_{\infty}} dx$$
$$= \frac{(ab; q)_{\infty}(q^c; q)_{\infty}(q^{1-c}; q)_{\infty}\pi}{(bq^c; q)_{\infty}(aq^{-c}; q)_{\infty}(q; q)_{\infty} \sin \pi c}.$$

30. Prove Theorem 10.9.2, that
$$\text{for } \max_{1 \leq i \leq 5} (|a_i|, |q|) < 1,$$
$$\int_{-1}^{1} \frac{h(x, 1)h(x, \sqrt{q})h(x, -1)h(x, -\sqrt{q})h(x, A)}{\prod_1^5 h(x, a_i)} \frac{dx}{\sqrt{1-x^2}}$$
$$= \frac{2\pi \prod_{i=1}^5 (A/a_i; q)_{\infty}}{(q; q)_{\infty} \prod_{1 \leq i < j \leq 5}(a_i; a_j; q)_{\infty}},$$

where $A = a_1 a_2 a_3 a_4 a_5$.

31. Prove the formulas (where $0 < q < 1$ and $\text{Re } a > 0$ and $\text{Re } b > 0$):

(a)
$$\int_0^{\infty} \frac{(-xq^b; q)_{\infty}(-q^{a+1}/x; q)_{\infty}}{(-x; q)_{\infty}(-x/q; q)_{\infty}} \frac{d_q x}{x} = \frac{\Gamma_q(a)\Gamma_q(b)}{\Gamma_q(a+b)},$$

(b)
$$\int_0^{\infty} \frac{(-xq^b; q)_{\infty}(-q^{a+1}/x; q)_{\infty}}{(-x; q)_{\infty}(-q/x; q)_{\infty}} \frac{dx}{x} = -\frac{\log q}{1-q} \frac{\Gamma_q(a)\Gamma_q(b)}{\Gamma_q(a+b)}.$$

(c) Extend the formula in (a) to

$$\int_0^\infty \frac{x^{c-1}(-xq^b; q)_\infty(-q^{a+1}/x; q)_\infty}{(-x; q)_\infty(-q/x; q)_\infty} d_q x$$

$$= \frac{(-q^c; q)_\infty(-q^{1-c}; q)_\infty}{(-1; q)_\infty(-q; q)_\infty} \frac{\Gamma_q(a+c)\Gamma_q(b-c)}{\Gamma_q(a+b)},$$

where $\mathrm{Re}(a+c) > 0$ and $\mathrm{Re}(b-c) > 0$ and $0 < q < 1$.

32. Prove that when $\max(|a_1|, |a_2|, |a_3|, |a_4|) < 1$,

$$\int_0^\pi \frac{\sin^2\theta d\theta}{\prod_{j=1}^4 (1 - 2a_j \cos\theta + a_j^2)} = \frac{\pi(1 - a_1 a_2 a_3 a_4)}{2\prod_{1 \le i < j \le 4}(1 - a_i a_j)}.$$

33. Use Theorem 10.12.1 to prove the following results:
    (a) If $0 < s < \min(a, b)$, then

$$\int_0^\infty x^{s-1} {}_2F_1(a, b; c; -x) dx = \frac{\Gamma(c)}{\Gamma(a)\Gamma(b)} \frac{\Gamma(s)\Gamma(a-s)\Gamma(b-s)}{\Gamma(c-s)}.$$

    (b) If $0 < s < 1$, then

$$\int_0^\infty x^{s-1}(1^{-a} - 2^{-a}x + 3^{-a}x^2 - \cdots) dx = \frac{\pi}{\sin s\pi}(1-s)^{-a}.$$

34. Show the formal equivalence of the formulas

    (a) $$\int_0^\infty x^{s-1}\{\phi(0) - x\phi(1) + x^2\phi(z) - \cdots\} dx = \frac{\pi}{\sin s\pi}\phi(-s),$$

    (b) $$\int_0^\infty x^{s-1}\left\{\lambda(0) - \frac{x}{1!}\lambda(1) + \frac{x^2}{2!}\lambda(z) - \cdots\right\} dx = \Gamma(s)\lambda(-s).$$

35. Suppose $s > 0$, and

$$\Lambda(x) = \sum_{n=0}^\infty (-1)^n \frac{\lambda(n)}{n!} x^n$$

converges for all x. Show the equivalence of Newton's difference formula

$$\lambda(-s) = \lambda(0) + \frac{s}{1!}\Delta\lambda(0) + \frac{s(s+1)}{2!}\Delta^2\lambda(0) + \cdots,$$

where $\Delta\lambda(n) = \lambda(n) - \lambda(n+1)$, to formula (b) in the previous problem.

36. Suppose $a > 0, b > 0$, and

$$e^{-ax} = \sum_{n=0}^\infty (-1)^n \frac{\lambda(n)}{n!}(xe^{bx})^n.$$

Use formula (b) of Exercise 34 to prove that $\lambda(n) = a(a+nb)^{n-1}$.
The results in Exercises 34–36 are due to Ramanujan. See Hardy [1940, Chapter 11].

37. Use Exercise 5 and the q-binomial to prove that

$$(aq; q^2) \sum_{n=0}^{\infty} \frac{q^{n^2} a^n}{(q^2; q^2)_n (aq^2, q^2)_n} = \sum_{n=0}^{\infty} \frac{q^{4n^2} a^{2n}}{(q^4; q^4)_n}.$$

38. (a) Use Heine's transformation in Theorem 10.9.1 to show that

$$_2\phi_1 \left(\begin{array}{c} b^2, b^2/c : q^2, cq/b^2 \\ c \end{array} \right)$$

$$= \frac{1}{2} \frac{(b^2; q^2)_\infty (q; q^2)_\infty}{(c; q^2)_\infty (cq/\beta^2; q^2)_\infty} \left(\frac{(c/b)_\infty}{(b)_\infty} + \frac{(-c/b)_\infty}{(-b)_\infty} \right).$$

(b) Take $b = q^{-n}$ and let $c \to \infty$ to obtain

$$\sum_{m=0}^{n} \frac{(q^2; q^2)_n q^m}{(q^2; q^2)_m (q^2; q^2)_{n-m}} = (-q; q)_n.$$

The last identity was obtained by Gauss [1808, §9] by a different method.

39. From Exercise 38(b) and the q-binomial theorem derive the following identities:

$$(-aq; q^2)_\infty \sum_{n=0}^{\infty} \frac{q^{n^2+n} a^n}{(q^2; q^2)_n (-aq; q^2)_n}$$

$$= (-aq^2; q^2)_\infty \sum_{n=0}^{\infty} \frac{q^{n^2} a^n}{(q^2; q^2)_n (-aq^2; q^2)_n} = \sum_{n=0}^{\infty} \frac{q^{n^2} a^n}{(q; q)_n}.$$

40. Show that the recurrence relation (10.11.3) follows from the generating function (10.11.2), and conversely.

41. (a) Prove the linearization formula (10.11.17) by first observing that for $x = \cos\theta$,

$$\sum_{k,m,n \geq 0} \frac{a^k b^m c^n}{(q; q)_k (q; q)_m (q; q)_n} \int_0^\pi H_k(x \mid q) H_m(x \mid q) H_n(x \mid q) |(e^{2i\theta}; q)|^2 d\theta$$

$$= \frac{2\pi}{(q; q)_\infty (ab; q)_\infty (ac; q)_\infty (bc; q)_\infty}$$

$$= \frac{2\pi}{(q; q)_\infty} \sum_{r,s,t \geq 0} \frac{a^{r+s} b^{r+t} c^{s+t}}{(q; q)_r (q; q)_s (q; q)_t}.$$

(b) Complete the evaluation of the integral in (10.11.19).

42. Prove the formula (10.11.18).

43. Take $x = \epsilon - \pi i m/2n$ in Theorem 10.12.6 and let $\epsilon \to 0$ to show that

$$\frac{1}{\sqrt{n}} \sum_{\ell=0}^{2n-1} e^{\pi i k^2 m/2n} = \frac{1+i}{\sqrt{m}} \sum_{\ell=0}^{m-1} e^{-2\pi i k^2 n/m}.$$

Deduce that for odd q,

$$\sum_{\ell=0}^{q-1} e^{2\pi i \ell^2/q} = \frac{1 - i^q}{1 - i} \sqrt{q}.$$

44. Consider a more general divided difference operator

$$Df(x) = \frac{f(y_2(x)) - f(y_1(x))}{y_2(x) - y_1(x)}.$$

If D takes polynomials of degree n to polynomials of degree $n - 1$, show that $y_1(x)$ and $y_2(x)$ satisfy a quadratic equation

$$Ay^2 + Bxy + Cx^2 + Dy + Ex + F = 0,$$

and show that two solutions of this equation can be used to define D so that it takes polynomials of degree n to polynomials of degree $n - 1$. Specific choices of $y_2(x)$ and $y_1(x)$ and limits of them give the operators used in this book, and the operators used in this book are essentially the standard forms to which $y_1(x)$ and $y_2(x)$ can be reduced. See Magnus [1988].

45. Show that

$$D_q T_n(x) = q^{(1-n)/2} \frac{(1 - q^n)}{(1 - q)} U_{n-1}(x).$$

46. If

$$\int_0^\pi f(\cos\theta)d\theta = 0$$

and f is smooth then

$$f(\cos\theta) = \sum_{k=1}^\infty a_k \cos k\theta.$$

If $f(\cos\theta)$ is a polynomial in $\cos\theta$ or the coefficients a_k decrease sufficiently rapidly, then

$$D_q f(x) = \sum_{k=1}^\infty a_k q^{(1-k)/2} \frac{(1 - q^k)}{(1 - q)} \frac{\sin k\theta}{\sin\theta} = g(x).$$

Define

$$I_q g(x) = f(x).$$

Show that

$$I_q g(x) = \frac{1}{\pi} \int_{-\pi}^\pi g(\cos\varphi) K(\theta + \varphi) \sin\varphi \, d\varphi,$$

where

$$K(\theta) = \frac{(1 - q)}{2q^{1/2}} \frac{d}{d\theta} \log(q^{1/2} e^{i\theta}; q)_\infty (q^{1/2} e^{-i\theta}; q)_\infty.$$

See Brown and Ismail [1995].

11

Partitions

The theory of partitions is a subject that, on the one hand, fits naturally into the subject of q-series, and on the other, is highly combinatorial in its methods. This provides for a variety of treatments of this subject. P. A. MacMahon, one of the pioneers in the study of partitions, titled his seminal two-volume work, *Combinatory Analysis*. It is clear that he saw a major role for analysis in the study of partitions. We shall follow his lead and examine partitions by means of the analytical technique he developed: partition analysis. This method is used to find the generating functions of various kinds of interesting partition functions. Examples of a few other ways of developing the theory of partitions will be given in passing.

11.1 Background on Partitions

The theory of partitions concerns representing integers as sums of positive integers. Thus there are five partitions of 4, namely 4, $3+1$, $2+2$, $2+1+1$, and $1+1+1+1$. Note that the order of the summands (or parts) is not considered: $1+2+1$ is the same partition of 4 as $2+1+1$.

One object of study is $p(n)$, the number of partitions of n. Other examples of interesting partition functions are $p_m(n)$, the numbers of partitions of n into $\leq m$ parts and $p_N^{(1)}(n)$, the number of partitions of n into distinct parts. Thus $p(4) = 5$, $p_2(4) = 3$, and $p_N^{(1)}(4) = 2$. An explanation for the last notation is given below.

The theory of partitions dates back to Euler. The generating function of a given partition function has turned out to be one of the most fundamental objects in the study of partitions. Euler's basic observation lay in the introduction of the geometric series in treating generating functions. Suppose A is some set of positive integers. A partition of n into elements of A is a representation of n as a sum of elements of A (where the order of summands can be disregarded.) Thus $n = a_1 + a_2 + \cdots + a_r$,

$a_i \in A$, and to make the representation unique we may require $a_1 \geq a_2 \geq \cdots \geq a_r$.

Let $p_A(n)$ denote the number of partitions of n into elements of A. The generating function for $p_A(n)$ is given by

$$\sum_{n=0}^{\infty} p_A(n)q^n = \prod_{a \in A}(1 + q^a + q^{2a} + q^{3a} + \cdots). \qquad (11.1.1)$$

The equality in (11.1.1) becomes clear once we multiply the terms together and see that the general exponent on q that arises is

$$f_1a_1 + f_2a_2 + \cdots + f_ja_j + \cdots.$$

This last expression is an arbitrary partition into elements of A, where a_1 appears f_1 times, a_2 appears f_2 times, and so on. Consequently,

$$\sum_{n=0}^{\infty} p_A(n)q^n = \prod_{a \in A} \frac{1}{1 - q^a}. \qquad (11.1.2)$$

If we were to require that each part appear $\leq s$ times and to define $p_A^{(s)}(n)$ to be the number of these partitions of n, then we would see as before

$$\sum_{n=0}^{\infty} p_A^{(s)}(n)q^n = \prod_{a \in A}(1 + q^a + q^{2a} + \cdots q^{sa})$$

$$= \prod_{a \in A} \frac{1 - q^{(s+1)a}}{1 - q^a}. \qquad (11.1.3)$$

Note that when $s = 1$, $p_A^{(1)}(n)$ is the number of partitions of n into distinct elements of A and

$$\sum_{n=0}^{\infty} p_A^{(1)}(n)q^n = \prod_{a \in A} \frac{1 - q^{2a}}{1 - q^a} = \prod_{a \in A}(1 + q^a). \qquad (11.1.4)$$

These observations on generating functions allow us to prove one of Euler's striking, albeit elementary, theorems.

Theorem 11.1.1   *The number of partitions of n into elements of $\mathbb{O}$ (the set of odd numbers) equals the number of partitions of n into distinct parts (i.e., the parts taken from $\mathbb{N}$, the set of all positive integers.) More succinctly,*

$$p_{\mathbb{O}}(n) = p_{\mathbb{N}}^{(1)}(n).$$

Proof. This is now easily seen because

$$
\sum_{n=0}^{\infty} p_{\mathbb{O}}(n) q^n = \frac{1}{(1-q)(1-q^3)(1-q^5)\cdots}
$$

$$
= \frac{(1-q^2)(1-q^4)\cdots}{(1-q)(1-q^2)(1-q^3)\cdots}
$$

$$
= \prod_{n=1}^{\infty} \frac{1-q^{2n}}{1-q^n} = \sum_{n=0}^{\infty} p_{\mathbb{N}}^{(1)}(n) q^n.
$$

Here (11.1.2) was used in the first step and (11.1.4) in the last step. The theorem is proved. ■

At this point the reader may try to prove the following result by the method of Theorem 11.1.1:

The number of partitions of n into summands not divisible by 3 is equal to the number of partitions of n where no summand occurs more than twice.

One of the difficulties encountered in studying the theory of partitions lies in the fact that each new result seems to require some new trick. Although this fact may seem charming to insiders it is somewhat discouraging for noncombinatorial outsiders. In this chapter we hope to present a systematic derivation of a variety of elementary results. Our focus will be MacMahon's partition analysis for obtaining generating functions for a number of interesting partition functions.

11.2 Partition Analysis

To illustrate this method, we start with the following problem: What is a closed form for the generating function $\sum_{n=0}^{\infty} p_m(n) q^n$? Recall that $p_m(n)$ is the number of partitions of n into $\leq m$ parts.

It is not difficult to see that we can write this generating function as a multidimensional sum. Thus

$$
\sum_{n=0}^{\infty} p_m(n) q^n = \sum_{n_1 \geq n_2 \geq \cdots \geq n_m \geq 0} q^{n_1+n_2+\cdots+n_m}.
$$

The requirement $n_1 \geq n_2 \geq \cdots \geq n_m \geq 0$ comes from the fact that order is disregarded in a partition. Thus each partition can be written uniquely as a sum of a nonincreasing sequence of numbers. MacMahon's idea is to introduce new variables $\lambda_1, \lambda_2, \ldots, \lambda_{m-1}$ that handle the inequalities satisfied by n_j while the n_j themselves become free. Consider the sum

$$
\sum_{n_1, n_2, \ldots, n_m \geq 0} q^{n_1+n_2+\cdots+n_m} \lambda_1^{n_1-n_2} \lambda_2^{n_2-n_3} \cdots \lambda_{m-1}^{n_{m-1}-n_m}.
$$

If we select only terms in this sum with nonnegative exponents on the λ, then the corresponding exponent will be a partition of n into $\leq m$ parts. For example, when $m = 2$ and $n = 4$ the exponents that result are $4 + 0$, $3 + 1$, and $2 + 2$. The method of partition analysis applies a linear operator $\Omega_\geq$ to such multiple Laurent series in $\lambda_1, \ldots, \lambda_{m-1}$. The operator annihilates terms with any negative exponents and in the remaining terms sets each $\lambda_i = 1$. Hence

$$\sum_{n=0}^{\infty} p_m(n)q^n = \Omega_\geq \sum_{n_1,\ldots,n_m \geq 0} q^{n_1+\cdots+n_m} \lambda_1^{n_1-n_2} \cdots \lambda_{m-1}^{n_{m-1}-n_m}$$

$$= \Omega_\geq \sum_{n_1 \geq 0} (q\lambda_1)^{n_1} \sum_{n_2 \geq 0} (q\lambda_2/\lambda_1)^{n_2} \cdots \sum_{n_m \geq 0} (q/\lambda_{m-1})^{n_m}$$

$$= \Omega_\geq \frac{1}{(1 - q\lambda_1)(1 - q\lambda_2/\lambda_1) \cdots (1 - q/\lambda_{m-1})}. \qquad (11.2.1)$$

The next step is to produce an algorithm to evaluate the effect of $\Omega_\geq$. To this end we state and prove the next lemma.

Lemma 11.2.1

$$\Omega_\geq \frac{1}{(1 - \lambda x)(1 - y/\lambda)} = \frac{1}{(1 - x)(1 - xy)}.$$

Proof.   The left side equals

$$\Omega_\geq \sum_{n,m \geq 0} \lambda^{n-m} x^n y^m = \sum_{n \geq m \geq 0} x^n y^m.$$

Set $k = n - m$ so that the last sum becomes

$$\sum_{k,m \geq 0} x^{m+k} y^m = \sum_{k \geq 0} x^k \sum_{m \geq 0} (xy)^m = \frac{1}{(1 - x)(1 - xy)},$$

and the lemma is proved.   ■

Repeated application of Lemma 11.2.1 gives the closed form for the generating function of $p_m(n)$.

Theorem 11.2.2

$$\sum_{n=0}^{\infty} p_m(n)q^n = 1/(1 - q)(1 - q^2) \cdots (1 - q^m).$$

Proof.   One application of Lemma 11.2.1 and (11.2.1) gives

$$\sum_{n=0}^{\infty} p_m(n)q^n = \Omega_\geq \frac{1}{(1 - q)(1 - \lambda_2 q^2)(1 - \lambda_3 q/\lambda_2) \cdots (1 - q/\lambda_{m-1})}.$$

A second application gives

$$\underset{\geq}{\Omega} \frac{1}{(1-q)(1-q^2)(1-\lambda_3 q^3)\cdots(1-q/\lambda_{m-1})}.$$

It is now clear that the result of the theorem follows after applying the lemma to each of $\lambda_1, \lambda_2, \ldots, \lambda_{m-1}$. The theorem is proved. ∎

Before developing partition analysis further, we consider a simple example to further illustrate the power of this method. We need an extension of Lemma 11.2.1, which will be useful for other purposes as well.

Lemma 11.2.3 *If α is a nonnegative integer,*

$$\underset{\geq}{\Omega} \frac{\lambda^{-\alpha}}{(1-\lambda x)(1-y/\lambda)} = \frac{x^\alpha}{(1-x)(1-xy)}.$$

The proof of Lemma 11.2.2 will work here as well. The reader may check this. The example is the following: Let $\Delta(n)$ denote the number of noncongruent triangles of perimeter n with positive integer sides. What is $\sum_{n=0}^{\infty} \Delta(n)q^n$?

Suppose n_1, n_2, n_3 are the sides of the triangle in decreasing order. We must have $n_2+n_3 \geq n_1+1$. MacMahon's partition analysis gives us the answer automatically:

$$\sum_{n=0}^{\infty} \Delta(n)q^n = \underset{\geq}{\Omega} \sum_{n_1,n_2,n_3 \geq 0} q^{n_1+n_2+n_3} \lambda_1^{n_1-n_2} \lambda_2^{n_2-n_3} \lambda_3^{n_2+n_3-n_1-1}$$

$$= \underset{\geq}{\Omega} \frac{\lambda_3^{-1}}{(1-q\lambda_1/\lambda_3)(1-q\lambda_2\lambda_3/\lambda_1)(1-q\lambda_3/\lambda_2)}$$

$$= \underset{\geq}{\Omega} \frac{\lambda_3^{-1}}{(1-q/\lambda_3)(1-q^2\lambda_2)(1-q\lambda_3/\lambda_2)}$$

$$= \underset{\geq}{\Omega} \frac{\lambda_3^{-1}}{(1-q/\lambda_3)(1-q^2)(1-q^3\lambda_3)}$$

$$= \frac{q^3}{(1-q^2)(1-q^3)(1-q^4)}.$$

Therefore, $\Delta(n) =$ the number of partitions of n into twos, threes, and fours with at least 1 three.

11.3 A Library for the Partition Analysis Algorithm

The examples of the preceding section illuminate the technique of partition analysis. From a variety of simple and simply proved evaluations of the operator $\Omega_{\geq}$ (such as Lemma 11.2.3), it is possible to apply the operator to numerous rational functions in several λ_i. We list a few more results similar to Lemma 11.2.3. These and others were originally given by MacMahon.

Proposition 11.3.1

(a)
$$\Omega \atop \geq \frac{1}{(1 - \lambda x)(1 - y_1/\lambda)(1 - y_2/\lambda) \cdots (1 - y_j/\lambda)}$$

$$= \frac{1}{(1 - x)(1 - xy_1) \cdots (1 - xy_j)}.$$

(b)
$$\Omega \atop \geq \frac{1}{(1 - \lambda x)(1 - \lambda y)(1 - z/\lambda)} = \frac{1 - xyz}{(1 - x)(1 - y)(1 - xz)(1 - yz)}.$$

(c)
$$\Omega \atop \geq \frac{1}{(1 - \lambda x)(1 - \lambda y)(1 - z/\lambda^2)} = \frac{1 + xyz - x^2yz - xy^2z}{(1 - x)(1 - y)(1 - x^2z)(1 - y^2z)}.$$

Proof. Each such result (and countless others) can be proved in several ways. The method of partial fractions will reduce most to applications of Lemma 11.2.3. The proofs of (a) and (b) are given below and (c) is left to the reader.

Note that for $j = 1$, the result in (a) is just Lemma 11.2.1. Using induction, suppose the result is true up to $j - 1$. Observe that

$$\frac{1}{(1 - y_{j-1}/\lambda)(1 - y_j/\lambda)} = \frac{1}{y_{j-1} - y_j} \left(\frac{y_{j-1}}{1 - y_{j-1}/\lambda} - \frac{y_j}{1 - y_j/\lambda} \right).$$

Then the expression in (a) can be written as

$$\frac{1}{y_{j-1} - y_j} \Omega \atop \geq \left[\frac{y_{j-1}}{(1 - \lambda x)(1 - y_1/\lambda) \cdots (1 - y_{j-1}/\lambda)} \right.$$

$$\left. - \frac{y_j}{(1 - \lambda x)(1 - y_1/\lambda) \cdots (1 - y_{j-2}/\lambda)(1 - y_{j/\lambda})} \right]$$

$$= \frac{1}{y_{j-1} - y_j} \left[\frac{y_{j-1}}{(1 - x)(1 - xy_1) \cdots (1 - xy_{j-1})} \right.$$

$$\left. - \frac{y_j}{(1 - x)(1 - xy_1) \cdots (1 - xy_{j-2})(1 - xy_j)} \right]$$

$$= \frac{1}{(1 - x)(1 - xy_1) \cdots (1 - xy_j)}.$$

The proof of (b) is very similar. We have

$$\Omega \atop \geq \frac{1}{(1 - x\lambda)(1 - y\lambda)(1 - z/\lambda)} = \frac{1}{x - y} \Omega \atop \geq \left(\frac{x}{1 - x\lambda} - \frac{y}{1 - y\lambda} \right) \frac{1}{1 - z/\lambda}$$

$$= \frac{x}{(x - y)(1 - x)(1 - xz)} - \frac{y}{(x - y)(1 - y)(1 - yz)}$$

$$= \frac{1 - xyz}{(1 - x)(1 - y)(1 - xz)(1 - yz)}.$$

This proves the proposition. ∎

11.4 Generating Functions

In this section we apply partition analysis to find the generating functions of some important partition functions.

Theorem 11.4.1 *Let $Q_m(n)$ denote the number of partitions of n into exactly m distinct parts. Then*

$$\sum_{n=0}^{\infty} Q_m(n)q^n = \frac{q^{m(m+1)/2}}{(1-q)(1-q^2)\cdots(1-q^m)}.$$

Proof. If $n = n_1 + n_2 + \cdots + n_m$, then we require $n_1 \geq n_2 + 1, n_2 \geq n_3 + 1$, $\ldots, n_m \geq 1$, because the parts are distinct and exactly m in number. So

$$\sum_{n=0}^{\infty} Q_m(n)q^n = \underset{\geq}{\Omega} \sum_{n_1,\ldots,n_m \geq 0} q^{n_1+n_2+\cdots+n_m} \lambda_1^{n_1-n_2-1} \lambda_2^{n_2-n_3-1} \cdots \lambda_{m-1}^{n_{m-1}-n_m-1} \lambda_m^{n_m-1}$$

$$= \underset{\geq}{\Omega} \frac{\lambda_1^{-1}\lambda_2^{-1}\cdots\lambda_m^{-1}}{(1-\lambda_1 q)(1-\lambda_2 q/\lambda_1)\cdots(1-\lambda_m q/\lambda_{m-1})}.$$

Apply Lemma 11.2.3 to $\lambda_1, \lambda_2, \ldots, \lambda_n$ with $\alpha = 1$ and obtain

$$\sum_{n=0}^{\infty} Q_m(n)q^n = \underset{\geq}{\Omega} \frac{q\lambda_2^{-1}\cdots\lambda_m^{-1}}{(1-q)(1-\lambda_2 q^2)(1-\lambda_3 q/\lambda_2)\cdots(1-\lambda_m q/\lambda_{m-1})}$$

$$= \underset{\geq}{\Omega} \frac{q \cdot q^2 \lambda_3^{-1}\cdots\lambda_m^{-1}}{(1-q)(1-q^2)(1-\lambda_3 q/\lambda_2)\cdots(1-\lambda_m q/\lambda_{m-1})}$$

$$= \frac{q \cdot q^2 \cdots q^m}{(1-q)(1-q^2)\cdots(1-q^m)},$$

which proves the theorem. ∎

In exactly the same way, we may consider $Q_m^{(k,\ell)}(n)$, the number of partitions of n into m parts where each part differs from the next by at least k and the smallest part is $\geq \ell$. The closed form for the generating function is given in the next theorem.

Theorem 11.4.2

$$\sum_{n\geq 0} Q_m^{(k,\ell)}(n)q^n = \frac{q^{\ell m+km(m-1)/2}}{(1-q)(1-q^2)\cdots(1-q^m)}.$$

Proof. Reasoning as before, we can see that

$$\sum_{n\geq 0} Q_m^{(k,\ell)}(n)q^n = \underset{\geq}{\Omega} \sum_{n_1,\ldots,n_m \geq 0} q^{n_1+n_2+\cdots+n_m} \lambda_1^{n_1-n_2-k} \cdots \lambda_{m-1}^{n_{m-1}-n_m-k} \lambda_m^{n_m-\ell}$$

$$= \underset{\geq}{\Omega} \frac{\lambda_1^{-k}\lambda_2^{-k}\cdots\lambda_{m-1}^{-k}\lambda_m^{-\ell}}{(1-\lambda_1 q)(1-\lambda_2 q/\lambda_1)\cdots(1-\lambda_m q/\lambda_{m-1})}$$

$$= \frac{q^{\ell m+km(m-1)/2}}{(1-q)(1-q^2)\cdots(1-q^m)}.$$

The only change is that in the first $m-1$ applications of $\Omega_\geq, \alpha = k$, and in the final (trivial) application, $\alpha = \ell$, $y = 0$ in Lemma 11.2.3. This proves the theorem. ∎

We can also introduce further variables that keep track of other facts about the partitions. For example, we have the following theorem.

Theorem 11.4.3   *Suppose $p_m(j,n)$ (respectively $Q_m(j,n)$) denotes the number of partitions of n into $\leq m$ parts (respectively exactly m distinct parts) with largest part j. Then*

$$\sum_{j,n\geq 0} p_m(j,n)z^j q^n = 1/(1-zq)(1-zq^2)\cdots(1-zq^m)$$

and

$$\sum_{j,n\geq 0} Q_m(j,n)z^j q^n = z^m q^{m(m+1)/2}/(1-zq)(1-zq^2)\cdots(1-zq^m).$$

Proof.   From the definition of $p_m(j,n)$ it is clear that

$$\sum_{j,n\geq 0} p_m(j,n)z^j q^n = \Omega_{\geq} \sum_{n_1,\ldots,n_m\geq 0} z^{n_1} q^{n_1+n_2+\cdots+n_m} \lambda_1^{n_1-n_2} \lambda_2^{n_2-n_3} \cdots \lambda_{m-1}^{n_{m-1}-n_m}$$

$$= \Omega_{\geq} \frac{1}{(1-zq\lambda_1)(1-q\lambda_2/\lambda_1)\cdots(1-q/\lambda_{m-1})}$$

$$= \Omega_{\geq} \frac{1}{(1-zq)(1-zq^2\lambda_2)(1-q\lambda_2/\lambda_3)\cdots(1-q/\lambda_{m-1})}$$

$$= \Omega_{\geq} \frac{1}{(1-zq)(1-zq^2)(1-zq^3\lambda_3)\cdots(1-q/\lambda_{m-1})}$$

$$= \frac{1}{(1-zq)(1-zq^2)\cdots(1-zq^m)}.$$

For the other generating function, the argument is similar. Thus,

$$\sum_{j,n\geq 0} Q_m(j,n)z^j q^n$$

$$= \Omega_{\geq} \sum_{n_1,\ldots,n_m\geq 0} z^{n_1} q^{n_1+\cdots+n_m}$$

$$\cdot \lambda_1^{n_1-n_2-1} \lambda_2^{n_2-n_3-1} \cdots \lambda_{m-1}^{n_{m-1}-n_m-1} \lambda_m^{n_m-1}$$

$$= \frac{z^m q^{m(m+1)/2}}{(1-zq)(1-zq^2)\cdots(1-zq^m)}.$$

The theorem is proved. ∎

In what follows we use the notation $[z^j]\sum_{n=0}^{\infty} a_n z^n = a_j$, that is, the operator $[z^j]$ applied to a power series gives the coefficient of z^j. An observation needed in the proof of the next theorem is that

$$\sum_{h=0}^{N} a_h = \sum_{h=0}^{N} [z^h] \sum_{n=0}^{\infty} a_n z^n$$

$$= [z^N][(1 + z + z^2 + \cdots)(a_0 + a_1 z + a_2 z^2 + \cdots)]$$

$$= [z^N]\frac{\sum_{0}^{\infty} a_n z^n}{1 - z}. \tag{11.4.1}$$

Theorem 11.4.4 *Suppose* $p(N, M, n)$ *(respectively* $Q(N, M, n)$*) denotes the number of partitions of n into* $\leq M$ *parts (respectively exactly M parts), each* $\leq N$*. Then*

$$\sum_{n=0}^{\infty} p(N, M, n)q^n = \begin{bmatrix} N + M \\ M \end{bmatrix}_q,$$

and

$$\sum_{n=0}^{\infty} Q(N, M, n)q^n = q^{M(M+1)/2} \begin{bmatrix} N \\ M \end{bmatrix}_q.$$

Here $\begin{bmatrix} n \\ m \end{bmatrix}_q$ *is the q-binomial coefficient defined by (10.0.5).*

Proof.   By Theorem 11.4.3 and the observation (11.4.1),

$$\sum_{n=0}^{\infty} p(N, M, n)q^n = \sum_{h=0}^{N} [z^h] \sum_{j,n\geq 0} p_M(j, n)z^j q^n$$

$$= \sum_{h=0}^{N} [z^h]\frac{1}{(1 - zq) \cdots (1 - zq^M)}$$

$$= [z^N]\frac{1}{(1 - z)(1 - zq) \cdots (1 - zq^M)}$$

$$= [z^N] \sum_{k=0}^{\infty} \begin{bmatrix} M + k \\ k \end{bmatrix}_q z^k = \begin{bmatrix} N + M \\ M \end{bmatrix}_q.$$

Note that Corollary 10.2.2(d) was used to get the second-to-last equation. For the second part, we have

$$\sum_{n=0}^{\infty} Q(N, M, n)q^n = \sum_{h=0}^{N} [z^h] \sum_{j,n\geq 0} Q_M(j, n)z^j q^n$$

$$= [z^N]z^M q^{M(M+1)/2}/[(1 - z)(1 - zq) \cdots (1 - zq^M)]$$

$$= [z^N]z^M q^{M(M+1)/2} \sum_{k=0}^{\infty} \begin{bmatrix} k + M \\ M \end{bmatrix}_q z^k = q^{M(M+1)/2} \begin{bmatrix} N \\ M \end{bmatrix}_q.$$

This proves the theorem.   ■

Corollary 11.4.5

$$\sum_{n,M\geq 0} Q(N, M, n)z^M q^n = (1 + zq)\cdots(1 + zq^N).$$

Proof. By Theorem 11.4.4 and Corollary 10.2.2(c)

$$\sum_{n,M\geq 0} Q(N, M, n)z^M q^n = \sum_{M\geq 0} z^M \left(\sum_{n\geq 0} Q(N, M, n)q^n \right)$$

$$= \sum_{M\geq 0} z^M q^{M(M+1)/2} \begin{bmatrix} N \\ M \end{bmatrix}_q,$$

$$= (1 + zq)(1 + zq^2)\cdots(1 + zq^N),$$

and the corollary is proved. ■

The next theorem gives limiting cases and other consequences of some previous results.

Theorem 11.4.6 *Let $p(n)$ denote the total number of partitions of n and let $p(m, n)$ denote the number of partitions of n into exactly m parts. Then*

(a) $$\sum_{n=0}^{\infty} p(n)q^n = 1/(q; q)_\infty,$$

(b) $$\sum_{n,m\geq 0} p(m, n)z^m q^n = 1/(zq; q)_\infty,$$

(c) $$\sum_{m,n\geq 0} Q_m(n)z^m q^n = (-zq; q)_\infty,$$

(d) $$\sum_{n,m\geq 0} Q_m^{(2,1)}(n)z^m q^n = \sum_{m=0}^{\infty} z^m q^{m^2}/(q; q)_m,$$

(e) $$\sum_{n,m\geq 0} Q_m^{(2,2)}(n)z^m q^n = \sum_{m=0}^{\infty} z^m q^{m(m+1)}/(q; q)_m.$$

Proof. By Theorem 11.2.2,

$$\sum_{n=0}^{\infty} p(n)q^n = \lim_{m\to\infty} \sum_{n=0}^{\infty} p_m(n)q^n = \lim_{m\to\infty} 1/[(1 - q)(1 - q^2)\cdots(1 - q^m)]$$

$$= 1/(q; q)_\infty.$$

To prove (b), first observe that $p(m, n) = p_m(n) - p_{m-1}(n)$. Then

$$\sum_{n,m\geq 0} p(m, n)z^m q^n = \sum_{m\geq 0} z^m \sum_{n\geq 0}(p_m(n) - p_{m-1}(n))q^n$$

$$= \sum_{m\geq 0} z^m \left(\frac{1}{(q; q)_m} - \frac{1}{(q; q)_{m-1}} \right)$$

$$= \sum_{m\geq 0} \frac{z^m q^m}{(q; q)_m} = \frac{1}{(zq; q)_\infty},$$

where the last step follows from Corollary 10.10.4. Formula (c) is obtained from Corollary 11.4.5 by letting $N \to \infty$. To derive (d), observe that, by Theorem 11.4.2,

$$\sum_{n,n\geq 0} Q_m^{(2,1)}(n)z^m q^n = \sum_{m\geq 0} z^m \sum_{n\geq 0} Q_m^{(2,1)}(n)q^n$$

$$= \sum_{m\geq 0} \frac{z^m q^{m^2}}{(q; q)_m}.$$

The final formula is obtained in a similar way. The theorem is proved. ∎

The series on the right side of (d) and (e) occur in the Rogers–Ramanujan formulas, which will be stated and proved in the next chapter.

11.5 Some Results on Partitions

In Section 11.1, we showed that the number of partitions of n into odd parts equals the number of partitions into distinct parts by showing that their generating functions are equal. This is a very powerful way of obtaining results on partitions. In this section, we give some applications of the theorems derived in the previous section. For example, a simple consequence of Theorem 11.4.4 is the next result.

Theorem 11.5.1 *The number of partitions of n into $\leq M$ parts each $\leq N$ equals the number of partitions of n into $\leq N$ parts each $\leq M$, that is,*

$$p(N, M, n) = p(M, N, n).$$

Proof. By Theorem 11.4.4, the generating function for $p(N, M, n)$ is

$$\begin{bmatrix} N + M \\ M \end{bmatrix}_q,$$

which is clearly the generating function for $p(M, N, n)$ as well. This proves the theorem. ∎

An immediate consequence of this theorem is the following:

Corollary 11.5.2   *The number of partitions of n into $\leq m$ parts equals the number of partitions of n in which each part $\leq m$.*

A direct proof of this corollary can also be given by showing that the generating function of $b(m, n)$, the number of partitions where the parts are $\leq m$, is also $1/(q; q)_m$. Let ℓ_k denote the number of times k occurs in some partition of n. Then

$$\sum_{n=0}^{\infty} b(m, n)q^n = \sum_{\ell_1, \ldots, \ell_m \geq 0} q^{\ell_1(1)+\ell_2(2)+\cdots+\ell_m(m)}$$

$$= \sum_{\ell_1} q^{\ell_1} \sum_{\ell_2} q^{2\ell_2} \cdots \sum_{\ell_m} q^{m\ell_m}$$

$$= \frac{1}{(1-q)(1-q^2)\cdots(1-q^m)}.$$

Here ℓ_k is the number of times k occurs in some partition of n.

The results of the previous section can also be used to give a different derivation of some identities obtained in Chapter 10 from the q-binomial theorem. For example, it is intuitively clear that

$$(1 + zq + z^2q^{1+1} + \cdots)(1 + zq^2 + z^2q^{2+2} + \cdots)(1 + zq^3 + z^2q^{3+3} + \cdots)\cdots$$
$$= \sum_{m,n\geq 0} p(m, n)z^m q^n,$$

where $p(m, n)$ is the number of partitions of n with exactly m parts. However, by partition analysis (Theorem 11.2.2)

$$\sum_{n\geq 0} p(m, n)q^n = \sum_{n\geq 0} (p_m(n) - p_{m-1}(n))q^n$$

$$= \frac{1}{(q; q)_m} - \frac{1}{(q; q)_{m-1}} = \frac{q^m}{(q; q)_m}.$$

Thus

$$\frac{1}{(zq; q)_\infty} = \sum_{m=0}^{\infty} \frac{q^m z^m}{(q; q)_m}$$

or

$$\sum_{m=0}^{\infty} \frac{z^m}{(q; q)_\infty} = \frac{1}{(z; q)_\infty},$$

a result of Euler (Corollary 10.2.2).

In a similar way,

$$(-zq; q)_\infty = (1 + zq)(1 + zq^2)(1 + zq^3) \cdots$$

$$= \sum_{m,n \geq 0} Q_m(n) z^m q^n = \sum_{m=0}^\infty \frac{q^{m(m+1)/2} z^m}{(q; q)_m},$$

where the last equality follows from partition analysis (Theorem 11.4.1). This implies the other result of Euler in Corollary 10.2.2, namely

$$\sum_{n=0}^\infty \frac{q^{\binom{n}{2}} x^n}{(q; q)_n} = (-x; q)_\infty.$$

We end this section with the statement of the Rogers–Ramanujan identities and their partition theoretic interpretation. A more complete discussion of these identities and their proofs is given in the next chapter. The identities are

$$\sum_{m=0}^\infty \frac{q^{m^2}}{(q; q)_m} = \prod_{n=0}^\infty (1 - q^{5n+1})^{-1} (1 - q^{5n+4})^{-1} \tag{11.5.1}$$

and

$$\sum_{m=0}^\infty \frac{q^{m(m+1)}}{(q; q)_m} = \prod_{n=0}^\infty (1 - q^{5n+2})^{-1} (1 - q^{5n+3})^{-1}. \tag{11.5.2}$$

The right-hand side of (11.5.1) is obviously the generating function for the number of partitions in which the parts are identical to 1 or 4 mod 5. The left-hand side, by Theorem 11.4.6(d), is the number of partitions where the parts differ by at least 2. A similar interpretation holds for (11.5.2).

Theorem 11.5.3

(a) *The number of partitions of n in which the difference between any two parts is at least 2 equals the number of partitions of n into parts $\equiv 1$ or $4 \pmod 5$.*

(b) *The number of partitions of n in which the least part ≥ 2 and the difference between any two parts is at least 2 equals the number of partitions of n into parts $\equiv 2$ or $3 \pmod 5$.*

This interpretation of (11.5.1) and (11.5.2) is due to MacMahon [1917–1918].

11.6 Graphical Methods

A powerful way of studying partitions is by representing them graphically. This method was discovered by N. M. Ferrers in the 1850s. It has since been used extensively in partition theory, unlike MacMahon's partition analysis, which has begun to gain some prominence only recently. This may partly be due to MacMahon's admission that partition analysis failed to give significant results in plane partitions

Figure 11.1

for which he had initially developed it. As a contrast, Sylvester, who published Ferrers's graphical method, gave this method much positive publicity. In a short paper titled, "Note on the graphical method in partitions," published in 1883, Sylvester wrote,

The discovery of this process is due to Dr. Ferrers, who informs me that he himself never published it but left it to me to do so in his name in the *London and Edinburgh Philosophical Magazine* for 1853. I may mention that I have never missed an opportunity of expressing my sense of the great importance of the discovery and bringing it under the notice of my pupils. . . .

Ferrers's graphical representation of a partition is a collection of lattice points where each row of points (or dots) corresponds to a part of the partition. For example, the graphical representation of $5 + 2 + 2 + 1$ is shown in Figure 11.1.

The conjugate of this partition is another partition obtained by reading the columns of the above partition. In this case the conjugate is $4 + 3 + 1 + 1 + 1$.

According to Sylvester, the result that Ferrers proved by this method is the following: The number of partitions of n into exactly m parts equals the number of partitions of n with maximum part m. This is intuitively obvious from Ferrers's graph. Each partition with exactly m parts has a conjugate with largest part m and conversely. In a footnote Sylvester [1853] makes the following interesting remark: "I learn from Mr. Ferrers that this theorem was brought under his cognizance through a Cambridge examination paper set by Mr. Adams of Neptune notability."

The reader should check that Theorem 11.5.1 and its corollary also follow immediately from an application of Ferrers's method. In this section, we consider a few applications of this method to see some of its scope and power.

Consider the result contained in Theorem 11.4.1:

$$\sum_{n=0}^{\infty} Q_m(n)q^n = \frac{q^{m(m+1)/2}}{(1-q)(1-q^2)\cdots(1-q^m)},$$

where $Q_m(n)$ is the number of partitions of n into exactly m distinct parts. We have seen that the generating function for the number of partitions with $\leq m$ parts is $1/(q; q)_m$. To understand the factor $m(m+1)/2$, take the partition $7 + 6 + 4 + 2 + 1$ where $m = 5$. Graphically this is depicted in Figure 11.2. There are $\frac{1}{2}(5)(6)$ dots

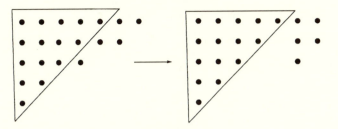

Figure 11.2

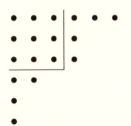

Figure 11.3

inside the triangle. The remaining nodes form the partition of some number with ≤ 5 parts.

Now consider Euler's identity contained in Corollary 10.10.4:

$$1 + \sum_{n=1}^{\infty} \frac{q^{n^2}}{(1-q)^2(1-q^2)^2 \cdots (1-q^n)^2} = \prod_{n=1}^{\infty} (1-q^n)^{-1}.$$

For each partition π we find the largest square of points (starting at the upper left-hand corner) contained in the Ferrers graph. Such a square is called a Durfee square, named after a student of Sylvester who used this idea. Suppose π is given by $6+4+4+2+1+1$; then its Ferrers graph and the 3×3 Durfee square are shown in Figure 11.3. In general each partition π of m has a Durfee square of side n, for some n, and we can write $\pi = n^2 + \pi_1 + \pi_2$, where π_1 is the partition made by the points below the square and π_2 is the conjugate partition of the points to the right of the square. Since the partitions with parts $\leq n$ are generated by $1/(q;q)_n$, it follows that the set of all partitions with Durfee square of side n is generated by

$$q^{n^2} \cdot \frac{1}{(q;q)_n} \cdot \frac{1}{(q;q)_n} = \frac{q^{n^2}}{(1-q)^2(1-q^2)^2 \cdots (1-q^n)^2}.$$

Now the generating function for $p(n)$ is $1/(q;q)_\infty$, and thus

$$\sum_{n=0}^{\infty} \frac{q^{n^2}}{(q;q)_n(q;q)_n} = \frac{1}{(q;q)_\infty} = \prod_{n=1}^{\infty} (1-q^n)^{-1}.$$

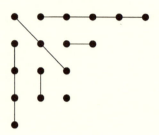

Figure 11.4

As an exercise the reader should prove the following identity:

$$\sum_{n=0}^{\infty} \frac{z^n q^{n^2}}{(q;q)_n (zq;q)_n} = 1/(zq;q)_\infty,$$

using the Durfee square.

As a final example, we take another look at the triple product identity,

$$\frac{1}{(q;q)_\infty} \sum_{n=-\infty}^{\infty} q^{\binom{n+1}{2}} x^n = (-xq;q)_\infty (-x^{-1};q)_\infty.$$

We have seen that the real difficulty in some proofs is to show that the term independent of x on the left-hand side is in fact $1/(q;q)_\infty$.

One way to show this fairly easily is to use the Frobenius symbol for a partition. To describe this idea, consider the partition $6+4+3+3+1$ whose Ferrers graph consists of the points in the graph shown in figure 11.4.

Associate the points as indicated in the Figure and read off the Frobenius symbol for π:

$$\begin{pmatrix} 5 & 2 & 0 \\ 4 & 2 & 1 \end{pmatrix}.$$

The top row represents the horizontal lines to the right of the diagonal and the bottom row the vertical lines below the diagonal. Clearly the sum of the numbers in the Frobenius symbol plus the number of columns gives the number being partitioned. More generally, every partition π of n can be represented by a Frobenius symbol

$$\begin{pmatrix} a_1 & a_2 \ldots a_r \\ b_1 & b_2 \ldots b_r \end{pmatrix},$$

where $a_1 > a_2 > \cdots > a_r \geq 0, b_1 > \cdots > b_r \geq 0$, and $n = r + \Sigma a_i + \Sigma b_i$. Now let us prove that the constant term in

$$(-xq;q)_\infty (-x^{-1};q)_\infty = \prod_{n=1}^{\infty} (1+xq^n) \prod_{m=1}^{\infty} (1+q^{m-1}/x)$$

is $1/(q; q)_\infty$, which will imply the result we want. Observe that a contribution to the constant term is obtained whenever r terms of the xq^n are multiplied with r terms of the form q^{m-1}/x, that is,

$$q^{a_1+a_2+\cdots+a_r} \cdot q^{b_1+b_2+\cdots+b_r},$$

where the as are positive and distinct and the bs are nonnegative and distinct. This is a partition represented by the Frobenius symbol

$$\begin{pmatrix} a_1 - 1 & a_2 - 1 & \cdots & a_r - 1 \\ b_1 & b_2 & \cdots & b_r \end{pmatrix},$$

Hence the constant term is the generating function for $p(n)$, which is $1/(q; q)_\infty$. The result is proved.

11.7 Congruence Properties of Partitions

Congruence properties of $p(n)$, the number of partitions of n, were first discovered by Ramanujan on studying the table of values of $p(n)$ constructed by MacMahon from $n = 1$ to 200. Ramanujan gave simple proofs of the theorems

$$p(5n + 4) \equiv 0 (\text{mod } 5), \tag{11.7.1}$$

$$p(7n + 5) \equiv 0 (\text{mod } 7). \tag{11.7.2}$$

He also found expressions for the generating functions of $p(5n+4)$ and $p(7n+5)$ as products (or a sum of two products). These are given by the formulas below:

$$p(4) + p(9)q + p(14)q^2 + \cdots = 5\frac{\{(1 - q^5)(1 - q^{10})(1 - q^{15}) \cdots\}^5}{\{(1 - q)(1 - q^2)(1 - q^3) \cdots\}^6}, \tag{11.7.3}$$

$$p(5) + p(12)q + p(19)q^2 + \cdots = 7\frac{\{(1 - q^7)(1 - q^{14})(1 - q^{21}) \cdots\}^3}{\{(1 - q)(1 - q^2)(1 - q^3) \cdots\}^4}$$

$$+ 49q\frac{\{(1 - q^7)(1 - q^{14})(1 - q^{21}) \cdots\}^7}{\{(1 - q)(1 - q^2)(1 - q^3) \cdots\}^8}. \tag{11.7.4}$$

Ramanujan [1927, paper 25] sketched a proof of (11.7.3) promising to give details in a later paper. A year later he died and the promised paper never appeared. However, he gave the necessary details in an unpublished manuscript. See Ramanujan [1988, p. 238]. We reproduce this proof here. It has also appeared in an interesting unpublished book manuscript by Thiruvenkatachar and Venkatachaliengar.

Observe that the congruences (11.7.1) and (11.7.2) follow immediately from (11.7.3) and (11.7.4) respectively. These generating functions can also be used to prove congruences modulo 5^2 and 7^2. First observe that for any prime p,

$$(1 - q)^p \equiv 1 - q^p \,(\mathrm{mod}\, p)$$

or

$$\frac{1 - q^p}{(1 - q)^p} \equiv 1 \,(\mathrm{mod}\, p). \tag{11.7.5}$$

Thus (11.7.3) and (11.7.5) with $p = 5$ imply that

$$
\frac{p(4)q + p(9)q^2 + \cdots}{5\{(1 - q^5)(1 - q^{10}) \cdots\}^4} = \frac{q}{(1 - q)(1 - q^2) \cdots} \frac{(1 - q^5)(1 - q^{10}) \cdots}{\{(1 - q)(1 - q^2) \cdots\}^5}
$$

$$
\equiv \frac{q}{(1 - q)(1 - q^2) \cdots} \,(\mathrm{mod}\, 5). \tag{11.7.6}
$$

Since

$$\frac{q}{(1 - q)(1 - q^2) \cdots} = \sum_{n=1}^{\infty} p(n - 1)q^n,$$

the coefficient of q^{5m} is divisible by 5. The coefficient of q^{5m} on the left of (11.7.6) is $p(25m - 1)$ and hence

$$p(25m - 1) \equiv 0 \,(\mathrm{mod}\, 25). \tag{11.7.7}$$

Similarly (11.7.4) implies

$$p(49m - 2) \equiv 0 \,(\mathrm{mod}\, 49). \tag{11.7.8}$$

To prove (11.7.3), start with Euler's pentagonal number theorem (Corollary 10.4.2(c)):

$$\prod_{n=1}^{\infty} (1 - q^{n/5}) = \sum_{n=-\infty}^{\infty} (-1)^n q^{n(3n+1)/10}.$$

Partition the series into five parts according to $n \equiv 0, \pm 1, \pm 2 \,(\mathrm{mod}\, 5)$. For example, the subseries of terms with $n = 5m - 1$ is

$$-\sum_{m=-\infty}^{\infty} (-1)^m q^{(5m-1)(15m-2)/10} = -q^{1/5} \sum_{m=-\infty}^{\infty} (-1)^m q^{5 \cdot m(3m-1)/2}$$

$$= -q^{1/5} \prod_{n=1}^{\infty} (1 - q^{5n}).$$

Thus

$$\prod_{n=1}^{\infty}(1 - q^{n/5}) = \sum_{m=-\infty}^{\infty}(-1)^m q^{m(5m+1)/2} + \sum_{m=-\infty}^{\infty}(-1)^m q^{(3m-1)(5m-2)/2}$$

$$+ q^{2/5}\left[\sum_{m=-\infty}^{\infty}(-1)^m q^{(3m+2)(5m+1)/2} - \sum_{m=-\infty}^{\infty}(-1)^m q^{m(15m+7)/2}\right]$$

$$- q^{1/5}\prod_{n=1}^{\infty}(1 - q^{5n}).$$

Divide by the infinite product $\prod_{n=1}^{\infty}(1 - q^{5n})$ to get

$$\prod_{n=1}^{\infty}\left(\frac{1 - q^{n/5}}{1 - q^{5n}}\right) = \xi_1 - q^{1/5} - \xi q^{2/5}, \tag{11.7.9}$$

where ξ and ξ_1 are power series in q. Our claim is that $\xi\xi_1 = 1$. We recall that by Corollary 10.4.2(e) and (11.7.9),

$$\frac{\sum_{n=-\infty}^{\infty}(-1)^n(2n + 1)q^{n(n+1)/10}}{\prod_{n=1}^{\infty}(1 - q^{5n})^3} = (\xi_1 - q^{1/5} - \xi q^{2/5})^3. \tag{11.7.10}$$

Since the power of q, given by $n(n+1)$, is either 0, 2, or 6 (mod 10), it follows that no power of q is of the form $2/5$ + an integer. So the term $3q^{2/5}\xi_1 - 3\xi_1^2\xi q^{2/5} = 3q^{2/5}\xi_1(1 - \xi\xi_1)$ on the right side of (11.7.10) must be zero. This implies that $\xi_1 = \xi^{-1}$. Therefore,

$$\prod_{n=1}^{\infty}\left(\frac{1 - q^{5n}}{1 - q^{n/5}}\right) = \frac{1}{\xi^{-1} - q^{1/5} - \xi q^{2/5}}. \tag{11.7.11}$$

The denominator on the right-hand side is

$$(\xi^{-1}q^{-1/5} - \xi q^{1/5} - 1)q^{1/5}.$$

Consider the expression $\lambda^{-1} - \lambda - 1$, where $\lambda = \xi q^{1/5}\omega$, and ω is a fifth root of unity. Observe that if $\lambda^{-1} - \lambda = 1$; then a simple calculation shows that $\lambda^{-5} + \lambda^5 = 11$. Thus

$$\xi^{-5} - 11q - \xi^5 q^2 = \prod_{k=0}^{4}(\xi^{-1} - q^{1/5}\omega^k - \xi q^{2/5}\omega^{2k}), \tag{11.7.12}$$

where $\omega = e^{2\pi i/5}$. It is now easy to check by long division that (11.7.11) can be written as

$$\prod_{n=1}^{\infty}\left(\frac{1-q^{5n}}{1-q^{n/5}}\right)$$

$$=\frac{\xi^{-4}-3q\xi+q^{1/5}(\xi^{-3}+2q\xi^2)+q^{2/5}(2\xi^{-2}-q\xi)+q^{3/5}(3\xi^{-1}+q\xi^4)+5q^{4/5}}{\xi^{-5}-11q-q^2\xi^5}.$$

$$(11.7.13)$$

(In fact, Ramanujan [1927, paper 25] starts the sketch of his proof of (11.7.3) by observing that (11.7.13) can be shown to be true.) Now multiply across by $q^{1/5}$ and replace $q^{1/5}$ with $q^{1/5}e^{2\pi ik/5}$, where $k=1,2,3,4$. Add the five identities after using the fact that

$$q^{1/5}\prod_{n=1}^{\infty}(1-q^{n/5})^{-1}=\sum_{n=1}^{\infty}p(n-1)q^{n/5}$$

to obtain

$$\prod_{n=1}^{\infty}(1-q^{5n})\sum_{n=0}^{\infty}p(5n+4)q^n=\frac{5}{\xi^{-5}-11q-q^2\xi^5}.\qquad(11.7.14)$$

Replace q with $qe^{2\pi ik}$, where $k=\pm1,\pm2$ in (11.7.11) and multiply the five equations. Note that

$$\prod_{k=0}^{4}(1-q^{n/5}e^{2\pi ikn/5})=\begin{cases}1-q^n, & n\not\equiv0(\mathrm{mod}\,5),\\(1-q^m)^5, & n=5m.\end{cases}$$

By (11.7.12), we now get

$$\prod_{n=1}^{\infty}\left(\frac{1-q^{5n}}{1-q^n}\right)^6=\frac{1}{\xi^{-5}-11q-\xi^5q^2}.$$

Combine this with (11.7.14) to get

$$\sum_{n=0}^{\infty}p(5n+4)q^n=5\frac{\{(1-q^5)(1-q^{10}(1-q^{15})\cdots\}^5}{\{(1-q)(1-q^2)(1-q^3)\cdots\}^6}.$$

About this identity, Hardy wrote that if he were to select one formula from Ramanujan's work for supreme beauty, he would agree with MacMahon in selecting this one.

Remark 11.7.1   Ramanujan remarked that

$$\xi^{-1}=\prod_{n=0}^{\infty}\frac{(1-q^{5n+2})(1-q^{5n+3})}{(1-q^{5n+1})(1-q^{5n+4})}.\qquad(11.7.15)$$

Note the connection of this function with the Rogers–Ramanujan identities, which Ramanujan used to express ξ^{-1} as a continued fraction. Formula (11.7.15) can be proved by first observing that (11.7.11) and the pentagonal number theorem, stated immediately after (11.7.8), imply that

$$\xi^{-1}\prod_{n=1}^{\infty}(1-q^{5n}) = \sum_{\substack{n\equiv 0,3 \\ (\mathrm{mod}\,5)}}(-1)^n q^{(n/10)(3n+1)}.$$

Now apply the quintuple product identity (Exercise 10.14) to the last sum and rearrange the resulting product to get (11.7.15). The continued fraction for ξ^{-1} can be found in Hardy [1940, p. 99].

Remark 11.7.2 Several proofs of (11.7.3) and (11.7.4) have been given. For a relatively simple one, which is similar to Ramanujan's proof in some respects, see Kolberg [1957]. For a proof that uses the machinery of modular functions, see Knopp [1971, §8.3].

Exercises

1. Verify the following formulas:

(a)
$$\Omega_{\geq}\frac{1}{(1-\lambda x)(1-y/\lambda^2)} = \frac{1}{(1-x)(1-x^2 y)},$$

(b)
$$\Omega_{\geq}\frac{1}{(1-\lambda^2 x)(1-y/\lambda)} = \frac{1+xy}{(1-x)(1-xy^2)},$$

(c)
$$\Omega_{\geq}\frac{1}{(1-\lambda x)(1-y/\lambda^s)} = \frac{1}{(1-x)(1-x^s y)},$$

(d)
$$\Omega_{\geq}\frac{1}{(1-\lambda^s x)(1-y/\lambda)} = \frac{1+xy(1-y^{s-1})/(1-y)}{(1-x)(1-xy^s)},$$

(e)
$$\Omega_{\geq}\frac{1}{(1-\lambda^2 x)(1-y/\lambda)(1-z/\lambda)} = \frac{1+xy+xz+xyz}{(1-x)(1-xy^2)(1-xz^2)},$$

(f)
$$\Omega_{\geq}\frac{1}{(1-\lambda^2 x)(1-\lambda y)(1-z/\lambda)} = \frac{1+xz-xyz-xyz^2}{(1-x)(1-y)(1-yz)(1-xz^2)},$$

(g)
$$\Omega_{\geq}\frac{1}{(1-\lambda x)(1-\lambda y)(1-\lambda z)(1-\omega/\lambda)}$$
$$= \frac{1-xy\omega-xz\omega-yz\omega+xyz\omega+xyz\omega^2}{(1-x)(1-y)(1-z)(1-x\omega)(1-y\omega)(1-z\omega)}.$$

2. Prove that the number of partitions of n into parts not divisible by 3 equals the number of partitions of n where no part occurs more than twice.

3. Show that the number of partitions of n in which only odd parts may be repeated equals the number of partitions of n in which no part appears more than three times.

4. Generalize Exercise 3 by showing that the number of partitions of n in which only parts $\not\equiv 0 \pmod{2^m}$ may be repeated equals the number of partitions of n in which no part appears more than $2^{m+1} - 1$ times.

5. Prove that the number of partitions of n in which each part appears 2, 3, or 5 times equals the number of partitions of n into parts congruent to 2, 3, 6, 9, or 10 modulo 12.

(This result is due to Subbarao; see Andrews [1976, p. 15].)

The next three exercises are from Ramanujan [1927, paper 25].

6. Prove that $p(5m + 4) \equiv 0 \pmod 5$ as follows:

(a) Show that

$$q \prod_{n=1}^{\infty} (1 - q^n)^4 = q \prod_{n=1}^{\infty}(1 - q^n) \prod_{n=1}^{\infty}(1 - q^n)^3$$

$$= \frac{1}{2} \sum_{\ell=-\infty}^{\infty} \sum_{m=-\infty}^{\infty} (-1)^{\ell+m}(2\ell + 1)q^{1+\ell(\ell+1)/2+m(3m+1)/2}.$$

(b) Show that if the exponent $1 + \ell(\ell + 1)/2 + m(3m + 1)/2$ is a multiple of 5, then the coefficient $2\ell + 1$ is also a multiple of 5.

(c) Show that $\frac{1}{(1-q)^5} \equiv \frac{1}{1-q^5} \pmod 5$.

(d) Use (c) to observe that

$$f(q) = q \frac{\prod_{n=1}^{\infty}(1 - q^{5n})}{\prod_{n=1}^{\infty}(1 - q^n)} = q \prod_{n=1}^{\infty}(1 - q^n)^4 \frac{\prod_{n=1}^{\infty}(1 - q^{5n})}{\prod_{n=1}^{\infty}(1 - q^n)^5}$$

$$\equiv q \prod_{n=1}^{\infty}(1 - q^n)^4 \pmod 5.$$

(e) Deduce that the coefficient of q^{5m+5} in $f(q)$ is a multiple of 5.

(f) Conclude that $p(5m + 4)$, the coefficient of q^{5m+5} in $q / \prod_{n=1}^{\infty}(1 - q^n)$, is a multiple of 5.

7. Show that $p(7m + 5) \equiv 0 \pmod 7$. Use the identity

$$q^2 \prod_{n=1}^{\infty}(1-q^n)^6 = \frac{1}{4} \sum_{\ell=-\infty}^{\infty} \sum_{m=-\infty}^{\infty} (-1)^{\ell+m}(2\ell+1)(2m+1)q^{2+[\ell(\ell+1)+m(m+1)]/2}$$

and the method employed in Exercise 6.

8. Use (11.7.4) to prove that $p(49m - 2) \equiv 0 \pmod{49}$.

9. Prove (11.7.4).

10. Show that the number of partitions of n that are self-conjugates, that is, identical with their conjugates, equals the number of partitions of n with distinct odd parts.

11. Let $M_1(n)$ denote the number of partitions of n into parts, each greater than 1, such that consecutive integers do not appear as parts. Let $M_2(n)$ denote the number of partitions of n in which no part appears exactly once. Show that $M_1(n) = M_2(n)$.

The result in Exercise 10 is due to Sylvester and that in Exercise 11 to MacMahon. See Andrews [1976, p. 14].

12. Use Euler's pentagonal number theorem, namely

$$\prod_{n=1}^{\infty}(1 - q^n) = \sum_{m=-\infty}^{\infty}(-1)^m q^{m(3m-1)/2},$$

to show that

$$P_E^{(1)}(n) - P_0^{(1)}(n) = \begin{cases} (-1)^m & \text{if } n = m(3m \pm 1)/2, \\ 0 & \text{otherwise.} \end{cases}$$

Here $P_E^{(1)}(n)$ (respectively $P_0^{(1)}(n)$) is the number of partitions of n into an even (respectively odd) number of distinct parts.

13. Prove the following relation that gives an efficient algorithm for computing $p(n)$:

$$p(n) - p(n - 1) - p(n - 2) + p(n - 5)$$
$$+ p(n - 7) + \cdots + (-1)^m(p(n - m(3m - 1)/2)$$
$$+ (-1)^m p(n - m(3m + 1)/2) + \cdots = 0,$$

where $p(M) = 0$ when M is negative. [*Hint:* $(q;q)_\infty \frac{1}{(q;q)_\infty} = 1$.]

14. Use the Durfee square discussed in Section 11.6 to prove that

$$\sum_{n=0}^{\infty} \frac{x^n q^{n^2}}{(q;q)_n(xq;q)_n} = (xq;q)_\infty.$$

15. Show that

$$\sum_{n=0}^{\infty} \frac{x^n y^n q^{n^2}}{(xq;q)_n(yq;q)_n} = \sum_{n=0}^{\infty} \frac{x^{2n} y^n q^{2n^2}}{(xq;q)_n(yq;q)_{2n}} + \sum_{n=0}^{\infty} \frac{x^{2n+1} y^{n+1} q^{(n+1)(2n+1)}}{(xq;q)_n(yq;q)_{2n+1}}.$$

For the left-hand side, use the idea of Exercise 14 to show that the coefficient of $y^m x^r q^n$ is the number of partitions of n into m parts with the largest part equal to r. Do the same for the right side, where instead of a Durfee square consider the largest rectangles that can be of size $n \times 2n$ or $(n + 1) \times (2n + 1)$. Note that the largest rectangles of these dimensions cover all the possibilities.

16. Use the fact that $n^2 = 1 + 3 + 5 + \cdots + 2n - 1$ to see that $\sum_{n=0}^{\infty} q^{n^2}/(q;q)_n$ is the generating function for the number of partitions in which the difference between the parts is at least 2.

The next six results are the partition theoretic interpretations of the six identities of Rogers given in Exercise 12.6. Prove them. For references see Andrews and Askey [1977].

17. The number of partitions $b_1 + b_2 + \cdots + b_r$ of n where $b_1 \geq b_2 \geq b_3 \geq \cdots$ and each b_i is odd or $\equiv \pm 4 \pmod{20}$ equals the number of partitions $c_1 + c_2 + \cdots$ of n where $c_1 > c_2 \geq c_3 > c_4 \geq c_5 > \cdots$.

Hint: Note that $n^2 = 0 + 1 + 1 + 2 + \cdots + (n-1) + n$ and that $1/(q;q)_{2n}$ is the generating function for partitions in which there are at most $2n$ parts.

(Gordon)

18. The number of partitions $b_1 + b_2 + \cdots$ of n where $b_1 \geq b_2 \geq b_3 \ldots$ and each b_i is odd or $\equiv \pm 8 \pmod{20}$ equals the number of partitions $c_1 + c_2 + \cdots + c_{2k+1}$ of n into an odd number of parts where $c_1 \geq c_2 \geq c_3 > c_4 \geq c_5 > c_6 \geq c_7 > \cdots$.

Hint: $n^2 + 2n = 1 + 1 + 2 + 2 + \cdots + (n-1) + (n-1) + n + n + n$.

(Connor)

19. The number of partitions $b_1 + b_2 + \cdots + b_r$ of n where $b_1 \geq b_2 \geq b_3 \geq \cdots$ and each b_i is $\neq \pm 1, \pm 8, \pm 9, 10 \pmod{20}$ equals the number of partitions $c_1 + c_2 + \cdots + c_{2k}$ of n into an even number of parts where $c_1 \geq c_2 > c_3 \geq c_4 > c_5 \geq \cdots$.

(Connor)

20. The number of partitions $b_1 + b_2 + \cdots + b_r$ of n where $b_1 \geq b_2 \geq b_3 \geq \cdots$ and each $b_i \neq \pm 3, \pm 4, \pm 7, 10 \pmod{20}$ equals the number of partitions $c_1 + c_2 + \cdots + c_k$ of n where $c_1 \geq c_2 > c_3 \geq c_4 > c_5 \geq c_6 > \cdots$.

(Connor)

21. The number of partitions of n with distinct parts and with each even part larger than twice the number of odd parts equals the number of partitions of n into parts $\equiv 1$ or $4 \pmod 5$.

Hint: The left-hand side of Exericise 12.6 (e) can be written as

$$\sum_{n=0}^{\infty} \frac{q^{1+3+5+\cdots+2n-1}}{(1-q^2)(1-q^4)\cdots(1-q^{2n})} \prod_{m=1}^{\infty}(1+q^{2n+2m}).$$

22. The number of partitions of n into distinct parts each larger than 1 in which each even part is larger than twice the number of odd parts equals the number of partitions of n into parts $\equiv 2$ or $3 \pmod 5$.

Hint: $3 + 5 + \cdots + (2n+1) = n^2 + 2n$.

12

Bailey Chains

L. J. Rogers is the pioneer of the work leading to the Rogers–Ramanujan identities and beyond. His idea, published in Rogers [1917], provides the starting point for the work of this chapter. We shall recount his seminal idea in Section 12.1. In the 1940s, W. N. Bailey began a systematic study of identities of the Rogers-Ramanujan type. See Bailey [1949]. He saw great generality in the methods introduced by Rogers. This greater level of generality provides for a wide variety of applications well beyond those considered by Rogers.

Motivation for the techniques presented here is scant. As you will see, Rogers's original idea seems almost magical in its construction. Since the advent of computer algebra we can better see how to make sense of Rogers's fortuitous discoveries. However, it is still not evident why one would initially expect that this method would bear fruit.

A systematic account of Bailey's ideas leading up to Bailey's lemma is given in Section 12.2. As an application of these ideas, the important $_8\phi_7$ transformation formula of Watson is derived in the next section. A few consequences of this formula are also included. The last section makes passing mention of other applications of the ideas in Section 12.2.

12.1 Rogers's Second Proof of the Rogers–Ramanujan Identities

The Rogers–Ramanujan identities were first discovered by Rogers [1894]. Rogers made considerable contributions to several areas of mathematics but surprisingly his work went largely unnoticed and did not have the influence it should have had. Part of the surprise comes from the fact that Rogers's early work in invariant theory was noticed by Sylvester and given a prominent place in his "Lectures on the Theory of Reciprocants" [1886]. The long neglect of invariant theory after Hilbert's discoveries may have played a role here. Rogers did not receive credit for his discovery of the Hölder inequality, and his papers containing the Rogers–Ramanujan identities, among the most beautiful formulas in mathematics, went unheralded.

These identities were later rediscovered by Ramanujan. Ramanujan's first letter to Hardy in 1913 contains some continued-fraction formulas that are consequences of these identities. Ramanujan did not have a proof of the identities and he posed them as a problem in 1914 in the *Journal of the Indian Mathematical Society*. MacMahon stated them without proof in the second volume of his *Combinatory Analysis*. He also noted the connection with partitions.

What happened next is best described in the words of Hardy [1940, p. 91]:

The mystery was solved, trebly, in 1917. In that year Ramanujan, looking through old volumes of the *Proceedings of the London Mathematical Society*, came accidentally across Rogers's paper. I can remember very well his surprise, and the admiration he expressed for Rogers's work. A correspondence followed in the course of which Rogers was led to a considerable simplification of the original proof. About the same time I. Schur, who was then cut off from England by war, rediscovered the identities again. Schur published two proofs, one of which is "combinatorial" and quite unlike any other proof known.

In this section we discuss Rogers's second proof published in Rogers [1917]. As noted before, the Rogers–Ramanujan identities are

$$\sum_{n=0}^{\infty} \frac{q^{n^2}}{(q;q)_n} = \frac{1}{(q;q^5)_\infty (q^4;q^5)_\infty}$$

$$= \frac{(q^2;q^5)_\infty (q^3;q^5)_\infty (q^5;q^5)_\infty}{(q;q)_\infty} \qquad (12.1.1)$$

and

$$\sum_{n=0}^{\infty} \frac{q^{n(n+1)}}{(q;q)_n} = \frac{1}{(q^2;q^5)_\infty (q^3;q^5)_\infty} = \frac{(q;q^5)_\infty (q^4;q^5)_\infty (q^5;q^5)_\infty}{(q;q)_\infty}. \qquad (12.1.2)$$

The products on the right side can be transformed by the triple product identity (Theorem 10.4.1),

$$(x;q)_\infty (q/x;q)_\infty (q;q)_\infty = \sum_{k=-\infty}^{\infty} (-1)^k q^{k(k-1)/2} x^k.$$

Replace q with q^5 and x with q^2 and then with q to get

$$(q^2;q^5)_\infty (q^3;q^5)_\infty (q^5;q^5)_\infty = \sum_{k=-\infty}^{\infty} (-1)^k q^{k(5k-1)/2}$$

and

$$(q;q^5)_\infty (q^4;q^5)_\infty (q^5;q^5)_\infty = \sum_{k=-\infty}^{\infty} (-1)^k q^{k(5k-3)/2}.$$

Thus it is sufficient to prove that

$$\sum_{n=0}^{\infty} \frac{q^{n^2}}{(q;q)_n} = \frac{\sum_{k=-\infty}^{\infty}(-1)^k q^{k(5k-1)/2}}{(q;q)_\infty} \tag{12.1.3}$$

and

$$\sum_{n=0}^{\infty} \frac{q^{n(n+1)}}{(q;q)_n} = \frac{\sum_{k=-\infty}^{\infty}(-1)^k q^{k(5k-3)/2}}{(q;q)_\infty}. \tag{12.1.4}$$

Rogers's second proof of the Rogers–Ramanujan identities depends on the following lemma.

Lemma 12.1.1   *Let*

$$S_{2\ell} = \sum_{k=-\ell}^{\ell} \frac{(-1)^k q^{k(3k-1)/2}}{(q)_{\ell+k}(q)_{\ell-k}} \quad and \quad S_{2\ell+1} = \sum_{k=-\ell}^{\ell+1} \frac{(-1)^k q^{k(3k-1)/2}}{(q)_{\ell+1-k}(q)_{\ell+k}}.$$

Then

$$S_{2\ell} = S_{2\ell+1} = \frac{1}{(q)_\ell}, \quad where \quad (q)_k = (q;q)_k.$$

Proof.   The proof consists in rearranging the terms in the sums. In $S_{2\ell}$, combine the terms corresponding to the indices $-k$ and $k+1$. (Note that when $k = \ell$, the term corresponding to $\ell + 1$ is 0.) We get

$$\frac{(-1)^k q^{k(3k+1)/2}}{(q)_{\ell+k}(q)_{\ell-k}} + \frac{(-1)^{k+1} q^{(k+1)(3k+2)/2}}{(q)_{\ell+k+1}(q)_{\ell-k-1}}$$

$$= \frac{(-1)^k q^{k(3k+1)/2}}{(q)_{\ell+k}(q)_{\ell-k}}\left[1 - \frac{q^{2k+1}(1-q^{\ell-k})}{(1-q^{\ell+k+1})}\right] = \frac{(-1)^k q^{k(3k+1)/2}}{(q)_{\ell+1+k}(q)_{\ell-k}}(1 - q^{2k+1}).$$

It is readily seen that this is also the expression obtained in combining the terms corresponding to indices $-k$ and $k+1$ in the sum $S_{2\ell+1}$. Thus $S_{2\ell} = S_{2\ell+1}$.

We now prove that

$$S_{2\ell+1} = (1 - q^{\ell+1})S_{2\ell+2}, \tag{12.1.5}$$

which together with $S_{2\ell} = S_{2\ell+1}$ will prove the lemma by induction. Consider the sum of the terms in $S_{2\ell+1}$ corresponding to $+k$ and $-k$. For $k \neq 0$,

$$\frac{(-1)^k q^{k(3k-1)/2}}{(q)_{\ell+1-k}(q)_{\ell+k}} + \frac{(-1)^k q^{k(3k+1)/2}}{(q)_{\ell+1+k}(q)_{\ell-k}} = \frac{(-1)^k q^{k(3k-1)/2}}{(q)_{\ell+k}(q)_{\ell+1-k}}\left[1 + \frac{q^k(1-q^{\ell+1-k})}{(1-q^{\ell+1+k})}\right]$$

$$= (1 - q^{\ell+1})\left[\frac{(-1)^k q^{k(3k-1)/2}}{(q)_{\ell+1+k}(q)_{\ell+1-k}}\right](1 + q^k).$$

This expression is $(1 - q^{\ell+1})$ times the sum of the terms in $S_{2\ell+2}$ corresponding to the indices $\pm k$, when $k \neq 0$. When $k = 0$, the corresponding terms in $S_{2\ell+1}$ and $S_{2\ell+2}$ are

$$\frac{(1 - q^{\ell+1})}{(q)_{\ell+1}(q)_{\ell+1}} \quad \text{and} \quad \frac{1}{(q)_{\ell+1}(q)_{\ell+1}}.$$

This proves (12.1.5) and the lemma. ∎

The idea of Rogers's second proof of (12.1.1) and (12.1.2) is to expand the function $(-\sqrt{q}e^{i\theta}; q)_\infty(-\sqrt{q}e^{-i\theta}; q)_\infty$ as a Fourier series in two different ways. One way is to use the triple product identity and the other is to apply the q-binomial identity to each of the products. Since the Fourier expansion is unique, the corresponding Fourier coefficients in the two expansions are identical. Hence the exponentials $e^{in\theta}$ can be replaced in the two expansions by any numbers as long as the series are convergent. Rogers shows that the Rogers–Ramanujan identities are obtained if this replacement is done appropriately. The details follow.

By Euler's result in Corollary 10.2.2 (b), we have

$$(-xe^{i\theta}q; q)_\infty = \sum_{n=0}^\infty \frac{q^{n(n+1)/2}}{(q; q)_n} x^n e^{in\theta}$$

and

$$(-xe^{-i\theta}q; q)_\infty = \sum_{m=0}^\infty \frac{q^{m(m+1)/2}}{(q; q)_m} x^m e^{-im\theta}.$$

So the product is

$$(-xe^{i\theta}q; q)_\infty(-xe^{-i\theta}q; q)_\infty = \sum_{m,n\geq 0} \frac{q^{[n(n+1)+m(m+1)]/2} x^{m+n} e^{i(n-m)\theta}}{(q; q)_m(q; q)_n}$$

$$= \sum_{p=0}^\infty x^p \sum_{n=0}^p \frac{q^{[n(n+1)+(p-n)(p-n+1)]/2}}{(q; q)_{p-n}(q; q)_n} e^{i(2n-p)\theta}.$$

Break up the last sum into two parts for p even ($p = 2\ell$) and p odd ($p = 2\ell + 1$) and set $n = \ell + k$. Then

$$(-xe^{i\theta}q; q)_\infty(-xe^{-i\theta}q; q)_\infty = \sum_{\ell=0}^\infty q^{\ell^2+\ell} x^{2\ell} \sum_{k=-\ell}^\ell \frac{q^{k^2} e^{2ki\theta}}{(q; q)_{\ell+k}(q; q)_{\ell-k}}$$

$$+ \sum_{\ell=0}^\infty q^{(\ell+1)^2} x^{2\ell+1} \sum_{k=-\ell}^{\ell+1} \frac{q^{k^2-k} e^{(2k-1)i\theta}}{(q; q)_{\ell+k}(q; q)_{\ell+1-k}}.$$

When $qx = q^{-1/2}$, this relation becomes

$$(-\sqrt{q}e^{i\theta};q)_\infty(-\sqrt{q}e^{-i\theta};q)_\infty$$

$$= \sum_{\ell=0}^{\infty} q^{\ell^2} \sum_{k=-\ell}^{\ell} \frac{q^{k^2}e^{2ki\theta}}{(q;q)_{\ell+k}(q;q)_{\ell-k}} + \sqrt{q}\sum_{\ell=0}^{\infty} q^{\ell(\ell+1)} \sum_{k=-\ell}^{\ell+1} \frac{q^{k^2-k}e^{(2k-1)i\theta}}{(q;q)_{\ell+k}(q;q)_{\ell+1-k}}$$

$$= \sum_{k=-\infty}^{\infty} q^{k^2}e^{2ki\theta} \sum_{\ell=k}^{\infty} \frac{q^{\ell^2}}{(q;q)_{\ell+k}(q;q)_{\ell-k}}$$

$$+ \sqrt{q}\sum_{k=-\infty}^{\infty} q^{k^2-k}e^{(2k-1)i\theta} \sum_{\ell=k-1}^{\infty} \frac{q^{\ell(\ell+1)}}{(q;q)_{\ell+k}(q;q)_{\ell+1-k}}. \tag{12.1.6}$$

This gives one Fourier expansion. To get the other, start with the triple product identity

$$(q,q)_\infty(x;q)_\infty(q/x;q)_\infty = \sum_{k=-\infty}^{\infty} (-1)^k q^{k(k-1)/2}x^k$$

and replace x with $-\sqrt{q}e^{i\theta}$ to get

$$(-\sqrt{q}e^{i\theta};q)_\infty(-\sqrt{q}e^{-i\theta};q)_\infty$$

$$= \frac{\sum_{k=-\infty}^{\infty} q^{k^2/2}e^{ik\theta}}{(q;q)_\infty} = \frac{\sum_{k=-\infty}^{\infty} q^{2k^2}e^{2ki\theta} + \sqrt{q}\sum_{k=-\infty}^{\infty} q^{2k(k-1)}e^{(2k-1)i\theta}}{(q;q)_\infty}.$$

$$\tag{12.1.7}$$

Thus the right sides of (12.1.6) and (12.1.7) are equal to each other. Now replace $e^{2ki\theta}$ with $(-1)^k q^{k(k-1)/2}$ and $e^{(2k-1)i\theta}$ with 0. Apply Lemma 12.1.1 to get (12.1.3), which is equivalent to the first Rogers–Ramanujan identity. The second identity is obtained by taking $e^{2ki\theta} = 0$ and $e^{(2k-1)i\theta} = (-1)^k q^{k(k+1)/2}$ and applying Lemma 12.1.1.

Remark 12.1.1   Rogers's argument provides one of the simplest proofs of (12.1.3) and (12.1.4). It appears to depend on formulas from the theory of theta functions. These formulas, however, are used only for motivation and are not necessary for the proof. Observe that, in the above argument, the equality of the coefficients of $e^{2ki\theta}$ in (12.1.6) and (12.1.7) are a consequence of the uniqueness of Fourier series. Thus we have

$$\sum_{\ell=k}^{\infty} \frac{q^{\ell^2+k^2}}{(q;q)_{\ell+k}(q;q)_{\ell-k}} = \frac{q^{2k^2}}{(q;q)_\infty}. \tag{12.1.8}$$

But this follows directly from Cauchy's formula in Corollary 10.9.4, which can be written

$$\sum_{n=0}^{\infty} \frac{q^{n^2}x^n}{(q;q)_n(xq;q)_n} = \frac{1}{(xq;q)_\infty}. \tag{12.1.9}$$

In fact, we can give a very simple graphical proof of (12.1.9), using Durfee squares, where the exponent of x counts the number of parts of a partition. Then (12.1.8) follows by taking $x = q^{2k}$. Next, (12.1.3) is obtained by multiplying both sides of (12.1.8) by $(-1)^k q^{k(k-1)/2}$, summing over all k, and applying Lemma 12.1.1. In the next section, we shall see that the scope of this simple argument can be greatly extended.

Remark 12.1.2   Note that we get Euler's pentagonal number theorem

$$\sum_{k=-\infty}^{\infty} (-1)^k q^{k(3k-1)/2} = (q; q)_\infty$$

when we let $\ell \to \infty$ in Lemma 12.1.1.

12.2 Bailey's Lemma

In a series of two papers in the 1940s, W. N. Bailey elucidated the underlying structure of Rogers's proof. He began with Rogers's replacement of $2 \cos n\theta = e^{in\theta} + e^{-in\theta}$ in certain Fourier-series expansions and quickly observed the following simplified version of the relevant expansion.

Lemma 12.2.1 (weak Bailey lemma)   *Suppose α_n and β_n are two sequences related by*

$$\beta_n = \sum_{r=0}^{n} \frac{\alpha_r}{(q; q)_{n-r} (aq; q)_{n+r}}; \tag{12.2.1}$$

then, subject to convergence conditions (which in most applications boil down to $|q| < 1$),

$$\sum_{n=0}^{\infty} a^n q^{n^2} \beta_n = \frac{1}{(aq; q)_\infty} \sum_{n=0}^{\infty} a^n q^{n^2} \alpha_n. \tag{12.2.2}$$

Proof.   The proof of this is quite short:

$$\sum_{n=0}^{\infty} a^n q^{n^2} \beta_n = \sum_{n=0}^{\infty} a^n q^{n^2} \sum_{r=0}^{n} \frac{\alpha_r}{(q; q)_{n-r} (aq; q)_{n+r}}$$

$$= \sum_{r=0}^{\infty} \sum_{n=r}^{\infty} \frac{a^n q^{n^2} \alpha_r}{(q; q)_{n-r} (aq; q)_{n+r}}$$

$$= \sum_{r=0}^{\infty} \alpha_r \sum_{n=0}^{\infty} \frac{q^{(n+r)^2} a^{n+r}}{(q; q)_n (aq; q)_{n+2r}}$$

$$= \sum_{r=0}^{\infty} \alpha_r a^r q^{r^2} \sum_{n=0}^{\infty} \frac{q^{n(n+2r)} a^n}{(q; q)_n (aq; q)_{n+2r}}$$

$$= \sum_{r=0}^{\infty} \alpha_r a^r q^{r^2} \frac{1}{(aq; q)_\infty},$$

where the last step follows from Corollary 10.9.4.   ∎

The proof given by Rogers can be subsumed by this result. For example, to get the first identity, take $a = 1$ and

$$\alpha_n = \begin{cases} 1, & n = 0, \\ (-1)^n q^{n(3n-1)/2}(1 + q^n), & n > 0. \end{cases}$$

By Lemma 12.1.1,

$$\beta_n = 1 + \sum_{r=1}^{n} \frac{(-1)^r q^{r(3r-1)/2}(1 + q^r)}{(q;q)_{n-r}(q;q)_{n+r}} = \sum_{-n}^{n} \frac{(-1)^r q^{r(3r-1)/2}}{(q;q)_{n-r}(q;q)_{n+r}} = \frac{1}{(q;q)_n},$$

so that Bailey's result implies

$$\sum_{n=0}^{\infty} \frac{q^{n^2}}{(q;q)_n} = \frac{1 + \sum_{n=1}^{\infty}(-1)^n q^{n^2}(q^{n(3n-1)/2} + q^{n(3n+1)/2})}{(q;q)_\infty}$$

$$= \frac{\sum_{n=-\infty}^{\infty}(-1)^n q^{n(5n-1)/2}}{(q;q)_\infty}.$$

The reader should work out the details for the second identity.

Seeing this, it is clear that the method may be greatly extended by using more general summations than the very special limiting case of the q-analog of Gauss's sum.

Bailey observed in his second paper that one could indeed invoke the full force of the q-Pfaff–Saalschütz summation. He carefully described the proof of such a generalization. However, seeing that it would look quite complicated he chose not to write it down. This omission caused him to miss the full power of what we now call Bailey's Lemma.

We start with the statement of "Bailey's transform."

Lemma 12.2.2 *Subject to suitable convergence conditions, if*

$$\beta_n = \sum_{r=0}^{n} \alpha_r U_{n-r} V_{n+r}$$

and

$$\gamma_n = \sum_{r=n}^{\infty} \delta_r U_{r-n} V_{r+n},$$

then

$$\sum_{n=0}^{\infty} \alpha_n \gamma_n = \sum_{n=0}^{\infty} \beta_n \delta_n.$$

Proof.

$$\sum_{n=0}^{\infty} \alpha_n \gamma_n = \sum_{n=0}^{\infty} \sum_{r=n}^{\infty} \alpha_n \delta_r U_{r-n} V_{r+n}$$

$$= \sum_{r=0}^{\infty} \sum_{n=0}^{r} \alpha_n \delta_r U_{r-n} V_{r+n}$$

$$= \sum_{r=0}^{\infty} \delta_r \beta_r,$$

and the lemma is proved. ■

The above argument is purely formal and the suitable convergence conditions are those necessary to make all the infinite series converge and to validate the change of order of summation. We now state and prove Bailey's Lemma.

Theorem 12.2.3 *If for $n \geq 0$*

$$\beta_n = \sum_{r=0}^{n} \frac{\alpha_r}{(q; q)_{n-r}(aq; q)_{n+r}},$$

then

$$\beta_n' = \sum_{r=0}^{n} \frac{\alpha_r'}{(q; q)_{n-r}(aq; q)_{n+r}}, \tag{12.2.3}$$

where

$$\alpha_r' = \frac{(\rho_1; q)_r (\rho_2; q)_r (aq/\rho_1 \rho_2)^r \alpha_r}{(aq/\rho_1; q)_r (aq/\rho_2; q)_r}$$

and

$$\beta_n' = \sum_{j \geq 0} \frac{(\rho_1; q)_j (\rho_2; q)_j (aq/\rho_1 \rho_2; q)_{n-j} (aq/\rho_1 \rho_2)^j \beta_j}{(q; q)_{n-j} (aq/\rho_1; q)_n (aq/\rho_2; q)_n}. \tag{12.2.4}$$

Proof. Take $U_n = 1/(q; q)_n$ and $V_n = 1/(aq; q)_n$ and

$$\delta_n = \frac{(\rho_1; q)_n (\rho_2; q)_n (q^{-N}; q)_n q^n}{(\rho_1 \rho_2 q^{-N}/a; q)_n}$$

in Lemma 12.2.2. To compute γ_n, we need the q-Pfaff–Saalschütz identity (10.11.3), which we restate here for the reader's convenience:

$$_3\phi_2 \left(\begin{matrix} q^{-n}, a, b \\ c, abq^{1-n}/c \end{matrix}; q, q \right) = \frac{(c/a; q)_n (c/b; q)_n}{(c; q)_n (c/ab; q)_n}.$$

Then

$$\gamma_n = \sum_{r=n}^{\infty} \delta_r U_{r-n} V_{r+n}$$

$$= \sum_{r=n}^{\infty} \frac{(\rho_1; q)_r (\rho_2; q)_r (q^{-N}; q)_r q^r}{(\rho_1 \rho_2 q^{-N}/a; q)_r (q; q)_{r-n} (aq; q)_{r+n}}$$

$$= \sum_{r=0}^{\infty} \frac{(\rho_1; q)_{r+n} (\rho_2; q)_{r+n} (q^{-N}; q)_{r+n} q^{r+n}}{(\rho_1 \rho_2 q^{-N}/a; q)_{r+n} (q; q)_r (aq; q)_{r+2n}}$$

$$= \frac{(\rho_1; q)_n (\rho_2; q)_n (q^{-N}; q)_n q^n}{(\rho_1 \rho_2 q^{-N}/a; q)_n (aq; q)_{2n}} \, _3\phi_2 \left(\begin{array}{c} \rho_1 q^n, \rho_2 q^n, q^{-(N-n)}; q, q \\ \rho_1 \rho_2 q^{n-N}/a, aq^{2n+1} \end{array} \right)$$

$$= \frac{(\rho_1; q)_n (\rho_2; q)_n (q^{-N}; q)_n q^n (aq^{n+1}/\rho_1; q)_{N-n} (aq^{n+1}/\rho_2; q)_{N-n}}{(\rho_1 \rho_2 q^{-N}/a; q)_n (aq; q)_{2n} (aq^{2n+1}; q)_{N-n} (aq/\rho_1 \rho_2; q)_{N-n}}$$

$$= \frac{(aq/\rho_1; q)_N (aq/\rho_2; q)_N}{(aq; q)_N (aq/\rho_1 \rho_2; q)_N} \frac{(-1)^n (\rho_1; q)_n (\rho_2; q)_n (q^{-N}; q)_n}{(aq/\rho_1; q)_n (aq/\rho_2; q)_n (aq^{N+1}; q)_n}$$

$$\times (aq/\rho_1 \rho_2)^n q^{nN-n(n-1)/2}.$$

We now turn to the proof of (12.2.3). Note that

$$\sum_{r=0}^{N} \frac{\alpha'_r}{(q; q)_{N-r} (aq; q)_{N+r}}$$

$$= \sum_{r=0}^{N} \frac{(\rho_1; q)_r (\rho_2; q)_r (aq/\rho_1 \rho_2)^r \alpha_r}{(aq/\rho_1; q)_r (aq/\rho_2; q)_r (q; q)_{N-r} (aq; q)_{N+r}}$$

$$= \sum_{r=0}^{N} \frac{(-1)^r (\rho_1; q)_r (\rho_2; q)_r (q^{-N}; q)_r}{(aq/\rho_1; q)_r (aq/\rho_2; q)_r (aq; q)_{N+r}} (aq/\rho_1 \rho_2)^r \cdot q^{rN-r(r-1)/2} \alpha_r$$

$$= \frac{(aq/\rho_1 \rho_2; q)_N}{(aq/\rho_1; q)_N (aq/\rho_2; q)_N} \sum_{r=0}^{N} \gamma_r \alpha_r$$

$$= \frac{(aq/\rho_1 \rho_2; q)_N}{(aq/\rho_1; q)_N (aq/\rho_2; q)_N} \sum_{r=0}^{N} \beta_r \delta_r \qquad \text{(by Lemma 12.2.2)}$$

$$= \frac{(aq/\rho_1 \rho_2; q)_N}{(aq/\rho_1; q)_N (aq/\rho_2; q)_N} \sum_{r=0}^{N} \frac{(\rho_1; q)_r (\rho_2; q)_r (q^{-N}; q)_r q^r \beta_r}{(\rho_1 \rho_2 q^{-N}/a; q)_r}$$

$$= \beta'_N,$$

where the last equation follows after an algebraic simplification that produces the expression given in (12.2.4). The theorem is proved. ∎

Remark 12.2.1   The Weak Bailey Lemma (Lemma 12.2.1) follows from the full
Bailey Lemma by letting $n, \rho_1, \rho_2 \to \infty$.

We call (α_n, β_n) a Bailey pair if they are related as in Bailey's lemma. The
power of the full Bailey Lemma is that, given a Bailey pair (α_n, β_n) a new Bailey
pair is produced. Consequently, from one such pair one can construct an infinite
sequence $(\alpha_n, \beta_n) \to (\alpha_n', \beta_n') \to (\alpha_n'', \beta_n'') \to \cdots$ of Bailey pairs by successive
application of Bailey's Lemma. This sequence is called a Bailey chain. The simplest
conceivable chain starts with

$$\beta_n = \delta_{n,0} = \begin{cases} 1 & \text{if } n = 0, \\ 0 & \text{if } n > 0 . \end{cases} \tag{12.2.5}$$

It is an simple exercise to show that the corresponding α_n is

$$\alpha_n = \frac{(1 - aq^{2n})(a; q)_n (-1)^n q^{\binom{n}{2}}}{(1 - a)(q; q)_n}. \tag{12.2.6}$$

Remark 12.2.2   The fact that (α_n, β_n) is a Bailey pair follows immediately from
a formula of Agarwal [1953],

$$\sum_{j=0}^{M} \frac{(1 - aq^{2j})(q^{-n}; q)_j (a; q)_j q^{nj}}{(1 - a)(aq^{n+1}; q)_j (q; q)_j} = \frac{(aq; q)_M q^{nM}(q^{1-n}; q)_M}{(q; q)_M (aq^{n+1}; q)_M}, \tag{12.2.7}$$

a result that can be proved directly by induction on M. Another way is to use the
"inversion" formula: If β_n is given by (12.2.1), then

$$\alpha_n = (1 - aq^{2n}) \sum_{j=0}^{n} \frac{(aq; q)_{n+j-1}(-1)^{n-j} q^{\binom{n-j}{2}} \beta_j}{(q; q)_{n-j}}. \tag{12.2.8}$$

12.3 Watson's Transformation Formula

In this section we use the Bailey chain to derive some important formulas. In
particular, we obtain a q-analog of Whipple's transformation due to Watson [1929].
This transforms a terminating very well poised $_8\phi_7$ to a terminating balanced $_4\phi_3$.
The phrase "very well poised" is defined below.

We start with a simpler result. Recall that (α_n, β_n) in (12.2.5) and (12.2.6) form
a Bailey pair. By Theorem 12.2.3, the next pair in the chain, namely (α_n', β_n'), is
given by

$$\beta_n' = \frac{(aq/\rho_1\rho_2; q)_n}{(q; q)_n (aq/\rho_1; q)_n (aq/\rho_2; q)_n},$$

$$\alpha_n' = \frac{(\rho_1; q)_n (\rho_2; q)_n (aq/\rho_1\rho_2)^n (1 - aq^{2n})(a; q)_n (-1)^n q^{\binom{n}{2}}}{(aq/\rho_1; q)_n (aq/\rho_2; q)_n (1 - a)(q; q)_n}.$$

And the relation

$$\beta'_n = \sum_{r=0}^{n} \frac{\alpha'_r}{(q;q)_{n-r}(aq;q)_{n+r}},$$

when written out in full, is

$$\frac{(aq/\rho_1\rho_2)_n}{(q)_n(aq/\rho_1)_n(aq/\rho_2)_n}$$
$$= \sum_{r=0}^{n} \frac{(\rho_1)_r(\rho_2)_r(aq/\rho_1\rho_2)^r(1-aq^{2r})(a)_r(-1)^r q^{\binom{r}{2}}}{(q)_{n-r}(aq)_{n+r}(aq/\rho_1)_r(aq/\rho_2)_r(1-a)(q)_r}, \quad (12.3.1)$$

where $(x)_r = (x;q)_r$. Note that

$$\frac{1-aq^{2r}}{1-a} = \frac{(1-\sqrt{a}q^r)(1+\sqrt{a}q^r)}{(1-\sqrt{a})(1+\sqrt{a})} = \frac{(\sqrt{aq};q)_r(-\sqrt{aq};q)_r}{(\sqrt{a},q)_r(-\sqrt{a};q)_r},$$

$$(q)_{n-r} = \frac{(-1)^r(q)_n}{(q^n)_r q^{r(2n-r+1)/2}},$$

and

$$(aq)_{n+r} = (aq)_n(aq^{n+1})_r.$$

Thus (12.3.1) is equivalent to the formula

$$_6\phi_5\left(\begin{array}{c} a, q\sqrt{a}, -q\sqrt{a}, \rho_1, \rho_2, q^{-n} \\ \sqrt{a}, -\sqrt{a}, aq/\rho_1, aq/\rho_2, aq^{n+1} \end{array}; q, \frac{aq^{n+1}}{\rho_1\rho_2}\right) = \frac{(aq)_n(aq/\rho_1\rho_2)_n}{(aq/\rho_1)_n(aq/\rho_2)_n}.$$
$$(12.3.2)$$

The $_6\phi_5$ is well poised because the product of a numerator parameter with the corresponding denominator parameter is aq. Additionally, the presence of the factor $(1-aq^{2r})/(1-a)$ now makes it a very well poised $_6\phi_5$.

Watson's transformation is obtained by moving up the Bailey chain to (α''_n, β''_n). The relation

$$\sum_{r=0}^{n} \frac{\alpha''_r}{(q)_{n-r}(aq)_{n+r}} = \beta''_n$$

is seen to be equivalent to

$$\sum_{r=0}^{n} \frac{(\lambda_1)_r(\lambda_2)_r(aq/\lambda_1\lambda_2)^r(\rho_1)_r(\rho_2)_r(aq/\rho_1\rho_2)^r(1-aq^{2r})(a)_r(-1)^r q^{\binom{r}{2}}}{(aq/\lambda_q)_r(aq/\lambda_2)_r(aq/\rho_1)_r(aq/\rho_2)_r(1-a)(q)_r(q)_{n-r}(aq)_{n+r}}$$
$$= \sum_{j\geq 0} \frac{(\lambda_1)_j(\lambda_2)_j(aq/\lambda_1\lambda_2)_{n-j}(aq/\lambda_1\lambda_2)^j(aq/\rho_1\rho_2)_j}{(q)_{n-j}(aq/\lambda_1)_j(aq/\lambda_2)_j(q)_j(aq/\rho_1)_j(aq/\rho_2)_j}. \quad (12.3.3)$$

By the formulas used to obtain (12.3.2), it is easy to show that (12.3.3) can be written as follows:

$$
{}_8\phi_7 \left(\begin{array}{c} a, q\sqrt{a}, -q\sqrt{a}, \lambda_1, \lambda_2, \rho_1, \rho_2, q^{-n} \\ \sqrt{a}, -\sqrt{a}, aq/\lambda_1, aq/\lambda_2, aq/\rho_1, aq/\rho_2, aq^{n+1} \end{array} ; q, \frac{a^2 q^{n+2}}{\lambda_1 \lambda_2 \rho_1 \rho_2} \right)
$$
$$
= \frac{(aq)_n (aq/\rho_1\rho_2)_n}{(aq/\rho_1)_n (aq/\rho_2)_n} {}_4\phi_3 \left(\begin{array}{c} aq/\lambda_1\lambda_2, \rho_1, \rho_2, q^{-n} \\ aq/\lambda_1, aq/\lambda_2, \rho_1\rho_2 q^{-n}/a \end{array} ; q, q \right). \quad (12.3.4)
$$

This is Watson's formula. When $aq/\lambda_1\lambda_2 = \rho_1\rho_2 q^{-n}/a$, the ${}_4\phi_3$ becomes a balanced ${}_3\phi_2$, which can be summed by the q-Pfaff–Saalschütz formula (10.11.3). The result in this case is the q-analog of Dougall's formula due to Jackson [1921]:

$$
{}_8\phi_7 \left(\begin{array}{c} a, q\sqrt{a}, -q\sqrt{a}, \lambda_1, \lambda_2, \rho_1, \rho_2, q^{-n} \\ \sqrt{a}, -\sqrt{a}, aq/\lambda_1, aq/\lambda_2, aq/\rho_1, aq/\rho_2, aq^{n+1} \end{array} ; q, q \right)
$$
$$
= \frac{(aq)_n (aq/\rho_1\rho_2)_n (aq/\lambda_1\rho_1)_n (aq/\lambda_1\lambda_2)_n}{(aq/\lambda_1)_n (aq/\lambda_2)_n (aq/\rho_1)_n (aq/\lambda_1\lambda_2\rho_1)_n}, \quad \text{when } a^2 q = \lambda_1\lambda_2\rho_1\rho_2 q^{-n}.
$$
$$
(12.3.5)
$$

Let $n \to \infty$ in (12.3.5). This gives

$$
{}_6\phi_5 \left(\begin{array}{c} a, q\sqrt{a}, -q\sqrt{a}, \lambda_1, \lambda_2, \rho_1 \\ \sqrt{a}, -\sqrt{a}, aq/\lambda_1, aq/\lambda_2, aq/\rho_1 \end{array} ; q, \frac{aq}{\lambda_1\lambda_2\rho_1} \right)
$$
$$
= \frac{(aq)_\infty (aq/\rho_1\rho_2)_\infty (aq/\lambda_1\rho_1)_\infty (aq/\lambda_1\lambda_2)_\infty}{(aq/\lambda_1)_\infty (aq/\lambda_2)_\infty (aq/\rho_1)_\infty (aq/\lambda_1\lambda_2\rho_1)_\infty}. \quad (12.3.6)
$$

This is a nonterminating form of (12.3.2).

Now observe that the ${}_8\phi_7$ in (12.3.4) is symmetric in $\lambda_1, \lambda_2, \rho_1, \rho_2$ and hence

$$
\frac{(aq/\rho_1\rho_2)_n}{(aq/\rho_1)_n (aq/\rho_2)_n} {}_4\phi_3 \left(\begin{array}{c} aq/\lambda_1\lambda_2, \rho_1, \rho_2, q^{-n} \\ aq/\lambda_1, aq/\lambda_2, \rho_1\rho_2 q^{-n}/a \end{array} ; q, q \right)
$$
$$
= \frac{(aq/\lambda_1\lambda_2)_n}{(aq/\lambda_1)_n (aq/\lambda_2)_n} {}_4\phi_3 \left(\begin{array}{c} aq/\rho_1\rho_2, \lambda_1, \lambda_2, q^{-n} \\ aq/\rho_1, aq/\rho_2, \lambda_1\lambda_2 q^{-n}/a \end{array} ; q, q \right). \quad (12.3.7)
$$

A q-analog of Dixon's formula for a well-poised ${}_3F_2$ is obtained by setting $\rho_1 = \sqrt{a}$ in (12.3.6):

$$
{}_4\phi_3 \left(\begin{array}{c} a, -q\sqrt{a}, \lambda_1, \lambda_2 \\ -\sqrt{a}, aq/\lambda_1, aq/\lambda_2 \end{array} ; q, \frac{q\sqrt{a}}{\lambda_1\lambda_2} \right)
$$
$$
= \frac{(aq)_\infty (aq/\lambda_1\lambda_2)_\infty (q\sqrt{a}/\lambda_1)_\infty (q\sqrt{a}/\lambda_2)_\infty}{(aq/\lambda_1)_\infty (aq/\lambda_2)_\infty (q\sqrt{a})_\infty (q\sqrt{a}/\lambda_1\lambda_2)_\infty}.
$$

The two Rogers–Ramanujan formulas can also be derived from Watson's transformation (12.3.4). Let $\lambda_1, \lambda_2, \rho_1, \rho_2 \to \infty$ to get

$$\sum_{j=0}^{n} \frac{(a;q)_j(1-aq^{2j})(q^{-n};q)_j(a^2q^{n+2j})^j}{(q;q)_j(1-a)(aq^{n+1};q)_j}$$

$$= (aq;q)_n \sum_{j=0}^{n} \frac{(-1)^j(q^{-n};q)_j}{(q;q)_j}(aq^{n+1+(j-1)/2})^j.$$

Now let $n \to \infty$ and apply the dominated convergence theorem. After simplification, the result is

$$1 + \sum_{j=1}^{\infty} \frac{(aq;q)_{j-1}(1-aq^{2j})}{(q;q)_j}(-1)^j a^{2j}q^{j(5j-1)/2} = (aq;q)_\infty \sum_{j=0}^{\infty} \frac{a^jq^{j^2}}{(q;q)_J}.$$

For $a = 1$, this is the first Rogers–Ramanujan identity and for $a = q$ it is the second one.

If one wants to generalize a q-series identity that has quadratic powers in it, it is natural to try to replace each $q^{\binom{k}{2}}$ with $(a;q)_k$ times something else that will give $q^{\binom{k}{2}}$ using

$$\lim_{a \to \infty} (a;q)_k/(-a)^k = q^{\binom{k}{2}}.$$

Since there are two Rogers–Ramanujan identities and five factors $q^{\binom{k}{2}}$, the minimum extension of the Rogers–Ramanujan identities of this type should have six free parameters. Formula (12.3.4) is this extension.

12.4 Other Applications

The summation formula for a very well poised $_6\phi_5$ and Watson's transformation were fairly straightforward consequences of Bailey's Lemma. Further interesting results are possible by starting with the Bailey pair (12.2.5) and (12.2.6) and moving further along the chain. For example,

$$\sum_{n_1 \geq n_2 \geq \cdots \geq n_k \geq 0} \frac{q^{n_1^2+n_2^2+\cdots n_k^2}}{(q)_{n_1-n_2}(q)_{n_2-n_3}\cdots(q)_{n_{k-1}-n_k}(q)_{n_k}}$$

$$= \prod_{\substack{n=1 \\ n\neq 0,\pm(k+1)(\bmod\, 2k+3)}}^{\infty} \frac{1}{1-q^n}.$$

This identity reduces to the first Rogers–Ramanujan identity when $k = 1$. As noted before, it can be proved by examining $(\alpha_n^{(k+1)}, \beta_n^{(k+1)})$ in the Bailey chain starting with the pair (12.2.5) and (12.2.6).

It should be stressed that this is not the only Bailey chain of interest. In a totally different context, another of Ramanujan's series,

$$S(q) := \sum_{n=0}^{\infty} \frac{q^{\binom{n+1}{2}}}{(-q;q)_n} =: \sum_{n=0}^{\infty} s_n q^n,$$

was studied. The relevant Bailey pair is

$$\beta_n = \frac{(-q)^n}{(q^2;q^2)_n}$$

and

$$\alpha_n = q^{n^2+n} \sum_{j=-n}^{n} (-1)^j q^{-j^2}.$$

From

$$\beta'_n = \sum_{r=0}^{n} \frac{\alpha'_r}{(q;q)_{n-r}(q;q)_{n+r}},$$

we can deduce

$$S(q) = \sum_{\substack{n \geq 0 \\ |j| \leq n}} (-1)^{n+j} q^{n(3n+1)/2-j^2}(1-q^{2n+1}).$$

From this result it can be shown that almost all s_n equal zero and that for any integer M (positive, negative, or zero), there exist infinitely many n so that $s_n = M$. For details, see Andrews, Dyson, and Hickerson [1988]. This paper also contains references to the work of Slater and Bressoud in Exercises 3, 4, and 5.

Exercises

1. Obtain the second Rogers–Ramanujan identity (12.1.2) from the weak Bailey Lemma 12.2.1.

2. Prove Agarwal's formula (12.2.7), namely

$$\sum_{j=0}^{M} \frac{(1-aq^{2j})(q^{-n};q)_j(a;q)_m q^{nj}}{(1-a)(aq^{n+1};q)_j(q;q)_j} = \frac{(aq;q)_M q^{nM}(q^{1-n};q)_M}{(q;q)_M(aq^{n+1};q)_M}.$$

3. Show that the (α_n, β_n) defined below is a Bailey pair with $a = 1$:

$$\alpha_m = \begin{cases} -q^{6n^2-5n+1}, & m = 3n-1 > 0, \\ q^{6n^2-n} + q^{6n^2+n}, & m = 3n > 0, \\ -q^{6n^2+5n+1}, & m = 3n+1 > 0, \\ 1, & m = 0, \end{cases}$$

$$\beta_n = 1/(q;q)_{2n}.$$

(Slater)

4. Suppose

$$
\alpha_m = \begin{cases}
-q^{6n^2-2n}, & m = 3n-1 > 0, \\
q^{6n^2-2n} + q^{6n^2+2n}, & m = 3n > 0, \\
-q^{6n^2+2n}, & m = 3n+1 > 0, \\
1, & m = 0,
\end{cases}
$$

and

$$
\beta_n = q^n/(q;q)_{2n}.
$$

Prove that (α_n, β_n) is a Bailey pair with $a = 1$.   (Slater)

5. Prove that

$$
\alpha_n = \begin{cases}
(-1)^m (q^{m(5m+1)/2} + q^{m(5m-1)/2}), & n = 2n, \\
1, & n = 0, \\
0, & n = \text{odd},
\end{cases}
$$

$$
\beta_n = \frac{1}{(q;q)_{2n}} \sum_{j=0}^{n} \begin{bmatrix} n \\ j \end{bmatrix}_q q^{j^2}
$$

is a Bailey pair with $a = 1$.   (Bressoud)

6. Apply (12.3.8), Exercises 10.37 and 10.39, and Jacobi's triple product formula to obtain the following six identities of Rogers:

(a) $$1 + \sum_{n=1}^{\infty} \frac{q^{n^2}}{(q;q)_{2n}} = 1/[(q;q^2)_{\infty}(q^4;q^{20})_{\infty}(q^{16};q^{20})_{\infty}].$$

(b) $$1 + \sum_{n=1}^{\infty} \frac{q^{n^2+2n}}{(q;q)_{2n+1}} = 1/[(q;q^2)_{\infty}(q^8;q^{20})_{\infty}(q^{12};q^{20})_{\infty}].$$

(c) $$1 + \sum_{n=1}^{\infty} \frac{q^{n^2+n}}{(q;q)_{2n}} = \prod_{\substack{n=1 \\ n \neq \pm 1, \pm 8, \pm 9, 10 (\mathrm{mod}\, 20)}}^{\infty} (1-q^n)^{-1}.$$

(d) $$1 + \sum_{n=1}^{\infty} \frac{q^{n^2+n}}{(q;q)_{2n+1}} = \prod_{\substack{n=1 \\ n \neq \pm 3, \pm 4, \pm 7, 10 (\mathrm{mod}\, 20)}}^{\infty} (1-q^n)^{-1}.$$

(e) $$(-q^2;q^2)_{\infty} \sum_{n=0}^{\infty} \frac{q^{n^2}}{(q^4;q^4)_n} = \prod_{n=0}^{\infty} \frac{1}{(1-q^{5n+1})(1-q^{5n+4})}.$$

(f) $$(-q^2;q^2)_{\infty} \sum_{n=0}^{\infty} \frac{q^{n^2+2n}}{(q^4;q^4)_n} = \prod_{n=0}^{\infty} \frac{1}{(1-q^{5n+2})(q-q^{5n+3})}.$$

Whipple's $_7F_6$ transformation was obtained in Chapter 7 as a byproduct of the solution of the connection coefficient problem for Jacobi polynomials. There are several q-analogs of the Jacobi polynomials. The next set of four problems shows how the little q-Jacobi polynomials can be used to derive Watson's $_8\phi_7$ transformation. See Andrews and Askey [1977].

7. Define the little q-Jacobi polynomials by

$$p_n(x; \alpha, \beta : q) = {}_2\phi_1 \left(\begin{matrix} q^{-n}, \alpha\beta q^{n+1}; \\ \alpha q \end{matrix} ; q, qx \right).$$

(a) Show that

$$\lim_{q \to 1^-} p_n(x; q^\alpha, q^\beta : q) = P_n^{(\alpha, \beta)}(1 - 2x) / P_n^{\alpha, \beta}(1).$$

(b) Derive the orthogonality relation

$$\sum_{i=0}^{\infty} \frac{\alpha^i q^i (q^{i+1}; q)_\infty}{(\beta q^{i+1}; q)_\infty} p_n(q^i; \alpha, \beta : q) p_m(q^i; \alpha, \beta : q)$$

$$= \begin{cases} 0 & \text{if } m \neq n, \\ \dfrac{\alpha^n q^n (q; q)_\infty (\alpha\beta q^{n+1}; q)_\infty (q; q)_n}{(\beta q^{n+1}; q)_\infty (\alpha q; q)_\infty (\alpha q; q)_n (1 - \alpha\beta q^{2n+1})} & \text{if } m = n. \end{cases}$$

[*Hint*: First obtain the following identity:

$$\sum_{i=0}^{\infty} \frac{\alpha^i q^i (q^{i+1}; q)_\infty}{(\beta q^{i+1}; q)_\infty} p_n(q^i; \alpha, \beta : q) q^{im}$$

$$= \begin{cases} 0 & \text{if } 0 \leq m < n, \\ \dfrac{(-\alpha)^n q^{n(n+1)/2} (q; q)_\infty (\alpha\beta q^{2n+2}; q)_\infty (q; q)_n}{(\beta q^{n+1}; q)_\infty (\alpha q; q)_\infty} & \text{if } m = n.] \end{cases}$$

8. Suppose that

$$p_n(x; \gamma, \delta : q) = \sum_{k=0}^{n} a_{kn} p_k(x; \alpha, \beta : q).$$

Show that

$$a_{kn} = \frac{(-1)^k q^{k(k+1)/2} (\gamma\delta q^{n+1}; q)_k (q^{-n}; q)_k (\alpha q; q)_k}{(\alpha\beta q^{k+1}; q)_k (q; q)_k (\gamma q; q)_k}$$

$$\cdot {}_3\phi_2 \left(\begin{matrix} q^{-n+k}, \gamma\delta q^{n+k+1}, \alpha q^{k+1}; \\ \gamma q^{k+1}, \alpha\beta q^{2k+2} \end{matrix} ; q, q \right).$$

9. Note that the identity in Exercise 8 is a polynomial identity in x. Deduce that with the same a_{kn}, we have

$$_{r+2}\phi_{r+1} \left(\begin{matrix} q^{-n}, \gamma\delta q^{n+1}, a_1, \ldots, a_r; \\ \gamma q, b_1, \ldots, b_r \end{matrix} ; q, qx \right)$$

$$= \sum_{k=0}^{n} a_{kn} \, {}_{r+2}\phi_{r+1} \left(\begin{matrix} q^{-k}, \alpha\beta q^{k+1}, a_1, \ldots, a_r; \\ \alpha q, b_1, \ldots b_r \end{matrix} ; q, qx \right).$$

10. Take $r = 2$, $\beta = \delta$, $a_1 = \alpha q$, $x = 1$, and $b_2 = q^2 \alpha\delta a_2 / b_1$ in the previous exercise. Observe that the $_3\phi_2$ in a_{kn} becomes balanced and may be computed by (10.10.3). Deduce that

$$_4\phi_3\left(\begin{array}{c} q^{-n}, \gamma\delta q^{n+1}, \alpha q, a_2 \\ \gamma q, b_1, q^2\alpha\delta a_2/b_1 \end{array}; q, q\right) = \frac{\alpha^n q^n (\delta q; q)_n (\gamma/\alpha; q)_n}{(\alpha\delta q^2; q)_n (\gamma q; q)_n}$$

$$\cdot \, _8\phi_7\left(\begin{array}{c} q^{-n}, q\sqrt{\alpha\delta q}, -q\sqrt{\alpha\delta q}, \alpha\delta q, \alpha q, \alpha\delta q^2/b_1, \gamma\delta q^{n+1}, b_1/a_2 \\ \sqrt{\alpha\delta q}, -\sqrt{\alpha\delta q}, \delta q, b_1, \alpha q^{-n+1}/\gamma, \alpha\delta a_2 q^2/b_1, \alpha\delta q^{n+2} \end{array}; q, a_2/\gamma\right).$$

The standard form of Watson's transformation is obtained from this formula after proper identification.

11. The big q-Jacobi polynomials are defined by

$$P_n^{(\alpha,\beta)}(x; c, d : q) = c^n \frac{q^{-(\alpha+1)n}(q^{\alpha+1}; q)_n(-q^{\alpha+1}d/c; q)_n}{(q; q)_n(-q; q)_n}$$

$$\cdot \, _3\phi_2\left(\begin{array}{c} q^{-n}, q^{n+\alpha+\beta+1}, xq^{\alpha+1}d/c \\ q^{\alpha+1}, -q^{\alpha+1}d/c \end{array}; q, q\right).$$

Prove that the orthogonality relation is

$$\int_{-d}^{c} P_n^{(\alpha,\beta)}(x; c, d : q) P_m^{(\alpha,\beta)}(x : c, d : q) \frac{(qx/c; q)_\infty(-qx/d; q)_\infty d_q x}{(q^{\alpha+1}x/c; q)_\infty(-q^{\beta+1}x/d; q)_\infty}$$

$$= \begin{cases} 0 & m \neq n, \\ \dfrac{(cd)^n q^{n(n-1)/2}(q^{\alpha+1}; q)_n(q^{\beta+1}; q)_n}{(q^{\alpha+\beta+1}; q)(q; q)_n(1-q^{2n+\alpha+\beta+1})} \\ \quad \cdot \dfrac{(-q^{\beta+1}c/d; q)_n(-q^{\alpha+1}d/c; q)_n(1-q^{\alpha+\beta+1})M}{(-q; q)_n(-q; q)_n}, & m = n, \end{cases}$$

where

$$M = \frac{cd(1-q)(q; q)_\infty(q^{\alpha+\beta+2}; q)_\infty(-d/c; q)_\infty(-c/d; q)_\infty}{(c+d)(q^{\alpha+1}; q)_\infty(q^{\beta+1}; q)_\infty(-q^{\alpha+1}d/c; q)_\infty(-q^{\beta+1}c/d; q)_\infty}.$$

Note that when $c = d = 1$, the weight function in the q-integral tends to $(1-x)^\alpha(1+x)^\beta$ as $q \to 1^-$.

12. A set of q-Laguerre polynomials is defined by

$$L_n^{(\alpha)}(x; q) = \frac{(q^{\alpha+1}; q)_n}{(q; q)_n} \sum_{k=0}^{n} \frac{(q^{-n}; q)_k q^{\binom{k}{2}}(1-q)^k(q^{n+\alpha+1}x)^k}{(q^{\alpha+1}; q)_k(q; q)_k}.$$

(a) Replace x with $-(1-q)q^{-(\beta+1)}x$ and let $\beta \to -\infty$ in the little q-Jacobi polynomial of degree n. Show that the result is

$$\frac{(q; q)_n}{(q^{\alpha+1}; q)_n} L_n^{(\alpha)}(x; q).$$

(b) Prove that $\lim_{q \to 1^-} L_n^{(\alpha)}(x; q) = L_n^\alpha(x)$.

(c) Prove the discrete orthogonality relation

$$\sum_{k=-\infty}^{\infty} L_n^{(\alpha)}(q^k; q)L_m^{(\alpha)}(q^k; q)\frac{q^{k\alpha+k}}{(-(1-q)q^k; q)_\infty}$$

$$= \begin{cases} 0, & m \neq n, \\ \dfrac{(q^{\alpha+1}; q)_n}{q^n(q; q)_n} \displaystyle\sum_{k=-\infty}^{\infty} \dfrac{q^{k\alpha+k}}{(-(1-q)q^k; q)_\infty} & m = n. \end{cases}$$

(Use Ramanujan's $_1\psi_1$ sum.)

(d) Prove the continuous orthogonality relation

$$\int_0^\infty L_m^{(\alpha)}(x; q)L_n^{(\alpha)}(x; q)\frac{x^\alpha}{(-(1-q)x; q)_\infty} dx$$

$$= \begin{cases} 0, & m \neq n, \\ \dfrac{\Gamma(\alpha+1)\Gamma(-\alpha)(q^{n+1}, q)_n}{\Gamma_q(-\alpha)(q; q)_n q^n}, & m = n. \end{cases}$$

(Moak)

13. A multiplicative shift of the variable x in Exercise 12 gives a set of q-Laguerre polynomials defined by

$$L_n^\alpha(x; q) = \sum_{k=0}^n \frac{(q^{-n}; q)_k q^{nk+(\alpha-\beta)k+k(k+1)/2} x^k}{(q^{\alpha+1}; q)_k(q; q)_k}.$$

(a) Show that

$$\int_0^\infty \frac{L_n^\alpha(x; q)L_m^\alpha(x; q)x^{\alpha-\beta}(-q^{\beta+1}x^{-1}; q)_\infty}{(-x; q)_\infty(-qx^{-1}; q)_\infty} dx$$

$$= \begin{cases} \dfrac{q^{-n}(q; q)_n(q; q)_\infty}{(q^{\alpha+1}; q)_n(q^{\alpha+1}; q)_n} \dfrac{\Gamma(\alpha+1-\beta)\Gamma(\beta-\alpha)}{\Gamma_q(\alpha+1-\beta)\Gamma_q(\beta-\alpha)}, & m = n, \\ 0, & m \neq n. \end{cases}$$

(b) Set $\alpha - \beta = c$ (fixed c) and let $\alpha \to \infty$ in $L_n^\alpha(x; q)$. Show that the result is the Stieltjes–Wigert polynomial

$$S_n(x; q) = \sum_{k=0}^n \frac{q^{k^2}(-q^c x)^k}{(q; q)_k(q; q)_{n-k}}.$$

(c) Prove that a weight function for the Stieltjes–Wigert polynomials is

$$\omega(x) = \frac{x^c}{(-x; q)_\infty(-q/x; q)_\infty}.$$

For Exercises 12 and 13, see Gasper and Rahman [1990, Chapter 7].

A

Infinite Products

A.1 Infinite Products

For readers unfamiliar with infinite products, a brief introduction is given here.

Definition A.1.1 *Let $p_{n,k} = \prod_{m=k}^{n}(1+a_m)$. If there is a k for which $p_{n,k}$ converges to a nonzero value p as $n \to \infty$, then we say that the infinite product $\prod_{n=1}^{\infty}(1+a_n)$ converges. We write it as $p = \prod_{n=k}^{\infty}(1 + a_n)$. The reason for not taking $k = 1$ is to allow an finite number of zero factors. The convergence of the product is said to be absolute if $\prod_{n=1}^{\infty}(1 + |a_n|)$ converges.*

The following basic theorem reduces the problem of convergence of a product to that of a series. For simplicity, assume that $\mathrm{Re}\, a_n > -1, n = 1, 2, \dots$. If not, start the product after this holds.

Theorem A.1.2 *The product $\prod_{n=1}^{\infty}(1 + a_n)$ converges if and only if the series $\sum_{n=1}^{\infty} \log(1 + a_n)$ converges.*

Proof. Suppose that $S_n = \sum_{m=1}^{n} \log(1 + a_m)$ converges to S. Since exp is a continuous function,

$$\prod_{m=1}^{n}(1 + a_m) = \exp(S_n) \quad \text{converges to } e^S = \prod_{n=1}^{\infty}(1 + a_n).$$

To prove the converse, let $1 + a_m = A_m e^{i\theta_m}$ and $\prod_{m=1}^{n}(1 + a_m) = B_n e^{i\phi_n}$. The convergence of $\Pi(1 + a_n)$ implies $a_m \to 0$ so that $\theta_m \to 0$ and ϕ_n can be chosen so that, say, $\phi_n \to \phi$. We also have

$$\theta_1 + \theta_2 + \cdots + \theta_n = \phi_n + 2\pi k_n,$$

where k_n is an integer. Therefore,

$$\theta_{n+1} = \phi_{n+1} - \phi_n + 2\pi(k_{n+1} - k_n).$$

Since $k_{n+1} - k_n$ is an integer, $\theta_{n+1} \to 0$, and since $\phi_{n+1} - \phi_n \to 0$, we must have $k_n = k$, a constant for sufficiently large n. Thus, for sufficiently large n,

$$S_n = \log p_n + 2\pi k i.$$

Let $n \to \infty$ to get

$$S = \log p + 2\pi k i.$$

This proves the theorem. For absolute convergence the condition is simpler and is contained in the next theorem. ∎

Theorem A.1.3   *The product $\prod_{n=1}^{\infty}(1 + a_n)$ converges absolutely if and only if $\sum_{n=1}^{\infty} a_n$ converges absolutely.*

Proof.   Convergence of the series or the product implies that $a_n \to 0$. For sufficiently large n we must have $|a_n| \leq 1/2$. Suppose $a_n \neq 0$ and n large. Then

$$\left| 1 - \frac{\log(1 + a_n)}{a_n} \right| = \left| \frac{a_n}{2} - \frac{a_n^2}{3} + \cdots \right|$$

$$\leq \frac{1}{2}[|a_n| + |a_n|^2 + \cdots]$$

$$= \frac{1}{2} \frac{|a_n|}{1 - |a_n|} \leq \frac{1}{2}.$$

So

$$-\frac{1}{2} \leq \left| \frac{\log(1 + a_n)}{a_n} \right| - 1 \leq \frac{1}{2}$$

or

$$\frac{1}{2}|a_n| \leq |\log(1 + a_n)| \leq \frac{3}{2}|a_n|.$$

These inequalities together with Theorem A.1.2 imply the result and also the fact that absolute convergence of a product implies its convergence. ∎

Definition A.1.4   *The infinite product*

$$\prod_{n=1}^{\infty}(1 + a_n(x)),$$

where x is a real or complex variable in a domain, is uniformly convergent if

$$p_n(x) = \prod_{m=k}^{n}(1 + a_m(x))$$

converges uniformly in that domain, for each k.

Theorem A.1.5 *If the series $\sum_{n=1}^{\infty} |a_n(x)|$ converges uniformly in some region, then the product $\prod_{n=1}^{\infty} (1 + a_n(x))$ also converges uniformly in that region.*

The proof of this result is left to the reader.

Corollary A.1.6 *If $a_n(x)$ is analytic in some region of the complex plane and $\Pi(1+a_n(x))$ converges uniformly in that region, then the infinite product represents an analytic function in that region.*

Exercises

1. Prove directly from the definition that the product $(1 - \frac{1}{2})(1 + \frac{1}{3})(1 - \frac{1}{4}) \cdots$ is convergent.

2. Prove that an absolutely convergence product is convergent.

3. Show that if $\Sigma a_n, \Sigma a_n^2, \ldots, \Sigma a_n^{k-1}, \Sigma |a_n|^k$ are all convergent, then $\Pi(1 + a_n)$ is convergent.

4. Discuss the convergence of the product

$$\prod_{n=1}^{\infty} \left\{ \left(1 + \frac{1}{n}\right)^x \left(1 - \frac{x}{n}\right) \right\}.$$

B

Summability and Fractional Integration

B.1 Abel and Cesàro Means

The following theorem, the first part of which was proved by Abel, is often encountered in a first course in real analysis.

Theorem B.1.1 *Suppose that the series $\sum_{n=0}^{\infty} b_n$ converges to B. Then $\sum_{n=0}^{\infty} b_n x^n$ converges uniformly on $[0, 1]$, and in particular*

$$\lim_{x \to 1^-} \sum_{n=0}^{\infty} b_n x^n = B. \tag{B.1.1}$$

Moreover, if $B_n = \sum_{k=0}^{n} b_k$, then

$$\lim_{n \to \infty} \frac{B_0 + B_1 + \cdots + B_n}{n + 1} = B. \tag{B.1.2}$$

The example of the series $1 - 1 + 1 - \cdots$ shows that the limits in (B.1.1) and (B.1.2) may exist even when the series does not converge. Thus (B.1.1) and (B.1.2) may be used to assign a sum to divergent series.

The series $\sum_{n=0}^{\infty} b_n$ is Abel summable or summable A to B if (B.1.1) holds; and it is Cesàro summable or summable $(C, 1)$ to B if (B.1.2) holds. We also speak of (B.1.1) as the Abel mean and (B.1.2) as the Cesàro mean of the series $\sum_{n=0}^{\infty} b_n$.

The Abel and Cesàro means of the series $1 - 1 + 1 - \cdots$ are both $1/2$. The next theorem shows that Cesàro summability is a stronger requirement than Abel summability.

Theorem B.1.2 *If the series $\sum_{n=0}^{\infty} b_n$ is $(C, 1)$ summable to B, then it is A summable to B.*

599

Proof.   Use summation by parts to get

$$\sum_{n=0}^{\infty} b_n x^n - B = (1-x) \sum_{n=0}^{\infty} B_n x^n - B \left(\text{where } B_n = \sum_{k=0}^{n} b_k \right)$$

$$= (1-x)^2 \sum_{n=0}^{\infty} (B_0 + \cdots + B_n) x^n - B$$

$$= (1-x)^2 \sum_{n=0}^{\infty} \{(B_0 + \cdots + B_n) - (n+1)B\} x^n. \quad \text{(B.1.3)}$$

Now (B.1.2) implies that, for $\epsilon > 0$, there exists an integer N such that

$$|(B_0 + \cdots + B_n) - (n+1)B| < \epsilon(n+1) \quad \text{for } n \geq N.$$

Use this in (B.1.3) to arrive at the necessary result. ∎

The example of the series $1 - 2 + 3 - \cdots$ shows that the Abel mean may exist (in this case it is $1/4$) but the Cesàro mean may not. The Abel and Cesàro means, together with other summability methods, are very useful in analysis and analytic number theory. As an elementary example, consider the following theorem of Abel on the product of two series:

Theorem B.1.3   *Suppose $\sum_{n=0}^{\infty} a_n$ and $\sum_{n=0}^{\infty} b_n$ are convergent series. Let $c_n = \sum_{k=0}^{n} a_k b_{n-k}$ and suppose that $\sum_{n=0}^{\infty} c_n$ is convergent. Then*

$$\sum_{n=0}^{\infty} c_n = \sum_{n=0}^{\infty} a_n \sum_{n=0}^{\infty} b_n. \quad \text{(B.1.4)}$$

Proof.   Since

$$c_n = \sum_{k=0}^{n} a_{n-k} b_k,$$

$$\sum_{k=0}^{\infty} c_n r^n = \sum_{j=0}^{\infty} a_j r^j \sum_{k=0}^{\infty} b_k r^k$$

and Theorem B.1.1 completes the proof. ∎

The Cesàro and Abel means play a very significant role in the theory of Fourier series and transforms. We take a very brief look at how they appear in Fourier series since these ideas will play a role elsewhere in the book.

Suppose $f(x)$ is an integrable function of period 2π. Let $\sum_{-\infty}^{\infty} a_n e^{inx}$ be its Fourier series. An important question is: When does the Fourier series of a function converge to the function? Once again it is easier to deal with the Cesàro mean or the Abel mean. The Abel sum is given by

$$\sum_{-\infty}^{\infty} a_n r^{|n|} e^{inx}, \quad |r| < 1. \tag{B.1.5}$$

Since

$$a_n = \frac{1}{2\pi} \int_0^{2\pi} f(t) e^{-int} dt,$$

the sum in (B.1.5) is equal to

$$\frac{1}{2\pi} \int_0^{2\pi} f(t) \sum_{-\infty}^{\infty} r^{|n|} e^{in(x-t)} dt.$$

The sum inside the integral is called the Poisson kernel and it is a straightforward calculation to show that it is equal to

$$\frac{1 - r^2}{1 - 2r \cos(x - t) + r^2} \equiv P_r(x - t). \tag{B.1.6}$$

The following properties of $P_r(x)$ are worth noting:

(i) $P_r(x) \geq 0$,

(ii) $\frac{1}{2\pi} \int_0^{2\pi} P_r(x) dx = 1$, and

(iii) for $\delta > 0$, $\max_{\delta \leq x \leq 2\pi - \delta} P_r(x) \to 0$ as $r \to 1^-$.

These properties can be used to give a proof of the following theorem:

Theorem B.1.4 *If f is periodic and integrable on $(0, 2\pi)$ then the Abel means of the Fourier series converge to $\frac{1}{2}\{f(x_0+) + f(x_0-)\}$ at every point x_0 where the right and left limits $f(x_0\pm)$ exist.*

A similar result exists for Cesàro means. The nth partial sum of the Fourier series is given by

$$s_n(x) = \sum_{k=-n}^{n} a_k e^{ikx} = \frac{1}{2\pi} \int_0^{2\pi} \sum_{k=-n}^{n} e^{ik(x-t)} f(t) dt.$$

The sum inside the integral is

$$1 + 2 \sum_{k=0}^{n} \cos k(x - t) = \frac{\sin(n + 1/2)(x - t)}{\sin((x - t)/2)}. \tag{B.1.7}$$

This expression is called the Dirichlet kernel. One drawback of the Dirichlet kernel is that it is not always positive. However, if we take the $(C, 1)$ mean of the Fourier series, then we get

$$\sigma_n(x) = \frac{s_0(x) + s_1(x) + \cdots + s_n(x)}{n+1}$$

$$= \frac{1}{2\pi(n+1)} \int_0^{2\pi} f(t) \sum_{k=0}^n \frac{\sin\left(n+\frac{1}{2}\right)(x-t)}{\sin((x-t)/2)} dt.$$

The sum in the integral denoted by $K_n(x-t)$ is called the Fejér kernel. The sum is equal to

$$\left\{ \frac{\sin\frac{1}{2}(n+1)(x-t)}{\sin\frac{1}{2}(x-t)} \right\}^2 \geq 0. \tag{B.1.8}$$

The Fejér kernel has the three properties of the Poisson kernel mentioned above. One may then deduce the following theorem:

Theorem B.1.5 *For f as in Theorem B.1.4, the Cesàro means of the Fourier series of f converge to $\frac{1}{2}\{f(x_0+) + f(x_0-)\}$ at every point x_0 where the right and left limits $f(x_0\pm)$ exist.*

If we assume f to be continuous on $[0, 2\pi]$, then the Abel and Cesàro means converge to f. An important consequence of Theorem B.1.5 is the following corollary:

Corollary B.1.6 *A continuous function on $[0, 2\pi]$ can be uniformly approximated by trigonometric polynomials, that is, polynomials of the form $\sum_{-n}^n a_k e^{ikx}$.*

Since e^{ikx} is uniformly approximated on $[0, 2\pi]$ by partial sums of its Taylor series, we get another proof (see Exercise 1.40) of the Weierstrass approximation theorem.

B.2 Cesàro Means (C, α)

In the previous section, we saw that the $(C, 1)$ means do not assign a value to the sum $1 - 2 + 3 - 4 + \cdots$. To handle this and some other situations, we define Cesàro means of higher order.

For the series $\sum_0^\infty b_n$, set

$$B_n^{(0)} = B_0 + B_1 + \cdots + B_n,$$

where

$$B_n = \sum_{k=0}^n b_k.$$

The limit of the sequence $\{\frac{B_n^{(0)}}{n+1}\}$ gives the $(C, 1)$ mean. Now, set

$$B_n^{(1)} = B_0^{(0)} + B_1^{(0)} + \cdots + B_n^{(0)} \tag{B.2.1}$$

and

$$E_n^{(1)} = 1 + 2 + \cdots + n + 1 = \frac{(n+1)(n+2)}{2}.$$

The limit of the sequence $\{\frac{B_n^{(1)}}{E_n^{(1)}}\}$ is the $(C, 2)$ mean of the series $\sum_0^\infty b_n$.
Similarly,

$$B_n^{(2)} = B_0^{(1)} + B_1^{(1)} + \cdots + B_n^{(1)}$$

and

$$E_n^{(2)} = 1 + 3 + \cdots + \frac{(n+1)(n+2)}{2} = \frac{(n+1)(n+2)(n+3)}{3!}.$$

Thus, the $(C, 3)$ mean is given by

$$\lim_{n \to \infty} \left(B_n^{(2)} / E_n^{(2)} \right).$$

Define $B_n^{(k)}$ and $E_n^{(k)}$ inductively by

$$B_n^{(k)} = B_0^{(k-1)} + B_1^{(k-1)} + \cdots + B_n^{(k-1)}$$

and

$$E_n^{(k)} = E_0^{(k-1)} + E_1^{(k-1)} + \cdots + E_n^{(k-1)}.$$

It is possible to express $B_n^{(k)}$ explicitly in terms of b_n as follows. Note that

$$\sum_{n=0}^\infty B_n^{(k)} x^n = (1-x)^{-1} \sum_{n=0}^\infty B_n^{(k-1)} x^n = \cdots = (1-x)^{-(k+1)} \sum_0^\infty b_n x^n$$

and

$$\sum_{n=0}^\infty E_n^{(k)} x^n = (1-x)^{-(k+1)}.$$

So

$$B_n^{(k)} = \sum_{\ell=0}^n \left(\frac{(k+1)_\ell}{\ell!} \right) b_{n-\ell} \tag{B.2.2}$$

and

$$E_n^{(k)} = \frac{(k+1)_n}{n!}. \tag{B.2.3}$$

We define the (C, k) mean of $\sum_0^\infty b_n$ as the limit of the $B_n^{(k)}/E_n^{(k)}$ as $n \to \infty$. Note that these quotients are meaningful even when k is not a positive integer. We take k to be real and greater than -1.

Definition B.2.1   *The series $\sum_0^\infty b_n$ is (C, α) summable to B (for $\alpha > -1$) if*

$$\lim_{n \to \infty} \frac{n!}{(\alpha + 1)_n} \sum_{\ell=0}^n \frac{(\alpha + 1)_k}{\ell!} b_{n-\ell} = B. \tag{B.2.4}$$

Remark B.2.1   Note that the limit in (B.2.4) is equal to

$$\lim_{n \to \infty} \frac{\Gamma(\alpha + 1)}{n^\alpha} B_n^{(\alpha)}.$$

B.3 Fractional Integrals

It is possible to extend the summability definitions to integrals. Suppose $f(t)$ is integrable on finite intervals $(0, x)$. We say, for example, that f is Abel integrable, that is, $\int_0^\infty f(t)dt$ exists in the Abel sense, if

$$\lim_{\lambda \to 0^+} \int_0^\infty e^{-\lambda t} f(t)dt \quad \text{exists.}$$

We also write this as

$$\int_0^\infty f(t)dt = \lim_{\lambda \to 0^+} \int_0^\infty e^{-\lambda t} f(t)dt \quad \text{(Abel)}.$$

To define Cesàro integrability (C, k), where k is an integer, we first have the integral analog of $B_n^{(k)}$ (see (B.2.2)) as

$$f_{(k)}(x) = \int_0^x f_{(k-1)}(t)dt = \int_0^x (x - t) f_{(k-2)}(t)dt = \cdots = \frac{1}{k!} \int_0^x (x - t)^k f(t)dt. \tag{B.3.1}$$

Following the remark after (B.2.4) we say that $\int_0^\infty f(t)dt$ is (C, k) integrable if

$$\lim_{x \to \infty} \frac{\Gamma(k + 1)}{x^k} f_k(x) = \lim_{x \to \infty} \int_0^x \left(1 - \frac{t}{x}\right)^k f(t)dt \quad \text{exists.}$$

Observe that the final expression for $f_k(x)$ in (B.3.1) is meaningful for all real $k > -1$. Thus we have a definition of (C, α) integrability for $\alpha > -1$.

Note also that the formula for $f_{(k)}$ in (B.3.1) expresses $f_{(k)}$ as a $(k + 1)$-fold integral of f. Thus, we define $I_\alpha f$, the α-fold integral of f for Re $\alpha > 0$, by the formula

$$I_\alpha f(x) = \frac{1}{\Gamma(\alpha)} \int_0^x (x - t)^{\alpha-1} f(t)dt. \tag{B.3.2}$$

It is easy to check that the operator I_α satisfies the relation

$$I_\alpha I_\beta = I_{\alpha+\beta}, \quad \text{Re}(\alpha, \beta) > 0.$$

The operator I_α is called a fractional integral of order α. Euler's integral expression for the $_2F_1$ hypergeometric series can now be interpreted as a fractional integral. Write the formula as

$$_2F_1\left(\begin{matrix} a, b \\ c \end{matrix}; x\right) = \frac{\Gamma(c)x^{1-c}}{\Gamma(b)\Gamma(c-b)} \int_0^x t^{b-1}(x-t)^{c-b-1}(1-t)^{-a}\,dt$$

$$= \frac{\Gamma(c)}{\Gamma(b)}x^{1-c}(I_{c-b}f)(x),$$

where $f(t) = t^{b-1}(1-t)^{-a}$.

Remark B.3.1 Abel and Cesàro integrability can be used to study Fourier integrals just as the corresponding summability is used to study Fourier series.

B.4 Historical Remarks

Euler had used Abel means and other methods for associating a sum to a divergent series. In particular, he obtained the functional equation of the zeta function in the form

$$\frac{1 - 2^{s-1} + 3^{s-1} - \cdots}{1 - 2^{-s} + 3^{-s} - \cdots} = -\frac{(s-1)!(2^s-1)}{(2^{s-1}-1)\pi^s} \cos\frac{1}{2}s\pi.$$

When Re $s > 1$, the series in the denominator converges but the series in the numerator is interpreted in the Abelian sense. It can be proved that

$$\lim_{x\to 1^-} \sum_1^\infty (-1)^{n-1}n^{s-1}x^n = \zeta(1-s)(2^s-1);$$

thus Euler had the functional equation correct. He verified it only for integer values of s and for $s = \frac{1}{2}$ and $\frac{3}{2}$. Abel's name is attached to this method of summation since he proved formula (B.1.1).

Leibniz used $(C, 1)$ means to evaluate $1 - 1 + 1 - \cdots$ and somewhat later D. Bernoulli employed the same method to consider the more general series $\sum_{n=0}^\infty b_n$, where $b_{n+p} = b_n$ for all n and $\sum_0^{p-1} b_k = 0$. Neither Leibniz nor Bernoulli explicitly stated that he was giving a new definition of convergence. This was done by Cesàro. For more on the history of Abel and Cesàro means the reader should consult the very interesting treatment given in Hardy [1949].

Abel was apparently the first mathematician to use fractional calculus, though the concept had been considered by others before him. Abel used fractional calculus in his solution to the problem of finding the tautochrone – the curve down which a particle slides freely under gravity with zero initial velocity and reaches the bottom in the same amount of time regardless of the starting point, provided that the starting and lowest points are distinct. This problem had already been solved by Huygens; the motivation was the construction of a clock in which the period of oscillation does not depend on the amplitude of the pendulum.

Suppose that the particle slides down from the point (a, b) and that the bottom of the curve is the origin of coordinates. If s is the length measured from the origin, then conservation of energy implies that

$$\text{velocity} = v = ds/dt = \sqrt{2g(b - y)}$$

at a point on the curve with ordinate y. Set $ds/dy = f(y)$ to see that the time taken to reach the bottom is given by

$$T(b) = \frac{1}{\sqrt{2g}} \int_0^b \frac{f(y)}{\sqrt{b - y}} dy.$$

Abel observed that

$$T(b) = \sqrt{\frac{\pi}{2g}} (I_{1/2} f)(b).$$

Since $T(b)$ is independent of b, use (B.3.2) to get

$$I_{1/2}(I_{1/2} f)(b) = \frac{\sqrt{2g} T}{\pi} \int_0^b \frac{dy}{\sqrt{b - y}} = \frac{2T}{\pi} \sqrt{2gb}.$$

Now take the derivative of both sides to obtain

$$f(b) = \frac{T}{\pi} \sqrt{\frac{2g}{b}}.$$

Since

$$f(y) = \frac{ds}{dy} = \left[1 + \left(\frac{dx}{dy}\right)^2\right]^{1/2},$$

we get a differential equation that can be solved. The tautochrone curve turns out to be a cycloid. An English translation of Abel's [1826] paper on this topic is also available.

The above calculations suggest that the concept of a fractional derivative would also be useful. For smooth enough functions, in particular analytic functions, it is possible to define fractional derivatives as follows:

$$D_\alpha f(x) = \frac{d^n}{dx^n} I_{n-\alpha} f(x), \tag{B.4.1}$$

where $0 < \operatorname{Re} \alpha < n$, and n is an integer. Observe that we took the integral first and then the derivative. If the calculations are done in the opposite order, a different function arises since the first $n - 1$ terms are removed by the nth derivative. For an analytic function

$$f(x) = \sum_{0}^{\infty} a_k x^k$$

the fractional integral has the form

$$I_\alpha f(x) = \sum_{0}^{\infty} \frac{k!}{\Gamma(k + \alpha + 1)} a_k x^{k+\alpha},$$

as is easily verified. This shows that the value of n in (B.4.1) may be arbitrarily chosen as long as $n > \operatorname{Re} \alpha$. In fact,

$$D_k I_\alpha = D_n I_{\alpha+n-k}.$$

However, the inclusion of fractional derivatives with fractional integrals does not give a group, for although

$$D_\alpha D_\beta = D_{\alpha+\beta}$$

when α, β and $\alpha + \beta$ are not integers,

$$D_{1/2} D \neq D_{3/2}$$

for the reason mentioned above.

Exercises

1. Prove Theorem B.1.1.
2. With the notation of Theorem B.1.3, prove that if $\sum_0^\infty a_n$ converges absolutely and $\sum_0^\infty b_n$ converges, then $\sum_0^\infty c_n$ converges.
3. Prove that if the series $\sum_0^\infty b_n$ is (C, α) summable to B, and $\beta > \alpha$, then the series is (C, β) summable to B.
4. Prove the following extension of Theorem B.1.2: If the series $\sum_0^\infty b_n$ is (C, α) summable to B, then it is Abel summable to B.
5. Show that

$$\sum_{n=1}^{\infty}(-1)^{n-1}n^m e^{-ny} = (-1)^m \frac{d^m}{dy^m}(1 + e^y)^{-1}.$$

Deduce that

$$1 - 2^{2k} + 3^{2k} = \cdots = 0,$$

$$1 - 2^{2k-1} + 3^{2k-1} - \cdots = (-1)^{k-1}\frac{2^{2k} - 1}{2k} B_k.$$

in the sense of Abel summability. This result is due to Euler. See the historical introduction in Hardy [1949].

6. A sequence S_n converges to S in the sense of Borel if

$$\lim_{t \to \infty} e^{-t} \sum_{n=0}^{\infty} \frac{S_n}{n!} t^n = S.$$

Show that if S_n converges to S then $S_n \to S$ in the Borel sense. Also prove that if $S_n(x) = 1 + x + \cdots + x^n$, then $S_n(x) \to 1/(1-x)$ in the Borel sense for $\operatorname{Re} x < 1$.

7. Show that if $\beta \geq \alpha \geq -1$ and the (C, β) means of Σa_n are nonnegative, then the (C, α) means of $\Sigma a_n r^n$ are nonnegative for $0 \leq r \leq (\alpha + 1)/(\beta + 1)$.

   Hint: Observe that

$$(1 - \omega)^{-\alpha - 1} \Sigma a_n r^n \omega^n = (1 - \omega)^{-\alpha - 1} (1 - rw)^{\beta + 1} (1 - rw)^{-\beta - 1} \Sigma a_n r^n \omega^n$$

   and show that the power series coefficients of $(1 - \omega)^{-\alpha - 1} (1 - rw)^{\beta + 1}$ and those of $(1 - rw)^{-\beta - 1} \Sigma a_n r^n \omega^n$ are nonnegative. This proof is due to Bustoz [1974].

8. A series $\sum_{1}^{\infty} a_n$ is summable (L) (Lambert summable) to S if

$$\lim_{x \to 1^-} (1 - x) \sum_{1}^{\infty} \frac{n a_n x^n}{1 - x^n} = S.$$

   Prove that if $\Sigma a_n = S(C, \alpha)$, for some α, then $\Sigma a_n = S(L)$.

9. Prove Dirichlet's theorem that if χ is a quadratic Dirichlet character, then $L(1, \chi) = \sum_{n=1}^{\infty} \chi(n)/n \neq 0$ as follows:

   (a) Show that $\sum_{d|n} \chi(d) \geq 0$.

   (b) Show that

$$\lim_{x \to 1^-} \sum_{n=1}^{\infty} \frac{\chi(n) x^n}{1 - x^n} = \lim_{x \to 1^-} \sum_{n=1}^{\infty} \left(\sum_{d|n} \chi(d) \right) x^n = \infty.$$

   (c) Let

$$f(x) = \sum_{n=1}^{\infty} \left(\frac{\chi(n)}{n(1 - x)} - \frac{\chi(n) x^n}{1 - x^n} \right) =: \sum_{n=1}^{\infty} \frac{\chi(n)}{1 - x} b_n.$$

   Show that $b_1 \geq b_2 \geq \cdots \geq b_n \geq b_{n+1} \geq \cdots$.

   (d) Note that $| \sum_{n=1}^{m} \chi(n) | \leq M$, where M is a constant that does not depend on m. Use summation by parts to prove that $f(x) \leq \frac{M b_1}{1 - x} = M$.

   (e) Deduce that $L(1, \chi) \neq 0$.

   This proof is taken from Monsky [1994].

10. Prove the following theorem of Hardy and Littlewood: If $a_n \geq 0$ and

$$\lim_{x \to 1^-} (1 - x) \sum_{n=0}^{\infty} a_n x^n = 1,$$

then

$$\lim_{n\to\infty} \frac{1}{n} \sum_{k=0}^{n} a_k = 1.$$

Karamata's proof is sketched below.

(a) Show that

$$\lim_{x\to 1^-} (1-x) \sum_{n=0}^{\infty} a_n x^n g(x^n) = \int_0^1 g(t)\,dt,$$

when $g(t)$

(i) is a polynomial,

(ii) is a continuous function, or

(iii) has a discontinuity of the first kind.

(b) Take $g(t) = 0$ $(0 \le t < 1/e) = 1/t(1/e \le t \le 1)$ and complete the proof of the theorem.

C

Asymptotic Expansions

C.1 Asymptotic Expansion

Let x be a real or complex variable in an unbounded region D and let $\sum_0^\infty a_n x^{-n}$ be a formal power series that may be convergent or divergent.

Definition C.1.1 *The series $\sum_0^\infty a_n x^{-n}$ is an asymptotic expansion of a function $f(x)$ if*

$$f(x) = \sum_0^{n-1} a_k x^{-k} + R_n(x), \tag{C.1.1}$$

where $R_n(x) = 0(x^{-n})$ as $x \to \infty$ in D. Usually (C.1.1) is written as

$$f(x) \sim a_0 + a_1 x^{-1} + a_2 x^{-2} + \cdots \qquad as\ x \to \infty\ in\ D. \tag{C.1.2}$$

This definition is due to Poincaré [1886]. As a simple example we have the following asymptotic expansion of $(1 + x)^{-1}$:

$$\frac{1}{1+x} \sim \frac{1}{x} - \frac{1}{x^2} + \frac{1}{x^3} - \cdots \qquad as\ x \to \infty. \tag{C.1.3}$$

In this case the series is convergent. The more interesting situation occurs when the series is divergent. Consider the complementary error function of a real variable $x \geq 0$,

$$\mathrm{erfc}\ x = \frac{2}{\sqrt{\pi}} \int_x^\infty e^{-t^2} dt. \tag{C.1.4}$$

Successive integration by parts gives

$$
\begin{aligned}
\mathrm{erfc}\ x &= \frac{2}{\sqrt{\pi}} \left[\frac{e^{-x^2}}{2x} - \int_x^\infty \frac{e^{-t^2}}{2t^2} dt \right] \\
&= \frac{e^{-x^2}}{\sqrt{\pi}x} \left[1 + \sum_{k=1}^n (-1)^k \frac{1 \cdots 3 \cdot (2k-1)}{(2x^2)^k} + R_n(x) \right],
\end{aligned}
$$

where

$$R_n(x) = (-1)^{n+1} \frac{1 \cdot 3 \cdots (2n+1)}{2^n} x e^{x^2} \int_x^\infty \frac{e^{-t^2}}{t^{2n+2}} dt.$$

It is easy to see that

$$|R_n(x)| \le \frac{1 \cdot 3 \cdots (2n+1)}{(2x^2)^{n+1}}. \qquad (C.1.5)$$

Hence

$$\frac{e^{-x^2}}{\sqrt{\pi x}} \left[1 + \sum_{k=1}^\infty (-1)^k \frac{1 \cdot 3 \cdots (2k-1)}{(2x^2)^k} \right]$$

is an asymptotic expansion of the complementary error function. It is clear that the series diverges for all $x > 0$. However, if a fixed number of terms is taken, then for large enough x a good approximation of erfc x is obtained. On the other hand, for a given x, taking more and more terms of the series does not improve the approximation, since the series diverges.

C.2 Properties of Asymptotic Expansions

The following theorem follows almost immediately from the definition.

Theorem C.2.1 *A function $f(x)$ has an asymptotic expansion $\Sigma a_n x^{-n}$ if and only if for each n,*

$$x^n \left[f(x) - \sum_{k=0}^{n-1} a_k x^{-k} \right] \to a_n \quad as\ x \to \infty\ in\ D, \qquad (C.2.1)$$

uniformly with respect to arg x when x is complex.

Proof. It is easy to see that (C.2.1) implies (C.1.2). To reason in the other direction, observe that

$$x^n R_n(x) = x^n [a_n x^{-n} + R_{n+1}(x)] \to a_n \quad as\ x \to \infty,$$

and the theorem is proved. ∎

A consequence of this theorem is that a function has at most one asymptotic expansion in D. In a different unbounded region the asymptotic expansion may be different. However, two different functions may have the same asymptotic expansion in some region. For example, the function e^{-x} in $|\arg x| \le \frac{1}{2}\pi - \delta < \frac{1}{2}\pi$ satisfies $x^n e^{-x} \to 0$ as $x \to \infty$. By (C.2.1),

$$e^{-x} \sim 0 + \frac{0}{x} + \frac{0}{x^2} + \cdots \quad |\arg x| \le \frac{1}{2}\pi - \delta, \quad x \to \infty.$$

So the zero function and e^{-x} have the same expansion in this region.

The next theorem gives some algebraic properties of asymptotic expansions.

Theorem C.2.2 *Suppose that $f(x) \sim \Sigma a_n x^{-n}$ in D_1 and $g(x) \sim \Sigma b_n x^{-n}$ in D_2. Then:*

(i) For constants λ and μ

$$\lambda f(x) + \mu g(x) \sim \Sigma (\lambda a_n + \mu b_n) x^{-n} \quad \text{in } D_1 \cap D_2.$$

(ii) $f(x) g(x) \sim \Sigma c_n x^{-n}$ in $D_1 \cap D_2$, where

$$c_n = a_0 b_0 + a_1 b_{n-1} + \cdots + a_n b_0.$$

(iii) If $a_0 \neq 0$, then

$$\frac{1}{f(x)} = \sum_{k=0}^{n-1} \frac{d_k}{x^k} + O(x^{-n}) \quad \text{as } x \to \infty \text{ in } D_1,$$

where $a_0^{k+1} d_k$ is a polynomial in $a_0, a_1, \ldots, a_k$. The d_k can be obtained from the relations

$$a_0 d_0 = 1, \qquad a_0 d_k = -(a_1 d_{k-1} + a_2 d_{k-2} + \cdots + a_k d_0), \quad k = 1, 2, \ldots.$$

The proof of this theorem is left to the reader.

Asymptotic series can be integrated over the interval $x \leq t < \infty$, if $a_0 = a_1 = 0$. In this case

$$f(x) \sim \frac{a_2}{x^2} + \frac{a_3}{x^3} + \cdots, \quad x \to \infty,$$

and

$$\int_x^\infty f(t)\, dt \sim \frac{a_2}{x} + \frac{a_3}{2x^2} + \cdots, \quad x \to \infty.$$

Integration is possible here because $f(t) = a_2 t^{-2} + O(t^{-2})$ for t large. If a_0 and a_1 are not zero, then

$$\int_x^\infty [f(t) - a_0 - a_1 t^{-1}]\, dt \sim \frac{a_2}{x} + \frac{a_3}{2x^3} + \cdots, \quad x \to \infty.$$

Differentiation of an asymptotic expansion may not always be valid. A standard example where differentiation fails is $f(x) = e^{-x} \sin e^x$ where $x > 0$. The derivative $f'(x) = \cos e^x - e^{-x} \sin e^x$ oscillates as $x \to \infty$ and hence does not have an asymptotic expansion by Theorem C.2.1. But

$$f(x) \sim 0 + \frac{0}{x} + \frac{0}{x^2} + \cdots, \quad x \to \infty.$$

If $f'(x)$ is continuous and has an asymptotic expansion, then it can be obtained from term-by-term differentiation of the expansion for $f(x)$. This follows from the result on integration and uniqueness of the asymptotic expansion.

C.3 Watson's Lemma

In this section, we discuss a method of obtaining the asymptotic expansion in some region of a function expressible as a Laplace integral. Section 1 has an example of an integral whose asymptotic expansion was obtained using integration by parts. The failure of this method in one case was discussed in Chapter 2. Also see Wong [1989, p. 18].

The next theorem, called Watson's lemma, gives the asymptotic expansion of $\int_0^\infty e^{-xt} f(t)dt$. See Watson [1918].

Theorem C.3.1 *Let $f(t)$ be analytic in $|t| \leq a + \delta$, where $a > 0$, $\delta > 0$, except possibly for a branch point at 0; and let*

$$f(t) = \sum_{m=1}^{\infty} a_m t^{(m/r)-1}, \tag{C.3.1}$$

when $|t| \leq a$ and $r > 0$. Suppose also that $|f(t)| < \kappa e^{bt}$, where κ and b are positive numbers independent of t, when $t \geq a$. Then

$$\int_0^\infty e^{-xt} f(t)dt \sim \sum_{m=1}^{\infty} a_m \Gamma(m/r) x^{-m/r} \quad \text{as } x \to \infty \tag{C.3.2}$$

for $|\arg x| \leq \frac{1}{2}\pi - \delta < \frac{1}{2}\pi$.

Proof. It is clear that for any fixed integer M, a constant C can be found such that for $t \geq 0$

$$\left| f(t) - \sum_{m=1}^{M-1} a_m t^{(m/r)-1} \right| \leq C t^{(M/r)-1} e^{bt}.$$

Hence

$$\int_0^\infty e^{-xt} f(t)dt = \sum_{m=1}^{M-1} \int_0^\infty e^{-xt} a_m t^{(m/r)-1} dt + R_M$$

$$= \sum_{m=1}^{M-1} a_m \Gamma(m/r) x^{-m/r} + R_M,$$

where

$$|R_M| \leq \int_0^\infty |e^{-xt}| C t^{(M/r)-1} e^{bt} dt$$

$$= C\Gamma(M/r)/[\text{Re } x - b]^{M/r},$$

provided Re $x > b$. Since $|\arg x| \leq \frac{1}{2}\pi - \delta$, it follows that Re $x > b$ if $|x| > b \csc \delta$. Thus,

$$|x^{M/r} R_M| < \frac{C\Gamma(M/r)|x|^{M/r}}{(|x|\sin \delta - b)^{M/r}} = O(1).$$

This proves the theorem. ∎

Watson's lemma can be generalized to the case in which the integral in (C.3.2) is a contour integral, $\int_\infty^{(0+)} e^{-xt} f(t)dt$ and (C.3.1) is replaced by an asymptotic expansion for $f(t)$. See Olver [1974, pp. 112–115] and Wong [1989, p. 22].

C.4 The Ratio of Two Gamma Functions

We give an application of Watson's lemma to obtain the asymptotic expansion of a ratio of gamma functions. This expansion is due to Tricomi and Erdélyi [1951].

Observe that for $\mathrm{Re}(x + a) > 0$ and $\mathrm{Re}(b - a) > 0$, we have

$$\frac{\Gamma(x + a)}{\Gamma(x + b)} = \frac{1}{\Gamma(b - a)} \int_0^\infty e^{-xt} e^{-at} (1 - e^{-t})^{b-a-1} dt. \tag{C.4.1}$$

Define the generalized Bernoulli polynomials $B_n^\sigma(u)$ by

$$\frac{t^\sigma e^{ut}}{(e^t - 1)^\sigma} = \sum_{n=0}^\infty B_n^\sigma(u) \frac{t^n}{n!}, \qquad |t| < 2\pi. \tag{C.4.2}$$

Then for $\sigma = a + 1 - b$,

$$\frac{e^{-at}}{(1 - e^{-t})^\sigma} = \sum_{n=0}^\infty \frac{(-1)^n}{n!} B_n^\sigma(a) t^{n-\sigma}, \qquad |t| < 2\pi.$$

Watson's lemma now implies that

$$\frac{\Gamma(x + a)}{\Gamma(x + b)} \sim \sum_{n=0}^\infty \frac{(-1)^n}{n!} B_n^\sigma(a) \frac{\Gamma(b - a + n)}{\Gamma(b - a)} \frac{1}{x^{b-a+n}},$$

$$|\arg x| \leq \frac{1}{2}\pi - \delta, \quad |x| \to \infty. \tag{C.4.3}$$

If instead of the integral representation (C.4.1), one were to use

$$\frac{\Gamma(x + a)}{\Gamma(x + b)} = \frac{\Gamma(1 + a - b)}{2\pi i} \int_{-\infty \cdot e^{i\alpha}}^{(0+)} e^{(x+a)t} (e^t - 1)^{b-a-1} dt, \tag{C.4.4}$$

$\mathrm{Re}[(x + a)e^{i\alpha}] > 0$, $|\alpha| < \pi/2$, and for small $|t|$,

$$\alpha - \pi \leq \arg(e^t - 1) \leq \alpha + \pi;$$

then one could extend (C.4.3) to $|\arg x| \leq \pi - \delta$.

There is an improvement of (C.4.3) from the computational standpoint due to Fields [1964]. Note that if u is replaced by $\sigma - u$ in (C.4.2), then $B_k^\sigma(\sigma - u) = (-1)^k B_k^\sigma(u)$. This implies that $B_{2k+1}^\sigma(\sigma/2) \equiv 0$. Now write the integrand in

(C.4.1) as

$$e^{-(x+a-\sigma/2)t}e^{-\sigma t/2}(1-e^{-t})^{-\sigma} = e^{-(x+a-\sigma/2)t}\sum_{n=0}^{\infty}\frac{(-1)^n}{n!}B_n^\sigma(\sigma/2)t^{n-\sigma}$$

$$= e^{-(x+a-\sigma/2)t}\sum_{n=0}^{\infty}\frac{B_{2n}^\sigma(\sigma/2)}{(2n)!}t^{2n-\sigma}.$$

This implies

$$\frac{\Gamma(x+a)}{\Gamma(x+b)} \sim \sum_{n=0}^{\infty}\frac{B_{2n}^\sigma(\sigma/2)}{(2n)!}\cdot\frac{\Gamma(b-a+2n)}{\Gamma(b-a)}\frac{1}{(x+a-\sigma/2)^{2n+b-a}}$$

for $|\arg(x+a)| \le \frac{1}{2}\pi - \delta, |x| \to \infty$.

Exercises

1. Let Re $x \ge 0$. Show that as $x \to 0$,

$$\int_0^\infty \frac{e^{-t}}{1+xt}dt \sim \sum_{n=0}^{\infty}(-1)^n n!x^n$$

by two methods: (a) Expand $1/(1+xt)$ as a series. (b) Integrate by parts repeatedly.

2. Suppose that

$$F(x) = \int_0^\infty e^{-xt}f(t)dt$$

converges for some $x = x_0$ and f has continuous derivatives of all orders in $0 \le t \le a$. Show that

$$F(x) \sim \Sigma f^{(n)}(0)x^{-n-1}$$

uniformly in arg x, as $x \to \infty$ in $|\arg x| < \pi/2 - \epsilon$ for $\epsilon > 0$.

3. Show that

$$x^{-a}e^x \int_x^\infty e^{-x}x^{a-1}dx \sim \frac{1}{x} + \frac{a-1}{x^2} + \frac{(a-1)(a-2)}{x^3} + \cdots.$$

4. Suppose θ_n is defined by

$$1 + \frac{n}{1!} + \frac{n^2}{2!} + \cdots + \frac{n^{n-1}}{(n-1)!} + \frac{n^n}{n!}\theta_n = \frac{1}{2}e^n.$$

Show that

$$\theta_n = 1 + \frac{n}{2}\left(\int_0^1 (xe^{1-x})^n dx - \int_1^\infty (xe^{1-x})^n dx\right)$$

and hence

$$\theta_n \sim \frac{1}{3} + \frac{4}{135n} - \frac{8}{2835n^2} + \cdots \qquad \text{as } n \to \infty.$$

For a discussion of this result of Ramanujan, see Berndt [1989, pp. 181–184].

5. Prove Theorem C.2.2.

D

Euler–Maclaurin Summation Formula

D.1 Introduction

Some consequences of the close connection between series and integrals are brought out even in a first course in calculus. The integral test, for example, states that for a decreasing continuous function on $[1, \infty)$ the series $\sum_1^\infty f(n)$ and integral $\int_1^\infty f(x)dx$ converge or diverge together. In the derivation of Stirling's approximation for $\Gamma(x)$ given in Chapter 1, we saw that for finite sums the function f need not be decreasing for the integral to provide a good approximation. The Euler–Mclaurin summation formula makes the connection between the sum and the integral explicit for sufficiently smooth functions. In this appendix we give a statement and proof of the formula and a few applications.

Start with a differentiable function defined on a set that contains the interval $[m, n]$, where m and n are integers. Let $m \le j < n$, where j is an integer. As a first approximation we consider

$$\int_j^{j+1} f(x)dx \approx \frac{1}{2}(f(j) + f(j+1)). \tag{D.1.1}$$

It is possible to express the error in this approximation as an integral. Observe that

$$\frac{1}{2}(f(j) + f(j+1)) = \int_j^{j+1} \frac{d}{dx}\left[\left(x - j - \frac{1}{2}\right)f(x)\right]dx$$

$$= \int_j^{j+1} f(x)dx + \int_j^{j+1} \left(x - j - \frac{1}{2}\right)f'(x)dx. \tag{D.1.2}$$

Set $j = m, m+1, \ldots, n-1$ to get

$$\frac{1}{2}(f(m) + f(m+1)) = \int_m^{m+1} f(x)dx + \int_m^{m+1} \left(x - m - \frac{1}{2}\right)f'(x)dx,$$

617

$$\frac{1}{2}(f(m+1)+f(m+2))=\int_{m+1}^{m+2}f(x)dx+\int_{m+1}^{m+2}\left(x-(m+1)-\frac{1}{2}\right)f'(x)dx,$$

$$\frac{1}{2}(f(n-1)+f(n))=\int_{n-1}^{n}f(x)dx+\int_{n-1}^{n}\left(x-(n-1)-\frac{1}{2}\right)f'(x)dx.$$

The second integrals on the right can be added together if we note that the integrand can be written as $(x-[x]-\frac{1}{2})f'(x)$. Addition gives

$$\sum_{m+1}^{n-1}f(k)+\frac{1}{2}(f(m)+f(n))=\int_{m}^{n}f(x)dx+\int_{m}^{n}\left(x-[x]-\frac{1}{2}\right)f'(x)dx.$$

$$(D.1.3)$$

This is a particular case of the Euler–Maclaurin formula. To put it in a slightly different form, set $B_1(x)=x-\frac{1}{2}$, so that $B_1(x-[x])=x-[x]-\frac{1}{2}$. Also let $B_1=B_1(0)=-1/2$. Now write (D.1.3) as

$$\sum_{m+1}^{n}f(k)+B_1\{f(n)-f(m)\}=\int_{m}^{n}f(x)dx+\int_{m}^{n}B_1(x-[x])f'(x)dx.$$

$$(D.1.4)$$

Recall that $B_1(x)$ is the first Bernoulli polynomial and B_1 the first Bernoulli number.

As an application take $f(x)=\log x$ and $m=1$ in (D.1.4) to get

$$\log n!-\left(n+\frac{1}{2}\right)\log n+n=1+\int_{1}^{n}\frac{B_1(x-[x])}{x}dx.\qquad(D.1.5)$$

Since $B_1(x-[x])$ is periodic of period 1 and

$$\int_{0}^{1}B_1(x-[x])dx=\int_{t}^{t+1}B_1(x-[x])dx=0\qquad(D.1.6)$$

for any $t\geq 0$, we see that the limit as $n\to\infty$ in (D.1.5) exists and is equal to

$$1+\int_{1}^{\infty}\frac{B_1(x-[x])}{x}dx.\qquad(D.1.7)$$

This is exactly Stirling's approximation, if we can show that the expression (D.1.7) equals $\frac{1}{2}\log 2\pi$. We prove this later by a method different from the one given in Chapter 1, which uses Wallis's formula.

Formula (D.1.4) was obtained by one integration by parts. With Stirling's formula in mind, it makes sense to repeat the process. This would give an expression with higher derivatives of $f(x)$. If $f(x)=\log x$, then $f'(x)=1/x$, $f''(x)=-1/x^2$, and so on. These quantities get small for large x and so an extension of (D.1.4) would be useful.

D.2 The Euler–Maclaurin Formula

Let $\tilde{B}_2(x)$ be a primitive (antiderivative) of $B_1(x - [x]) = \tilde{B}_1(x)$. It follows from (D.1.6) that $\tilde{B}_2$ is periodic with period one. In particular, this implies that $\tilde{B}_2(0) = \tilde{B}_2(1) = \cdots = \tilde{B}_2(j) = \tilde{B}_2(j+1) = \cdots$. Now, integration by parts gives

$$\int_j^{j+1} \tilde{B}_1(x) f'(x) dx = \tilde{B}_2(0)(f'(j+1) - f'(j)) - \int_j^{j+1} \tilde{B}_2(x) f''(x) dx.$$

We assume that f has continuous derivatives of as high an order as necessary.

Sum from $j = m$ to $j = n$ to get

$$\int_m^n \tilde{B}_1(x) f'(x) dx = \tilde{B}_2(0)(f'(n) - f'(m)) - \int_m^n \tilde{B}_2(x) f''(x) dx. \quad \text{(D.2.1)}$$

Note that it was the periodicity of $\tilde{B}_2(x)$ that gave us the simple expression on the right side of Equation (D.2.1). This suggests that we should choose the constant of integration in $\tilde{B}_2(x)$ such that

$$\int_0^1 \tilde{B}_2(x) dx = \int_t^{t+1} \tilde{B}_2(x) dx = 0. \quad \text{(D.2.2)}$$

Let $\frac{1}{2} B_2(x) = \tilde{B}_2(x)$ for $0 \le x \le 1$, so that $\tilde{B}_2(x) = \frac{1}{2} B_2(x - [x])$. From the definition of $\tilde{B}_2(x)$, it follows that

$$B_2(x) = x^2 - x + \frac{1}{6},$$

which is the second Bernoulli polynomial. We state the general Euler–Maclaurin formula in the next theorem:

Theorem D.2.1 *Suppose f has continuous derivatives up to order s. Then*

$$\sum_{m+1}^n f(x) = \int_m^n f(x) dx + \sum_{\ell=1}^s (-1)^\ell \frac{B_\ell}{\ell!} \{ f^{(\ell-1)}(n) - f^{(\ell-1)}(m) \}$$

$$+ \frac{(-1)^{s-1}}{s!} \int_m^n B_s(x - [x]) f^{(s)}(x) dx,$$

where $B_s(x)$ are the Bernoulli polynomials and $B_\ell = B_\ell(0)$, the Bernoulli numbers.

Proof. As in the derivation of (D.2.1), apply integration by parts successively to obtain a sequence of periodic functions $\tilde{B}_n(x)$ such that $\tilde{B}_n'(x) = \tilde{B}_{n-1}(x)$ $(0 < x < 1, \ n \ge 1)$ and $\int_0^1 \tilde{B}_n(x) dx = 0, n \ge 1$. With respect to these functions, we obtain the formula

$$\sum_{m+1}^{n} f(k) = \int_{m}^{n} f(x)dx + \sum_{\ell=1}^{s}(-1)^{\ell}\tilde{B}_{\ell}(0)\{f^{(\ell-1)}(n) - f^{(\ell-1)}(m)\}$$

$$+ (-1)^{s-1}\int_{m}^{n}\tilde{B}_{s}(x)f^{(s)}(x)dx.$$

To show the relation of $\tilde{B}_{\ell}(x)$ to Bernoulli polynomials, consider the generating function

$$G(x, t) := \sum_{0}^{\infty}\tilde{B}_{n}(x)t^{n}, \quad \text{where } \tilde{B}_{0}(x) = 1, \tag{D.2.3}$$

of the sequence $\{\tilde{B}_{\ell}(x)\}$. Observe that for $0 < x < 1$,

$$\frac{\partial G}{\partial x} = \sum_{n=1}^{\infty}\tilde{B}_{n}'(x)t^{n} = \sum_{n=1}^{\infty}\tilde{B}_{n-1}(x)t^{n} = t\sum_{n=0}^{\infty}\tilde{B}_{n}(x)t^{n} = tG.$$

We then have

$$G(x, t) = A(t)e^{xt}. \tag{D.2.4}$$

Use $\int_{0}^{1}\tilde{B}_{n}(x)dx = 0$ in (D.2.3) and (D.2.4) to obtain

$$1 = A(t)\frac{e^{t} - 1}{t} \quad \text{or} \quad A(t) = \frac{t}{e^{t} - 1}.$$

Thus

$$G(x, t) = \frac{te^{xt}}{e^{t} - 1} = \sum_{0}^{\infty}B_{n}(x)\frac{t^{n}}{n!}, \tag{D.2.5}$$

and

$$\tilde{B}_{n}(x) = \frac{B_{n}(x)}{n!} \quad \text{for } 0 < x < 1.$$

By the periodicity of $\tilde{B}_{n}(x)$ we arrive at

$$\tilde{B}_{n}(x) = \frac{1}{n!}B_{n}(x - [x]) \quad \text{for all } x.$$

Although this formal argument has not been justified, it can be done easily since the generating function (D.2.5) is analytic in the disk $|t| < 2\pi$. This proves the theorem. ∎

D.3 Applications

Take $f(x) = x^{-s} \log x$, $m = 1$, and $n = N$ in (D.1.3) to get

$$\sum_{n=2}^{N} \frac{\log n}{n^s} = \frac{N^{1-s} \log N}{1-s} + \frac{1}{(s-1)^2} - \frac{N^{1-s}}{(s-1)^2}$$

$$+ \frac{1}{2}(N^{-s} \log N) + \int_1^N \frac{x - [x] - \frac{1}{2}}{x^{s+1}}(1 - s \log x)dx.$$

Let Re $s > 1$ and $N \to \infty$ to obtain

$$\zeta'(s) = \frac{-1}{(s-1)^2} - \int_1^\infty \frac{x - [x] - \frac{1}{2}}{x^{s+1}}(1 - s \log x)dx.$$

The integral on the right side of the equality converges for Re $s > -1$ by Abel's test. So

$$\zeta'(0) = -1 - \int_1^\infty \frac{x - [x] - \frac{1}{2}}{x}dx. \tag{D.3.1}$$

By Corollary 1.3.3 we have $\zeta'(0) = -\frac{1}{2} \log 2\pi$. Thus the constant (D.1.7) in Stirling's formula is $\frac{1}{2} \log 2\pi$.

As another application, we prove the following useful theorem on the order of $\zeta(s)$ and $\zeta'(s)$ for large Im s.

Theorem D.3.1 *Let $s = \sigma + it$. For $1 - \frac{A}{\log |t|} \leq \sigma \leq 2$, where A is any positive constant, and $|t|$ large enough,*

$$\zeta(s) = O(\log |t|)$$

and

$$\zeta'(s) = O(\log^2 |t|).$$

Proof. In (D.1.3), set $f(x) = x^{-s}$, $m = N$, and let $n \to \infty$ to arrive at

$$\zeta(s) = \sum_{n=1}^{N} \frac{1}{n^s} + \frac{N^{1-s}}{s-1} - \frac{1}{2}N^{-s} - s \int_N^\infty \frac{x - [x] - \frac{1}{2}}{x^{s+1}}dx.$$

The integral is $O(|t|/\sigma N^\sigma)$. Note that

$$|n^{-s}| = n^{-\sigma} \leq n^{-(1-A/\log |t|)} = O(1/n),$$

where the last equality holds for $n \leq |t|$. Thus, take $N = [|t|]$ to get

$$\zeta(s) = \sum_{n=1}^{N} O\left(\frac{1}{n}\right) + O(1) = O(\log |t|).$$

To derive the result for $\zeta'(s)$, take $f(x) = x^{-s} \log x$ and do a similar calculation. See Rademacher [1973, p. 100], if necessary. ∎

We complete this section with a proof of the following theorem stated in Chapter 1.

Theorem D.3.2   *Let $z \in \mathbb{C} - (-\infty, 0]$. Then*

$$\log \Gamma(z) = \left(z - \frac{1}{2}\right) \log z - z + \sum_{j=1}^{m} \frac{B_{2j}}{2j(2j-1)} \frac{1}{z^{2j-1}}$$
$$+ \frac{1}{2} \log 2\pi - \frac{1}{2m} \int_0^{\infty} \frac{B_{2m}(x - [x])}{(x+z)^{2m}} dx.$$

Proof.   Start with the following expression of the gamma function:

$$\Gamma(z) = \lim_{n \to \infty} \prod_{\ell=1}^{n} \frac{\ell}{z+\ell-1} \left(\frac{\ell+1}{\ell}\right)^{z-1}.$$

Then

$$\log \Gamma(z) = \lim_{n \to \infty} \left[(z-1) \log(n+1) - \sum_{\ell=1}^{n} \log \frac{z+\ell-1}{\ell} \right], \qquad \text{(D.3.2)}$$

where the principal branch of the log function in $\mathbb{C} - (-\infty, 0]$ is chosen. In Theorem D.2.1, take

$$f(x) = \log \frac{x+z-1}{x} = \log(x+z-1) - \log x$$

to get

$$\sum_{\ell=1}^{n} \log \frac{z+\ell-1}{\ell} = \log z + \sum_{\ell=2}^{n} \log \frac{z+\ell-1}{\ell}$$

$$= \log z + \int_1^n [\log(x+z-1) - \log x] dx$$

$$+ \sum_{j=1}^{m} \frac{B_{2j}}{2j(2j-1)} \left[\frac{1}{(n+z-1)^{2j-1}} - \frac{1}{n^{2j-1}} - \frac{1}{z^{2j-1}} + 1 \right]$$

$$+ \frac{1}{2} [\log(n+z-1) - \log n - \log z]$$

$$+ \frac{1}{2m} \int_0^n B_{2m}(x - [x]) \left[\frac{1}{(z+x-1)^{2m}} - \frac{1}{x^{2m}} \right] dx.$$

Here we have used the fact that $B_1 = -1/2$ and $B_{2j+1} = 0$ for $j \geq 1$. Compute the first of the above integrals and observe that after some cancellation the terms that involve n are

$$(n + z - 1) \log(n + z - 1) - n \log n + \frac{1}{2} \log \frac{n + z - 1}{n}$$

$$+ \sum_{j=1}^{m} \frac{B_{2j}}{2j(2j-1)} \left[\frac{1}{(n + z - 1)^{2j-1}} - \frac{1}{n^{2j-1}} \right].$$

Subtract this from $(z-1) \log(n+1)$ and let $n \to \infty$ to compute the limit in (D.3.2). The result is

$$\log \Gamma(z) = \left(z - \frac{1}{2} \right) \log z - z + 1 + \sum_{j=1}^{m} \frac{B_{2j}}{2j(2j-1)} \left(\frac{1}{z^{2j-1}} - 1 \right)$$

$$- \frac{1}{2m} \int_1^\infty B_{2m}(x - [x]) \left[\frac{1}{(x + z - 1)^{2m}} - \frac{1}{x^{2m}} \right] dx. \quad \text{(D.3.3)}$$

From (D.1.5) and (D.3.1) we know that

$$\lim_{z \to \infty} [\log \Gamma(z) - (z - 1/2) \log z + z] = \frac{1}{2} \log 2\pi.$$

So let $z \to \infty$ in (D.3.3) to see that

$$1 - \sum_{j=1}^{m} \frac{B_{2j}}{2j(2j-1)} + \frac{1}{2m} \int_1^\infty \frac{B_{2m}(x - [x])}{x^{2m}} dx = \frac{1}{2} \log 2\pi.$$

This result combined with (D.3.3) gives the formula in Theorem D.3.2. ∎

D.4 The Poisson Summation Formula

In this section we state and prove an important and useful formula from the theory of Fourier series. This is the Poisson summation formula. It has numerous applications though we mention only a few. One consequence of Poisson's formula is the Euler–Maclaurin summation formula.

Start with a result on Fourier series attributed to Jordan. Recall that a function $f(x)$ is said to be of bounded variation on an interval $[a, b]$ (which may be infinite), if there is a constant $C > 0$ such that for any set of points $x_0 < x_1 < \cdots < x_n$ in the interval

$$\sum_{k=1}^{n} |f(x_i) - f(x_{i-1})| < C.$$

If the interval is the whole real line, then we say that f is of bounded total variation. An important though easily proved result is that if f is of bounded variation on $[a, b]$, then $f = g - h$, where g and h are increasing functions on $[a, b]$.

Jordan's theorem is the following: Suppose f is integrable on $[0, 1]$ and periodic with period one. If f is of bounded variation on $[0, 1]$, then

$$\frac{f(x+) + f(x-)}{2} = \lim_{N \to \infty} \sum_{|n| \le N} \left(\int_0^1 f(t) e^{-2\pi i n t} dt \right) e^{2\pi i n x}.$$

Poisson's summation formula is contained in the next theorem.

Theorem D.4.1   *Suppose g is integrable on $(-\infty, \infty)$ and of bounded total variation. Suppose h is an even, positive, integrable function that is decreasing on $[0, \infty)$ and such that $|g(x)| \le h(x)$. Then*

$$\sum_{n=-\infty}^{\infty} \frac{g(x + n+) + g(x + n-)}{2} = \lim_{N \to \infty} \sum_{|n| \le N} \int_{-\infty}^{\infty} g(t) e^{-2\pi i n t} dt \, e^{2\pi i n x}.$$

$$(D.4.1)$$

Proof.   The inequality

$$\sum_{n=-M}^{N-1} |g(x + n)| \le |g(x)| + \int_{-M}^{N} h(x + t) dt$$

implies that $\sum_{-\infty}^{\infty} g(x+n)$ is an absolutely convergent series and defines a periodic function $f(x)$ with period one. Moreover, since g is of bounded total variation, f is of bounded variation on $[0, 1]$. By Jordan's theorem

$$\frac{f(x+) + f(x-)}{2} = \lim_{N \to \infty} \sum_{|n| \le N} \int_0^1 f(t) e^{-2\pi i n t} dt \, e^{2\pi i n x}.$$

Now

$$\int_0^1 f(t) e^{-2\pi i n t} dt = \int_0^1 \sum_{-\infty}^{\infty} g(m + t) e^{-2\pi i n t} dt$$

$$= \sum_{-\infty}^{\infty} \int_0^1 g(m + t) e^{-2\pi i n t} dt$$

$$= \sum_{-\infty}^{\infty} \int_m^{m+1} g(u) e^{-2\pi i n u} du$$

$$= \int_{-\infty}^{\infty} g(u) e^{-2\pi i n u} du.$$

The summation in the second line can be taken outside the integral by the dominated convergence theorem.   ∎

Remark D.4.1   If the series on the right-hand side in Theorem D.4.1 is absolutely convergent, then we can write

$$\sum_{n=-\infty}^{\infty} \frac{g(x+n+) + g(x+n-)}{2} = \sum_{n=-\infty}^{\infty} \left(\int_{-\infty}^{\infty} g(t)e^{-2\pi int} dt \ e^{2\pi inx} \right).$$

An interesting example of Poisson's formula connects the Poisson kernel for the upper half plane given by $g(x) = y/(x^2 + y^2)$, $y > 0$, with the kernel for the unit disk given by $(1 - r^2)/(1 - 2r \cos x + r^2)$. Observe that with g as defined here,

$$\hat{g}(t) = \int_{-\infty}^{\infty} \frac{ye^{-2\pi itx}}{x^2 + y^2} dx = \pi e^{-2\pi y|t|}.$$

The reader may verify this by contour integration or by other means. Substitution of this in Theorem D.4.1 gives

$$\sum_{n=-\infty}^{\infty} \frac{y}{y^2 + (x+n)^2} = \sum_{n=-\infty}^{\infty} e^{-2\pi y|n|} e^{2\pi inx} = \frac{1 - e^{-4\pi y}}{1 - 2e^{-2\pi y} \cos 2\pi x + e^{-4\pi y}}.$$

The partial fraction expansions for the trigonometric functions can also be derived from Theorem D.4.1. Take $s > 0$ and set

$$g(t) = \begin{cases} e^{-st}, & t \geq 0, \\ 0, & t < 0. \end{cases}$$

The left side of Poisson's formula becomes

$$\sum_{n>-x} e^{-s(n+x)} + \begin{cases} 1/2, & x \text{ integer}, \\ 0, & \text{otherwise}. \end{cases}$$

The right-hand side is given by

$$\sum_{n=-\infty}^{\infty} \frac{e^{2\pi inx}}{s + 2\pi in}.$$

When $x = 0$, we get

$$\frac{e^s + 1}{e^s - 1} = \frac{1}{s} + \sum_{n=1}^{\infty} \left(\frac{2}{s + 2\pi in} + \frac{1}{s - 2\pi in} \right).$$

By analytic continuation, we can take $s = 2\pi iy$. The result is

$$\pi \cot \pi y = \frac{1}{y} + \sum_{n=1}^{\infty} \left(\frac{1}{y+n} + \frac{1}{y-n} \right).$$

When $x = 1/2$, we obtain

$$\frac{e^{s/2}}{e^s - 1} = \frac{1}{s} + \sum_{n=1}^{\infty} \left(\frac{(-1)^n}{s + 2\pi in} + \frac{(-1)^n}{s - 2\pi in} \right).$$

This time, $s = 2\pi i y$ gives

$$\pi \csc \pi y = \frac{1}{y} + \sum_{n=1}^{\infty} (-1)^n \left(\frac{1}{y+n} + \frac{1}{y-n} \right).$$

An important transformation formula for theta functions follows from the Poisson summation formula by taking $g(x) = e^{-s\pi x^2}$. The formula is

$$\sum_{n=-\infty}^{\infty} e^{-s\pi(n+x)^2} = s^{-1/2} \sum_{n=-\infty}^{\infty} e^{-\pi n^2/s} e^{2\pi inx}. \qquad \text{(D.4.2)}$$

The Euler–Maclaurin formula follows from the Poisson summation formula by taking

$$g(x) = \begin{cases} f(x), & \text{for } a \le x \le b, \\ 0, & \text{otherwise.} \end{cases}$$

Here a and b are assumed to be nonnegative integers and $f(x)$ is continuously differentiable in $[a, b]$. This implies that f is of bounded variation on $[a, b]$ and hence $g(x)$ is of bounded total variation. Use this $g(x)$ in (D.4.1). The result is

$$\frac{1}{2}(f(a) + f(b)) + \sum_{k=a+1}^{b-1} f(k) = \lim_{N\to\infty} \sum_{|n|\le N} \int_a^b f(t) e^{-2\pi int} dt$$

$$= \int_a^b f(t)dt + 2\sum_{n=1}^{\infty} \int_a^b f(t) \cos 2\pi nt\, dt. \qquad \text{(D.4.3)}$$

Integration by parts gives

$$\int_a^b f(t) \cos 2\pi nt\, dt = -\int_a^b f'(t) \frac{\sin 2n\pi t}{2n\pi} dt.$$

Also,

$$2\sum_{n=1}^{\infty} \frac{\sin 2n\pi x}{2n\pi} = -B_1(x - [x]),$$

except when x is an integer (see Exercise 44 in Chapter 1). Thus, by the dominated convergence theorem, we can write (D.4.2) as

$$\sum_{a<n\le b} f(n) = \int_a^b f(t)dt - B_1(f(b) - f(a)) + \int_a^b f'(x) B_1(x - [x])dx.$$

If we assume that $f^{(m)}(x)$ is continuously differentiable, then the application of

$$\frac{d B_n(x)}{dx} = n B_{n-1}(x), \qquad 0 < x < 1,$$

and successive integration by parts gives Theorem D.2.1. For other applications and extensions of the summation formulas studied here, see Berndt [1975].

Exercises

1. Prove that with the notation and conditions of Theorem D.3.1 $\zeta'(s) = O(\log^2 |t|)$.

2. Define $\Delta f(0) = f(1) - f(0)$. Show that if $f(x)$ is a polynomial of degree m, then

$$f(0) = \int_0^1 f(x)dx + \sum_{j=1}^m \frac{B_j}{j} \Delta f^{(j-1)}(0).$$

3. Use the Euler–Maclaurin formula to obtain the relation

$$\sum_{k=1}^n \frac{1}{k} = \log n + \gamma + \frac{1}{2n} - \sum_{k=1}^{p-1} \frac{B_{2k}}{2kn^{2k}} - \frac{\theta B_{2p}}{2pn^{2p}},$$

where $0 < \theta < 1$ and γ is Euler's constant.

4. Show that there is a constant C such that

$$\sum_{2 \le n \le x} \frac{1}{n \log n} = \log \log x + C + O\left(\frac{1}{x \log x}\right).$$

5. Use the Poisson summation formula to prove that

$$\sum_{-\infty}^{\infty} e^{-s(n+x)^2} = \frac{1}{\sqrt{s}} \sum_{-\infty}^{\infty} e^{-\pi n^2/s} e^{2\pi inx}.$$

6. Use Poisson summation to obtain the following theorem of Lipschitz: Suppose $\operatorname{Re} x > 0, 0 < \alpha \le 1$, and $\operatorname{Re} s > 1$. Then

$$\frac{(2\pi)^s}{\Gamma(s)} \sum_{n=0}^{\infty} (n+\alpha)^{s-1} e^{-2\pi x(n+\alpha)} = \sum_{n=-\infty}^{\infty} \frac{e^{2\pi in\alpha}}{(x+ni)^s}.$$

(For a different proof where $\alpha = 1$, see Exercise 37 in Chapter 2.)

7. Suppose a, b, and c are integers such that $ac+b$ is even. Use Poisson summation to prove the reciprocity for Gaussian sums

$$\sum_{n=0}^{|c|-1} e^{\pi i(an^2+bn)/c} = \sqrt{\left|\frac{c}{a}\right|} \, e^{\pi i(|ac|-b^2)/(4ac)} \sum_{n=0}^{|a|-1} e^{-\pi i(cn^2+bn)/a}.$$

E

Lagrange Inversion Formula

E.1 Reversion of Series

There are situations in analysis in which one knows a series expansion for $y(x)$ but would like to obtain the series for x in terms of y. Newton, for example, encountered such a problem when he had the series for $\sin^{-1} x$ by integrating the expansion of $(1 - x^2)^{-1/2}$ term by term and he wanted the series for $\sin x$. Some results of Newton on reversion of series can be found in Newton [1960, p. 147].

Suppose a series for x in powers of y is required when $x = y\phi(x)$. Assume that ϕ is analytic in a neighborhood of $x = 0$ with $\phi(0) \neq 0$. Then

$$y = x/\phi(x) = \sum_{n=1}^{\infty} a_n x^n, \quad a_1 \neq 0. \tag{E.1.1}$$

We shall see below that Lagrange's inversion formula gives

$$x = \sum_{n=1}^{\infty} b_n y^n,$$

where

$$b_n = \frac{1}{n!} \frac{d^{n-1}}{dx^{n-1}} [\phi(x)^n]_{x=0}.$$

This means that $n b_n$ is the coefficient of x^{n-1} in the expansion of $\phi^n(x)$ or the coefficient of x^{-1} in the expansion of $1/y^n$.

More generally, suppose that (E.1.1) holds and that $f(x)$ is an analytic function in a neighborhood of $x = 0$. Then Lagrange's formula is

$$f(x) = f(0) + \sum_{n=1}^{\infty} \frac{y^n}{n!} \left[\frac{d^{n-1}}{dx^{n-1}} (f'(x)\phi^n(x)) \right]_{x=0}. \tag{E.1.2}$$

E.2 A Basic Lemma

The proof of Lagrange's formula (E.1.2) depends on a simple lemma given below. The treatment here follows that of Gessel and Stanton [1982], which is based on the work of Jacobi [1830]. See also Bromwich [1926]. An extension of Lagrange inversion to several variables is in Good [1960] and q-analogs are discussed in Stanton [1988]. A history of this formula can be found in W. Johnson's unpublished manuscript, *Notes on the Lagrange inversion formula*.

Define the residue of a formal Laurent series $f(x) = \sum_{j=-m}^{\infty} a_j x^j$ by $\mathrm{Res}[f(x)] = a_{-1}$. The next lemma states that the residue does not change under a certain change of variables.

Lemma E.2.1   *Let* $G(x) = \sum_{j=-m}^{\infty} a_j x^j$ *and* $h(x) = \sum_{i=1}^{\infty} b_i x^i$, *where* $b_1 \neq 0$. *Then* $\mathrm{Res}[G(h(x))h'(x)] = \mathrm{Res}[G(x)]$.

Proof.   Since both sides of the equation are linear in G, it is sufficient to prove it for $G(x) = x^m$. Since $\mathrm{Res}[g'(x)] = 0$ for any Laurent series $g(x)$, it follows that, for $m \neq -1$,

$$\mathrm{Res}[h^m(x)h'(x)] = \frac{1}{m+1} \mathrm{Res}[\{h^{m+1}(x)\}'] = 0.$$

For $m = -1$, let $h(x) = b_1 x f(x)$. Then

$$\mathrm{Res}\left[\frac{h'(x)}{h(x)}\right] = \mathrm{Res}\left[\frac{1}{x} + \frac{f'(x)}{f(x)}\right] = 1 + \mathrm{Res}[\{\log f(x)\}'] = 1.$$

The lemma is proved.   ∎

To prove (E.1.2) note that $y = a_1 x + a_2 x^2 + \cdots$ (x a complex variable) is conformal in a neighborhood of 0, because $a_1 \neq 0$, and so x can be expanded in powers of y. Since $f(x)$ is analytic, it can be expanded in powers of y. Suppose that

$$f(x) = f(0) + \sum_{n=1}^{\infty} c_n y^n.$$

Then

$$f'(x) = \sum_{n=1}^{\infty} n c_n y^{n-1} \frac{dy}{dx}.$$

Let $G(x) = \sum_{n=1}^{\infty} n c_n x^{n-1}$. By Lemma E.2.1,

$$\mathrm{Res}\left[\frac{f'(x)\phi^r(x)}{x^r}\right] = \mathrm{Res}\left[\frac{f'(x)}{y^r}\right] = \mathrm{Res}\left[\frac{G(y)dy/dx}{y^r}\right] = \mathrm{Res}\left[\frac{G(x)}{x^r}\right] = r c_r.$$

But the Taylor series of $f'(x)\phi^r(x)$ is

$$\sum_{n=0}^{\infty} \frac{x^n}{n!} \left[\frac{d^n}{dx^n} (f'(x)\phi^r(x)) \right]_{x=0},$$

and so

$$\text{Res}\left[\frac{f'(x)\phi^r(x)}{x^r} \right] = \frac{1}{(r-1)!} \left[\frac{d^{r-1}}{dx^{r-1}} (f'(x)\phi^r(x)) \right]_{x=0}.$$

Thus

$$c_r = \frac{1}{r!} \left[\frac{d^{r-1}}{dx^{r-1}} (f'(x)\phi^r(x)) \right]_{x=0},$$

which proves Lagrange's formula (E.1.2).

A variant of (E.1.2) is the formula

$$\frac{f(x)}{1 - y\phi'(x)} = \sum_{n=0}^{\infty} \frac{y^n}{n!} \left[\frac{d^n}{dx^n} (f(x)\phi^n(x)) \right]_{x=0}, \tag{E.2.1}$$

where y is defined by (E.1.1). To prove (E.2.1), note that since $x - y\phi(x) = 0$, we have

$$1 - \phi(x) \frac{dy}{dx} - y\phi'(x) = 0,$$

or

$$\frac{dy}{dx} = \frac{1 - y\phi'(x)}{\phi(x)}. \tag{E.2.2}$$

Take the derivative of (E.1.2) with respect to x and use (E.2.2) to get

$$\frac{f'(x)\phi(x)}{1 - y\phi'(x)} = \sum_{n=0}^{\infty} \frac{y^n}{n!} \left[\frac{d^n}{dx^n} (f'(x)\phi(x)\phi^n(x)) \right]_{x=0}.$$

Set $F(x) = f'(x)\phi(x)$ to obtain (E.2.1).

E.3 Lambert's Identity

Lagrange's inversion formula can be used to obtain a number of interesting series expansions for specific functions. A few examples are given in the exercises. Here we derive an identity of Lambert. This is given by

$$(1 + x)^\alpha = 1 + \sum_{n=1}^{\infty} \frac{\alpha}{\alpha + n\beta} \binom{\alpha + \beta n}{n} y^n, \tag{E.3.1}$$

where $y = x(1 + x)^{-\beta}$. Take $f(x) = (1 + x)^\alpha$ and $\phi(x) = (1 + x)^\beta$ in (E.1.2). Then

$$(1 + x)^\alpha = 1 + \sum_{n=1}^{\infty} \frac{y^n}{n!} \left[\frac{d^{n-1}}{dx^{n-1}} (\alpha(1 + x)^{n\beta + \alpha - 1}) \right]_{x=0}.$$

The expression in square brackets equals

$$\alpha(n\beta + \alpha - 1)\cdots(n\beta + \alpha - n + 1) = \alpha(n-1)!\binom{\alpha + n\beta - 1}{n - 1},$$

and this gives (E.3.1). The identity corresponding to (E.2.1) is

$$\frac{(1 + x)^{\alpha+1}}{1 + x(1 - \beta)} = 1 + \sum_{n=1}^{\infty}\binom{\alpha + \beta n}{n}y^{n}. \tag{E.3.2}$$

An interesting binomial identity comes from (E.3.1) and (E.3.2). Equate the coefficient of y^n on both sides of

$$(1 + x)^{\gamma}\frac{(1 + x)^{\alpha+1}}{1 + x(1 - \beta)} = \frac{(1 + x)^{\alpha+\gamma+1}}{1 + x(1 - \beta)}.$$

The result is

$$\sum_{k=0}^{n}\frac{\gamma}{\gamma + \beta k}\binom{\gamma + \beta k}{k}\binom{\alpha + \beta(n - k)}{n - k} = \binom{\alpha + \gamma + \beta n}{n}. \tag{E.3.3}$$

E.4 Whipple's Transformation

Gessel and Stanton showed how to derive Whipple's transformation, which we obtained in different ways in Chapters 3 and 7, from Lemma E.2.1.

Theorem E.4.1 *Suppose $A(x)$, $B(x)$, $C(x)$, and $D(x)$ are power series whose coefficients of x^k are A_k, B_k, C_k, and D_k respectively. Suppose*

$$(1 - x)^{-\alpha}A(x/(1 - x)^{\beta+1}) = B(x), \tag{E.4.1}$$

$$(1 + \beta x)(1 - x)^{-\gamma}C(x/(1 - x)^{\beta+1}) = D(x). \tag{E.4.2}$$

If $n(\beta + 1) = 1 - \alpha - \gamma$, then

$$\sum_{k=0}^{n}B_k D_{n-k} = \sum_{k=0}^{n}A_k C_{n-k}.$$

Proof. First note that

$$\mathrm{Res}\left[\frac{B(x)D(x)}{x^{n+1}}\right] = \text{coefficient of } x^n \text{ in } B(x)D(x) = \sum_{k=0}^{n}B_k D_{n-k}.$$

Now write $h(x) = x(1 - x)^{-\beta-1}$ so that $h'(x) = (1 + \beta x)(1 - x)^{-\beta-2}$ and

$$\begin{aligned}
\frac{B(x)D(x)}{x^{n+1}} &= \frac{(1 + \beta x)(1 - x)^{-\alpha-\gamma}A(x/(1 - x)^{\beta+1})C(x/(1 - x)^{\beta+1})}{x^{n+1}} \\
&= \frac{(1 + \beta x)(1 - x)^{n(\beta+1)-1}A(h(x))C(h(x))}{x^{n+1}} \\
&= \frac{A(h(x))C(h(x))h'(x)}{[h(x)]^{n+1}}.
\end{aligned}$$

By Lemma E.2.1,

$$\text{Res}\left[\frac{B(x)D(x)}{x^{n+1}}\right] = \text{Res}\left[\frac{A(x)C(x)}{x^{n+1}}\right].$$

By the remark at the beginning of the proof, this implies that

$$\sum_{k=0}^{n} B_k D_{n-k} = \sum_{k=0}^{n} A_k C_{n-k}.$$

This proves the theorem. ∎

To derive Whipple's transformation for a $_7F_6$, recall Whipple's quadratic transformation noted in Chapter 2, namely,

$$(1-x)^{-a}{}_3F_2\left(\begin{matrix} a/2, (1+a)/2, 1+a-b-c \\ 1+a-b, 1+a-c \end{matrix}; -\frac{4x}{(1-x)^2}\right)$$

$$= {}_3F_2\left(\begin{matrix} a, b, c \\ 1+a-b, 1+a-c \end{matrix}; x\right). \tag{E.4.3}$$

This corresponds to (E.4.1) with $\alpha = a$ and $\beta = 1$. The transformation corresponding to (E.4.3) is a result of Bailey [1929]:

$$(1+x)(1-x)^{-a-1}{}_3F_2\left[\begin{matrix} (a+1)/2, 1+a/2, a+1-b-c \\ 1+a-b, 1+a-c \end{matrix}; -\frac{4x}{(1-x)^2}\right]$$

$$= {}_4F_3\left[\begin{matrix} a, 1+a/2, b, c \\ a/2, 1+a-b, 1+a-c \end{matrix}; x\right]. \tag{E.4.4}$$

This can be obtained from (E.4.3) by differentiation or directly by equating the coefficients of x^n. To apply Theorem E.4.1, note that from (E.4.3),

$$A_k = (a/2)_k((1+a)/2)_k(1+a-b-c)_k(-4)^k/[k!(1+a-b)_k(1+a-c)_k],$$
$$B_k = (a)_k(b)_k(c)_k/[k!(1+a-b)_k(1+a-c)_k],$$

and from (E.4.4) after renaming the parameters,

$$C_k = ((1+d)/2)_k((2+d)/2)_k(1+d-e-f)_k(-4)^k/$$
$$[k!(1+d-e)_k(1+d-f)_k],$$
$$D_k = (d)_k(1+d/2)_k(e)_k(f)_k/[k!\,(d/2)_k(1+d-e)_k(1+d-f)_k].$$

So

$$\sum_{k=0}^{n} A_k C_{n-k}$$

$$= \frac{(-4)^n((1+d)/2)_n((2+d)/2)_n(1+d-e-f)_n}{n!(1+d-e)_n(1+D-f)_n}$$

$$\cdot \sum_{k=0}^{n} \frac{(a/2)_k((1+a)/2)_k(1+a-b-c)_k(-n)_k(e-d-n)_k(f-d-n)_k}{k!(1+a-b)_k(1+a-c)_k((1-d)/2-n)_k(-n-d/2)_k(e+f-n-d)_k}.$$

Theorem E.4.1 is applicable when $2n = -a - d$. This simplifies the expression in the sum. For example, $-n - d/2 = a/2$, and the terms involving these expressions cancel. After simplification the sum becomes

$$_4F_3\left(\begin{matrix} -n, 1+a-b-c, e+a+n, f+a+n \\ 1+a-b, 1+a-c, e+f+a+n \end{matrix}; 1\right).$$

The sum $\sum_{k=0}^n B_k D_{n-k}$ leads to a $_7F_6$ and the equality

$$\sum_{k=0}^n B_k D_{n-k} = \sum_{k=0}^n A_k C_{n-k}$$

results in Whipple's theorem

$$_7F_6\left(\begin{matrix} a, 1+a/2, b, c, e+a+n, f+a+n, -n \\ a/2, 1+a-b, 1+a-c, 1-e-n, 1-f-n, 1+a+n \end{matrix}; 1\right)$$

$$= \frac{(a+1)_n(a+c+f+n)_n}{(e)_n(f)_n}$$

$$\times {}_4F_3\left(\begin{matrix} 1+a-b-c, e+a+n, f+a+n, -n \\ 1+a-b, 1+a-c, e+f+a+n \end{matrix}; 1\right).$$

Exercises

1. Show that

$$e^{\alpha x} = \sum_{n=0}^{\infty} \frac{\alpha(\alpha+bn)^{n-1}}{n!}(xe^{-bx})^n$$

and that

$$x = \sum_{n=1}^{\infty} \frac{n^{n-1}}{n!}(xe^{-x})^n.$$

2. Show formally that

$$f(x+a) = \sum_{n=0}^{\infty} \frac{f^{(n)}(a-nb)}{n!} x(x+nb)^{n-1}.$$

Give sufficient conditions for this to be correct.

3. Show that

$$(a+x)^{\alpha} = \sum_{k=0}^{\infty} x(x-kb)^{k-1} \frac{(-\alpha)_k}{k!}(-1)^k(a+kb)^{\alpha-k}.$$

4. Show that if $f^{-1}(x)$ is the inverse of $f(x)$ and $f(0) = 0$, then assuming the necessary analyticity of the functions,

$$f^{-1}(x) = x\left(\frac{x}{f(x)}\right)_{x=0} + \frac{x^2}{2!}\left(\frac{d}{dx}\left(\frac{x^2}{f(x)}\right)\right)_{x=0} + \cdots.$$

5. Show that

$$\log(1+x) = \sum_{n=1}^{\infty} \binom{n\beta-1}{n-1} \frac{y^n}{n}$$

when $y = x(1+x)^{-\beta}$.

6. Suppose that $x^{m+1} + ax - b = 0$. Show that

$$x = \frac{b}{a} - \frac{b^{m+1}}{a^{m+2}} + \frac{2m+2}{2!} \frac{b^{2m+1}}{a^{2m+3}} - \frac{(3m+2)(3m+3)}{3!} \frac{b^{3m+1}}{a^{3m+4}} + \cdots.$$

Use this formula to find a solution of $x^5 + 4x + 2 = 0$ to four decimal places of accuracy. When $m = 0$, this series reduces to the geometric series. Write this sum as a hypergeometric series.

7. Use the Lagrange inversion formula to derive the generating function for Laguerre polynomials.

F

Series Solutions of Differential Equations

F.1 Ordinary Points

For readers not familiar with series solutions of differential equations, we give a few basic definitions so that the discussions of the hypergeometric equation in Chapter 2 and of the confluent and Bessel's equations in Chapter 4 are intelligible.

Consider the differential equation

$$a(x)\frac{d^2y}{dx^2} + b(x)\frac{dy}{dx} + c(x)y = 0 \tag{F.1.1}$$

with $a(x)$, $b(x)$, and $c(x)$ analytic in the neighborhood of $x = x_0$. We take x to be a complex variable in this discussion. The simplest situation occurs when $a(x_0) \neq 0$. In this case, x_0 is called an ordinary point of the Equation (F.1.1).

It is not difficult to show that if x_0 is an ordinary point of (F.1.1), then (F.1.1) has a unique solution $f(x)$ analytic in a neighborhood of x_0 with prescribed values $f(x_0) = f_0$ and $f'(x_0) = f_1$. This implies that there are exactly two linearly independent solutions in a neighborhood of x_0. To prove this result, it is convenient to divide (F.1.1) by $a(x)$ and rewrite the equation as

$$y'' = B(x)y' + C(x)y. \tag{F.1.2}$$

Again for convenience, take $x_0 = 0$. Then

$$B(x) = \sum_{n=0}^{\infty} b_n x^n \quad \text{and} \quad C(x) = \sum_{n=0}^{\infty} c_n x^n.$$

Suppose that (F.1.2) has an analytic solution $f(x)$ with

$$f(x_0) = f_0, \quad f'(x_0) = f_1.$$

Then

$$f(x) = \sum_{n=0}^{\infty} f_n x^n. \tag{F.1.3}$$

Substitute this series and the series for $B(x)$ and $C(x)$ in (F.1.2); equate the coefficients of x^n. The result is

$$(n+2)(n+1)f_{n+2} = \sum_{k=0}^{n}(n+1-k)f_{n+1-k}b_k + \sum_{k=0}^{n}f_{n-k}c_k, \qquad \text{(F.1.4)}$$

$n = 0, 1, 2, \ldots$. This shows that f_0 and f_1 uniquely determine f_n for $n \geq 2$. It is now enough to prove that the series (F.1.3) has a positive radius of convergence. Since the series for $B(x)$ and $C(x)$ have a positive radius of convergence, there exist constants M and R such that

$$|b_n| \leq M/R^n \quad \text{and} \quad |c_n| \leq M/R^n.$$

We show by induction that there exist suitable positive numbers M_1 and r such that

$$|f_n| \leq M_1/r^n. \qquad \text{(F.1.5)}$$

This implies the needed result. Take M_1 such that $|f_1| \leq M_1$ and choose r such that $r < R, |f_1| \leq M_1/r$, and $M(r/2 + r^2) \leq 1$. Suppose $n \geq 2$ and assume (F.1.5) true up to $n - 1$. By (F.1.4), the inequality (F.1.5) follows after a small calculation.

F.2 Singular Points

When $a(x_0) = 0$ and $b(x_0)$ and/or $c(x_0)$ is not zero, then $x = x_0$ is called a singular point of (F.1.1). Divide (F.1.1) by $a(x)$ and write it as

$$y'' + d(x)y' + e(x)y = 0. \qquad \text{(F.2.1)}$$

Consider first the simplest case where x_0 is a singular point and $a(x)$ has a simple zero at x_0. Then $d(x)$ and $e(x)$ have at most simple poles. Take $x_0 = 0$. Then

$$d(x) = \sum_{n=-1}^{\infty} d_n x^n \quad \text{and} \quad e(x) = \sum_{n=-1}^{\infty} e_n x^n. \qquad \text{(F.2.2)}$$

Substitute the series (F.1.3) in (F.2.1) and equate coefficients as before. The reader may check that in this case we get only one solution, since the value of f_0 determines f_n for $n \geq 1$. Moreover, if d_{-1} is a nonnegative integer, then this method may fail to produce a solution.

Now consider the case where $d(x)$ has at most a simple pole and $e(x)$ at most a double pole. The simplest special case of (F.1.1) that leads to this is when

$$a(x) = a_2(x - x_0)^2, \quad b(x) = b_1(x - x_0), \quad c(x) = c_0. \qquad \text{(F.2.3)}$$

Two linearly independent solutions exist, at least one of the form $y = (x - x_0)^\mu$. To determine μ, substitute this expression for y in (F.1.1) to get

$$a_2\mu(\mu - 1) + b_1\mu + c_0 = 0. \tag{F.2.4}$$

If this quadratic has two unequal roots, these roots give two independent solutions of (F.1.1). If the two roots equal μ_1 (say), then we have the solution $(x - x_0)^{\mu_1}$. To find the other independent solution, set $y = (x - x_0)^{\mu_1} w$. The differential equation for w has $\log(x - x_0)$ as a solution. Thus the second independent solution is $(x - x_0)^{\mu_1} \log(x - x_0)$.

F.3 Regular Singular Points

A point $x = x_0$ is called a regular singular point of (F.1.1) if

$$\lim_{x \to x_0} \frac{(x - x_0)b(x)}{a(x)} \quad \text{and} \quad \lim_{x \to x_0} \frac{(x - x_0)^2 c(x)}{a(x)}$$

both exist. If one of these limits does not exist, the singular point is irregular.

To see the difference between a regular and an irregular singular point, the easiest case to consider is the first-order analog of (F.1.1). A regular singular point occurs when

$$b(x)y' + c(x)y = 0 \tag{F.3.1}$$

with $b(x)$ and $c(x)$ analytic in a neighborhood of $x = x_0$, and

$$\lim_{x \to x_0} (x - x_0)c(x)/b(x)$$

exists.

When $b(x) = (x - x_0)^2$ and $c(x) = c_0$, then

$$y(x) = Ae^{-c_0/(x-x_0)}$$

is the solution, and it has an essential singularity at $x = x_0$.

The simple case considered in the previous section, where $a(x), b(x), c(x)$ are as in (F.2.3), shows that, at a regular singular point, solutions may involve noninteger powers of $x - x_0$. Moreover, the second solution may have a logarithmic singularity.

Now suppose that x_0 is a regular singular point of (F.1.1) such that $a(x)$ has a double zero at x_0. Then

$$a(x) = \sum_{n=2}^{\infty} a_n(x - x_0)^n, \quad b(x) = \sum_{n=1}^{\infty} b_n(x - x_0)^n, \quad \text{and} \quad c(x) = \sum_{n=0}^{\infty} c_n(x - x_0)^n.$$

Suppose $y = (x - x_0)^\mu f(x)$ is a solution of (F.1.1). The equation for $f(x)$ can be seen to be

$$a(x) f'' + \left(2\mu \sum_{n=1}^{\infty} a_{n+1}(x - x_0)^n + b(x) \right) f'$$

$$+ \left(c(x) + \mu \sum_{n=0}^{\infty} b_{n+1}(x - x_0)^n + \mu(\mu - 1) \sum_{n=0}^{\infty} a_{n+2}(x - x_0)^n \right) f = 0.$$

$$\text{(F.3.2)}$$

Now if μ is chosen so that

$$\mu(\mu - 1)a_2 + \mu b_1 + c_0 = 0, \tag{F.3.3}$$

then (F.3.1) is an equation of the same type as the one considered at the beginning of Section F.2. In that case a solution of the form $f(x) = \sum_{n=0}^{\infty} f_n(x - x_0)^n$ is always possible provided $2\mu + b_1/a_2$ is not a nonnegative integer.

Note that the quadratic (F.3.2) has two solutions μ_1 and μ_2. Thus it may be possible to obtain two solutions of the form $(x - x_0)^\mu f(x)$. It follows from (F.3.2) that

$$\mu_1 - \mu_2 = 2\mu_1 - (\mu_1 + \mu_2)$$
$$= 2\mu_1 - (1 - b_1/a_2)$$
$$= 2\mu_1 + b_1/a_2 - 1.$$

The remarks in the previous paragraph now imply that if the difference of the two roots of (F.3.2) is not an integer, then (F.1.1) has two independent solutions of the form $\sum_{n=0}^{\infty} f_n(x - x_0)^{n+\mu}$. However, if $\mu_1 - \mu_2 \geq 0$ is an integer, then a solution of this form with $\mu = \mu_1$ exists. The other solution may involve a logarithmic singularity.

Equation (F.3.2) is called the indicial equation. In practice it is obtained by substituting the series $\Sigma f_n(x - x_0)^{n+\mu}$ in (F.1.1) and equating the coefficients of the lowest power of $x - x_0$.

For the hypergeometric equation (2.3.5), there are three regular singular points 0, 1, and infinity. For the confluent equation (4.1.3) and the Bessel equation (4.5.1), 0 is a regular singular point, and infinity is an irregular singular point.

Bibliography

Abel, N. H. [1826]. Auflösung einer mechanischen Aufgabe, *J. Reine Ang. Math.*, **1**, 153–157. English translation in *Source Book in Mathematics*, D. E. Smith, ed., McGraw-Hill, New York, 656–662.

Adams, J. C. [1877]. On the expression of the product of any two Legendre coefficients by means of a series of Legendre's coefficients, *Proc. R. Soc. London*, **27**, 63–71.

Agarwal, R. P. [1953]. On the partial sums of series of hypergeometric type, *Proc. Cambridge Phil. Soc.*, **49**, 441–445.

Ahern, P. and Rudin, W. [1996]. Geometric properties of the gamma function, *Am. Math. Monthly*, **103**, 678–681.

Al-Salam, W. A. and Ismail, M. E. H. [1988]. q-Beta integrals and the q-Hermite polynomials, *Pac. J. Math.*, 209–221.

Amend, B. [1996]. Fox Trot, Comic strip, June 2.

Anastassiadis, J. [1964]. *Définition des Fonctions Eulériennes par des Équations Functionnelles*, Gauthier-Villars, Paris.

Anderson, G. W. [1990]. The evaluation of Selberg sums, *Comp. Rend. Acad. Sci., Paris*, **311**, Série 1, 469–472.

Anderson, G. W. [1991]. A short proof of Selberg's generalized beta formula, *Forum Math.*, **3**, 415–417.

Andrews, G. E. [1976]. *The Theory of Partitions*, Addison-Wesley, Reading, MA.

Andrews, G. E. [1986]. *q-Series: Their Development and Application in Analysis, Number Theory, Combinatorics, Physics, and Computer Algebra*, Amer. Math. Soc., Providence, RI.

Andrews, G. E. [1994]. The death of proof? Semi-rigorous mathematics? You've got to be kidding!, *Math. Intelligencer*, **16** (4), 16–18.

Andrews, G. E. and Askey, R. [1977]. Enumeration of partitions: The role of Eulerian series and q-orthogonal polynomials, in *Higher Combinatorics*, M. Aigner, ed., Reidel, Dordrecht, 3–26.

Andrews, G. E. and Burge, W. H. [1993]. Determinant identities, *Pac. J. Math.*, **158**, 1–14.

Andrews, G. E., Dyson, F. J., and Hickerson, D. [1988]. Partitions and indefinite quadratic forms, *Invent. Math.*, **91**, 391–407.

Anno, M. and Mori, T. [1986]. *Socrates and the Three Little Pigs*, Philomel Books, New York.

Aomoto, K. [1987]. Jacobi polynomials associated with Selberg integrals, *SIAM J. Math. Anal.*, **18**, 545–549.

Artin, E. [1964]. *The Gamma Function*, Holt, Rinehart and Winston, New York.

Askey, R. [1970]. Orthogonal polynomials and positivity, *Studies in Applied Mathematics* 6, in *Wave Propagation and Special Functions*, D. Ludwig and F. W. J. Olver, eds., SIAM, Philadelphia, 68–85.

Askey, R. [1975]. *Orthogonal Polynomials and Special Functions*, SIAM, Philadelphia.

Askey, R. [1987]. Ramanujan's $_1\psi_1$ and formal Laurent series, *Indian J. Math.*, **19**, 101–105.

Askey, R. [1988]. Beta integrals and q-extensions, *Papers of the Ramanujan Centennial International Conference, Ramanujan Math. Soc.*, 1987, 85–102.

Askey, R. [1994]. A look at the Bateman project, *Contemporary Math.*, **169**, 29–43.

Askey, R. and Gasper, G. [1971]. Jacobi polynomial expansions of Jacobi polynomials with non-negative coefficients, *Proc. Cambridge Phil. Soc.*, **70**, 243–255.

Askey, R. and Gasper, G. [1977]. Convolution structures for Laguerre polynomials, *J. Analyse Math.*, **31**, 48–68.

Askey, R., Gasper, G., and Ismail, M. [1975]. A positive sum from summability theory, *J. Approx. Theory*, **13**, 413–420.

Askey, R., Koornwinder, T., and Rahman, M. [1986]. An integral of products of ultraspherical functions and a q-extension, *J. London Math. Soc.*, **33** (2), 133–148.

Askey, R. and Steinig, J. [1974]. Some positive trigonometric sums, *Trans. Am. Math. Soc.*, **187**, 295–307.

Askey, R. and Wilson, J. A. [1984]. A recurrence relation generalizing those of Apéry, *J. Austral. Math. Soc.* (Series A) **36**, 267–278.

Askey, R. and Wilson, J. A. [1985]. *Some Basic Hypergeometric Orthogonal Polynomials that Generalize Jacobi Polynomials*, Am. Math. Soc., Providence, RI.

Auslander, L. and Tolimieri, R. [1979]. Is computing with finite Fourier transform pure or applied mathematics?, *Bull. Am. Math. Soc.* (New Series), **11**, 847–897.

Azor, R., Gillis, J., and Victor, J. D. [1982]. Combinatorial applications of Hermite polynomials, *SIAM J. Math. Anal.*, **13**, 879–890.

Baernstein II, A., Drasin, D., Duren, P., and Marden, A. [1986]. *The Bieberbach Conjecture*, Amer. Math. Soc. Surveys and Monographs No. 21, Providence, RI.

Bailey, W. N. [1929]. Some identities involving generalized hypergeometric series, *Proc. London Math. Soc.*, **26** (2), 503–516.

Bailey, W. N. [1931]. The partial sum of the coefficients of the hypergeometric series, *J. London Math. Soc.*, **6**, 40–41.

Bailey, W. N. [1932]. On one of Ramanujan's theorems, *J. London Math. Soc.*, **7**, 34–36.

Bailey, W. N. [1933]. On the product of two Legendre polynomials, *Proc. Cambridge Phil. Soc.*, **29**, 173–177.

Bailey, W. N. [1935]. *Generalized Hypergeometric Series*, Cambridge University Press, Reprinted by Hafner Pub. Co., New York, 1972.

Bailey, W. N. [1938]. The generating function of Jacobi polynomials, *J. London Math. Soc.*, **13**, 8–11.

Bailey, W. N. [1941]. A note on certain q-identities, *Quart. J. Math.* (Oxford), **18**, 157–166.

Bailey, W. N. [1949]. Identities of the Rogers–Ramanujan type, *Proc. London Math. Soc.*, **50** (2), 1–10.

Bailey, W. N. [1954]. Contiguous hypergeometric functions of the type $_3F_2(1)$, *Proc. Glasgow Math. Assoc.*, **2**, 62–65.

Bak, J. and Newman, D. J. [1982]. *Complex Analysis*, Springer-Verlag, New York.

Barnes, E. W. [1908]. A new development of the theory of the hypergeometric functions, *Proc. London Math. Soc.*, **6** (2), 141–177.

Barnes, E. W. [1910]. A transformation of generalized hypergeometric series, *Quart. J. Math.*, **41**, 136–140.

Bateman, H. [1909]. The solution of linear differential equations by means of definite integrals, *Trans. Cambridge Phil. Soc.*, **21**, 171–196.

Bateman, H. [1932]. *Partial Differential Equations of Mathematical Physics*, Cambridge University Press, Cambridge, UK.

Bellman, R. [1961]. *A Brief Introduction to Theta Functions*, Holt, Rinehart, and Winston, Inc., New York.

Berggren, L., Borwein, J., and Borwein, P. [1997]. *Pi: A Source Book*, Springer-Verlag, New York.

Berndt, B. [1975]. Periodic Bernoulli numbers, summation formulas and applications, in *Theory and Application of Special Functions*, R. Askey, ed., Academic Press, New York.

Berndt, B. [1985]. *Ramanujan's Notebooks, Part I*, Springer-Verlag, New York.

Berndt, B. [1989]. *Ramanujan's Notebooks, Part II*, Springer-Verlag, New York.

Beukers, F. [1979]. A note on the irrationality of $\zeta(2)$ and $\zeta(3)$, *Bull. London Math. Soc.*, **11**, 2268–2272.

Boas, R. P. [1954]. *Entire Functions*, Academic Press, New York.

Bochner, S. [1952]. Remarks on Gaussian sums and Tauberian theorems, *J. Indian Math. Soc.*, **15**, 99–106.

Bochner, S. [1955]. *Harmonic Analysis and the Theory of Probability*, University of California Press, Berkeley.

Bochner, S. [1979]. Review of Gesammelte Schriften by Gustav Herglotz, *Bull. Am. Math. Soc.* (New Series), **1**, 1020–1022.

Bohr, H. and Mollerup, J. [1922]. *Laerebog i Matematisk Analyse*, Vol. III, J. Gjellerup, Kopenhagen.

Borwein, J. and Borwein, P. [1987]. *Pi and the AGM*, Wiley, New York.

Boyarsky, M. [1980]. p-adic gamma functions and Dwork cohomology, *Trans. Am. Math. Soc.*, **257**, 359–369.

Brafman, F. [1951]. Generating functions of Jacobi and related polynomials, *Proc. Am. Math. Soc.*, **2**, 924–949.

Brent, R. P. [1976]. Fast multiple-precision evaluation of elementary functions, *J. Assoc. Comp. Mech.*, **23**, 242–251.

Bressoud, D. M. [1987]. Almost poised basic hypergeometric series, *Proc. Indian Acad. Sci.* (Math./Sci.), **97**, 61–66.

Bromwich, T. J. I'A. [1926]. *An Introduction to the Theory of Infinite Series*, 2nd ed., Macmillan, London.

Brown, G. and Hewitt, E. [1984]. A class of positive trigonometric sums, *Math. Ann.*, **268**, 91–122.

Brown, M. and Ismail, M. E. H. [1995]. A right inverse of the Askey–Wilson operator, *Proc. Amer. Math. Soc.*, **123**, 2071–2079.

Brualdi, R. A. and Ryser, H. J. [1991]. *Combinatorial Matrix Theory*, Cambridge University Press, New York.

Bustoz, J. [1974]. Note on "Positive Cesàro means of numerical series," *Proc. Amer. Math. Soc.*, **45**, 69.

Caratheodory, C. [1954]. *Funktionentheorie*, English translation, *Theory of Functions*, Vols. I and II, [1958, 1960], Chelsea, New York.

Cartier, P. and Foata, D. [1969]. *Problémes Combinatoires de Commutation et Réarrangements*, Lecture Notes in Math., No. 85, Springer-Verlag, Berlin.

Cauchy, A.-L. [1843]. Mémoire sur les fonctions dont plusieurs valeurs . . . , *C.R. Acad. Sci. Paris*, **17**, 523, reprinted in *Oeuvres de Cauchy*, [1893], Ser. 1, **8**, 42–50.

Cauchy, A.-L. [1843a]. Deuxième Mémoir sur les fonctions dont plusieurs valeurs . . . , *C.R. Acad. Sci. Paris*, reprinted in *Oeuvres de Cauchy*, [1893], **8**, 50–55.

Chebyshev, P. L. [1870]. Sur les fonctions analogues a celles de Legendre, in *Oeuvres de P. L. Tchebychef*, Vol. 2, A. Markoff and N. Sonin, eds., St. Petersburg [1907], 61–68, reprinted Chelsea, New York [1961].

Chihara, J. S. [1978]. *An Introduction to Orthogonal Polynomials*, Gordon & Breach, New York.

Clausen, T. [1828]. Ueber die Fälle, wenn die Reihe von der Form $y = 1 + \frac{\alpha}{1} \cdot \frac{\beta}{\gamma} x + \frac{\alpha \cdot \alpha + 1}{1 \cdot 2}$ $\cdot \frac{\beta \cdot \beta + 1}{\gamma \cdot \gamma + 1} x^2+$ etc. ein quadrat von der Form $z = 1 + \frac{\alpha'}{1} \cdot \frac{\beta'}{\gamma'} \cdot \frac{\delta'}{\epsilon'} x + \frac{\alpha' \cdot \alpha' + 1}{1 \cdot 2} \cdot \frac{\beta' \cdot \beta' + 1}{\gamma' \cdot \gamma' + 1}$ $\cdot \frac{\delta' \cdot \delta' + 1}{\epsilon' \cdot \epsilon' + 1} x^2+$ etc. hat, *J. Reine Ang. Math.*, **3**, 89–91.

Cooley, J. W. and Tukey, J. W. [1965]. An algorithm for the machine calculation of complex Fourier series, *Math. Comp.*, **19**, 297–301.

Copson, E. T. [1935]. *Theory of Functions of a Complex Variable*, Oxford Univ. Press, London.

Cox, D. [1984]. The arithmetic-geometric mean of Gauss, *Enseign. Math.*, **30**, 270–330.

Davenport, H. and Hasse, H. [1934]. Die Nullstellen der Kongruenzzetafunktion in gewissen zyklischen Fällen, *J. Reine Ang. Math.*, **172**, 151–182.

de Boor, C. [1980]. FFT as nested multiplication, with a twist, *SIAM J. Sci. Stat. Comp.*, **1**, 173–178.

de Branges, L. [1972]. Gauss spaces of entire functions, *J. Math. Anal. Appl.*, **37**, 1–41.

de Branges, L. [1972a]. Tensor product spaces, *J. Math. Anal. App.*, **38**, 109–148.

de Branges, L. [1985]. A proof of the Bieberbach conjecture, *Acta Math.*, **154**, 137–152.

de Bruijn, N. G. [1967]. Uncertainty principles in Fourier analysis, in *Inequalities*, O. Shisha, ed., Academic Press, New York.

de Bruijn, N. G., Saff, E. B., and Varga, R. S. [1981]. On the zeros of generalized Bessel polynomials, *Proc. K. Ned. Akad. Wet.*, Series A, **84** (1), 1–25.

de Moivre, A. [1730]. *Miscellanea Analytica de Seriebus et Quadraturis*, Tonson and Watts, London.

Dedekind, R. [1853]. Über ein Eulersches Integral, *J. Reine Ang. Math.*

DeSainte-Catherine, M. and Viennot, G. [1983]. Combinatorial interpretation of integrals of products of Hermite, Laguerre and Tchebycheff polynomials, *Polynômes Orthogonaux et Applications*, C. Brezinski et al., eds., Lecture Notes in Math. 1171, Springer-Verlag, 120–128.

Din, A. M. [1981]. *Lett. Math. Phys.*, **5**, 207.

Dixon, A. C. [1903]. Summation of a certain series, *Proc. London Math. Soc.*, **35** (1), 285–289.

Dougall, J. [1907]. On Vandermonde's theorem and some more general expansions, *Proc. Edinburgh Math. Soc.*, **25**, 114–132.

Dougall, J. [1919]. A theorem of Sonine in Bessel functions, with two extensions to spherical harmonics, *Proc. Edinburgh Math. Soc.*, **37**, 33–47.

Edwards, C. H. [1979]. *The Historical Development of Calculus*, Springer-Verlag, New York.

Edwards, A. W. F. [1987]. *Pascal's Arithmetical Triangle*, Charles Griffin and Co., London.

Eisenstein, G. [1847]. Genaue Untersuchung der unendlichen Doppelproducte, *J. Reine Ang. Math.*, **35**, 153–274.

Elliot, E. B. [1904]. A formula including Legendre's $EK' + KE' - KK' = \frac{1}{2}\pi$, *Messenger of Math.*, **33**, 31–40.

Erdélyi, A. [1937]. Der Zusammenhang zwischen verschiedenen Integraldarstellungen hypergeometrischer Funktionen, *Quart. J. Math.* (Oxford), **8**, 200–213.

Erdélyi, A. [1938]. Note on the transformation of Eulerian hypergeometric functions, *Quart. J. Math.* (Oxford), **9**, 129–134.

Erdélyi, A. [1939]. Transformation of hypergeometric integrals by means of fractional integration by parts, *Quart. J. Math.* (Oxford), **9**, 176–189.

Erdélyi, A. [1953]. *Higher Transcendental Functions*, Vols. I, II, III, A. Erdélyi, ed., McGraw-Hill. Reprinted by Krieger Publishing Co., Malabar, FL [1981].

Erdélyi, A. [1956]. *Asymptotic Expansions*, Dover, New York.

Erdélyi, A. [1960]. Asymptotic solutions of differential equations with transition points or singularities, *J. Mathematical Phys.*, **1**, 16–26.

Euler, L. [1730]. De progressionibus transcendentibus sen quaroum termini generales algebrare dari nequeunt, *Comm. Acad. Sci. Petropolitanae*, **5**, 36–57.

Euler, L. [1739]. De productis ex infinitis factoribus ortis, *Comm. Acad. Sci. Petropolitanae*, **11**, 3–31. Reprinted in *Opera Omnia*, **14** [1924], 260–290.

Euler, L. [1748]. *Introductio in Analysin Infinitorum*, Marcum-Michaelem Bousquet, Lausannae. English translation published by Springer-Verlag, [1988].

Euler, L. [1769]. *Institutiones Calculi Integralis*, II, *Opera Omnia*, Ser. 1, Vols. 11–13.

Euler, L. [1794]. Specimen transformationis singularis serierum, *Nova Acta Acad. Sci. Petropolitanae*, **12**, 58–70. Reprinted in *Opera Omnia*, Ser. 1, Vol. 16, Part 2, 41–55.

Evans, R. [1981]. Identities for Gauss sums over finite fields, *Enseign. Math.*, **27** (2), 197–209.

Evans, R. [1991]. The evaluation of Selberg character sums, *Enseign. Math.*, **37** (2), 235–248.

Fejér, L. [1925]. Abschätzungen für die Legendreschen und verwandte Polynome, *Math. Zeit.*, **24**, 267–284.

Feldheim, E. [1941]. Contribution à la theorie des polynomes de Jacobi (Hungarian, French summary), *Mat. Fiz. Lapik*, **48**, 453–504.

Feldheim, E. [1941a]. Sur les polynomes généralisés de Legendre, *Izv. Akad. Nauk. SSSR Ser. Math.*, **5**, 241–248.

Feldheim, E. [1943]. Contributi alla teoria della funzioni ipergeometriche di piú variabili, *Annali della Scuola Norm. Super. di Pisa*, Ser. II, **12**, 17–60.

Feldheim, E. [1963]. Appendix by G. Szegö. On the positivity of certain sums of ultraspherical polynomials, *J. Anal. Math.*, **11**, 275–284.

Ferrers, N. M. [1877]. *An Elementary Treatise on Spherical Harmonics and Subjects Connected with Them*, Macmillan, London.

Fields, J. L. [1964]. A note on the asymptotic expansion of a ratio of gamma functions, *Proc. Edinburgh Math. Soc.*, **15**, 43–45.

Fine, N. J. [1988]. *Basic Hypergeometric Series and Applications*, Amer. Math. Soc., Providence, RI.

Foata, D. [1965]. Etude algébrique de certains problémes d'analyse combinatoire et du calcul des probabilités, *Publ. Inst. Stat. Univ. Paris*, **14**, 81–241.

Forrester, P. J. and Rogers, J. B. [1986]. Electrostatics and the zeros of the classical polynomials, *SIAM J. Math. Anal.*, **17**, 461–468.

Frobenius, F. G. [1871]. Über die Entwicklung analytischer Functionen in Reihen, die nach gegebenen Functionen fortschreiten, *J. Reine Ang. Math.*, **73**, 1–30. Reprinted in Gesammelte Abhandlungen, Band I, Springer-Verlag, [1968], 35–64.

Funk, P. [1916]. Beiträge zur Theorie der Kugelfunktionen, *Math. Ann.*, **77**, 136–152.

Fuss, N. [1843]. *Correspondance Mathématique et Physique de Quelques Célèbres Géomètres du XVIIIème siècle*, 1, Saint-Pétersbourg.

Gasper, G. [1975]. Positive integrals of Bessel functions, *SIAM J. Math. Anal.*, **6**, 868–881.

Gasper, G. [1977]. Positive sums of the classical orthogonal polynomials, *SIAM J. Math. Anal.*, (8), 423–447.

Gasper, G. [1985]. Rogers' linearization formula for the continuous q-ultraspherical polynomials and quadratic transformation formulas, *SIAM J. Math. Anal.*, **16**, 1061–1071.

Gasper, G. and Rahman, M. [1990]. *Basic Hypergeometric Series*, Cambridge University Press, Cambridge, UK.

Gauss, C. F. [1808]. Summatio quarumdam serierum singularium, *Commen. Soc. Reg. Sci. Götting. Rec.*, vol. I. German translation in *Arithmetische Untersuchungen*, Chelsea, New York [1981].

Gauss, C. F. [1812]. Disquisitiones generales circa seriem infinitam, *Comm. Soc. Reg Gött. II, Werke*, **3**, 123–162.

Gauss, C. F. [1866]. *Zur Theorie der neuen Transscendenten*, II, *Werke*, **3**, 436–445.

Gauss, C. F. [1866a]. *Hundert Theoreme über die neuen Transscendenten, Werke*, **3**, 461–469.

Gauss, C. F. [1866b]. *Arithmetisch Geometrisches Mittel, Werke*, vol. **3**, 361–403.

Gautschi, W. [1967]. Computational aspects of three-term recurrence relations, *SIAM Review*, **9**, 24–82.

Gegenbauer, L. [1875]. Ueber einige Bestimmte Integrale, *Sitz. Math. Natur. Klasse Akad. Wiss. Wien*, **70**, 433–443.

Gessel, I. and Stanton, D. [1982]. Strange evaluations of hypergeometric series, *SIAM J. Math. Anal.*, **11**, 295–308.

Godsil, C. D. [1981]. Hermite polynomials and a duality relation for the matching polynomial, *Combinatorica*, **1**, 257–262.

Godsil, C. D. [1993]. *Algebraic Combinatorics*, Chapman and Hall, New York.

Good, I. J. [1960]. Generalization to severable variables of Lagrange's expansion, with applications to stochastic processes, *Proc. Cambridge Phil. Soc.*, **56**, 367–380.

Good, I. J. [1962]. A short proof of MacMahon's "master theorem," *Proc. Cambridge Phil. Soc.*, **58**, 160.

Good, I. J. [1970]. Short proof of a conjecture of Dyson, *J. Math. Phys.*, **11**, 1884.

Gosper, R. W., Jr. [1978]. Decision procedure for indefinite hypergeometric summation, *Proc. Natl. Acad. Sci. USA*, **75**, 40–42.

Gould, H. W. and Hsu, L. C. [1973]. Some new inverse series relations, *Duke Math. J.*, **40**, 885–891.

Gray, J. [1986]. *Linear Differential Equations and Group Theory from Riemann to Poincaré*, Birkhäuser, Boston.

Gronwall, T. H. [1912]. Ueber die Gibbssche Erscheinung und die trigonometrischen Summen $\sin x + 1/2 \sin 2x + 1/3 \sin 3x + \cdots + 1/n \sin nx$, *Math. Ann.*, **72**, 228–243.

Gross, B. and Koblitz, N. [1979]. Gauss sums and the p-adic gamma function, *Ann. Math.*, **109**, 569–581.

Halmos, P. [1950]. *Measure Theory*, Van Nostrand, Princeton.

Hamburger, H. [1922]. Ueber einige Beziehungen, die mit der Funktionalgleichung der Riemannschen ζ-Funktion aequivalent sind, *Math. Ann.*, **85**, 129–140.

Hankel, H. [1875]. Bestimmte Integrale mit Cylinderfunctionen, *Math. Ann.*, **8**, 453–470.

Hardy, G. H. [1922]. A new proof of the functional equation for the zeta function, *Mat. Tidsskrift*, **B**, 71–73.

Hardy, G. H. [1933]. A theorem concerning Fourier transforms, *J. London Math. Soc.*, **8**, 227–231.

Hardy, G. H. [1940]. *Ramanujan*, Cambridge University Press, Cambridge, UK.

Hardy, G. H. [1949]. *Divergent Series*, Cambridge University Press, Cambridge, UK.

Hecke, E. [1918]. Ueber orthogonal-invariante Integralgleichungen, *Math. Ann.*, **78**, 398–404.

Heckman, G. and Schlicktkrull, H. [1994]. *Harmonic Analysis and Special Functions on Symmetric Spaces*, Academic Press, San Diego.

Heine, E. [1847]. Untersuchungen über die Reihe..., *J. Reine Ang. Math.*, **34**, 285–328.

Helversen-Pasotto, A. [1978]. L'identité de Barnes pour les corps finis, *C. R. Acad. Sci. Paris*, Sér. A–B, **286**, A297–A300.

Helversen-Pasotto, A. and Solé, P. [1993]. Barnes' first lemma and its finite analogue, *Canadian Math. Bull.*, **36**, 273–282.

Hermite, C. [1890]. Sur les Polynomes de Legendre, *Rend. Circ. Mat. Palermo*, **IV**, 146–152. Reprinted in *Oeuvres*, **4**, 314–320.

Hewitt, E. [1954]. Remark on orthonormal sets in $L_2(a, b)$, *Amer. Math. Monthly*, **61**, 249–250.

Hill, M. J. M. [1908]. On a formula for the sum of a finite number of terms of the hypergeometric series when the fourth element is equal to unity, *Proc. London Math. Soc.*, (2), **6**, 339–348.

Hille, E. [1962]. *Analytic Function Theory*, Vol. II, Ginn and Co., Republished, Chelsea, New York [1977].

Hölder, O. [1889]. Über einen Mittelwertsatz, *Goettinger Nach.*, 38–47.

Hsü, H.-Y. [1938]. Certain integrals and infinite series involving ultraspherical polynomials and Bessel functions, *Duke Math. J.*, **4**, 374–383.

Hylleraas, E. [1962]. Linearization of products of Jacobi polynomials, *Math. Scand.*, **10**, 189–200.

Ingham, A. E. [1932]. *The Distribution of Prime Numbers*, Cambridge University Press, London.

Ireland, K. and Rosen, M. [1991]. *A Classical Introduction to Modern Number Theory*, 2nd Ed., Springer-Verlag, New York.

Ismail, M. E. H. [1977]. A simple proof of Ramanujan's $_1\psi_1$ sum, *Proc. Amer. Math. Soc.*, **63**, 185–186.

Ismail, M. E. H. and Stanton, D. [1988], On the Askey–Wilson and Rogers polynomials, *Canadian J. Math.*, **40**, 1025–1045.

Ismail, M. E. H. and Tamhankar, M. V. [1979]. A combinatorial approach to some positivity problems, *SIAM J. Math. Anal.*, **10**, 478–485.

Jackson, D. [1911]. Ueber eine trigonometrische Summe, *Rend. Circ. Mat. Palermo*, **32**, 257–262.

Jackson, F. H. [1910]. On q-definite integrals, *Quart. J. Pure Appl. Math.*, **41**, 193–203.

Jackson, F. H. [1921]. Summation of a q-hypergeometric series, *Messenger of Math.*, **50**, 101–112.

Jacobi, C. G. [1826]. Über Gauss' neue Methode, die Werthe der Integrale näherungsweise zu finden, *J. Reine Ang. Math.*, **1**, 301–308.

Jacobi, C. G. [1829]. *Fundamenta Nova*, Regiomontis, fratrum Borntraeger. Reprinted in *Werke*, Vol. 1, Chelsea, New York [1969], 49–239.

Jacobi, C. G. [1830]. De resolutione aequationum per series infinitam, *J. Reine Ang. Math.*, **6**, 257–286.

Jacobi, C. G. [1834]. Demonstratio Formulae. . ., *J. Reine Ang. Math.*, **11**, 307.

Jacobi, C. G. [1859]. Untersuchung über die Differentialgleichung der hypergeometrischen Reihe, *J. Reine Ang. Math.*, **56**, 149–165.

Jensen, J. L. W. V. [1915–1916]. An elementary exposition of the theory of the gamma function, *Ann. Math.*, **17**, 124–166.

John, F. [1938]. Special solutions of certain difference equations, *Acta Math.*, **71**, 175–189.

Karlin, S. and McGregor, J. L. [1957]. The differential equations of birth-and-death processes, *Trans. Amer. Math. Soc.*, **85**, 489–546.

Kirillov, A. N. [1994]. *Dilogarithm Identities*, Preprint Series, Dept. of Math. Sci., University of Tokyo.

Klein, F. [1894]. *Vorlesungen über die Hypergeometrische Funktion*, Göttingen.

Knopp, M. [1971]. *Modular Forms and Analytic Number Theory*, Markham, New York.

Koblitz, N. [1977]. *P-Adic Numbers, p-adic Analysis, and Zeta-Functions*, Springer-Verlag, New York.

Koblitz, N. [1984]. *Introduction to Elliptic Curves and Modular Forms*, Springer-Verlag, New York.

Koekoek, R. and Swarttouw, R. F. [1994]. The Askey-scheme of hypergeometric orthogonal polynomials and its q-analogue, *Reports of the Faculty of Technical Mathematics and Informatics*, no. 98–117, Delft.

Kogbetliantz, E. [1924]. Recherches sur la sommabilité des séries ultrasphériques par la méthode des moyennes arithmetiques, *J. Math. Pures Appl.*, (9), **3**, 107–187.

Kolberg, O. [1957]. Some identities involving the partition functions, *Math. Scand.*, **5**, 77–92.

Koornwinder, T. H. [1972]. The addition formula for Jacobi polynomials, I, Summary of results, *Indag. Math.*, **34**, 188–191.

Koornwinder, T. H. [1974]. Jacobi polynomials, II. An analytic proof of the product formula, *SIAM J. Math. Anal.*, **5**, 125–137.

Koornwinder, T. H. [1975]. Jacobi polynomials, III. An analytic proof of the addition formula, *SIAM J. Math. Anal.*, **6**, 533–543.

Koornwinder, T. H. [1978]. Positivity proofs for linearization and connection coefficients of orthogonal polynomials satisfying an addition formula, *J. London Math. Soc.* (2), **18**, 101–114.

Koornwinder, T. H. [1990]. Jacobi functions as limit cases of q-ultraspherical polynomials, *J. Math. Anal. Appl.*, **148**, 44–54.

Kummer, E. [1836]. Ueber die hypergeometrische Reihe, *J. Reine Ang. Math.*, **15**, 39–83, 127–172.

Kummer, E. [1847]. Beitrag zur Theorie der Function $\Gamma(x)$, *J. Reine Ang. Math.*, **35**, 1–4.

Lang, S. [1980]. *Cyclotomic Fields II*, Springer-Verlag, New York.

Lanzewizky, I. L. [1941]. Ueber die Orthogonalität der Fejér-Szegöschen Polynome, *C. R. (Dokl.) Acad. Sci. USSR*, **31**, 199–200.

Lewin, L. [1981]. *Polylogarithms and Associated Functions*, North-Holland, New York.

Lewin, L. [1981a]. *Structural Properties of Polylogarithms*, Amer. Math. Soc., Providence, RI.

Liouville, J. [1839]. Sur quelques intégrales définies, *J. Math. Pures Appl.*, Sér. 1, **4**, 229–235.

Liouville, J. [1855]. Sur un théorème relatif à 1' intégrale eulérienne de seconde espèce, *J. Math. Pure Appl.*, Sér. 1, **20**, 157–160.

Littlewood, J. E. [1986]. *Littlewood's Miscellany* (B. Bollobás, ed.), Cambridge University Press, Cambridge, UK.

Lorch, L., Muldoon, M. E., and Szego, P. [1970]. Higher monotonicity properties of certain Sturm–Liouville functions, III, *Canadian J. Math.*, **22**, 1238–1265.

Lorch, L., Muldoon, M. E., and Szego, P. [1972]. Higher monotonicity properties of certain Sturm–Liouville functions, IV, *Canadian J. Math.*, **24**, 349–368.

Lorch, L. and Szego, P. [1963]. Higher monotonicity properties of certain Sturm–Liouville functions, *Acta Math.*, **109**, 55–73.

Lorentz, G. G. and Zeller, K. [1964]. Abschnittlimitierbarkeit und der Satz von Hardy–Bohr, *Arch. Math.* (Basel), **15**, 208–213.

Lorentzen, L. and Waadeland, H. [1992]. *Continued Fractions with Applications*, North-Holland, Amsterdam.

Macdonald, I. G. [1995]. *Symmetric Functions and Hall Polynomials*, 2nd ed., Oxford University Press, Oxford.

MacMahon, P. A. [1917–1918]. *Combinatory Analysis*, Cambridge University Press, Cambridge, UK. Reprinted by Chelsea, New York [1984].

Magnus, A. [1988]. Associated Askey–Wilson polynomials as Laguerre–Hahn orthogonal polynomials, in *Orthogonal Polynomials and Their Applications* (Segovia, 1986), M. Alfaro et al., eds., *Lecture Notes in Math.* no. 1329, Springer-Verlag, New York, 261–378.

Makai, E. [1974]. An integral inequality satisfied by Bessel functions, *Acta Math. Acad. Sci. Hungaricae*, **25**, 387–390.

Mehta, M. L. [1991]. *Random Matrices*, 2nd Ed., Academic Press, Boston.

Miller, W., Jr. [1972]. *Symmetry Groups and Their Applications*, Academic Press, New York.

Milne, S. C. [1988]. Multiple q-series and $U(n)$ generalizations of Ramanujan's $_1\psi_1$ sum, in *Ramanujan Revisited*, G. Andrews et al., eds., Academic Press, New York, 473–524.

Milne, S. C. and Lilly, G. M. [1995]. Consequences of the A_ℓ and C_ℓ Bailey transform and Bailey lemma, *Discrete Math.*, **139**, 319–346.

Moak, D. S. [1980]. The q-gamma function for $q > 1$, *Aequat. Math.*, **20**, 278–285.

Monsky, P. [1994]. Simplifying the proof of Dirichlet's theorem, *Amer. Math. Monthly*, **100**, 861–862.

Morita, Y. [1975]. A p-adic analog of the Γ-function, *J. Fac. Sci. Tokyo*, Sec. 1A, **22**, 255–266.

Morris, W. G. [1984]. *Constant term identities for finite and affine root systems: Conjectures and theorems*, Ph.D. Thesis, University of Wisconsin–Madison.

Müller, C. [1966]. *Spherical Harmonics, Lecture Notes in Mathematics*, **7**, Springer-Verlag, New York.

Murphy, R. [1835]. Second memoir on the inverse method of definite integrals, *Trans. Cambridge Phil. Soc.*, **5**, 113–148.

Nassarallah, B. and Rahman, M. [1985]. Projection formulas, a reproducing kernel and a generating function for q-Wilson polynomials, *SIAM J. Math. Anal.*, **16**, 186–197.

Natanson, I. P. [1965]. *Constructive Function Theory*, Vol. III, Frederick Ungar Publishing Co., New York.

Nemes, I., Petkovšek, M., Wilf, H., and Zeilberger, D. [1997]. How to do Monthly Problems with your computer, *Amer. Math. Monthly*, **104**, 505–519.

Nevai, P. [1979]. *Orthogonal Polynomials*, Amer. Math. Soc., Providence, RI.

Nevai, P. [1990]. *Orthogonal Polynomials: Theory and Practice*, P. Nevai, ed., Kluwer Academic Publishers, Dordrecht.

Newton, I. [1960]. *The Correspondence of Isaac Newton*, **2**, 1676–1687, Cambridge University Press, Cambridge, UK.

Nielsen, N. [1906]. *Handbuch der Theorie der Gamma Funktion*, B. B. Teubner, Leipzig.

Olver, F. W. J. [1974]. *Asymptotics and Special Functions*, Academic Press, New York.

Papperitz, E. [1889]. Ueber die Darstellung der hypergeometrischen Transcendenten durch eindeutige Functionen, *Math. Ann.*, **34**, 247–296.

Petkovšek, M., Wilf, H., and Zeilberger, D. [1996]. $A = B$, A. K. Peters, Wellesley, MA.

Pfaff, J. F. [1797]. *Disquisitiones Analyticae, I*, Helmstadt.

Pfaff, J. F. [1797a]. Observationes analyticae ad L. Euleri Institutiones Calculi Integralis, Vol. IV, Supplem. II et IV, Historie de 1793, *Nova Acata Acad. Scie. Petropolitanae*, **XI**, 38–57. (Note: The history section is paged separately from the scientific section of this journal.)

Poincaré, H. [1886]. Sur les intégrales irrégulières des équations linéares, *Acta Math.*, **8**, 295–344.

Poisson, S. D. [1823]. Suite du Mémoire sur les intégrales définies et sur la sommation des séries, *Paris Jour. de l'École Polytechnique*, **19**, 404–509, especially, pp. 477–478.

Pólya, G. [1984]. *Collected Papers*, Vols. I–IV, MIT Press, Cambridge, MA.

Pólya, G. and Szegö, G. [1972]. *Problems and Theorems in Analysis*, Vols. I and II, Springer-Verlag, New York.

Rademacher, H. [1955]. On the transformation of $\log \eta(\tau)$, *J. Indian Math. Soc.*, **19**, 25–30.

Rademacher, H. [1973]. *Topics in Analytic Number Theory*, Springer-Verlag, New York.

Rahman, M. [1981]. A non-negative representation of the linearization coefficients of the product of Jacobi polynomials, *Canadian J. Math.*, **33**, 915–928.

Rahman, M. [1986]. q-Wilson functions of the second kind, *SIAM J. Math. Anal.*, **17**, 1280–1286.

Ramanujan, S. [1927]. *Collected Papers*, Cambridge University Press. Reprinted by Chelsea, New York [1962].

Ramanujan, S. [1988]. *The Lost Notebook*, Narosa Publishing House, New Delhi.

Raynal, J. [1979]. On the definition and properties of generalized 6-j symbols, *J. Math. Phys.*, **20**, 2398–2415.

Riemann, B. [1857]. Beiträge zur Theorie der durch Gauss'sche Reihe $F(\alpha, \beta, \gamma, x)$ darstellbaren Functionen, *K. Gess. Wiss. Göttingen*, **7**, 1–24.

Riemann, B. [1859]. Über die Anzahl der Primzahlen unter einer gegebene Grösse, *Monatsb. Berliner Akad.*, 671–680. Reprinted in Gesammelte Mathematische Werke, paper 7, Springer-Verlag, [1990].

Rogers, L. J. [1888]. An extension of a certain theorem in inequalities, *Messenger of Math.*, **17**, 145–150.

Rogers, L. J. [1894]. Second memoir on the expansion of certain infinite products, *Proc. London Math. Soc.*, **25**, 318–343.

Rogers, L. J. [1895]. Third memoir on the expansion of certain infinite products, *Proc. London Math. Soc.*, **26**, 15–32.

Rogers, L. J. [1907]. On function sum theorems connected with the series $\sum_{n=1}^{\infty} x^n/n^2$, *Proc. London Math. Soc.*, **4**, 169–189.

Rogers, L. J. [1917]. On two theorems of combinatory analysis and allied identities, *Proc. London Math. Soc.*, (2), **16**, 315–336.

Roosenraad, C. T. [1969]. *Inequalities with orthogonal polynomials*, Ph.D. Thesis, University of Wisconsin–Madison.

Rothe, H. A. [1811]. *Systematisches Lehrbuch der Arithmetic*, Leipzig.

Roy, R. [1990]. The discovery of the series formula for π by Leibniz, Gregory and Nilakantha, *Math. Mag.*, **63**, 291–306.

Roy, R. [1993]. The work of Chebyshev on orthogonal polynomials, in *Topics in Polynomials of One and Several Variables and Their Applications*, Th. Rassias, H. M. Srivastava, A. Yanushauskas, eds., World Scientific, Singapore, 495–512.

Rudin, W. [1976]. *Principles of Mathmatical Analysis*, 3rd ed., McGraw-Hill, New York.

Saalschütz, L. [1890]. Eine Summationsformel, *Zeitschrift Math. Phys.*, **35**, 186–188.

Saff, E. B. and Varga, R. S. [1976]. Zero-free parabolic regions for sequences of polynomials, *SIAM J. Math. Anal.*, 7, 344–357.

Salamin, E. [1976]. Computation of π using arithmetic-geometric mean, *Math. Comput.*, **30**, 565–570.

Săpiro, R. L. [1968]. Special functions related to representations of the group $SU(n)$, of class I with respect to $SU(n-1)$ ($n \geq 3$), *Izv. Vyssh. Uchebn. Zaved. Matematika*, **71** (4), 97–107 (in Russian).

Sarmanov, I. O. [1968]. A generalized symmetric gamma correlation, *Dokl. Akad. Nauk SSSR*, **179**, 1276–1278; *Soviet Math. Dokl.*, **9**, 547–550.

Scharlau, W. and Opolka, H. [1985]. *From Fermat to Minkowski*, Springer-Verlag, New York.

Schiff, L. I. [1947]. *Quantum Mechanics*, Addison-Wesley, New York.

Schur, I. [1918]. Über die Verteilung der Wurzeln bei gewissen algebraischen Gleichungen mit ganzzahligen Koeffizienten, *Math. Zeit.*, **1**, 377–402.

Schur, I. [1921]. Über die Gaussschen Summen, *Nach. Gessel. Göttingen, Math-Phys. Klasse*, 147–153.

Schützenberger, M.–P. [1953]. Une interprétation de certaines solutions de l'équation fonctionelle: $F(x+y) = F(x)F(y)$, *C. R. Acad. Sci. Paris*, **236**, 352–353.

Sears, D. B. [1951]. Transformations of basic hypergeometric functions of special type, *Proc. London Math. Soc.*, (2), **53**, 138–157.

Selberg, A. [1944]. Bemerkninger om et multipelt integral, *Norske Mat. Tidsskr.*, **26**, 71–78.

Sheppard, W. F. [1912]. Summation of the coefficients of some terminating hypergeometric series, *Proc. London Math. Soc.*, (2), **10**, 469–478.

Siegel, C. L. [1945]. The trace of totally positive and real algebraic integers, *Ann. Math.*, **46**, 302–313.

Siegel, C. L. [1954]. A simple proof of $\eta(-1/\tau) = \eta(\tau)\sqrt{\tau/2}$, *Mathematica*, **1**, 4.

Simpson, T. [1759]. The invention of a general method for determining the sum of every second, third, fourth, or fifth, etc. term of a series, taken in order; the sum of the whole series being known, *Phil. Trans. Royal Soc. London*, **50**, 757–769.

Smith, H. J. S. [1859–1865]. *Report on the Theory of Numbers*, Reprinted by Chelsea, New York [1965].

Stanton, D. [1988]. Recent results for the q-Lagrange inversion formula, in *Ramanujan Revisited*, G. E. Andrews, R. Askey, B. Berndt, K. G. Ramanathan, and R. A. Rankin, eds., Academic Press, San Diego, 525–536.

Stieltjes, T. J. [1993]. *Collected Papers*, Vols. I and II, Springer-Verlag, New York.

Stirling, J. [1730]. *Methodus Differentialis*, London.

Stone, M. H. [1962]. A generalized Weierstrass approximation theorem, in *Studies in Modern Analysis*, R. C. Buck, ed., Math. Assoc. America,

Sylvester, J. J. [1853]. On Mr. Cayley's impromptu demonstration of the rule for determining at sight the degree of any symmetrical function of the roots of an equation expressed in terms of the coefficients, *Phil. Mag.*, **5**, 199–202. *Collected Papers*, Vol. 1, 594–598.

Sylvester, J. J. [1883]. Note on the graphical method in partitions of n, *Proc. Cambridge Phil. Soc.*, **19**, 207–210.

Sylvester, J. J. [1886]. Lectures on the theory of reciprocants, *Amer. J. Math.*, **8**, 196–260; **9**, 1–37, 113–11661, 297–352; **10**, 1–116.

Szász, O. [1950]. On the relative extrema of ultraspherical polynomials, *Bollettino della Unione Matematica Italiana*, **5** (3), 125–127.

Szász, O. [1951]. On the relative extrema of the Hermite orthogonal functions, *J. Indian Math. Soc.*, **25**, 1340–1345.

Szegö, G. [1933]. Ueber gewisse Potenzreihen mit lauter positiven Koeffizienten, *Math. Zeit.*, **37**, 674–688.

Szegö, G. [1936]. Inequalities for the zeros of Legendre polynomials and related functions, *Trans. Amer. Math. Soc.*, **39**, 1–17.

Szegö, G. [1948]. On an inequality of Turán concerning Legendre polynomials, *Bull. Amer. Math. Soc.*, **54**, 401–405.

Szegö, G. [1950]. On the relative extrema of Legendre polynomials, *Bollettino della Unione Matematica Italiana*, **5** (4), 120–121.

Szegö, G. [1975]. *Orthogonal Polynomials*, Amer. Math. Soc., Providence, RI.

Szwarc, R. [1992]. Orthogonal polynomials and a discrete boundary value problem II, *SIAM J. Math. Anal*, **23**, 965–969.

Takács, L. [1973]. On an identity of Shih-Chieh Chu, *Acta Sci. Math.*, (Szeged), **34**, 383–391.

Thomae, J. [1869]. Beiträge zur Theorie der durch die Heinesche Reihe. . . , *J. Reine Ang. Math.*, **70**, 258–281.

Thomae, J. [1879]. Über die Funktionen welche durch Reihen von der Form dargestellt werden: $1 + \frac{pp'p''}{1q'q''} + \cdots$, *Journal Math.*, **87**, 26–73.

Titchmarsh, E. C. [1937]. *Introduction to the Theory of Fourier Integrals*, Oxford University Press, London.

Todd, J. [1950]. On the relative extrema of Laguerre orthogonal functions, *Bollettino della Unione Matematica Italiana*, **5** (3), 120–125.

Tricomi, F. G. and Erdélyi, A. [1951]. The asymptotic expansion of a ratio of gamma functions, *Pac. J. Math.*, **1**, 133–142.

Turán, P. [1952]. On a trigonometrical sum, *Ann. Soc. Polonaise Math.*, **25**, 155–161.

Tweddle, I. [1988]. *James Stirling*, Scottish Academic Press, Edinburgh.

van der Poorten, A. [1979]. A proof that Euler missed...Apéry's proof of the irrationality of $\zeta(3)$, *Math. Intelligencer*, **2**, 195–203.

Van Loan, C. [1992]. *Computational Frameworks for the Fast Fourier Transform*, SIAM, Philadelphia.

Viennot, G. [1983]. *Une Theorie Combinatoire des Polynômes Orthogonaux Généraux*, Lecture Notes, UQAM.

Vietoris, L. [1958]. Ueber das Vorzeichen gewisser trigonometrischer Summen, *Sitzungsber. Oest. Akad. Wiss.*, **167**, 125–135.

Vietoris, L. [1959]. Ueber das Vorzeichen gewisser trigonometrischer Summen, II, *Sitzungsber. Oest. Akad. Wiss.*, **167**, 192–193.

Vilenkin, N. J. [1958]. Some relations for Gegenbauer functions, *Uspekhi Matem. Nauk (N.S.)*, **13** (3), 167–172.

Vilenkin, N. J. [1968]. *Special Functions and the Theory of Group Representations*, Translations of Math. Monographs 22, Amer. Math. Soc., Providence, RI.

Vilenkin, N. J. and Klimyk, A. [1992]. *Representation of Lie Groups, Special Functions and Integral Transforms*, Kluwer Academic, Amsterdam.

Wallis, J. [1656]. *Arithmetica Infinitorum*, Oxford.

Wang, Z. X. and Guo, D. R. [1989]. *Special Functions*, World Scientific, Singapore.

Watson, G. N. [1918]. Asymptotic expansions of Hypergeometric functions, *Trans. Cambridge Phil. Soc.*, **22**, 277–308.

Watson, G. N. [1925]. A note on generalized hypergeometric series, *Proc. London Math. Soc.*, (2), **23**, xiii–xv (Records for 8 Nov. 1923).

Watson, G. N. [1929]. A new proof of the Rogers–Ramanujan identities, *J. London Math. Soc.*, **4**, 4–9.

Watson, G. N. [1944]. *A Treatise on the Theory of Bessel Functions*, 2nd Ed., Cambridge University Press, Cambridge, UK.

Weil, A. [1949]. Number of solutions of equations in a finite field, *Bull. Amer. Math. Soc.*, **55**, 497–508.

Weil, A. [1968]. Sur une formule classique, *J. Math. Soc. Japan*, **20**, 400–402.

Weil, A. [1974]. Two lectures on number theory, past and present, *Collected Papers*, Vol. III, Springer-Verlag, New York.

Weil, A. [1976]. *Elliptic Functions According to Eisenstein and Kronecker*, Springer-Verlag, New York.

Weil, A. [1983]. *Number Theory, from Hammurabi to Legendre*, Birkhäuser, Boston.

Whipple, F. J. W. [1926]. On well-poised series, generalized hypergeometric series having parameters in pairs each pair with the same sum, *Proc. London Math. Soc.*, (2), **24**, 247–263.

Whittaker, E. T. [1904]. An expression of certain known functions as generalized hypergeometric series, *Bull. Amer. Math. Soc.*, **10**, 125–134.

Whittaker, E. T. and Watson, G. N. [1940]. *A Course of Modern Analysis*, 4th Ed., Cambridge University Press, London.

Wiener, N. [1933]. *The Fourier Integral and Certain of Its Applications*, Cambridge. Reprinted by Dover, New York [1958].

Wilkins, J. E. [1948]. Nicholson's integral for $J_n^2(z) + Y_n^2(z)$, *Bull. Amer. Math. Soc.*, **54**, 232–234.

Wilson, J. A. [1977]. Three-term contiguous relations and some new orthogonal polynomials, *Padé and Rational Approximations*, E. B. Saff and R. S. Varga, eds., Academic Press, New York.

Wilson, J. A. [1978]. *Hypergeometric series, recurrence relations and some new orthogonal polynomials*, Ph.D. Thesis, University of Wisconsin, Madison.

Wilson, J. A. [1991]. Orthogonal functions for Gram determinants, *SIAM J. Math. Anal.*, **22**, 1147–1155.

Wong, R. [1989]. *Asymptotic Approximations of Integrals*, Academic Press, New York.

Zagier, D. [1989]. The dilogarithm in geometry and number theory, in *Number Theory and Related Topics*, R. Askey et al., eds., Oxford University Press, Oxford, UK, 231–249.

Zeilberger, D. [1982]. Sister Celine's technique and its generalizations, *J. Math. Anal. Appl.*, **85**, 114–145.

Zeilberger, D. [1994]. Theorems for a price: Tomorrow's semi-rigorous mathematical culture, *The Math. Intelligencer*, **16** (4), 11–14.

Index

Subject Index

Symbol Index

663